Springer Series in Materials Science

Volume 167

For further volumes:
http://www.springer.com/series/856

The Springer Series in Materials Science covers the complete spectrum of materials physics, including fundamental principles, physical properties, materials theory and design. Recognizing the increasing importance of materials science in future device technologies, the book titles in this series reflect the state-of-the-art in understanding and controlling the structure and properties of all important classes of materials.

Sitangshu Bhattacharya
Kamakhya Prasad Ghatak

Effective Electron Mass in Low-Dimensional Semiconductors

Springer

Sitangshu Bhattacharya
Department of Electronics Systems
Engineering, Nano Scale Device
Research Laboratory
Indian Institute of Science
Bangalore
India

Kamakhya Prasad Ghatak
Department of Electronics
and Communication Engineering
National Institute of Technology
Agartala, Tripura West
India

ISSN 0933-033X
ISBN 978-3-642-31247-2 ISBN 978-3-642-31248-9 (eBook)
DOI 10.1007/978-3-642-31248-9
Springer Heidelberg New York Dordrecht London

Library of Congress Control Number: 2012945124

Printed on acid-free paper

Springer is part of Springer Science+Business Media (www.springer.com)

The first author dedicates this book to his beautiful and wonderful wife Rekha Verma, the source of inspiration to complete this monograph
The second author dedicates this monograph to the Jadavpur University, Kolkata for accepting him as the first recipient of the Degree of Doctor of Engineering in 1991 since the University inception in 1955

Preface

The merging of the concept of introduction of asymmetry of the wave vector space of the charge carriers in semiconductors with the modern techniques of fabricating nanostructured materials such as MBE, MOCVD, and FLL in one, two, and three dimensions (such as UFs, nipi structures, inversion, and accumulation layers, quantum wire superlattices, carbon nanotubes, nanowires, quantum dots, magneto inversion and accumulation layers, quantum dot superlattices, etc.) spawns not only useful quantum effect devices but also unearths new concepts in the realm of low-dimensional materials science and related disciplines. It is worth remarking that these semiconductor nanostructures occupy a paramount position in the entire arena of nanoscience and technology by their own right and find extensive applications in quantum registers, resonant tunneling diodes and transistors, quantum switches, quantum sensors, quantum logic gates, hetero-junction field-effect transistors, quantum well and nanowire transistors, high-speed digital networks, high-frequency microwave circuits, quantum cascade lasers, high-resolution terahertz spectroscopy, superlattice photo-oscillator, advanced integrated circuits, superlattice photocathodes, thermoelectric devices, superlattice coolers, intermediate-band solar cells, micro-optical systems, high performance infrared imaging systems, band-pass filters, thermal sensors, optical modulators, optical switching systems, single electron/molecule electronics, nanotube-based diodes, and other nano-electronic devices. Knowledge regarding these quantized structures may be gained from original research contributions in scientific journals, proceedings of various international conferences, and different review articles respectively. Mathematician Simmons rightfully tells us [1] that the mathematical knowledge is said to be doubling in every 10 years and in this context we can also envision the extrapolation of the Moore's law by projecting it in the perspective of the advancement of new research and analyses, in turn, generating novel concepts particularly in the area of nanoscience and technology [2]. In this context, it may be noted that the available books on solid-state and allied sciences cannot afford to cover even an entire chapter excluding few pages on the Effective Electron Mass (EEM) in Low-Dimensional Semiconductors.

The effective mass of the carriers in semiconductors, being connected with the mobility, is known to be one of the most important physical quantities, used for the analysis of electron devices under different operating conditions [3]. The carrier degeneracy in semiconductors influences the effective mass when it is energy dependent. Under degenerate conditions, only the electrons at the Fermi surface of n-type semiconductors participate in the conduction process and hence, the effective mass of the electrons corresponding to the Fermi level would be of interest in electron transport under such conditions. The Fermi energy is again determined by the electron energy spectrum and the carrier statistics and therefore, these two features would determine the dependence of the EEM in degenerate n-type semiconductors under the degree of carrier degeneracy. In recent years, various energy wave vector dispersion relations have been proposed [4–10] which have created the interest in studying the effective mass in such materials under external conditions. It has, therefore, different values in different materials and varies with electron concentration, with the magnitude of the reciprocal quantising magnetic field under magnetic quantization, with the quantizing electric field as in inversion layers, with the nano-thickness as in UFs and nanowires and with superlattice period as in the quantum confined superlattices of small gap semiconductors with graded interfaces having various carrier energy spectra [11–57].

This book, divided into three parts which contain nine chapters and three Appendices, is partially based on our ongoing researches on the effective mass from 1980 and an attempt has been made to present a cross section of the effective mass for a wide range of low-dimensional semiconductors with varying carrier energy spectra under various physical conditions. The first part deals with the influence of quantum confinement on the EEM in non-parabolic semiconductors. Chapter 1 investigates the EEM in UFs of nonlinear optical materials on the basis of a generalized electron dispersion law introducing the anisotropies of the effective electron masses and the spin orbit splitting constants respectively together with the inclusion of the crystal field splitting within the framework of the *k.p* formalism. The results of III–V (e.g. InAs, InSb, GaAs, etc.), ternary (e.g. $Hg_{1-x}Cd_xTe$), quaternary (e.g. $In_{1-x}Ga_xAs_{1-y}P_y$ lattice matched to InP) compounds form a special case of our generalized analysis under certain limiting conditions. The EEM in UFs of II–VI, Bi, IV–VI, stressed Kane-type semiconductors, Te, GaP, $PtSb_2$, Bi_2Te_3, Ge and GaSb compounds have also been investigated by using the appropriate energy band structures for these materials. The importance of the aforementioned semiconductors has also been described in the same chapter. It is well known that the semiconductor superlattices find extensive applications in avalanche photodiodes, photo-detectors, electro-optic modulators, etc. In Chap. 2 the EEM in nipi structures of nonlinear optical, III–V, II–VI, IV–VI, and stressed Kane-type semiconductors has been studied.

In recent years, there has been considerable interest in the study of the inversion layers which are formed at the surfaces of semiconductors in metal–oxide–semiconductor field-effect transistors (MOSFET) under the influence of a sufficiently strong electric field applied perpendicular to the surface by means of a large gate bias. In such layers, the carriers form a two-dimensional gas and are free to move parallel to the surface while their motion is quantized perpendicular to it leading to the formation of electric subbands [58]. In Chap. 3, the EEM in *n*-channel inversion layers of nonlinear optical, III–V, II–VI, IV–VI stressed Kane-type semiconductors, Ge and GaSb has been investigated.

The effects of quantizing magnetic field on the band structures of compound semiconductors are more striking than that of the parabolic one and are easily observed in experiments. A number of interesting physical features originate from the significant changes in the basic energy wave vector relation of the carriers caused by the magnetic field. The valuable information could also be obtained from experiments under magnetic quantization regarding the important physical properties such as Fermi energy and effective masses of the carriers, which affect almost all the transport properties of the electron devices [59–63] of various materials having different carrier dispersion relations [64]. In Chap. 4, the EEM in nonlinear optical, III–V. II–VI, Bi. IV–VI, stressed Kane-type semiconductors, Te, GaP, $PtSb_2$, Bi_2Te_3, Ge, GaSb and II–V compounds have also been studied under magnetic quantization. Since Iijima's discovery [65], carbon nanotubes (CNTs) have been recognized as fascinating materials with nanometer dimensions uncovering new phenomena in different areas of nanoscience and technology. The remarkable physical properties of these quantum materials make them ideal candidates to reveal new phenomena in nano-electronics. Chapter 5 contains the study of the EEM in nanowires of the nonlinear optical, III–V, II–VI, Bi, IV–VI, stressed Kane-type semiconductors, Te, GaP, $PtSb_2$, Bi_2Te_3, Ge, GaSb and II–V semiconductors together with CNTs respectively.

With the advent of nanophotonics, there has been considerable interest in studying the optical processes in semiconductors and their nanostructures [66–67]. It appears from the literature that investigations have been carried out on the assumption that the carrier energy spectra are invariant quantities in the presence of intense light waves, which is not fundamentally true. The physical properties of semiconductors in the presence of light waves which change the basic dispersion relation have been relatively less investigated in the literature [68, 69]. The second part of this book studies the influence of light waves of the EEM in opto-electronic semiconductors and Chap. 6 investigates the influence of light waves on the EEM in quantum confined III–V, ternary, and quaternary semiconductors. Under external photo excitation the electron dispersion relation changes profoundly and the EEM has been studied by formulating a new electron dispersion law on the basis of *k.p* formalism. In the same chapter the influence of magnetic quantization on the EEM has been investigated. The same chapter also explores the effect of light waves on the EEM for 2D systems (e.g. UFs, nipi structures, and inversion layers), 1D systems (such as quantum wire effective mass superlattices, and quantum wire superlattices with graded interfaces) and the influence of quantizing magnetic field on the EEM for effective mass superlattices, and superlattices with graded interfaces respectively.

With the advent of nanodevices, the inbuilt electric field becomes so large that the electron energy spectrum changes fundamentally and the single Chap. 7 of the third part investigates the influence of intense electric field on the EEM in II–V, ternary and quaternary semiconductors. The same chapter also explores the influence of electric field on the 2D systems (e.g. UFs, nipi structures and inversion layers) and 1D systems (such as, nano wire effective mass superlattices, and nano wire superlattices with graded interfaces) in this context. Chapter 8 contains the applications and brief review of experimental results. Chapter 9 contains the conclusion and the scope for future research.

It may be noted that the influence of crossed electric and quantizing magnetic fields on the transport properties of semiconductors having various band structures are relatively less investigated as compared with the corresponding magnetic quantization, although, the cross-fields are fundamental with respect to the addition of new physics and the related experimental findings. It is well known that in the presence of electric field (E_o) along x-axis and the quantizing magnetic field (B) along z-axis, the dispersion relations of the conduction electrons in semiconductors become modified and for which the electron moves in both the z and y directions. The motion along y-direction is purely due to the presence of E_O along x-axis and in the absence of electric field, the EEM along y-axis tends to infinity which indicates the fact that the electron motion along y-axis is forbidden. The EEM of the isotropic, bulk semiconductors having parabolic energy bands exhibits mass anisotropy in the presence of cross fields and this anisotropy depends on the electron energy, the magnetic quantum number, the electric and the magnetic fields respectively, although, the EEM along z-axis is a constant quantity. In 1966, Zawadzki and Lax [70] formulated the electron dispersion law for III-V semiconductors in accordance with the two-band model of Kane under cross fields configuration which generates the interest to study this particular topic of solid state science in general [71–77].

Appendix A investigates the EEM under cross field configuration in nonlinear optical, III–V, II–VI, Bi, IV–VI, and stressed Kane-type semiconductors and ultra thin films of the aforementioned materials. It is an amazing fact that though heavily doped semiconductors have been deeply studied in the literature but the study of the carrier transport in heavily doped materials through proper formulation of the Boltzmann transport equation which needs in turn, the corresponding heavily doped carrier energy spectra is still one of the open research problems [78–81]. Appendix B attempts to touch the enormous field of active research with respect to EEM of heavily doped compound semiconductors in a nutshell. Appendix C deals with the EEM in III–V, II–VI, IV–VI, HgTe/CdTe, and strained layer heavily doped superlattices with graded interfaces and effective mass superlattices of the said constituent materials. In these appendices no graphs together with results and discussions are being presented since we feel that the readers will enjoy the complex computer algorithm to investigate the EEM in the respective case generating new physics and thereby transforming each appendix into a short monograph by considering various materials having different dispersion relations. Since there is no existing book devoted totally to the EEM in low-dimensional semiconductors to the best of our knowledge, we hope that this book will be a useful reference source for the present and the next generation of readers and researchers of materials and allied sciences in general. In spite of our joint efforts, the production of error-free first edition of any book from every point of view enjoys permanently the domain of impossibility theorems and the same stands very true for this monograph also. Various expressions of this book have been appearing for the first time in printed form. The suggestions of the readers for the development of this book will be highly appreciated for the purpose of future edition, if any.

In this book, from Chap. 1 till the end, we have presented **250 open research problems** in this particular topic. The problems presented here are the integral part of this book and will be useful for the readers to initiate their own contributions on the effective mass. This aspect is also important for Ph.D. aspirants and researchers. Each chapter ends with a table containing the main results excluding the last two and the Appendices.

In this monograph, we have investigated various dispersion relations of different quantized structures and the corresponding electron statistics to study effective mass. Our theoretical formulation of the density-of-states effective mass of tetragonal materials based on our generalized electron dispersion relation agrees well with the available experimental data as given elsewhere [82]. Thus, in this book, the readers will get a lot of information regarding quantum confined low-dimensional materials having different band structures. Although the name of the book is extremely specific, from the content, one can infer that it should be useful in graduate courses on materials science, nanoscience and technology, solid-state science, semiconductor physics, and nanostructured devices in many universities and institutes. Last but not the least, we do hope that our humble effort will kindle the desire to delve deeper into this fascinating topic by anyone engaged in materials research and device development either in academics or in industries.

Acknowledgments

Acknowledgment by Sitangshu Bhattacharya

In spite of many hurdles, the completion of this monograph is owed to my teacher S. Mahapatra, in the Department of Electronic Systems Engineering at the Indian Institute of Science, Bangalore, with whom I have learned to perform quality research. My sincere gratitude is also to my friend R. C. Mallik of Thermoelectric Materials and Device Laboratory, Department of Physics of the same Institute for his constant inspiration. I offer special thanks to all my friends at my department and institute for standing by my side at difficult times of my research life. I am indebted to the Department of Science and Technology, India, for sanctioning the project and the fellowship under "SERC Fast Track Proposal of Young Scientist" scheme-2008-2009 (SR/FTP/ETA-37/08) under which this monograph has been completed. As forever, I am immensely grateful to the second author, my friend, mentor, and Ph.D. thesis advisor.

Acknowledgment by Kamakhya Prasad Ghatak

Like the first author, I am ever grateful to Professor A. N. Chakravarti, my Ph.D. thesis advisor and mentor who convinced a 21-year-old Circuit theorist that theoretical semiconductor science is the hidden dual dance of quantum mechanics and statistical mechanics, and even to appreciate the incredible beauty, he placed a stiff note for me to understand deeply the Course of Theoretical Physics, the Classics of Landau–Lifshitz together with the two-volume classics of Morse-Feshbach 35 years ago. I am grateful to Professor P. K. Bose, Director, National Institute of Technology, Agartala, Tripura my present mentor at the last phase of my academic life and a very pivotal person in my academic career, for instigating me to carry out extensive research by ignoring all the difficulties. I express my gratitude to Professors R. K. Poddar and R. N. Basu, Ex-Vice Chancellors of the University of Calcutta, Professors S. K. Sen and A. N. Basu, Ex-Vice Chancellors of Jadavpur University, and Professor D. K. Basu, Ex-Vice Chancellor of Burdwan and Tripura Universities, the five pivotal persons in the research growth of my career. I am grateful to Professors S. C. Dasgupta, P. K. Choudhury, M. Mitra and S. Sarkar of the Department of Mathematics of the then Bengal Engineering College (presently Bengal Engineering and Science University), Shibpur, Howrah for creating the interest in various topics of Engineering Mathematics when I was pursuing the bachelor degree in the branch of Electronics and Telecommunication Engineering 40 years ago. I am indebted to Late Professor C. K. Majumdar of the Department of Physics of the University of Calcutta for lighting the fire for Theoretical Physics. I express my gratitude to Professors H. L. Hartnagel, D. Bimberg, W. L. Freeman, and W. Schommers for various academic interactions spanning the last two decades. The well-known scientist Late Professor P. N. Butcher has been a steady hidden force since 1987, before his demise, with respect to our scripting the series in band structure-dependent properties of nanostructured materials. He insisted to me repeatedly regarding it and to tune with his high rigorous academic standard, my colleagues and I wrote the Einstein Relation in Compound Semiconductors and their Nanostructures, Springer Series in Materials Science, Vol. 116, 2009 as the first one, Photoemission from Optoelectronic Materials and their Nanostructures, Springer Series in Nanostructure Science and Technology, 2009 as the second one, "Thermoelectric Power in Nano-Structured Materials Under Strong Magnetic Fields", Springer Series in Materials Science, Vol. 137, 2010 as the third one, "Fowler-Nordheim Field Emission :Effects in Semiconductor Nanostructures", Springer Series in Solid State Sciences, Vol 170, 2012 as the fourth one, and the present monograph as the fifth one.

I offer special thanks to Late Mr. N. Guhachoudhury of Jadavpur University for instilling in me the thought that the *academic output* = ((*desire* χ *determination* χ *dedication*) – (*false enhanced self-ego pretending like a true friend although a real unrecognizable foe*)). I must not allow even a thank you to my beloved better half for really forming the backbone of my long unperturbed research career, since in accordance with Sanatan Hindu Dharma, the fusion of marriage has transformed us to form a single entity, where the individuality is being lost. I am grateful to all the members of my research team for not only quantum confining me in the infinitely deep quantum wells of Ramanujan and Rabindranath but also inspiring me in the real sense of the term to teach quantum mechanics and related topics from eight volume classics of Greiner et. al. I must express my gratitude to Mr. N. Paitya, one of the strong member of my research group, for critically reading the manuscript and offering important suggestions for the betterment of the book. I offered my special thanks to the Assistant Professors and Ph.D. scholars of my present Department for overall supervision of the book in its last phase before sending it to Dr. C. Ascheron, Executive Editor Physics, Springer Verlag. Myself and Dr. D. De of the Department of Computer Science and Engineering, West Bengal University of Technology are grateful to the University Grant Commission for sanctioning the research project No-F 40-469/2011(SR) and Department of Science and Technology for further sanctioning the project SERC/ET-0213/2011 respectively, under which this book has been completed. Last but not the least, I offer special thanks to my life long time tested friend Mr. B. Nag of Applied Physics Department for motivating me during rather turbulent moments of my academic career.

Joint acknowledgments

As always. we are grateful to Dr. C. Ascheron in the real sense of the term for his inspiration and priceless technical assistance from the very start of our first book from Springer. We owe a lot to Ms. A. Duhm, Associate Editor Physics, Springer and Mrs. E. Suer, assistant to Dr. Ascheron. Naturally, the authors are responsible for non-imaginative shortcomings. We firmly believe that our Mother Nature has propelled this joint collaboration in her own unseen way in spite of several insurmountable obstacles.

Bangalore, India S. Bhattyacharya
Tripura, India K. P. Ghatak

References

1. G.E. Simmons, *Differential Equations with Application and Historical Notes, International Series in Pure and Applied Mathematics* (McGraw-Hill, New York, 1991)
2. H. Huff (ed.), *Into the Nano Era—Moore's Law beyond Planar Silicon CMOS*, Springer Series in Materials Science, vol. 106 (Springer, Berlin, 2009)
3. S. Adachi, J. Appl. Phys. **58**, R11 (1985)
4. S. Bhattacharya, K.P. Ghatak, *Fowler-Nordheim Field Emission: Effects in Semiconductor Nanostructures.* Series in Solid State Sciences, vol. **170**, (Springer, Berlin, 2012)
5. K.P. Ghatak, D. De, S. Bhattacharya, *Photoemission from Optoelectronic materials and Their Nanostructures.* Series in Nanostructure Science and Technology (Springer, New York, 2009a)
6. S. Choudhury, L.J. Singh, K.P. Ghatak, Nanotechnology **15**, 180 (2004)
7. K.P. Ghatak, J.P. Banerjee, D. Bhattacharya, Nanotechnology **7**, 110 (1996)

8. S. Bhattacharya, S. Choudhury. K.P. Ghatak. Superlatt. Microstruct. **48,** 257 (2010)
9. K.P. Ghatak, S. Bhattacharya, S. Pahari, D. De, S. Ghosh, M. Mitra, Ann. Phys. **17**, 195 (2008)
10. S. Pahari, S. Bhattacharya, K.P. Ghatak, J. Comput. Theor. Nanosci. (Invited Paper) **6**, 2088 (2009)
11. R.W. Cunningham, Phys. Rev. **167**, 761 (1968)
12. M. Krieehbaum, P. Kocevar, H. Pascher, G. Bauer, IEEE QE **24**, 1727 (1988)
13. M.S. Lundstrom, J. Guo, *Nanoscale Transistors, Device Physics, Modeling and Simulation* (Springer, New York, 2006)
14. R. Saito, G. Dresselhaus, M.S. Dresselhaus, *Physical Properties of Carbon Nanotubes* (Imperial College Press, London, 1998)
15. X. Yang, J. Ni, Phys. Rev. B **72**, 195426 (2005)
16. W. Mintmire, C.T. White, Phys. Rev. Lett. **81**, 2506 (1998)
17. D.G. Seiler, B.D. Bajaj, A.E. Stephens, Phys. Rev. B **16**, 2822 (1977)
18. A.V. Germaneko, G.M. Minkov, Phys. Stat. Sol. **184**, 9 (1994)
19. G.L. Bir, G.E. Pikus, *Symmetry and Strain—Induced effects in Semiconductors* (Nauka, Moscow, 1972). (in Russian)
20. M. Mondal, K.P. Ghatak, Phys. Stat. Sol. **135**, K21 (1986)
21. C.C. Wu, C.J. Lin, J. Low, Temp. Phys. **57**, 469 (1984)
22. G.P. Chuiko, Sov. Phys. Semi. **19**, 1381 (1985)
23. Y. Yamada, J. Phys. Soc. Jpn. **35**, 1600 (1973)
24. D.G. Seiler, W.M. Beeker, K.M. Roth, Phys. Rev. **1**, 764 (1970)
25. S. Bhattacharya, D. De, S.M. Adhikari, K.P. Ghatak, Superlatt. Microstruc. **51**, 203 (2012)
26. K.P. Ghatak, M. Mondal, Z.F. Naturforschung **41A**, 821 (1986)
27. A.N. Chakravarti, A.K. Choudhury, K.P. Ghatak, S. Ghosh, A. Dhar, Appl. Phys. **25**, 105 (1981)
28. P.K. Chakraborty, G.C. Datta, K.P. Ghatak, Phys Scrip **68**, 368 (2003)
29. K.P. Ghatak, S. Bhattacharya, S.K. Biswas, A. Dey, A.K. Dasgupta, Phys. Scrip. **75**, 820–836 (2007)
30. K.P. Ghatak, M. Mondal, Z.F. Physik B **B69**, 471 (1988)
31. A.N. Chakravarti, K.P. Ghatak, K.K. Ghosh, S. Ghosh, A. Dhar, Z.F, Physik B. **47**, 149 (1982)
32. P.K. Bose, N. Paitya, S. Bhattacharya, D. De, S. Saha, K.M. Chatterjee, S. Pahari, K.P. Ghatak, *Quantum Matter* (Invited Paper, 2012)
33. H.A. Lyden, Phys. Rev. **135**, A514 (1964)
34. E.D. Palik, G.B. Wright, in *Semiconductors and Semimetals*, ed. by R.K. Willardson, A.C Beer, **3**, (Academic Press, New York, 1967), p. 421
35. H.I. Zhang, Phys. Rev **1B**, 3450 (1970)
36. M. Mondal, K.P. Ghatak, Phys. Letts. **131 A**, 529 (1988)
37. K.P. Ghatak, B. Mitra, Int. J. Electron. **72**, 541 (1992)
38. B. Mitra, A. Ghoshal, K.P. Ghatak, Nouvo Cimento D **12D**, 891 (1990)
39. K.P. Ghatak, S.N. Biswas, Nonlinear Opt. Quan. Opt. **4**, 347 (1993)
40. K.P. Ghatak, S.N. Biswas, Nonlinear Opt. Quan. Opt. **12**, 83 (1995)

41. K.P. Ghatak, A. Ghoshal, B. Mitra. Nouvo Cimento **14D**, 903 (1992)
42. K.P. Ghatak, A. Ghoshal, B. Mitra. Nouvo Cimento **13D**, 867 (1991)
43. B. Mitra, K.P. Ghatak, Solid State Electron. **32**, 177 (1989)
44. M. Mondal, N. Chattapadhyay, K.P. Ghatak, J. Low Temp. Phys. **66**, 131 (1987)
45. P.N. Hai, W.M. Chen, I.A. Buyanova, H.P. Xin, CWTu Appl. Phys. Lett. **77**, 1843 (2000)
46. D.P. DiVincenzo, E.J. Mele, Phys. Rev. B **29**, 1685 (1984)
47. P. Perlin, E. Litwin-Staszewska, B. Suchanek, W. Knap, J. Camassel, T. Suski, R. Piotrzkowski, I. Grzegory, S. Porowski, E. Kaminska, J.C. Chervin, Appl. Phys. Lett. **68**, 1114 (1996)
48. G.E. Smith, Phys. Rev. Lett. **9**, 487 (1962)
49. D. Schneider, D. Rurup, A. Plichta, H.-U. Grubert, A. Schlachetzki, K. Hansen, Z. Phys. B **95**, 281 (1994)
50. F. Masia, G. Pettinari, A. Polimeni, M. Felici, A. Miriametro, M. Capizzi, A. Lindsay, S.B. Healy, E.P. O'Reilly, A. Cristofoli, G. Bais, M. Piccin, S. Rubini, F. Martelli, A. Franciosi, P.J. Klar, K. Volz, W. Stolz, Phys. Rev. B **73**, 073201 (2006)
51. V.K. Arora, H. Jeafarian, Phys. Rev. B. **13**, 4457 (1976)
52. S.E. Ostapov, V.V. Zhikharevich, V.G. Deibuk, Semicond. Phys. Quan. Electron. Optoelectron. **9**, 29 (2006)
53. M.J. Aubin, L.G. Caron, J.-P. Jay-Gerin, Phys. Rev. B **15**, 3872 (1977)
54. S.L. Sewall, R.R. Cooney, P. Kambhampati, Appl. Phys. Lett. **94**, 243116 (2009)
55. K. Tanaka, N. Kotera, *in 20th International Conference on Indium Phosphide and Related Materials, 25–29 May 2008,* Versailles, France, 2008, pp. 1–4
56. M. Singh, P.R. Wallace, S.D. Jog, J. Erushanov, J. Phys. Chem. Solids **45**, 409 (1984)
57. W. Zawadzki, Adv. Phys. **23**, 435 (1974)
58. T. Ando, H. Fowler, F. Stern, Rev. Mod. Phys. **54**, 437 (1982)
59. K.H.J. Buschow, F.R. de Boer, *Physics of Magnetism and Magnetic Materials* (Springer, New York, 2003)
60. D. Sellmyer, R. Skomski (eds.), *Advanced Magnetic Nanostructures* (Springer, New York, 2005)
61. J.A.C. Bland, B. Heinrich (eds.), *Ultrathin Magnetic Structures III: Fundamentals of Nanomagnetism (Pt. 3)* (Springer, Berlin, 2005)
62. N. Miura, *Physics of Semiconductors in High Magnetic Fields, Series on Semiconductor Science and Technology* (Oxford University Press, New York, 2007)
63. S. Blundell, *Magnetism in Condensed Matter, Oxford Master Series in Condensed Matter Physics* (Oxford University Press, New York, 2001)
64. K.P. Ghatak, S. Bhattacharya, D. De, *Einstein Relatin in Compound Semiconductors and Their Nanostructures. Series in Materials Science*, vol. 116 (Springer, Berlin, 2009b)
65. S. Iijima, Nature **354**, 56 (1991)

66. K.P. Ghatak, S Bhattacharya, *Thermo Electric Power In Nano structured Materials Strong Magnetic Fields.* Series in Materials Science Vol 137 (Springer, Berlin, 2010)
67. P.K. Basu, *Theory of Optical Process in Semiconductors, Bulk and Microstructures* (Oxford University Press, Oxford, 1997)
68. K.P. Ghatak, S. Bhattacharya, J. Appl. Phys. **102**, 073704 (2007)
69. K.P. Ghatk, S. Bhattacharya, S. Bhowmik, R. Benedictus, S. Chowdhury, J. Appl. Phys. **103**, 094314 (2008)
70. W. Zawadzki, B. Lax, Phys. Rev. Lett. **16**, 1001 (1966)
71. K.P. Ghatak, J.P. Banerjee, B. Goswami, B. Nag, Nonlinear Opt. Quan. Opt. **16**, 241 (1996b)
72. M. Mondal, K.P. Ghatak, Phys. Stat. Sol. **133**, K67 (1986)
73. M. Mondal, K.P. Ghatak, Phys. Stat. Sol. **147**, K179 (1988)
74. B. Mitra, A. Ghoshal, K.P. Ghatak, Phys. Stat. Sol. **154**, K147 (1989)
75. B. Mitra, K.P. Ghatak, Phys. Stat. Sol. **164**, K13 (1991)
76. K.P. Ghatak, B. Mitra, Int. J. Electron. **70**, 345 (1991)
77. K.P. Ghatak, B. Goswami, M. Mitra, B. Nag, Nonlinear Opt. Quan. Opt. **16**, 9 (1996)
78. P.K. Chakraborty, A. Sinha, S. Bhattacharya, K.P. Ghatak, Physica B **390**, 325 (2007)
79. P.K. Chakraborty, K.P. Ghatak, J. Phys. Chem. Solids **62**, 1061 (2001a)
80. P.K. Chakraborty, K.P. Ghatak, Phys. Letts. A **288**, 335 (2001b)
81. P.K. Chakraborty, K.P. Ghatak, Phys. D. Appl. Phys. **32**, 2438 (1999)
82. E.A. Arushanov, A.F. Knyazev, A.N. Natepov, S.T. Radautsan, Sov. Phys. Semicond. **15**, 828 (1981)

Contents

Part I
Influence of Quantum Confinement on the Effective Electron Mass (EEM) in Non-Parabolic Semiconductors

Chapter 1
The EEM in Ultrathin Films (UFs) of Nonparabolic Semiconductors

1.1 Introduction

The concept of the effective mass of the carriers in semiconductors is one of the basic pillars in the realm of solid state and related sciences [1]. It must be noted that among the various definitions of the effective electron mass (e.g effective acceleration mass, density-of-state effective mass, concentration effective mass, conductivity effective mass, Faraday rotation effective mass, etc) [2], it is the effective momentum mass that should be regarded as the basic quantity [3]. This is due to the fact that it is this mass which appears in the description of transport phenomena and all other properties of the conduction electrons in a semiconductor with arbitrary band nonparabolicity [3]. It can be shown that it is the effective momentum mass which enters in various transport coefficients and plays the most dominant role in explaining the experimental results of different scattering mechanisms through Boltzmann's transport equation [4, 5]. The carrier degeneracy in semiconductors influences the effective mass when it is energy dependent. Under degenerate conditions, only the electrons at the Fermi surface of n-type semiconductors participate in the conduction process and hence, the effective momentum mass of the electrons (EEM) corresponding to the Fermi level would be of interest in electron transport under such conditions. The Fermi energy is again determined by the carrier energy spectrum and the electron statistics and therefore, these two features would determine the dependence of the EEM in degenerate n-type semiconductors under the degree of carrier degeneracy. In recent years, various energy wave vector dispersion relations have been proposed [6–38] which have created the interest in studying the EEM in such materials under external conditions. The nature of these variations has been investigated in the literature [39–85]. Some of the significant features, which have emerged from these studies, are:

(a) The EEM increases monotonically with electron concentration.
(b) The EEM increases with doping in heavily doped materials in the presence of band tails.

S. Bhattacharya and K. P. Ghatak, *Effective Electron Mass in Low-Dimensional Semiconductors*, Springer Series in Materials Science 167,
DOI: 10.1007/978-3-642-31248-9_1, © Springer-Verlag Berlin Heidelberg 2013

(c) The nature of variations is significantly influenced by the energy band constants of various materials having different band structures.
(d) The EEM oscillates with inverse quantizing magnetic field due to SdH effect. The EEM in Bismuth under magnetic quantization depends both on the Fermi energy and on the magnetic quantum number due to the presence of band nonparabolicity only.
(e) The EEM increases with the magnitude of the quantizing electric field in n-channel inversion layers of III-V semiconductors and depend on the subband index for both low and high electric field limits.
(f) The EEM in ultrathin films of nonlinear optical materials depends on the Fermi energy and size quantum numbers due to the specific dispersion relations.
(g) The EEM has significantly different values in superlattices and also in the presence of quantum confined superlattices of small gap semiconductors with graded interfaces.

In recent years, with the advent of fine lithographical methods [86, 87] molecular beam epitaxy [88], organometallic vapor-phase epitaxy [89], and other experimental techniques, the restriction of the motion of the carriers of bulk materials in one (ultrathin films, NIPI structures, inversion, and accumulation layers), two (nanowires) and three (quantum dots, magnetosize quantized systems, magneto accumulation layers, magneto inversion layers, quantum dot superlattices, magneto ultrathin film superlattices, and magneto NIPI structures) dimensions have in the last few years, attracted much attention not only for their potential in uncovering new phenomena in nanoscience but also for their interesting quantum device applications [90–93]. In ultrathin films (UFs), the restriction of the motion of the carriers in the direction normal to the film (say, the z direction) may be viewed as carrier confinement in an infinitely deep 1D rectangular potential well, leading to quantization [known as quantum size effect (QSE)] of the wave vector of the carrier along the direction of the potential well, allowing 2D carrier transport parallel to the surface of the film representing new physical features not exhibited in bulk semiconductors [94–98]. The low-dimensional heterostructures based on various materials are widely investigated because of the enhancement of carrier mobility [99].These properties make such structures suitable for applications in ultrathin film lasers [100], heterojunction FETs [101, 102], high-speed digital networks [103–106], high-frequency microwave circuits [107], optical modulators [108], optical switching systems [109], and other devices. The constant energy 3D wave-vector space of bulk semiconductors becomes 2D wave-vector surface in UFs due to dimensional quantization. Thus, the concept of reduction of symmetry of the wave-vector space and its consequence can unlock the physics of low-dimensional structures.

In this chapter, we study the EEM in UFs of nonparabolic semiconductors having different band structures. At first we shall investigate the EEM in UFs of nonlinear optical compounds which are being used in nonlinear optics and light emitting diodes [110]. The quasi-cubic model can be used to investigate the symmetric properties of both the bands at the zone center of wave vector space of the same compound. Including the anisotropic crystal potential in the Hamiltonian, and special features of the nonlinear optical compounds, Kildal [111] formulated the electron dispersion law under the assumptions of isotropic momentum matrix element and the isotropic spin-orbit splitting constant, respectively, although the anisotropies in the two aforementioned band constants are the significant physical features of the said materials [112–114]. In Sect. 1.2.1, the EEM in UFs of nonlinear optical semiconductors has been investigated by considering the combined influence of the anisotropies of the said energy band constants together with the inclusion of the crystal field splitting respectively within the framework of *k.p* formalism. The III-V compounds find applications in infrared detectors [115], quantum dot light emitting diodes [116], quantum cascade lasers [117], ultrathin film wires [118], optoelectronic sensors [119], high electron mobility transistors [120], etc. The electron energy spectrum of III-V semiconductors can be described by the three- and two-band models of Kane [121, 122], together with the models of Stillman et al. [123], Newson and Kurobe [124] and, Palik et al. [125] respectively. In this context it may be noted that the ternary and quaternary compounds enjoy the singular position in the entire spectrum of optoelectronic materials. The ternary alloy $Hg_{1-x}Cd_xTe$ is a classic narrow gap compound. The band gap of this ternary alloy can be varied to cover the spectral range from 0.8 to over 30 μm [126] by adjusting the alloy composition. $Hg_{1-x}Cd_xTe$ finds extensive applications in infrared detector materials and photovoltaic detector arrays in the 8–12 μm wave bands [127]. The above uses have generated the $Hg_{1-x}Cd_xTe$ technology for the experimental realization of high mobility single crystal with specially prepared surfaces. The same compound has emerged to be the optimum choice for illuminating the narrow subband physics because the relevant material constants can easily be experimentally measured [128]. Besides, the quaternary alloy $In_{1-x}Ga_xAs_yP_{1-y}$ lattice matched to InP, also finds wide use in the fabrication of avalanche photodetectors [129], hetero-junction lasers [130], light emitting diodes [131] and avalanche photodiodes[132], field effect transistors, detectors, switches, modulators, solar cells, filters, and new types of integrated optical devices are made from the quaternary systems [133]. It may be noted that all types of band models as discussed for III-V semiconductors are also applicable for ternary and quaternary compounds. In Sect. 1.2.2, the EEM in UFs of III-V, ternary and quaternary semiconductors has been studied in accordance with the said band models and the simplified results for wide gap materials having parabolic energy bands under certain limiting conditions have further been demonstrated as a special case and thus confirming the compatibility test.

The II-VI semiconductors are being used in nanoribbons, blue green diode lasers, photosensitive thin films, infrared detectors, ultra high-speed bipolar transistors, fiber optic communications, microwave devices, solar cells, semiconductor gamma-ray detector arrays, semiconductor detector gamma camera and allow for a greater density of data storage on optically addressed compact discs [134–141]. The carrier energy spectra in II-VI compounds are defined by the Hopfield model [142] where the splitting of the two-spin states by the spin-orbit coupling and the crystalline field has been taken into account. The Sect. 1.2.3 contains the investigation of the EEM in UFs of II-VI compounds.

In recent years, Bismuth (Bi) nanolines have been fabricated and Bi also finds use in array of antennas which leads to the interaction of electromagnetic waves with such Bi-nanowires [143, 144]. Several dispersion relations of the carriers have been proposed for Bi. Shoenberg [145, 146] experimentally verified that the de Haas-Van Alphen and cyclotron resonance experiments supported the ellipsoidal parabolic model of Bi, although, the magnetic field dependence of many physical properties of Bi supports the two-band model [147].The experimental investigations on the magneto-optical and the ultrasonic quantum oscillations support the Lax ellipsoidal nonparabolic model [147]. Kao [148], Dinger and Lawson [149] and Koch and Jensen [150] demonstrated that the Cohen model [151] is in conformity with the experimental results in a better way. Besides, the hybrid model of bismuth, as developed by Takoka et al., also finds use in the literature [152]. McClure and Choi [153] derived a new model of Bi and they showed that it can explain the data for a large number of magneto-oscillatory and resonance experiments.

In Sect. 1.2.4, the EEM in UFs of Bi has been formulated in accordance with the aforementioned energy band models for the purpose of relative assessment. Besides, under certain limiting conditions all the results for all the models of 2D systems are reduced to the well-known result of the EEM in UFs of wide gap materials. This above statement exhibits the compatibility test of our theoretical analysis.

Lead chalcogenides (PbTe, PbSe, and PbS) are IV-VI nonparabolic semiconductors whose studies over several decades have been motivated by their importance in infrared IR detectors, lasers, light-emitting devices, photovoltaics, and high temperature thermoelectrics [154–158]. PbTe, in particular, is the end compound of several ternary and quaternary high performance high temperature thermoelectric materials [159–163]. It has been used not only as bulk but also as films [164–167], ultrathin films [168] superlattices [169, 170] nanowires [171] and colloidal and embedded nanocrystals [172–175], and PbTe films doped with various impurities have also been investigated [176–183] These studies revealed some of the interesting features that had been seen in bulk PbTe, such as Fermi level pinning and, in the case of superconductivity [184]. In Sect. 1.2.5, the EEM in UFs of IV-VI semiconductors has been studied taking PbTe, PbSe, and PbS as examples.

The stressed semiconductors are being investigated for strained silicon transistors, quantum cascade lasers, semiconductor strain gages, thermal detectors, and strained-layer structures [185–188]. The EEM in UFs of stressed compounds (taking stressed n-InSb as an example) has been investigated in Sect. 1.2.6 The vacuum deposited Tellurium (Te) has been used as the semiconductor layer in thin-body transistors (TFT) [189] which is being used in CO_2 laser detectors [190], electronic imaging, strain sensitive devices [191, 192], and multichannel Bragg cell [193]. Section 1.2.7 contains the investigation of EEM in UFs of Tellurium.

The n-Gallium Phosphide (n-GaP) is being used in quantum dot light emitting diode [194], high efficiency yellow solid state lamps, light sources, high peak current pulse for high gain tubes. The green and yellow light emitting diodes made of nitrogen-doped n-GaP possess a longer device life at high drive currents [195–197]. In Sect. 1.2.8, the EEM in UFs of n-GaP has been studied. The Platinum Antimonide ($PtSb_2$) finds application in device miniaturization, colloidal nanoparticle synthesis, sensors and detector materials and thermo-photovoltaic devices [198–200]. Section 1.2.9 explores the EEM in UFs of $PtSb_2$.Bismuth telluride

(Bi_2Te_3) was first identified as a material for thermoelectric refrigeration in 1954 [201] and its physical properties were later improved by the addition of bismuth selenide and antimony telluride to form solid solutions [202–206]. The alloys of Bi_2Te_3 are useful compounds for the thermoelectric industry and have been investigated in the literature [202–206]. In Sect. 1.2.10, the EEM in UFs of Bi_2Te_3 has been considered.

The usefulness of elemental semiconductor Germanium is already well known since the inception of transistor technology and, it is also being used in memory circuits, single photon detectors, single photon avalanche diode, ultrafast optical switch, THz lasers and THz spectrometers [207–210]. In Sect. 1.2.11, the EEM has been studied in UFs of Ge. Gallium Antimonide (GaSb) finds applications in the fiber optic transmission window, heterojunctions, and ultrathin films. A complementary heterojunction field effect transistor in which the channels for the p-FET device and the n-FET device forming the complementary FET are formed from GaSb. The band gap energy of GaSb makes it suitable for low power operation [211–216]. In Sect. 1.2.12, the EEM in UFs of GaSb has been studied. Section 1.3 contains the result and discussions pertaining to this chapter. The last Sect. 1.4 contains open research problems.

1.2 Theoretical Background

1.2.1 The EEM in UFs of Nonlinear Optical Semiconductors

The form of *k.p* matrix for nonlinear optical compounds can be expressed extending Bodnar [112] as

$$H = \begin{bmatrix} H_1 & H_2 \\ H_2^+ & H_1 \end{bmatrix} \tag{1.1}$$

where,

$$H_1 \equiv \begin{bmatrix} E_{g0} & 0 & P_\parallel k_z & 0 \\ 0 & (-2\Delta_\parallel/3) & \left(\sqrt{2}\Delta_\perp/3\right) & 0 \\ P_\parallel k_z & \left(\sqrt{2}\Delta_\perp/3\right) & -\left(\delta + \frac{1}{3}\Delta_\parallel\right) & 0 \\ 0 & 0 & 0 & 0 \end{bmatrix}$$

$$H_2 \equiv \begin{bmatrix} 0 & -f_{,+} & 0 & f_{,-} \\ f_{,+} & 0 & 0 & 0 \\ 0 & 0 & 0 & 0 \\ f_{,+} & 0 & 0 & 0 \end{bmatrix}$$

in which E_{g_0} is the band gap in the absence of any field, $P_\parallel$ and $P_\perp$ are the momentum matrix elements parallel and perpendicular to the direction of crystal axis respectively,

δ is the crystal-field splitting constant, $\Delta_\parallel$ and $\Delta_\perp$ are the spin-orbit splitting constants parallel and perpendicular to the C-axis respectively, $f_{,\pm} \equiv \left(P_\perp/\sqrt{2}\right)\left(k_x \pm i k_y\right)$ and $i = \sqrt{-1}$. Thus, neglecting the contribution of the higher bands and the free electron term, the diagonalization of the above matrix leads to the dispersion relation of the conduction electrons in bulk specimens of nonlinear optical semiconductors as

$$\gamma(E) = f_1(E)k_s^2 + f_2(E)k_z^2 \tag{1.2}$$

where,

$$\gamma(E) \equiv E(E+E_{g0})\left[(E+E_{g0})(E+E_{g0}+\Delta_\parallel) + \delta\left(E+E_{g_o}+\frac{2}{3}\Delta_\parallel\right) + \frac{2}{9}(\Delta_\parallel^2 - \Delta_\perp^2)\right],$$

E is the total energy of the electron as measured from the edge of the conduction band in the vertically upward direction in the absence of any quantization, $k_s^2 = k_x^2 + k_y^2$,

$$f_1(E) \equiv \frac{\hbar^2 E_{g0}\left(E_{g0}+\Delta_\perp\right)}{\left[2m_\perp^*\left(E_{g0}+\frac{2}{3}\Delta_\parallel\right)\right]} \times \left[\delta\left(E+E_{g0}+\frac{1}{3}\Delta_\parallel\right) + (E+E_{g0})\left(E+E_{g0}+\frac{2}{3}\Delta_\parallel\right) + \frac{1}{9}(\Delta_\parallel^2 - \Delta_\perp^2)\right]$$

$$f_2(E) \equiv \frac{\hbar^2 E_{g0}\left(E_{g0}+\Delta_\parallel\right)}{\left[2m_\parallel^*\left(E_{g0}+\frac{2}{3}\Delta_\parallel\right)\right]}\left[(E+E_{g0})\left(E+E_{g0}+\frac{2}{3}\Delta_\parallel\right)\right], \quad \hbar = h/2\pi,$$

h is *Planck's constant* and $m_\parallel^*$ and $m_\perp^*$ are the *longitudinal and transverse effective electron masses* at the edge of the *conduction band respectively*.

For dimensional quantization along z-direction, the dispersion relation of the 2D electrons in this case can be written following (1.2) as

$$\psi_1(E) = \psi_2(E)k_s^2 + \psi_3(E)(n_z\pi/d_z)^2 \tag{1.3}$$

where $\psi_1(E) = \gamma(E)$, $\psi_2(E) = f_1(E)$, $\psi_3(E) = f_2(E)$, $n_z(= 1, 2, 3, \ldots)$ and d_z are the size quantum number and the nano-thickness along the z-direction respectively.

The EEM is defined as the ratio of the electron momentum to the group velocity. The EEM at the Fermi level in the xy-plane can be written as

$$m^*(E_F, n_z) = \hbar^2 k_s \left.\frac{\partial k_s}{\partial E}\right|_{E=E_{Fs}} \tag{1.4}$$

where E_{Fs} is the Fermi energy in the presence of size quantization as measured from the edge of the conduction band in the vertically upward direction in the absence of any quantization. From (1.3) and (1.4), the EEM in this case can be written as

$$m^*(E_{Fs}, n_z) = \left(\frac{\hbar^2}{2}\right)[\psi_2(E_{Fs})]^{-2}\left[\psi_2(E_{Fs})\left\{\{\psi_1(E_{Fs})\}' - \{\psi_3(E_{Fs})\}'\left(\frac{n_z\pi}{d_z}\right)^2\right\}\right. \\ \left. - \left\{\psi_1(E_{Fs}) - \psi_3(E_{Fs})\left(\frac{n_z\pi}{d_z}\right)^2\right\}\{\psi_2(E_{Fs})\}'\right] \tag{1.5}$$

where, the primes denote the differentiation of the differentiable functions with respect to Fermi energy. Thus, we observe that the EEM is the function of size quantum number and the Fermi energy due to the combined influence of the crystal-field splitting constant and the anisotropic spin-orbit splitting constants respectively. To study the dependence of the EEM as a function of electron concentration per unit area we have to formulate the corresponding density-of-states function (DOS).

The general expression of the total 2D DOS ($N_{2DT}(E)$)in this case is given by

$$N_{2DT}(E) = \frac{2g_v}{(2\pi)^2}\sum_{n_z=1}^{n_{zmax}} \frac{\partial A(E, n_z)}{\partial E} H\left(E - E_{n_z}\right) \tag{1.6}$$

where, g_v is the valley degeneracy, $A(E, n_z)$ is the area of the constant energy 2D wave vector space for UFs, $H(E - E_{n_z})$is the Heaviside step function and (E_{n_z}) is the corresponding subband energy. Using (1.3) and (1.6), the expression of the $N_{2DT}(E)$ for UFs of nonlinear optical semiconductors can be written as

$$N_{2DT}(E) = \left(\frac{g_v}{2\pi}\right)\sum_{n_z=1}^{n_{zmax}} [\psi_2(E)]^{-2}\left[\psi_2(E)\left\{\{\psi_1(E)\}' - \{\psi_3(E)\}'\left(\frac{n_z\pi}{d_z}\right)^2\right\}\right. \\ \left. - \left\{\psi_1(E) - \psi_3(E)\left(\frac{n_z\pi}{d_z}\right)^2\right\}\{\psi_2(E)\}'\right] H(E - E_{n_{z1}}) \tag{1.7}$$

where, the subband energies ($E_{n_{z1}}$) in this case is given by

$$\psi_1(E_{n_{z1}}) = \psi_2(E_{n_{z1}})(n_z\pi/d_z)^2 \tag{1.8}$$

Combining (1.7) with the Fermi-Dirac occupation probability factor, integrating between $E_{n_{z1}}$ to infinity and applying the generalized Somerfeld's lemma, the 2D carrier statistics in this case assumes the form

$$n_{2D} = \frac{g_v}{2\pi}\sum_{n_x=1}^{n_{xmax}} \left[T_{51}(E_{Fs}, n_z) + T_{52}(E_{Fs}, n_z)\right] \tag{1.9}$$

where,

$$T_{51}(E_{Fs}, n_z) \equiv \left[\frac{\psi_1(E_{Fs}) - \psi_3(E_{Fs})(n_z\pi/d_z)^2}{\psi_2(E_{Fs})}\right],$$

$$T_{52}(E_{Fs}, n_z) \equiv \sum_{r=1}^{s} L(r)[T_{51}(E_{Fs}, n_z)],$$

$L(r) = 2(k_B T)^{2r}(1 - 2^{1-2r})\xi(2r)\frac{\partial^{2r}}{\partial E_f^{2r}}$, k_B is the Boltzmann constant, T is the temperature, r is the set of real positive integers whose upper limit is s, $\xi(2r)$ is the Zeta function of order 2r [217].

1.2.2 The EEM in UFs of III–V Semiconductors

The dispersion relation of the conduction electrons of III-V compounds are described by the models of Kane (both three and two bands) [121, 122], Stillman et al. [123], Newson and Kurobe [124] and Palik et al. [125] respectively. For the purpose of complete and coherent presentation, the EEM in UFs of III-V semiconductors have also been investigated in accordance with the aforementioned different dispersion relations for the purpose of relative comparison as follows:

(a) The three-band model of Kane

Under the conditions, $\delta = 0$, $\Delta_\parallel = \Delta_\perp = \Delta$ (isotropic spin orbit splitting constant) and $m_\perp^* = m_\perp^* = m_c$ (isotropic effective electron mass at the edge of the conduction band), (1.2) gets simplified into the form

$$\frac{\hbar^2 k^2}{2m_c} = I_{11}(E), I_{11}(E) \equiv \frac{E(E + E_{g_0})(E + E_{g_0} + \Delta)(E_{g_0} + \frac{2}{3}\Delta)}{E_{g_0}(E_{g_0} + \Delta)(E + E_{g_0} + \frac{2}{3}\Delta)} \tag{1.10}$$

which is known as the three-band model of Kane [121, 122] and is often used to study the electronic properties of III-V materials.

Thus, under the conditions $\delta = 0$, $\Delta_\parallel = \Delta_\perp = \Delta$ and $m_\parallel^* = m_\perp^* = m_c$, (1.3) assumes the form

$$\frac{\hbar^2 k_s^2}{2m_c} + \frac{\hbar^2}{2m_c}(n_z\pi/d_z)^2 = I_{11}(E) \tag{1.11}$$

Using (1.11) and (1.4), the EEM in *x–y* plane for this case can be written as

$$m^*(E_{Fs}) = m_c\{I_{11}(E_{Fs})\}' \tag{1.12}$$

It is worth noting that the EEM in this case is a function of Fermi energy alone and is independent of size quantum number.

The total 2D DOS function can be written as

$$N_{2DT}(E) = \left(\frac{m_c g_v}{\pi \hbar^2}\right) \sum_{n_z=1}^{n_{zmax}} \left\{[I_{11}(E)]' H\left(E - E_{n_{z2}}\right)\right\} \tag{1.13}$$

where, the subband energies $E_{n_{z2}}$ can be expressed as

$$I_{11}(E_{n_{z2}}) = \frac{\hbar^2}{2m_c}(n_z\pi/d_z)^2 \tag{1.14}$$

The 2D carrier concentration assumes the form

$$n_{2D} = \frac{m_c g_v}{\pi \hbar^2} \sum_{n_z=1}^{n_{zmax}} [T_{53}(E_{Fs}, n_z) + T_{54}(E_{Fs}, n_z)] \tag{1.15}$$

where

$$T_{53}(E_{Fs}, n_z) \equiv \left[I_{11}(E_{Fs}) - \frac{\hbar^2}{2m_c}\left(\frac{n_z\pi}{d_z}\right)^2\right] \text{ and}$$

$$T_{54}(E_{Fs}, n_z) \equiv \sum_{r=1}^{s} L(r) T_{53}(E_{Fs}, n_z).$$

Under the inequalities $\Delta \gg E_{g_0}$ or $\Delta \ll E_{g_0}$ (1.10) can be expressed as

$$E(1 + \alpha E) = \frac{\hbar^2 k^2}{2m_c} \tag{1.16}$$

where, $\alpha \equiv 1/E_{g_0}$ and is known as band nonparabolicity.
It may be noted that (1.16) is the well-known two-band model of Kane and is used in the literature to study the physical properties of those III-V and optoelectronic materials whose energy band structures obey the aforementioned inequalities.
Under the said inequalities (1.11) assumes the form

$$E(1 + \alpha E) = \frac{\hbar^2 k_s^2}{2m_c} + \frac{\hbar^2}{2m_c}\left(\frac{n_z\pi}{d_z}\right)^2 \tag{1.17}$$

The EEM in this case can be written as

$$m^*(E_{Fs}) = m_c(1 + 2\alpha E_{Fs}) \tag{1.18}$$

Thus, we observe that the EEM in the present case is a function of Fermi energy only due to the presence of band nonparabolicity.
The total 2D DOS function assumes the form

$$N_{2DT}(E) = \frac{m_c g_v}{\pi \hbar^2} \sum_{n_z=1}^{n_{zmax}} (1 + 2\alpha E) H\left(E - E_{n_{z3}}\right) \tag{1.19}$$

where, the subband energy ($E_{n_{z3}}$) can be expressed as

$$\frac{\hbar^2}{2m_c}(n_z \pi / d_z)^2 = E_{n_{z3}}\left(1 + \alpha E_{n_{z3}}\right) \tag{1.20}$$

The 2D electron statistics can be written as

$$n_{2D} = \frac{m_c g_v}{\pi \hbar^2} \sum_{n_z=1}^{n_{zmax}} \int_{E_{n_{z3}}}^{\infty} \frac{(1 + 2\alpha E) dE}{1 + exp\left(\frac{E - E_{Fs}}{k_B T}\right)}$$

$$= \frac{m_c k_B T g_v}{\pi \hbar^2} \sum_{n_z=1}^{n_{zmax}} \left[(1 + 2\alpha E_{n_{z3}}) F_0(\eta_{n_1}) + 2\alpha k_B T F_1(\eta_{n_1})\right] \tag{1.21}$$

where, $\eta_{n_1} \equiv (E_{Fs} - E_{n_{z3}})/k_B T$ and $F_j(\eta)$ is the one-parameter Fermi-Dirac integral of order j which can be written [218, 219] as

$$F_j(\eta) = \left(\frac{1}{\Gamma(j+1)}\right) \int_0^{\infty} \frac{x^j dx}{1 + \exp(x - \eta)}, \quad j > -1 \tag{1.22}$$

or for all j, analytically continued as a complex contour integral around the negative *x-axis*

$$F_j(\eta) = \left(\frac{\Gamma(-j)}{2\pi\sqrt{-1}}\right) \int_{-\infty}^{+0} \frac{x^j dx}{1 + \exp(-x - \eta)} \tag{1.23}$$

where η is the dimensionless x independent variable.
Under the condition $\alpha \to 0$, the expressions of total 2D DOS, for UFs whose bulk electrons are defined by the isotropic parabolic energy bands can, be written as

$$N_{2DT}(E) = \frac{m_c g_v}{\pi \hbar^2} \sum_{n_z=1}^{n_{zmax}} H\left(E - E_{n_{zp}}\right) \tag{1.24}$$

The subband energy ($E_{n_{zp}}$), the EEM, and the n_{2D} can respectively be expressed as

$$E_{n_{zp}} = \frac{\hbar^2}{2m_c}\left(\frac{n_z \pi}{d_z}\right)^2 \tag{1.25}$$

$$m^*(E_{Fs}) = m_c \tag{1.26}$$

and

$$n_{2D} = \frac{m_c k_b T g_v}{\pi \hbar^2} \sum_{n_z=1}^{n_{zmax}} F_0(\eta_{n_2}) \tag{1.27}$$

where, $\eta_{n_2} \equiv \frac{1}{k_B T}\left[E_{Fs} - \frac{\hbar^2}{2m_c}\left(\frac{n_z \pi}{d_z}\right)^2\right]$

It may be noted that the results of this section are already well known in the literature [220].

(b) The model of Stillman et al.

In accordance with the model of Stillman et al. [123], the electron dispersion law of III-V materials assumes the form

$$E = \bar{t}_{11} k^2 - \bar{t}_{12} k^4 \tag{1.28}$$

where,

$$\bar{t}_{11} \equiv \frac{\hbar^2}{2m_c}; \bar{t}_{12} \equiv \left(1 - \frac{m_c}{m_0}\right)^2 \left(\frac{\hbar^2}{2m_c}\right)^2 \times \left[\left(3E_{g0} + 4\Delta + \frac{2\Delta^2}{E_{g0}}\right).\{(E_{g0} + \Delta)(2\Delta + 3E_{g0})\}^{-1}\right]$$

and m_0 is the free electron mass.

Equation (1.28) can be expressed as

$$\frac{\hbar^2 k^2}{2m_c} = I_{12}(E) \tag{1.29}$$

where, $I_{12}(E) \equiv a_{11}\left[1 - (1 - a_{12}E)^{1/2}\right]$, $a_{11} \equiv \left(\frac{\hbar^2 \bar{t}_{11}}{4m_c \bar{t}_{12}}\right)$ and $a_{12} \equiv \frac{4\bar{t}_{12}}{\bar{t}_{11}^2}$.

The 2D electron dispersion relation in this case assumes the form

$$\frac{\hbar^2 k_s^2}{2m_c} + \frac{\hbar^2}{2m_c}(n_z \pi / d_z)^2 = I_{12}(E) \tag{1.30}$$

Using (1.30) and (1.4), the EEM in x–y plane for this case can be written as

$$m^*(E_{Fs}) = m_c\{I_{12}(E_{Fs})\}' \tag{1.31}$$

It appears that the EEM in this case is a function of Fermi energy alone and is independent of size quantum number.

The total 2D DOS function can be written as

$$N_{2DT}(E) = \left(\frac{m_c g_v}{\pi\hbar^2}\right) \sum_{n_z=1}^{n_{zmax}} \left\{[I_{12}(E)]' H(E - E_{n_{z3}})\right\} \tag{1.32}$$

where, the subband energies $E_{n_{z3}}$ can be expressed as

$$I_{12}(E_{n_{z3}}) = \frac{\hbar}{2m_c}(n_z\pi/d_z)^2 \tag{1.33}$$

The 2D carrier concentration assumes the form

$$n_{2D} = \frac{m_c g_v}{\pi\hbar^2} \sum_{n_z=1}^{n_{zmax}} [T_{55}(E_{Fs}, n_z) + T_{56}(E_{Fs}, n_z)] \tag{1.34}$$

where

$$T_{55}(E_{Fs}, n_z) \equiv \left[I_{12}(E_{Fs}) - \frac{\hbar^2}{2m_c}\left(\frac{n_z\pi}{d_z}\right)^2\right] \text{ and}$$

$$T_{56}(E_{Fs}, n_z) \equiv \sum_{r=1}^{s} L(r) T_{55}(E_{Fs}, n_z)$$

(c) Model of Palik et al.

The energy spectrum of the conduction electrons in III-V semiconductors up to the fourth order in effective mass theory, taking into account the interactions of heavy hole, light hole and the split-off holes can be expressed in accordance with the model of Palik et al. [125] as

$$E = \frac{\hbar^2 k^2}{2m_c} - \bar{B}_{11} k^4 \tag{1.35}$$

where

$$\bar{B}_{11} = \left[\frac{\hbar^4}{4E_{g0}(m_c)^2}\right]\left[\frac{1 + \frac{x_{11}^2}{2}}{1 + \frac{x_{11}}{2}}\right](1 - y_{11})^2,$$

$$x_{11} = \left[1 + \left(\frac{\Delta}{E_{g0}}\right)\right]^{-1} \text{ and } y_{11} = \frac{m_c}{m_o}$$

The (1.35) gets simplified as

$$\frac{\hbar^2 k^2}{2m_c} = I_{13}(E) \tag{1.36}$$

where

$$I_{13}(E) = \bar{b}_{12}\left[\bar{a}_{12} - ((\bar{a}_{12})^2 - 4E\bar{B}_{11})^{1/2}\right],$$

$$\bar{a}_{12} = \left(\frac{\hbar^2}{2m_c}\right) \text{ and } \bar{b}_{12} = \left[\frac{\bar{a}_{12}}{2\bar{B}_{11}}\right]$$

The 2D electron dispersion relation in this case assumes the form

$$\frac{\hbar^2 k_s^2}{2m_c} + \frac{\hbar^2}{2m_c}(n_z\pi/d_z)^2 = I_{13}(E) \tag{1.37}$$

Using (1.37) and (1.4), the EEM in x–y plane for this case can be written as

$$m^*(E_{Fs}) = m_c\{I_{13}(E_{Fs})\}' \tag{1.38}$$

It appears that the EEM in this case is a function of Fermi energy alone and is independent of size quantum number.
The total 2D DOS function can be written as

$$N_{2DT}(E) = \left(\frac{m_c g_v}{\pi\hbar^2}\right)\sum_{n_z=1}^{n_{zmax}}\left\{[I_{13}(E)]' H(E - E_{n_{z4}})\right\} \tag{1.39}$$

where, the subband energies $E_{n_{z4}}$ can be expressed as

$$I_{13}(E_{n_{z4}}) = \frac{\hbar}{2m_c}(n_z\pi/d_z)^2 \tag{1.40}$$

The 2D carrier concentration assumes the form

$$n_{2D} = \frac{m_c g_v}{\pi\hbar^2}\sum_{n_z=1}^{n_{zmax}}\left[T_{57}(E_{Fs}, n_z) + T_{58}(E_{Fs}, n_z)\right] \tag{1.41}$$

where

$$T_{57}(E_{Fs}, n_z) \equiv \left[I_{13}(E_{Fs}) - \frac{\hbar^2}{2m_c}\left(\frac{n_z\pi}{d_z}\right)^2\right] \text{ and}$$

$$T_{58}(E_{Fs}, n_z) \equiv \sum_{r=1}^{s} L(r) T_{57}(E_{Fs}, n_z)$$

1.2.3 The EEM in UFs of II–VI Semiconductors

The carrier energy spectra in bulk specimens of II-VI compounds in accordance with Hopfield model [142] can be written as

$$E = a_o' k_s^2 + b_o' k_z^2 \pm \bar{\lambda}_o k_s \tag{1.42}$$

where $a_o' \equiv \hbar^2/2m_\perp^*$, $b_o' \equiv \hbar^2/2m_\perp^*$, and $\bar{\lambda}_o$ represents the splitting of the two-spin states by the spin-orbit coupling and the crystalline field.

The dispersion relation of the conduction electrons of UFs of II-VI materials for dimensional quantization along z-direction can be written following (1.42) as

$$E = a_o' k_s^2 + b_o' \left(\frac{n_z \pi}{d_z}\right)^2 \pm \bar{\lambda}_o k_s \tag{1.43}$$

Using (1.43), the EEM in this case can be written as

$$m^*(E_{Fs}, n_z) = m_\perp^* \left[1 \mp \frac{(\bar{\lambda}_o)}{\left[(\bar{\lambda}_o)^2 - 4a_o' b_o' \left(\frac{n_z \pi}{d_z}\right)^2 + 4a_o' E_{Fs}\right]^{1/2}} \right] \tag{1.44}$$

Thus, we can infer that the EEM in the UFs of II-VI compounds is a function of both the size quantum number and the Fermi energy due to the presence of the term $\bar{\lambda}_o$.

The subband energy $E_{n_{z5}}$ assumes the form

$$E_{n_{z5}} = b_o' (n_z \pi / d_z)^2 \tag{1.45}$$

The area of constant energy 2D quantized surface in this case is given by

$$A_\pm (E, n_z) = \left[\frac{\pi}{2 (a_0')^2} \left[(\overline{\lambda_0})^2 + 2a_0' \left(E - E_{n_{z5}}\right) \pm \overline{\lambda_0} \left[(\overline{\lambda_0})^2 + 4a_0' \left(E - E_{n_{z5}}\right) \right]^{1/2} \right] \right] \tag{1.46}$$

The surface electron concentration under the condition of extreme carrier degeneracy can be expressed in this case as

$$n_{2D} = \frac{2g_v}{2 (2\pi)^2} \sum_{n_z=1}^{n_{z\max}} \left[A_+ \left(E_{Fs}, n_z\right) + A_- \left(E_{Fs}, n_z\right) \right] \tag{1.47}$$

Using (1.46) and (1.47) we get

$$n_{2D} = \frac{g_v m_\perp^*}{\pi \hbar^2} \sum_{n_z=1}^{n_{z\max}} \left(E_{Fs} - E_{n_{z5}} + \left(\overline{\lambda}\right)^2 m_\perp^* \hbar^{-2} \right) \tag{1.48}$$

1.2.4 The EEM in UFs of Bismuth

(a) The McClure and Choi model

The dispersion relation of the carriers in Bi can be written, following the McClure and Choi [153], as

$$E(1+\alpha E) = \frac{p_x^2}{2m_1} + \frac{p_y^2}{2m_2} + \frac{p_z^2}{2m_3} + \frac{p_y^2}{2m_2}\alpha E \left\{1 - \left(\frac{m_2}{m_2'}\right)\right\} + \frac{p_y^4 \alpha}{4m_2 m_2'} - \frac{\alpha p_x^2 p_y^2}{4m_1 m_2} - \frac{\alpha p_y^2 p_z^2}{4m_2 m_3} \tag{1.49}$$

where $p_i \equiv \hbar k_i, i = x, y, z, m_1, m_2$ and m_3 are the effective carrier masses at the band-edge along x, y and z directions respectively and m_2' is the effective-mass tensor component at the top of the valence band (for electrons) or at the bottom of the conduction band (for holes).

The dispersion relation of the conduction electrons in UFs of Bi for dimensional quantization along k_z direction can be written following (1.49) for this model as

$$E(1+\alpha E) = \frac{p_x^2}{2m_1} + \frac{p_y^2}{2m_2} + \frac{\hbar^2}{2m_3}\left(\frac{n_z\pi}{d_z}\right)^2 + \frac{p_y^2}{2m_2}\alpha E \left\{1 - \left(\frac{m_2}{m_2'}\right)\right\} + \frac{p_y^4 \alpha}{4m_2 m_2'} - \frac{\alpha p_x^2 p_y^2}{4m_1 m_2} - \frac{\alpha p_y^2 \hbar^2}{4m_2 m_3}\left(\frac{n_z\pi}{d_z}\right)^2 \tag{1.50}$$

Equation (1.50) can, approximately, be expressed as

$$\gamma_1(E, n_z) = p_1 k_x^2 + q_1(E) k_y^2 + R_1(E, n_z) k_y^4 \tag{1.51}$$

where,

$$\gamma_1(E, n_z) \equiv \left[E(1+\alpha E) - \frac{\hbar^2}{2m_3}\left(\frac{n_z\pi}{d_z}\right)^2 \right], \quad p_1 \equiv \frac{\hbar^2}{2m_1},$$

$$q_1(E) \equiv \frac{\hbar^2}{2m_2}\left[1 + \alpha E \left(1 - \frac{m_2}{m_2'}\right) - \alpha E(1+\alpha E) \right]$$

and

$$R_1(E,n_z) \equiv \left[\frac{\alpha\hbar^4}{4m_2m_2'} + \alpha\left(\frac{\hbar^2}{2m_2}\right)^2\left\{1+\alpha E\left(1-\frac{m_2}{m_2'}\right)-\frac{\alpha\hbar^2}{2m_3}\left(\frac{n_z\pi}{d_z}\right)^2\right\}\right]$$

The area enclosed by (1.51) is defined by the following integral:

$$A(E,n_z) = 4\left[\frac{R_1(E,n_z)}{p_1}\right]^{1/2} \cdot J_1(E,n_z) \tag{1.52}$$

where,

$$J_1(E,n_z) \equiv \int_0^{u_o(E,n_z)} \left[\frac{\gamma_1(E,n_z)}{R_1(E,n_z)} - \frac{q_1(E)k_y^2}{R_1(E,n_z)} - k_y^4\right]^{1/2} dk_y$$

and

$$u_0(E,n_z) \equiv \left[\sqrt{\frac{q_1^2(E)}{4R_1^2(E,n_z)} + \frac{\gamma_1(E,n_z)}{R_1(E,n_z)}} - q_1(E)\right]^{1/2}$$

Thus, the area enclosed can be written as

$$A(E,n_z) = \frac{4}{3}\left[\frac{R_1(E,n_z)}{p_1}\right]^{1/2}\left[a^2(E,n_z)+b^2(E,n_z)\right]^{1/2}$$

$$\left[a^2(E,n_z)F\left[\frac{\pi}{2},l(E,n_z)\right] - \left[a^2(E,n_z)-b^2(E,n_z)\right]E\left[\frac{\pi}{2},l(E,n_z)\right]\right] \tag{1.53}$$

where,

$$a^2(E,n_z) \equiv \frac{q_1(E)}{2R_1(E,n_z)} + \frac{1}{2}\left[\frac{q_1^2(E)}{R_1^2(E,n_z)} + \frac{4\gamma_1(E,n_z)}{R_1(E,n_z)}\right]^{1/2},$$

$$b^2(E,n_z) \equiv \frac{1}{2}\left[\frac{q_1^2(E)}{R_1^2(E,n_z)} + \frac{4\gamma_1(E,n_z)}{R_1(E,n_z)}\right]^{1/2} - \left(\frac{q_1(E)}{2R_1(E,n_z)}\right),$$

$$l(E,n_z) \equiv \frac{b(E,n_z)}{\sqrt{a^2(E,n_z)+b^2(E,n_z)}},\ F\left[\frac{\pi}{2},l(E,n_z)\right] \text{ and } E\left[\frac{\pi}{2},l(E,n_z)\right]$$

are the complete elliptic integral of the first and second kinds respectively [217] Using (1.53), the EEM can be written as

$$m^*(E_{Fs}, n_z) = \left(\frac{2\hbar^2}{3\pi\sqrt{p_1}}\right)[R_3(E, n_z)]|_{E=E_{Fs}} \tag{1.54}$$

where,

$$\begin{aligned}
&R_3(E_{Fs}, n_z)\\
&\equiv \frac{1}{2}\left[R_1(E_{Fs}, n_z)\right]^{-1/2}[R_1(E_{Fs}, n_z)]'[a^2(E_{Fs}, n_z) + p^2(E_{Fs}, n_z)]^{1/2}\\
&\quad \times \left[a^2(E_{Fs}, n_z)F\left(\frac{\pi}{2}, l(E_{Fs}, n_z)\right) - [a^2(E_{Fs}, n_z)\right.\\
&\qquad \left.- b^2(E_{Fs}, n_z)]E\left(\frac{\pi}{2}, l(E_{Fs}, n_z)\right)\right]\\
&\quad + \sqrt{R_1(E_{Fs}, n_z)}\left[a^2(E_{Fs}, n_z) + b^2(E_{Fs}, n_z)\right]^{-1/2}\\
&\quad \times \left[a(E_{Fs}, n_z)(a(E_{Fs}, n_z))' + b(E_{Fs}, n_z)(b(E_{Fs}, n_z))'\right]\\
&\quad \times \left[a^2(E_{Fs}, n_z)F\left(\frac{\pi}{2}, l(E_{Fs}, n_z)\right) - [a^2(E_{Fs}, n_z)\right.\\
&\qquad \left.-b^2(E_{Fs}, n_z)]E\left(\frac{\pi}{2}, l(E_{Fs}, n_z)\right)\right]\\
&\quad + \sqrt{R_1(E_{Fs}, n_z)}[a^2(E_{Fs}, n_z) + b^2(E_{Fs}, n_z)]^{1/2}\\
&\quad \times \left[2a(E_{Fs}, n_z)(a(E_{Fs}, n_z))'F\left(\frac{\pi}{2}, l(E_{Fs}, n_z)\right)\right.\\
&\quad +a^2(E_{Fs}, n_z)\left\{F\left(\frac{\pi}{2}, l(E_{Fs}, n_z)\right)\right\}' - [2a(E_{Fs}, n_z)[a(E_{Fs}, n_z)]'\\
&\quad -2b(E_{Fs}, n_z)(b(E_{Fs}, n_z))']E\left(\frac{\pi}{2}, l(E_{Fs}, n_z)\right)\\
&\quad \left.-\left[a^2(E_{Fs}, n_z) - b^2(E_{Fs}, n_z)\right]\left(E\left(\frac{\pi}{2}, l(E_{Fs}, n_z)\right)\right)'\right]
\end{aligned}$$

Thus, the EEM in this case is a function of both the Fermi energy and the size quantum number due to the presence of band nonparabolicity only.
The total 2D DOS function can be written following (1.53), as

$$N_{2DT}(E) = \left(\frac{2g_v}{3\pi^2\sqrt{p_1}}\right)\sum_{n_z=1}^{n_{zmax}} R_3(E, n_z)H(E - E_{n_{z6}}) \tag{1.55}$$

where, the subband energies $E_{n_{z6}}$ assume the form

$$E_{n_{z6}}(1 + \alpha E_{n_{z6}}) = \frac{\hbar^2}{2m_3}\left(\frac{n_2\pi}{d_z}\right)' \tag{1.56}$$

Combining (1.55) with the Fermi-Dirac occupation probability factor, the 2D electron statistics in UFs of Bi in accordance with the McClure and Choi model can be expressed as

$$n_{2D} = \left(\frac{2g_v}{3\pi^2\sqrt{p_1}}\right)\sum_{n_z=1}^{n_{zmax}}[\theta_1(E_{Fs}, n_z) + \theta_2(E_{Fs}, n_z) \tag{1.57}$$

where,

$$\theta_1(E_{Fs}, n_z) \equiv \left\{\sqrt{R_1(E_{Fs}, n_z)}[a^2(E_{Fs}, n_z) + b^2(E_{Fs}, n_z)]^{1/2} \times \left[a^2(E_{Fs}, n_z)F\left(\frac{\pi}{2}, l(E_{Fs}, n_z)\right) - [a^2(E_{Fs}, n_z) - b^2(E_{Fs}, n_z)]F\left(\frac{\pi}{2}, l(E_{Fs}, n_z)\right)\right]\right\}$$

and $\theta_2(E_{Fs}, n_z) \equiv \sum_{r=1}^{s} L(r)[\theta_1(E_{Fs}, n_z)]$.

(b) The Hybrid Model

The dispersion relation of the carriers in bulk specimens of Bi in accordance with the Hybrid model can be represented as [152]

$$E(1 + \alpha E) = \frac{\theta_0(E)(\hbar k_y^2)}{2M_2} + \frac{\alpha\gamma_0\hbar^4 k_y^4}{4M_2^2} + \frac{\hbar^2 k_x^2}{2m_1} + \frac{\hbar^2 k_z^2}{2m_3} \tag{1.58}$$

in which $\theta_0(E) \equiv [1 + \alpha E(1 - \gamma_0) + \bar{\delta}_0]$, $\gamma_0 \equiv \frac{M_2}{m_2}$, $\bar{\delta}_0 \equiv \frac{M_2}{M_2'}$ and the other notations are defined in [152].

In the presence of size quantization along y-direction, the 2D electron dispersion relation can be written as

$$\frac{\hbar^2 k_x^2}{2m_1} + \frac{\hbar^2 k_z^2}{2m_3} = E(1 + \alpha E) - \frac{\theta_0(E)\hbar^2}{2M_2}\left(\frac{\pi n_y}{d_y}\right)^2 - \frac{\alpha\gamma_0\hbar^4}{4M_2^2}\left(\frac{\pi n_y}{d_y}\right)^4 \tag{1.59}$$

The 2D area is given by

$$A(E, n_y) = \frac{2\pi\sqrt{m_1 m_3}}{\hbar_2} t_{29}(E, n_y) \tag{1.60}$$

$$t_{29}(E, n_y) = \left[E(1 + \alpha E) - \frac{\theta_0(E)\hbar^2}{2M_2}\left(\frac{\pi n_y}{d_y}\right)^2 - \frac{\alpha\gamma_0\hbar^4}{4M_2^4}\left(\frac{\pi n_y}{d_y}\right)^4\right]$$

The effective mass in the X–Z plane can be written as

$$m^*(E_{Fs}, n_y) = [\sqrt{m_1 m_3}]t_{29}'(E_{Fs}, n_y) \tag{1.61}$$

Therefore, the effective mass in UFs of Bi in accordance with Hybrid model is a function of Fermi energy and the size quantum number due to the presence of band nonparabolicity only.

The subband energy are given as

$$E_{n_y}(1+\alpha E) - \frac{\theta_0(E_{n_y})\hbar^2}{2M_2}\left(\frac{\pi n_y}{d_y}\right)^2 - \frac{\alpha\gamma_0\hbar^4}{4M_2^2}\left(\frac{\pi n_y}{d_y}\right)^4 = 0 \tag{1.62}$$

The total DOS function in this case can be written as

$$N_{2DT}(E) = \frac{g_v\sqrt{m_1 m_3}}{\pi\hbar^2}\sum_{n_y=1}^{n_{ymax}}\{t_{29}(E,n_y)\}' H(E-E_{n_y}) \tag{1.63}$$

The use of (1.63) leads to the 2D electron statistics in UFs of Bi in this case as

$$N_{2D} = \frac{g_v\sqrt{m_1 m_3}}{\pi\hbar^2}\sum_{n_y=1}^{n_{ymax}}\left[t_{29}(E_{Fs},n_y) + t_{30}(E_{Fs},n_y)\right] \tag{1.64}$$

in which $t_{30}(E_{Fs},n_y) = \sum_{r=1}^{s_o} L(r)[t_{29}(E_{Fs},n_y)]$

(c) The Cohen model

In accordance with the Cohen model [151], the dispersion law of the carriers in Bi is given by

$$E(1+\alpha E) = \frac{p_x^2}{2m_1} + \frac{p_z^2}{2m_3} - \frac{\alpha E p_y^2}{2m_2'} + \frac{p_y^2(1+\alpha E)}{2m_2} + \frac{\alpha p_y^4}{4m_2 m_2'} \tag{1.65}$$

The 2D electron dispersion law in UFs of Bi in accordance with this model can be written following (1.65) as

$$E(1+\alpha E) = \frac{p_x^2}{2m_1} + \frac{\hbar^2}{2m_3}\left(\frac{n_z\pi}{d_z}\right)^2 - \frac{\alpha E p_y^2}{2m_2'} + \left(\frac{\alpha p_y^4}{4m_2 m_2'}\right) + \frac{p_y^2}{2m_2}(1+\alpha E) \tag{1.66}$$

The (1.66) can be written as

$$\gamma_1(E,n_z) = p_1 k_x^z + q_2(E)k_y^2 + R_2 k_y^4 \tag{1.67}$$

where, $q_2(E) \equiv \left[\frac{\hbar^2}{2m_2}(1+\alpha E) - \frac{\alpha E\hbar^2}{2m_2'}\right]$ and $R_2 \equiv \left(\frac{\alpha\hbar^4}{4m_2 m_2'}\right)$.

The EEM in this case can be written as

$$m^*(E_{Fs},n_z) = \left(\frac{2\hbar^2}{3\pi\sqrt{p_1}}\right)\left[R_4(E,n_z)\right]|_{E=E_{Fs}} \tag{1.68}$$

in which,

$$\begin{aligned}
&R_4(E_{Fs}, n_z) \\
&\equiv \sqrt{R_2}[a_1^2(E_{Fs}, n_z) + b^2(E_{Fs}, n_z)]^{-1/2}[a_1(E_{Fs}, n_z)(a_1(E_{Fs}, n_z))' \\
&\quad + b_1(E_{Fs}, n_z)(b_1(E_{Fs}, n_z))'] \left[a_1^2(E_{Fs}, n_z) F\left(\frac{\pi}{2}, l(E_{Fs}, n_z)\right)\right. \\
&\quad \left. - [a_1^2(E_{Fs}, n_z) - b_1^2(E_{Fs}, n_z)] E\left(\frac{\pi}{2}, l_1(E_{Fs}, n_z)\right)\right] \\
&\quad + \sqrt{R_2}\left[a_1^2(E_{Fs}, n_z) + b_1^2(E_{Fs}, n_z)\right]^{1/2}\left[2a_1(E_{Fs}, n_z)(a_1(E_{Fs}, n_z))'\right. \\
&\quad \times F\left(\frac{\pi}{2}, l_1(E_{Fs}, n_z)\right) + a_1^2(E_{Fs}, n_z)\left\{F\left(\frac{\pi}{2}, l_1(E_{Fs}, n_z)\right)\right\}' \\
&\quad \times \left[2a_1(E_{Fs}, n_z)(a_1(E_{Fs}, n_z))' - 2b_1(E_{Fs}, n_z)(b_1(E_{Fs}, n_z))'\right] E\left(\frac{\pi}{2}, l_1(E_{Fs}, n_z)\right) \\
&\quad \left. - \left[a_1^2(E_{Fs}, n_z) - b_1^2(E_{Fs}, n_z)\right]\left(E\left(\frac{\pi}{2}, l_1(E_{Fs}, n_z)\right)\right)'\right]\Bigg],
\end{aligned}$$

$$a_1^2(E_{Fs}, n_z) \equiv \frac{q_2(E_{Fs})}{2R_2} + \frac{1}{2}\left[\frac{q_2^2(E_{Fs})}{R_2^2} + \frac{4\gamma_1(E_{FS}, n_z)}{R_2}\right]^{1/2},$$

$$b_1^2(E_{Fs}, n_z \equiv \frac{1}{2}\left[\frac{q_2^2(E_{Fs})}{R_2^2} + \frac{4\gamma_1(E_{Fs}, n_z)}{R_2}\right]^{1/2} - \left(\frac{q_2(E_{Fs})}{2R_2}\right)$$

$$\text{and } l_1(E_{Fs}, n_z) \equiv \frac{b_1(E_{Fs}, n_z)}{\sqrt{a_1^2(E_{Fs}, n_z) + b_1^2(E_{Fs}, n_z)}}.$$

which shows that the EEM in this present case is again a function of both the size quantum number and the Fermi energy due to the presence of the band nonparabolicity only.
The total DOS is given by

$$N_{2DT}(E) = \left(\frac{2g_v}{3\pi^2\sqrt{p_1}}\right)\sum_{n_z=1}^{n_{zmax}} R_4(E, n_z)H(E - E_{nz7}) \tag{1.69a}$$

where, $E_{n_{z7}}$ is the lowest positive root of the equation

$$\gamma_1\left(E_{n_{z7}}, n_z\right) = 0 \tag{1.69b}$$

Combining (1.69a) with the Fermi-Dirac occupation probability factor, the 2D electron statistics in UFs of Bi in accordance with the Cohen model can be written as

$$n_{2D} = \left(\frac{2g_v}{3\pi^2\sqrt{p_1}}\right)\sum_{n_z=1}^{n_{zmax}} [\theta_3(E_{Fs}, n_z) + \theta_4(E_{Fs}, n_z)] \tag{1.70}$$

where,

$$\theta_3(E_{Fs}, n_z) \equiv \left\{\sqrt{R_2}\left[a_1^2(E_{Fs}, n_z) - b_1^2(E_{Fs}, n_z)\right]^{1/2}\left[a_1^2(E_{Fs}, n_z)F\left(\frac{\pi}{2}, l_1(E_{Fs}, n_z)\right)\right.\right.$$
$$\left.\left. - \left[a_1^2(E_{Fs}, n_z) - b_1^2(E_{Fs}, n_z)\right]F\left(\frac{\pi}{2}, l_1(E_{Fs}, n_z)\right)\right]\right\},$$

and $\theta_4(E_{Fs}, n_z) \equiv \sum_{r=1}^{s} L(r)\left[\theta_3(E_{Fs}, n_z)\right]$.

(d) The Lax model

The electron energy spectra in bulk specimens of Bi in accordance with the Lax model can be written as [147]

$$E(1+\alpha E) = \frac{p_x^2}{2m_1} + \frac{p_y^2}{2m_2} + \frac{p_z^2}{2m_3} \tag{1.71}$$

The 2D electron dispersion law in this case can be written as

$$E(1+\alpha E) = \frac{\hbar^2 k_x^2}{2m_1} + \frac{\hbar^2 k_y}{2m_2} + \frac{\hbar^2}{2m_3}\left(\frac{n_z \pi}{d_z}\right)^2 \tag{1.72}$$

The EEM in this case assumes the form

$$m^*(E_{Fs}) = \sqrt{m_1 m_2}(1 + 2\alpha E_{Fs}) \tag{1.73}$$

Thus, we see that the EEM for the Lax model is a function of the Fermi energy alone due to the band nonparabolicity.

The subband energy, the total DOS function and the 2D electron statistics for this model can, respectively, be expressed as

$$E_{n_{z8}}(1+\alpha E_{n_{z8}}) = \frac{\hbar^2}{2m_3}(n_z \pi / d_z)^2 \tag{1.74}$$

$$N_{2DT}(E) = \frac{g_v \sqrt{m_1 m_2}}{\pi \hbar^2} \sum_{n_z=1}^{n_{zmax}} (1+2\alpha E) H(E - E_{n_{z8}}) \tag{1.75}$$

$$n_{2D} = \frac{g_v \sqrt{m_1 m_2} k_B T}{\pi \hbar^2} \sum_{n_z=1}^{n_{zmax}} \left[(1+2\alpha E_{n_{z8}}) F_0(\eta_{y2}) + 2\alpha k_B T F_1(\eta_{y2})\right] \tag{1.76}$$

where, $\eta_{y2} = \dfrac{E_{Fs} - E_{n_{z8}}}{k_B T}$.

(e) The ellipsoidal parabolic model

The 2D dispersion relation, the EEM, the subband energy ($E_{n_{z9}}$), the total DOS, and the 2D electron statistics for this model can respectively be written as

$$E = \left(\frac{\hbar^2 k_x^2}{2m_1}\right) + \left(\frac{\hbar^2 k_y^2}{2m_2}\right) + \left(\frac{\hbar^2}{2m_3}\right)\left(\frac{n_z \pi}{d_z}\right)^2 \tag{1.77}$$

$$m^*(E_{Fs}) = \left(\sqrt{m_1 m_2}\right) \tag{1.78}$$

$$N_{2DT}(E) = \frac{g_v \sqrt{m_1 m_2}}{\pi \hbar^2} \sum_{n_z=1}^{n_{zmax}} H(E - E_{n_{z9}}) \tag{1.79}$$

$$E_{n_{z9}} = \left(\frac{\hbar^2}{2m_3}\right)\left(\frac{n_z \pi}{d_z}\right)^2 \tag{1.80}$$

$$N_{2D} = \left[\frac{k_B T g_v \sqrt{m_1 m_2}}{\pi \hbar^2}\right] \sum_{n_z=1}^{n_{zmax}} F_0(\eta_{y3}) \tag{1.81}$$

where, $\eta_{y3} \equiv (k_B T)^{-1}\left[E_{Fs} - E_{n_{z9}}\right]$

1.2.5 The EEM in UFs of IV–VI Semiconductors

The dispersion relation of the conduction electrons in IV-VI semiconductors can be expressed in accordance with Dimmock [221] as

$$\left[\overline{\varepsilon} - \frac{E_{g0}}{2} - \frac{\hbar^2 k_s^2}{2m_t^-} - \frac{\hbar^2 k_z^2}{2m_l^-}\right]\left[\overline{\varepsilon} + \frac{E_{g0}}{2} + \frac{\hbar^2 k_s^2}{2m_t^-} + \frac{\hbar^2 k_z^2}{2m_l^-}\right] = P_\perp^2 k_s^2 + P_\parallel^2 k_z^2 \tag{1.82}$$

where $\overline{\varepsilon}$ is the energy as measured from the center of the band gap E_{g0}, $m_t^\pm$ and $m_l^\pm$ represent the contributions to the transverse and longitudinal effective masses of the external L_6^+ and L_6^- bands arising from the $\vec{k}.\vec{p}$ perturbations with the other bands taken to the second order. Using $\varepsilon = E + \left(E_{g0}/2\right)$, $P_\perp^2 = \frac{\hbar^2 E_{g0}}{2m_t^*}$, $P_\parallel^2 = \frac{\hbar^2 E_{g0}}{2m_l^*}$ (m_t^* and m_l^* are the transverse and longitudinal effective electron masses at $k = 0$) in (1.82), we can write

$$\left[E - \frac{\hbar^2 k_s^2}{2m_t^-} - \frac{\hbar^2 k_z^2}{2m_l^-}\right]\left[1 + \alpha E + \alpha + \frac{\hbar^2 k_s^2}{2m_t^+} + \alpha + \frac{\hbar^2 k_z^2}{2m_l^+}\right] = \frac{\hbar^2 k_s^2}{2m_t^*} + \frac{\hbar^2 k_z^2}{2m_l^*} \tag{1.83}$$

The 2D dispersion relation of the conduction electrons in IV-VI materials in UFs for the dimensional quantization along z direction can be expressed as

$$E(1+\alpha E)+\alpha E\left(\frac{\hbar^2k_x^2}{2x_4}+\frac{\hbar^2k_y^2}{2x_5}\right)+\alpha E\frac{\hbar^2}{2x_6}\left(\frac{n_z\pi}{d_z}\right)^2-(1+\alpha E)\left(\frac{\hbar^2k_x^2}{2x_1}+\frac{\hbar^2k_y^2}{2x_2}\right)$$
$$-\alpha\left(\frac{\hbar^2k_x^2}{2x_1}+\frac{\hbar^2k_y^2}{2x_2}\right)\left(\frac{\hbar^2k_x^2}{2x_4}+\frac{\hbar^2k_y^2}{2x_5}\right)-\alpha\left(\frac{\hbar^2k_x^2}{2x_1}+\frac{\hbar^2k_y^2}{2x_2}\right)\frac{\hbar^2}{2x_6}\left(\frac{n_z\pi}{d_z}\right)^2$$
$$-(1+\alpha E)\frac{\hbar^2}{2x_3}\left(\frac{n_z\pi}{d_z}\right)^2-\alpha\frac{\hbar^2}{2x_3}\left(\frac{n_z\pi}{d_z}\right)^2\left(\frac{\hbar^2k_x^2}{2x_4}+\frac{\hbar^2k_y^2}{2x_5}\right)-\alpha\frac{\hbar^2}{2x_3}\left(\frac{n_z\pi}{d_z}\right)^2\frac{\hbar^2}{2x_6}\left(\frac{n_z\pi}{d_z}\right)^2$$
$$=\frac{\hbar^2k_x^2}{2m_1}+\frac{\hbar^2k_y^2}{2m_2}+\frac{\hbar^2}{2m_3}\left(\frac{n_z\pi}{d_z}\right)^2 \tag{1.84}$$

where

$$x_4=m_t^+,\ x_5=\frac{m_t^++2m_l^+}{3},\ x_6=\frac{3m_t^+m_l^+}{2m_l^++m_t^+},\ x_1=m_t^-,\ x_2=\frac{m_t^-+2m_l^-}{3},$$

$$x_3=\frac{3m_t^-m_l^-}{2m_l^-+m_t^-},\ m_1=m_t^*,\ m_2=\frac{m_t^++2m_l^*}{3}\text{ and }m_3=\frac{3m_l^*m_t^*}{m_t^*+2m_l^*}.$$

Substituting $k_x=r\text{Cos}\theta$ and $k_y=r\text{Sin}\theta$ (where r and θ are 2D polar coordinates in 2D wave vector space) in (1.84), we can write

$$r4\left[\alpha\frac{1}{4}\left(\frac{\hbar^2Cos^2\theta}{x_1}+\frac{\hbar^2Sin^2\theta}{x_2}\right)\left(\frac{\hbar^2Cos^2\theta}{x_4}+\frac{\hbar^2Sin^2\theta}{x_5}\right)\right]+r^2\frac{1}{2}\left[\left(\frac{\hbar^2Cos^2\theta}{m_1}+\frac{\hbar^2Sin^2\theta}{m_2}\right)\right.$$
$$+\alpha\frac{\hbar^2}{2x_3}\left(\frac{n_z\pi}{d_z}\right)^2\left(\frac{\hbar^2Cos^2\theta}{x_4}+\frac{\hbar^2Sin^2\theta}{x_5}\right)+\alpha\left(\frac{\hbar^2Cos^2\theta}{x_1}+\frac{\hbar^2Sin^2\theta}{x_2}\right)\frac{\hbar^2}{2x_6}\left(\frac{n_z\pi}{d_z}\right)^2$$
$$\left.+\hbar^2(1+\alpha E)\left(\frac{Cos^2\theta}{x_1}+\frac{Sin^2\theta}{x_2}\right)-\hbar^2\alpha E\left(\frac{Cos^2\theta}{x_4}+\frac{Sin^2\theta}{x_5}\right)\right]-[E(1+\alpha E)$$
$$\left.+\alpha E\frac{\hbar^2}{2x_6}\left(\frac{n_z\pi}{d_z}\right)^2-(1+\alpha E)\frac{\hbar^2}{2x_3}\left(\frac{n_z\pi}{d_z}\right)^2-\alpha\left(\frac{\hbar^4}{4x_3x_6}\left(\frac{n_z\pi}{d_z}^4\right)\right)\right]=0 \tag{1.85}$$

The area $A(E,n_z)$ of the 2D wave vector space can be expressed as

$$A(E,n_z)=\overline{J}_1-\overline{J}_2 \tag{1.86}$$

where

$$\overline{J}_1\equiv 2\int_0^{\pi/2}\frac{c}{b}d\theta \tag{1.87}$$

and

$$\overline{J}_2\equiv 2\int_0^{\pi/2}\frac{ac^2}{b^3}d\theta \tag{1.88}$$

in which

$$\alpha \equiv \left[\alpha\left(\frac{\hbar^4}{4}\right)\left(\frac{Cos^2\theta}{x_1}+\frac{Sin^2\theta}{x_2}\right)\left(\frac{Cos^2\theta}{x_4}+\frac{Sin^2\theta}{x_3}\right)\right],$$

$$\begin{aligned}b \equiv \left(\frac{\hbar^2}{2}\right)&\left[\left(\frac{Cos^2\theta}{m_1}+\frac{Sin^2\theta}{m_2}\right)+\alpha\left(\frac{\hbar^2}{2x_3}\right)\left(\frac{n_z\pi}{d_z}\right)^2\left(\frac{Cos^2\theta}{x_4}+\frac{Sin^2\theta}{x_5}\right)\right.\\&+\alpha\left(\frac{\hbar^2}{2x_6}\right)\left(\frac{n_z\pi}{d_z}\right)^2\left(\frac{Cos^2\theta}{m_1}+\frac{Sin^2\theta}{m_2}\right)+(1+\alpha E)\left(\frac{Cos^2\theta}{x_1}+\frac{Sin^2\theta}{x_2}\right)\\&\left.-\alpha E\left(\frac{Cos^2\theta}{x_4}+\frac{Sin^2\theta}{x_5}\right)\right]\end{aligned}$$

and

$$\begin{aligned}c \equiv &\left[E(1+\alpha E)+\alpha E\left(\frac{\hbar^2}{2x_6}\right)\left(\frac{n_z\pi}{d_z}\right)^2-(1+\alpha E)\left(\frac{\hbar^2}{2x_3}\right)\left(\frac{n_z\pi}{d_z}\right)^2\right.\\&\left.-\alpha\left(\frac{\hbar^4}{4x_3x_6}\right)\left(\frac{n_z\pi}{d_z}^4\right)\right]\end{aligned}$$

(1.87) can be expressed as $\overline{J}_1 = 2\int_0^{\pi/2}\frac{t_3(E,n_z)d\theta}{A_1(E,n_z)Cos^2\theta+B_1(E,n_z)Sin^2\theta}$ where, $t_3(E,n_z) \equiv c$, $A_1(E,n_z) \equiv \frac{\hbar^2}{2m_1}t_1(E,n_z)$,

$$t_1(E,n_z) \equiv \left[1+m_1\left[\frac{1}{x_4}\frac{\alpha\hbar^2}{2x_3}\left(\frac{n_z\pi}{d_z}\right)^2+\frac{\alpha\hbar^2}{2x_1x_6}\left(\frac{n_z\pi}{d_z}\right)^2+\frac{1+\alpha E}{x_1}-\frac{\alpha E}{x_4}\right]\right]$$

$$B_1(E,n_z) \equiv \frac{\hbar^2}{2m_2}t_2(E,n_z) \text{ and}$$

$$t_2(E,n_z) \equiv \left[1+m_2\left[\frac{\alpha\hbar^2}{2x_3x_5}\left(\frac{n_z\pi}{d_z}\right)^2+\frac{\alpha\hbar^2}{2x_2x_6}\left(\frac{n_z\pi}{d_z}\right)^2+\frac{1+\alpha E}{d_z}-\frac{\alpha E}{x_5}\right]\right].$$

Performing the integration, we get

$$\overline{J}_1 = \pi t_3(E,n_z)[A_1(E,n_z)B_1(E,n_z)]^{-1/2} \tag{1.89}$$

From (1.88) we can write

$$\overline{J}_2 = \frac{\alpha t_3^2(E,n_z)\hbar^4}{2B_1^3(E,n_z)}I \tag{1.90}$$

where

$$I \equiv \int_0^\infty \frac{(a_1 + a_2 z^2)(a_3 + a_4 z^2)dz}{[(\overline{a})^2 + z^2]^3} \tag{1.91}$$

in which $a_1 \equiv \frac{1}{x_1}, a_2 \equiv \frac{1}{x_2}, z = \tan\theta, \theta$ is a new variable, $a_3 \equiv \frac{1}{x_4}, a_4 \equiv \frac{1}{x_5}$ and $(\overline{a})^2 \equiv \left(\frac{A_1(E,n_z)}{B_1(E,n_z)}\right)$. The use of the Residue theorem leads to the evaluation of the integral in (1.91) as

$$I \equiv \frac{\pi}{4\overline{a}}[a_1 a_4 + 3a_2 a_4] \tag{1.92}$$

Therefore, the 2D area of the 2D wave vector space can be written as

$$A(E,n_z) = \frac{\pi t_3(E,n_z)}{\sqrt{A_1(E,n_z)B_1(E,n_z)}}\left[1 - \frac{1}{x_5}\left(\frac{1}{x_1} + \frac{3}{x_2}\right)\frac{\alpha t_3(E,n_z)\hbar^4}{8B_1^2(E,n_z)}\right] \tag{1.93}$$

The EEM for the UFs of IV-VI materials can thus be written as

$$m^*(E,n_z) = \frac{\hbar^2}{2}[\theta_5(E,n_z)]\Big|_{E=E_{Fs}} \tag{1.94}$$

where,

$$\begin{aligned}
\theta_5(E,n_z) \equiv & \left[1 - \frac{1}{x_5}\left(\frac{1}{x_1} + \frac{3}{x_2}\right)\frac{\alpha t_3(E,n_z)\hbar^4}{8[B_1(E,n_z)]^2}\right][A_1(E,n_z)B_1(E,n_z)]^{-1} \\
& \times \left[\sqrt{A_1(E,n_z)B_1(E,n_z)}\{t_3(E,n_z)\}' - t_3(E,n_z)\right. \\
& \times \left.\left\{\frac{1}{2}\{A_1(E,n_z)\}'\left[\frac{B_1(E,n_z)}{A_1(E,n_z)}\right]^{-1/2} + \frac{1}{2}\{B_1(E,n_z)\}'\left[\frac{A_1(E,n_z)}{B_1(E,n_z)}\right]^{-1/2}\right\}\right] \\
& - \frac{1}{8}\frac{t_3(E,n_z)\alpha\hbar^4}{\sqrt{A_1(E,n_z)B_1(E,n_z)}}\frac{1}{x_5}\left(\frac{1}{x_1} + \frac{3}{x_2}\right)[B_1(E,n_z)]^{-4} \\
& \times \left[\{B_1(E,n_z)\}^2\{t_3(E,n_z)\}' - 2B_1(E,n_z)\{B_1(E,n_z)\}' t_3(E,n_z)\right]
\end{aligned}$$

Thus, the EEM is a function of Fermi energy and the quantum number due to the band nonparabolicity.

The total DOS function can be written as

$$N_{2DT}(E) = \left(\frac{g_v}{2\pi}\right)\sum_{n_z=1}^{n_{zmax}} \theta_5(E,n_z)H(E - E_{n_{z10}}) \tag{1.95}$$

where the subband energy $(E_{n_{z10}})$ in this case can be written as

$$E_{n_{z10}}\left(1+\alpha E_{n_{z10}}\right)+\alpha E_{n_{z10}}\frac{\hbar^2}{2x_6}\left(\frac{n_z\pi}{d_z}\right)^2-(1+\alpha E_{n_{z10}})\frac{\hbar^2}{2x_3}\left(\frac{n_z\pi}{d_z}\right)^2$$
$$-\alpha\frac{\hbar^2}{2x_3}\left(\frac{n_z\pi}{d_z}\right)^2\frac{\hbar^2}{2x_6}\left(\frac{n_z\pi}{d_z}\right)^2-\left[\frac{\hbar^2}{2m_3}\left(\frac{n_z\pi}{d_z}\right)^2\right]=0 \quad (1.96)$$

The use of (1.95) leads to the expression of 2D electron statistics as

$$n_{2D}=\frac{g_v}{2\pi}\sum_{n_z=1}^{n_{zmax}}[T_{59}(E_{Fs},n_z)+T_{60}(E_{Fs},n_z)] \quad (1.97)$$

where $T_{59}(E_{Fs},n_z)\equiv\frac{A(E_{Fs},n_z)}{\pi}$ and $T_{60}(E_{Fs},n_z)\equiv\sum_{r=1}^{s}L(r)T_{59}(E_{Fs},n_z)$.

1.2.6 The EEM in UFs of Stressed Semiconductors

The electron energy spectrum in stressed Kane-type semiconductors can be written [222–225] as

$$\left(\frac{k_x}{\bar{a}_0(E)}\right)^2+\left(\frac{k_y}{\bar{b}_0(E)}\right)^2+\left(\frac{k_z}{\bar{c}_0(E)}\right)^2=1 \quad (1.98)$$

where

$$[\bar{a}_0(E)]^2\equiv\frac{\overline{K}_0(E)}{\overline{A}_0(E)+\frac{1}{2}\overline{D}_0(E)},\ \overline{K}_0(E)\equiv\left[E-C_1\varepsilon-\frac{2C_2^2\varepsilon_{xy}^2}{3E_g'}\right]\left(\frac{3E_g'}{2B_2^2}\right),$$

C_1 is the conduction band deformation potential, ε is the trace of the strain tensor $\hat{\varepsilon}$ which can be written as $\hat{\varepsilon}=\begin{bmatrix}\varepsilon_{xx} & \varepsilon_{xy} & 0\\ \varepsilon_{xy} & \varepsilon_{yy} & 0\\ 0 & 0 & \varepsilon_{zz}\end{bmatrix}$, C_2 is a constant which describes the strain interaction between the conduction and valance bands, $E_g'\equiv E_g+E-C_1\varepsilon$, B_2 is the momentum matrix element,

$$\overline{A}_0(E)\equiv\left[1-\frac{(\bar{a}_0+C_1)}{E_g'}+\frac{3\bar{b}_0\varepsilon_{xx}}{2E_g'}-\frac{\bar{b}_0\varepsilon}{2E_g'}\right],$$
$$\bar{a}_0\equiv-\frac{1}{3}(\bar{b}_0+2\bar{m}),\ \bar{b}_0\equiv\frac{1}{3}(\bar{l}-\bar{m}),\ \bar{d}_0\equiv\frac{2\bar{n}}{\sqrt{3}},$$

$\bar{l},\bar{m},\bar{n}$ are the matrix elements of the strain perturbation operator, $\overline{D}_0(E)\equiv(\bar{d}_0\sqrt{3})\frac{\varepsilon_{xy}}{E_g'}$,

$$[\bar{b}_0(E)]^2 \equiv \frac{\overline{K}_0(E)}{\overline{A}_0(E) - \frac{1}{2}\overline{D}_0(E)}, \quad [\bar{c}_0(E)]^2 \equiv \frac{\overline{K}_0(E)}{\overline{L}_0(E)},$$

$$\text{and } \overline{L}_0(E) \equiv \left[1 - \frac{(\bar{a}_0 + C_1)}{E'_g} + \frac{3\bar{b}_0\varepsilon_{zz}}{E'_g} - \frac{\bar{b}_0\varepsilon}{2E'_g}\right]$$

The 2D electron energy spectrum in UFs of stressed materials assumes the form

$$\frac{K_x^2}{[\bar{a}_0(E)]^2} + \frac{K_y^2}{[\bar{b}_0(E)]^2} + \frac{1}{[\bar{c}_0(E)]^2}(n_z\pi/d_z) = 1 \tag{1.99}$$

The area of 2D wave vector space enclosed by (1.99) can be written as

$$A(E, n_z) = \pi P^2(E, n_z)\bar{a}_0(E)\bar{b}_0(E) \tag{1.100}$$

where $P^2(E, n_z) = [1 - [n_z\pi/d_z\bar{c}_0(E)]^2]$.

The expression of the surface EEM in this case can be written as

$$m^*(E_{Fs}, n_z) = \frac{\hbar^2}{2}[\theta_6(E, n_z)]\bigg|_{E=E_{Fs}} \tag{1.101}$$

in which,

$$\theta_6(E, n_z) = \left[2P(E, n_z)\{P(E, n_z)\}'\bar{a}_0(E)\bar{b}_0(E) + \{P(E, n_z)\}^2\{\bar{a}_0(E)\}'\bar{b}_0(E) + \{P(E, n_z)\}^2\{\bar{b}_0(E)\}'\bar{a}_0(E)\right]$$

The EEM in this case is the function of Fermi energy and the size quantization number due to the presence of stress only.

Thus, the total 2D DOS function can be expressed as

$$N_{2DT}(E) = \left(\frac{g_v}{2\pi}\right)\sum_{n_z=1}^{n_{zmax}} \theta_6(E, n_z)H(E - E_{n_{z11}}) \tag{1.102}$$

The subband energies ($E_{n_{z11}}$) are given by

$$\bar{c}_0(E_{n_{z11}}) = n_z\pi/d_z \tag{1.103}$$

The 2D surface electron concentration per unit area for UFs of stressed Kane-type compounds can be written as

$$n_{2D} = \frac{g_v}{2\pi}\sum_{n_z=1}^{n_{zmax}} [T_{61}(E_{Fs}, n_z) + T_{62}(E_{Fs}, n_z)] \tag{1.104}$$

where

$$T_{61}(E_{Fs}, n_z) \equiv [P^2(E_{Fs}, n_z)\bar{a}_0(E_{Fs})\bar{b}_0(E_{Fs})]$$

$$\text{and } T_{62}(E_{Fs}, n_z) \equiv \sum_{r=1}^{s} L(r)T_{61}(E_{Fs}, n_z).$$

In the absence of stress together with the substitution, $B_2^2 \equiv 3\hbar^2(E_g/4m_c)$, (1.98) assumes the same form as given by (1.16).

1.2.7 The EEM in UFs of Tellurium

The dispersion relation of the conduction electrons in Te can be expressed as [226]

$$E = \psi_1 k_z^2 + \psi_2 k_s^2 \pm [\psi_3^2 k_s^2 + \psi_2^4 k_s^2]^{1/2} \tag{1.105}$$

where, $\psi_1 = 6.7 \times 10^{-16}$ mev.m^2, $\psi_2 = 4.2 \times 10^{-16}$ mev.m^2, $\psi_3 = 6 \times 10^{-8}$, mev.m and $\psi_4 = 3.6 \times 10^{-8}$ mev.m

The 2D electron energy spectrum in ultrathin films of Te assumes the form

$$k_s^2 = \psi_5(E) - \psi_6\left(\frac{\pi n_z}{d_z}\right)^2 \pm \psi_7\left[\psi_8^2(E) - \left(\frac{\pi n_z}{d_z}\right)^2\right]^{\frac{1}{2}} \tag{1.106}$$

where, $\psi_5(E) = \left[\frac{E}{\psi_2} + \frac{\psi_4^2}{2\psi_2^2}\right]$, $\psi_6 = \frac{\psi_1}{\psi_2}$, $\psi_7 = \frac{\psi_4\sqrt{\psi_1}}{\psi_2^{\frac{3}{2}}}$, $\psi_8^2(E) = \frac{\psi_4^4 + 4E\psi_2\psi_4^2 + 4\psi_2^2\psi_3^2}{4\psi_1\psi_2\psi_4^2}$

The EEM in this case is given by

$$m^*(E_{Fs}, n_z) = \frac{\hbar^2}{2}\left[t'_{40}(E, n_z)\right]\Big|_{E=E_{Fs}} \tag{1.107}$$

where, $t_{40}(E, n_z) = \left[\psi_5(E) - \psi_6\left(\frac{\pi n_z}{d_z}\right)^2 \pm \psi_7\left[\psi_8^2(E) - \left(\frac{\pi n_z}{d_z}\right)^2\right]^{1/2}\right]^{1/2}$

It appears that the EEM in UFs of Te is a function of Fermi energy and size quantum number which are the characteristics of such systems.

Thus, the total 2D DOS function can be expressed as

$$N_{2DT}(E) = \left(\frac{g_v}{\pi}\right)\sum_{n_z=1}^{n_{zmax}} t'_{40}(E, n_2)H(E - E_{n_{z12}}) \tag{1.108}$$

The subband energies $(E_{n_{z12}})$ are given by

$$E_{n_{z12}} = \psi_1(n_z\pi/d_z)^2 \pm \psi_3(n_z\pi/d_z) \quad (1.109)$$

The 2D surface electron concentration per unit area for UFs of Te can be written as

$$n_{2D} = \frac{g_v}{\pi}\sum_{n_z=1}^{n_{zmax}}[t_{40}(E_{Fs},n_2) + t_{41}(E_{Fs},n_2)] \quad (1.110)$$

where $t_{41}(E_{Fs},n_2) \equiv \sum_{r=1}^{s} L(r)t_{40}(E_{Fs},n_z)$.

1.2.8 The EEM in UFs of Gallium Phosphide

The energy spectrum of the conduction electrons in n-GaP can be written as [227]

$$E = \frac{\hbar^2 k_s^2}{2m_\perp^*} + \frac{\hbar^2}{2m_{||}^*}[\overline{A}' k_s^2 + k_z^2] - \left[\frac{\hbar^4 k_0^2}{m_{||}^{*2}}(k_s^2 + k_z^2) + |V_G|^2\right]^{1/2} + |V_G| \quad (1.111)$$

where, K_0 and $|V_G|$ are constants of the energy spectrum and $\overline{A}' = 1$.

The 2D electron dispersion relation in size quantized n-GaP can be expressed as

$$E = ak_s^2 + C(n_z\pi/d_z)^2 + |V_G| - \left[Dk_s^2 + |V_G|^2 + D(n_z\pi/d_z)^2\right]^{1/2} \quad (1.112)$$

in which, $a \equiv \frac{\hbar^2}{2m_\perp^*} + \frac{\hbar^2}{2m_{||}^*}$, $C \equiv \frac{\hbar^2}{2m_{||}^*}$ and $D \equiv (\hbar^2 k_0/m_{||}^*)^2$

The subband energy $(E_{n_{z13}})$ are given by

$$E_{n_{z13}} = C(\pi n_z/d_z)^2 + |V_G| - \left[|V_G|^2 + D(\pi n_z/d_z)^2\right]^{1/2} \quad (1.113)$$

Equation (1.112) can be expressed as

$$k_s^2 = t_{42}(E,n_z) \quad (1.114)$$

in which, $t_{42}(E,n_z) \equiv [\{2a(E-t_1)+D\} - \{[2a(E-t_1)+D]^2 - 4a^2[(E-t_1)^2 - t_2]\}^{1/2}]$, $t_1 \equiv |V_G| + C(\pi n_z/d_z)^2$ and $t_2 \equiv |V_G|^2 + D(\pi n_z/d_z)^2$

The EEM can be expressed from (1.114) as

$$m^*(E_{Fs},n_z) = \frac{\hbar^2}{2}t'_{42}(E_{Fs},n_z) \quad (1.115)$$

It appears that the EEM in UFs of GaP is a function of Fermi energy and size quantum number due to the presence of the system constant k_0 .

The total DOS function is given by

$$N_{2DT}(E) = \frac{g_v}{4\pi a^2} \sum_{n_z=1}^{n_{zmax}} \left[t'_{42}(E, n_z)\right] H(E - E_{n_{z13}}) \tag{1.116}$$

The electron statistics in UFs in n-GaP assumes the form

$$n_{2D} = \frac{g_v}{4\pi a^2} \sum_{n_z=1}^{n_{zmax}} \left[t_{42}(E_{Fs}, n_z) + t_{43}(E_{Fs}, n_z)\right] \tag{1.117}$$

where, $t_{43}(E_{Fs}, n_z) \equiv \sum_{r=1}^{s} L(r) \left[t_{42}(E_{Fs}, n_z)\right]$

1.2.9 The EEM in UFs of Platinum Antimonide

The dispersion relation for the n-type $PtSb_2$ can be written as [228]

$$\left(E + \lambda_0 \frac{(\bar{a})^2}{4} k^2 - l k_s^2 \frac{(\bar{a})^2}{4}\right) \left(E + \delta_0 - \nu \frac{(\bar{a})^2}{4} k^2 - \bar{\bar{n}} \frac{(\bar{a})^2}{4} k_s^2\right) = I \frac{(\bar{a})^4}{16} k^4 \tag{1.118}$$

where $\omega_1 \equiv \left(\lambda_0 \frac{(\bar{a})^2}{4} - l \frac{(\bar{a})^2}{4}\right)$, $\omega_2 \equiv \lambda_0 \frac{(\bar{a})^2}{4}$, $\omega_3 \equiv \left((\bar{\bar{n}}) \frac{(\bar{a})^2}{4} + \nu \frac{(\bar{a})^2}{4}\right)$, $\omega_4 \equiv \nu \frac{(\bar{a})^2}{4}$, $I_1 \equiv I \left(\frac{(\bar{a})^2}{4}\right)^2$, $\lambda_0, l, \delta_0, \nu$ and $\bar{\bar{n}}$ are the band constants and $\bar{a}$ is the lattice constant.

The (1.118) can be expressed as

$$\left[E + \omega_1 k_s^2 + \omega_2 k_z^2\right] \left[E + \delta_0 - \omega_3 k_s^2 - \omega_4 k_z^2\right] = I_1 (k_z^2 + k_s^2)^2 \tag{1.119}$$

The use of (1.119) leads to the expression of the 2D dispersion law in UFs of n-$PtSb_2$ as

$$k_s^2 = t_{44}(E, n_z) \tag{1.120}$$

where,

$$t_{44}(E, n_z) = [2A_9]^{-1} \left[-A_{10}(E, n_z) + \sqrt{A_{10}^2(E, n_z) + 4A_9 A_{11}(E, n_z)}\right] \tag{1.121}$$

$A_9 = [I_1 + \omega_1 \omega_3]$, $A_{10}(E, n_z)$

$$= \left[\omega_3 E + \omega_1 \left\{ E + \delta_0 - \omega_4 \left(\frac{\pi n_z}{d_z} \right)^2 \right\} + \omega_2 \omega_3 \left(\frac{\pi n_z}{d_z} \right)^2 + 2I_1 \left(\frac{\pi n_z}{d_z} \right)^2 \right]$$

and

$$A_{11}(E, n_z) \equiv \left[E \left[E + \delta_0 - \omega_4 \left(\frac{\pi n_z}{d_z} \right)^2 \right] \right.$$
$$\left. + \omega_2 \left(\frac{\pi n_z}{d_z} \right)^2 \left[E + \delta_0 - \omega_4 \left(\frac{\pi n_z}{d_z} \right)^2 \right] - I_1 \left(\frac{\pi n_z}{d_z} \right)^4 \right]$$

The area of k_s space can be expressed as

$$A(E, n_z) = \frac{\pi}{2A_9} t_{44}(E, n_z) \tag{1.122}$$

The EEM can be written as

$$m^*(E_{Fs}, n_z) = \frac{\hbar^2}{4A_9} t'_{44}(E_{Fs}, n_z) \tag{1.123}$$

It appears that the EEM in UFs of $PtSb_2$ is a function of Fermi energy and size quantum number which is the characteristic features of such systems.

The total DOS function assumes the form

$$N_{2DT}(E) = \frac{g_v}{4\pi A_9} \sum_{n_z=1}^{n_{zmax}} [t'_{44}(E, n_z)] H(E - E_{n_{z14}}) \tag{1.124}$$

where the quantized levels $E_{n_{z14}}$ can be expressed through the equation

$$E_{n_{z14}} = (2)^{-1} \left[- \left[\omega_2 \left(\frac{\pi n_z}{d_z} \right)^2 + \delta_0 - \omega_4 \left(\frac{\pi n_z}{d_z} \right)^2 \right] \right.$$
$$+ \left\{ \left[\omega_2 \left(\frac{\pi n_z}{d_z} \right)^2 + \delta_0 - \omega_4 \left(\frac{\pi n_z}{d_z} \right)^2 \right]^2 \right.$$
$$\left. \left. +4 \left[I_1 \left(\frac{\pi n_z}{d_z} \right)^4 + \omega_2 \omega_4 \left(\frac{\pi n_z}{d_z} \right)^4 - \omega_2 \delta_0 \left(\frac{\pi n_z}{d_z} \right)^2 \right] \right\}^{1/2} \right] \tag{1.125}$$

The electron statistics can be written as

$$n_{2D} = \frac{2g_v}{(2\pi)^2} \frac{\pi}{2A_9} \sum_{n_z=1}^{n_{zmax}} \int_{E_{nz}}^{\infty} \frac{\partial}{\partial E} [t_{44}(E, n_z)] f(E) dE$$

$$n_{2D} = \frac{g_v}{4\pi A_9} \sum_{n_z=1}^{n_{zmax}} [t_{44}(E_{Fs}, n_z) + t_{45}(E_{Fs}, n_z)] \tag{1.126}$$

where $t_{45}(E_{Fs}, n_z) \equiv \sum_{r=1}^{s} L(r)[t_{44}(E_{Fs}, n_z)]$

1.2.10 The EEM in UFs of Bismuth Telluride

The dispersion relation of the conduction electron Bi_2Te_3 can be written as [229–231]

$$E(1 + \alpha E) = \overline{\omega}_1 k_x^2 + \overline{\omega}_2 k_y^2 + \overline{\omega}_3 k_z^2 + 2\overline{\omega}_4 k_z k_y \tag{1.127}$$

where

$$\overline{\omega}_1 = \frac{\hbar^2}{2m_0} \overline{\alpha}_{11}, \overline{\omega}_2 = \frac{\hbar 2}{2m_0} \overline{\alpha}_{22}, \overline{\omega}_3 = \frac{\hbar^2}{2m_0} \overline{\alpha}_{33}, \overline{\omega}_4 = \frac{\hbar^2}{2m_0} \overline{\alpha}_{23}$$

in which $\overline{\alpha}_{11}, \overline{\alpha}_{22}, \overline{\alpha}_{33}$ and $\overline{\alpha}_{23}$ are system constants.

The 2D electron dispersion law in UFs of $Bi_2\,Te_3$ assumes the form

$$E(1 + \alpha E) = \overline{\omega}_1 (\frac{n_x \pi}{d_x})^2 + \overline{\omega}_2 k_y^2 + \overline{\omega}_3 k_z^2 + 2\overline{\omega}_4 k_z k_y \tag{1.128}$$

The area of the ellipse is given by

$$A_n(E, n_z) = \frac{\pi}{\sqrt{\overline{\alpha}_{22}\overline{\alpha}_{33} - 4\overline{\alpha}_{23}^2}} \left[\frac{2m_0 E(1 + \alpha E)}{\hbar^2} - \overline{\alpha}_{11} (\frac{n_x \pi}{d_x})^2 \right] \tag{1.129}$$

The EEM can be expressed as

$$m^*(E_{Fs}) = \frac{m_0(1 + 2\alpha E_{Fs})}{\sqrt{\overline{\alpha}_{22}\overline{\alpha}_{33} - 4\overline{\alpha}_{23}^2}} \tag{1.130}$$

It appears that the EEM in UFs of Bi_2Te_3 is a function of Fermi energy due to the presence of the band nonparabolicity.

The total DOS function assumes the form

$$N_{2DT}(E) = \frac{g_v m_0}{\pi \hbar^2 \sqrt{\bar{\alpha}_{22}\bar{\alpha}_{33} - 4\bar{\alpha}_{23}^2}} \sum_{n_z=1}^{n_x max} (1 + 2\alpha E) H(E - E_{n_{z15}}) \quad (1.131)$$

where, $(E_{n_{z15}})$ can be expressed through the equation

$$E_{n_{z15}}(1 + \alpha E_{n_{z15}}) = \bar{\omega}_1 \left(\frac{n_x \pi}{d_x}\right)^2 \quad (1.132)$$

The electron concentration can be written as

$$n_{2D} = \frac{k_B T g_v}{\pi \hbar^2} \left(\frac{m_0}{\sqrt{\bar{\alpha}_{22}\bar{\alpha}_{33} - 4\bar{\alpha}_{23}^2}}\right) \sum_{n_z=1}^{n_{zmax}} [(1 + 2\alpha E_{n_{z15}}) F_0(\eta_{n_{15}}) + 2\alpha k_B T F_1(\eta_{n15})] \quad (1.133)$$

where, $\eta_{n_{15}} = \frac{E_{Fs} - E_{n_{z15}}}{k_B T}$.

1.2.11 The EEM in UFs of Germanium

It is well known that the conduction electrons $n-Ge$ obey two different types of dispersion laws since band nonparabolicity has been included in two different ways as given in the literature [232, 234].

a. The energy spectrum of the conduction electrons in bulk specimens of $n-Ge$ can be expressed in accordance with Cardona et al. [232, 233] as

$$E = -\frac{E_{g0}}{2} + \frac{\hbar^2 k_z^2}{2m_\parallel^*} + \left[\frac{E_{g0}^2}{4} + E_{g0} k_s^2 \left(\frac{\hbar^2}{2m_\perp^*}\right)\right]^{\frac{1}{2}} \quad (1.134)$$

where in this case $m_\parallel^*$ and $m_\perp^*$ are the longitudinal and transverse effective masses along <111> direction at the edge of the conduction band respectively. Equation (1.134) can be written as

$$\frac{\hbar^2 k_s^2}{2m_\perp^*} = E(1 + \alpha E) + \alpha \left(\frac{\hbar^2 k_z^2}{2m_\parallel^*}\right) - (1 + 2\alpha E)\left(\frac{\hbar^2 k_z^2}{2m_\parallel^*}\right) \quad (1.135)$$

In the presence of size quantization along k_z direction, the 2D dispersion relation of the conduction relations in UFs of n-Ge can be written by extending the method as given in [235] as

$$\frac{\hbar^2 k_x^2}{2m_1^*} + \frac{\hbar^2 k_y^2}{2m_2^*} = \gamma(E, n_z) \quad (1.136)$$

where, $m_1^* \equiv m_\perp^*$, $m_2^* = \frac{m_\perp^* + 2m_\parallel^*}{3}$,

$$\gamma(E, n_z) \equiv \left[E(1+\alpha E) - (1+2\alpha E)\frac{\hbar^2}{2m_3^*}\left(\frac{n_z\pi}{d_z}\right)^2 + \alpha\left[\frac{\hbar^2}{2m_3^*}\left(\frac{n_z\pi}{d_z}\right)^2\right]^2\right]$$

and $m_3^* = \frac{3m_\parallel^* m_\perp^*}{2m_\parallel^* + m_\perp^*}$

The area of ellipse of the 2D surface as given by (1.136) can be written as

$$A(E, n_z) = \frac{2\pi\sqrt{m_1^* m_2^*}}{\hbar^2}\gamma(E, n_z) \tag{1.137}$$

The EEM can be expressed using (1.137) as

$$m^*(E_{Fs}, n_z) \equiv \sqrt{m_1^* m_2^*}\left[(1+2\alpha E_{Fs}) - (2\alpha)\frac{\hbar^2}{2m_3^*}\left(\frac{n_z\pi}{d_z}\right)^2\right] \tag{1.138}$$

Therefore, the EEM is a function of Fermi energy and size quantum number due to the presence of band nonparabolicity only.
The DOS function per subband can be expressed as

$$N_{2D}(E) = \frac{4\sqrt{m_1^* m_2^*}}{\pi\hbar^2}\left[1 + 2\alpha E - 2\alpha\left(\frac{\hbar^2}{2m_3^*}\left(\frac{\pi n_z}{d_z}\right)^2\right)\right] \tag{1.139}$$

The total DOS function is given by

$$N_{2DT}(E) = \frac{4}{\pi\hbar^2}\sqrt{m_1^* m_2^*}\sum_{n_z=1}^{n_{zmax}} \times\left[1 + 2\alpha E - 2\alpha\left(\frac{\hbar^2}{2m_3^*}\left(\frac{\pi n_z}{d_z}\right)^2\right)\right] H(E - E_{n_z 16}) \tag{1.140}$$

where, $E_{n_{z16}}$ is the positive root of the following equation

$$E_{n_{z16}}(1+\alpha E_{n_{z16}}) - (1+2\alpha E_{n_{z16}})\left(\frac{\hbar^2}{2m_3^*}\left(\frac{\pi n_z}{d_z}\right)^2\right) + \alpha\left(\frac{\hbar^2}{2m_3^*}\left(\frac{\pi n_z}{d_z}\right)^2\right)^2 = 0 \tag{1.141}$$

Thus combining (1.140) with the Fermi Dirac occupation probability factor, the electron statistics in this case can be written as

$$n_{2D} = \frac{4\sqrt{m_1^* m_2^*} k_B T}{\pi \hbar^2} \sum_{n_z=1}^{n_{zmax}} \left[(A_1(n_z) + 2\alpha E_{n_{z16}}) F_0(E_{n_{z16}}) + 2\alpha k_B T F_1(E_{n_{z16}}) \right] \tag{1.142}$$

where $A_1(n_z) \equiv \left[1 + 2\alpha(\hbar^2/2m_3^*)(\pi n_z/d_z)^2\right]$ and $\eta_{n_{z16}} \equiv \frac{1}{K_B T}[E_{F2D} - E_{n_{z16}}]$

b. The dispersion relation of the conduction electron in bulk specimens of $n-Ge$ can be expressed in accordance with the model of Wang and Ressier [234] and can be written as

$$E = \frac{\hbar^2 k_z^2}{2m_\parallel^*} + \frac{\hbar^2 k_s^2}{2m_\perp^*} - \bar{c}_1 \left(\frac{\hbar^2 k_s^2}{2m_\perp^*}\right)^2 - \bar{d}_1 \left(\frac{\hbar^2 k_s^2}{2m_\perp^*}\right)\left(\frac{\hbar^2 k_z^2}{2m_\parallel^*}\right) - \bar{e}_1 \left(\frac{\hbar^2 k_z^2}{2m_\parallel^*}\right)^2 \tag{1.143}$$

where $\bar{c}_1 = \overline{C}(2m_\perp^*/\hbar^2)^2, \overline{C} = 1.4\overline{A}, \overline{A} = \frac{1}{4}(\hbar^4/E_{g0}m_\perp^{*2})\left(1 - \frac{m_\perp^*}{m_0}\right)^2, \bar{d}_1 = \bar{d}\left(\frac{4m_\perp^* m_\parallel^*}{\hbar^4}\right), \bar{d} = 0.8\overline{A}, \bar{e}_1 = \bar{e}_0(2m_\parallel^*/\hbar^2)^2$ and $\bar{e}_0 = 0.005\overline{A}$.

Therefore the 2D dispersion law can be expressed as

$$E = A_5(n_z) + A_6(n_z)\beta - \bar{c}_1\beta^2 \tag{1.144}$$

where $A_5(n_z) \equiv \frac{\hbar^2}{2m_3^*}\left(\frac{\pi n_z}{d_z}\right)^2\left[1 - \bar{e}_1\left(\frac{\hbar^2}{2m_3^*}\right)\left(\frac{\pi n_z}{d_z}\right)^2\right]$,

$A_6(n_z) \equiv \left[1 - \bar{d}_1\left(\frac{\hbar^2}{2m_3^*}\right)\left(\frac{n_z\pi}{d_z}\right)^2\right]$ and $\beta \equiv \frac{\hbar^2 k_x^2}{2m_1^*} + \frac{\hbar^2 k_y^2}{2m_2^*}$.

Equation (1.144) can be written as

$$\frac{\hbar^2 k_x^2}{2m_1^*} + \frac{\hbar^2 k_y^2}{2m_2^*} = I_1(E, n_z) \tag{1.145}$$

where $I_1(E, n_z) \equiv (2\bar{c}_1)^{-1}\left[A_6(n_z) - \left[A_6^2(n_z) - 4\bar{c}_1 E + 4\bar{c}_1 A_5(n_z)\right]^{1/2}\right]$

From (1.145), the area of the 2D k_s -space is given by

$$A(E, n_z) = \frac{2\pi\sqrt{m_1^* m_2^*}}{\hbar^2} I_1(E, n_z) \tag{1.146}$$

The EEm can be expressed using (1.146) as

$$m^*(E_{Fs}, n_z) \equiv \sqrt{m_1^* m_2^*}[I_1(E_{Fs}, n_z)]' \tag{1.147}$$

where $\{I_1(E, n_z)\}' \equiv \frac{\partial}{\partial E}[I_1(E, n_z)]$

Therefore, the EEM according to this model is a function of Fermi energy and size quantum number due to the presence of band nonparabolicity only

The DOS function per subband can be written as

$$N_{2D}(E) = \frac{4}{\pi}\frac{\sqrt{m_1^* m_2^*}}{\hbar^2}\{I_1(E, n_z)\}' \quad (1.148)$$

The total DOS function assumes the form

$$N_{2DT}(E) = \frac{4\sqrt{m_1^* m_2^*}}{\pi\hbar^2}\sum_{n_z=1}^{n_{zmax}}\{I_1(E, n_z)\}' H(E - E_{n_{z17}}) \quad (1.149)$$

where, the subband energy $E_{n_{z17}}$ are given by

$$E_{n_{z17}} = \frac{\hbar^2}{2m_3^*}\left(\frac{\pi n_z}{d_z}\right)^2\left[1 - \bar{e}_1\left(\frac{\hbar^2}{2m_3^*}\right)\left(\frac{\pi n_z}{d_z}\right)^2\right] \quad (1.150)$$

The electron statistics can be written as

$$n_{2D} = \frac{4\sqrt{m_1^* m_2^*}}{\pi\hbar^2}\sum_{n_z=1}^{n_{zmax}}[t_{46}(E_{Fs}, n_z) + t_{47}(E_{Fs}, n_z)] \quad (1.151)$$

where $t_{46}(E_{Fs}, n_z) \equiv I_1(E_{Fs}, n_z)$, $t_{47}(E_{Fs}, n_z) \equiv \sum_{r=1}^{S} L(r)(t_{46}(E_{Fs}, n_z))$.

1.2.12 The EEM in UFs of Gallium Antimonide

The dispersion relation of the conduction electrons in n-GaSb can be written as [236]

$$E = \frac{\hbar^2 k^2}{2m_0} - \frac{\bar{E}'_{g0}}{2} + \frac{\bar{E}'_{g0}}{2}\left[1 + \frac{2\hbar^2 k^2}{\bar{E}'_{g0}}\left(\frac{1}{m_c} - \frac{1}{m_0}\right)\right]^{\frac{1}{2}} \quad (1.152)$$

where $\overline{E}'_{g0} = \left[E_{g0} + \frac{5.10^{-5}T^2}{2(112 + T)}\right]$ eV

Equation (1.152) can be expressed as

$$\frac{\hbar^2 k^2}{2m_c} = I_{36}(E) \quad (1.153)$$

where

$$I_{36}(E) = [E + \bar{E}'_{g0} - (m_c/m_o)(\bar{E}'_{g0}/2) - [(\bar{E}'_{g0}/2)^2 + [((\bar{E}'_{g0})^2/2)(1 - (m_c/m_0))] + [(\bar{E}'_{g0}/2)(1 - (m_c/m_0))]^2 + E\bar{E}'_{g0}(1 - (m_c/m_0))]^{1/2}]$$

The 2D electron dispersion relation in this case assumes the form

$$\frac{\hbar^2 k_s^2}{2m_c} + \frac{\hbar^2}{2m_c}(n_z\pi/d_z)^2 = I_{36}(E) \tag{1.154}$$

Using (1.154) , the EEM in x–y plane for this case can be written as

$$m^*(E_{Fs}) = m_c\{I_{36}(E_{Fs})\}' \tag{1.155}$$

It appears that the EEM in this case is a function of Fermi energy alone and is independent of size quantum number.

The total 2D DOS function can be written as

$$N_{2DT}(E) = \left(\frac{m_c g_v}{\pi\hbar^2}\right)\sum_{n_z=1}^{n_{zmax}}\left\{[I_{36}(E)]' H(E - E_{n_{z17}})\right\} \tag{1.156}$$

where, the subband energies $E_{n_{z17}}$ can be expressed as $\mathring{u}$

$$I_{36}(E_{n_{z17}}) = \frac{\hbar^2}{2m_c}(n_z\pi/d_z)^2 \tag{1.157}$$

The 2D carrier concentration assumes the form

$$n_{2D} = \left(\frac{m_c g_v}{\pi\hbar^2}\right)\sum_{n_z=1}^{n_{zmax}}[t_{55}(E_{Fs}, n_z) + t_{56}(E_{Fs}, n_z)] \tag{1.158}$$

where $t_{55}(E_{Fs}, n_z) \equiv \left[I_{36}(E_{Fs}) - \frac{\hbar^2}{2m_c}\left(\frac{n_z\pi}{d_z}\right)^2\right]$

and $t_{56}(E_{Fs}, n_z) \equiv \sum_{r=1}^{s} L(r) t_{55}(E_{Fs}, n_z)$

1.3 Results and Discussions

Using (1.5) and (1.9) and taking the energy band constants as given in Table 1.1, we have plotted Fig. 1.1. The EEM in UFs of Cd_3As_2 as a function of film thickness and have been shown in Fig 1.1. For comparison, we have also plotted the EEM in the absence of the crystal-field splitting for the three- and the two-band models of Kane. Figure 1.1 exhibits the effect of size quantization on the EEM in general, and bears a good amount of discussion. It appears that the effect of van Hove singularity makes the EEM to suffer severe discontinuities. Assuming a carrier degeneracy of 10^{15} m^{-2}, Fig. 1.1a shows that the EEM can reach upto about 10% of its free

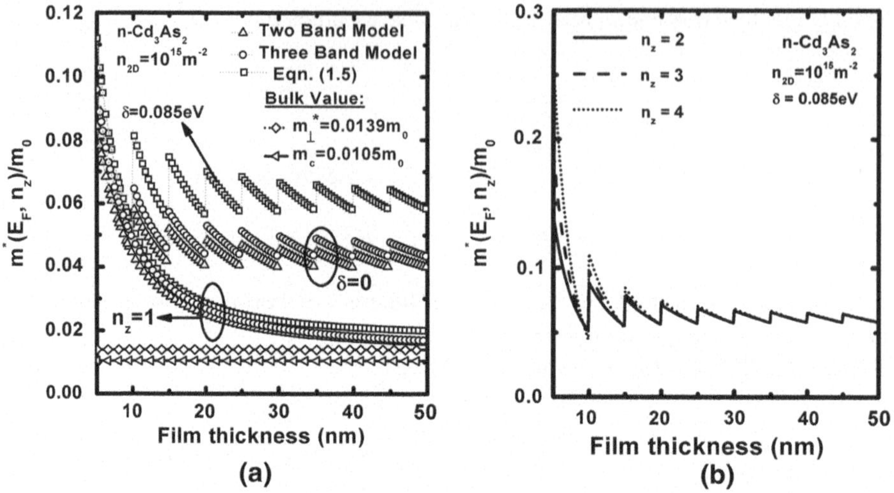

Fig. 1.1 **a** Plot of the EEM as function of film thickness for UFs of n-Cd_3As_2 considering (1.5). The *plots* for three-and two-band models of Kane have also been exhibited in which, $m_\perp^* = 0.0139m_0$ and $m_c = \frac{1}{2}(m_{11}^* + m_\perp^*)m_0 = 0.0105m_0$ are the corresponding bulk values. **b** Plot of the EEM as function of film thickness for UFs of n-Cd_3As_2 for all cases of Fig. 1.1a at different three subband levels

mass at a film thickness of 5 nm, which is quite high from its bulk value and may degrade the carrier mobility to a great extent. In the same figure, we have also demonstrated the effect of assuming only the lowest level subband. It appears that with this approximation, the EEM approaches to the bulk value $m_\perp^* = 0.0139m_0$ more quickly than that by considering the subbands. With this, it is now more obvious to note that the assumption of a single subband occupancy throughout leads to the practical approach to the determination of EEM. All the models of the single subband occupied curves tend to merge with the bulk value near 50 nm thickness. The increase in the EEM with the reduction of film thickness is due to the increased Fermi energy of the material. It must be noted that with such a highly doped system, the Fermi energy is determined by the carrier statistics equation. It is this Fermi energy which should be used in the determination of the EEM. This is not the case in an intrinsic material. In such a case, the Fermi energy coincides with the intrinsic energy level, which is very near to the energy band gap of the material and thus the variation of the energy band gap with the film thickness needs great concern. The variation of the energy band gap however is significant at extremely narrow film thickness, more in the region below sub-4 nm, a context which shall be highlighted in Chap. 8, where Applications and brief review of experimental results have been discussed. Thus, all the curves below such thickness are expected to suffer deviation with our existing theoretical model, if plotted. In all the subsequent geometry dependent curves, we have restricted ourselves above sub-4 nm regime. Since Cd_3As_2 crystals are usually grown as degenerate n-type specimens, the Fermi level mass will be the effective mass of consideration for transport in Cd_3As_2. Hence in the quantum limit, the effective mass at the Fermi level corresponding to the lowest electric sub-band will be the effective conductivity mass for electron transport in Cd_3As_2. It appears from these figures that the Fermi level mass is significantly influenced by the effects of size quantization particularly in tetragonal semiconductors like n- Cd_3As_2 having crystalline field effects and energy-dependent anisotropy of the effective mass. It has been found that the effective mass at the Fermi level depends on the size quantum number due to the combined influence of crystal-field splitting and the anisotropic spin-orbit splitting constant, resulting in different effective masses at the Fermi level corresponding to different electric subbands (the different effective masses being the same in the absence of field splitting as can be seen from Fig. 1.1a and b). It has further been observed that the different effective masses corresponding to different electric subbands closely approach each other, for a given film thickness, with increasing electron concentration and for a given electron concentration, with increasing film thickness. These are in conformity with expectations since both with increasing electron concentration at a given film thickness and with increasing film thickness for a given electron concentration, the effects of size quantization gradually become less and less significant. As in bulk specimens, the Fermi level mass increases with increasing carrier concentration at a given value of the film thickness. Besides, for particular values of the film thickness and electron concentration, the combined effect of $\delta \neq 0$ and $\Delta_{11} \neq \Delta_{\perp}$effect of crystal-field splitting is to reduce the effective mass corresponding to any particular subband. It may further be noted that if the direction normal to the film is taken as one of the transverse directions of the single ellipsoid at the zone center and not as the longitudinal direction as assumed in the present chapter, the effective mass at the Fermi level corresponding to any given subband would be somewhat different. Nevertheless, since the mass anisotropy in $Cd_3As_2$is indeed small as can be seen from the values of $P_{11}$and $P_{\perp}$ which are very close to each other, the arbitrary choice of the direction normal to the film with respect to the major axis of the ellipsoid would not result in a significant change in the effective mass at the Fermi level corresponding to a particular subband. The Fermi level mass should gradually become closer to that of bulk specimens with increasing film thickness since, for such thicknesses, the effects of size quantization are greatly diminished. This has also been confirmed in our present work. Furthermore, the general features of the effects of size quantization on the effective mass as discussed here would also be valid with the only exception that the effective mass at the Fermi level will be independent of the size quantum number in the absence of crystal-field splitting and anisotropic spin-orbit splitting constant for the III-V small-gap semiconductors since these semiconductors have nonparabolic energy bands obeying Kane's dispersion relation and the present chapter is based on the generalized Kane's model.

Figure 1.2 exhibits the plot EEM in UFs of n-$CdGeAs_2$ as a function of film thickness in accordance with the three- and two-band models of Kane together with the incorporation of the crystal-field parameter. It appears that the effect of the crystal-field splitting increases the EEM sharply below sub-20 nm. The EEM also increases about 7 % at 5 nm and converges to its bulk value beyond 20 nm at the same value of electron degeneracy. The effect of film thickness on the EEM of III-V semiconduc-

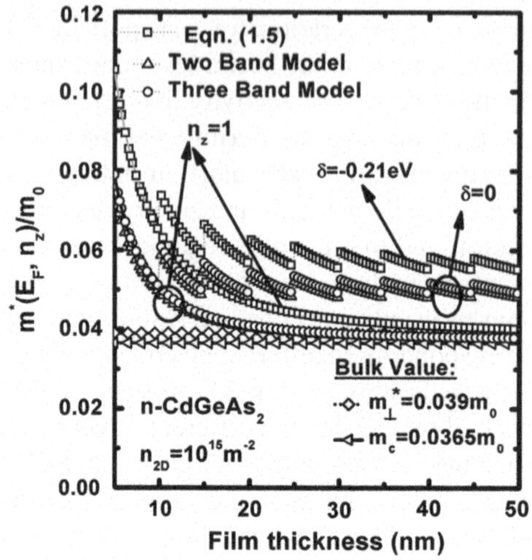

Fig. 1.2 Plot of the EEM as function of film thickness for UFs of n-$CdGeAs_2$ for all the cases of Fig. 1.1 in which, $m_{\perp}^{*} = 0.039m_0$ and $m_c = \frac{1}{2}\left(m_{11}^{*} + m_{\perp}^{*}\right)m_0 = 0.0105m_0$ are the corresponding bulk values

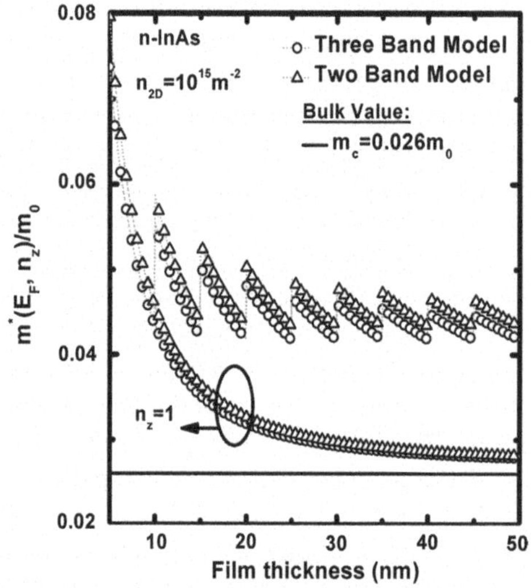

Fig. 1.3 Plot of the EEM as function of film thickness for UFs of n-InAs in accordance with the three- and the two-band model of Kane

tors, most important with respect to extremely low field high mobility of which are n-InAs, n-InSb, and n-GaAs has been exhibited in Figs. 1.3,1.4 and 1.5, respectively.

The effect of nonlinearity of the energy band structure on the respective EEMs has been clearly indicated. It appears that in the determination of the EEM, it is sufficient to take the two band model of Kane to explain the variation of the EEM over a wide range of thickness. The deviation from the three-band model of Kane is much less indicating that the complexity in the energy band model can be reduced to a large extent by considering only the two-band model of Kane. This is extremely important with respect to the numerical computation in device analyses performance where sufficient longer computation time affects the efficiency in characterizing the

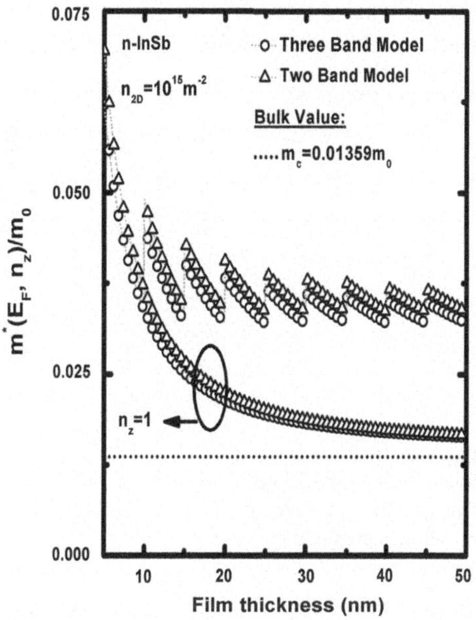

Fig. 1.4 Plot of EEM as function of film thickness for UFs of n-InSb for all the cases of Fig. 1.3

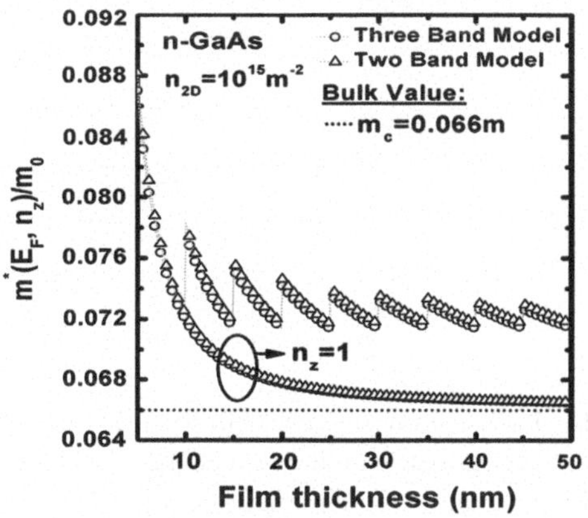

Fig. 1.5 Plot of the EEM as function of film thickness for UFs of n-GaAs for all the cases of Fig. 1.3

compact model with respect to the said materials. In all the Figs. 1.3, 1.4, 1.5, we have demonstrated the effect of two widely known models viz. the three and the two band models. Figures 1.6 and 1.7 exhibits the variation of the EEM with respect to the film thickness for the ternary and quaternary materials at same carrier degeneracy level. It appears that at an alloy composition $x = 0.3$, the EEM in both the cases tends to about 0.1 times the rest mass at film thickness of 5 nm. The effect of variation of EEM on the alloy composition for these two materials has been exhibited in Fig. 1.8.

The effect of increasing the alloy composition increases the EEM for the said two materials. For the purpose of comparison, we have also plotted the variation of

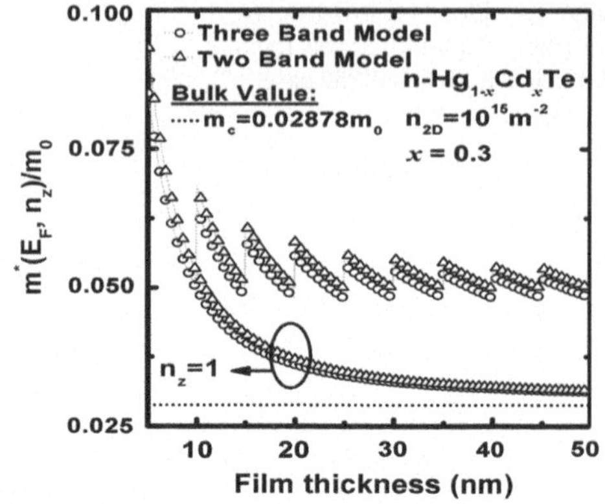

Fig. 1.6 Plot of the EEM as function of film thickness for UFs of n-$Hg_{0.3}Cd_{0.7}Te$ for all the cases of Fig. 1.3

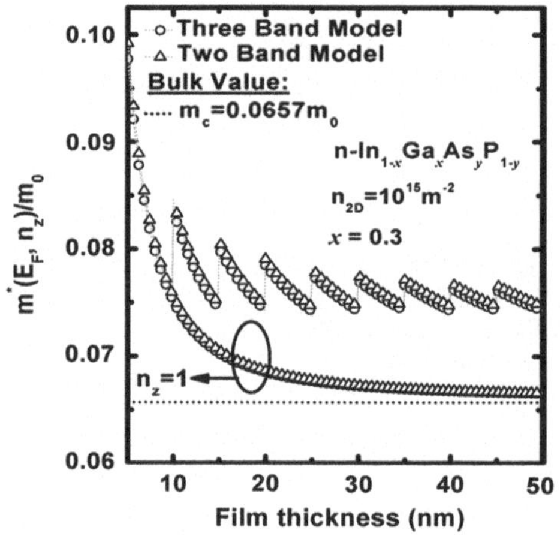

Fig. 1.7 Plot of the EEM as function of film thickness for UFs of n-$In_{1-x}Ga_xAs_yP_{1-y}$ for all the cases of Fig. 1.3

the bulk effective mass with the alloy composition. For the quaternary material, the difference between the two energy band models is not much, as can also be seen from Figs. 1.6 and 1.7. The increment in EEM is rather linear in case of InGaAsP than that of HgCdTe. This also exhibits the variation of the electron mobility in these systems as the alloy composition changes. It appears that with increase in x, the mobility falls down assuming a constant relaxation rate.

The effect of carrier degeneracy on the EEM in nonlinear optical, III-V, ternary and quaternary materials have been exhibited in Figs. 1.9, 1.10, 1.11, 1.12, 1.13, 1.14, 1.15. It appears that the EEM for all the aforementioned materials at 10 nm film thickness are almost invariant below sub 10^{15} m^{-2}. The effect of inclusion of both the higher order subbands and the lowest subband has been exhibited. From all the curves, it appears that the EEM bears almost exponential relation with the carrier degeneracy. This notion comes straightforward from the carrier concentration relation (1.27).

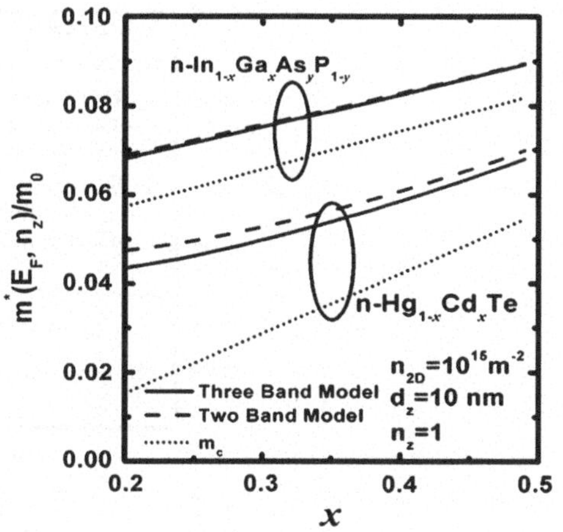

Fig. 1.8 Plot of the EEM at the lowest subband as function of alloy composition in UFs of n-$Hg_{1-x}Cd_x$ Te and n-$In_{1-x}Ga_xAs_yP_{1-y}$ for the three and the two band models of Kane respectively

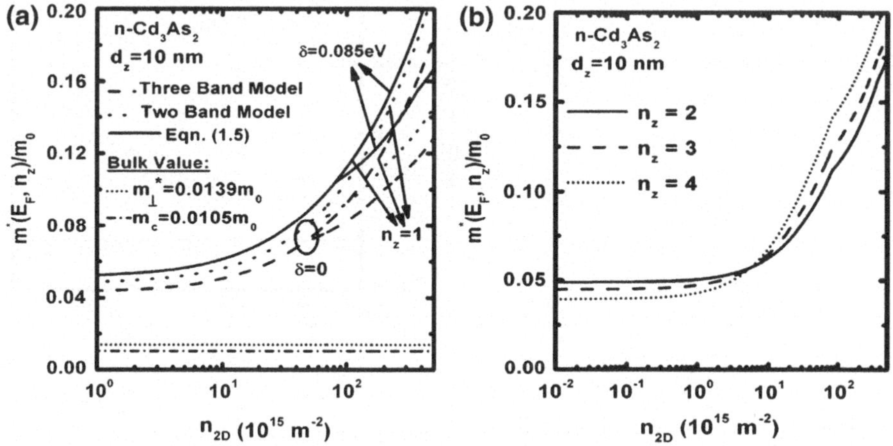

Fig. 1.9 **a** Plot of the EEM as function of surface electron concentration in UFs of n-Cd_3As_2. The plots for three- and two-band models of Kane have also been exhibited in which, $m_{\perp}^* = 0.0139m_0$ and $m_c = \frac{1}{2}\left(m_{11}^* + m_{\perp}^*\right) m_0 = 0.0105m_0$ are the corresponding bulk values. **b** Plot of the EEM as function of surface electron concentration in UFs of n-Cd_3As_2 at different subband levels for all cases of Fig. 1.9

The variation of the EEM in II-VI materials like p-CdS has been exhibited in Figs. 1.16 and 1.17 as functions of film thickness and Fermi energy respectively. In these two figures, instead of obtaining the Fermi energy from the corresponding carrier statistics, we have followed the opposite route, i.e., what values of the Fermi energy makes the EEM to be very low or very high. A corresponding concentration of that order can then be evaluated. A decision of this kind aids a good amount of estimation in the optimization. Using this approach, we estimate that the EEM can

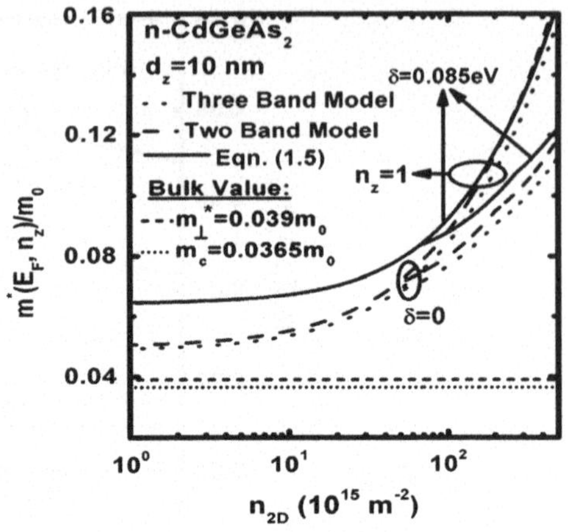

Fig. 1.10 Plot of the EEM as function of surface electron concentration in UFs of n-$CdGeAs_2$ for all the cases of Fig. 1.9

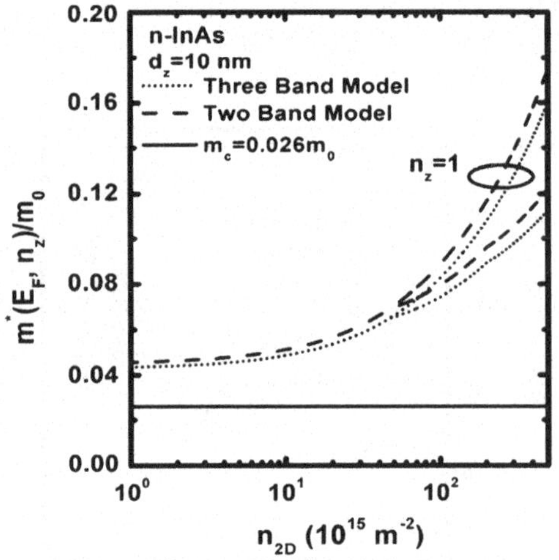

Fig. 1.11 Plot of the EEM as function of surface electron concentration in UFs of n-InAs

soar up to 0.77 times rest mass in the higher valley, while for the lower valley, it may plunge upto about 0.55 times rest mass.

The effect of valley degeneracy as we see from these two curves expresses much in understanding the electron transport direction.

It appears from the two curves that the channel oriented along the lower valley direction will most probably result in an increased value of current due to the low EEM. It would have been of much interest to figure out how the energy band gap at the two valleys changes with respect to the thickness and is left as an exercise to the reader.

Figures 1.18 and 1.19 exhibit the effect of film thickness and the carrier concentration on the EEM of UFs of Bismuth. The effect of increasing the carrier degeneracy has also been exhibited in Fig. 1.18. It appears that the EEM increases from its cor-

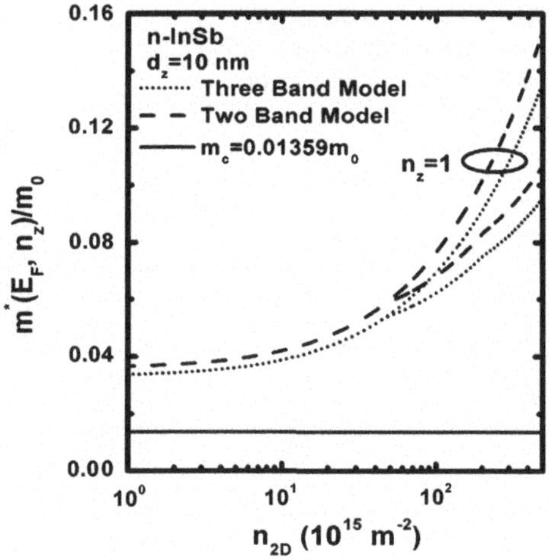

Fig. 1.12 Plot of the EEM as function of surface electron concentration in UFs of n-InSb

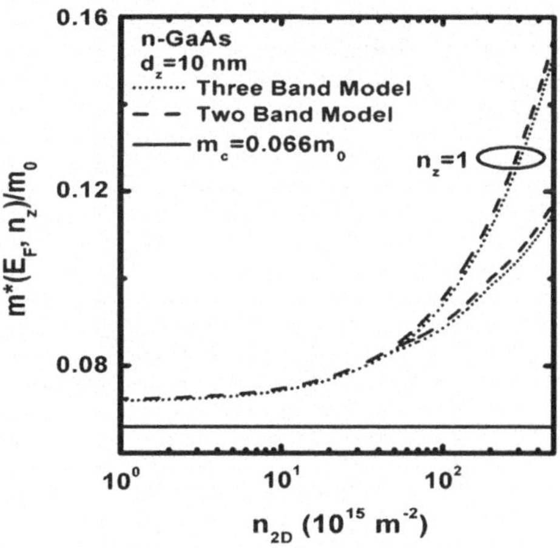

Fig. 1.13 Plot of the EEM as function of surface electron concentration in UFs of n-GaAs

responding bulk value sharply at the 5 nm film thickness implying a tremendous decrease in the carrier mobility.

Figure 1.19 exhibits the effect of different energy band model of Bi on EEM for a varying surface electron concentration. It appears that at the lowest subband energy level, there is almost no difference between the Mc Clure and Cohen model extracted EEM, however there is a significant change in the Hybrid and Lax ellipsoidal model. Figures 1.20, 1.21, and 1.22 exhibit the variation of the EEM at the lowest subband level for QWs of IV-VI, strained InSb and Ge. The effective mass in IV-VI materials exhibits strong variation for PbTe, an excellent thermoelectric material, whereas least for PbSnSe. It also appears that the EEM of PbTe is higher than that of PbSnSe and PbSnTe. With the advent of strained quantum effect devices, the analysis of EEM in strained quantum wells becomes very much important. It appears that the

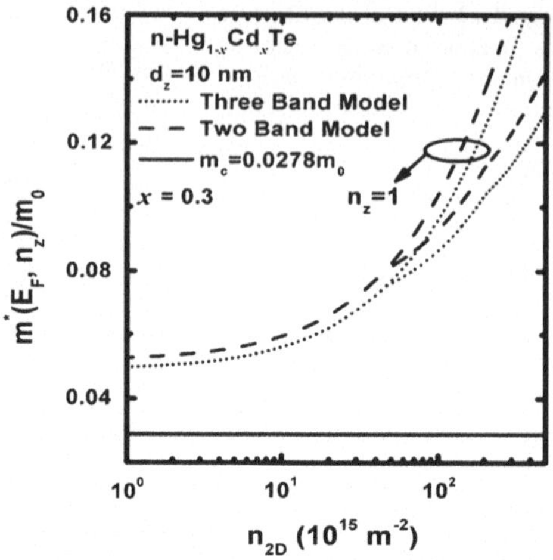

Fig. 1.14 Plot of the EEM as function of surface electron concentration in UFs of n-HgCdT

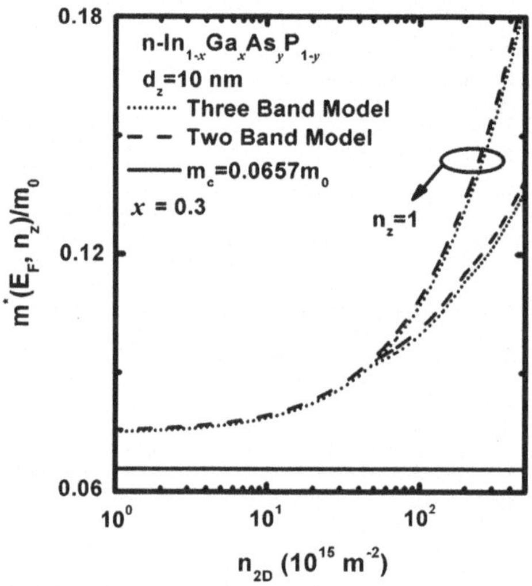

Fig. 1.15 Plot of the EEM as function of surface electron concentration in UFs of n-InGaAsP

compressive and tensile strain does not tend to modify the respective magnitude of the EEM in strained quantum wells of InSb. It should be noted that the EEM in Fig. 1.21 has been evaluated by considering the momentum matrix element $B_2 = 0.9$ eVnm. This is a bulk value. However, an arbitrary increase in this geometry dependent parameter sufficiently reduces the EEM and thus finds extensive use in strained film transitors. In Chap. 8, we shall be presenting a much detailed explanation of the effect of uniaxial and biaxial strain on Si nanowires and the effect on energy band gap. The variation of the EEM in Ge has been exhibited in Fig. 1.22 as function of film thickness for the model of Cardona et al. The general trend of increase in the EEM has also been exhibited here at least 6 times the bulk value $\sqrt{m_1^* m_2^*}$ for three different carrier concentration levels.

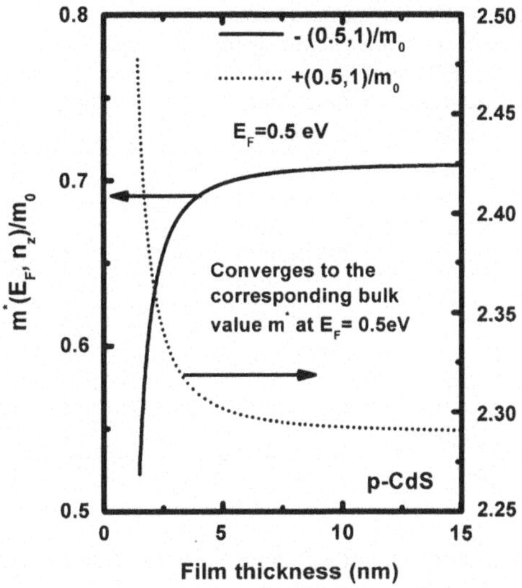

Fig. 1.16 Plot of the EEM as function of film thicknesss in UFs of p-CdS in two different conduction band valleys

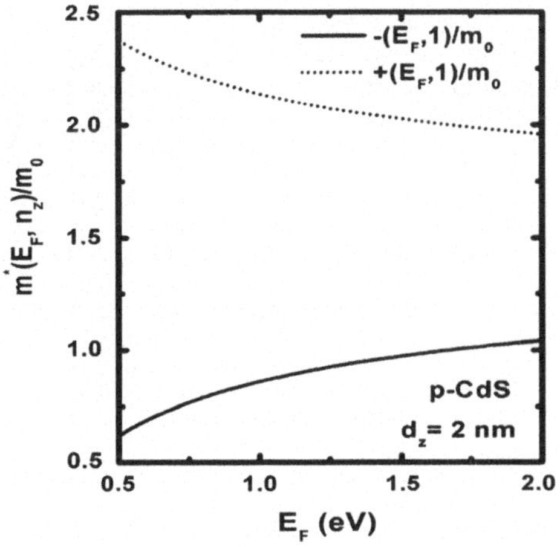

Fig. 1.17 Plot of the EEM as function of Fermi energy for all the cases of Fig. 1.16

We observe that considering the various subband levels, the EEM exhibits a step-functional decreasing dependence with increase in film thickness for UFs of all the single valley materials. The combined influence of the anisotropies of the energy band constants and the crystal-field splitting is to enhance the EEM as compared with the corresponding which is based on two band model of Kane in the whole range of thicknesses as considered in Fig. 1.1. The periodicity with respect to the film thickness is the same in both the cases and is invariant of the energy band constants.

The influence of quantum confinement is immediately apparent from Figs. 1.1, 1.2, 1.3, 1.4, 1.5, 1.6, 1.7 and 1.16, 1.18, 1.20, 1.21, 1.22 since the EEM depends

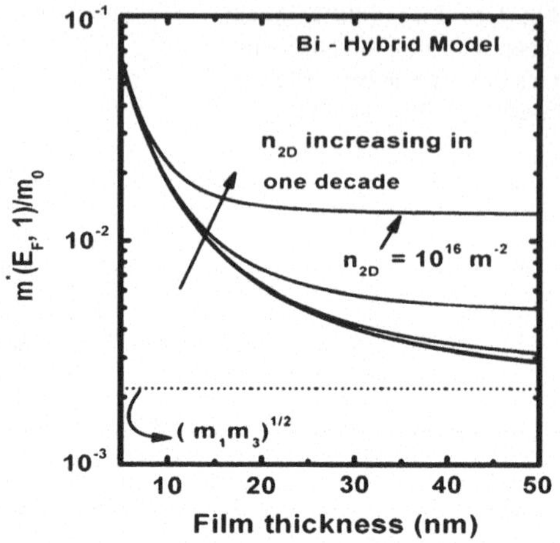

Fig. 1.18 Plot of the EEM as function of film thicknesss in UFs of Bismuth for different carrier concentration values using the Hybrid model

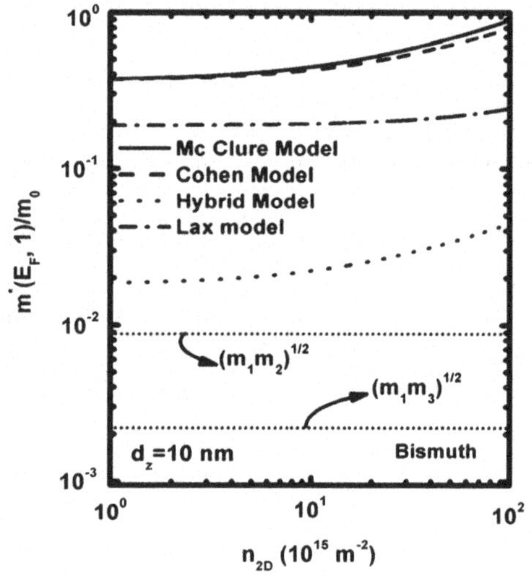

Fig. 1.19 Plot of the EEM at the lowest subband as function of surface electron coencentration in UFs of Bismuth using the Mc Clure, Cohen, Hybrid, and Lax energy band models

strongly on the thickness of the quantum-confined materials in contrast with the corresponding bulk specimens. The EEM changes with increasing carrier concentration suffering discontinuities with different numerical magnitudes. It appears from the aforementioned figures that the EEM exhibits spikes for particular values of film thickness which, in turn, depends on the particular band structure of the specific semiconductor. Moreover, the EEM from QWs of different compounds can be smaller than bulk specimens of the same materials having multi valley conduction band like in case of p-CdS, which is also a direct signature of quantum confinement. This effect of the discontinuity on the EEM will be less and less prominent with increasing film thickness. For bulk specimens of the same material, the EEM will be found to increase continuously with increasing electron degeneracy in a non-

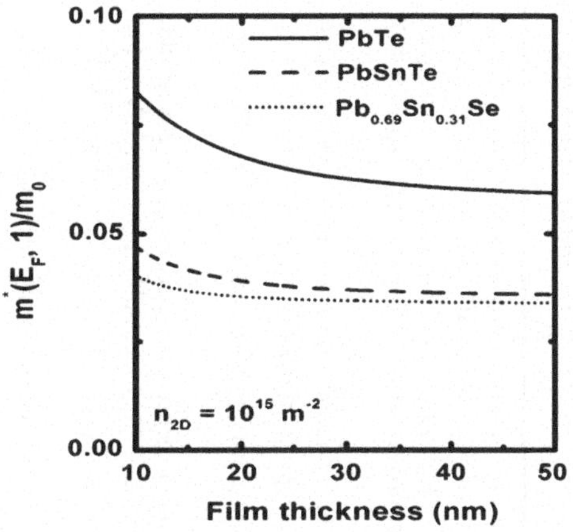

Fig. 1.20 Plot of the EEM at the lowest subband as function of film thickness in UFs of IV-VI materials

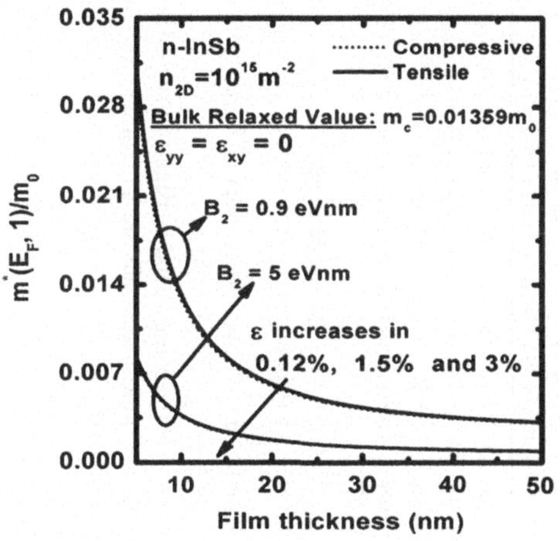

Fig. 1.21 Plot of the EEM at the lowest subband as function of film thickness in UFs of uniaxial strained InSb

oscillatory manner. The appearance of the discrete jumps in the respective figures is due to the redistribution of the electrons among the quantized energy levels when the size quantum number corresponding to the highest occupied level changes from one fixed value to the others.

With varying electron degeneracy, a change is reflected in the EEM through the redistribution of the electrons among the size-quantized levels. It may be noted that at the transition zone from one subband to another, the height of the peaks between any two subbands decreases with the increasing in the degree of quantum confinement and is clearly shown in the respective figures. It should be noted that although the EEM changes in various manners with all the variables as is evident from all the figures, the rates of variations are totally band-structure dependent.

It is imperative to state that the present investigation excludes the many-body, hot electron, broadening and the allied effects in the simplified theoretical formalism due to the absence of proper analytical techniques for including them for generalized

Table 1.1 The numerical values of the energy band constants of few materials

	Materials	Numerical values of the energy band constants
1	(a) The conduction electrons of n-cadmium germanium arsenide can be described by three types of band models	1. The values of the energy band constants in accordance with the generalized electron dispersion relation of nonlinear optical materials are as follows: $E_{g_0} = 0.57\,\text{eV}$, $\Delta_{\parallel} = 0.30\,\text{eV}$, $\Delta\perp = 0.36\,\text{eV}$, $m_{\parallel}^{*} = 0.034m_0$, $m_{\perp}^{*} = 0.039m_0$, $t = 4k$, $\delta = -0.21\,\text{eV}$, $g_v = 1$ [118, 119], $\varepsilon_{sc} = 18.4\varepsilon_0$ [120] (ε_{sc} and ε_0 are the permittivity of the semiconductor material and free space respectively) and W(electron affinity) = 4 eV [121, 122] 2. In accordance with the three-band model of Kane the spectrum constants are given by $\Delta = (\Delta_{\parallel} + \Delta_{\perp})/2 = 0.33\,\text{eV}$, $E_{g_0} = 0.57\,\text{eV}$, $m_c = \left(m_{\parallel}^{*} + m_{\perp}^{*}\right)/2 = 0.0365m_0$ and $\delta = 0\,\text{eV}$. 3. In accordance with two-band model of Kane, the spectrum constants are given by $E_{g_0} = 0.57\,\text{eV}$ and $m_c = (m_{\parallel}^{*} + m_{\perp}^{*})/2 = 0.0365m_0$
	(b) The conduction electrons of n-cadmium arsenide can be described by three types of band models	1. The values of the energy band constants in accordance with the generalized electron dispersion relation of nonlinear optical materials are as follows $\lvert E_{g_0}\rvert = 0.095\,\text{eV}$, $\Delta_{\parallel} = 0.27\,\text{eV}$, $\Delta_{\perp} = 0.25\,\text{eV}$, $m_{\parallel}^{*} = 0.00697m_o$, $m_{\perp}^{*} = 0.013933m_0$, $T = 4K$, $\delta = 0.085\,\text{eV}$, $g_v = 1$ [118, 119] and $\varepsilon_{sc} = 16\varepsilon_0$ [121, 122] 2. In accordance with the three-band model of Kane, the spectrum constants are given by $\Delta = (\Delta_{\parallel} + \Delta_{\perp})/2 = 0.26\,\text{eV}$, $\lvert E_{g_0}\rvert = 0.095\,\text{eV}$, $m_c = \left(m_{\parallel}^{*} + m_{\perp}^{*}\right)/2 = 0.0105m_0$ and $\delta = 0\,\text{eV}$ 3. In accordance with two-band model of Kane, the spectrum constants are given by $\lvert E_{g_0}\rvert = 0.095\,\text{eV}$, and $m_c = \left(m_{\parallel}^{*} + m_{\perp}^{*}\right)/2 = 0.0105m_0$
2	n-Indium arsenide	The values $E_{g_0} = 0.36\,\text{eV}$, $\Delta = 0.43\,\text{eV}$, $m_c = 0.026m_0$, $g_v = 1$, $\varepsilon_{sc} = 12.25\varepsilon_0$ [111] and $W = 5.06\,\text{eV}$ [123] are valid for three-band model of Kane
3	n-Gallium arsenide	The values $E_{g_0} = 1.55\,\text{eV}$, $\Delta = 0.35\,\text{eV}$, $m_c = 0.066m_0$, $g_v = 1$, $\varepsilon_{sc} = 12.9\varepsilon_0$ [111] and $W = 4.07\,\text{eV}$ [124] are valid for three-band model of Kane. The values $a_1 3 = -1.97 \times 10^{-37}\,\text{eV}m^4$ and $a_{15} = -2.3 \times 10^{-34}\,\text{eV}m^4$ [125] are valid for the Newson and Kurobe model [125]
4	n-Gallium aluminium arsenide	$E_{g_0} = (1.424 + 1.266x + 0.26x^2)\,\text{eV}$, $\Delta = (0.34 - 0.5x)\,\text{eV}$, $g_v = 1$, $m_c = (0.066 + 0.088x)m_0$, $\varepsilon_{sc} = [13.18 - 3.12x]\varepsilon_0$[1] and $W = (3.64 - 0.14x)\,\text{eV}$ [126]
5	n-Mercury cadmium telluride	$E_{g_0} = (-0.302 + 1.93x + 5.35 \times 10^{-4}(1 - 2x)T - 0.810x^2 + 0.832x^3)\,\text{eV}$, $\Delta = (0.63 + 0.24x - 0.27x^2)\,\text{eV}$, $m_c = 0.1m_0 E_{g0}(\text{eV})^{-}1$, $g_v = 1$, $\varepsilon_{sc} = [20.262 - 14.812x + 5.22795x^2]\varepsilon_0$ [108] and $W = (4.23 - 0.183(E_{g_0} - 0.083))\,\text{eV}$ [127]
	n-Indium gallium arsenide	$E_{g_0} = (1.337 - 0.73y + 0.13y^2)\,\text{eV}$, $\delta = (0.114 + 0.26y - 0.22y^2)\,\text{eV}$

(continued)

Table 1.1 (continued)

	Materials	Numerical values of the energy band constants
6	Phosphide lattice matched to indium phosphide	$y = (0.1896\text{–}0.4052x)/(0.1896\text{–}0.0123x)$, $m_c = (0.08\text{–}0.039y)m_0$, $g_v = 1$, $\varepsilon_{sc} = [10.65+0.1320y]\varepsilon_0$ and [128] $W(x, y) = [5.06(1-x)y + 4.38(1-x)(1-y) + 3.64xy + 3.75\{x(1-y)\}]\,\text{eV}$
7	n-Indium antimonide	$E_{g_0} = 0.2352\,\text{eV}$, $\Delta = 0.81\,\text{eV}$, $m_c = 0.01359m_0$, $g_v = 1$, $\varepsilon_{sc} = 15.56\varepsilon_0$ [121, 122] and $W = 4.72\,\text{eV}$ [123]
8	n-Gallium antimonide	The values of $E_{g_0} = 0.81\,\text{eV}$, $\Delta = 0.80\,\text{eV}$, $p = 9.48 \times 10^{-10}\,\text{eVm}$, $\bar{\zeta}_0 = -2.1$, $\bar{v}_0 = -1.49$, $\bar{\omega} = 0.42$, $g_v = 1$ [129] and [130] are valid for the model of Seiler et al. [129]
9	n-Cadmium sulphide	$m_{\parallel}^* = 0.7m_0$, $m_{\perp}^* = 1.5m_0$, $C_0 = 1.4 \times 10^{-8}\,\text{eV}m$, $g_v = 1$ [121, 122], $\varepsilon_{sc} = 15.5\varepsilon_0$ [131] and $W = 4.5\,\text{eV}$ [123]
10	n-Lead telluride	The values $m_1^- = 0.070m_0$, $m_1^- = 0.54m_0$, $m_1^+ = 0.010m_0$, $m_1^+ = 1.4m_0$, $P_{\parallel} = 141\,\text{meVnm}$, $P_{\perp} = 486\,\text{meVnm}$, $E_{g_0} = 190meV$, $g_v = 4$ [121, 122], $\varepsilon_{sc} = 33\varepsilon_0$ [121, 122, 132] and $W = 4.6\,\text{eV}$ [133] are valid for the Dimmock model [134]. The values $m_1 = 0.0239m_0$, $m_2 = 0.024m_0$, $m_2' = 0.31m_0$, $m_3 = 0.24m_0$ [135] are valid for the Cohen model [136–141].
11	Stressed n-indium antimonide	The values $m_c = 0.01359m_0$, $E_{g_0} = 0.081\,\text{eV}$, $B_2 = 9 \times 10^{-10}\,\text{eVm}$, $C_1 = 3\,\text{eV}$, $C_2 = 2\,\text{eV}$, $\bar{a}_0 = -10\,\text{eV}$, $\bar{b}_0 = -1.7\,\text{eV}$, $\bar{d}_0 = -4.4\,\text{eV}$, $S_{xx} = 0.6 \times 10^{-3}(kbar)^{-1}$, $S_{yy} = 0.42 \times 10^{-3}(kbar)^{-1}$, $S_{zz} = 0.39 \times 10^{-3}(kbar)^{-1}$, $S_{xy} = 0.5 \times 10^{-3}(kbar)^{-1}$, $\varepsilon_{xx} = \sigma S_{xx}$, $\varepsilon_{yy} = \sigma Syy$, $\varepsilon_{zz} = \sigma S_{zz}$, $\varepsilon_{xy} = \sigma Sxy$, σ is the stress in kilobar, $g_v = 1$ [142] are valid for the model of Seiler et al. [142]
12	Bismuth	$E_{g_0} = 0.0153\,\text{eV}$, $m_1 = 0.00194m_0$, $m_2 = 0.313m_0$, $m_3 = 0.00246m_0$, $m_2' = 0.36m_0$, $g_v = 3$ [143, 144], $M_2 = 1.25m_0$, $M_2' = 0.36m_0$ [145, 146] and $W = 4.34\,\text{eV}$.
13	Mercury telluride	$m_v^* = 0.028m_0$, $g_v = 1$, $\varepsilon_\infty = 15.2\varepsilon_0$ [147] and $W = 5.5\,\text{eV}$ [148]
14	Platinum antimonide	For valence bands, along <100> direction, $\bar{\lambda}_0 = (0.02/4)\,\text{eV}$, $\bar{l} = (-0.32/4)\,\text{eV}$, $\bar{v} = (0.39/4)\,\text{eV}$, $\bar{n} = (-0.65/4)\,\text{eV}$, $\bar{a} = 0.643\,\text{nm}$, $I = 0.30(\text{eV})^2$, $\bar{\delta}_0 = 0.02\,\text{eV}$, $g_v = 6$ [151], $\varepsilon_{sc} = 30\varepsilon_0$ [149] and $\phi_w \approx 3.0\,\text{eV}$ [150, 151] For conduction bands, along <111> direction, $g_v = 8$ [149, 151], $\bar{\lambda}_0 = (0.33/4)\,\text{eV}$, $\bar{l} = (1.09/4)\,\text{eV}$, $\bar{v} = (0.17/4)\,\text{eV}$ and $\bar{n} = (0.22/4)\,\text{eV}$
15	n-Gallium phosphide	$m_{\parallel}^* = 0.92m_0$, $m_{\perp}^* = 0.25m_0$, $k_0 = 1.7 \times 10^{15}m^{-1}$, $\|V_G\| = 0.21\,\text{eV}$, $g_v = 6$ [152] and $W = 3.75\,\text{eV}$ [123]
16	Germanium	$E_{g_0} = 0.785\,\text{eV}$, $m_{\parallel}^* = 1.57m_0$, $m_{\perp}^* = 0.0807m_0$ [123] and $W = 4.14\,\text{eV}$ [153], $g_v = 4$
17	Tellurium	The values $\psi_1 = 6.7 \times 10^{-16}meVm^2$, $\psi_2 = 4.2 \times 10^{-16}meVm^2$, $\psi_3 = 6 \times 10^{-8}\,\text{meVm}$ and $\psi_4 = (3.6 \times 10^{-8}\,\text{meVm})$ [154–158] are valid for the model of Bouat et al. [154–158]

(continued)

Table 1.1 (continued)

	Materials	Numerical values of the energy band constants
18	Graphite	The values $\bar{\Delta} = -0.0002\,\text{eV}, \bar{\gamma}_1 = 0.392\,\text{eV}, \bar{\gamma}_5 = 0.194\,\text{eV}, c_6 = 0.674\,\text{nm}, \bar{\gamma}_2 = -0.019\,\text{eV}, a_6 = 0.246nm, \bar{\gamma}_0 = 3\,\text{eV}, \bar{\gamma}_4 = 0.193\,\text{eV}, \bar{\gamma}_3 = 0.21\,\text{eV}$ [159–163] and $W = 4.6\,\text{eV}$ [164–167] are valid for the model of Brandt et al. [159–163]
19	Lead germanium telluride	The values $g_v = 4$ [168] and $\phi_w \approx 6\,\text{eV}$ [169, 170] are valid for the model of Vassilev [208] as given by (R.1.18)
20	Cadmium antimonide	The values $a_1 = -32.3 \times 10^{-20}\,\text{eVm}^2, b_1 = -60.7 \times 10^{-20}\,\text{eVm}^2, a_2 = -16.3 \times 10^{-20}\,\text{eVm}^2, b_2 = -24.4 \times 10^{-20}\,\text{eVm}^2, a_3 = -91.9 \times 10^{-20}\,\text{eVm}^2, b_3 = -105 \times 10^{-20}\text{eVm}^2 A = 2.92 \times 10^{-10}\,\text{eVm}, B = -3.47 \times 10^{-10}\,\text{eVm}, G_3 = 1.3 \times 10^{-10}\,\text{eVm}, \Delta_3 = 0.070eV$ [171] and $\phi_w \approx 2\,\text{eV}$ [172–175]
21	Cadmium diphosphide	The values $\beta_1 = 8.6 \times 10^{-21}\,\text{eVm}^2, \beta_2 = 1.8 \times 10^{-21}\,\text{eVm}^2, \beta_4 = 0.0825\,\text{eV}, \beta_5 = -1.9 \times 10^{-19}\,\text{eVm}^2$ [176–183] and $\phi_w \approx 5\,\text{eV}$ [184] are valid for the model of Chuiko [190]
22	Zinc diphosphide	The values $\beta_1 = 8.7 \times 10^{-21}\,\text{eVm}^2, \beta_2 = 1.9 \times 10^{-21}\,\text{eVm}^2, \beta_4 = 0.0875\,\text{eV}, \beta_5 = -1.9 \times 10^{-19}\,\text{eVm}^2$ [176–183] and $W \approx 3.9\,\text{eV}$ [184] are valid for the model of Chuiko [190]
23	Bismuth telluride	The values $E_{g0} = 0.145\,\text{eV}, \bar{\alpha}_{11} = 4.9, \bar{\alpha}_{22} = 5.92, \bar{\alpha}_{33} = 9.5, \bar{\alpha}_{23} = 4.22, g_v = 6$ [185–188] and $\phi_w = 5.3\,\text{eV}$ [189]
24	Carbon nanotube	The values $a_c = 0.144\,\text{nm}$ [190], $t_c = 2.7\,\text{eV}$ [191, 192], $\bar{r}_0 = 0.7nm$ [193] and $W = 3.2\,\text{eV}$ [96] are valid for graphene band structure realization of carbon nanotube [195]
25	Antimony	The values $\alpha_{11} = 16.7, \alpha_{22} = 5.98, \alpha_{33} = 11.61, \alpha_{23} = 7.54$ [195–197] and $W = 4.63\,\text{eV}$ are valid for the model of Ketterson [195–197]
26	Zinc selenide	$m_{c2} = 0.16m_0, \Delta = 0.42\,\text{eV}, E_{g02} = 2.82\,\text{eV}$ [153] and $W = 3.2\,\text{eV}$ [198–200]
27	Lead selenide	$m_t^- = 0.23m_0, m_l^- = 0.32m_0, m_t^+ = 0.115m_0, m_l^+ = 0.303m_0, P_\parallel \approx 138\,\text{meVnm}, P_\perp = 471\,\text{meVnm}, E_{g0} = 0.28\,\text{eV}$ [201], $\varepsilon_{sc} = 21.0\varepsilon_0$ [153] and $W = 4.2\,\text{eV}$ [202–206]

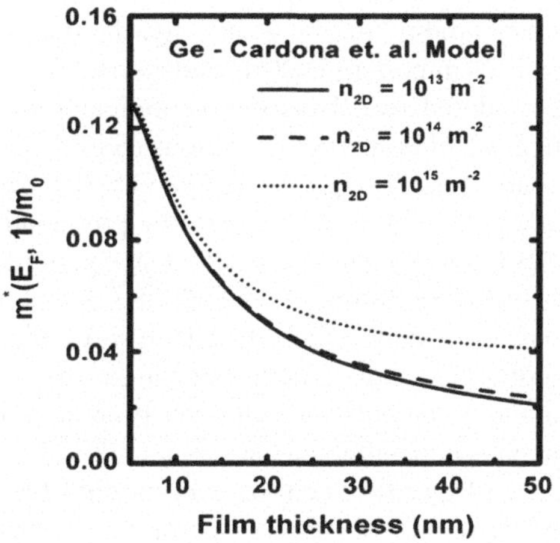

Fig. 1.22 Plot of the EEM at the lowest subband as function of film thickness in UFs of Ge

systems as considered here. We have also approximated the variation of value of the work function from its bulk value in the present system. Our simplified approach will be appropriate for the purpose of comparison when the methods of tackling the formidable problems after inclusion of the said effects for the generalized systems emerge. The results of this simplified approach get transformed to the well-known formulation of the EEM for wide gap materials having parabolic energy bands. This indirect test not only exhibits the mathematical compatibility of the formulation but also shows the fact that this simple analysis is a more generalized one, since one can obtain the corresponding results for materials having parabolic energy bands under certain limiting conditions from the present derivation. For the purpose of computer simulations for obtaining the plots of EEM versus various external variables, we have taken very low temperatures since the quantization effects are basically low temperature phenomena together with the fact that the temperature dependence of all the energy band constants of all the semiconductors and their nanostructures as considered in this chapter are not available in the literature. Our results as formulated in this chapter are valid for finite temperatures and are useful in comparing the results for temperature variations of EEM after the availability of the temperature dependences of such constants of various dispersion relations in this context. It is worth noting that the nature of the curves of EEM with various physical variables based on our simplified formulations as presented here would be useful to analyze the experimental results when they materialize. The inclusion of the said effects would certainly increase the accuracy of the results although the qualitative features of EEM would not change in the presence of the aforementioned effects.

It can be noted that on the basis of the dispersion relations of the various quantized structures as discussed above the low field carrier mobility, drive currents in field effect transistors, Fowler Nordhiem field current, the Debye screening length, the plasma frequency, the activity coefficient, the carrier contribution to the elastic constants, the diffusion coefficient of minority carriers, the third-order nonlinear optical susceptibility, the heat capacity, the dia and paramagnetic susceptibilities and the various important DC/AC transport coefficients can be probed for all types of UFs as considered here. Thus, our theoretical formulation comprises the dispersion relation-dependent properties of various technologically important quantum-confined semiconductors having different band structures. We have not considered other types of compounds in order to keep the presentation concise and succinct. With different sets of energy band parameters, one gets different numerical values of the EEM. The nature of variations of the EEM as shown here would be similar for the other types of materials and the simplified analysis of this chapter exhibits the basic qualitative features of the EEM. The reader can also explore the EEM for the leftover 2D materials to enjoy the intricate computer programming and the 2D physics in this context. It may be noted that the basic aim of this chapter is not solely to demonstrate the influence of quantum confinement on the EEM for a wide class of quantized materials but also to formulate the appropriate carrier statistics in the most generalized form, since the transport and other phenomena in modern nano-structured devices having different band structures and the derivation of the expressions of many important carrier properties are based on the temperature-dependent carrier statistics in such systems.

For the purpose of condensed presentation, the carrier statistics and the EEM in different materials as considered in this chapter have been presented in Table 1.2.

1.4 Open Research Problems

The problems under these sections of this monograph are by far the most important part and a few open research problems from this chapter till the end are being presented. The numerical values of the energy band constants for various semiconductors are given in Table 1.1 *for the related computer simulations.*

R.1.1. Investigate the effective acceleration mass (EAM), density-of-state effective mass (DEM), concentration effective mass (CEM), conductivity effective mass (CoEM), Faraday rotation effective mass (FREM), and Optical effective mass (OEM) from all the bulk semiconductors whose respective dispersion relations of the carriers are given in this chapter.

R.1.2. Repeat R.1.1 for the bulk semiconductors whose respective dispersion relations of the carriers in the absence of any field are given below:

(a) The electron dispersion law in n-GaP can be written as [237]

$$E = \frac{\hbar^2 k_s^2}{2m_{\|}^*} + \frac{\hbar^2 k_s^2}{2m_{\perp}^*} \mp \frac{\overline{\Delta}}{2} \pm \left[\left(\frac{\overline{\Delta}}{2} \right)^2 + P_1 k_z^2 + D_1 k_x^2 k_y^2 \right]^{1/2} \tag{R1.1}$$

Table 1.2 The EEM and the electron statistics in ultrathin films of nonparabolic semiconductors

Type of materials	EEM	The electron statistics
1. Non-linear optical semiconductors	$m^*(E_{Fs}, n_z) = \left(\frac{\hbar^2}{2}\right)[\psi_2(E_{Fs})]^{-2} \left[\psi_2(E_{Fs})\left\{\{\psi_1(E_{Fs})\}' - \{\psi_3(E_{Fs})\}'\left(\frac{n_z\pi}{d_z}\right)^2\right\} - \left\{\psi_1(E_{Fs}) - \psi_3(E_{Fs})\left(\frac{n_z\pi}{d_z}\right)^2\right\}\{\psi_2(E_{Fs})\}'\right]$ (1.5)	$n_{2D} = \frac{g_v}{2\pi}\sum_{n_x=1}^{n_{xmax}}[T_{51}(E_{Fs}, n_z) + T_{52}(E_{Fs}, n_z)]$ (1.9)
2. III-V semiconductors i.		
(a) Three band model of Kane	$m^*(E_{Fs}) = m_c\{I_{11}(E_{Fs})\}'$ (1.12)	$n_{2D} = \frac{m_c g_v}{\pi\hbar^2}\sum_{n_z=1}^{n_{zmax}}[T_{53}(E_{Fs}, n_z) + T_{54}(E_{Fs}, n_z)]$ (1.15)
(b) Two band model of Kane	$m^*(E_{Fs}) = m_c(1 + 2\alpha E_{Fs})$ (1.18)	$n_{2D} = \frac{m_c k_B T g_v}{\pi\hbar^2}\sum_{n_z=1}^{n_{zmax}}\left[(1 + 2\alpha E_{n_{z3}} F_0(\eta_{n_1}) + 2\alpha k_B T F_1(\eta_{n_1})\right]$ (1.21)
(c) Parabolic energy bands	$m^*(E_{Fs}) = m_c$ (1.26)	
ii. Stillman et al.	$m^*(E_{Fs}) = m_c\{I_{12}(E_{Fs})\}'$ (1.31)	$n_{2DT} = \frac{m_c k_b T g_v}{\pi\hbar^2}\sum_{n_z=1}^{n_{zmax}} F_0(\eta_{n_2})$ (1.27)
iii. Palik et al.	$m^*(E_{Fs}) = m_c\{I_{13}(E_{Fs})\}'$ (1.38)	$n_{2D} = \frac{m_c g_v}{\pi\hbar^2}\sum_{n_z=1}^{n_{zmax}}\left[T_{55}(E_{Fs}, n_z) + T_{56}(E_{Fs}, n_z)\right]$ (1.34)
		$n_{2D} = \frac{m_c g_v}{\pi\hbar^2}\sum_{n_z=1}^{n_{zmax}}\left[T_{57}(E_{Fs}, n_z) + T_{58}(E_{Fs}, n_z)\right]$ (1.41)
3. II-VI semiconductors	$m^*(E_{Fs}, n_z) = m_\perp^* \left[1 \mp \frac{(\bar{\lambda}_o)}{\left[(\bar{\lambda}_o)^2 - 4a_o' b_o'\left(\frac{n_z\pi}{d_z}\right)^2 + 4a_o' E_{Fs}\right]^{1/2}}\right]$ (1.44)	$n_{2D} = \frac{g_v m_\perp^*}{\pi\hbar^2}\sum_{n_z=1}^{n_{z\max}}\left(E_{Fs} - E_{n_{z5}} + \left(\bar{\lambda}\right)^2 m_\perp^* \hbar^{-2}\right)$ (1.48)

(continued)

Table 1.2 (continued)

Type of materials	EEM	The electron statistics
4. Bismuth		
i. McClure and Choi model	$m^*(E_{Fs}, n_z) = \left(\frac{2\hbar^2}{3\pi\sqrt{p_1}}\right)[R_3(E, n_z)]\Big\vert_{E=E_{Fs}}$ (1.54)	$n_{2D}(E) = \left(\frac{2g_v}{3\pi^2\sqrt{p_1}}\right)\sum_{n_z=1}^{n_{max}}[\theta_1(E_{Fs}, n_z) + \theta_2(E_{Fs}, n_z)$ (1.57)
ii. Hybrid model	$m^*(E_{Fs}, n_y) = [\sqrt{m_1 m_3}]t'_{29}(E, n_y)$ (1.61)	$n_{2D}(E) = \frac{g_v\sqrt{m_1 m_3}}{\pi\hbar^2}\sum_{n_y=1}^{n_{y_{max}}}\left[t_{29}(E_{Fs}, n_y) + t_{30}(E_{Fs}, n_y)\right]$ (1.64)
iii. Cohen model	$m^*(E_{Fs}, n_z) = \left(\frac{2\hbar^2}{3\pi\sqrt{p_1}}\right)\left[R_4(E, n_z)\right]\Big\vert_{E=E_{Fs}}$ (1.68)	$n_{2D} = \left(\frac{2g_v}{3\pi^2\sqrt{p_1}}\right)\sum_{n_z=1}^{n_{z_{max}}}\left[\theta_3(E_{Fs}, n_z) + \theta_4(E_{Fs}, n_z)\right]$ (1.70)
iv. Lax model	$m^*(E_{Fs}) = \sqrt{m_1 m_2}(1 + 2\alpha E_{Fs})$ (1.73)	$n_{2D} = \frac{g_v\sqrt{m_1 m_2}k_B T}{\pi\hbar^2} \times \sum_{n_z=1}^{n_{z_{max}}}\left[(1 + 2\alpha E_{n_z 8})F_0(\eta_{y2}) + 2\alpha k_B T F_1(\eta_{y2})\right]$ (1.76)
v. Ellipsoidal parabolic model	$m^*(E_{Fs}) = \left(\sqrt{m_1 m_2}\right)$ (1.78)	$n_{2D} = \left[\frac{k_B T g_v\sqrt{m_1 m_2}}{\pi\hbar^2}\right]\sum_{n_z=1}^{n_{z_{max}}}F_0(\eta_{y_3})$ (1.81)
5. IV-VI semiconductors	$m^*(E, n_z) = \frac{\hbar^2}{2}[\theta_5(E, n_z)]\Big\vert_{E=E_{Fs}}$ (1.94)	$n_{2D} = \frac{g_v}{2\pi}\sum_{n_z=1}^{n_{z_{max}}}[T_{59}(E_{Fs}, n_z) + T_{60}(E_{Fs}, n_z)]$ (1.97)
6. Stressed materials	$m^*(E_{Fs}, n_z) = \frac{\hbar^2}{2}[\theta_6(E, n_z)]\Big\vert_{E=E_{Fs}}$ (1.101)	$n_{2D} = \frac{g_v}{2\pi}\sum_{n_z=1}^{n_{z_{max}}}[T_{61}(E_{Fs}, n_z) + T_{62}(E_{Fs}, n_z)]$ (1.104)

Table 1.2 (continued)

Type of materials	EEM	The electron statistics
7. Tellurium	$m^*(E_{Fs}, n_z) = \frac{\hbar^2}{2}\left[t'_{40}(E, n_z)\right]\Big\vert_{E=E_{Fs}}$ (1.107)	$n_{2D} = \frac{g_v}{\pi} \sum_{n_z=1}^{n_{zmax}} [t_{40}(E_{Fs}, n_z) + t_{41}(E_{Fs}, n_z)]$ (1.110)
8. n-GaP	$m^*(E_{Fs}, n_z) = \frac{\hbar^2}{2} t'_{42}(E_{Fs}, n_z)$ (1.115)	$n_{2D} = \frac{g_v}{4\pi a^2} \sum_{n_z=1}^{n_{zmax}} \left[t_{42}(E_{Fs}, n_z) + t_{43}(E_{Fs}, n_z)\right]$ (1.117)
9. n type $PtSb_2$	$m^*(E_{Fs}, n_z) = \frac{\hbar^2}{4A_9} t'_{44}(E_{Fs}, n_z)$ (1.123)	$n_{2D} = \frac{g_v}{4\pi A_9} \sum_{n_z=1}^{n_{zmax}} [t_{44}(E_{Fs}, n_z) + t_{45}(E_{Fs}, n_z)]$ (1.126)
10. Bi_2Te_3	$m^*(E_{Fs}) = \frac{m_0(1+2\bar{\alpha}E_{Fs})}{\sqrt{\bar{\alpha}_{22}\bar{\alpha}_{33} - 4\bar{\alpha}_{23}^2}}$ (1.130)	$n_{2D} = \frac{k_B T g_v}{\pi\hbar^2}\left(\frac{m_0}{\sqrt{\bar{\alpha}_{22}\bar{\alpha}_{33} - 4\bar{\alpha}_{23}^2}}\right) \sum_{n_z=1}^{n_{zmax}} [(1+2\alpha E_{n_{z15}})F_0(\eta_{n_{n15}})$ $+2\alpha k_B T F_1(\eta_{n15})]$ (1.133)
11. n- Ge:		
i. Model by Cardona et al.	$m^*(E_{Fs}, n_z) \equiv \sqrt{m_1^* m_2^*} \times \left[(1 + 2\alpha E_{Fs}) - (2\alpha)\frac{\hbar^2}{2m_3^*}\left(\frac{n_z\pi}{d_z}\right)^2\right]$ (1.138)	$n_{2D} = \frac{4\sqrt{m_1^* m_2^*} k_B T}{\pi\hbar^2} \sum_{n_z=1}^{n_{zmax}} \times \left[(A_1(n_z) + 2\alpha E_{n_{z16}})F_0(E_{n_{z16}}) + 2\alpha k_B T F_1(E_{n_{z16}})\right]$ (1.142)
ii. Model by Wang and Ressler	$m^*(E_{Fs,n_z}) \equiv \sqrt{m_1^* m_2^*}\left[\{I_1(E_{Fs}, n_z)\}'\right]$ (1.147)	$n_{2D} = \frac{4\sqrt{m_1^* m_2^*}}{\pi\hbar^2} \sum_{n_z=1}^{n_{zmax}} [t_{46}(E_{Fs}, n_z) + t_{47}(E_{Fs}, n_z)]$ (1.151)
12. n- GaSb	$m^*(E_{Fs}) = m_c\{I_{36}(E_{Fs})\}'$ (1.155)	$n_{2D} = \left(\frac{m_c g_v}{\pi\hbar^2}\right) \sum_{n_z=1}^{n_{zmax}} [t_{55}(E_{Fs}, n_z) + t_{56}(E_{Fs}, n_z)]$ (1.158)

where, $\overline{\Delta} = 335\,\text{meV}$, $P_1 = 2 \times 10^{-10}\text{eVm}$, $D_1 = p_1 a_1$ and $a_1 = 5.4 \times 10^{-10}\text{m}$.

(b) In addition to the Cohen model, the dispersion relation for the conduction electrons for IV-VI semiconductors can also be described by the models of Bangert et al. [238] and Foley et al. [239], respectively.

(i) In accordance with Bangert et al. [238], the dispersion relation is given by

$$\Gamma(E) = F_1(E)k_s^2 + F_2(E)k_z^2 \tag{R1.2}$$

where, $\Gamma(E) = 2E$, $F_1(E) = \frac{R_1^2}{E+E_g} + \frac{S_1^2}{E+\Delta_c'} + \frac{Q_1^2}{E+E_g}$, $F_2(E) = \frac{2C_5^2}{E+E_g} + \frac{(S_1+Q_1)^2}{E+\Delta_c''}$ $R_1^2 = 2.3\times10^{-19}(\text{eVm})^2$, $C_5^2 = 0.83\times10^{-19}(\text{eVm})^2$, $Q_1^2 = 1.3R_1^2$, $S_1^2 = 4.6R_1^2$, $\Delta_c' = 3.07\,\text{eV}$, $\Delta_c'' = 3.28\,\text{eV}$ and $g_v = 4$. It may be noted that under the substitution $S_1 = 0$, $Q_1 = 0$, $R_1^2 \equiv \frac{\hbar^2 E_g}{m_\perp^*}$, $C_5^2 \equiv \frac{\hbar^2 E_g}{m_\parallel^*}$, (R1.2) assumes the form $E(1+\alpha E) = \frac{\hbar^2 k_s^2}{2m_\perp^*} + \frac{\hbar^2 k_z^2}{2m_\parallel^*}$ which is the simplified Lax model.

(ii) The carrier energy spectrum of IV-VI semiconductors in accordance with Foley et al. [239] can be written as

$$E + \frac{E_g}{2} = E_-(k) + \left[\left[E_+(k) + \frac{E_g}{2}\right]^2 + P_\perp^2 k_s^2 + P_\parallel^2 k_z^2\right]^{1/2} \tag{R1.3}$$

where, $E_+(k) = \frac{\hbar^2 k_s^2}{2m_\perp^+} + \frac{\hbar^2 k_z^2}{2m_\parallel^+}$, $E_-(k) = \frac{\hbar^2 k_s^2}{2m_\perp^-} + \frac{\hbar^2 k_z^2}{2m_\parallel^-}$ represents the contribution from the interaction of the conduction and the valance band edge states with the more distant bands and the free electron term, $\frac{1}{m_\perp^\pm} = \frac{1}{2}\left[\frac{1}{m_{tc}} \pm \frac{1}{m_{tc}}\right]$, $\frac{1}{m_\parallel^\pm} = \frac{1}{2}\left[\frac{1}{m_{lc}} \pm \frac{1}{m_{lv}}\right]$, For n-PbTe, $P_\perp = 4.61 \times 10^{-10}\,\text{eVm}$, $P_\parallel = 1.48 \times 10^{-10}\,\text{eVm}$, $\frac{m_0}{m_{tv}} = 10.36$, $\frac{m_0}{m_{tv}} = 0.75$, $\frac{m_0}{m_{tv}} = 11.36$, $\frac{m_0}{m_{lv}} = 1.20$ and $g_v = 4$.

(c) The hole energy spectrum of p-type zero-gap semiconductors (e.g. HgTe) is given by [240]

$$E = \frac{\hbar^2 k^2}{2m_v^*} + \frac{3e^2}{128\varepsilon_\infty}k - \left(\frac{2E_B}{\pi}\right)\ln\left|\frac{k}{k_0}\right| \tag{R1.4}$$

where m_v^* is the effective mass of the hole at the top of the valence band, ε_∞ is the semiconductor permittivity in the high frequency limit, $E_B \equiv \frac{m_0 e^2}{2\hbar^2 \varepsilon_\infty^2}$ and $k_0 \equiv \frac{m_0 e^2}{\hbar^2 \varepsilon_\infty}$.

(d) The conduction electrons of n-GaSb obey the following two dispersion relations:

(i) In accordance with the model of Seiler et al. [253]

$$E = \left[-\frac{E_g}{2} + \frac{E_g}{2}[1+\alpha_4 k^2]^{1/2} + \frac{\bar{\varsigma}_0 \hbar^2 k^2}{2m_0} + \frac{\bar{v}_0 f_1(k)\hbar^2}{2m_0} \pm \frac{\bar{\omega}_0 f_2(k)\hbar^2}{2m_0} \right] \tag{R1.5}$$

where $\alpha_4 \equiv 4p^2(E_g + \frac{2}{3}\Delta)[E_g^2(E_g+\Delta)]^{-1}$, P is the isotropic momentum matrix element, $f_1(k) \equiv k^{-2}[k_x^2 k_y^2 + k_y^2 k_z^2 + k_z^2 k_x^2]$ represents the warping of the Fermi surface, $f_2(k) \equiv [\{k^2(k_x^2 k_y^2 + k_y^2 k_z^2 + k_z^2 k_x^2) - 9k_x^2 k_y^2 k_z^2\}^{1/2} k^{-1}]$ represents the inversion asymmetry splitting of the conduction band, and $\bar{\varsigma}_0$, $\bar{v}_0$, and $\bar{\omega}_0$ represent the constants of the electron spectrum in this case.

(ii) In accordance with the model of Zhang et al. [241]

$$E = [E_2^{(1)} + E_2^{(2)} K_{4,1}]k^2 + [E_4^{(1)} + E_4^{(2)} K_{4,1}]k^4 + k^6[E_6^{(1)} + E_6^{(2)} K_{4,1} + E_6^{(3)} K_{6,1}] \tag{R1.6}$$

where $K_{4,1} \equiv \frac{5}{4}\sqrt{21}\left[\frac{k_x^4+k_y^4+k_z^4}{k^4} - \frac{3}{5}\right]$,

$K_{6,1} \equiv \sqrt{\frac{639639}{32}}\left[\frac{k_x^2 k_y^2 k_z^2}{k^6} + \frac{1}{22}\left(\frac{k_x^4+k_y^4+k_z^4}{k^4} - \frac{3}{5}\right) - \frac{1}{105}\right]$, the coefficients are in eV, the values of k are $10\left(\frac{a}{2\pi}\right)$ times those of k in atomic units (α is the lattice constant), $E_2^{(1)} = 1.0239620$, $E_2^{(2)} = 0$, $E_4^{(1)} = -1.1320772$, $E_4^{(2)} = 0.05658$, $E_6^{(1)} = 1.1072073$, $E_6^{(2)} = -0.1134024$ and
$E_6^{(3)} = -0.0072275$.

(e) In addition to the well-known band models as discussed in this monograph, the conduction electrons of III-V semiconductors obey the following three dispersion relations:

(i) In accordance with the model of Rossler [242]

$$E = \frac{\hbar^2 k^2}{2m^*} + \bar{\alpha}_{10} k^4 + \bar{\beta}_{10}\left[k_x^2 k_y^2 + k_y^2 k_z^2 + k_z^2 k_x^2\right] \pm \bar{\gamma}_{10}\left[k^2(k_x^2 k_y^2 + k_y^2 k_z^2 + k_z^2 k_x^2) - 9k_x^2 k_y^2 k_z^2\right]^{1/2} \tag{R1.7}$$

where $\bar{\alpha}_{10} = \bar{\alpha}_{11} + \bar{\alpha}_{12}k$, $\bar{\beta}_{10} = \bar{\beta}_{11} + \bar{\beta}_{12}k$ and $\bar{\gamma}_{10} = \bar{\gamma}_{11} + \bar{\gamma}_{12}k$, in which, $\bar{\alpha}_{11} = -2132 \times 10^{-40}\,\text{eVm}^4$, $\bar{\alpha}_{12} = 9030 \times 10^{-50}\,\text{eVm}^5$, $\bar{\beta}_{11} = -2493 \times 10^{-40}\,\text{eVm}^4$, $\bar{\beta}_{12} = 12594 \times 10^{-50}\,\text{eVm}^5$, $\bar{\gamma}_{11} = 30 \times 10^{-30}\,\text{eVm}^3$ and $\bar{\gamma}_{12} = -154 \times 10^{-42}\,\text{eVm}^4$.

(ii) In accordance with Johnson and Dickey [243], the electron energy spectrum assumes the form

$$E = -\frac{E_g}{2} + \frac{\hbar^2 k^2}{2}\left[\frac{1}{m_0} + \frac{1}{m_{\gamma b}}\right] + \frac{E_g}{2}\left[1 + 4\frac{\hbar^2 k^2}{2m_c'}\frac{\bar{f}_1(E)}{E_g}\right]^{1/2} \tag{R1.8}$$

where, $\frac{m_0}{m'_c} \equiv P^2\left[\frac{\left(E_g+\frac{2\Delta}{3}\right)}{E_g(E_g+\Delta)}\right]$, $\bar{f}_1(E) \equiv \frac{(E_g)+\Delta)\left(E+E_g+\frac{2\Delta}{3}\right)}{\left(E_g+\frac{2\Delta}{3}\right)(E+E_g+\Delta)}$, $m'_c =$ $0.139m_0$ and $m_{\gamma b} = \left[\frac{1}{m'_c} - \frac{2}{m_0}\right]^{-1}$.

(iii) In accordance with Agafonov et al. [244], the electron energy spectrum can be written as

$$E = \frac{\bar{\eta} - E_g}{2}\left[1 - \frac{\hbar^2 k^2}{2\bar{\eta} m^*}\left\{\frac{D\sqrt{3} - 3\bar{B}}{2\left(\frac{\hbar^2}{2m^*}\right)}\right\}\left[\frac{k_x^4 + k_y^4 + k_z^4}{k^4}\right]\right] \tag{R1.9}$$

where, $\bar{\eta} \equiv \left(E_g^2 + \frac{8}{3}P^2k^2\right)$ and $\bar{B} \equiv -21\frac{\hbar^2}{2m_0}$ and $D \equiv -40\left(\frac{\hbar^2}{2m_0}\right)$.

(f) The dispersion relation of the carriers in n-type $Pb_{1-x}Ga_x$Te with x=0.01 can be written [246] as

$$\left[E - 0.606k_s^2 - 0.0722k_z^2\right]\left[E + \bar{E}_g + 0.411k_s^2 + 0.0377k_z^2\right]$$
$$= 0.23k_s^2 + 0.02k_z^2 \pm \left[0.06\bar{E}_g + 0.061k_s^2 + 0.0066k_z^2\right]k_s \tag{R1.10}$$

where, $\bar{E}_g(= 0.21\,\text{eV})$ is the energy gap for the transition point, the zero of the energy E is at the edge of the conduction band of the Γ point of the Brillouin zone and is measured positively upwards, k_x, k_y, and k_z are in the units of $10^9\,\text{m}^{-1}$.

(g) The energy spectrum of the carriers in the two higher valence bands and the single lower valence band of Te can, respectively, be expressed as [245]

$$\bar{E} = A_{10}k_z^2 + B_{10}k_s^2 \pm \left[\Delta_{10}^2 + (\beta_{10}k_z)^2\right]^{1/2}$$
$$\text{and } \bar{E} = \Delta_{\parallel} + A_{10}k_z^2 + B_{10}k_s^2 \pm \beta_{10}k_z \tag{R1.11}$$

where $\bar{E}$ is the energy of the hole as measured from the top of the valence and within it, $A_{10} = 3.77 \times 10^{-19}\,\text{eVm}^2$, $B_{10} = 3.57 \times 10^{-19}\,\text{eVm}^2$, $\Delta_{10} = 0.628\,\text{eV}$, $(\beta_{10})^2 = 6 \times 10^-20(\text{eVm})^2$ and $\Delta_{\parallel} = 1004 \times 10^{-5}\,\text{eV}$ are the spectrum constants.

(h) The dispersion relation for the electrons in graphite can be written following Brandt [252] as

$$E = \frac{1}{2}[E_2 + E_3] \pm \left[\frac{1}{4}(E_2 - E_3)^2 + \eta_2^2 k^2\right]^{1/2} \tag{R1.12}$$

where, $E_2 \equiv \bar{\Delta} - 2\bar{\gamma}_1\cos\phi_0 + 2\bar{\gamma}_5\cos^2\phi_0$, $\phi_0 \equiv \frac{c_6 k_z}{2}$, $E_3 \equiv 2\bar{\gamma}_2\cos^2\phi_0$ and $\eta_2 \equiv \left(\frac{\sqrt{3}}{2}\right)a_6(\bar{\gamma}_0 + 2\bar{\gamma}_4\cos\phi_0)$ in which the band constants are $\bar{\Delta}, \bar{\gamma}_0, \bar{\gamma}_1, \bar{\gamma}_2, \bar{\gamma}_4, \bar{\gamma}_5, a_6$ and c_6 respectively.

(i) The dispersion relation of the conduction electrons in Antimony (Sb) in accordance with Ketterson [251] can be written as

$$2m_0E = \alpha_{11}P_x^2 + \alpha_{22}P_y^2 + \alpha_{33}P_z^2 + 2\alpha_{23}P_yP_z \quad \text{(R1.13)}$$

and

$$2m_0E = a_1P_x^2 + a_2P_y^2 + a_3P_z^2 + a_4P_yP_z \pm a_5P_xP_z \pm a_6P_xP_y \quad \text{(R1.14)}$$

where, $a_1 = \frac{1}{4}(\alpha_{11} + 3\alpha_{22}), a_2 = \frac{1}{4}(\alpha_{22} + 3\alpha_{11}), a_3 = \alpha_{33}, a_4 = \alpha_{33}, a_5 = \sqrt{3}$ and $a_6 = \sqrt{3}(\alpha_{22} - \alpha_{11})$ in which $\alpha_{11}, \alpha_{22}, \alpha_{33}$ and α_{23} are the system constants.

(j) The dispersion relation of the holes in p-InSb can be written in accordance with Cunningham [247] as

$$\bar{E} = c_4(1 + \gamma_4 f_4)k^2 \pm \frac{1}{3}[2\sqrt{2}\sqrt{c_4}\sqrt{16 + 5\gamma_4}\sqrt{E_4 g_4 k}] \quad \text{(R1.15)}$$

where $c_4 \equiv \frac{\hbar^2}{2m_0} + \theta_4, \theta_4 \equiv 4.7\frac{\hbar^2}{2m_0}, \gamma_4 \equiv \frac{b_4}{c_4}, b_4 \equiv \frac{3}{2}b_5 + 2\theta_4, b_5 \equiv 2.4\frac{\hbar^2}{2m_0}, f_4 \equiv \frac{1}{4}[\sin^2 2\theta + \sin^4\theta \sin^2 2\phi]$, θ is measured from the positive z-axis, ϕ is measured from positive x-axis, $g_4 \equiv \sin\theta\left[\cos^2\theta + \frac{1}{4}sin^4\theta \sin^2 2\phi\right]$ and $E_4 = 5 \times 10^{-4}$ eV.

(k) The energy spectrum of the valence bands of CuCl in accordance with Yekimov et al. [248] can be written as

$$E_h = (\gamma_6 - 2\gamma_7)\frac{\hbar^2k^2}{2m_0} \quad \text{(R1.16)}$$

and

$$E_{l,s} = (\gamma_6 + \gamma_7)\frac{\hbar^2k^2}{2m_0} - \frac{\Delta_1}{2} \pm \left[\frac{\Delta_1^2}{4} + \gamma_7\Delta_1\frac{\hbar^2k^2}{2m_0} + 9\left(\frac{\gamma_7\hbar^2k^2}{2m_0}\right)^2\right]^{1/2} \quad \text{(R1.17)}$$

where, $\gamma_6 = 0.53, \gamma_7 = 0.07, \Delta_1 = 70$ meV.

(l) In the presence of stress, χ_6 along <001> and <111> directions, the energy spectra of the holes in semiconductors having diamond structure valence bands can be respectively expressed following Roman [249] et al. as

$$E = A_6k^2 \pm \left[\bar{B}_7^2k^4 + \delta_6^2 + B_7\delta_6(2k_z^2 - k_s^2)\right]^{1/2} \quad \text{(R1.18)}$$

and

$$E = A_6 k^2 \pm \left[\bar{B}_7^2 k^4 + \delta_7^2 + \frac{D_6}{\sqrt{3}} \delta_7 (2k_z^2 - k_s^2) \right]^{1/2} \quad \text{(R1.19)}$$

where A_6, B_7, D_6, and C_6 are inverse mass band parameters in which $\delta_6 \equiv l_7(\bar{S}_{11} - \bar{S}_{12})\chi_6$, $\bar{S}_{ij}$ are the usual elastic compliance constants, $\bar{B}_7^2 \equiv \left(B_7^2 + \frac{c_6^2}{5} \right)$ and $\delta_7 \equiv \left(\frac{d_g S_{44}}{2\sqrt{3}} \right) \chi_6$. For gray tin, $d_g = -4.1\,\text{eV}$, $l_7 = -2.3\,\text{eV}$,
$A_6 = 19.2 \frac{\hbar^2}{2m_o}$, $B_7 = 26.3 \frac{\hbar^2}{2m_o}$, $D_6 = 31 \frac{\hbar^2}{2m_o}$ and $c_6^2 = -1112 \frac{\hbar^2}{2m_o}$.

(m) The dispersion relation of the carriers of cadmium and zinc diphosphides are given by [250]

$$E = \left[\beta_1 + \frac{\beta_2 \beta_3(k)}{8\beta_4} \right] k^2 \pm \left\{ \left[\beta_4 \beta_3(k) x \left(\beta_5 - \frac{\beta_2 \beta_3(k)}{8\beta_4} \right) k^2 \right] + 8\beta_4^2 \left(1 - \frac{\beta_3^2(k)}{4} \right) - \beta_2 \left(1 - \frac{\beta_3^2(k)}{4} \right) k^2 \right\}^{1/2} \quad \text{(R1.20)}$$

where β_1, β_2, β_4 and β_5 are system constants, and $\beta_3(k) = \frac{k_x^2 + k_y^2 - 2k_z^2}{k^2}$

R1.3. Investigate the EEM, EAM, DEM, CEM, CoEM, FREM, and OEM for ultrathin films, wires and dots of all the semiconductors as considered in R1.1 and R1.2, respectively.

R1.4. Investigate the same set of masses as defined in (R1.3) for bulk specimens of the heavily–doped semiconductors in the presences of Gaussian, exponential, Kane, Halperian, Lax and Bonch-Burevich types of Band tails [121, 121] for all systems whose unperturbed carrier energy spectra are defined in R1.1 and R1.2, respectively.

R1.5. Investigate the same set of masses as defined in (R1.3) for ultrathin films, wires, and dot of all the heavily doped semiconductors as considered in R1.4.

R1.6. Investigate the same set of masses as defined in (R1.3) for bulk specimens of the negative refractive index, organic, magnetic, and other advanced optical materials in the presence of an arbitrarily oriented alternating electric field.

R1.7. Investigate the same set of masses as defined in (R1.3) for ultrathin films, wires and dot of the negative refractive index, organic, magnetic and other advanced optical materials in the presence of an arbitrarily oriented alternating electric field.

R1.8. Investigate the same set of masses as defined in (R1.3) for the multiple ultrathin films, wires, and dots of semiconductors whose unperturbed carrier energy spectra are defined in R1.1, R1.2 and heavily doped semiconductors in the presences of Gaussian, exponential, Kane, Halperian, Lax, and Bonch-Burevich types of Band tails [121, 122] for all systems whose unperturbed carrier energy spectra are defined in the same problems respectively.

R1.9. Investigate the same set of masses as defined in (R1.3) for all the appropriate low-dimensional systems of this chapter in the presence of finite potential wells.

R1.10. Investigate the same set of masses as defined in (R1.3) for all the appropriate low-dimensional systems of this chapter in the presence of parabolic potential wells.

R1.11. Investigate the same set of masses as defined in (R1.3) for all the appropriate systems of this chapter forming quantum rings.

R1.12. Investigate the same set of masses as defined in (R1.3) for all the above appropriate problems in the presence of elliptical Hill and quantum square rings.

R1.13. Investigate the same set of masses as defined in (R1.3) for the appropriate accumulation layres for all the materials whose unperturbed carrier energy spectra are defined in R1.1 and R1.2, respectively.

R1.14. Investigate the same set of masses as defined in (R1.3) for parabolic cylindrical quantum dots in the presence of an arbitrarily oriented alternating electric field for all the materials whose unperturbed carrier energy spectra are defined in R1.1 and R1.2, respectively.

R1.15. Investigate the same set of masses as defined in (R1.3) for wedge shaped, cylindrical quantum dots of the negative refractive index, and other advanced optical materials in the presence of an arbitrarily oriented alternating electric field and non-uniform lightwaves.

R1.16. Investigate the same set of masses as defined in (R1.3) for triangular, cylindrical quantum dots of the negative refractive index, organic, magnetic and other advanced optical materials in the presence of an arbitrarily oriented alternating electric field in the presence of strain.

R1.17. Investigate the same set of masses as defined in (R1.3) for conical quantum dots of the negative refractive index, organic, magnetic, and other advanced optical materials in the presence of an arbitrarily oriented alternating electric field.

R1.18. (a) Investigate the same set of masses as defined in (R1.3) for conical quantum dots of the negative refractive index, organic, magnetic, and other advanced optical materials in the presence of an arbitrarily oriented alternating electric field considering many-body effects.
(b) Investigate all the appropriate problems of this chapter for a Dirac electron.
(c) Investigate all the appropriate problems of this chapter by including the many-body, image force, broadening and hot carrier effects respectively.

R1.19. Investigate all the appropriate problems of this chapter by removing all the mathematical approximations and establishing the respective appropriate uniqueness conditions.

References

1. S. Adachi, J. Appl. Phys. **58**, R11 (1985)
2. R. Dornhaus, G. Nimtz, *Springer Tracts in Modern Physics*, vol. 78 (Springer, Berlin, 1976), p. 1
3. W. Zawadzki, *Handbook of Semiconductor Physics*, vol. 1, ed. by W. Paul (North Holland, Amsterdam, 1982), p 719
4. I.M. Tsidilkovski, Cand. Thesis Leningrad University SSR (1955)
5. F.G. Bass, I.M. Tsidilkovski, Ivz. Acad. Nauk Azerb SSR **10**, 3 (1966)
6. B. Mitra, K.P. Ghatak, Phys. Scr. **40**, 776 (1989)
7. S.K. Biswas, A.R. Ghatak, A. Neogi, A. Sharma, S. Bhattacharya, K.P. Ghatak, Phys. E **36**, 163 (2007)
8. M. Mondal, S. Banik, K.P. Ghatak, J. Low Temp. Phys. **74**, 423 (1989)
9. K.P. Ghatak, S. Bhattacharya, H. Saikia, A. Sinha, J. Comp. Theor. Nanosci. **3**, 1 (2006)
10. K.P. Ghatak, S.N. Biswas, J. Vac. Sci. Tech. **7B**, 104 (1989)
11. P.K. Chakraborty, S. Choudhury, K.P. Ghatak, Phys. B **387**, 333 (2007)
12. S. Bhattacharya, S. Pahari, D.K. Basu, K.P. Ghatak, J. Comp. Theor. Nanosci. **3**, 280 (2006)
13. A. Sinha, A.K. Sharma, R. Barui, A.R. Ghatak, S. Bhattacharya, K.P. Ghatak, Phys. B **391**, 141 (2007)
14. S. Choudhury, L.J. Singh, K.P. Ghatak, Nanotechnology **15**, 180 (2004)
15. S. Chowdhary, L.J. Singh, K.P. Ghatak, Phys. B **B365**, 5 (2005)
16. P.K. Chakraborty, A. Sinha, S. Bhattacharya, K.P. Ghatak, Phys. B **390**, 325 (2007)
17. K.P. Ghatak, J.P. Banerjee, D. Bhattacharyya, Nanotechnology **7**, 110 (1996)
18. P.K. Chakraborty, B. Nag, K.P. Ghatak, J. Phys. Chem. Solids **64**, 2191 (2003)
19. K.P. Ghatak, J.P. Banerjee, B. Nag, J. Appl. Phys. **83**, 1420 (1998)
20. B. Nag, K.P. Ghatak, J. Phys. Chem. Solids **59**, 713 (1998)
21. P.K. Chakraborty, B. Nag, K.P. Ghatak, J. Phys. Chem. Solids **64**, 2191 (2003)
22. P.K. Chakraborty, K.P. Ghatak, Phys. Lett. **288**, 335 (2001)
23. P.K. Chakraborty, K.P. Ghatak, J. Phys. Chem. Sol. **62**, 1061 (2001)
24. P.K. Chakraborty, K.P. Ghatak, J. Phys. D Appl. Phys. **32**, 2438 (1999)
25. B. Nag, K.P. Ghatak, J. Phys. Chem. Solids **58**, 427 (1997)
26. K.P. Ghatak, D.K. Basu, B. Nag, J. Phys. Chem. Solids **58**, 133 (1997)
27. B. Nag, K.P. Ghatak, Nonlinear Opt. Quantum Opt. **19**, 1 (1998)
28. K.P. Ghatak, B. Nag, Nanostruct. Mat. **10**, 923 (1998)
29. K.P. Ghatak, B. Nag, D. Bhattacharyya, J. Low Temp. Phys. **14**, 1 (1995)
30. M. Mondal, K.P. Ghatak, Phys. Scr. **30**, 217 (1984)
31. M. Mondal, K.P. Ghatak, J. Magn. Magn. Mat. **62**, 115 (1986)
32. M. Mondal, K.P. Ghatak, Thin Solid Films **148**, 219 (1987)
33. K.P. Ghatak, A.K. Choudhury, S. Ghosh, A.N. Chakravarti, Appl. Phys. **23**, 241 (1980)
34. M. Mondal, N. Chattopadhyay, K.P. Ghatak, J. Low Temp, J. Low Temp. Phys. **73**, 321 (1988)
35. K.P. Ghatak, N. Chattopadhyay, M. Mondal, Appl. Phys. A **44**, 305 (1987)
36. B. Mitra, K.P. Ghatak, Phys. Lett. **135A**, 397 (1989)
37. B. Mitra, K.P. Ghatak, Solid State Electron. **32**, 810 (1989)
38. B. Mitra, K.P. Ghatak, Phys. Scr. **42**, 103 (1990)
39. K.P. Ghatak, M. Mondal, Z.F. Naturforschung **41A**, 821 (1986)
40. A.N. Chakravarti, A.K. Choudhury, K.P. Ghatak, S. Ghosh, A. Dhar, Appl. Phys. **25**, 105 (1981)
41. K.P. Ghatak, M. Mondal, Z.F. Phys. B **B69**, 471 (1988)
42. A.N. Chakravarti, K.P. Ghatak, K.K. Ghosh, S. Ghosh, A. Dhar, Z.F. Phys. B **47**, 149 (1982)
43. H.A. Lyden, Phys. Rev. **135**, A514 (1964)
44. E.D. Palik, G.B. Wright, in *Semiconductors and Semimetals*, ed. by R.K. Willardson, A.C. Beer, vol. 3. (Academic Press, New York, 1967), p. 421
45. H.L. Stomer, R. Dingle, A.C. Gossard, W. Wiegmann, M.D. Sturge, Solid State Commun. **29**, 705 (1979)

46. D.C. Rogers, J. Singleton, R.J. Nicholas, C.T. Foxon, in *High Magnetic Fields in Semiconductor Physics*, ed. by G. Landwehr, Springer Series in Solid-State Sciences, vol. 71. (Springer, Berlin, 1987), p. 223
47. M. Potemski, J.C. Maan, K. Pllog, G. Weimann, Solid State Commun. **75**, 185 (1990)
48. B.C. Cavenett, E.J. Pakulis, Phys. Rev. B **32**, 8449 (1985)
49. C. Wetzel, Al.L. Efros, A. Moll, B.K. Meyer, P. Omling, P. Sobkowicz. Phys. Rev. B **45**, 14052 (1992)
50. G. Hendorfer, M. Seto, H. Ruckser, W. Jantsch, M. Helm, G. Brunthaler, W. Jost, H. Obloh, K. Kohler, D.J. As, Phys. Rev. B **48**, 2328 (1993)
51. Q.X. Zhao, P.O. Holtz, B. Monemar, T. Lundstrom, J. Wallin, G. Landgren, Phys. Rev. B **48**(11), 890 (1983)
52. U. Ekenberg, Phys. Rev. B **36**, 6152 (1987)
53. H.I. Zhang, Phys. Rev. B **1**, 3450 (1970)
54. Y. Zhang, A. Mascarenhas, H.P. Xin, C.W. Tu, Phys. Rev. B **61**, 7479 (2000)
55. P.N. Hai, W.M. Chen, I.A. Buyanova, H.P. Xin, C.W. Tu, Appl. Phys. Lett. **77**, 1843 (2000)
56. J. Wu, W. Shan, W. Walukiewicz, K.M. Yu, J.W. Ager III, E.E. Haller, H.P. Xin, C.W. Tu, Phys. Rev. B **64**, 085320 (2001)
57. D.L. Young, D.L. Geisz, T.J. Coutts, Appl. Phys. Lett. **82**, 1236 (2003)
58. Y.J. Wang, X. Wei, Y. Zhang, A. Mascarenhas, H.P. Xin, Y.G. Hong, C.W. Tu, Appl. Phys. Lett. **82**, 4453 (2003)
59. F. Masia, A. Polimeni, G. Baldassarri Höger von Högersthal, M. Bissiri, M. Capizzi, P.J. Klar, W. Stolz, Appl. Phys. Lett. **82**, 4474 (2003)
60. A. Polimeni, G. Baldassarri Höger von Högersthal, F. Masia, A. Frova, M. Capizzi, S. Sanna, V. Fiorentini, P.J. Klar, W. Stolz, Phys. Rev. B **69**, 041201(R) (2004)
61. G. Baldassarri Höger von Högersthal, A. Polimeni, F. Masia, M. Bissiri, M. Capizzi, D. Gollub, M. Fischer, A. Forchel, Phys. Rev. B **67**, 233304 (2003)
62. M. Mondal, K.P. Ghatak, Phys. Lett. **131 A**, 529 (1988)
63. K.P. Ghatak, B. Mitra, Int. J. Electron. **72**, 541 (1992)
64. B. Mitra, A. Ghoshal, K.P. Ghatak, Nouvo Cimento D **12D**, 891 (1990)
65. K.P. Ghatak, S.N. Biswas, Nonlinear Opt. Quantum Opt. **4**, 347 (1993)
66. K.P. Ghatak, S.N. Biswas, Nonlinear Opt. Quantum Opt. **12**, 83 (1995)
67. V. Milanovic, D. Tjapkin, Phys. B and C **114**(3), 375(1982)
68. Z.S. Gribnikov, K. Hess, G.A. Kosinovsky, J. Appl. Phys. **7**(4), 1337 (1995)
69. M. Dyakonov, M.S. Shur, J. Appl. Phys. **84**(7), 3726 (1998)
70. K.P. Ghatak, A. Ghoshal, B. Mitra, Nouvo Cimento **14D**, 903 (1992)
71. K.P. Ghatak, A. Ghoshal, B. Mitra, Nouvo Cimento **13D**, 867 (1991)
72. B. Mitra, K.P. Ghatak, Solid State Electron. **32**, 177 (1989)
73. M. Mondal, N. Chattapadhyay, K.P. Ghatak, J. Low Temp. Phys. **66**, 131 (1987)
74. D.P. DiVincenzo, E.J. Mele, Phys. Rev. B **29**, 1685 (1984)
75. P. Perlin, E. Litwin-Staszewska, B. Suchanek, W. Knap, J. Camassel, T. Suski, R. Piotrzkowski, I. Grzegory, S. Porowski, E. Kaminska, J.C. Chervin, Appl. Phys. Lett. **68**, 1114 (1996)
76. G.E. Smith, Phys. Rev. Lett. **9**, 487 (1962)
77. D. Schneider, D. Rurup, A. Plichta, H.-U. Grubert, A. Schlachetzki, K. Hansen, Z. Phys. B **95**, 281 (1994)
78. F. Masia, G. Pettinari, A. Polimeni, M. Felici, A. Miriametro, M. Capizzi, A. Lindsay, S.B. Healy, E.P. O'Reilly, A. Cristofoli, G. Bais, M. Piccin, S. Rubini, F. Martelli, A. Franciosi, P.J. Klar, K. Volz, W. Stolz, Phys. Rev. B **73**, 073201 (2006)
79. V.K. Arora, H. Jeafarian, Phys. Rev. B **13**, 4457 (1976)
80. S.E. Ostapov, V.V. Zhikharevich, V.G. Deibuk, Semicond. Phys. Quantum Electron. Optoelectron. **9**, 29 (2006)
81. M.J. Aubin, L.G. Caron, J.-P. Jay- Gerin, Phys. Rev. B **15**, 3872 (1977)
82. S.L. Sewall, R.R. Cooney, P. Kambhampati, Appl. Phys. Lett. **94**, 243116 (2009)

83. K. Tanaka, N. Kotera, in *20th International Conference on Indium Phosphide and Related Materials*, Versailles, France, 25–29 May 2008, pp. 1–4
84. M. Singh, P.R. Wallace, S.D. Jog, J. Erushanov, J. Phys. Chem. Solids **45**, 409 (1984)
85. W. Zawadski, Adv. Phys. **23**, 435 (1974)
86. P.M. Petroff, A.C. Gossard, W. Wiegmann, Appl. Phys. Lett. **45**, 620 (1984)
87. J.M. Gaines, P.M. Petroff, H. Kroemar, R.J. Simes, R.S. Geels, J.H. English, J. Vac. Sci. Technol. B **6**, 1378 (1988)
88. J. Cilbert, P.M. Petroff, G.J. Dolan, S.J. Pearton, A.C. Gossard, J.H. English, Appl. Phys. Lett. **49**, 1275 (1986)
89. T. Fujui, H. Saito, Appl. Phys. Lett. **50**, 824 (1987)
90. H. Sasaki, Jpn. J. Appl. Phys. **19**, 94 (1980)
91. P.M. Petroff, A.C. Gossard, R.A. Logan, W. Weigmann, Appl. Phys. Lett. **41**, 635 (1982)
92. H. Temkin, G.J. Dolan, M.B. Panish, S.N.G. Chu, Appl. Phys. Lett. **50**, 413 (1988)
93. I. Miller, A. Miller, A. Shahar, U. Koren, P.J. Corvini, Appl. Phys. Lett. **54**, 188 (1989)
94. L.L. Chang, H. Esaki, C.A. Chang, L. Esaki, Phys. Rev. Lett. **38**, 1489 (1977)
95. K. Less, M.S. Shur, J.J. Drunnond, H. Morkoc, IEEE Trans. Electron. Devices **ED-30**, 07 (1983)
96. G. Bastard, *Wave Mechanics Applied to Semiconductor Heterostructures* (Halsted, Les Ulis, Les Editions de Physique, New York, 1988)
97. M.J. Kelly, *Low dimensional semiconductors: materials, physics, technology, devices* (Oxford University Press, Oxford, 1995)
98. C. Weisbuch, B. Vinter, *Quantum Semiconductor Structures* (Boston Academic Press, Boston, 1991)
99. N.T. Linch, Festkorperprobleme **23**, 27 (1985)
100. D.R. Sciferes, C. Lindstrom, R.D. Burnham, W. Streifer, T.L. Paoli, Electron. Lett. **19**, 170 (1983)
101. P.M. Solomon, Proc. IEEE **70**, 489 (1982)
102. T.E. Schlesinger, T. Kuech, Appl. Phys. Lett. **49**, 519 (1986)
103. D. Kasemet, C.S. Hong, N.B. Patel, P.D. Dapkus, Appl. Phys. Lett. **41**, 912 (1982)
104. K. Woodbridge, P. Blood, E.D. Pletcher, P.J. Hulyer, Appl. Phys. Lett. **45**, 16 (1984)
105. S. Tarucha, H.O. Okamoto, Appl. Phys. Lett. **45**, 16 (1984)
106. H. Heiblum, D.C. Thomas, C.M. Knoedler, M.I. Nathan, Appl. Phys. Lett. **47**, 1105 (1985)
107. O. Aina, M. Mattingly, F.Y. Juan, P.K. Bhattacharyya, Appl. Phys. Lett. **50**, 43 (1987)
108. I. Suemune, L.A. Coldren, IEEE J. Quant. Electron. **24**, 1178 (1988)
109. D.A.B. Miller, D.S. Chemla, T.C. Damen, J.H. Wood, A.C. Burrus, A.C. Gossard, W. Weigmann, IEEE J. Quant. Electron. **21**, 1462 (1985)
110. J.W. Rowe, J.L. Shay, Phys. Rev. B **3**, 451 (1973)
111. H. Kildal, Phys. Rev. B **10**, 5082 (1974)
112. J. Bodnar, in *Proceedings of the International Conference on Physics of Narrow-gap Semiconductors* (Polish Science Publishers, Warsaw, 1978)
113. G.P. Chuiko, N.N. Chuiko, Sov. Phys. Semicond. **15**, 739 (1981)
114. K.P. Ghatak, S.N. Biswas, Proc. SPIE **1484**, 149 (1991)
115. A. Rogalski, J. Alloys Comp. **371**, 53 (2004)
116. A. Baumgartner, A. Chaggar, A. Patanè, L. Eaves, M. Henini, Appl. Phys. Lett. **92**, 091121 (2008)
117. J. Devenson, R. Teissier, O. Cathabard, A.N. Baranov, Proc. SPIE **6909**, 69090U (2008)
118. B.S. Passmore, J. Wu, M.O. Manasreh, G.J. Salamo, Appl. Phys. Lett. **91**, 233508 (2007)
119. M. Mikhailova, N. Stoyanov, I. Andreev, B. Zhurtanov, S. Kizhaev, E. Kunitsyna, K. Salikhov, Y. Yakovlev, Proc. SPIE **6585**, 658526 (2007)
120. W. Kruppa, J.B. Boos, B.R. Bennett, N.A. Papanicolaou, D. Park, R. Bass, Electron. Lett. **42**, 688 (2006)
121. E.O. Kane, in *Semiconductors and Semimetals*, vol. 1, ed. by R.K. Willardson, A.C. Beer (Academic Press, New York, 1966), p. 75
122. B.R. Nag, *Electron Transport in Compound Semiconductors* (Springer, Heidelberg, 1980)

123. G.E. Stillman, C.M. Wolfe, J.O. Dimmock, in *Semiconductors and Semimetals*, vol. 12, ed. by R.K. Willardon, A.C. Beer (Academic Press, New York, 1977), p. 169
124. D.J. Newson, A. Karobe, Semicond. Sci. Tech. **3**, 786 (1988)
125. E.D. Palik, G.S. Picus, S. Teither, R.E. Wallis, Phys. Rev. **122**, 475 (1961)
126. P.Y. Lu, C.H. Wung, C.M. Williams, S.N.G. Chu, C.M. Stiles, Appl. Phys. Lett. **49**, 1372 (1986)
127. N.R. Taskar, I.B. Bhat, K.K. Prat, D. Terry, H. Ehasani, S.K. Ghandhi, J. Vac. Sci. Tech. **7A**, 281 (1989)
128. F. Koch, *Springer Series in Solid States Sciences* (Springer, Germany, 1984)
129. L.R. Tomasetta, H.D. Law, R.C. Eden, I. Reyhimy, K. Nakano, IEEE J. Quant. Electron. **14**, 800 (1978)
130. T. Yamato, K. Sakai, S. Akiba, Y. Suematsu, IEEE J. Quantum Electron. **14**, 95 (1978)
131. T.P. Pearsall, B.I. Miller, R.J. Capik, Appl. Phys. Lett. **28**, 499 (1976)
132. M.A. Washington, R.E. Nahory, M.A. Pollack, E.D. Beeke, Appl. Phys. Lett. **33**, 854 (1978)
133. M.I. Timmons, S.M. Bedair, R.J. Markunas, J.A. Hutchby, in *Proceedings of the 16th IEEE Photovoltaic Specialist Conference* (IEEE, San Diego, California 666, 1982)
134. J.A. Zapien, Y.K. Liu, Y.Y. Shan, H. Tang, C.S. Lee, S.T. Lee, Appl. Phys. Lett. **90**, 213114 (2007)
135. M. Park, Proc. SPIE **2524**, 142 (1995)
136. S.-G. Hur, E.T. -Kim, J. H. -Lee, G.H. -Kim, S.G. -Yoon, Electrochem. Solid-State Lett. **11**, H176 (2008)
137. H. Kroemer, Rev. Mod. Phys. **73**, 783 (2001)
138. T. Nguyen Duy, J. Meslage, G. Pichard, J. Crys. Growth **72**, 490 (1985)
139. T. Aramoto, F. Adurodija, Y. Nishiyama, T. Arita, A. Hanafusa, K. Omura, A. Morita, Solar Energy Mater. Solar Cells **75**, 211 (2003)
140. H.B. Barber, J. Electron. Mater. **25**, 1232 (1996)
141. S. Taniguchi, T. Hino, S. Itoh, K. Nakano, N. Nakayama, A. Ishibashi, M. Ikeda, Electron. Lett. **32**, 552 (1996)
142. J.J. Hopfield, J. Appl. Phys. **32**, 2277 (1961)
143. R.V. Belosludov, A.A. Farajian, H. Mizuseki, K. Miki, Y. Kawazoe, Phys. Rev. B **75**, 113411 (2007)
144. J. Heremans, C.M. Thrush, Y.-M Lin, S. Cronin, Z. Zhang, M.S. Dresselhaus, J.F. Mansfield, Phys. Rev. B **61**, 2921 (2000)
145. D. Shoenberg, Proc. R. Soc. (London) **170**, 341 (1939)
146. B. Abeles, S. Meiboom, Phys. Rev. **101**, 544 (1956)
147. B. Lax, J.G. Mavroides, H.J. Zieger, R.J. Keyes, Phys. Rev. Lett. **5**, 241 (1960)
148. Y.-H. Kao, Phys. Rev. **129**, 1122 (1963)
149. R.J. Dinger, A.W. Lawson, Phys. Rev. B **3**, 253 (1971)
150. J.F. Koch, J.D. Jensen, Phys. Rev. **184**, 643 (1969)
151. M.H. Cohen, Phys. Rev. **121**, 387 (1961)
152. S. Takaoka, H. Kawamura, K. Murase, S. Takano, Phys. Rev. B **13**, 1428 (1976)
153. J.W. McClure, K.H. Choi, Solid State Commun. **21**, 1015 (1977)
154. G.P. Agrawal, N.K. Dutta, *Semicond. Lasers* (Van Nostrand Reinhold, New York, 1993)
155. S. Chatterjee, U. Pal, Opt. Eng. (Bellingham), **32**, 2923 (1993)
156. T.K. Chaudhuri, Int. J. Energy Res. **16**, 481 (1992)
157. J.H. Dughaish, Phys. B **322**, 205 (2002)
158. C. Wood, Rep. Prog. Phys. **51**, 459 (1988)
159. K.-F. Hsu, S. Loo, F. Guo, W. Chen, J.S. Dyck, C. Uher, T. Hogan, E.K. Polychroniadis, M.G. Kanatzidis, Science **303**, 818 (2004)
160. J. Androulakis, K.F. Hsu, R. Pcionek, H. Kong, C. Uher, J.J. D'Angelo, A. Downey, T. Hogan, M.G. Kanatzidis, Adv. Mater. **18**, 1170 (2006)
161. P.F.P. Poudeu, J.D'Angelo, A.D. Downey, J.L. Short, T.P. Hogan, M.G. Kanatzidis, Angew. Chem. Int. Ed. **45**, 3835 (2006)

162. P.F. Poudeu, J. D'Angelo, H. Kong, A. Downey, J.L. Short, R. Pcionek, T.P. Hogan, C. Uher, M.G. Kanatzidis, J. Am. Chem. Soc. **128**, 14347 (2006)
163. J.R. Sootsman, R.J. Pcionek, H. Kong, C. Uher, M.G. Kanatzidis, Chem. Mater. **18**, 4993 (2006)
164. A.J. Mountvala, G. Abowitz, J. Am. Ceram. Soc. **48**, 651 (1965)
165. E.I. Rogacheva, I.M. Krivulkin, O.N. Nashchekina, AYu. Sipatov, V.A. Volobuev, M.S. Dresselhaus, Appl. Phys. Lett. **78**, 3238 (2001)
166. H.S. Lee, B. Cheong, T.S. Lee, K.S. Lee, W.M. Kim, J.W. Lee, S.H. Cho, J.Y. Huh, Appl. Phys. Lett. **85**, 2782 (2004)
167. K. Kishimoto, M. Tsukamoto, T. Koyanagi, J. Appl. Phys. **92**, 5331 (2002)
168. E.I. Rogacheva, O.N. Nashchekina, S.N. Grigorov, M.A. Us, M.S. Dresselhaus, S.B. Cronin, Nanotechnology **14**, 53 (2003)
169. E.I. Rogacheva, O.N. Nashchekina, A.V. Meriuts, S.G. Lyubchenko, M.S. Dresselhaus, G. Dresselhaus, Appl. Phys. Lett. **86**, 063103 (2005)
170. E.I. Rogacheva, S.N. Grigorov, O.N. Nashchekina, T.V. Tavrina, S.G. Lyubchenko, AYu. Sipatov, V.V. Volobuev, A.G. Fedorov, M.S. Dresselhaus, Thin Solid Bodies **493**, 41 (2005)
171. X. Qiu, Y. Lou, A.C.S. Samia, A. Devadoss, J.D. Burgess, S. Dayal, C. Burda, Angew. Chem. Int. Ed. **44**, 5855 (2005)
172. C. Wang, G. Zhang, S. Fan, Y. Li, J. Phys. Chem. Solids **62**, 1957 (2001)
173. B. Poudel, W.Z. Wang, D.Z. Wang, J.Y. Huang, Z.F. Ren, J. Nanosci. Nanotechnol. **6**, 1050 (2006)
174. B. Zhang, J. He, T.M. Tritt, Appl. Phys. Lett. **88**, 043119 (2006)
175. W. Heiss, H. Groiss, E. Kaufmann, G. Hesser, M. Böberl, G. Springholz, F. Schäffler, K. Koike, H. Harada, M. Yano, Appl. Phys. Lett. **88**, 192109 (2006)
176. B.A. Akimov, V.A. Bogoyavlenskiy, L.I. Ryabova, V.N. Vasil'kov, Phys. Rev. B **61**, 16045 (2000)
177. Ya. A. Ugai, A.M. Samoilov, M.K. Sharov, O.B. Yatsenko, B.A. Akimov, Inorg. Mater. **38**, 12 (2002)
178. Ya. A. Ugai, A.M. Samoilov, S.A. Buchnev, Yu. V. Synorov, M.K. Sharov, Inorg. Mater. **38**, 450 (2002)
179. A.M. Samoilov, S.A. Buchnev, YuV Synorov, B.L. Agapov, A.M. Khoviv. Inorg. Mater. **39**, 1132 (2003)
180. A.M. Samoilov, S.A. Buchnev, E.A. Dolgopolova, YuV Synorov, A.M. Khoviv, Inorg. Mater. **40**, 349 (2004)
181. H. Murakami, W. Hattori, R. Aoki, Phys. C **269**, 83 (1996)
182. H. Murakami, W. Hattori, Y. Mizomata, R. Aoki, Phys. C **273**, 41 (1996)
183. H. Murakami, R. Aoki, K. Sakai, Thin Solid Bodies **27**, 343 (1999)
184. B.A. Volkov, L.I. Ryabova, D.R. Khokhlov, Phys. Usp. **45**, 819 (2002), and references therein
185. F. Hüe, M. Hÿtch, H. Bender, F. Houdellier, A. Claverie, Phys. Rev. Lett. **100**, 156602 (2008)
186. S. Banerjee, K.A. Shore, C.J. Mitchell, J.L. Sly, M. Missous, IEE Proc. Circuits Devices Syst. **152**, 497 (2005)
187. M. Razeghi, A. Evans, S. Slivken, J.S. Yu, J.G. Zheng, V.P. Dravid, Proc. SPIE **5840**, 54 (2005)
188. R.A. Stradling, Semicond. Sci. Technol. **6**, C52 (1991)
189. P.K. Weimer, Proc. IEEE **52**, 608 (1964)
190. G. Ribakovs, A.A. Gundjian, IEEE J. Quant. Electron. **QE-14**, 42 (1978)
191. S.K. Dey, J. Vac. Sci. Technol. **10**, 227 (1973)
192. S.J. Lynch, Thin Solid Bodies **102**, 47 (1983)
193. V.V. Kudzin, V.S. Kulakov, D.R. Pape', S.V. Kulakov, V.V. Molotok, IEEE. Ultrason. Symp. **1**, 749 (1997)
194. F. Hatami, V. Lordi, J.S. Harris, H. Kostial, W.T. Masselink, J. Appl. Phys. **97**, 096106 (2005)
195. B.W. Wessels, J. Electrochem. Soc. **122**, 402 (1975)
196. D.W.L. Tolfree, J. Sci. Instrum. **41**, 788 (1964)
197. P.B. Hart, Proc. IEEE **61**, 880 (1973)

198. M.A. Hines, G.D. Scholes, Adv. Mater. **15**, 1844 (2003)
199. C.A. Wang, R.K. Huang, D.A. Shiau, M.K. Connors, P.G. Murphy, P.W. O'Brien, A.C. Anderson, D.M. DePoy, G. Nichols, M.N. Palmisiano, Appl. Phys. Lett. **83**, 1286 (2003)
200. C.W. Hitchcock, R.J. Gutmann, J.M. Borrego, I.B. Bhat, G.W. Charache, IEEE Trans. Electron. Devices **46**, 2154 (1999)
201. H.J. Goldsmid, R.W. Douglas, Br. J. Appl. Phys. **5**, 386 (1954)
202. F.D. Rosi, B. Abeles, R.V. Jensen, J. Phys. Chem. Sol. **10**, 191 (1959)
203. T.M. Tritt (ed.), *Semiconductors and Semimetals*, vol. 69, 70 and 71: Recent Trends in Thermoelectric Materials Research I, II and III (Academic Press, New York, 2000)
204. D.M. Rowe (ed.), *CRC Handbook of Thermoelectrics* (CRC Press, Boca Raton, 1995)
205. D.M. Rowe, C.M. Bhandari, *Modern Thermoelectrics* (Reston Publishing Company, Virginia, 1983)
206. D.M. Rowe (ed.), *Thermoelectrics Handbook: Macro to Nano* (CRC Press, Boca Raton, 2006)
207. H. Choi, M. Chang, M. Jo, S.J. Jung, H. Hwang, Electrochem. Solid-State Lett. **11**, H154 (2008)
208. S. Cova, M. Ghioni, A. Lacaita, C. Samori, F. Zappa, Appl. Opt. **35**, 1956 (1996)
209. H.W.H. Lee, B.R. Taylor, S.M. Kauzlarich, Nonlinear Optics: Materials, Fundamentals, and Applications (Technical Digest, 12, 2000)
210. E. Brundermann, U. Heugen, A. Bergner, R. Schiwon, G.W. Schwaab, S. Ebbinghaus, D.R. Chamberlin, E.E. Haller, M. Havenith, IN *29th International Conference on Infrared and Millimeter Waves and 12th International Conference on Terahertz, Electronics*, vol 283 (2004)
211. A.N. Baranov, T.I. Voronina, N.S. Zimogorova, L.M. Kauskaya, Y.P. Yakoviev, Sov. Phys. Semicond. **19**, 1676 (1985)
212. M. Yano, Y. Suzuki, T. Ishii, Y. Matsushima, M. Kimata, Jpn. J. Appl. Phys. **17**, 2091 (1978)
213. F.S. Yuang, Y.K. Su, N.Y. Li, Jpn. J. Appl. Phys. **30**, 207 (1991)
214. F.S. Yuang, Y.K. Su, N.Y. Li, K.J. Gan, J. Appl. Phys. **68**, 6383 (1990)
215. Y.K. Su, S.M. Chen, J. Appl. Phys. **73**, 8349 (1993)
216. S.K. Haywood, A.B. Henriques, N.J. Mason, R.J. Nicholas, P.J. Walker, Semicond. Sci. Technol. **3**, 315 (1988)
217. M. Abramowitz, I.A. Stegun, *Handbook of Mathematical Functions* (Dover Publications, New York, 1965)
218. J.S. Blakemore, *Semiconductor Statistics* (Dover Publications, New York, 1987)
219. K.P. Ghatak, S. Bhattacharya, S.K. Biswas, A. Dey, A.K. Dasgupta, Phys. Scr. **75**, 820 (2007)
220. K.P. Ghatak, S. Bhattachaya, D. De, *Einstein Relation in Compound Semiconductors and Nanostructures*, Springer Series in Material Sciences, vol. 116 (Springer, Germany, 2009)
221. J.O. Dimmock, in *The Physics of Semimetals and Narrowgap Semiconductors*, ed. by D.L. Carter, R.T. Bates (Pergamon Press, Oxford, 1971)
222. D.G. Seiler, B.D. Bajaj, A.E. Stephens, Phys. Rev. B **16**, 2822 (1977)
223. A.V. Germaneko, G.M. Minkov, Phys. Stat. Sol. (b) **184**, 9 (1994)
224. G.L. Bir, G.E. Pikus, *Symmetry and Strain -Induced effects in Semiconductors* (Nauka, Russia, 1972)
225. M. Mondal, K.P. Ghatak, Phys. Stat. Sol. (b) **135**, K21 (1986)
226. J. Bouat, J.C. Thuillier, Surf. Sci. **73**, 528 (1978)
227. G.J. Rees, Physics of Compounds, in *Proceedings of the 13th International Conference* ed. by F.G. Fumi (North Holland Company, 1976), p. 1166
228. P.R. Emtage, Phys. Rev. **138**, A246 (1965)
229. M. Stordeur, W. Kuhnberger, Phys. Stat. Sol. (b) **69**, 377 (1975)
230. D.R. Lovett, *Semimetals and Narrow-Bandgap Semiconductor* (Pion Limited, UK, 1977)
231. H. Kohler, Phys. Stat. Sol. (b) **74**, 591 (1976)
232. M. Cardona, W. Paul, H. Brooks Helv, Acta Phys. **33**, 329 (1960)
233. A.F. Gibson in *Proceeding of International School of Physics, ENRICO FERMI, course XIII*, ed. by R.A. Smith (Academic Press, New York, 1963), p. 171
234. C.C. Wang, N.W. Ressler, Phys. Rev. **2**, 1827 (1970)
235. M. Zalazny, Phys. B **124**, 352 (1984)

236. P.C. Mathur, S. Jain, Phys. Rev. **19**, 1359 (1979)
237. E.L. Ivchenko, G.E. Pikus Sov, Phys. Semicond. **13**, 579 (1979)
238. E. Bangert, P. Kastner, Phys. Stat. Sol (b) **61**, 503 (1974)
239. G.M.T. Foley, P.N. Langenberg, Phys. Rev. B **15B**, 4850 (1977)
240. V.I. Ivanov-Omskii, ASh Mekhtisev, S.A. Rustambekova, E.N. Ukraintsev, Phys. Stat. Sol. (b) **119**, 159 (1983)
241. H.I. Zhang, Phys. Rev. B **1**, 3450 (1970)
242. U. Rossler, Solid State Commun. **49**, 943 (1984)
243. J. Johnson, D.H. Dickey, Phys. Rev. **1**, 2676 (1970)
244. V.G. Agafonov, P.M. Valov, B.S. Ryvkin, I.D. Yarashetskin, Sov. Phys. Semiconduct. **12**, 1182 (1978)
245. L.A. Vassilev, Phys. State sol(b), **121**, 203 (1984)
246. N.S. Averkiev, V.M. Asnin, A.A. Bakun, A.M. Danishevskii, E.L. Ivchenko, G.E. Pikus, A.A. Rogachev, Sov. Phys. Semicond. **18**, (1984) pp. 379, 402
247. R.W. Cunningham, Phys. Rev. **167**, 761 (1968)
248. A.I. Yekimov, A.A. Onushchenko, A.G. Plyukhin, AI.L. Efros, J. Expt. Theor. Phys. **88**, 1490 (1985)
249. B.J. Roman, A.W. Ewald, Phys. Rev. **B5**, 3914 (1972)
250. G.P. Chuiko, Sov. Phys. Semiconduct. **19**(12), 1381 (1985)
251. J.B. Ketterson, Phys. Rev. **129**, 18 (1963)
252. N.B. Brandt, V.N. Davydov, V.A. Kulbachinskii, O.M. Nikitina, Sov. Phys. Sol. Stat. **29**, 1014 (1987)
253. D.G. Seiler, W.M. Beeker, L.M. Roth, Phys. Rev. **1**, 764 (1970)

Chapter 2
The EEM in Nipi Structures of Nonparabolic Semiconductors

2.1 Introduction

The concept of doping superlattices (SLs) was introduced by Esaki and Tsu [1] and extensive work in this subject was initiated by Dohler [2–15]. In the compositional SL the periodic potential is due to a change in the band gap of two materials. In doping SLs, the periodicity is space-charge induced and in addition a homogeneous material is used. With the advent of modern experimental techniques of fabricating nanomaterials, it is possible to grow semiconductor SLs composed of alternative layers of two different degenerate layers with controlled thickness. These structures have found wide applications in many new devices such as photodiodes, photoresistors [16], transistors [17], light emitters [18], tunneling devices [19], etc [20–33]. The investigations of the physical properties of narrow gap SLs have increased extensively, since they are important for optoelectronic devices and also since the quality of heterostructures involving narrow gap materials has been greatly improved. It may be noted that the nipi structures, also called the doping superlattices as mentioned above, are crystals with a periodic sequence of ultrathin film layers [19, 20] of the same semiconductor with the intrinsic layer in between together with the opposite sign of doping. All the donors will be positively charged and all the acceptors negatively. This periodic space charge causes a periodic space charge potential which quantizes the motions of the carriers in the z-direction together with the formation of the subband energies.

In Fig. 2.1a, the layers and the impurity types in different layers are shown. Electrons from neutral donors recombine with neutral acceptors, leaving behind a net space charge associated with ionized impurities. The concentration of the impurities is shown in Fig. 2.1b. The periodic potential is due to three terms:

$$V(z) = V_{\text{imp}}(z) + V_{\text{H}}(z) + V_{\text{xc}}(z),$$

where, $V_{\text{H}}(z)$ is the Hartree potential of electrons and holes and $V_{\text{xc}}(z)$ is the exchange potential. The potential due to ionized impurities, $V_{\text{imp}}(z)$ is obtained from Poisson's equation:

S. Bhattacharya and K. P. Ghatak, *Effective Electron Mass in Low-Dimensional Semiconductors*, Springer Series in Materials Science 167,
DOI: 10.1007/978-3-642-31248-9_2, © Springer-Verlag Berlin Heidelberg 2013

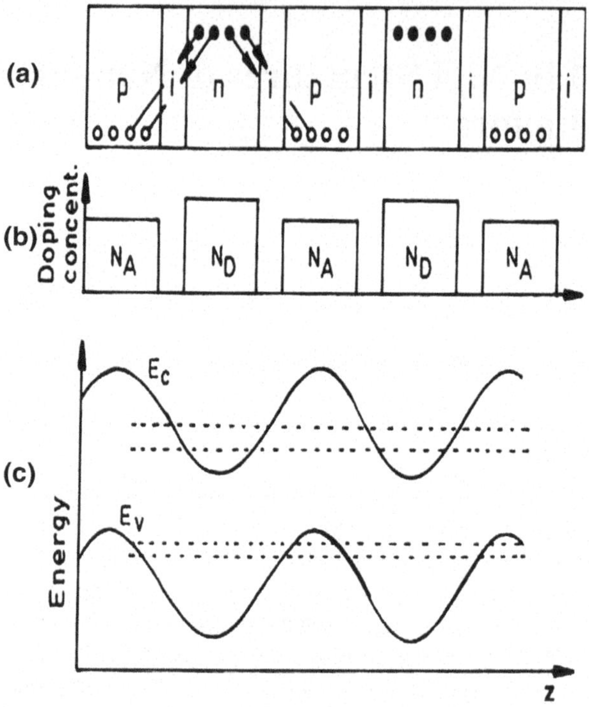

Fig. 2.1 Periodic band edge modulation in an NIPI SL: **a** structure; **b** doping profile; and **c** spatial variation of conduction and valence band edges showing the development of the 1D periodic potential

The Poisson's equation in this case is given by, [34]

$$\frac{d^2 V_{imp}}{dz^2} = \frac{e^2}{\varepsilon_{sc}}[N_D(z) - N_A(z)] \tag{2.1a}$$

where, ε_{sc} is the semiconductor permittivity,$N_D(z)$is the donor concentration along z-axis and $N_A(z)$is the acceptor concentration along z-axis.

Energy levels for the z quantized motion for electrons are to be calculated self-consistently from Schrodinger equation

$$\left[-\frac{\hbar^2}{2m_e}\frac{\partial^2}{\partial z^2} + V(z)\right]\phi(z) = E\phi(z) \tag{2.1b}$$

where m_e is the effective electron mass, $\phi(z)$is the electron wave function, and E is the energy eigenvalue. The envelope function $\phi(z)$ is given by

$$\phi(z) = \sum_m \exp(iqmd) f(z - md) \tag{2.1c}$$

where d is the period.

Some special features of this SL are stated below [34]:

1. When there are equal numbers of donors and acceptors in a period, i.e.,

$$\int_{-d/2}^{d/2} N_D(z)dz = \int_{-d/2}^{d/2} N_A(z)dz \tag{2.1d}$$

there are no free carriers in the unexcited sample at low temperatures.

2. Assume that the thickness of the doping layers are equal: $d_n = d_p$, the doping levels are uniform: $N_D = N_A$, and also there is no undoped layer: $d_i = 0$. The periodic space charge layer is then due to impurity ions only. $V_{imp}(z)$ is parabolic in nature and has amplitude

$$V_0 = \frac{e^2}{\varepsilon_s} N_D d_n^2/8. \tag{2.1e}$$

The potential variation is sketched in Fig 2.1c. The effective energy gap becomes

$$E_g(\text{eff}) = E_g - 2V_0 + E_{e0} + E_{h0} \tag{2.1f}$$

where, E_{e0} and E_{h0} are the energies of the first subbands. The envelope functions $\phi(z)$in the tight binding approximation are harmonic oscillator functions and the eigenvalues are expressed as

$$E_{en} = \hbar \left(\frac{e^2 N_D}{\varepsilon_{sc} m_e} \right)^{1/2} \left[n + 1/2 \right]. \tag{2.1g}$$

One may therefore conclude from (2.1g) and calculated values that the effective gap may be reduced from that in bulk material.

3. If the thickness of and/or the doping concentration in the n-layer is increased, there will be a finite, two-dimensional electron concentration in the n-layers. Therefore, it appears that the electronic structures of the nipi's differ radically from the corresponding bulk semiconductors as stated below:

   (a) Each band is split into mini bands.
   (b) The magnitude and spacing of these mini bands may be designed by the choice of the superlattices parameters and
   (c) The electron energy spectrum of the nipi crystal becomes two-dimensional leading to the step functional dependence of the density-of-states function.

In Sect. 2.2.1, of the theoretical background, the EEM in nipi structures of non-linear optical materials has been investigated. Section 2.2.2 contains the results for nipi structures of III–V, ternary and quaternary compounds in accordance with the three- and two-band models of Kane together with parabolic energy bands and they form the special cases of Sect. 2.2.1. Sections 2.2.3–2.2.5 contain the study of the EEM for nipi's of II–VI, IV–VI, and stressed Kane-type semiconductors respectively. The Sect. 2.3 contains the results and discussion of this chapter. The last Sect. 2.4 presents the open research problems pertaining to this chapter.

2.2 Theoretical Background

2.2.1 Formulation of the EEM in Nipi Structures of Nonlinear Optical Materials

The dispersion relation of the conduction electrons in nipi structures of nonlinear optical materials can be expressed using (1.2) and following the method as given in [34, 35] as

$$\psi_1(E) = \psi_2(E)k_s^2 + \psi_3(E)\left(n_i + \frac{1}{2}\right)\frac{2m_{\parallel}^*}{\hbar}\omega_8(E) \tag{2.1h}$$

where

$$\omega_8(E) \equiv \left(\frac{n_0\,|e|^2}{\varepsilon_{sc}\,[\theta_1(E)]}\right)^{1/2} \quad \text{and } \theta_1(E) \equiv \frac{\hbar^2}{2}\left\{\frac{\psi_3(E)\,[\psi_1(E)]' - \psi_1(E)\,[\psi_3(E)]'}{[\psi_3(E)]^2}\right\}$$

and $n_i(= 0, 1, 2\ldots)$ is the miniband index for nipi structures.

The EEM in this case assumes the form

$$m^*(E_{\mathrm{Fn}}, n_i) = \left(\frac{\hbar^2}{2}\right) R_{81}(E, n_i)\Big|_{E=\bar{E}_{\mathrm{Fn}}} \tag{2.2}$$

where,

$$\begin{aligned} R_{81}(E, n_i) \equiv [\psi_2(E)]^{-2}\Bigg[\psi_2(E)\Bigg\{[\psi_1(E)]' - \left(\frac{2m_{\parallel}^*}{\hbar}\right)[\psi_3(E)]'\left(n_i + \frac{1}{2}\right)[\omega_8(E)] \\ - \left(\frac{2m_{\parallel}^*}{\hbar}\right)[\psi_3(E)]\left(n_i + \frac{1}{2}\right)[\omega_8(E)]'\Bigg\} \\ - \left\{[\psi_1(E)] - \left(\frac{2m_{\parallel}^*}{\hbar}\right)[\psi_3(E)]\left(n_i + \frac{1}{2}\right)[\omega_8(E)]\right\}[\psi_2(E)]'\Bigg] \end{aligned}$$

and $\bar{E}_{\mathrm{Fn}}$ is the Fermi energy in the present case as measured from the edge of the conduction band in vertically upward direction in the absence of any quantization.

From (2.2), we observe that the EEM is a function of the Fermi energy, nipi subband index, and the other material constants which is the characteristic feature of nipi structures of nonlinear optical materials.

The subband energy ($E_{1\mathrm{ni}}$) can be written as

$$\psi_1(E_{1\mathrm{ni}}) = \psi_3(E_{1\mathrm{ni}})\left(n_i + \frac{1}{2}\right)\frac{2m_{||}^*}{\hbar}\omega_8(E_{1\mathrm{ni}}) \tag{2.3}$$

The density-of-states function for nipi structures of non-linear optical materials can be expressed as

$$N_{\mathrm{nipi}}(E) = \frac{g_v}{2\pi d_0}\sum_{n_i=0}^{n_{i\,\max}} R_{81}(E, n_i)H(E - E_{1\mathrm{ni}}) \tag{2.4}$$

in which d_0 is the superlattice period.

The electron concentration, can be written as

$$n_0 = \frac{g_v}{2\pi d_0}\sum_{n_i=0}^{n_{i\,\max}}\left[T_{81}(\bar{E}_{\mathrm{Fn}}, n_i) + T_{82}(\bar{E}_{\mathrm{Fn}}, n_i)\right] \tag{2.5}$$

where, $T_{81}(\bar{E}_{\mathrm{Fn}}, n_i) \equiv \left[\psi_1(\bar{E}_{\mathrm{Fn}}) - \psi_3(\bar{E}_{\mathrm{Fn}})\left(n_i + \frac{1}{2}\right)\frac{2m_{||}^*}{\hbar}\omega_8(\bar{E}_{\mathrm{Fn}})\right]\left[\psi_2(\bar{E}_{\mathrm{Fn}})\right]^{-1}$

and $T_{82}(\bar{E}_{\mathrm{Fn}}, n_i) \equiv \sum_{r=1}^{s} L(r)T_{81}(\bar{E}_{\mathrm{Fn}}, n_i)$.

2.2.2 *EEM in the Nipi Structures of III–V, Ternary and Quaternary Semiconductors*

(a) The electron energy spectrum in nipi structures of III–V, ternary and quaternary materials can be expressed from (2.1) under the conditions $\Delta_{||} = \Delta_{\perp} = \Delta$, $\delta = 0$ and $m_{||}^* = m_{\perp}^* = m_c$, as

$$I_{11}(E) = \left(n_i + \frac{1}{2}\right)\hbar\omega_9(E) + \frac{\hbar^2 k_s^2}{2m_c} \tag{2.6}$$

where $\omega_9(E) \equiv \left(\frac{n_0\,|e|^2}{\varepsilon_{\mathrm{sc}} I'(E) m_c}\right)^{1/2}$.

The EEM in this case can be written as

$$m^*(E_{\mathrm{Fn}}, n_i) = m_c R_{82}(E, n_i)|_{E=E_{\mathrm{Fn}}} \tag{2.7}$$

in which, $R_{82}(E, n_i) \equiv \{[I_{11}(E)]' - (n_i + \frac{1}{2})\,\hbar\,[\omega_9(E)]'\}$.
From (2.7) we observe that the EEM in this case is a function of the Fermi energy, nipi subband index and the other material constants which is the characteristic feature of nipi structures of III–V, ternary and quaternary compounds whose bulk dispersion relations is defined by the three-band model of Kane.
The subband energies (E_{2ni})can be written as

$$I_{11}(E_{2ni}) = \left(n_i + \frac{1}{2}\right)\hbar\omega_9(E_{2ni}). \tag{2.8}$$

The density-of-states function in this case can be expressed as

$$N_{nipi}(E) = \frac{m_c g_v}{\pi \hbar^2 d_0} \sum_{n_i=0}^{n_{i\,\max}} R_{82}(E, n_i) H(E - E_{2ni}). \tag{2.9}$$

The use of (2.9) leads to the expression of the electron concentration as

$$n_0 = \frac{m_c g_v}{\pi \hbar^2 d_0} \sum_{n_i=0}^{n_{i\,\max}} \left[T_{83}(\bar{E}_{Fn}, n_i) + T_{84}(\bar{E}_{Fn}, n_i)\right] \tag{2.10}$$

where $T_{83}(\bar{E}_{Fn}, n_i) \equiv \left[I_{11}(\bar{E}_{Fn}) - \left(n_i + \frac{1}{2}\right)\hbar\omega_9(\bar{E}_{Fn})\right]$ and $T_{84}(\bar{E}_{Fn}, n_i) \equiv \sum_{r=1}^{s} L(r) T_{83}(\bar{E}_{Fn}, n_i)$.

(b) For the two-band model of Kane, the expressions of the dispersion relation, the EEM, the subband energies, the density-of-states function, and n_0 remain the same where

$$I_{11}(E) = E(1 + \alpha E), \quad \{I_{11}(E)\}' = (1 + 2\alpha E) \quad \text{and} \quad \{I_{11}(E)\}'' = 2\alpha.$$

The EEM in this case can be written as

$$m^*(E_{Fn}, n_i) = m_c \left\{(1 + 2\alpha E_{Fn}) + \left(n_i + \frac{1}{2}\right)\hbar\,[\omega_9(E_{Fn})]\frac{\alpha}{(1 + 2\alpha E_{Fn})}\right\} \tag{2.11}$$

From 2.11 we observe that the EEM in this case is a function of the Fermi energy, nipi subband index and the other material constants due to the band nonparabolicity only.

(c) For parabolic energy bands, the forms of the expressions of dispersion relation, the EEM, the subband energies, the density-of-states function ,and n_0 remain the same, where $I_{11}(E) = E$,

$$\{I_{11}(E)\}' = 1 \quad \text{and} \quad \{I_{11}(E)\}'' = 0.$$

The EEM can be written as

$$m^*(E_{\mathrm{Fn}}, n_i) = m_c. \tag{2.12}$$

From (2.12) we observe that the EEM in this case is a constant quantity.

2.2.3 EEM in the Nipi Structures of II–VI Semiconductors

The carrier dispersion law in nipi structures of II–VI compounds can be expressed as

$$E = a_0' k_s^2 + \left(n_i + \frac{1}{2}\right)\hbar\omega_{10} \pm \bar{\lambda}_0 k_s, \quad \omega_{10} \equiv \left(\frac{n_0 \, |e|^2}{\varepsilon_{sc} m_{||}^*}\right)^{1/2}. \tag{2.13}$$

Using (2.13), the EEM in this case can be written as

$$m^*(E_{\mathrm{Fn}}, n_i) = m_\perp^* \left\{1 - \bar{\lambda}_0 \left[(\bar{\lambda}_0)^2 + 4a_0' E_{\mathrm{Fn}} - 4a_0' \left(n_i + \frac{1}{2}\right)\hbar\omega_{10}\right]^{-1/2}\right\}. \tag{2.14}$$

Thus, the EEM in this case is a function of the Fermi energy, the nipi subband index number and the energy spectrum constants due to the presence of only $\bar{\lambda}_0$.

The subband energies ($E_{3\mathrm{ni}}$)can be written as

$$E_{3\mathrm{ni}} = \left(n_i + \frac{1}{2}\right)\hbar\omega_{10} \tag{2.15}$$

The density-of-states function in this case can be expressed as

$$N_{\mathrm{nipi}}(E) = \frac{m_\perp^* g_v}{\pi \hbar^2 d_0} \sum_{n_i=0}^{n_{i\,\max}} \left[1 - \frac{a_{81}}{\sqrt{E + b_{81}(n_i)}}\right] H(E - E_{3\mathrm{ni}}) \tag{2.16}$$

in which, $a_{81} \equiv \frac{\bar{\lambda}_0}{2\sqrt{a_0'}}$ and $b_{81}(n_i) \equiv \left[\frac{1}{4a_0'}\left[(\bar{\lambda}_0)^2 - 4a_0'\left(n_i + \frac{1}{2}\right)\hbar\omega_{10}\right]\right]$.

The use of the (2.16) leads to the electron concentration under the condition of extreme degeneracy as

$$n_0 = \frac{g_v m_\perp^*}{\pi \hbar^2} \sum_{ni=0}^{ni_{\max}} \left(E_{\mathrm{Fn}} - E_{3\mathrm{ni}} + (\overline{\lambda_0})^2 m_\perp^* \hbar^{-2}\right) \tag{2.17}$$

2.2.4 EEM in the Nipi Structures of IV–VI Semiconductors

The carrier energy spectrum in nipi structures of IV–VI compounds can be written as

$$k_s^2 = (\hbar^2 S_{19})^{-1}\left[-S_{20}(E,n_i) + \sqrt{S_{20}^2(E,n_i) + 4S_{19}S_{21}(E,n_i)}\right] \tag{2.18}$$

in which,

$$S_{19} \equiv \left(\frac{\alpha}{m_t^+ m_t^-}\right), \quad S_{20}(E,n_i) \equiv \left\{\frac{1}{m_t^*} - \left(\frac{\alpha E}{m_t^+}\right) + \frac{1+\alpha E}{m_t^-} + \frac{\alpha\hbar^2}{2m_l^+ m_t^-}\left(n_i + \frac{1}{2}\right)\right.$$
$$\left. T(E) + \frac{\alpha\hbar^2}{2m_l^- m_t^+}\left(n_i + \frac{1}{2}\right)T(E)\right\}$$

$$T(E) \equiv \frac{2m^*(0)}{\hbar}\omega_{11}(E), \ m^*(0) \equiv \left(\frac{m_l^* m_l^-}{m_l^* + m_l^-}\right), \quad \omega_{11}(E) \equiv \left(\frac{n_0\,|e|^2}{\varepsilon_{sc} m^*(E)}\right)^{1/2},$$

$$m^*(E) \equiv \frac{1}{4t_1}\left[-(t_2(E))' + \frac{t_2(E)(t_2(E))' + 2t_1(1+2\alpha E)}{\sqrt{t_2^2(E) + 4Et_1(1+\alpha E)}}\right],$$

$$t_1 \equiv \left(\frac{\alpha}{4m_l^+ m_l^-}\right), \quad t_2(E) \equiv \frac{1}{2}\left[\left(\frac{1}{m_l^*}\right) - \left(\frac{\alpha E}{m_l^+}\right) + \left(\frac{1+\alpha E}{m_l^-}\right)\right],$$

$$(t_2(E))' \equiv \frac{\alpha}{2}\left(\frac{1}{m_l^-} - \left(\frac{1}{m_l^+}\right)\right) \text{ and}$$

$$S_{21}(E,n_i) \equiv \left[E(1+\alpha E) + \frac{\alpha E\hbar^2}{2m_l^+}\left(n_i + \frac{1}{2}\right)T(E) + \frac{\hbar^2}{2m_l^-}\left(n_i + \frac{1}{2}\right)T(E)(1+\alpha E)\right.$$
$$\left. + \frac{\hbar^4}{4m_l^- m_l^+}\left(n_i + \frac{1}{2}\right)T(E) - \left(\frac{\hbar^2}{2m_l^*}\right)T(E)\left(n_i + \frac{1}{2}\right)\right].$$

Using (2.18), the EEM in this case can be written as

$$m^*(E_{\mathrm{Fn}}, n_i) = R_{84}(E,n_i)|_{E=E_{\mathrm{Fn}}} \tag{2.19}$$

where,

$$R_{84}(E,n_i) \equiv (2S_{19})^{-1} \times \left[-(S_{20}(E,n_i))' + \frac{S_{20}(E,n_i)\left[S_{20}(E,n_i)\right]' + 2S_{19}\left[S_{21}(E,n_i)\right]'}{\left[\left\{\left[S_{20}(E,n_i)\right]'\right\}^2 + 4S_{19}S_{21}(E,n_i)\right]^{1/2}} \right].$$

Thus, one can observe that the EEM in this case is a function of both the Fermi energy and the nipi subband index number together with the spectrum constants of the system due to the presence of band nonparabolicity.

The subband energies ($E_{4\text{ni}}$) can be written as

$$\left[E_{4\text{ni}} - \frac{\hbar^2}{2m_l^-}T(E_{4\text{ni}})\left(n_i + \frac{1}{2}\right)\right] \times \left[1 + \alpha E_{4\text{ni}} + +\alpha\frac{\hbar^2}{2m_l^+}T(E_{4\text{ni}})\left(n_i + \frac{1}{2}\right)\right] = \left[\frac{\hbar^2}{2m_l^*}T(E_{4\text{ni}})\left(n_i + \frac{1}{2}\right)\right]. \tag{2.20}$$

The density-of-states function in this case assumes the form as

$$N_{\text{nipi}}(E) = \frac{g_v}{\pi\hbar^2 d_0}\sum_{n_i=0}^{n_{i\,\max}} R_{84}(E,n_i)H(E - E_{4\text{ni}}) \tag{2.21}$$

The use of (2.21) leads to the expression of the electron concentration as

$$n_0 = \frac{g_v}{2\pi\hbar^2 S_{19} d_0}\sum_{n_i=0}^{n_{i\,\max}}\left[T_{85}(\bar{E}_{\text{Fn}},n_i) + T_{86}(\bar{E}_{Fn},n_i)\right] \tag{2.22}$$

where, $T_{85}(\bar{E}_{\text{Fn}},n_i) \equiv \left[-S_{20}(E_{\text{Fn}},n_i) + \sqrt{[S_{20}(E_{\text{Fn}},n_i)]^2 + 4S_{19}S_{21}(E_{\text{Fn}},n_i)}\right]$
and $T_{86}(\bar{E}_{\text{Fn}},n_i) \equiv \sum_{r=1}^{s} L(r)T_{85}(\bar{E}_{\text{Fn}},n_i)$.

2.2.5 *EEM in the Nipi Structures of Stressed Semiconductors*

The electron dispersion law in the nipi structures of stressed semiconductors can be written as

$$\frac{k_x^2}{[\bar{a}_0(E)]^2}+\frac{k_y^2}{[\bar{b}_0(E)]^2}+\frac{1}{[\bar{c}_0(E)]^2}\frac{2m_z^*(0)}{\hbar}\left(n_i+\frac{1}{2}\right)\omega_{12}(E)=1 \tag{2.23}$$

where $\omega_{12}(E)\equiv\left(\frac{n_0\,|e|^2}{\varepsilon_{sc}m_z^*(E)}\right)^{1/2}$ and $m_z^*(E)\equiv\hbar^2\bar{c}_0(E)\frac{\partial}{\partial E}[\bar{c}_0(E)]$.

The use of (2.23) leads to the expression of the EEM as

$$m^*(E_{Fn},n_i)=\left(\frac{\hbar^2}{2}\right)R_{85}(E,n_i)\Big|_{E=E_{Fn}} \tag{2.24}$$

where,

$$\begin{aligned}R_{85}(E,n_i)\equiv&\Big[\big[(\bar{a}_0(E))'b_0(E)+(\bar{b}_0(E))'\bar{a}_0(E)\big]\\&\times\left[1-\frac{1}{[\bar{c}_0(E)]^2}\frac{2m_z^*(0)}{\hbar}\left(n_i+\frac{1}{2}\right)\omega_{12}(E)\right]\\&-\left[\frac{\bar{a}_0(E)\bar{b}_0(E)}{[\bar{c}_0(E)]^2}\frac{2m_z^*(0)}{\hbar}\left(n_i+\frac{1}{2}\right)[\omega_{12}(E)]'\right]\\&+\left[\frac{\bar{a}_0(E)\bar{b}_0(E)\,[\bar{c}_0(E)]'}{[\bar{c}_0(E)]^3}\frac{4m_z^*(0)}{\hbar}\left(n_i+\frac{1}{2}\right)[\omega_{12}(E)]\right]\Big].\end{aligned}$$

Thus, the EEM is a function of the Fermi energy and the nipi subband index due to the presence of stress and band nonparabolicity only.

The subband energies (E_{5ni})can be written as

$$\frac{1}{[\bar{c}_0(E_{4\mathrm{ni}})]^2}\frac{2m_z^*(0)}{\hbar}\left(n_i+\frac{1}{2}\right)\omega_{12}(E_{4\mathrm{ni}})=1. \tag{2.25}$$

The density-of-states function can be expressed as

$$N_{\mathrm{nipi}}(E)=\frac{g_v}{\pi\hbar^2 d_0}\sum_{n_i=0}^{n_{i\,\max}}R_{85}(E,n_i)H(E-E_{5\mathrm{ni}}). \tag{2.26}$$

Thus, using (2.26), the electron concentration in nipi structures of stressed compounds can be written as

$$n_0=\frac{g_v}{2\pi d_0}\sum_{n_i=0}^{n_{i\,\max}}\big[C_3(\bar{E}_{\mathrm{Fn}},n_i)+C_4(\bar{E}_{\mathrm{Fn}},n_i)\big]. \tag{2.27}$$

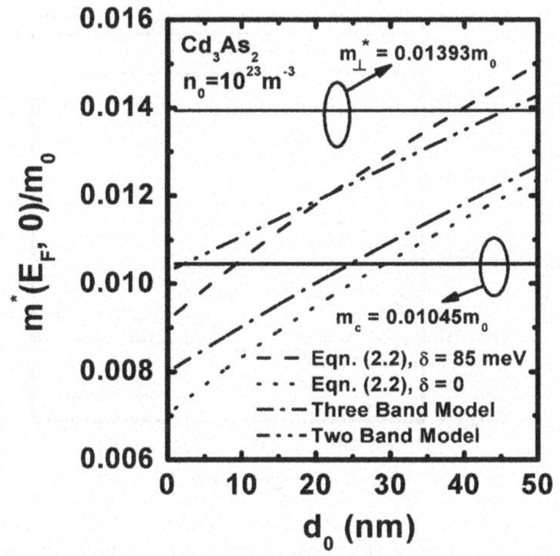

Fig. 2.2 Plot of the EEM as function of superlattice period for n-Cd_3As_2 considering Eq. (2.2) The plots for three- and two-band models of Kane have also been exhibited with their corresponding bulk values

where

$$C_3(\bar{E}_{\mathrm{Fn}}, n_i) \equiv \bar{a}_0(\bar{E}_{\mathrm{Fn}})\bar{b}_0(E_{\mathrm{Fn}})\left[1 - \frac{2m_z^*(0)}{\hbar}\left(n_i + \frac{1}{2}\right)\frac{\omega_{12}(\bar{E}_{\mathrm{Fn}})}{(\bar{c}_0(\bar{E}_{\mathrm{Fn}}))^2}\right]$$

and

$$C_4(\bar{E}_{\mathrm{Fn}}, n_i) \equiv \sum_{r=1}^{s} L(r)C_3(\bar{E}_{\mathrm{Fn}}, n_i).$$

2.3 Results and Discussion

The effect of nipi superlattice period on the EEM has been exhibited in Figs. 2.2, 2.3, 2.4, 2.5, 2.6, 2.7, 2.8, 2.9, 2.10 for different materials. Using (2.2) and (2.5) together with the energy band constants as given in Table 1.1, we have plotted the EEM in nipi structures of nonlinear optical materials taking Cd_3As_2 and $CdGeAs_2$ as examples in Figs. 2.2 and 2.3

From both Figs. 2.2 and 2.3, it appears that the effect of increment of the superlattice period increases the EEM in the presence of extreme carrier degeneracy of the order of 10^{23} m^{-3}. For comparison with the bulk anisotropic effective masses, we have also exhibited the same in the said figures. It appears that the EEM can be much less than that of the corresponding bulk values below 10 nm period for Cd_3As_2. Thus in such condition, one can expect an increase in the carrier mobility to a great extent,

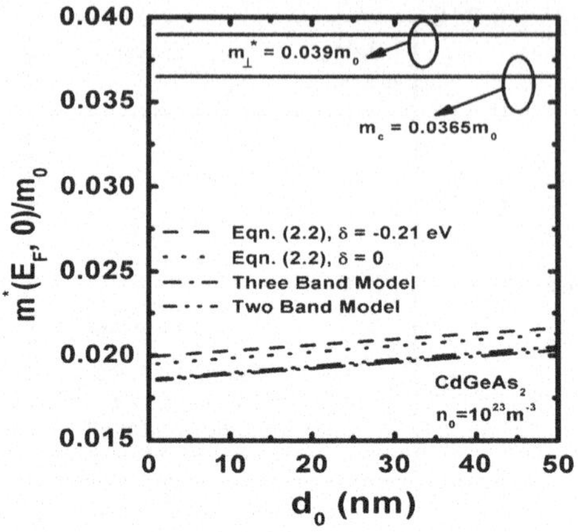

Fig. 2.3 Plot of the EEM as function of superlattice period for n-$CdGeAs_2$ for all cases of Fig. 2.2

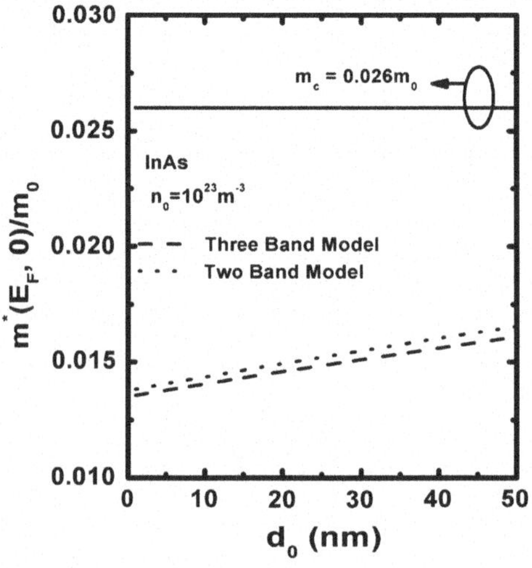

Fig. 2.4 Plot of the EEM as function of superlattice period for n-InAs considering the three- and two-band model of Kane with the corresponding bulk value

in fact almost double. The effect of crystal field splitting has also been exhibited in the same Figs. 2.2 and 2.3.

It appears that the effect of δ on the EEM is the largest in case of Cd_3As_2. The approximation in the energy band structure also makes a significant deviation of the EEM in case of Cd_3As_2. However, for $CdGeAs_2$, the EEM exhibits a slow variation over the superlattice period as compared with Fig. 2.2 At this point it should be noted that with the increase in the superlattice period, the EEM in Cd_3As_2 by considering the energy dispersion relation with the absence of the crystal field splitting and the three-band model of Kane actually tends to the anisotropic bulk value $0.01393m_0$.

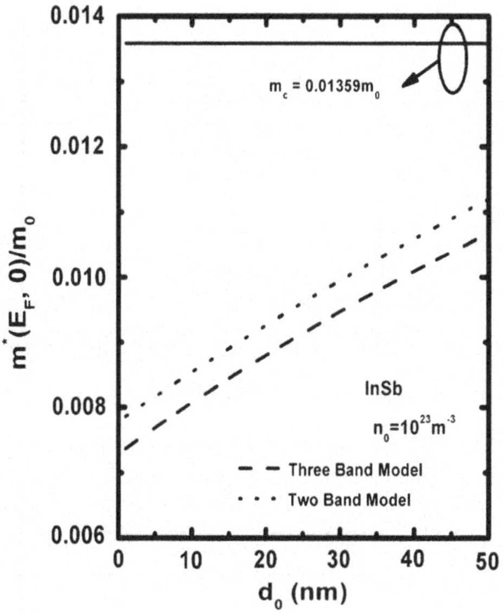

Fig. 2.5 Plot of the EEM as function of superlattice period for n-InSb considering the three- and two-band model of Kane with the corresponding bulk value

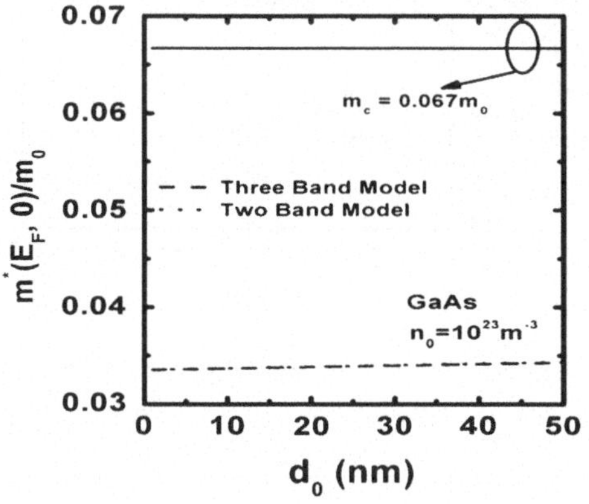

Fig. 2.6 Plot of the EEM as function of superlattice period for n-GaAs considering the three- and two-band model of Kane with the corresponding bulk value

This is not with the case when the effect of crystal field splitting and the two-band equivalent model is considered. In these two cases, the EEM is overestimated against the bulk value. This is not with the case of Fig. 2.3 of $CdGeAs_2$, where the EEM converges to the bulk anisotropic value at larger superlattice period.

The effect of superlattice period on the EEM in the ground state subband in III–V materials has been evaluated using the three- and the two-band model of Kane in Figs. 2.4, 2.5, 2.6 for InAs, InSb and GaAs respectively.

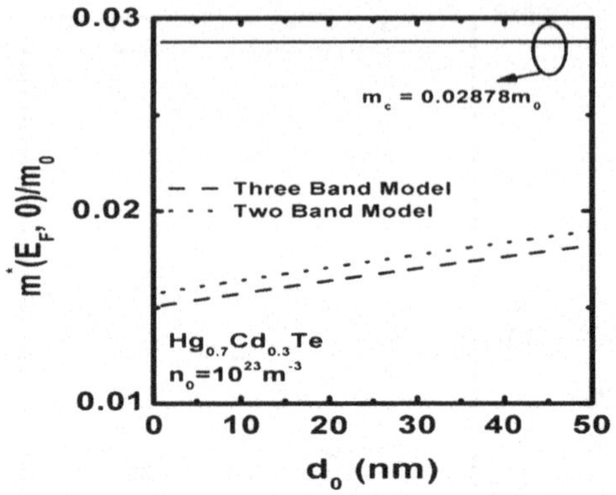

Fig. 2.7 Plot of the EEM as function of superlattice period for n-$Hg_{1-x}Cd_xTe$ considering the three- and two-band model of Kane with the corresponding bulk value at $x = 0.3$

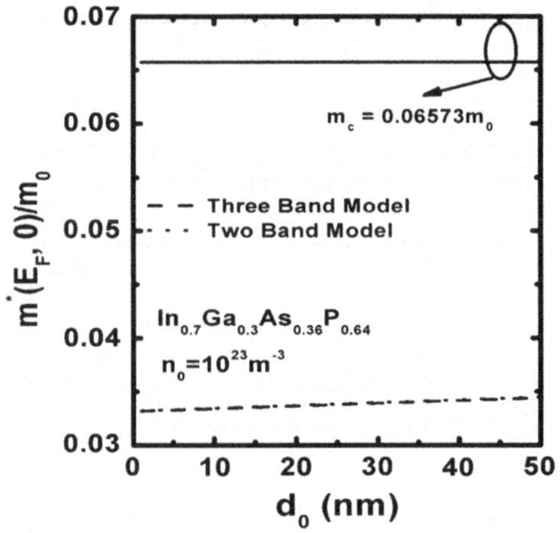

Fig. 2.8 Plot of the EEM as function of superlattice period for n-$In_{1-x}Ga_xAs_{1-y}P_y$ considering the three- and two-band model of Kane with the corresponding bulk value at $x = 0.3$

It appears from these figures that the effect of the variation of the energy dispersion relation model on EEM is almost insignificant for GaAs nipi structures, whereas for InSb, the EEM exhibits a significant deviation. In almost all the cases of about 1 nm period, the EEM approximately becomes half of the respective isotropic effective bulk masses indicating the mobility rise of up to 200 %.

In Figs. 2.7 and 2.8, the EEM as function of the periods has been further evaluated for the ternary and quaternary materials like $Hg_{1-x}Cd_xTe$ and $In_{1-x}Ga_xAs_{1-y}P_y$, where the energy band gap in these materials can be modulated by changing the alloy fraction x. We see that the EEM in case of $In_{1-x}Ga_xAs_{1-y}P_y$ almost exhibits no significant variation and approaches quickly its bulk normalized value 0.0287 at $x = 0.3$ as compared to $Hg_{1-x}Cd_xTe$.

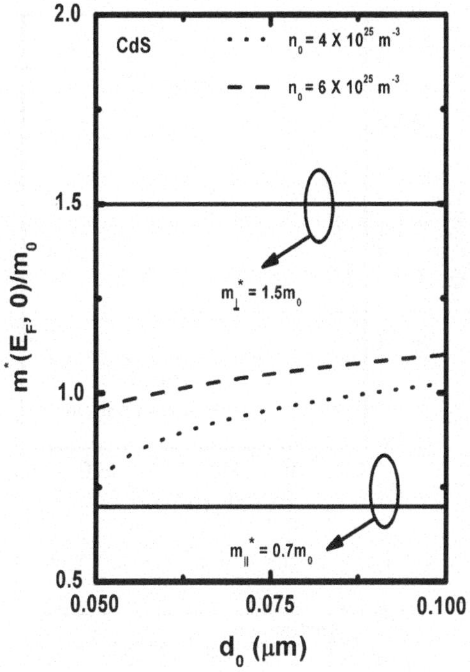

Fig. 2.9 Plot of the EEM as function of superlattice period for p-CdS

Figures 2.9 and 2.10 exhibit the EEM at the lowest subband in II–VI and IV–VI nipi structures of CdS and PbTe respectively.

The effect of increasing the doping concentration from 4×10^{25} to $6\times10^{25}\,m^{-3}$ on EEM in CdS has been also exhibited in Fig. 2.9 for a period bandwidth of 50–100 μm. It appears that with the increase in the doping concentration, the EEM in CdS increases and approaches the bulk longitudinal normalized value 1.5. However, in case of PbTe, we see that the EEM saturates above superlattice period of about 20 μm.

The effect of doping concentration on the EEM in the lowest subband level in all the aforementioned materials has been exhibited in Figs. 2.11, 2.12, 2.13, 2.14, 2.15, 2.16, 2.17, 2.18. It appears that the EEM increases with the increases in carrier degeneracy for all the cases and may become even larger than that of their corresponding bulk value along the proper transport direction. From Fig. 2.12 we see that the EEM is almost constant below the degeneracy of about $10^{23}\,m^{-3}$. The effect of different models of energy band structures has been exhibited to present the dependency of the EEM on the same. It appears from Figs. 2.15 and 2.17 that the second and third order Kane model almost exhibits no differences of the EEM from the two. The respective saturation of the EEM with the decrease in the degeneracy is different for the materials as this depends on the Fermi energy which is a function of the energy band parameters.

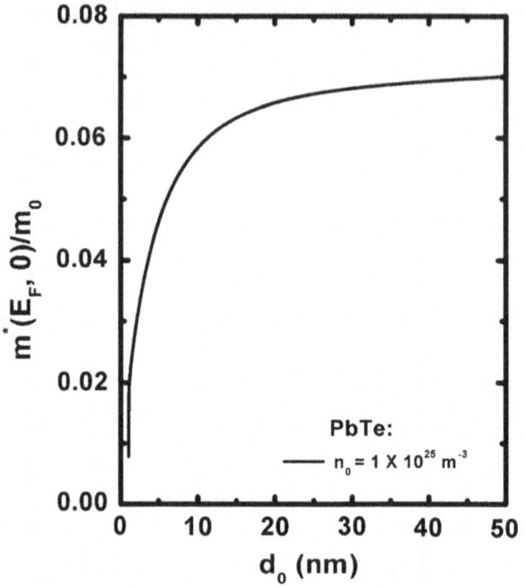

Fig. 2.10 Plot of the EEM as function of superlattice period for PbTe

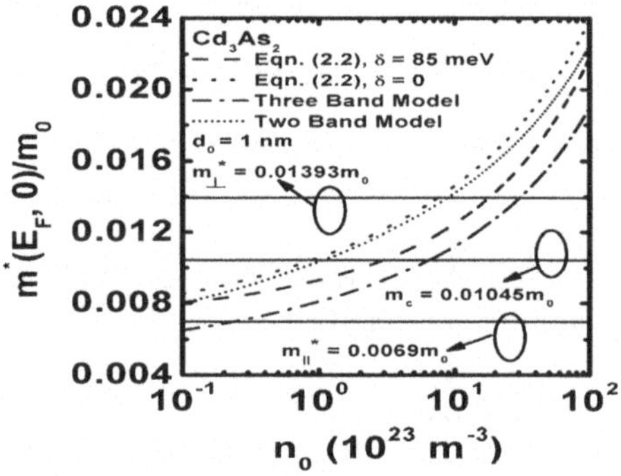

Fig. 2.11 Plot of the EEM as function of doping concentration for n-Cd_3As_2 considering all cases of Fig. 2.2

Figure 2.18 exhibits the variation of the EEM with increasing degeneracy for PbTe nipi. Anomalous behavior in the variation of the EEM has been exhibited as one increases the degeneracy. It appears that above $2 \times 10^{22}\ \text{m}^{-3}$, the EEM decreases. This should not be in general confused with other plots since an increase in the degeneracy increases the Fermi energy which increases the EEM. However, in this case, the effect of the different spectrum constants defines the variation of the EEM.

The variation of the EEM with alloy composition for the ternary and quaternary materials has been exhibited in Fig. 2.19 for the three- and two-energy band model of Kane. Almost no difference in the two-energy band model in this case is exhibited. The variation of the EEM for the quaternary is slower than that of the ternary which is due to the variation of the energy band gap through the alloy composition. The

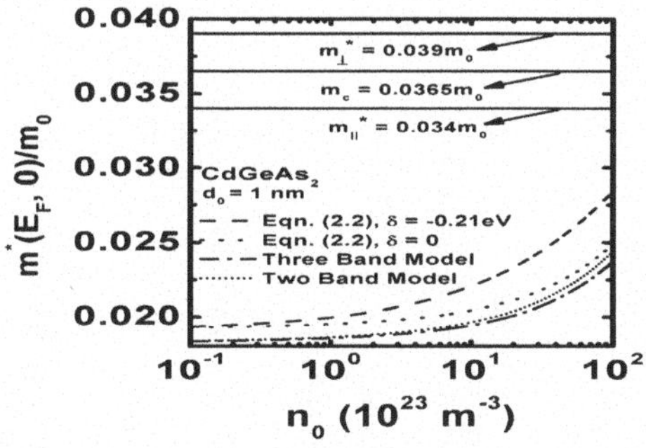

Fig. 2.12 Plot of the EEM as function of doping concentration for n-$CdGeAs_2$ considering all cases of Fig. 2.3

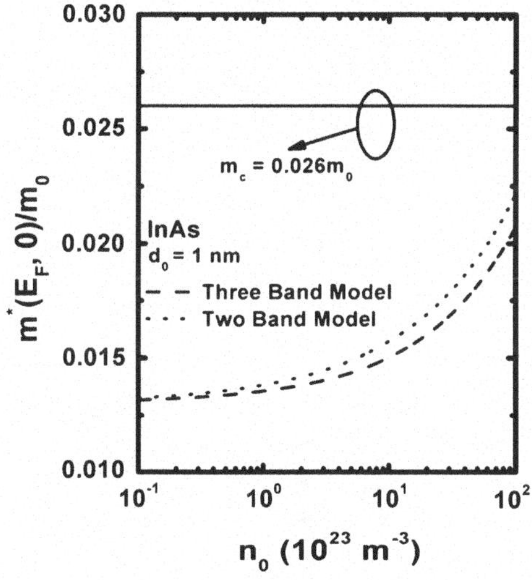

Fig. 2.13 Plot of the EEM as function of doping concentration for n-InAs considering all cases of Fig. 2.4

joy of executing the intricate computer programming for the variation of EEM in case of stressed InSb nipi structure whose energy band parameters have been given in Table 1.1 of Chap. 1 has been left to the reader. The summary of this chapter is presented in Table 2.1.

2.4 Open Research Problems

R.2.1 Investigate the EEM, EAM, DEM, CEM, CoEM, FREM, and OEM in the presence of an arbitrarily oriented nonquantizing magnetic field for nipi

Table 2.1 The electron statistics and the EEM in the nipi structures of nonlinear optical, III–V, ternary, quaternary, II–VI, IV–VI and stressed materials

Type of materials	The EEM	The electron statistics
1. Non-linear optical semiconductors	$m^*(E_{\mathrm{Fn}}, n_i) = \left(\frac{\hbar^2}{2}\right) R_{81}(E, n_i)\Big\vert_{E=\bar{E}_{\mathrm{Fn}}}$ (2.2)	$n_0 = \frac{g_v}{2\pi d_0} \sum_{n_i=0}^{n_{i\,\max}} \left[T_{81}(\bar{E}_{\mathrm{Fn}}, n_i) + T_{82}(\bar{E}_{\mathrm{Fn}}, n_i)\right]$ (2.5)
2. III–V semiconductors		
i. (a) Three band model of Kane	$m^*(E_{\mathrm{Fn}}, n_i) = m_c R_{82}(E, n_i)\vert_{E=E_{\mathrm{Fn}}}$ (2.7)	$n_0 = \frac{m_c g_v}{\pi \hbar^2 d_0} \sum_{n_i=0}^{n_{i\,\max}} \left[T_{83}(\bar{E}_{\mathrm{Fn}}, n_i) + T_{84}(\bar{E}_{\mathrm{Fn}}, n_i)\right]$ (2.10)
(b) Two band model of Kane	$m^*(E_{\mathrm{Fn}}, n_i) = m_c \left\{(1 + 2\alpha E_{\mathrm{Fn}}) + \left(n_i + \frac{1}{2}\right) \times \hbar\left[\omega_9(E_{Fn})\right] \frac{\alpha}{(1 + 2\alpha E_{Fn})}\right\}$ (2.11)	(2.10) remains same where $I(E) = E(1 + \alpha E)$, $\{I(E)\}' = (1 + 2\alpha E)$ and $\{I(E)\}'' = 2\alpha$
3. II–VI semiconductors	$m^*(E_{\mathrm{Fn}}, n_i) = m_{\perp}^* \left\{1 - \bar{\lambda}_0[(\bar{\lambda}_0)^2 + 4a_0' E_{\mathrm{Fn}} - 4a_0' \left(n_i + \frac{1}{2}\right) \hbar\omega_{10}]^{-1/2}\right\}$ (2.14)	$n_0 = \frac{g_v m_{\perp}^*}{\pi \hbar^2} \sum_{ni=0}^{ni_{\max}} \left(E_{\mathrm{Fn}} - E_{3ni} + (\overline{\lambda_0})^2 m_{\perp}^* \hbar^{-2}\right)$ (2.17)
4. IV–VI semiconductors	$m^*(E_{\mathrm{Fn}}, n_i) = R_{84}(E, n_i)\vert_{E=E_{\mathrm{Fn}}}$ (2.19)	$n_0 = \frac{g_v}{2\pi \hbar^2 S_{19} d_0} \sum_{n_i=0}^{n_{i\,\max}} \left[T_{85}(\bar{E}_{\mathrm{Fn}}, n_i) + T_{86}(\bar{E}_{\mathrm{Fn}}, n_i)\right]$ (2.22)
5. Stressed materials	$m^*(E_{\mathrm{Fn}}, n_i) = \left(\frac{\hbar^2}{2}\right) R_{85}(E, n_i)\Big\vert_{E=E_{\mathrm{Fn}}}$ (2.24)	$n_0 = \frac{g_v}{2\pi d_0} \sum_{n_i=0}^{n_{i\,\max}} \left[C_3(\bar{E}_{\mathrm{Fn}}, n_i) + C_4(\bar{E}_{\mathrm{Fn}}, n_i)\right]$ (2.27)

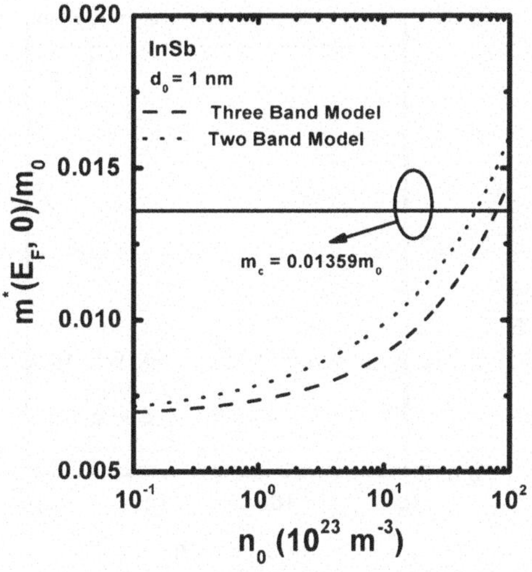

Fig. 2.14 Plot of the EEM as function of doping concentration for n-InSb considering all cases of Fig. 2.5

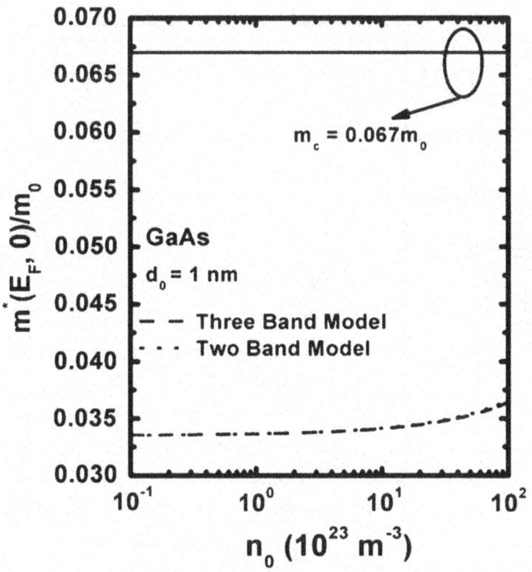

Fig. 2.15 Plot of the EEM as function of doping concentration for n-GaAs considering all cases of Fig. 2.6

structures of nonlinear optical semiconductors by including the electron spin. Study all the special cases for III–V, ternary and quaternary materials in this context.

R.2.2 Investigate the same set of masses as defined in (R 2.1) in nipi structures of IV–VI, II–VI and stressed Kane-type compounds in the presence of an

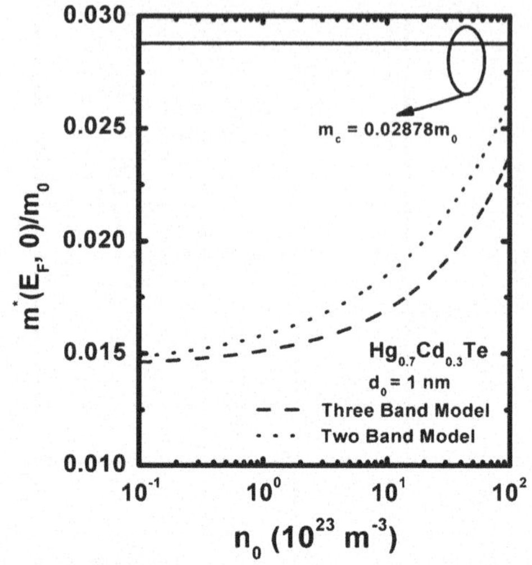

Fig. 2.16 Plot of the EEM as function of doping concentration for n-$Hg_{1-x}Cd_xTe$ considering all cases of Fig. 2.7

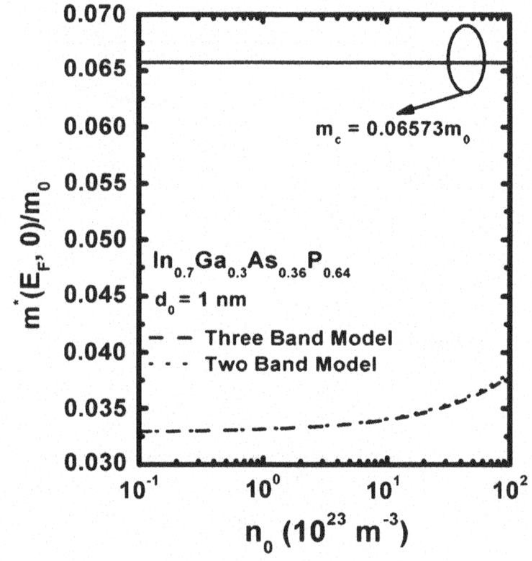

Fig. 2.17 Plot of the effective electron mass as function of doping concentration for n-$In_{1-x}Ga_xAs_{1-y}P_y$

arbitrarily oriented nonquantizing magnetic field by including the electron spin.

R.2.3 Investigate the same set of masses as defined in (R 2.1) for nipi structures of all the materials as stated in R.1.1 of chapter 1.

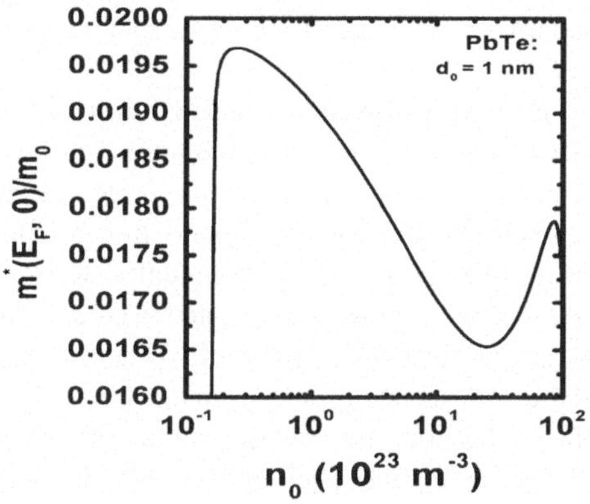

Fig. 2.18 Plot of the effective electron mass as function of doping concentration for PbTe

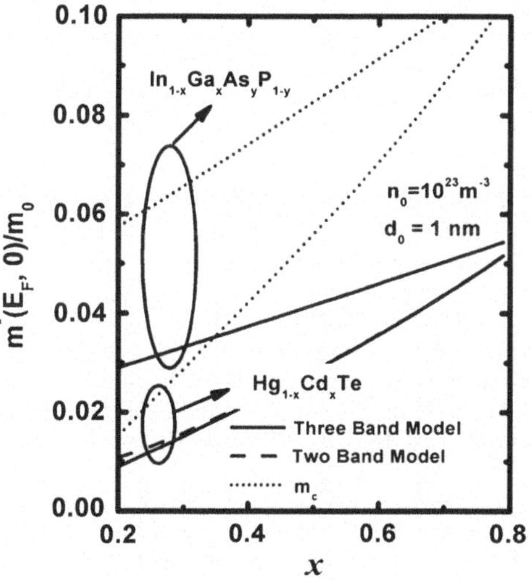

Fig. 2.19 Plot of the EEM as function of alloy composition for $Hg_{1-x}Cd_xTe$ and n-$In_{1-x}Ga_xAs_{1-y}P_y$ considering the three- and two-band model of Kane with the corresponding bulk value at different x

R.2.4 Investigate the same set of masses as defined in (R 2.1) for all the problems from R.2.1 to R.2.3 in the presence of an additional arbitrarily oriented electric field.

R.2.5 Investigate the same set of masses as defined in (R 2.1) for all the problems from R.2.1 to R.2.3 in the presence of arbitrarily oriented crossed electric and magnetic fields.

R 2.6 Investigate the same set of masses as defined in (R 2.1) for nipi structures of the heavily doped semiconductors in the presences of Gaussian, exponential, Kane, Halperian, Lax, and Bonch-Burevich types of Band tails for all systems whose unperturbed carrier energy spectra are defined in R1.1 and R1.2 respectively.

R 2.7 Investigate the same set of masses as defined in (R 2.1) for nipi structures of the negative refractive index, organic, magnetic, and other advanced optical materials in the presence of an arbitrarily oriented alternating electric field.

R 2.8 Investigate the same set of masses as defined in (R 2.1) for all the nipi systems of this chapter in the presence of finite potential wells.

R 2.9 Investigate the same set of masses as defined in (R 2.1) for all the nipi systems of this chapter in the presence of parabolic potential wells.

R 2.10 Investigate all the appropriate problems of this chapter by including the many body, image force, broadening, and hot carrier effects respectively.

R 2.11 Investigate all the appropriate problems of this chapter by removing all the mathematical approximations and establishing the respective appropriate uniqueness conditions.

References

1. L. Esaki, R. Tsu, IBM J. Res. Dev. **14**, 61 (1970)
2. G.H. Döhler, Phys. Status Solidi B **52**, 79 (1972)
3. G.H. Döhler, Phys. Status Solidi B **52**, 533 (1972)
4. G.H. Döhler, Surf. Sci. **73**, 97 (1978)
5. G.H. Döhler, J. Vac. Sci. Technol. **16**, 851 (1979)
6. G.H. Döhler, K. Ploog, Prog. Cryst. Growth Charact. **2**, 145 (1979)
7. G.H. Döhler, H. Künzel, K. Ploog, Phys. Rev. B **25**, 2616 (1982)
8. H. Künzel, G.H. Döhler, A. Fischer, K. Ploog, Appl. Phys. Lett. **38**, 171 (1980)
9. G.H. Döhler, H. Künzel, D. Olego, K. Ploog, P. Ruden, H.J. Stolz, Phys. Rev. Lett. **47**, 864 (1981)
10. K. Ploog, H. Künzel, J. Knecht, A. Fischer, G.H. Döhler, Appl. Phys. Lett. **38**, 870 (1981)
11. K. Ploog, A. Fischer, H. Künzel, J. Electrochem. Soc. **128**, 400 (1981)
12. H. Jung, G.H. Döhler, H. Kunzel, K. Ploog, P. Ruden, H.J. Stolz, Solid State Commun. **43**, 291 (1982)
13. H. Künzel, G.H. Döhler, P. Ruden, K. Ploog, Appl. Phys. Lett. **41**, 852 (1982)
14. N.G. Anderson, W.D. Laidig, R.M. Kolbas, Y.C. Lo, J. Appl. Phys. **60**, 2361 (1986)
15. F. Capasso, Semiconduct. Semimet. **22**, 2 (1985)
16. F. Capasso, K. Mohammed, A.Y. Cho, R. Hull, A.L. Hutchinson, Appl. Phys. Lett. **47**, 420 (1985)
17. F. Capasso, R.A. Kiehl, J. Appl. Phys. **58**, 1366 (1985)
18. K. Ploog, G.H. Doheler, Adv. Phys. **32**, 285 (1983)
19. F. Capasso, K. Mohammed, A.Y. Cho, Appl. Phys. Lett. **478**, (1986)
20. R. Grill, C. Metzner, G.H. Döhler, Phys. Rev. B **63**, 235316 (2001)
21. J.Z. Wang, Z.G. Wang, Z.M. Wang, S.L. Feng, Z. Yang, Phys. Rev. B **62**, 6956 (2000)
22. J.Z. Wang, Z.G. Wang, Z.M. Wang, S.L. Feng, Z. Yang, Phys. Rev. B **61**, 15614 (2000)
23. A.R. Kost, M.H. Jupina, T.C. Hasenberg, E.M. Garmire, J. Appl. Phys. **99**, 023501 (2006)
24. A.G. Smirnov, D.V. Ushakov, V.K. Kononenko, Proc. SPIE **4706**, 70 (2002)
25. D.V. Ushakov, V.K. Kononenko, I.S. Manak, Proc. SPIE **4358**, 171 (2001)
26. J.Z. Wang, Z.G. Wang, Z.M. Wang, S.L. Feng, Z. Yang, Phys. Rev. B **62**, 6956 (2000)
27. A.R. Kost, L. West, T.C. Hasenberg, J.O. White, M. Matloubian, G.C. Valley, Appl. Phys. Lett. **63**, 3494 (1993)

28. S. Bastola, S.J. Chua, S.J. Xu, J. Appl. Phys. **83**, 1476 (1998)
29. Z.J. Yang, E.M. Garmire, D. Doctor, J. Appl. Phys. **82**, 3874 (1997)
30. G.H. Avetisyan, V.B. Kulikov, I.D. Zalevsky, P.V. Bulaev, Proc. SPIE **2694**, 216 (1996)
31. U. Pfeiffer, M. Kneissl, B. Knüpfer, N. Müller, P. Kiesel, G.H. Döhler, J.S. Smith, Appl. Phys. Lett. **68**, 1838 (1996)
32. H.L. Vaghjiani, E.A. Johnson, M.J. Kane, R. Grey, C.C. Phillips, J. Appl. Phys. **76**, 4407 (1994)
33. P. Kiesel, K.H. Gulden, A. Hoefler, M. Kneissl, B. Knuepfer, S.U. Dankowski, P. Riel, X.X. Wu, J.S. Smith, G.H. Doehler, Proc. SPIE **1985**, 278 (1993)
34. P.K. Basu, Theory of Optical Process in Semiconductors: Bulk and Microstructures (Oxford University Press, Oxford 1997)
35. G.H. Doheler, Phys. Scr. **24**, 430 (1981)
36. S. Mukherjee, S.N. Mitra, P.K. Bose, A.R. Ghatak, A. Neoigi, J.P. Banerjee, A. Sinha, M. Pal, S. Bhattacharya, K.P. Ghatak, J. Comput. Theor. Nanosci. **4**, 550 (2007)

Chapter 3
The EEM in Inversion Layers of Non-Parabolic Semiconductors

3.1 Introduction

It is well known that the electrons in bulk semiconductors in general, have three dimensional freedom of motion. When, these electrons are confined in a one dimensional potential well whose width is of the order of the carrier wavelength, the motion in that particular direction gets quantized while that along the other two directions remains as free. Thus, the energy spectrum appears in the shape of discrete levels for the one dimensional quantization, each of which has a continuum for the two dimensional free motion. The transport phenomena of such one dimensional confined carriers have recently studied [1–22] with great interest. For the metal-oxide-semiconductor (MOS) structures, the work functions of the metal and the semiconductor substrate are different and the application of an external voltage at the metal-gate causes the change in the charge density at the oxide semiconductor interface leading to a bending of the energy bands of the semiconductor near the surface. As a result, a one dimensional potential well is formed at the semiconductor interface. The spatial variation of the potential profile is so sharp that for considerable large values of the electric field, the width of the potential well becomes of the order of the de Broglie wavelength of the carriers. The Fermi energy, which is near the edge of the conduction band in the bulk, becomes nearer to the edge of the valance band at the surface creating inversion layers. The energy levels of the carriers bound with in the potential well get quantized and form electric subbands. Each of the subband corresponds to a quantized level in a plane perpendicular to the surface leading to a quasi two dimensional electron gas. Thus, the extreme band bending at low temperature allows us to observe the quantum effects at the surface.

In Sect. 3.2.1, of the theoretical background, the EEM in n-channel inversion layers of tetragonal materials has been investigated for both weak and strong electric field limits. The Sect. 3.2.2 contains the results for n-channel inversion layers of III–V, ternary and quaternary compounds for both the electric field limits whose bulk electrons obey the three and the two band models of Kane together with parabolic energy bands and they form the special cases of Sect. 3.2.1. The Sect. 3.2.3

S. Bhattacharya and K. P. Ghatak, *Effective Electron Mass in Low-Dimensional Semiconductors*, Springer Series in Materials Science 167,
DOI: 10.1007/978-3-642-31248-9_3,

contains the study of the EEM for n-channel inversion layers of II–VI materials. The Sects. 3.2.4 and 3.2.5 contains the study of the EEM in n-channel inversion layers of IV–VI and stressed semiconductors for both the limits respectively. The Sects. 3.2.6 and 3.2.7 contains the study of the EEM in n-channel inversion layers of Ge and GaSb for both the limits respectively. The Sect. 3.3 contains the results and discussion of this chapter. The last Sect. 3.4 presents the open research problems for this chapter.

3.2 Theoretical Background

3.2.1 Formulation of the EEM in n-Channel Inversion Layers of Non-Linear Optical Semiconductors

In the presence of a surface electric field F_s along z direction and perpendicular to the surface, (1.2) assumes the form

$$\psi_1(E - |e|\, F_s z) = \psi_2(E - |e|\, F_s z)k_s^2 + \psi_3(E - |e|\, F_s z)k_z^2 \tag{3.1}$$

where, for this chapter, E represents the electron energy as measured from the edge of the conduction band at the surface in the vertically upward direction.

The quantization rule for inversion layers is given by [5]

$$\int_0^{z_t} k_z \mathrm{d}z = \frac{2}{3}(S_i)^{3/2} \tag{3.2}$$

where, z_t is the classical turning point and S_i is the zeros of the Airy function $(Ai(-S_i) = 0)$.

Using (3.1) and (3.2), under the weak electric field limit, one can write,

$$\int_0^{z_t} \sqrt{A_7(E) - |e|\, F_s z D_7(E)}\mathrm{d}z = \frac{2}{3}(S_i)^{3/2} \tag{3.3}$$

in which, $z_t \equiv \dfrac{A_7(E)}{[|e|\, F_s D_7(E)]}$, $A_7(E) \equiv \left[\dfrac{\psi_1(E) - \psi_2(E)k_s^2}{\psi_3(E)}\right]$, $D_7(E) \equiv [B_7(E) - A_7(E)C_7(E)]$, $B_7(E) \equiv \left[\dfrac{(\psi_1(E))' - (\psi_2(E))'k_s^2}{\psi_3(E)}\right]$ and $C_7(E) \equiv \left[\dfrac{(\psi_3(E))'}{\psi_3(E)}\right]$.

Thus, the 2D electron dispersion law in n-channel inversion layers of tetragonal materials under the weak electric field limit can approximately be written as

$$\psi_1(E) = P_7(E, i)k_s^2 + Q_7(E, i) \tag{3.4}$$

where,

$$P_7(E,i) \equiv \left[\psi_2(E) - \left(\frac{2t_2(E)}{3\,[t_1(E)]^{1/3}}\right)\psi_3(E)S_i(|e|\,F_s)^{2/3}\right],$$

$$t_2(E) \equiv \left[\frac{[\psi_2(E)]'}{\psi_3(E)} - \left(\frac{\psi_2(E)\,[\psi_3(E)]'}{[\psi_3(E)]^2}\right)\right],$$

$$t_1(E) \equiv \left[\frac{[\psi_1(E)]'}{\psi_3(E)} - \left(\frac{\psi_1(E)\,[\psi_3(E)]'}{[\psi_3(E)]^2}\right)\right]$$

and $Q_7(E,i) \equiv S_i\psi_3(E)\,[|e|\,F_s t_1(E)]^{2/3}$.

The EEM in the $x-y$ plane can be expressed as

$$m^*(E_{\mathrm{Fiw}},i) = \left(\frac{\hbar^2}{2}\right)G_7(E,i)\Big|_{E=E_{\mathrm{Fiw}}} \tag{3.5}$$

where, $G_7(E,i) \equiv [P_7(E,i)]^{-2}\left[P_7(E,i)\left\{(\psi_1(E))' - (Q_7(E,i))'\right\} - \{\psi_1(E) - (Q_7(E,i))\}\,(P_7(E,i))'\right]$ and E_{Fiw} is the Fermi energy under the weak electric field limit as measured from the edge of the conduction band at the surface in the vertically upward direction. Thus, we observe that the EEM is the function of subband index, the Fermi energy and other band constants due to the combined influence of the crystal filed splitting constant and the anisotropic spin-orbit splitting constants respectively.

The subband energy ($E_{n_{\mathrm{iw1}}}$) in this case can be obtained from (3.4) as

$$\psi_1\left(E_{n_{\mathrm{iw1}}}\right) = Q_7\left(E_{n_{\mathrm{iw1}}}, i\right) \tag{3.6}$$

The general expression of the 2D total density-of-states function in this case can be written as

$$N_{2D_i}(E) = \frac{2g_v}{(2\pi)^2}\sum_{i=0}^{i_{\max}}\frac{\partial}{\partial E}\left[A(E,i)H(E-E_{n_i})\right] \tag{3.7}$$

where, $A(E,i)$ is the area of the constant energy 2D wave vector space for inversion layers and E_{n_i} is the corresponding subband energy.

Using (3.4) and (3.7), the total 2D density-of-states function under the weak electric field limit can be expressed as

$$N_{2D_i}(E) = \frac{g_v}{(2\pi)}\sum_{i=0}^{i_{\max}}\left[G_7(E,i)H(E-E_{n_{\mathrm{iw1}}})\right] \tag{3.8}$$

where, $E_{n_{\mathrm{iw1}}}$ is the subband energy for the weak electric field limit in this case.

Using (3.8) and the Fermi-Dirac occupation probability factor, the 2D surface electron concentration in n-channel inversion of tetragonal materials under the weak electric field limit ($n_{2\mathrm{Dw}}$) can be written as

$$n_{2\mathrm{Dw}} = g_v (2\pi)^{-1} \sum_{i=0}^{i_{\max}} \left[P_{7\mathrm{w}}(E_{\mathrm{Fiw}}, i) + Q_{7\mathrm{w}}(E_{\mathrm{Fiw}}, i) \right] \tag{3.9}$$

where, $P_{7\mathrm{w}}(E_{\mathrm{Fiw}}, i) \equiv [\psi_1(E_{\mathrm{Fiw}}, i) - Q_7(E_{\mathrm{Fiw}}, i)] \{P_7(E_{\mathrm{Fiw}}, i)\}^{-1}$, $Q_{7\mathrm{w}}(E_{\mathrm{Fiw}}, i) \equiv \sum_{r=1}^{s} \{L(r) [P_7(E_{\mathrm{Fiw}}, i)]\}$ and $F_s \equiv \left(\frac{(|e|\, n_{2\mathrm{Dw}})}{\varepsilon_{\mathrm{sc}}} \right)$.

Using (3.1) and (3.2), the 2D electron dispersion law in n-channel inversion layers of tetragonal materials under the strong electric field limit can be written as

$$k_s^2 = P_2(E, i) \tag{3.10}$$

$$P_2(E, i) \equiv [F_7(E)]^{-1} [F_6(E) - F_8(E, i)], \quad F_6(E) \equiv \left[\frac{\psi_1(E)}{\psi_3(E)} \right] \left[1 + \frac{[\psi_1(E)]' [\psi_3(E)]'}{\psi_3(E) [\psi_1(E)]''} + \frac{\psi_1(E) [\psi_3(E)]''}{2 [\psi_1(E)]'' [\psi_3(E)]} \right],$$

$$F_7(E) \equiv \left[\left(\frac{\psi_2(E)}{\psi_3(E)} \right) - \left(\frac{\psi_1(E) [\psi_2(E)]''}{2 [\psi_1(E)]'' [\psi_3(E)]} \right) + \left(\frac{\psi_2(E)}{\psi_3(E)} \right) \left[\frac{[\psi_1(E)]' [\psi_3(E)]'}{\psi_3(E) [\psi_1(E)]''} + \frac{\psi_1(E) [\psi_3(E)]''}{2 [\psi_1(E)]'' [\psi_3(E)]} \right] + \left(\frac{\psi_1(E)}{\psi_3(E)} \right) \left[\frac{[\psi_2(E)]' [\psi_3(E)]'}{\psi_3(E) [\psi_1(E)]''} + \frac{\psi_2(E) [\psi_3(E)]''}{2 [\psi_1(E)]'' [\psi_3(E)]} \right] \right]$$

$$\text{and} \quad F_8(E, i) \equiv \left[\frac{2\sqrt{2}}{3} (S_i)^{3/2} (|e|\, F_s) \sqrt{[\psi_1(E)]''} \right].$$

The EEM in the $x - y$ plane can be written in this case as

$$m^*(E_{\mathrm{Fis}}, i) = \left(\frac{\hbar^2}{2} \right) [P_2(E, i)]' \Big|_{E = E_{\mathrm{Fis}}} \tag{3.11}$$

where, E_{Fis} is the Fermi energy under the strong electric field limit as measured from the edge of the conduction band at the surface. Thus, we note that the EEM is the function of subband index and the Fermi energy due to the combined influence of the crystal filed splitting constant and the anisotropic spin-orbit splitting constants respectively.

The subband energy ($E_{n_{\mathrm{is1}}}$) in this case can be obtained from the (3.10) as

$$P_2(E_{n_{is1}}, i) = 0 \tag{3.12}$$

The total 2D density-of-states function under the strong electric field limit can be written as

$$N_{2D_i}(E) = \frac{g_v}{(2\pi)} \sum_{i=0}^{i_{max}} \left[(P_2(E, i))' H(E - E_{n_{is1}}) \right] \tag{3.13}$$

Using (3.13) and the Fermi-Dirac occupation probability factor, the 2D surface electron concentration (n_{2Ds}) in this case can be expressed as

$$n_{2Ds} = g_v (2\pi)^{-1} \sum_{i=0}^{i_{max}} \left[P_2(E_{Fis}, i) + Q_2(E_{Fis}, i) \right] \tag{3.14}$$

where, $Q_2(E_{Fis}, i) \equiv \sum_{r=1}^{s} \{L(r) \left[P_2(E_{Fis}, i) \right]\}$.

3.2.2 Formulation of the EEM in n-Channel Inversion Layers of III–V, Ternary and Quaternary Semiconductors

Using the substitutions $\delta = 0$, $\Delta_{\parallel} = \Delta_{\perp} = \Delta$ and $m_{\parallel}^* = m_{\perp}^* = m_c$, (3.4) under the condition of weak electric field limit, assumes the form

$$I_{11}(E) = \frac{\hbar^2 k_s^2}{2m_c} + S_i \left[\frac{\hbar\, |e|\, F_s\, [I_{11}(E)]'}{\sqrt{2m_c}} \right]^{2/3} \tag{3.15}$$

(3.15) represents the dispersion relation of the 2D electrons in n-channel inversion layers of III–V, ternary and quaternary materials under the weak electric field limit whose bulk electrons obey the three band model of Kane.

The EEM can be expressed as

$$m^*(E_{Fiw}, i) = m_c \left[P_3(E, i) \right] |_{E = E_{Fiw}} \tag{3.16}$$

where, $P_3(E, i) \equiv \left\{ [I_{11}(E)]' - \left\{ \frac{2}{3} S_i \left[\frac{\hbar |e| F_s}{\sqrt{2m^*}} \right]^{2/3} \{[I_{11}(E)]'\}^{-1/3} [I_{11}(E)]'' \right\} \right\}$.

Thus, one can observe that the EEM is a function of the subband index, surface electric field, the Fermi energy and the other spectrum constants due to the combined influence of E_g and Δ.

The subband energy ($E_{n_{iw2}}$) in this case can be obtained from (3.15) as

$$I_{11}\left(E_{n_{\mathrm{iw2}}}\right) = S_i \left[\frac{\hbar \left|e\right| F_s \left[I_{11}\left(E_{n_{\mathrm{iw2}}}\right)\right]'}{\sqrt{2m_c}} \right]^{2/3} \tag{3.17}$$

Using (3.15) and (3.7), the 2D total density-of-states function in weak electric field limit can be expressed as

$$N_{2\mathrm{D}_i}(E) = \frac{m_c g_v}{\pi \hbar^2} \sum_{i=0}^{i_{\max}} \left[P_3(E,i) H(E - E_{n_{\mathrm{iw2}}})\right] \tag{3.18}$$

Using (3.18) and the occupation probability, the $n_{2\mathrm{Dw}}$ in the present case can be written as

$$n_{2\mathrm{Dw}} = \frac{g_v m_c}{\pi \hbar^2} \sum_{i=0}^{i_{\max}} \left[P_{4\mathrm{w}}(E_{\mathrm{Fiw}}, i) + Q_{4\mathrm{w}}(E_{\mathrm{Fiw}}, i)\right] \tag{3.19}$$

where, $P_{4\mathrm{w}}(E_{\mathrm{Fiw}}, i) \equiv \left\{ I_{11}(E_{\mathrm{Fiw}}) - S_i \left[\frac{\hbar e F_s [I_{11}(E_{\mathrm{Fiw}})]'}{\sqrt{2m_c}} \right]^{2/3} \right\}$ and $Q_4(E_{\mathrm{Fiw}}, i) \equiv \sum_{r=1}^{s} \{L(r)\,[P_4(E_{\mathrm{Fiw}}, i)]\}$.

Using the substitutions $\delta = 0$, $\Delta_{||} = \Delta_{\perp} = \Delta$ and $m_{||}^* = m_{\perp}^* = m_c$, (3.10) under the condition of strong electric field limit, assumes the form

$$\left[I_{11}(E) - \left\{ \frac{|e|\, F_s \hbar}{\sqrt{2m_c}} \left(\frac{2\sqrt{2}(S_i)^{3/2}}{3} \right) \sqrt{[I_{11}(E)]''} \right\} \right] = \frac{\hbar^2 k_s^2}{2m_c} \tag{3.20}$$

(3.20) represents the dispersion relation of the 2D electrons in n-channel inversion layers of III–V, ternary and quaternary materials under the strong electric field limit whose bulk conduction electrons are defined by the three band model of Kane.

The EEM can be expressed as

$$m^*(E_{\mathrm{Fis}}, i) = m_c\, [P_5(E, i)]|_{E = E_{\mathrm{Fis}}} \tag{3.21}$$

where, $P_5(E, i) \equiv \left\{ \{I_{11}(E)\}' - \left\{ \frac{|e|\, F_s \hbar}{\sqrt{2m_c}} \left(\frac{\sqrt{2}(S_i)^{3/2}}{3} \right) ([I_{11}(E)]'')^{-1/2} [I_{11}(E)]''' \right\} \right\}$.

Thus, one can observe that the EEM is a function of the subband index, surface electric field, the Fermi energy and the other spectrum constants due to the combined influence of E_g and Δ.

The subband energy ($E_{n_{\mathrm{is2}}}$) in this case can be obtained from (3.20) as

$$I_{11}(E_{n_{\mathrm{is2}}}) - \left\{ \frac{|e|\, F_s \hbar}{\sqrt{2m_c}} \left(\frac{2\sqrt{2}(S_i)^{3/2}}{3} \right) \sqrt{\left[I_{11}(E_{n_{\mathrm{is2}}})\right]''} \right\} = 0 \tag{3.22}$$

Using (3.20) and (3.7), the total 2D density-of-states function under the strong electric field limit can be expressed as

$$N_{2D_i}(E) = \frac{m_c g_v}{\pi \hbar^2} \sum_{i=0}^{i_{\max}} [P_5(E,i) H(E - E_{n_{is2}})] \quad (3.23)$$

Using (3.23) and the Fermi-Dirac occupation probability factor, the n_{2Ds} in the present case under the strong electric field can be written as

$$n_{2Ds} = \frac{g_v m_c}{\pi \hbar^2} \sum_{i=0}^{i_{\max}} [P_{6s}(E_{Fis}, i) + Q_{6s}(E_{Fis}, i)] \quad (3.24)$$

where, $P_{6s}(E_{Fis}, i) \equiv \left\{ I_{11}(E_{Fis}) - \left[\frac{2\sqrt{2}}{3}(S_i)^{3/2} \frac{\hbar |e| F_s [I_{11}(E_{Fis})]''}{\sqrt{2m_c}} \right] \right\}$ and

$Q_{6s}(E_{Fis}, i) \equiv \sum_{r=1}^{s} \{L(r)[P_6(E_{Fis}, i)]\}$

Using the constraints $\Delta \gg E_{g_0}$ or $\Delta \ll E_{g_0}$, (3.15) under the low electric field limit assumes the form

$$E(1+\alpha E) = \frac{\hbar^2 k_s^2}{2m_c} + S_i \left[\frac{\hbar |e| F_s (1+2\alpha E)}{\sqrt{2m_c}} \right]^{2/3} \quad (3.25)$$

For large values of i, $S_i \to \left[\frac{3\pi}{2}\left(i + \frac{3}{4}\right)\right]^{2/3}$ [5], and (3.25) gets simplified as

$$E(1+\alpha E) = \frac{\hbar^2 k_s^2}{2m_c} + \left[\frac{3\pi \hbar |e| F_s}{2} \left(i + \frac{3}{4}\right) \frac{(1+2\alpha E)}{\sqrt{2m_c}} \right]^{2/3} \quad (3.26)$$

(3.26) was derived for the first time by Antcliffe et al. [3].

The EEM in this case is given by

$$m^*(E_{Fiw}, i) = m_c\, [P_6(E, i)]|_{E=E_{Fiw}} \quad (3.27)$$

where, $P_6(E, i) \equiv \left\{ 1 + 2\alpha E - \frac{4\alpha}{3} S_i \left[\frac{\hbar |e| F_s}{\sqrt{2m_c}} \right]^{2/3} \{1+2\alpha E\}^{-1/3} \right\}$.

Thus, one can observe that the EEM is a function of the subband index, surface electric field and the Fermi energy due to the presence of band nonparabolicity only.

The subband energies ($E_{n_{iw3}}$) are given by

$$E_{n_{iw3}}(1+\alpha E_{n_{iw3}}) = S_i \left[\frac{\hbar |e| F_s (1+2\alpha E_{n_{iw3}})}{\sqrt{2m_c}} \right]^{2/3} \quad (3.28)$$

The total 2D density-of-states function can be written as

$$N_{2D}(E)=\frac{m_c g_v}{\pi\hbar^2}\sum_{i=0}^{i_{max}}\left\{\left[\left[1+2\alpha E-\frac{4\alpha}{3}S_i\left[\frac{\hbar|e|F_s}{\sqrt{2m_c}}\right]^{2/3}(1+2\alpha E)^{-1/3}\right]H(E-E_{n_{iw3}})\right\}\tag{3.29}$$

Under the condition $\alpha E \ll 1$, the use of (3.29) and the Fermi-Dirac integral leads to the expression of n_{2Dw} as

$$n_{2Dw}=\left(\frac{g_v m_c k_B T}{\pi\hbar^2}\right)\sum_{i=0}^{i_{max}}\{[1+D_i+2\alpha E_{n_{iw3}}]F_0(\eta_{iw})+2\alpha k_B T F_1(\eta_{iw})\}\tag{3.30}$$

where, $D_i\equiv\frac{4\alpha S_i}{3}\left(\frac{\hbar|e|F_s}{\sqrt{2m_c}}\right)^{2/3}$ and $\eta_{iw}\equiv\left[\frac{E_{Fiw}-E_{n_{iw3}}}{k_BT}\right]$.

For all values of αE_{Fiw}, the n_{2Dw} can be written as

$$n_{2Dw}=\left(\frac{g_v m_c}{\pi\hbar^2}\right)\sum_{i=0}^{i_{max}}[P_{5w}(E_{Fiw},i)+Q_{5w}(E_{Fiw},i)]\tag{3.31}$$

where, $P_{5w}(E_{Fiw},i)\equiv\left[E_{Fiw}(1+\alpha E_{Fiw})-S_i\left[\frac{\hbar|e|F_s}{\sqrt{2m_c}}(1+2\alpha E_{Fiw})\right]^{2/3}\right]$ and $Q_{5w}(E_{Fiw},i)\equiv\sum_{r=1}^{s}L(r)P_{5w}(E_{Fiw},i)$.

For $\alpha\to 0$, as for inversion layers, whose bulk electrons are defined by the parabolic energy bands, we can write,

$$E=\frac{\hbar^2k_s^2}{2m_c}+S_i\left[\frac{\hbar|e|F_s}{\sqrt{2m_c}}\right]^{2/3}\tag{3.32}$$

The (3.32) is valid for all values of the surface electric field [1].

The electric subband energy ($E_{n_{i4}}$) assumes the form, from (3.32) as

$$E_{n_{i4}}=S_i\left[\frac{\hbar|e|F_s}{\sqrt{2m_c}}\right]^{2/3}\tag{3.33}$$

The total density-of-states function can be written using (3.33) as

$$N_{2D}(E)=\frac{m_c g_v}{\pi\hbar^2}\sum_{i=0}^{i_{max}}H(E-E_{n_{i4}})\tag{3.34}$$

The use of (3.34) leads to the expression of n_{2Di} as [1]

$$n_{2\mathrm{Di}} = \frac{g_v m_c k_B T}{\pi \hbar^2} \sum_{i=0}^{i_{\max}} F_0(\eta_i) \tag{3.35}$$

where, $\eta_i \equiv (k_B T)^{-1} \left[E_{\mathrm{Fi}} - S_i \left[\frac{\hbar |e| F_s}{\sqrt{2m_c}} \right]^{2/3} \right]$, E_{Fi} is the Fermi energy as measured from the edge of the conduction band at the surface.

Using the constraints $\Delta \gg E_g$ or $\Delta \ll E_g$, the (3.20) under the strong electric field limit assumes the form

$$E(1 + \alpha E) = \frac{\hbar^2 k_s^2}{2m_c} + \frac{2\sqrt{2}}{3} (S_i)^{3/2} \left[\frac{\hbar |e| F_s}{\sqrt{2m_c E_g}} \right] \tag{3.36}$$

For large values of i, $S_i \to \left[\frac{3\pi}{2} \left(i + \frac{3}{4} \right) \right]^{2/3}$ [5] and (3.36) gets simplified as

$$E(1 + \alpha E) = \frac{\hbar^2 k_s^2}{2m_c} + \left[\frac{\pi \hbar |e| F_s \sqrt{2}}{\sqrt{m_c E_g}} \left(i + \frac{3}{4} \right) \right] \tag{3.37}$$

The (3.37) was derived for the first time by Antcliffe et al. [3].

From (3.36), we observe that under the condition $E_g \to 0$, one cannot obtain the corresponding parabolic case, since under high electric field limit, the band becomes permanently nonparabolic.

The EEM is given by

$$m^*(E_{\mathrm{Fis}}, i) = m_c \, (1 + 2\alpha E_{\mathrm{Fis}})|_{i=0} \tag{3.38}$$

Thus, in the high electric field limit, the EEM is a function of Fermi energy due to the presence of band nonparabolicity only and is independent of the subband index.

The electric subband energy ($E_{n_{\mathrm{iw5}}}$) in the high electric field limit is given by

$$E_{n_{\mathrm{iw5}}}(1 + \alpha E_{n_{\mathrm{iw5}}}) = \left[\frac{\pi \hbar |e| F_s \sqrt{2}}{\sqrt{m_c E_g}} \left(i + \frac{3}{4} \right) \right] \tag{3.39}$$

The 2D total density-of-states function in this case can be written as

$$N_{2\mathrm{D}_i}(E) = \frac{m_c g_v}{\pi \hbar^2} \sum_{i=0}^{i_{\max}} \{ [1 + 2\alpha E] \, H \left(E - E_{n_{\mathrm{iw5}}} \right) \} \tag{3.40}$$

The surface electron concentration for all values of αE_{Fis} in this case assumes the form

$$n_{2\mathrm{Dw}} = \left(\frac{g_v m_c k_B T}{\pi \hbar^2}\right) \sum_{i=0}^{i_{\max}} \left\{[1 + 2\alpha k_B T]\, F_0(\eta_{\mathrm{is}}) + 2\alpha k_B T E_{n_{\mathrm{iw5}}} F_1(\eta_{\mathrm{is}})\right\} \quad (3.41)$$

where, $\eta_{\mathrm{is}} \equiv \left[\frac{E_{\mathrm{Fis}} - E_{n_{\mathrm{iw5}}}}{k_B T}\right]$.

3.2.3 Formulation of the EEM in n-Channel Inversion Layers of II–VI Semiconductors

The use of (1.42) and (3.2) leads to the quantization integral as

$$\frac{\sqrt{2m_{\parallel}^*}}{\hbar} \int_0^{z_t} \left[E - |e|\, F_s z - a_0' k_s^2 \mp (\bar{\lambda}_0) k_s\right]^{1/2} \mathrm{d}z = \frac{2}{3}(S_i)^{3/2} \quad (3.42)$$

where, $z_t \equiv (|e|\, F_s)^{-1}\left[E - a_0' k_s^2 \mp (\bar{\lambda}_0) k_s\right]$.

Therefore, the 2D electron dispersion law for n-channel inversion layers of II–VI semiconductors can be expressed for all values of F_s as

$$E = a_0' k_s^2 \pm (\bar{\lambda}_0)\, k_s + S_i \left(\frac{\hbar\, |e|\, F_s}{\sqrt{2m_{\parallel}^*}}\right)^{2/3} \quad (3.43)$$

The area of the 2D surface as enclosed by (3.43) can be expressed as

$$A(E,i) = \frac{\pi (m_{\perp}^*)^2}{\hbar^4} \left[\left\{2(\bar{\lambda}_0)^2 - \frac{2\hbar^2}{m_{\perp}^*} S_i \left(\frac{\hbar\, |e|\, F_s}{\sqrt{2m_{\parallel}^*}}\right)^{2/3} + \frac{2\hbar^2 E}{m_{\perp}^*}\right\}\right.$$
$$\left. -2(\bar{\lambda}_0) \left[(\bar{\lambda}_0)^2 - \frac{2\hbar^2}{m_{\perp}^*} S_i \left(\frac{\hbar\, |e|\, F_s}{\sqrt{2m_{\parallel}^*}}\right)^{2/3} + \frac{2\hbar^2 E}{m_{\perp}^*}\right]^{1/2}\right] \quad (3.44)$$

The EEM is given by

$$m^*(E_{\mathrm{Fi}}, i) = m_{\perp}^* \left[1 - \frac{\rho_{71}}{\sqrt{E_{\mathrm{Fi}} + \rho_{72}}}\right] \quad (3.45)$$

where, E_{Fi} is the Fermi energy in this case,

$\rho_{71} \equiv \frac{\bar{\lambda}_0}{2\sqrt{a_0'}}$ and $\rho_{72} \equiv \left[(\rho_{71})^2 - \left(\frac{\hbar |e| F_s}{\sqrt{2m_{\parallel}^*}}\right)^{2/3}\right]$.

Thus, the EEM depends on both the Fermi energy and the subband index due to the presence of the term $\bar{\lambda}_0$.

The subband energy ($E_{n_{i6}}$) can be written as

$$E_{n_{i6}} = S_i \left(\frac{\hbar \, |e| \, F_s}{\sqrt{2m_{||}^*}} \right)^{2/3} \tag{3.46}$$

The total 2D density-of-states function can be written as

$$N_{2D_i}(E) = \frac{m_{\perp}^* g_v}{\pi \hbar^2} \sum_{i=0}^{i_{\max}} \left\{ \left[1 - \frac{\rho_{71}}{\sqrt{E + \rho_{72}}} \right] H(E - E_{n_{i6}}) \right\} \tag{3.47}$$

The surface electron concentration under the condition of extreme degeneracy assumes the form

$$n_{2D} = \frac{g_v m_{\perp}^*}{\pi \hbar^2} \sum_{n_i=1}^{n_{i\max}} \left(E_{Fi} - E_{n_{i6}} + (\bar{\lambda})^2 m_{\perp}^* \hbar^{-2} \right) \tag{3.48}$$

3.2.4 Formulation of the EEM in n-Channel Inversion Layers of IV–VI Semiconductors

In the low electric field limit (1.83) assumes the form

$$\begin{aligned} &E(1 + \alpha E) - |e| \, F_s z (1 + 2\alpha E) \\ &= \frac{p_x^2}{2M_1} + \frac{p_z^2}{2M_3} + \left(\frac{\alpha p_y^4}{4M_2 M_2'} \right) + \frac{p_y^2}{2M_2} + \frac{p_y^2}{2M_2} \alpha (E - |e| \, F_s z) \left(1 - \frac{M_2}{M_2'} \right) \end{aligned} \tag{3.49}$$

where, $M_1 = m_{\perp c}$, $m_{\perp c}$ is the transverse effective electron mass at the edge of the conduction band at $k = 0$, $M_2 = \left(\frac{m_{\perp c} + 2m_{||c}}{3} \right)$, $m_{||c}$ is the longitudinal effective electron mass at the edge of the conduction band at $k = 0$, $M_3 = \left(\frac{3m_{\perp c} m_{||c}}{2m_{||c} + m_{\perp c}} \right)$, $M_2' = \left(\frac{m_{\perp v} + 2m_{||v}}{3} \right)$, $m_{\perp v}$ and $m_{||v}$ are the effective transverse and longitudinal hole masses at the edge of the valance band at $k = 0$.

The use of (3.49) and (3.2) leads to the simplified expression of the 2D electron dispersion law in n-channel inversion layers of IV–VI materials under the weak electric field limit as

$$\gamma_{71}(E,i) = p_{71}k_x^2 + q_{71}(E,i)k_y^2 + r_{71}k_y^4 \tag{3.50}$$

where,

$$\gamma_{71}(E,i) \equiv \left[E(1+\alpha E) - S_i\left(\frac{\hbar\,|e|\,F_s}{\sqrt{2M_3}}\right)^{2/3}\left(1+\frac{4}{3}\alpha E\right)\right],$$

$$p_{71} \equiv \frac{\hbar^2}{2M_1},$$

$$q_{71}(E,i) \equiv \left(\frac{\hbar^2}{2M_2}\right)\left[1+\alpha E\left(1-\frac{M_2}{M_2'}\right) - \frac{2\alpha S_i}{3}\left(\frac{\hbar\,|e|\,F_s}{\sqrt{2M_3}}\right)^{2/3}\left(1-\frac{M_2}{M_2'}\right)\right]$$

and $r_{71} \equiv \left(\frac{\alpha\hbar^4}{4M_2M_2'}\right)$.

The area enclosed by (3.50) is given by

$$A(E,i) = \frac{4}{3}\left(\frac{r_{71}}{p_{71}}\right)^{1/2}\left[\{a_{71\mathrm{w}}(E,i)\}^2 + \{b_{71\mathrm{w}}(E,i)\}^2\right]^{1/2}$$

$$\left[\{a_{71\mathrm{w}}(E,i)\}^2 F\left[\frac{\pi}{2},\ell_{71\mathrm{w}}(E,i)\right] - (\{a_{71\mathrm{w}}(E,i)\}^2 - \{b_{71\mathrm{w}}(E,i)\}^2)E\left[\frac{\pi}{2},\ell_{71\mathrm{w}}(E,i)\right]\right] \tag{3.51}$$

in which, $\{a_{71\mathrm{w}}(E,i)\}^2 \equiv \left[\frac{q_{71}(E,i)}{2r_{71}} + \frac{1}{2}\left[\frac{\{q_{71}(E,i)\}^2}{(r_{71})^2} + \frac{4\gamma_{71}(E,i)}{r_{71}}\right]^{1/2}\right],$

$$\{b_{71\mathrm{w}}(E,i)\}^2 \equiv \left[\frac{1}{2}\left[\frac{\{q_{71}(E,i)\}^2}{(r_{71})^2} + \frac{4\gamma_{71}(E,i)}{r_{71}}\right]^{1/2} - \left(\frac{q_1(E,i)}{2r_{71}}\right)\right],$$

$\ell_{71\mathrm{w}}(E,i) \equiv \frac{b_{71\mathrm{w}}(E,i)}{\sqrt{\{a_{71\mathrm{w}}(E,i)\}^2 + \{b_{71\mathrm{w}}(E,i)\}^2}}$, $F\left[\frac{\pi}{2},\ell_{71\mathrm{w}}(E,i)\right]$ and $E\left[\frac{\pi}{2},\ell_{71\mathrm{w}}(E,i)\right]$ are the complete elliptic integral of the first and second kinds respectively.

Using (3.51), the EEM in this case can be expressed as

$$m^*(E_{\mathrm{Fiw}},i) = \frac{\hbar^2}{2\pi}R_{71}(E,i)|_{E=E_{\mathrm{Fiw}}} \tag{3.52}$$

where,

$$R_{71}(E,i) \equiv \left[\frac{4}{3}\left(\frac{r_{71}}{p_{71}}\right)^{1/2}\left[\{a_{71w}(E,i)\{a_{71w}(E,i)\}' + b_{71w}(E,i)\{b_{71w}(E,i)\}'\}\right.\right.$$
$$\times\left[\{a_{71w}(E,i)\}^2+\{b_{71w}(E,i)\}^2\right]^{-1/2}\left[\{a_{71w}(E,i)\}^2 F\left[\frac{\pi}{2},\ell_{71w}(E,i)\right]\right.$$
$$\left.-(\{a_{71w}(E,i)\}^2-\{b_{71w}(E,i)\}^2)E\left[\frac{\pi}{2},\ell_{71w}(E,i)\right]\right]+\frac{4}{3}\left(\frac{r_{71}}{p_{71}}\right)^{1/2}$$
$$\times(\{a_{71w}(E,i)\}^2+\{b_{71w}(E,i)\}^2)^{1/2}2a_{71w}(E,i)\,\{a_{71w}(E,i)\}'$$
$$F\left[\frac{\pi}{2},\ell_{71w}(E,i)\right]+\{a_{71w}(E,i)\}^2\left\{F\left[\frac{\pi}{2},\ell_{71w}(E,i)\right]\right\}'$$
$$-\left\{E\left[\frac{\pi}{2},\ell_{71w}(E,i)\right]\right\}'(\{a_{71w}(E,i)\}^2-\{b_{71w}(E,i)\}^2)$$
$$E\left[\frac{\pi}{2},\ell_{71w}(E,i)\right][2a_{71w}(E,i)\,\{a_{71w}(E,i)\}'$$
$$\left.-2b_{71w}(E,i)\,\{a_{71w}(E,i)\}']\right].$$

Thus, the EEM is a function of the subband index number and the Fermi energy due to the presence of band nonparabolicity only.

The subband energies ($E_{n_{iw7}}$) are given by

$$\left[E_{n_{iw7}}\left(1+\alpha E_{n_{iw7}}\right)-S_i\left(\frac{\hbar|e|F_s}{\sqrt{2M_3}}\right)^{2/3}\left(1+\frac{4}{3}\alpha E_{n_{iw7}}\right)\right]=0 \tag{3.53}$$

The total 2D density-of-states function can be written as

$$N_{2D_i}(E)=\frac{g_v}{2\pi^2}\sum_{i=0}^{i_{max}}\left\{R_{71}(E,i)\,H\left(E-E_{n_{iw7}}\right)\right\} \tag{3.54}$$

The surface electron concentration assumes the form

$$n_{2Dw}=\frac{2g_v}{3\pi^2}\left(\frac{r_{71}}{p_{71}}\right)^{1/2}\left\{\sum_{i=0}^{i_{max}}[P_{7w}(E_{Fwi},i)+Q_{7w}(E_{Fwi},i)]\right\} \tag{3.55}$$

where, $P_{7w}(E_{Fwi},i)\equiv\left[\{a_{71w}(E_{Fwi},i)\}^2+\{b_{71w}(E_{Fwi},i)\}^2\right]^{1/2}$

$$\left[\{a_{71w}(E,i)\}^2 F\left[\frac{\pi}{2},\ell_{71w}(E_{Fwi},i)\right]\right.$$
$$\left.-(\{a_{71w}(E_{Fwi},i)\}^2-\{b_{71w}(E_{Fwi},i)\}^2)E\left[\frac{\pi}{2},\ell_{71w}(E_{Fwi},i)\right]\right]$$

and $Q_{7w}(E_{Fiw},i)\equiv\sum_{r=1}^{s}L(r)P_{7w}(E_{Fiw},i)$.

Under the strong electric field limit, the dispersion relation assumes the form

$$\gamma_{72}(E,i) = p_{72}p_x^2 + q_{72}(E)k_y^2 + r_{72}k_y^4 \tag{3.56}$$

where,

$$\gamma_{72}(E,i) \equiv \left[E(1+\alpha E) - \frac{2\sqrt{2\alpha}}{3}\frac{\hbar\,|e|\,F_s}{\sqrt{2M_3}}(S_i)^{3/2}\right],$$

$$p_{72} \equiv \left(\frac{\hbar^2}{2M_1}\right),$$

$$q_{72}(E) \equiv \left(\frac{\hbar^2}{2M_2}\right)\left[1+\alpha E\left(1-\frac{M_2}{M_2'}\right)\right]$$

and $r_{72} \equiv \left[\dfrac{\alpha\hbar^4}{4M_2M_2'}\right]$.

Comparing (3.56) with (3.50), we observe that the forms of the (3.52) and (3.55) remain unchanged provided, $\gamma_{71}(E,i)$, p_{71}, $q_{71}(E,i)$ and r_{71} are being replaced by the corresponding quantities $\gamma_{72}(E,i)$, p_{72}, $q_{72}(E)$ and r_{72} respectively.

3.2.5 Formulation of the EEM in n-Channel Inversion Layers of Stressed Semiconductors

The use of (1.98) and (3.2) leads to the expression of the dispersion relation of the 2D electrons in n-channel inversion layers of stressed III–V materials under the low electric field limit as

$$[T_{57}(E,i)]\,k_x^2 + [T_{67}(E,i)]\,k_y^2 = T_{77}(E,i) \tag{3.57}$$

where,

$$T_{57}(E,i) \equiv \left[E - \alpha_1 + \frac{2}{3}S_i\left(\frac{|e|^2}{\varepsilon_{sc}}\right)^{2/3}(n_{2Dw})^{2/3}L_{17}(E)\right],$$

$$L_{17}(E) \equiv \left[\frac{(E-\alpha_1)}{(E-\alpha_3)^{2/3}\left[\bar{T}_{47}(E)\right]^{1/3}} - (E-\alpha_3)^{1/3}\left[\bar{T}_{47}(E)\right]^{-1/3}\right],$$

$$\left[\bar{T}_{47}(E)\right] \equiv \left[\{\rho_5(E)\}' - \left(\frac{\rho_5(E)}{E-\alpha_3}\right)\right],$$

$$T_{67}(E,i) \equiv \left[E - T_2 + \frac{2}{3}S_i\left(\frac{|e|^2}{\varepsilon_{sc}}\right)^{2/3}(n_{2Dw})^{2/3}L_{27}(E)\right],$$

$$L_{27}(E) \equiv \left[\frac{(E-\alpha_2)}{(E-\alpha_3)^{2/3}\left[\bar{T}_{47}(E)\right]^{1/3}} - \left(\frac{(E-\alpha_3)^{1/3}}{\left[\bar{T}_{47}(E)\right]^{1/3}} \right) \right],$$

$$T_{77}(E,i) \equiv \left[\rho_5(E) - S_i \left(\frac{|e|^2}{\varepsilon_{sc}} \right)^{2/3} (n_{2Dw})^{2/3} L_{37}(E) \right]$$

and $L_{37}(E) \equiv (E-\alpha_3)^{1/3}\left[\bar{T}_{47}(E)\right]^{2/3}$.

The area of the 2D surface under the weak electric field limit can be written as

$$A(E,i) = \frac{\pi T_{77}(E,i)}{\sqrt{T_{57}(E,i)T_{67}(E,i)}} \tag{3.58}$$

The subband energies ($E_{n_{iw8}}$) in this case are defined by

$$T_{47}(E_{n_{iw8}}) = S_i \left(\frac{|e|^2}{\varepsilon_{sc}} \right)^{2/3} (n_{2Dw})^{2/3} L_{37}(E_{n_{iw8}}) \tag{3.59}$$

The expression of the EEM in this case can be written as

$$m^*(E_{Fiw},i) = \frac{\hbar^2}{2} L_{47}(E,i)|_{E=E_{Fiw}} \tag{3.60}$$

where,

$$L_{47}(E,i) \equiv \left[\frac{1}{T_{57}(E,i)T_{67}(E,i)} \right] \left[\{T_{77}(E,i)\}' \left[T_{57}(E,i)T_{67}(E,i)\right]^{1/2} - \left(\frac{T_{77}(E,i)}{2} \right) \times \left\{ \{T_{57}(E,i)\}' \left[\frac{T_{67}(E,i)}{T_{57}(E,i)} \right]^{1/2} + \{T_{67}(E,i)\}' \left[\frac{T_{57}(E,i)}{T_{67}(E,i)} \right]^{1/2} \right\} \right].$$

The total 2D density-of-states function can be expressed as

$$N_{2D}(E) = \frac{g_v}{2\pi} \sum_{i=0}^{i_{max}} \left\{ L_{47}(E,i) H\left(E - E_{n_{iw8}}\right) \right\} \tag{3.61a}$$

The surface electron concentration under the weak electric field limit assumes the form

$$n_{2Dw} = \frac{g_v}{(2\pi)} \left\{ \sum_{i=0}^{i_{max}} \left[P_{8w}(E_{Fwi},i) + Q_{8w}(E_{Fwi},i) \right] \right\} \tag{3.61b}$$

where, $P_{8w}(E_{Fwi}, i) \equiv \frac{T_{77}(E_{Fwi}, i)}{\sqrt{T_{57}(E_{Fwi}, i) T_{67}(E_{Fwi}, i)}}$ and $Q_{8w}(E_{Fiw}, i) \equiv \sum_{r=1}^{s} L(r) P_{8w}$ (E_{Fiw}, i).

The use of (1.98) and (3.2) leads to the simplified dispersion relation of the 2D electrons in n-channel inversion layers of stressed III–V materials under the high electric field limit can be expressed as

$$[T_{117}(E, i)] k_x^2 + [T_{127}(E, i)] k_y^2 = T_{137}(E, i) \tag{3.62}$$

where,

$$T_{117}(E, i) \equiv \left[E - \alpha_1 + \frac{4}{3} |e| F_s (S_i)^{2/3} a_{77}(E)\right],$$

$$a_{77}(E) \equiv \frac{1}{2}\left[\frac{1}{\sqrt{(E - \alpha_3)} [T_{97}(E)]} - \left\{\frac{(E - \alpha_1)^2}{\sqrt{T_{97}(E)} (E - \alpha_3)^{3/2}}\right\}\right],$$

$$[T_{97}(E)] \equiv \left[\frac{\{\rho_5(E)\}''}{2} - \left(\frac{(\rho_5(E))'}{E - \alpha_3}\right) + \left(\frac{\rho_5(E)}{(E - \alpha_3)^2}\right)\right],$$

$$T_{127}(E, i) \equiv \left[E - \alpha_2 + \frac{4}{3} |e| F_s (S_i)^{2/3} a_{87}(E)\right],$$

$$a_{87}(E) \equiv \frac{1}{2}\left[\frac{1}{\sqrt{(E - a_3)} T_{97}(E)} - \left(\frac{(E - \alpha_2)^{1/3}}{\sqrt{T_{97}(E)} (E - T_3)^{3/2}}\right)\right],$$

$$T_{137}(E, i) \equiv \left[\rho_5(E) - (S_i)^{3/2} F_s a_{97}(E)\right]$$

and $a_{97}(E) \equiv \frac{4}{3}\sqrt{T_{97}(E)(E - \alpha_3)}$.

The area of the 2D surface in this case is given by

$$A(E, i) = \frac{\pi T_{137}(E, i)}{\sqrt{T_{117}(E, i) T_{127}(E, i)}} \tag{3.63}$$

The subband energies ($E_{n_{is8}}$) in this case can be written as

$$T_{137}(E_{n_{is8}}, i) = 0 \tag{3.64}$$

The EEM in this case assumes the form

$$m^*(E_{Fis}, i) = \frac{\hbar^2}{2} T_{147}(E, i)|_{E=E_{Fis}} \tag{3.65}$$

where,

$$T_{147}(E,i) \equiv \left[\frac{1}{T_{117}(E,i)T_{127}(E,i)}\right]\left[\{T_{137}(E,i)\}'[T_{117}(E,i)T_{127}(E,i)]^{1/2} - \left(\frac{T_{137}(E,i)}{2}\right)\cdot\left\{\{T_{117}(E,i)\}'\left[\frac{T_{127}(E,i)}{T_{117}(E,i)}\right]^{1/2} + \{T_{127}(E,i)\}'\left[\frac{T_{117}(E,i)}{T_{127}(E,i)}\right]^{1/2}\right\}\right].$$

The expression of the total 2D density-of-states function is given by

$$N_{2D}(E) = \frac{g_v}{2\pi}\sum_{i=0}^{i_{max}}\{T_{147}(E,i)H(E - E_{n_{is8}})\} \tag{3.66}$$

The surface electron concentration in the strong electric field limit can be expressed as

$$n_{2Ds} = \frac{g_v}{(2\pi)}\left\{\sum_{i=0}^{i_{max}}[P_{9s}(E_{Fis},i) + Q_{9s}(E_{Fis},i)]\right\} \tag{3.67}$$

where, $P_{9s}(E_{Fis},i) \equiv \frac{T_{137}(E_{Fis},i)}{\sqrt{T_{117}(E_{Fis},i)T_{127}(E_{Fis},i)}}$ and $Q_{9s}(E_{Fis},i) \equiv \sum_{r=1}^{s} L(r)P_{9s}(E_{Fis},i)$.

3.2.6 Formulation of the EEM in n-Channel Inversion Layers of Germanium

Using (3.2) and (1.134), the 2D electron dispersion law in n-channel inversion layers of Ge can be expressed as

$$\frac{\hbar^2 k_x^2}{2m_1^*} + \frac{\hbar^2 k_y^2}{2m_2^*} = \left[E(1+\alpha E) + \alpha E_{i2}^2 - E_{i2}(1+2\alpha E)\right] \tag{3.68}$$

where, $E_{i2} = S_i\left(\frac{\hbar e F_s}{\sqrt{2m_3^*}}\right)^{2/3}$

The area of 2D space is

$$A = \frac{2\pi\sqrt{m_1^* m_2^*}}{\hbar^2}\left[E(1+\alpha E) + \alpha E_{i2}^2 - E_{i2}(1+2\alpha E)\right] \tag{3.69}$$

The EEM assumes the form

$$m^*(E_{\mathrm{Fiw}}, i) = \sqrt{m_1^* m_2^*}\,[1 + 2\alpha E_{\mathrm{Fiw}} - E_{i2} 2\alpha] \tag{3.70}$$

Thus the EEM is the function of both Fermi energy and quantum number due to band nonparabolicity.

The density-of-states function is given by

$$N_{2\mathrm{D}}(E) = \frac{2g_v}{(2\pi)^2} \cdot \frac{2\pi\sqrt{m_1^* m_2^*}}{\hbar^2} \sum_{i=0}^{i_{\max}} [1 + 2\alpha E - 2\alpha E_{i2}] H(E - E_{i2}) \tag{3.71}$$

The surface electron concentration is given by

$$n_{\mathrm{Ds}} = \frac{g_v k_B T}{\pi} \cdot \frac{\sqrt{m_1^* m_2^*}}{\hbar^2} \sum_{i=0}^{i_{\max}} [\tau_{71} F_0(\eta_4) + 2\alpha k_B T F_1(\eta_4)] \tag{3.72}$$

where, $\eta_4 = (k_B T)^{-1}[E_{\mathrm{Fw}} - E_{i2}]$ and, $\tau_{71} = [1 - 2\alpha E_{i2}]$

3.2.7 Formulation of the EEM in n-Channel Inversion Layers of GaSb

Using (3.2) and (1.153), the 2D electron dispersion law in n-channel inversion layers of GaSb under weak electric field limit can be expressed as

$$I_{36}(E) = \frac{\hbar^2 k_s^2}{2m_c} + S_i \left[\frac{\hbar\,|e|\,F_s\,[I_{36}(E)]'}{\sqrt{2m_c}}\right]^{2/3} \tag{3.73}$$

(3.73) represents the dispersion relation of the 2D electrons in n-channel inversion layers of III–V, ternary and quaternary materials under the weak electric field limit whose bulk electrons obey the three band model of Kane.

The EEM can be expressed as

$$m^*(E_{\mathrm{Fiw}}, i) = m_c\,[P_{36}(E, i)]|_{E=E_{\mathrm{Fiw}}} \tag{3.74}$$

where, $P_{36}\,(E, i) \equiv \left\{[I_{36}(E)]' - \left\{\frac{2}{3} S_i \left[\frac{\hbar\,|e|\,F_s}{\sqrt{2m^*}}\right]^{2/3} \{[I_{36}(E)]'\}^{-1/3}\,[I_{36}(E)]''\right\}\right\}$.

Thus, one can observe that the EEM is a function of the subband index, surface electric field, the Fermi energy and the other spectrum constants due to the combined influence of E_g and Δ.

The subband energy ($E_{n_{\mathrm{iw2}}}$) in this case can be obtained from (3.73) as

$$I_{36}(E_{n_{\mathrm{iw2}}}) = S_i \left[\frac{\hbar |e| F_s \left[I_{36}(E_{n_{\mathrm{iw2}}}) \right]'}{\sqrt{2m_c}} \right]^{2/3} \tag{3.75}$$

The 2D total density-of-states function in weak electric field limit can be expressed as

$$N_{2\mathrm{D}_i}(E) = \frac{m_c g_v}{\pi \hbar^2} \sum_{i=0}^{i_{\max}} \left[P_{36}(E,i) H(E_{36} - E_{n_{\mathrm{iw2}}}) \right] \tag{3.76}$$

Using (3.76) and the occupation probability, the $n_{2\mathrm{Dw}}$ in the present case can be written as

$$n_{2\mathrm{Dw}} = \frac{g_v m_c}{\pi \hbar^2} \sum_{i=0}^{i_{\max}} \left[P_{46\mathrm{w}}(E_{\mathrm{Fiw}}, i) + Q_{46\mathrm{w}}(E_{\mathrm{Fiw}}, i) \right] \tag{3.77}$$

where, $P_{46\mathrm{w}}(E_{\mathrm{Fiw}}, i) \equiv \left\{ I_{36}(E_{\mathrm{Fiw}}) - S_i \left[\frac{\hbar e F_s \left[I_{36}(E_{\mathrm{Fiw}}) \right]'}{\sqrt{2m_c}} \right]^{2/3} \right\}$ and $Q_{46}(E_{\mathrm{Fiw}}, i) \equiv \sum_{r=1}^{s} \{ L(r) [P_{46}(E_{\mathrm{Fiw}}, i)] \}$.

Under the condition of strong electric field limit, assumes the form

$$\left[I_{36}(E) - \left\{ \frac{|e| F_s \hbar}{\sqrt{2m_c}} \left(\frac{2\sqrt{2}(S_i)^{3/2}}{3} \right) \sqrt{[I_{36}(E)]''} \right\} \right] = \frac{\hbar^2 k_s^2}{2m_c} \tag{3.78}$$

(3.78) represents the dispersion relation of the 2D electrons in n-channel inversion layers of III–V, ternary and quaternary materials under the strong electric field limit whose bulk conduction electrons are defined by the three band model of Kane.

The EEM can be expressed as

$$m^*(E_{\mathrm{Fis}}, i) = m_c \ [P_{56}(E, i)] |_{E = E_{\mathrm{Fis}}} \tag{3.79}$$

where,

$$P_{56}(E, i) \equiv \left\{ \{I_{36}(E)\}' - \left\{ \frac{|e| F_s \hbar}{\sqrt{2m_c}} \left(\frac{\sqrt{2}(S_i)^{3/2}}{3} \right) ([I_{36}(E)]'')^{-1/2} [I_{36}(E)]''' \right\} \right\}.$$

Thus, one can observe that the EEM is a function of the subband index, surface electric field, the Fermi energy and the other spectrum constants due to the combined influence of E_g and Δ.

The subband energy $(E_{n_{\mathrm{is2}}})$ in this case can be obtained from (3.78) as

$$I_{36}(E_{n_{\mathrm{is2}}}) - \left\{ \frac{|e| F_s \hbar}{\sqrt{2m_c}} \left(\frac{2\sqrt{2}(S_i)^{3/2}}{3} \right) \sqrt{\left[I_{36}(E_{n_{\mathrm{is2}}}) \right]''} \right\} = 0 \tag{3.80}$$

The total 2D density-of-states function under the strong electric field limit can be expressed as

$$N_{2D_i}(E) = \frac{m_c g_v}{\pi \hbar^2} \sum_{i=0}^{i_{\max}} \left[P_{56}(E,i) H(E - E_{n_{is2}}) \right] \tag{3.81}$$

The n_{2Ds} in the present case under the strong electric field can be written as

$$n_{2Ds} = \frac{g_v m_c}{\pi \hbar^2} \sum_{i=0}^{i_{\max}} [P_{65s}(E_{Fis}, i) + Q_{65s}(E_{Fis}, i)] \tag{3.82}$$

where, $P_{65s}(E_{Fis}, i) \equiv \left\{ I_{36}(E_{Fis}) - \left[\frac{2\sqrt{2}}{3} (S_i)^{3/2} \frac{\hbar |e| F_s [I_{36}(E_{Fis})]''}{\sqrt{2m_c}} \right] \right\}$

and $Q_{56s}(E_{Fis}, i) \equiv \sum_{r=1}^{s} \{L(r) [P_{56s}(E_{Fis}, i)]\}$

3.3 Results and Discussion

The effect of surface electric field on the EEM at the quantum limit in n-channel inversion layers of Cd_3As_2 and $CdGeAs_2$ has been exhibited in Figs. 3.1, 3.2, 3.3 and 3.4. In Figs. 3.1 and 3.2, we have demonstrated the variation of the EEM with electric field in the weak inversion regime which was extended upto $10^5\ \text{Vm}^{-1}$. It appears that with the increase in the electric field, the EEM in Fig. 3.1 increases considering the generalized energy band model (3.5) and the three and the two band models of Kane which are the special cases of our generalized analysis. It appears that in the weak field regime, the deviation between the three and two band models of Kane is less however significant difference is with the consideration of the crystal field. It should be noted that it is these two models which tends to the isotropic bulk effective mass value $0.0105m_0$, rather than the generalized model. In the high field regime, Fig. 3.2, the difference in the three and two band models of Kane appears which marks a significant variation in the value of the EEM. The effect of crystal field splitting tends to decrease the EEM considering the generalized energy band model.

A closer look at the two figures reveal more interesting features of the continuity of the weak inversion energy band model in the strong field and strong inversion energy band model in the weak field. It is due to this non-convergence there is a slight mismatch of the EEM at the boundary of $10^5\ \text{Vm}^{-1}$ in both the Figs. 3.1 and 3.2 and needs more attention towards the development of the generalized theory valid for all values of electric field is still a formidable problem in this case.

Figure 3.3 and 3.4 exhibits the EEM in n-$CdGeAs_2$ for all the cases of Figs. 3.1 and 3.2 respectively. It appears that in the weak inversion regime the EEM is almost invariant of the electric field; however with the increase in the field, the EMM sharply

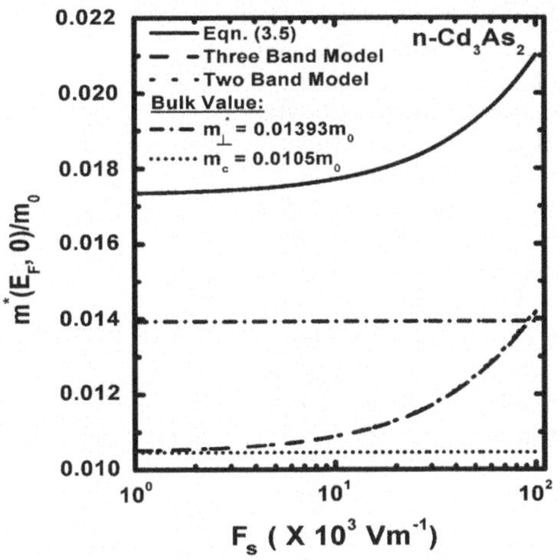

Fig. 3.1 Plot of the EEM at weak inversion as the function of surface electric field for n-channel inversion layers of Cd_3As_2 considering (3.5) in accordance with the generalized theory. The simplified results for three and two band models of Kane have also been exhibited in which, $m_{\perp}^{*} = 0.0139 m_0$ and $m_c = \frac{1}{2}(m_{\perp}^{*} + m_{\parallel}^{*}) = 0.0105 m_0$ are the corresponding bulk values

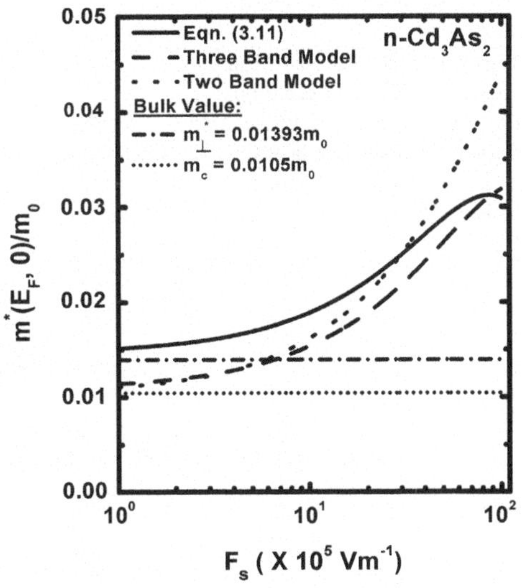

Fig. 3.2 Plot of the EEM at strong inversion as function of surface electric field for n-channel inversion layers of Cd_3As_2 for all cases of Fig. 3.1

decreases and tends to take negative values which challenge the applicability of the quantization condition (3.2) at strong electric field for n-channel $CdGeAs_2$.

The effect of electric field on the EEM of n-channel InAs has been exhibited in Figs. 3.5 and 3.6. Almost no variation of the EEM in weak field appears for n-InAs channel while for higher fields, the EEM tends to decrease. The effect of surface electric field on n-channel GaAs and InSb at weak and strong electric field has been exhibited in Fig. 3.7. Same trend as InAs in weak field again follows for GaAs, where the difference in the energy band model in determining the EEM is vanishing small. With the increase in the electric field at high value, the EEM in n-channel InSb tends

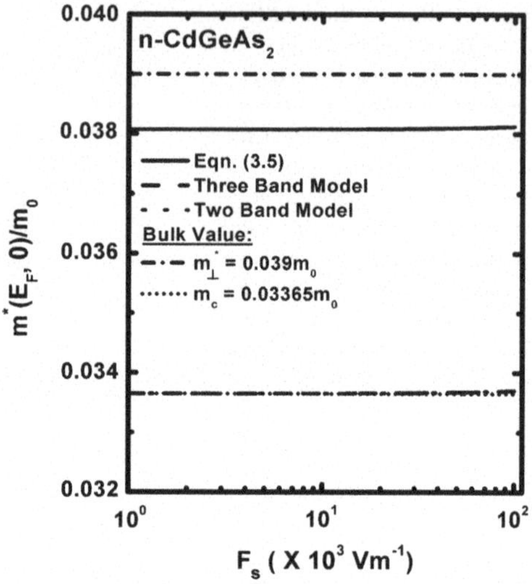

Fig. 3.3 Plot of the EEM at weak inversion as function of surface electric field for n-channel inversion layers of $CdGeAs_2$ considering (3.5) in accordance with the generalized theory. The simplified results for three and two band models of Kane have also been exhibited in which, $m_{\perp}^{*} = 0.039m_0$ and $m_c = \frac{1}{2}(m_{\perp}^{*} + m_{\parallel}^{*}) = 0.03365m_0$ are the corresponding bulk values

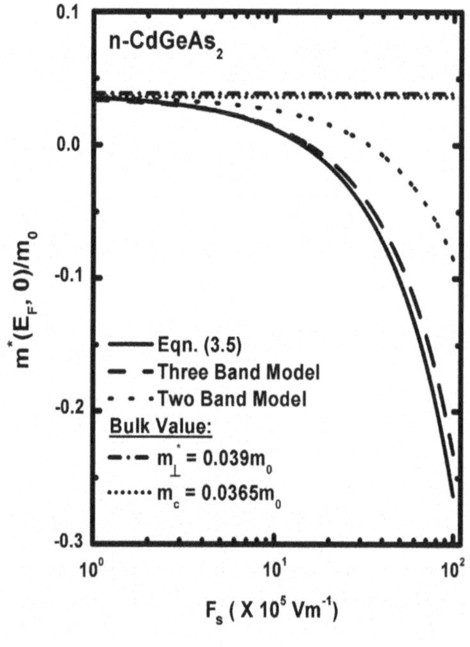

Fig. 3.4 Plot of the EEM at strong inversion as function of surface electric field for n-channel inversion layers of $CdGeAs_2$ for all cases of Fig. 3.3

to fall down. This is not with the case of CdS in Fig. 3.8 where the effect of increasing the electric field increases the EEM monotonically presenting a significant change.

Finally, in Fig. 3.9, we present the variation of the EEM in n-channel inversion layers of Ge following Cardona model under weak inversion regime and n-channel inversion layers of GaSb under strong inversion regime. It appears that the EEM

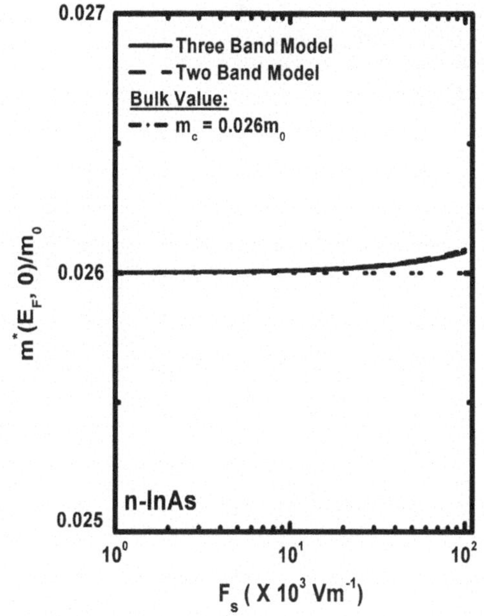

Fig. 3.5 Plot of the EEM at low electric field limit as function of surface electric field for n-InAs

in case of GaSb increases within the regime of $10^8\,\mathrm{Vm}^{-1}$, while the EEM in Ge stays almost constant. The study of the effect of surface electric field on the EEM in n-channel inversion layers of IV–VI and stressed InSb materials has been left as an exercise to the reader. For the purpose of condensed presentation, the carrier statistics and the EEM in different materials as considered in this chapter have been presented in Table 3.1.

3.4 Open Research Problems

R.3.1 Investigate the EEM, EAM, DEM, CEM, CoEM, FREM and OEM in the presence of an arbitrarily oriented electric quantization for n-channel inversion layers of non-linear optical materials. Study all the special cases for III–V, ternary and quaternary compounds in this context.

R.3.2 Investigate the same set of masses as defined in (R.3.1) in n-channel inversion layers of IV–VI, II–VI and stressed Kane type compounds in the presence of an arbitrarily oriented quantizing electric field.

R.3.3 Investigate the same set of masses as defined in (R.3.1) in n-channel inversion layers of all the materials as stated in R.1.1 of Chap. 1 in the presence of an arbitrarily oriented quantizing electric field.

R.3.4 Investigate the same set of masses as defined in (R.3.1) in the presence of an arbitrarily oriented non-quantizing magnetic field in n-channel inversion layers of non-linear optical semiconductors by including the electron spin.

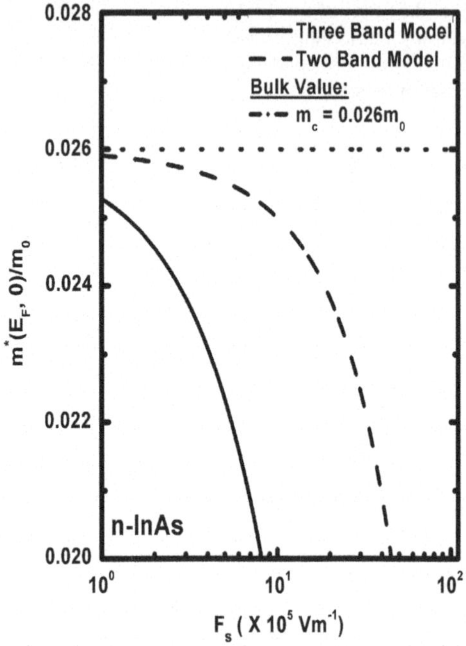

Fig. 3.6 Plot of the EEM at high electric field limit as function of surface electric field for n-InAs

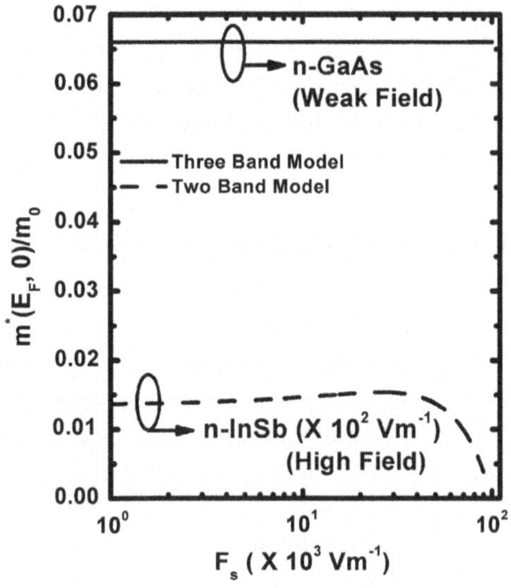

Fig. 3.7 Plot of the EEM at low and high electric field limits as function of surface electric field for n-channel inversion layers of GaAs and InSb respectively

Study all the special cases for III–V, ternary and quaternary materials in this context.

R.3.5 Investigate the same set of masses as defined in (R.3.1) in n-channel inversion layers of IV–VI, II–VI and stressed Kane type compounds in the presence

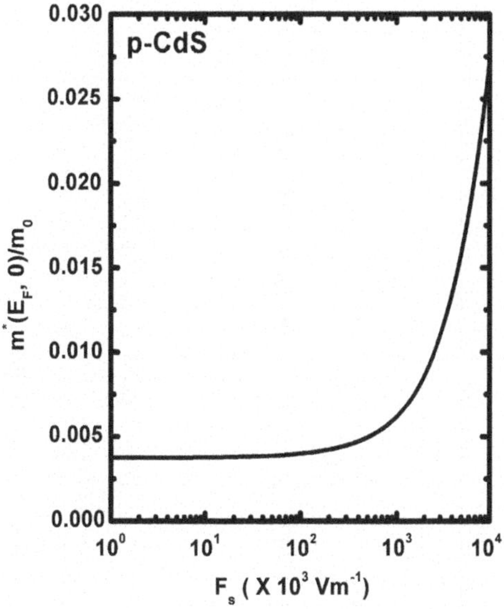

Fig. 3.8 Plot of the EEM as function of surface electric field for p-channel inversion layers of CdS

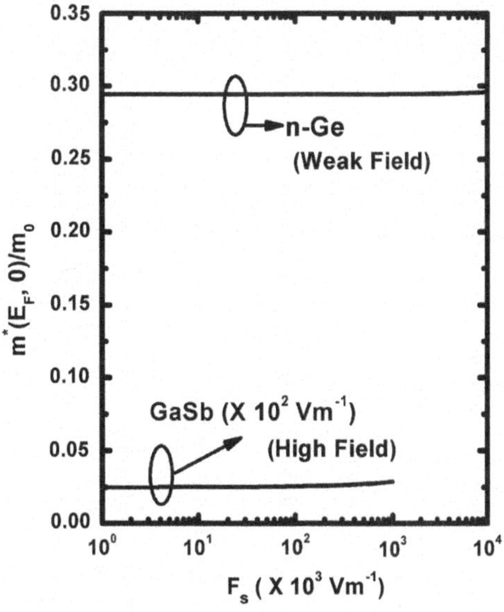

Fig. 3.9 Plot of the EEM at weak and strong electric field limits as function of surface electric field for n-channel inversion layers of Ge and GaSb respectively

of an arbitrarily oriented non-quantizing magnetic field by including the electron spin.

R.3.6 Investigate the same set of masses as defined in (R.3.1) in n-channel inversion layers of all the materials as stated in R.1.1 of Chap. 1 in the presence of an arbitrarily oriented non-quantizing magnetic field by including electron spin.

Table 3.1 The EEM and the electron statistics in n-channel inversion layers of non-parabolic semiconductors

Type of materials	EEM	The electron statistics
1. Non-linear optical semiconductors		
(a) Weak electric field	$m^*(E_{\text{Fiw}}, i) = \left(\frac{\hbar^2}{2}\right) G_7(E, i)\Big\vert_{E=E_{\text{Fiw}}}$ (3.5)	$n_{2\text{Dw}} = g_v(2\pi)^{-1} \sum_{i=0}^{i_{\max}} [P_{7\text{w}}(E_{\text{Fiw}}, i) + Q_{7\text{w}}(E_{\text{Fiw}}, i)]$ (3.9)
(b) Strong electric field	$m^*(E_{\text{Fis}}, i) = \left(\frac{\hbar^2}{2}\right) [P_2(E, i)]'\Big\vert_{E=E_{\text{Fis}}}$ (3.11)	$n_{2\text{Ds}} = g_v(2\pi)^{-1} \sum_{i=0}^{i_{\max}} [P_2(E_{\text{Fis}}, i) + Q_2(E_{\text{Fis}}, i)]$ (3.14)
2. III–V semiconductors under:		
(a) Weak electric field		
(i) Three band model of Kane	$m^*(E_{\text{Fiw}}, i) = m_c[P_3(E, i)]\vert_{E=E_{\text{Fiw}}}$ (3.16)	$n_{2\text{Dw}} = \frac{g_v m_c}{\pi\hbar^2} \sum_{i=0}^{i_{\max}} [P_{4\text{w}}(E_{\text{Fiw}}, i) + Q_{4\text{w}}(E_{\text{Fiw}}, i)]$ (3.19)
(ii) Two band model of Kane	$m^*(E_{\text{Fiw}}, i) = m_c\ [P_6(E, i)]\vert_{E=E_{\text{Fiw}}}$ (3.27)	$n_{2\text{Dw}} = \left(\frac{g_v m_c k_B T}{\pi\hbar^2}\right) \sum_{i=0}^{i_{\max}} \{[1 + D_i + 2\alpha E_{n_{\text{iw3}}}]\, F_0(\eta_{\text{iw}})$ $+ 2\alpha k_B T F_1(\eta_{\text{iw}})\}$ (3.30)
(b) Strong electric field		
(i) Three band model of Kane	$m^*(E_{\text{Fis}}, i) = m_c\ [P_5(E, i)]\vert_{E=E_{\text{Fis}}}$ (3.21)	$n_{2\text{Ds}} = \frac{g_v m_c}{\pi\hbar^2} \sum_{i=0}^{i_{\max}} [P_{6s}(E_{\text{Fis}}, i) + Q_{6s}(E_{\text{Fis}}, i)]$ (3.24)
(ii) Two band model of Kane	$m^*(E_{\text{Fis}}, i) = m_c\ (1 + 2\alpha E_{\text{Fis}})\vert_{i=0}$ (3.38)	$n_{2\text{Dw}} = \left(\frac{g_v m_c k_B T}{\pi\hbar^2}\right) \sum_{i=0}^{i_{\max}} \{[1 + 2\alpha k_B T]\, F_0(\eta_{\text{is}})$ $+ 2\alpha k_B T E_{n_{\text{iw5}}} F_1(\eta_{\text{is}})\}$ (3.41)
3. II–VI semiconductors	$m^*(E_{\text{Fi}}, i) = m_\perp^* \left[1 - \frac{\rho_{71}}{\sqrt{E_{\text{Fi}} + \rho_{72}}}\right]$ (3.45)	$n_{2\text{D}} = \frac{g_v m_\perp^*}{\pi\hbar^2} \sum_{n_i=1}^{n_{i\max}} (E_{\text{Fi}} - E_{n_{i6}} + (\bar{\lambda})^2 m_\perp^* \hbar^{-2})$ (3.48)
4. IV–VI semiconductors	$m^*(E_{\text{Fiw}}, i) = \frac{\hbar^2}{2\pi} R_{71}(E, i)\vert_{E=E_{\text{Fiw}}}$ (3.52)	$n_{2\text{Dw}} = \frac{2g_v}{3\pi^2} (\frac{r_{71}}{p_{71}})^{1/2} \left\{ \sum_{i=0}^{i_{\max}} [P_{7\text{w}}(E_{\text{Fwi}}, i) + Q_{7\text{w}}(E_{\text{Fwi}}, i)] \right\}$ (3.55)

(continued)

Table 3.1 (continued)

Type of materials	EEM	The electron statistics
5. Stressed materials		
(a) Weak electric field limit	$m^*(E_{\mathrm{Fiw}}, i) = \frac{\hbar^2}{2} L_{47}(E, i)\vert_{E=E_{\mathrm{Fiw}}}$ (3.60)	$n_{2\mathrm{Dw}} = \frac{g_v}{(2\pi)} \left\{ \sum_{i=0}^{i_{\max}} [P_{8\mathrm{w}}(E_{\mathrm{Fwi}}, i) + Q_{8\mathrm{w}}(E_{\mathrm{Fwi}}, i)] \right\}$ (3.61b)
(b) Strong electric field limit	$m^*(E_{\mathrm{Fis}}, i) = \frac{\hbar^2}{2} T_{147}(E, i)\vert_{E=E_{\mathrm{Fis}}}$ (3.65)	$n_{2\mathrm{Ds}} = \frac{g_v}{(2\pi)} \left\{ \sum_{i=0}^{i_{\max}} [P_{9s}(E_{\mathrm{Fis}}, i) + Q_{9s}(E_{\mathrm{Fis}}, i)] \right\}$ (3.67)
6. n-Ge	$m^*(E_{\mathrm{Fiw}}, i) = \sqrt{m_1^* m_2^*}$ $[1 + 2\alpha E_{\mathrm{Fiw}} - E_{i2} 2\alpha]$ (3.70)	$n_{\mathrm{Ds}} = \frac{g_v k_B T}{\pi} \cdot \frac{\sqrt{m_1^* m_2^*}}{\hbar^2} \sum_{i=0}^{i_{\max}} [\tau_{71} F_0(\eta_4) + 2\alpha k_B T F_1(\eta_4)]$ (3.72)
7. n-GaSb under:		
(a) Weak electric field	$m^*(E_{\mathrm{Fiw}}, i) = m_c\ [P_{36}(E, i)]\vert_{E=E_{\mathrm{Fiw}}}$ (3.74)	$n_{2\mathrm{Dw}} = \frac{g_v m_c}{\pi \hbar^2} \sum_{i=0}^{i_{\max}} [P_{46\mathrm{w}}(E_{\mathrm{Fiw}}, i) + Q_{46\mathrm{w}}(E_{\mathrm{Fiw}}, i)]$ (3.77)
(b) Strong electric field	$m^*(E_{\mathrm{Fis}}, i) = m_c\ [P_{56}(E, i)]\vert_{E=E_{\mathrm{Fis}}}$ (3.79)	$n_{2\mathrm{Ds}} = \frac{g_v m_c}{\pi \hbar^2} \sum_{i=0}^{i_{\max}} [P_{65s}(E_{\mathrm{Fis}}, i) + Q_{65s}(E_{\mathrm{Fis}}, i)]$ (3.82)

R.3.7 Investigate the same set of masses as defined in (R.3.1) in n-channel inversion layers for all the problems from R.3.1 to R.3.6 in the presence of an additional arbitrarily oriented electric field.

R.3.8 Investigate the same set of masses as defined in (R.3.1) in n-channel inversion layers for all the problems from R.3.1 to R.3.3 in the presence of arbitrarily oriented crossed electric and magnetic fields.

R.3.9 Investigate the same set of masses as defined in (R.3.1) in n-channel inversion layers for all the problems from R.3.1 to R.3.8 in the presence of surface states.

R.3.10 Investigate the same set of masses as defined in (R.3.1) in n-channel inversion layers for all the problems from R.3.1 to R.3.8 in the presence of hot electron effects.

R.3.11 Investigate the problems from R.3.1 to R.3.10 for the appropriate p-channel inversion layers.

R.3.12 Investigate all the appropriate problems of this chapter by including the many body, image force, broadening and hot carrier effects respectively.

R.3.13 Investigate all the appropriate problems of this chapter by removing all the mathematical approximations and establishing the respective appropriate uniqueness conditions.

References

1. T. Ando, H. Fowler, F. Stern, Rev. Mod. Phys. **54**, 437 (1982)
2. J.J. Quinn, P.J. Styles (eds.), *Electronic Properties of Quasi Two Dimensional Systems* (North Holland, Amsterdam, 1976)
3. G.A. Antcliffe, R.T. Bate, R.A. Reynolds, *Proceedings of the International Conference, Physics of Semi-Metals and Narrow-Gap Semiconductors* ed. by D.L. Carter, R.T. Bate (Pergamon Press, Oxford, 1971), p. 499
4. Z.A. Weinberg, Sol. Stat. Electron. **20**, 11 (1977)
5. G. Paasch, T. Fiedler, M. Kolar, I. Bartos, Phys. Stat. Sol. (b) **118**, 641 (1983)
6. S. Lamari, Phys. Rev. B **64**, 245340 (2001)
7. T. Matsuyama, R. Kürsten, C. Meißner, U. Merkt, Phys. Rev. B **61**, 15588 (2000)
8. P.V. Santos, M. Cardona, Phys. Rev. Lett. **72**, 432 (1994)
9. L. Bu, Y. Zhang, B.A. Mason, R.E. Doezema, J.A. Slinkman, Phys. Rev. B **45**, 11336 (1992)
10. P.D. Dresselhaus, C.M. Papavassiliou, R.G. Wheeler, R.N. Sacks, Phys. Rev. Lett. **68**, 106 (1992)
11. U. Kunze, Phys. Rev. B **41**, 1707 (1990)
12. E. Yamaguchi, Phys. Rev. B **32**, 5280 (1985)
13. Th. Lindner, G. Paasch, J. Appl. Phys. **102**, 054514 (2007)
14. S. Lamari, J. Appl. Phys. **91**, 1698 (2002)
15. K.P. Ghatak, M. Mondal, J. Appl. Phys. **70**, 299 (1991)
16. K.P. Ghatak, S.N. Biswas, J. Vac. Sci. Tech. **7B**, 104 (1989)
17. B. Mitra, K.P. Ghatak, Sol. State Electron. **32**, 177 (1989)
18. K.P. Ghatak, M. Mondal, J. Appl. Phys. **62**, 922 (1987)
19. M. Mondal, K. P. Ghatak, J. Magnet. Magn. Mat. **62**, 115 (1986)
20. M. Mondal, K.P. Ghatak, Phys. Script. **31**, 613 (1985)
21. K.P. Ghatak, M. Mondal, Z. fur Physik B **64**, 223 (1986)
22. K.P. Ghatak, S.N. Biswas, Sol. State Electron. **37**, 1437 (1994)

Chapter 4
The EEM in Nonparabolic Semiconductors Under Magnetic Quantization

4.1 Introduction

It is well known that the band structure of semiconductors can be dramatically changed by applying the external fields [1–68]. The effects of the quantizing magnetic field on the band structure of compound semiconductors are more striking and can be observed easily in experiments. Under magnetic quantization, the motion of the electron parallel to the magnetic field remains unaltered while the area of the wave vector space perpendicular to the direction of the magnetic field gets quantized in accordance with the Landau's rule of area quantization in the wave vector space [40–68].The energy levels of the carriers in a magnetic field(with the component of the wave-vector parallel to the direction of magnetic field be equated with zero) are termed as the Landau levels and the quantized energies are known as the Landau subbands. It is important to note that the same conclusion may be arrived either by solving the single-particle time-independent Schrödinger differential equation in the presence of a quantizing magnetic field or by using the operator method. The quantizing magnetic field tends to remove the degeneracy and increases the band gap. A semiconductor, placed in a magnetic field B, can absorb radiative energy with the frequency ($\omega_0 = (|e|\, B \big/ m_c)$). This phenomenon is known as cyclotron or diamagnetic resonance. The effect of energy quantization is experimentally noticeable when the separation between any two consecutive Landau levels is greater than $k_B T$. A number of interesting transport phenomena originate from the change in the basic band structure of the semiconductor in the presence of quantizing magnetic field. These have been widely been investigated and also served as diagnostic tools for characterizing the different materials having various band structures. The discreteness in the Landau levels leads to a whole crop of magneto-oscillatory phenomena, important among which are (i) Shubnikov-de Haas oscillations in magneto-resistance; (ii) de Haas-Van Alphen oscillations in magnetic susceptibility; (iii) magneto-phonon oscillations in thermoelectric power, etc.

In this chapter in Sect. 4.2.1, of the theoretical background, the EEM has been investigated in nonlinear optical semiconductors in the presence of an arbitrarily

S. Bhattacharya and K. P. Ghatak, *Effective Electron Mass in Low-Dimensional Semiconductors*, Springer Series in Materials Science 167,
DOI: 10.1007/978-3-642-31248-9_4, © Springer-Verlag Berlin Heidelberg 2013

oriented quantizing magnetic field. The Sect. 4.2.2 contains the results of III–V, ternary and quaternary compounds in accordance with the three- and the two-band models of Kane and forms the special case of Sect. 4.2.1. In the same section the well-known result of EEM from semiconductors having parabolic energy bands has been presented. In the same section the EEM in accordance with the models of Stillman et al. and Palik et al. have also been investigated for the purpose of relative comparison. The Sect. 4.2.3 contains the study of the EEM for the II–VI semiconductors under magnetic quantization. In Sect. 4.2.4, the magneto-EEM for Bismuth has been investigated in accordance with the models of the McClure and Choi, the Cohen and the Lax nonparabolic ellipsoidal respectively. In Sect. 4.2.5, the EEM in IV–VI materials has been discussed in accordance with the model of Dimmock, Bangert and Kastner and Foley and Landenberg respectively. In Sect. 4.2.6, the magneto-EEM for the stressed Kane type semiconductors has been investigated. In Sect. 4.2.7, the EEM in Te has been studied under magnetic quantization. In Sect. 4.2.8, the magneto-EEM in n-GaP has been studied. In Sect. 4.2.9, the EEM in $PtSb_2$ has been investigated under magnetic quantization. In Sect. 4.2.10, the magneto-EEM in Bi_2Te_3 has been studied. In Sect. 4.2.11, the EEM in Ge has been studied under magnetic quantization in accordance with the models of Cardona et al. and Wang and Ressler respectively. In Sects. 4.2.12 and 4.2.13, the magneto-EEM in n-GaSb and II–V compounds has respectively been studied. In Sect. 4.3 contains the result and discussions in this context. The last Sect. 3.4 contains open research problems for this chapter.

4.2 Theoritical Background

4.2.1 *The EEM in Non-Linear Optical Semiconductors Under Magnetic Quantization*

In the presence of an arbitrarily oriented quantizing magnetic field B along k_{z1} direction which makes an angle θ with k_z axis and lies in the $k_x - k_z$ plane, the magneto-dispersion law of the conduction electrons in nonlinear optical semiconductors can be expressed extending the method as given by Wallace [69] as

$$\psi_1(E) = \bar{A}_{\pm}(n, E, \theta) + a_0(E, \theta)(k_{z1})^2, \tag{4.1a}$$

where

$$\bar{A}_{\pm}(n, E, \theta) \equiv \frac{2|e|B}{\hbar}\left(n + \frac{1}{2}\right)\left[\psi_2(E)\left\{\psi_2(E)\cos^2\theta + \psi_3(E)\sin^2\theta\right\}\right]^{\frac{1}{2}}$$
$$\pm \left[\frac{|e|B\hbar E_g}{6}\left\{\frac{(E_g + \Delta_\perp)}{m_\perp^*\left(E_g + \frac{2}{3}\Delta_\perp\right)}\right\}^{\frac{1}{2}}\right]$$

$$\times\left[\left(E+E_g+\delta+\left[\frac{\Delta_{\|}^2-\Delta_{\perp}^2}{3\Delta_{\|}}\right]\right)^2\left\{\frac{\Delta_{\|}^2(E_g+\Delta_{\perp})\cos^2\theta}{m_{\perp}^*\left(E_g+\frac{2}{3}\Delta_{\perp}\right)}\right\}\right.$$
$$\left.+\left\{\frac{(E+E_g)^2(E_g+\Delta_{\|})\Delta_{\perp}^2\sin^2\theta}{m_{\|}^*\left(E_g+\frac{2}{3}\Delta_{\|}\right)}\right\}\right]^{\frac{1}{2}},$$

$n(=0,1,2,3,\ldots)$ Is the Landau quantum number and
$a_0(E,\theta)\equiv\dfrac{(\psi_2(E)\psi_3(E))}{(\psi_2(E)\cos^2\theta+\psi_3(E)\sin^2\theta)}$.

The EEM at the Fermi level along the direction of the quantizing magnetic field, can be expressed as

$$m_{k_{z1}}^*(E_{\text{FB},n,\theta})=\hbar^2 k_{z1}\left.\frac{\partial k_{z1}}{\partial E}\right|_{E=E_{\text{FB}}},\tag{4.1b}$$

where E_{FB} is the Fermi energy in the presence of magnetic quantization as measured from the edge of the conduction band in the vertically upward direction in the absence of any field. Using (4.1a) and (4.1b) we can write

$$m_{k_{z1}}^*(E_{\text{FB}},n,\theta)=\left(\frac{\hbar^2}{2}\right)\left\{\left[\frac{[\psi_1(E_{\text{FB}})]'-\left[\bar{A}_{\pm}(n,E_{\text{FB}},\theta)\right]'}{a_0(E_{\text{FB}},\theta)}\right]\right.$$
$$\left.-\left[\frac{[a_0(E_{\text{FB}},\theta)]'}{a_0^2(E_{\text{FB}},\theta)}\right]\left[\psi_1(E_{\text{FB}})-\bar{A}_{\pm}(n,E_{\text{FB}},\theta)\right]\right\}.\tag{4.2}$$

From (4.2), it appears that EEM is a function of the Fermi energy, the angle of orientation of the quantizing magnetic field, the magnetic quantum number, and the electron spin for tetragonal materials due to the combined influence of the crystal field splitting and the anisotropic spin orbit splitting constant. The dependence of the oscillatory mobility on the spin-dependent EEM in addition to Fermi energy is an important physical feature of tetragonal compounds.

To investigate the dependence of EEM on the electron concentration we have to determine the density of state function in the present case ($D_B(E)$) which can be written, including spin and extending the method as given in Nag [70] as

$$D_B(E)=\frac{g_v\,|e|\,B}{2\,\pi^2\,\hbar}\left(\sum_{n=0}^{n_{\max}}\frac{\partial k_{z1}}{\partial E}H(E-E_n)\right),\tag{4.3}$$

where E_n is the Landau energy. Using (4.1a) and (4.3), one obtains,

$$D_B(E) = \left(\frac{g_v\,|e|\,B}{4\,\pi^2\,\hbar}\right)\sum_{n=0}^{n_{\max}}\left[\frac{\psi_1(E) - \bar{A}_{\pm}(n,E,\theta)}{a_0(E,\theta)}\right]^{\frac{-1}{2}}[a_0(E,\theta)]^{-2}$$
$$\times\left[a_0(E,\theta)\left\{\{\psi_1(E)\}' - [\bar{A}_{\pm}(n,E,\theta)]'\right\}\right.$$
$$\left. - \left\{\psi_1(E) - \bar{A}_{\pm}(n,E,\theta)\right\}\{a_0(E,\theta)\}'\right] H(E - E_n). \quad (4.4)$$

Thus, combining (4.4) with the Fermi–Dirac occupation probability factor and using the generalized Sommerfeld's lemma [71], the electron concentration assumes the form

$$n_0 = \frac{g_v\,|e|\,B}{2\,\pi^2\,\hbar}\sum_{n=0}^{n_{\max}}[T_{33}(n,E_{FB}) + T_{34}(n,E_{FB})], \quad (4.5)$$

where

$$T_{33}(n,E_{FB}) = \left[\frac{\psi_1(E_{FB}) - \bar{A}_{\pm}(n,E_{FB},\theta)}{a_0(E_{FB},\theta)}\right]^{\frac{1}{2}},$$

$T_{34}(n,E_{FB}) \equiv \sum_{r=1}^{s} L_B(r)\,[T_{33}(n,E_{FB})]$ and $L_B(r) = 2(k_BT)^{2r}(1 - 2^{1-2r})$ $\xi(2r)\frac{\partial^{2r}}{\partial E_{FB}^{2r}}$

4.2.2 The EEM in Kane type III–V Semiconductors Under Magnetic Quantization

(a) Three band model of Kane:

Under the conditions $\delta = 0$, $\Delta_{||} = \Delta_{\perp} = \Delta$ and $m_{||}^* = m_{\perp}^* = m_c$ (4.1a) assumes the form

$$I_{11}(E) = \left(n + \frac{1}{2}\right)\hbar\omega_0 + \frac{\hbar^2 k_z^2}{2m_c} \pm |e|\,B\hbar\Delta\left[6m_c\left(E + E_g + \frac{2}{3}\Delta\right)\right]^{-1}. \quad (4.6)$$

Equation (4.6) is the dispersion relation of the conduction electrons of III–V, ternary and quaternary materials in the presence of a quantizing magnetic field B along z-direction [70].

From (4.6), the EEM along the direction of magnetic quantization can be written as

$$m_{k_z}^*(E_{FB}) = m_c\left[\{I_{11}(E_{FB})\}' \pm \frac{|e|\,B\hbar\Delta}{6m_c}\left[\left(E_{FB} + E_g + \frac{2}{3}\Delta\right)\right]^{-2}\right]. \quad (4.7)$$

Thus, the EEM is a function of the Fermi energy and the electron spin under magnetic quantization. The dependence of the EEM on the electron spin is due to the presence of the spin orbit splitting constant, excluding the dependence on $\{I_{11}(E_{FB})\}'$, is a special property of the three-band model of Kane.

Using (4.6) and (4.3), the density-of-states function in this case can be expressed as

$$D_B(E) = \frac{g_v\,|e|\,B\sqrt{2m_c}}{4\,\pi^2\,\hbar^2}\sum_{n=0}^{n_{max}}\left[I_{11}(E) - \left(n+\frac{1}{2}\right)\hbar\omega_0 \mp \frac{|e|\,B\hbar\Delta}{6m_c\left(E+E_g+\frac{2}{3}\Delta\right)}\right]^{-1/2} \times \left[\{I_{11}(E_{FB})\}' \pm \frac{|e|\,B\hbar\Delta}{6m_c\left(E+E_g+\frac{2}{3}\Delta\right)^2}\right] H(E-E_{n1}). \tag{4.8}$$

Thus, the electron concentration assumes the form

$$n_0 = \frac{g_v\,|e|\,B\sqrt{2m_c}}{2\,\pi^2\,\hbar^2}\sum_{n=0}^{n_{max}}[T_{35}(n, E_{FB}) + T_{36}(n, E_{FB})], \tag{4.9}$$

where

$$T_{35}(n, E_{FB}) \equiv \left[I_{11}(E_{FB}) - \left(n+\frac{1}{2}\right)\hbar\omega_0 \mp \frac{|e|\,B\hbar\Delta}{6m_c\left(E_{FB}+E_g+\frac{2}{3}\Delta\right)}\right]^{\frac{1}{2}}$$

and $T_{36}(n, E_{FB}) \equiv \sum_{r=1}^{s} L_B(r) T_{35}(n, E_{FB})$.

In the absence of spin, the electron concentration assumes the form

$$n_0 = \frac{g_v\,|e|\,B\sqrt{2m_c}}{\pi^2\,\hbar^2}\sum_{n=0}^{n_{max}}[T_{37}(n, E_{FB}) + T_{38}(n, E_{FB})], \tag{4.10}$$

where

$$T_{37}(n, E_{FB}) \equiv \left[I_{11}(E_{FB}) - \left(n+\frac{1}{2}\right)\hbar\omega_0\right]^{\frac{1}{2}} \quad \text{and}$$

$$T_{38}(n, E_{FB}) \equiv \sum_{r=1}^{s} L_B(r) T_{37}(n, E_{FB}).$$

(b) Two band model of Kane:

Under the condition $\Delta \gg E_{g0}$, (4.6) can be expressed as

$$E(1+\alpha E) = \left(n+\frac{1}{2}\right)\hbar\omega_0 + (\hbar^2 k_z^2/2m_c) \pm \frac{1}{2}\mu_0 g^* B, \quad (4.11)$$

where $\mu_0 = (|e|\,\hbar/2m_0)$ is known as the Bohr magnetron, g^* is the magnitude of the band edge g-factor and is equal to (m_0/m_c) in accordance with the two-band model of Kane.

From (4.11), the EEM along the direction of magnetic quantization can be expressed as

$$m_{k_z}^*(E_{FB}) = m_c\,[1 + 2\alpha E_{FB}]\,. \quad (4.12)$$

Thus, the EEM is a function of Fermi energy only due to the presence of band nonparabolicity factor α and is independent of the electron spin under magnetic quantization.

In accordance with the two-band model of Kane, the density-of-states function assumes the form

$$D_B(E) = \frac{g_v\,|e|\,B\sqrt{2m_c}}{4\,\pi^2\,\hbar^2}\sum_{n=0}^{n_{max}}[1+2\alpha E]\left[E(1+\alpha E) - \left(n+\frac{1}{2}\right)\hbar\omega_0 \mp \frac{1}{2}g^*\mu_0 B\right]^{-\frac{1}{2}} H(E - E_{n_2}). \quad (4.13)$$

Thus, the electron concentration can be written as

$$n_0 = \frac{g_v\,|e|\,B\sqrt{2m_c}}{2\,\pi^2\,\hbar^2}\sum_{n=0}^{n_{max}}[T_{39}(n, E_{FB}) + T_{310}(n, E_{FB})], \quad (4.14)$$

where

$$T_{39}(n, E_{FB}) \equiv \left[E_{FB}(1+\alpha E_{FB}) - \left(n+\frac{1}{2}\right)\hbar\omega_0 \pm \frac{1}{2}g^*\mu_0 B\right]^{\frac{1}{2}}$$

and $T_{310}(n, E_{FB}) \equiv \sum_{r=1}^{s} L_B(r) T_{39}(n, E_{FB})$.

In the absence of spin, the electron concentration assumes the form

$$n_0 = \frac{g_v\,|e|\,B\sqrt{2m_c}}{\pi^2\,\hbar^2}\sum_{n=0}^{n_{max}}[T_{311}(n, E_{FB}) + T_{312}(n, E_{FB})], \quad (4.15)$$

where

$$T_{311}(n, E_{FB}) \equiv \left[E_{FB}(1+\alpha E_{FB}) - \left(n+\frac{1}{2}\right)\hbar\omega_0\right]^{\frac{1}{2}}$$

and $T_{312}(n, E_{FB}) \equiv \sum_{r=1}^{s} L_B(r) T_{311}(n, E_{FB})$.

From (4.13), under the condition $\alpha E \ll 1$, the density-of-states function can be written as

$$D_B(E) = \frac{g_v |e| B\sqrt{2m_c}}{4\pi^2\hbar^2} \sum_{n=0}^{n_{max}} \left[\left(1+\frac{3}{2}\alpha E\right) \times \left[E - \left\{\frac{\left(n+\frac{1}{2}\right)\hbar\omega_0 \mp \frac{1}{2}g^*\mu_0 B}{1+\alpha E}\right\}\right]^{-\frac{1}{2}} \right] H(E - E_{n_2}). \quad (4.16)$$

Therefore, the electron concentration is given by

$$n_0 = \frac{g_v |e| B\sqrt{2m_c}}{4\pi^2\hbar^2} \sum_{n=0}^{n_{max}} \int_{E_{n2}}^{\infty} \left[E - \left\{\frac{\left(n+\frac{1}{2}\right)\hbar\omega_0 \mp \frac{1}{2}g^*\mu_0 B}{1+\alpha E}\right\}\right]^{-\frac{1}{2}} \left(1+\frac{3}{2}\alpha E\right) f_0 \mathrm{d}E. \quad (4.17)$$

Let us substitute,

$$y = E - \left\{\left[\left(n+\frac{1}{2}\right)\hbar\omega_0 \mp \frac{1}{2}g^*\mu_0 B\right](1+\alpha E)^{-1}\right\}, \quad (4.18)$$

where y is a new variable.

Since, E_{n2} is the root of (4.18), we can write $y(1+\alpha E_{n_2}) = 0$, since, $(1+\alpha E_{n_2}) \neq 0$, therefore, $y = 0$. Again when, $E \to \infty$, $y \to \infty$. Therefore, from (4.18), after binomial expansion and neglecting the terms in the order of $(\alpha E)^2$, we can write

$$E = \frac{y}{a_{01}} + b_{01}, \quad (4.19)$$

where

$$a_{01} \equiv \left[1 + \alpha\left(n+\frac{1}{2}\right)\hbar\omega_0 \pm \frac{1}{2}g^*\mu_0 B\right]$$

and

$$b_{01} \equiv (a_{01})^{-1}\left[\left(n+\frac{1}{2}\right)\hbar\omega_0 \pm \frac{1}{2}g^*\mu_0 B\right]$$

Therefore, combining (4.17) and (4.19) we get,

$$n_0 = \frac{g_v\,|e|\,B\sqrt{2m_c}}{4\,\pi^2\,\hbar^2}\sum_{n=0}^{n_{\max}}\frac{1}{a_{01}}\int_0^\infty (y)^{-1/2}\left[1+\frac{3}{2}\alpha\left(\frac{y}{a_{01}}+b_{01}\right)\right]$$
$$\times\left[1+e^{\frac{\frac{y}{a_{01}}+b_{01}-E_{FB}}{k_BT}}\right]^{-1}dy. \tag{4.20}$$

Let us substitute, $\beta_{01} = \dfrac{y}{a_{01}k_BT}$ and $\eta_B = \dfrac{E_{FB}-b_{01}}{k_BT}$.

Thus, using (4.20) and the Fermi–Dirac integrals, the electron concentration in this case assumes the form

$$n_0 = \frac{g_v N_C \theta_{B1}}{2}\left[\sum_{n=0}^{n_{\max}}\frac{1}{\sqrt{a_{01}}}\left[\left(1+\frac{3}{2}\alpha b_{01}\right)F_{\frac{-1}{2}}(\eta_B)+\frac{3}{4}\alpha k_BT F_{\frac{1}{2}}(\eta_B)\right]\right], \tag{4.21}$$

where

$$\theta_{B1}\equiv\frac{\hbar\omega_0}{k_BT},\ \text{and}\quad N_c = 2\left(\frac{2\pi m_c k_BT}{h^2}\right)^{\frac{3}{2}}.$$

In the absence of spin (4.21) assume the form [71],

$$n_0 = g_v N_C \theta_{B1}\left[\sum_{n=0}^{n_{\max}}\frac{1}{\sqrt{a_{01}^*}}\left[\left(1+\frac{3}{2}\alpha b_{01}^*\right)F_{\frac{-1}{2}}(\bar{\eta}_{B_1})+\frac{3}{4}\alpha k_BT F_{\frac{1}{2}}(\bar{\eta}_{B_1})\right]\right], \tag{4.22}$$

where

$$a_{01}^*\equiv 1+\alpha\left(n+\frac{1}{2}\right)\hbar\omega_0,\ b_{01}^*\equiv\left[\left(n+\frac{1}{2}\right)\hbar\omega_0\right](a_{01}^*)^{-1}$$

and

$$\bar{\eta}_{B_1}\equiv\frac{E_{FB}-b_{01}^*}{k_BT}.$$

(c) Parabolic energy bands:

Under the condition $\alpha\to 0$ (4.11) becomes

$$E = \left(n+\frac{1}{2}\right)\hbar\omega_0 + (\hbar^2k_z^2/2m_c)\pm\frac{1}{2}g^*\mu_0 B. \tag{4.23}$$

From (4.12), the EEM along the direction of quantizing magnetic field can be expressed as

$$m_{k_z}^*(E_{FB}) = m_c. \tag{4.24a}$$

The electron concentration in this case assumes the form

$$n_0 = g_v N_C \theta_{B1} \sum_{n=0}^{n_{\max}} F_{\frac{-1}{2}}(\eta_{B1}), \tag{4.24b}$$

where

$$\eta_{B1} \equiv (k_B T)^{-1} \left[E_{\mathrm{FB}} - \left(n + \frac{1}{2} \right) \hbar\omega_0 \right].$$

Thus, the quantizing magnetic field cannot influence the EEM in relatively wide gap semiconductors having parabolic energy bands.

(d) The model of Stillman, et al.

In accordance with the model, the electron energy spectrum in III–V semiconductors in the presence of the quantizing magnetic field $\vec{B}$ along z-direction can be written following (1.29) as

$$k_z^2 = \frac{2m_c}{\hbar^2} \left[I_{12}(E) - \left(n + \frac{1}{2} \right) \hbar\omega_0 \right]. \tag{4.25}$$

Therefore,

$$k_z^2 = \frac{2m_c}{\hbar^2} A_{33}(E, n), \tag{4.26}$$

where

$$A_{33}(E, n) = I_{12}(E) - (n + \frac{1}{2})\hbar\omega_0.$$

The EEM for this model can be expressed as

$$m^*_{k_z}(E_{\mathrm{FB}}) = m_c[I'_{11}(E_{\mathrm{FB}})]. \tag{4.27}$$

The electron concentration is given by

$$n_0 = \frac{\sqrt{2m_c} e B g_v}{\pi^2 \hbar^2} \sum_{n=0}^{n_{\max}} [Y_{33}(E_{\mathrm{FB}}, n) + Z_{33}(E_{\mathrm{FB}}, n)], \tag{4.28}$$

where

$$Y_{33}(E_{\mathrm{FB}}, n) = [\sqrt{A_{33}(E_{\mathrm{FB}}, n)}]$$

and

$$Z_{33}(E_{\mathrm{FB}}, n) = \sum_{r=1}^{s} L_B(r)[Y_{33}(E_{\mathrm{FB}}, n)].$$

(e) The model of Palik et al.

To the fourth order in effective mass theory and taking into account the interactions of the conduction, light hole, heavy-hole, and split-off hole bands, the electron energy spectrum in III–V semiconductors in the presence of a quantizing magnetic field $\vec{B}$ can be written in accordance with the present model extending (1.35) as

$$E = J_{31} + \left(n+\frac{1}{2}\right)\hbar\omega_0 + \frac{\hbar^2 k_z^2}{2m_c} \pm \frac{1}{4}\left(\frac{m_c}{m_0}\right)\hbar\omega_0 g_0^* \pm k_{30}\alpha\left(n+\frac{1}{2}\right)(\hbar\omega_0)^2$$
$$\pm k_{31}\alpha\hbar\omega_0\left(\frac{\hbar^2 k_z^2}{2m_c}\right) + k_{32}\alpha\left[\hbar\omega_0\left(n+\frac{1}{2}\right) + \frac{\hbar^2 k_z^2}{2m_c}\right]^2, \quad (4.29)$$

where

$$J_{31} = -\frac{1}{2}\alpha\hbar\omega_0\left[(1-y_{11})/(2+x_{11})^2\right].J_{32},$$

$$J_{32} = \left\{\left[\frac{1}{3}(1-x_{11})^2 - (2+x_{11}^2)\right](2+x_{11}).y_{11} + \frac{1}{2}(1-x_{11}^2)(1+x_{11})(1+y_{11})\right\},$$

$$g_0^* = 2\left\{1 - \left[\frac{(1-x_{11})}{(2+x_{11})}\right]\left[\frac{(1-y_{11})}{y_{11}}\right]\right\},$$

$$k_{30} = (1-y_{11})(1-x_{11})\left\{\left[\left(2+\frac{3}{2}x_{11}+x_{11}^2\right)\cdot\frac{(1-y_{11})}{(2+x_{11})^2}\right] - \frac{2}{3}y_{11}\right\},$$

$$k_{31} = (1-y_{11})\left[\frac{(1-x_{11})}{(2+x_{11})}\right]\cdot\left\{\left[\left(2+\frac{3}{2}x_{11}+x_{11}^2\right)\cdot\frac{(1-y_{11})}{(2+x_{11})}\right] - \frac{2}{3}(1-x_{11})y_{11}\right\}$$

and

$$k_{32} = -\left[\left(1+\frac{1}{2}x_{11}^2\right)\Big/\left(1+\frac{1}{2}x_{11}\right)\right](1-y_{11})^2.$$

The (4.29) assumes the form

$$J_{34}k_z^4 + J_{35,\pm}(n)k_z^2 + J_{36,\pm}(n) - E = 0, \quad (4.30)$$

where

$$J_{34} = \alpha k_{32}\left(\hbar^2/2m_c\right)^2, \quad J_{35,\pm}(n) = \left[\frac{\hbar^2}{2m_c} \pm \alpha k_{31}\hbar\omega_0.\frac{\hbar^2}{2m_c} + \alpha k_{32}\hbar\omega_0.\frac{\hbar^2}{2m_c}\left(n+\frac{1}{2}\right)\right],$$
$$J_{36,\pm}(n) = \left[J_{31} \pm \frac{1}{4}\left(\frac{m_c}{m_0}\right)\hbar\omega_0 g_0^* \pm k_{30}\alpha(\hbar\omega_0)^2\left(n+\frac{1}{2}\right) + k_{32}\alpha[(\hbar\omega_0)\left(n+\frac{1}{2}\right)]^2\right]$$

from (4.30) we get

$$k_z^2 = A_{35,\pm}(E, n), \tag{4.31}$$

where

$$A_{35,\pm}(E, n) = (2J_{34})^{-1}\left[-J_{35,\pm}(n) + \sqrt{\left(J_{35,\pm(n)}\right)^2 - 4J_{34}\left[J_{36,\pm}(n) - E\right]}\right]$$

The EEM for this model can be expressed as

$$m^*_{k_z}(E_{\mathrm{FB}}) = (\hbar^2 / 2)[A'_{35,\pm}(E_{\mathrm{FB}}, n)]. \tag{4.32}$$

Thus, EEM in accordance with this model is a function of Fermi energy, Landau quantum number, and the electron spin due to the presence of band nonparabolicity only.

The electron concentration is given by

$$n_0 = \frac{eBg_v}{2\pi^2 h}\sum_{n=0}^{n_{\max}}[Y_{34}(E_{\mathrm{FB}}, n) + Z_{34}(E_{\mathrm{FB}}, n)], \tag{4.33}$$

where

$$Y_{34}(E_{\mathrm{FB}}, n) = [\sqrt{A_{35,+}(E_{\mathrm{FB}}, n)} + \sqrt{A_{35,-}(E_{\mathrm{FB}}, n)}]$$

and

$$Z_{34}(E_{\mathrm{FB}}, n) = \sum_{r=1}^{s_0} L_B(r)[Y_{34}(E_{\mathrm{FB}}, n)]$$

4.2.3 The EEM in II–VI Semiconductors Under Magnetic Quantization

The Hamiltonian of the conduction electron of II–VI semiconductors in the presence of a quantizing magnetic field B along z-direction assumes the form

$$\hat{H}_B = \frac{(\hat{p}_x)^2}{2m^*_\perp} + \frac{(\hat{p}_y - |e|\,B\hat{x})^2}{2m^*_\perp} \pm \frac{\bar{\lambda}_0}{\hbar}\left[(\hat{p}_x)^2 + (\hat{p}_y - |e|\,B\hat{x})^2\right]^{1/2} + \frac{(\hat{p}_z)^2}{2m^*_{||}}, \tag{4.34}$$

where the "hats" denote the respective operators. The application of the operator method leads to the magneto-dispersion relation of the carriers of II–VI semiconductors, including spin, as

$$E = \frac{\hbar |e| B}{m_{\perp}^{*}} \left(n + \frac{1}{2}\right) + \frac{\hbar^2 k_z^2}{2m_{\parallel}^{*}} \pm \bar{\lambda}_0 \left[\frac{2|e| B}{\hbar}\left(n + \frac{1}{2}\right)\right]^{1/2} \pm \frac{1}{2} g^{*} \mu_0 B. \quad (4.35)$$

From (4.35), the EEM along the direction of the magnetic quantization can be expressed as

$$m_{k_z}^{*}(E_{\mathrm{FB}}) = m_{\parallel}^{*}. \quad (4.36)$$

Thus, the EEM in this case is a constants quantity and is not affected by magnetic field.

4.2.4 The EEM in Bismuth Under Magnetic Quantization

(a) The McClure and Choi Model

The Hamiltonian in the presence of a quantizing magnetic field B along the z-direction in accordance with this model can be written as

$$\begin{aligned} \hat{H}_B = & \frac{(\hat{p}_x)^2}{2m_1} + \frac{(\hat{p}_y - |e| B\hat{x})^2}{2m_2}\left[1 + \alpha E\left(1 - \frac{m_2}{m_2'}\right)\right] + \frac{(\hat{p}_z)^2}{2m_3} \\ & + \frac{\alpha(\hat{p}_y - |e| B\hat{x})^4}{4m_2 m_2'} - \alpha(\hat{p}_y - |e| B\hat{x})^2\left[\frac{(\hat{p}_x)^2}{4m_1 m_2}\right] \\ & + \frac{\alpha(\hat{p}_y - |e| B\hat{x})^4}{4m_2 m_3} - \alpha(\hat{p}_y - |e| B\hat{x})^2\left[\frac{(\hat{p}_x)^2}{4m_1 m_2} + \frac{(\hat{p}_z)^2}{4m_2 m_3}\right]. \end{aligned} \quad (4.37)$$

Thus, the modified carrier energy spectrum in accordance with McClure and Choi model up to the first order by including spin effects can be expressed as [72, 73]

$$\begin{aligned} E(1 + \alpha E) = & \left(n + \frac{1}{2}\right)\hbar\omega(E) + (n^2 + 1 + n)\frac{\alpha\hbar^2\omega^2(E)}{4} \\ & + \frac{\hbar^2 k_z^2}{2m_3}\left[1 - \frac{\alpha\left(n + \frac{1}{2}\right)\hbar\omega(E)}{2}\right] \pm \frac{1}{2}|g^{*}|\mu_0 B, \end{aligned} \quad (4.38)$$

where

$$\omega(E) \equiv \frac{|e| B}{\sqrt{m_1 m_2}}\left[1 + \alpha E\left(1 - \frac{m_2}{m_2'}\right)\right]^{1/2}.$$

From (4.38), the EEM along the direction of magnetic quantization assumes the form as

$$m^*_{k_z}(n, E_{FB}) = m_3 \left[\left[1 - \frac{\alpha}{2}\left(n + \frac{1}{2}\right)\hbar\omega(E_{FB}) \right]^{-1} \left[1 + 2\alpha E_{FB} - \left(n + \frac{1}{2}\right)\hbar\omega'(E_{FB}) \right. \right.$$
$$\left. - \frac{1}{2}(n^2 + n + 1)\alpha\hbar^2\omega(E_{FB})\omega'(E_{FB}) \right] + \frac{\alpha\left(n + \frac{1}{2}\right)\hbar\omega'(E_{FB})}{2}$$
$$\times \left[1 - \frac{\alpha\left(n + \frac{1}{2}\right)\hbar\omega(E_{FB})}{2} \right]^{-2} \left[E_{FB}(1 + \alpha E_{FB}) - \left(n + \frac{1}{2}\right)\hbar\omega(E_{FB}) \right.$$
$$\left. \left. - \frac{\alpha\hbar^2\omega^2(E_{FB})}{4}(n^2 + 1 + n) \pm \frac{1}{2}g^*\mu_0 B \right] \right]. \tag{4.39}$$

In the absence of band nonparabolicity, from (4.39) we get

$$m^*_{k_z}(n, E_{FB}) = m_3. \tag{4.40}$$

It is interesting to note that for the two-band model of Kane, the band nonparabolicity alone explains the dependence of the EEM on Fermi energy, and the EEM is independent of magnetic quantum number and the electron spin. In the case of McClure and Choi model of Bi under magnetic quantization, the same band nonparabolicity again alone explains the dependence of the EEM on the magnetic quantum number, electron spin, and the Fermi energy respectively. The density-of-states function for this model under magnetic quantization is given by

$$D_B(E) = \frac{g_v|e|B\sqrt{2m_3}}{4\pi^2\hbar^2} \sum_{n=0}^{n_{max}} \left[\left[1 - \frac{\alpha\left(n + \frac{1}{2}\right)\hbar\omega(E)}{2} \right]^{-3/2} \left(\frac{1}{2}\right)\alpha\left(n + \frac{1}{2}\right)\hbar[\omega(E)]' \right.$$
$$\times \left[E(1 + \alpha E) - \left(n + \frac{1}{2}\right)\hbar\omega(E) - (n^2 + 1 + n)\frac{\alpha\hbar^2\omega^2(E)}{4} \mp \frac{1}{2}g^*\mu_0 B \right]^{1/2}$$
$$+ \left[E(1+\alpha E) - \left(n + \frac{1}{2}\right)\hbar\omega(E) - (n^2 + 1 + n)\frac{\alpha\hbar^2\omega^2(E)}{4} \pm \frac{1}{2}|g^*|\mu_0 B \right]^{-1/2}$$
$$\times \left[1 + 2\alpha E - \left(n + \frac{1}{2}\right)\hbar\{\omega(E)\}' - (n^2 + 1 + n)\frac{\alpha\hbar^2\omega(E)\{\omega(E)\}'}{2} \right]$$
$$\left. \times \left[1 - \frac{\alpha\left(n + \frac{1}{2}\right)\hbar\omega(E)}{2} \right]^{-1/2} \right] H(E - E_{n_5}). \tag{4.41}$$

Combining (4.41) with the Fermi–Dirac occupation probability and using the generalized Sommerfeld's lemma [71], the electron concentration in this case assumes the form

$$n_0 = \frac{g_v|e|B\sqrt{2m_3}}{2\pi^2\hbar^2} \sum_{n=0}^{n_{max}} [T_{313}(n, E_{FB}) + T_{314}(n, E_{FB})], \tag{4.42}$$

where

$$T_{313}(n, E_{FB}) \equiv \left[1 - \frac{\alpha\left(n+\frac{1}{2}\right)\hbar\omega(E_{FB})}{2}\right]^{-1/2} \times \left[E_{FB}(1+\alpha E_{FB}) - \left(n+\frac{1}{2}\right)\hbar\omega(E_{FB}) - (n^2+n+1)\frac{\alpha\hbar^2\omega^2(E_{FB})}{4} \mp \frac{1}{2}g^*\mu_0 B\right]^{1/2}$$

and

$$T_{314}(n, E_{FB}) \equiv \sum_{r=1}^{s} L_B(r)\,[T_{313}(n, E_{FB})].$$

Under the condition $\alpha \to 0$, (4.42) get simplified as

$$n_0 = \frac{g_v N_{C_2}\theta_{B3}}{2} \sum_{n=0}^{n_{max}} F_{\frac{-1}{2}}(\eta_{B_3}), \tag{4.43}$$

where

$$N_{C_2} \equiv 2\left(\frac{2\pi m_{D3}^* k_B T}{h^2}\right)^{3/2}, \quad m_{D3}^* \equiv (m_1 m_2 m_3)^{1/3},$$

$$\theta_{B3} \equiv \frac{\hbar\omega_{03}}{k_B T}, \quad \omega_{03} \equiv (|e|\,B)/\sqrt{m_1 m_2}$$

and

$$\eta_{B_3} \equiv (k_B T)^{-1}\left[E_{FB} - \left(n+\frac{1}{2}\right)\hbar\omega_{03} \mp \frac{1}{2}g^*\mu_0 B\right].$$

In the absence of the spin, the electron concentration for McClure and Choi model can be written as

$$n_0 = \frac{g_v\,|e|\,B\sqrt{2m_3}}{\pi^2\,\hbar^2} \sum_{n=0}^{n_{max}} [T_{315}(n, E_{FB}) + T_{316}(n, E_{FB})], \tag{4.44}$$

where

$$T_{315}(n, E_{FB}) \equiv \left[1 - \frac{\alpha(n+\frac{1}{2})\hbar\omega(E_{FB})}{2}\right]^{-1/2} \left[E_{FB}(1+\alpha E_{FB}) - \left(n+\frac{1}{2}\right)\hbar\omega(E_{FB}) - (n^2+n+1)\frac{\alpha\hbar^2\omega^2(E_{FB})}{4}\right]^{1/2}$$

and

$$T_{316}(n, E_{FB}) \equiv \sum_{r=1}^{s} L_B(r)\,[T_{315}(n, E_{\mathrm{FB}})].$$

It should be noted that in the presence of a quantizing magnetic field B along y direction, the dispersion relation of the conduction electrons of Bi in accordance with the McClure and Choi model can be expressed, neglecting spin and using operator method as,

$$E(1+\alpha E) = \left(n+\frac{1}{2}\right)\hbar\omega_4 + \frac{p_y^2}{2m_2}\left(1+\alpha E\left(1-\frac{m_2}{m_2'}\right)\right) + \frac{\alpha p_y^4}{4m_2 m_2'} - \frac{\alpha p_y^2}{2m_2}\left(n+\frac{1}{2}\right)\hbar\omega_4, \quad (4.45)$$

where

$$\omega_4 \equiv \frac{|e|\,B}{\sqrt{m_1 m_3}}.$$

The electron concentration in this case can be written as

$$n_0 = \frac{g_v\,|e|\,B}{\sqrt{2}\,\pi^2\,\hbar^2}\sum_{n=0}^{n_{\max}}[T_{317}(n, E_{\mathrm{FB}}) + T_{318}(n, E_{\mathrm{FB}})], \quad (4.46)$$

where

$$T_{317}(n, E_{\mathrm{FB}}) \equiv \left[-q_1(n, E_{\mathrm{FB}}) + \sqrt{[q_1(n, E_{\mathrm{FB}})]^2 + 4q_2(n, E_{\mathrm{FB}})}\right]^{1/2},$$

$$q_1(n, E_{\mathrm{FB}}) \equiv \left(\frac{2m_2'}{\alpha}\right)\left[1+\alpha E_{\mathrm{FB}}\left(1-\frac{m_2}{m_2'}\right) - \alpha\left(n+\frac{1}{2}\right)\hbar\omega_4\right],$$

$$q_2(n, E_{\mathrm{FB}}) \equiv \left(\frac{4m_2 m_2'}{\alpha}\right)\left[E_{\mathrm{FB}}\,(1+\alpha E_{\mathrm{FB}}) - (n+\frac{1}{2})\hbar\omega_4\right]$$

and

$$T_{318}(n, E_{\mathrm{FB}}) \equiv \sum_{r=1}^{s} L_B(r)\,[T_{317}(n, E_{\mathrm{FB}})].$$

(b) The Cohen Model

The application of the above method in Cohen model leads to the electron energy spectrum in Bi in the presence of quantizing magnetic field B along z-direction as [72, 73]

$$E(1+\alpha E)=\left(n+\frac{1}{2}\right)\hbar\omega(E)\pm\frac{1}{2}g^{*}\mu_{0}B+\frac{3}{8}\alpha\left(n^{2}+n+\frac{1}{2}\right)\hbar^{2}\omega^{2}(E)+\frac{\hbar^{2}k_{z}^{2}}{2m_{3}}. \tag{4.47}$$

From (4.47), the EEM along the direction of the quantizing magnetic field can be expressed as

$$m_{k_z}^{*}(n,E_{\mathrm{FB}})=m_{3}\left[2\alpha E_{\mathrm{FB}}+1-\left(n+\frac{1}{2}\right)\hbar\omega'(E_{\mathrm{FB}})\right.$$
$$\left.-\frac{3}{4}\alpha\hbar^{2}\omega(E_{\mathrm{FB}})\omega'(E_{\mathrm{FB}})\left(n^{2}+n+\frac{1}{2}\right)\right]. \tag{4.48}$$

In the absence of band nonparabolicity, the (4.48) gets transformed into the well-known (4.40) and the mass becomes independent of Fermi energy and magnetic quantum number.

By comparing (4.48) and (4.39), it is important to note that the band nonparabolicity has been introduced between the McClure and Choi model and the Cohen model in two different ways so that in the first case, the band nonparabolicity alone explains the dependence of the EEM on the Fermi energy, magnetic quantum number and the electron spin, whereas for the Cohen model, the same band nonparabolicity alone explains the independence of the EEM on the electron spin excluding the other two dependences. In the absence of band nonparabolicity for both the models of Bi, the mass along the direction of the magnetic field is not perturbed by the magnetic quantization.

The density-of-states function under magnetic quantization in accordance with the Cohen model is given by

$$D_{B}(E)=\frac{g_{v}\,|e|\,B\sqrt{2m_{3}}}{4\,\pi^{2}\,\hbar^{2}}\sum_{n=0}^{n_{\max}}\left[\left[E(1+\alpha E)-\left(n+\frac{1}{2}\right)\hbar\omega(E)\right.\right.$$
$$\left.-\left(n^{2}+\frac{1}{2}+n\right)\frac{3\alpha\hbar^{2}\omega^{2}(E)}{8}\mp\frac{1}{2}g^{*}\mu_{0}B\right]^{-1/2}$$
$$\times\left[1+2\alpha E-\left(n+\frac{1}{2}\right)\hbar\{\omega(E)\}'\right.$$
$$\left.\left.-\left(n^{2}+\frac{1}{2}+n\right)\frac{3\alpha\hbar^{2}\omega(E)\{\omega(E)\}'}{4}\right]\right]H(E-E_{n_6}). \tag{4.49}$$

Thus, the electron concentration assumes the form

$$n_{0}=\frac{g_{v}\,|e|\,B\sqrt{2m_{3}}}{2\,\pi^{2}\,\hbar^{2}}\sum_{n=0}^{n_{\max}}[T_{319}(n,E_{\mathrm{FB}})+T_{320}(n,E_{\mathrm{FB}})], \tag{4.50}$$

where

$$T_{319}(n, E_{FB}) \equiv \left[E_{FB}(1+\alpha E_{FB}) - \left(n+\frac{1}{2}\right)\hbar\omega(E_{FB}) \mp \frac{1}{2}g^*\mu_0 B - \frac{3}{8}\alpha\left(n^2+n+\frac{1}{2}\right)\hbar^2\omega^2(E_{FB})\right]^{1/2},$$

$$\omega(E_{FB}) \equiv \frac{|e|\,B}{\sqrt{m_1 m_2}}\left[1+\alpha E_{FB}\left(1-\frac{m_2}{m_2'}\right)\right]^{1/2}$$

and

$$T_{320}(n, E_{FB}) \equiv \sum_{r=1}^{s} L_B(r)\,[T_{319}(n, E_{FB})].$$

In the presence of a quantizing magnetic field B along y direction, the magneto-Cohen model can be expressed by neglecting spin as

$$E(1+\alpha E) = \left(n+\frac{1}{2}\right)\hbar\omega_4 - \frac{\alpha E p_y^2}{2m_2'} + \frac{p_y^2}{2m_2}(1+\alpha E) + \frac{\alpha p_y^4}{4m_2 m_2'}. \tag{4.51}$$

The electron concentration in this case can be expressed as

$$n_0 = \frac{g_v\,|e|\,B}{\sqrt{2}\,\pi^2\,\hbar^2}\sum_{n=0}^{n_{max}}[T_{319}(n, E_{FB}) + T_{320}(n, E_{FB})], \tag{4.52}$$

where

$$T_{321}(n, E_{FB}) \equiv \left[-q_3(n, E_{FB}) + \sqrt{[q_3(n, E_{FB})]^2 + 4q_4(n, E_{FB})}\right]^{1/2}$$

and

$$T_{321}(n, E_{FB}) \equiv \sum_{r=1}^{s} L_B(r)\,[T_{320}(n, E_{FB})]$$

in which,

$$q_3(n, E_{FB}) \equiv \left(\frac{4m_2 m_2'}{\alpha}\right)\left[\frac{-\alpha E_{FB}}{2m_2'} + \frac{1}{2m_2}(1+\alpha E_{FB})\right]$$

and

$$q_4(n, E_{FB}) \equiv \left(\frac{4m_2 m_2'}{\alpha}\right)\left[E_{FB}(1+\alpha E_{FB}) - \left(n+\frac{1}{2}\right)\hbar\omega_4\right].$$

(c) The Lax Model

In accordance with this model, the magneto dispersion relation assumes the form [72, 73]

$$E(1+\alpha E) = \left(n+\frac{1}{2}\right)\hbar\omega_{03} + \frac{\hbar^2 k_z^2}{2m_3} \pm \frac{1}{2}\mu_0 g^* B, \tag{4.53}$$

where

$$\omega_{03} = \frac{eB}{\sqrt{m_1 m_2}}$$

$$\text{Therefore,} \quad k_z^2 = \frac{2m_3}{\hbar^2}[A_{40,\pm}(E,n)], \tag{4.54}$$

where

$$A_{40,\pm}(E,n) = E(1+\alpha E) - \left(n+\frac{1}{2}\right)\hbar\omega_{03} \mp \frac{1}{2}\mu_0 g^* B$$

The EEM assumes the form

$$m^*(E_{FB}) = m_3(1+2\alpha E_{FB}). \tag{4.55}$$

The electron concentration is given by

$$n_0 = \frac{eBg_v\sqrt{2m_3}}{2\pi^2\hbar^2}\sum_{n=0}^{n_{max}}[Y_{40}(E_{FB},n) + Z_{40}(E_{FB},n)], \tag{4.56}$$

where

$$Y_{40}(E_{FB},n) = [\sqrt{A_{40,+}(E_{FB},n)} + \sqrt{A_{40,-}(E_{FB},n)}]$$

and

$$Z_{40}(E_{FB},n) = \sum_{r=1}^{s} L_B(r)[Y_{40}(E_{FB},n)].$$

(d) Ellipsoidal parabolic energy bands

For this model, the magneto-dispersion relation can be written as

$$E = \left(n+\frac{1}{2}\right)\hbar\omega_{03} + (\hbar^2 k_z^2/2m_3) \pm \frac{1}{2}g^*\mu_0 B. \tag{4.57}$$

The expressions of the electron concentration for this model are the special cases of the models of the McClure and Choi, the Cohen and the Lax respectively.

4.2.5 The EEM in IV–VI Semiconductors Under Magnetic Quantization

(a) The Dimmock Model

In accordance with Dimmock model, the electron energy spectrum in IV–VI semiconductors in the presence of a quantizing magnetic field $\vec{B}$ along z-direction can be written following (1.83) as

$$\left[E-\frac{\hbar^2}{2m_t^-}\cdot\frac{2eB}{\hbar}\left(n+\frac{1}{2}\right)-\frac{\hbar^2k_z^2}{2m_l^-}\right]\left[1+\alpha E+\alpha\frac{\hbar^2}{2m_t^+}\cdot\frac{2eB}{\hbar}\left(n+\frac{1}{2}\right)+\frac{\alpha\hbar^2k_z^2}{2m_l^+}\right]$$
$$=\frac{\hbar eB}{m_t^*}\left(n+\frac{1}{2}\right)+\frac{\hbar^2k_z^2}{2m_l^*}, \tag{4.58}$$

Thus, (4.58) assumes the form

$$k_z^2=A_{42}(E,n), \tag{4.59}$$

where

$$A_{42}(E,n)=[2C_{31}]^{-1}\left[-C_{32}(E,n)+\left[C_{32}^2(E,n)-4C_{31}\left\{C_{33}(E,n)-E(1+\alpha E)\right\}\right]^{\frac{1}{2}}\right]$$

$$C_{31}=\frac{\alpha\hbar^4}{4m_l^+m_l^-},\quad C_{32}(E,n)=\left[\frac{-\alpha E\hbar^2}{2m_l^+}+\frac{\alpha\hbar^3eB}{2m_l^+m_t^-}\left(n+\frac{1}{2}\right)+\frac{(1+\alpha E)\hbar^2}{2m_l^-}\right.$$
$$\left.+\frac{\alpha\hbar^3eB}{2m_l^-m_t^+}\left(n+\frac{1}{2}\right)+\frac{\hbar^2}{2m_l^*}\right],$$
$$C_{33}(E,n)=\left[\frac{\hbar eB}{m_t^*}\left(n+\frac{1}{2}\right)-\frac{\alpha E\hbar eB}{m_t^+}\left(n+\frac{1}{2}\right)+\frac{(1+\alpha E)\hbar eB}{m_t^-}\left(n+\frac{1}{2}\right)\right.$$
$$\left.+\frac{\alpha(\hbar eB)^2}{m_t^+m_t^-}\left(n+\frac{1}{2}\right)^2\right].$$

The EEM for this model can be expressed as

$$m^*(E_{\mathrm{FB}},n)=\frac{\hbar^2}{2}A'_{42}(E_{\mathrm{FB}},n), \tag{4.60}$$

Therefore, the EEM is a function of Fermi energy and Landau quantum number due to the presence of band nonparabolicity only.

The electron concentration can be written as

$$n_0 = \left(\frac{eBg_v}{\pi^2 \hbar}\right) \sum_{n=0}^{n_{\max}} [Y_{41}(E_{FB}, n) + Z_{41}(E_{FB}, n)], \tag{4.61}$$

where

$$Y_{41}(E_{FB}, n) = [\sqrt{A_{42}(E_{FB}, n)}]$$

and

$$Z_{41}(E_{FB}, n) = \sum_{r=1}^{s_0} L_B(r)[Y_{41}(E_{FB}, n)].$$

(b) The Model of Bangert and Kastner

The electron energy spectrum of IV–VI materials in accordance with the model of Bangert and Kastner can be written as [74]

$$\Gamma(E) = \bar{F}_1(E)k_s^2 + \bar{F}_2(E)k_z^2, \tag{4.62}$$

where

$$\Gamma(E) = 2E, \quad \bar{F}_1(E) = \left[\frac{(\bar{R})^2}{E + E_{g0}} + \frac{(\bar{S})^2}{E + \Delta_l'} + \frac{(\bar{Q})^2}{E + \Delta_l''}\right],$$

$$\bar{F}_2(E) = \left[\frac{2(\bar{A})^2}{E + E_{g0}} + \frac{(\bar{S} + \bar{Q})^2}{E + \Delta_l''}\right]$$

and $\bar{R}$, $\bar{S}$, $\bar{Q}$, $\bar{A}$, Δ_l', Δ_l'' are the electron energy spectrum constants.

In the presence of a quantizing magnetic field $\vec{B}$ along z-direction, (4.62) assumes the form

$$\Gamma(E) = \bar{F}_1(E)\frac{2eB}{\hbar}\left(n + \frac{1}{2}\right) + \bar{F}_2(E)k_z^2 \tag{4.63}$$

$$\text{Therefore,} \quad k_z^2 = A_{44}(E, n), \tag{4.64}$$

where

$$A_{44}(E, n) = \frac{\Gamma(E) - \bar{F}_1(E)\left(\frac{2eB}{\hbar}\right)\left(n + \frac{1}{2}\right)}{\bar{F}_2(E)}.$$

The EEM for this model can be expressed as

$$m^*(E_{FB}, n) = \frac{\hbar^2}{2} A_{44}'(E_{FB}, n). \tag{4.65}$$

Therefore, the EEM is a function of Fermi energy and Landau quantum number which is the characteristic feature of this model.

The electron concentration can be expressed as

$$n_0 = \left(\frac{eBg_v}{\pi^2\hbar}\right)\sum_{n=0}^{n_{\max}}[Y_{42}(E_{FB},n)+Z_{42}(E_{FB},n)], \tag{4.66}$$

where

$$Y_{42}(E_{FB},n) = [\sqrt{A_{44}(E_{FB},n)}]$$

and

$$Z_{42}(E_{FB},n) = \sum_{r=1}^{s_0} L_B(r)[Y_{42}(E_{FB},n)].$$

(c) The Model of Foley and Landenberg

In accordance with the model of Foley and Landenberg, the electron energy spectrum in IV–VI semiconductors assumes the form [75]

$$E+\frac{E_{g0}}{2} = \frac{\hbar^2k_s^2}{2m_\perp^-}+\frac{\hbar^2k_z^2}{2m_\parallel^-}+\left[\left[\frac{\hbar^2k_s^2}{2m_\perp^+}+\frac{\hbar^2k_z^2}{2m_\parallel^+}+\frac{E_{g0}}{2}\right]^2+P_\parallel^2k_z^2+P_\perp^2k_s^2\right]^{\frac{1}{2}}, \tag{4.67}$$

where $\frac{1}{m_\perp^\pm} = \frac{1}{2}\left[\frac{1}{m_{tc}}\pm\frac{1}{m_{t2}}\right]$, $\frac{1}{m_\parallel^\pm} = \frac{1}{2}\left[\frac{1}{m_{lc}}\pm\frac{1}{m_{l2}}\right]$, m_{tc} and m_{lc} are the transverse and longitudinal effective electron masses of the conduction electrons at the edge of the conduction band and m_{t2} and m_{l2} are the transverse and longitudinal effective hole masses at the edge of the valence band. In the presence of magnetic quantization $\vec{B}$ along z-direction (4.67) assumes the form

$$\therefore k_z^2 = A_{46}(E,n), \tag{4.68}$$

where

$$A_{46}(E,n) = (2D_{31})^{-1}\left[-D_{32}(E,n)+[D_{32}^2(E,n)+4[E(E+E_{g0})-D_{33}(E,n)]D_{31}]^{\frac{1}{2}}\right],$$

$$D_{31} = \left[\frac{\hbar^4}{4(m_\parallel^+)^2}-\frac{\hbar^4}{4(m_\parallel^-)^2}\right],$$

$$D_{32}(E,n) = \left[\frac{\hbar^2}{2m_\parallel^+}\left\{E_{g0}+\frac{2\hbar eB}{m_\perp^+}\left(n+\frac{1}{2}\right)\right\}+P_\parallel^2 - \frac{\hbar^3eB}{m_\perp^-m_\parallel^-}\left(n+\frac{1}{2}\right)+(E_{g0}+2E)\frac{\hbar^2}{2m_\parallel^-}\right],$$

$$D_{33}(E,n)=\left[-\left\{\frac{\hbar eB}{m_{\perp}^{-}}(n+\frac{1}{2})\right\}^{2}+(E_{g0+2E})\frac{\hbar eB}{m_{\perp}^{-}}\left(n+\frac{1}{2}\right)+\left\{\frac{\hbar eB}{m_{\perp}^{+}}\left(n+\frac{1}{2}\right)\right\}^{2}\right.$$
$$\left.+E_{g0}\frac{\hbar eB}{m_{\perp}^{+}}\left(n+\frac{1}{2}\right)+P_{\perp}^{2}\frac{2eB}{\hbar}\left(n+\frac{1}{2}\right)\right].$$

The EEM for this model can be expressed as

$$m^{*}(E_{\mathrm{FB}},n)=\frac{\hbar^{2}}{2}A_{46}'(E_{\mathrm{FB}},n), \tag{4.69}$$

Therefore, the EEM is a function of Fermi energy and Landau quantum number which is the characteristic feature of this model.

The electron concentration can be expressed as

$$n_{0}=\left(\frac{eBg_{v}}{\pi^{2}\hbar}\right)\sum_{n=0}^{n_{\max}}[Y_{43}(E_{\mathrm{FB}},n)+Z_{43}(E_{\mathrm{FB}},n)], \tag{4.70}$$

where

$$Y_{43}(E_{\mathrm{FB}},n)=[\sqrt{A_{46}(E_{\mathrm{FB}},n)}]$$

and

$$Z_{43}(E_{\mathrm{FB}},n)=\sum_{r=1}^{s_{0}}L_{B}(r)[Y_{43}(E_{\mathrm{FB}},n)]$$

4.2.6 The EEM in Stressed Semiconductors Under Magnetic Quantization

The simplified expression of the electron energy spectrum in stressed Kane -type semiconductors in the presence of an arbitrarily oriented quantizing magnetic field B, which makes angles α_1, β_1 and γ_1 with k_x, k_y and k_z axes respectively can be written following (1.99) as

$$1-\left[k_{z}'\right]^{2}[I_{2}(E)]^{-1}=I_{3}(n,E), \tag{4.71}$$

where

$$I_{2}(E)\equiv[\bar{a}_{0}(E)]^{2}\cos^{2}\alpha_{1}+\left[\bar{b}_{0}(E)\right]^{2}\cos^{2}\beta_{1}+[\bar{c}_{0}(E)]^{2}\cos^{2}\gamma_{1}$$

and

$$I_3(n,E) \equiv \left(\frac{2|e|B}{\hbar}\right)\left(n+\frac{1}{2}\right)\left[[\bar{a}_0(E)]\left[\bar{b}_0(E)\right][\bar{c}_0(E)]\right]^{-1}[I_2(E)]^{1/2}.$$

The use of (3.71) leads to the expression of the EEM as

$$m^*_{k'_z}(n, E_{\mathrm{FB}}) = \frac{\hbar^2}{2}\left[-\{I_3(n, E_{\mathrm{FB}})\}' I_2(n, E_{\mathrm{FB}}) + (1 - I_3(n, E_{\mathrm{FB}}))\{I_2(n, E_{\mathrm{FB}})\}'\right]. \tag{4.72}$$

In the absence of stress, together with the substitution $B_2^2 \equiv \dfrac{3\hbar^2 E_g}{4m_c}$, the (4.72) gets simplified into (4.12).

By comparing (4.72) and (4.12), one can observe that the stress makes the EEM quantum number dependent in stressed Kane-type compounds under magnetic quantization, in addition to Fermi energy.

The density-of-states function in this case is given by

$$D_B(E) = \frac{g_v|e|B}{2\pi^2\hbar}\sum_{n=0}^{n_{\max}}\left\{\frac{\{I_2(E)\}'}{\sqrt{I_2(E)}}[1 - I_3(n,E)]^{1/2} - [1 - I_3(n,E)]^{-1/2} \times \{I_3(n,E)\}'\sqrt{I_2(E)}\right\} H(E - E_{n_8}). \tag{4.73}$$

The use of (4.73) leads to the expression of electron concentration as

$$n_0 = \frac{g_v|e|B}{\pi^2\hbar}\sum_{n=0}^{n_{\max}}[T_{327}(n, E_{\mathrm{FB}}) + T_{328}(n, E_{\mathrm{FB}})], \tag{4.74}$$

where

$$T_{327}(n, E_{\mathrm{FB}}) \equiv \sqrt{I_2(E_{\mathrm{FB}})}\left[\sqrt{1 - [I_3(n, E_{\mathrm{FB}})]}\right]$$

and

$$T_{328}(n, E_{\mathrm{FB}}) \equiv \sum_{r=1}^{s} L(r) T_{327}(n, E_{\mathrm{FB}}).$$

4.2.7 The EEM in Tellurium Under Magnetic Quantization

The dispersion under magnetic quantization can be written following (1.105) as

$$E = \Psi_1 k_z^2 + \Psi_2\frac{2eB}{\hbar}\left(n+\frac{1}{2}\right) \pm \left[\Psi_3^2 k_z^2 + \Psi_4^2\frac{2eB}{\hbar}\left(n+\frac{1}{2}\right)\right]^{\frac{1}{2}} \tag{4.75}$$

$$\text{Therefore,}\quad k_z^2 = A_{50,\pm}(E, n), \tag{4.76}$$

where

$$A_{50,\pm}(E,n) = (2\Psi_1^2)^{-1}\left[\Psi_5(E,n) \pm \left[\Psi_5^2(E,n) - 4\Psi_1^2\Psi_6(E,n)\right]^{\frac{1}{2}}\right],$$

$$\Psi_5(E,n) = \left[2\Psi_1\left[E - \Psi_2\frac{2eB}{\hbar}\left(n+\frac{1}{2}\right)\right] + \Psi_3^2\right],$$

$$\Psi_6(E,n) = \left[\left[E - \Psi_2\frac{2eB}{\hbar}\left(n+\frac{1}{2}\right)\right]^2 - \Psi_4^2\frac{2eB}{\hbar}\left(n+\frac{1}{2}\right)\right].$$

The expression for EEM can be written as

$$m^*(E_{FB},n) = \frac{\hbar^2}{2}A'_{50,\pm}(E_{FB},n). \tag{4.77}$$

The presence of the term Ψ_3 in (4.75) makes the mass both quantum number and the Fermi energy dependent in this case.

The electron concentration can be expressed as

$$n_0 = \left(\frac{eBg_v}{2\,\pi^2\,\hbar}\right)\sum_{n=0}^{n_{max}}[Y_{45}(E_{FB},n) + Z_{45}(E_{FB},n)], \tag{4.78}$$

where

$$Y_{45}(E_{FB},n) = [\sqrt{A_{50,+}(E_{FB},n)} + \sqrt{A_{50,-}(E_{FB},n)}]$$

and

$$Z_{45}(E_{FB},n) = \sum_{r=1}^{s_0} L_B(r)[Y_{45}(E_{FB},n)]$$

4.2.8 The EEM in n-Gallium Phosphide Under Magnetic Quantization

The magneto electron energy spectrum can be written following (1.111) as

$$E = a_0\frac{2eB}{\hbar}\left(n+\frac{1}{2}\right) + b_0k_z^2 - \left[\left[C\frac{2eB}{\hbar}\left(n+\frac{1}{2}\right) + |V_G|^2\,Ck_z^2\right]^{\frac{1}{2}}\right] + |V_G|, \tag{4.79}$$

where $a_0 = \frac{\hbar^2}{2m_\perp^*} + \frac{A\hbar^2}{2m_\parallel^*}$, $b_0 = \frac{\hbar^2}{2m_\parallel^*}$, $C = \frac{\hbar^4 k_0^2}{(m_\parallel^*)^2}$

$$\text{Therefore,} \quad k_z^2 = A_{52,\pm}(E, n), \tag{4.80}$$

where

$$A_{52,\pm}(E,n) = (2b_0^2)^{-1}\left[\Psi_{11}(E,n) \pm \left[\Psi_{11}^2(E,n) - 4b_0^2\Psi_{12}(E,n)\right]^{\frac{1}{2}}\right],$$

$$\Psi_{11}(E,n) = [2b_0\,[E - \Psi_9(n)] + C],$$

$$\Psi_{12}(E,n) = \left[[E - \Psi_9(n)]^2 - \Psi_{10}(n)\right],$$

$$\Psi_9(n) = |V_G| + a_0\frac{2eB}{\hbar}\left(n + \frac{1}{2}\right)$$

and

$$\Psi_{10}(n) = C\frac{2eB}{\hbar}\left(n + \frac{1}{2}\right) + |V_G|^2$$

The expression for EEM can be written as

$$m^*(E_{FB}, n) = \frac{\hbar^2}{2}A'_{52,\pm}(E_{FB}, n). \tag{4.81}$$

The presence of the term $|V_G|$ in (4.79) makes the mass both quantum number and the Fermi energy dependent in this case.

The electron concentration can be expressed as

$$n_0 = \left(\frac{eBg_v}{2\,\pi^2\,\hbar}\right)\sum_{n=0}^{n_{max}}[Y_{46}(E_{FB}, n) + Z_{46}(E_{FB}, n)], \tag{4.82}$$

where

$$Y_{46}(E_{FB}, n) = [\sqrt{A_{52,+}(E_{FB}, n)} + \sqrt{A_{52,-}(E_{FB}, n)}]$$

and

$$Z_{46}(E_{FB}, n) = \sum_{r=1}^{s_0} L_B(r)[Y_{46}(E_{FB}, n)].$$

4.2.9 The EEM in Platinum Antimonide Under Magnetic Quantization

The magneto-dispersion relation can be written following (1.118) as

$$\left[E + \frac{\bar{\lambda}_0(\bar{a})^2 eB}{2\hbar}\left(n+\frac{1}{2}\right) + \frac{\bar{\lambda}_0(\bar{a})^2}{4}k_z^2 - \frac{\bar{l}(\bar{a})^2 eB}{2\hbar}\left(n+\frac{1}{2}\right)\right]$$
$$\times\left[E + \bar{\delta}_0 - \frac{\bar{\upsilon}(\bar{a})^2 eB}{2\hbar}\left(n+\frac{1}{2}\right) - \frac{\bar{\upsilon}(\bar{a})^2}{4}k_z^2 - \frac{\bar{n}(\bar{a})^2 eB}{2\hbar}\left(n+\frac{1}{2}\right)\right]$$
$$= \frac{I(\bar{a})^4}{16}\left[k_z^2 + \frac{2eB}{\hbar}\left(n+\frac{1}{2}\right)\right]^2 \tag{4.83}$$

$$\text{Therefore,}\quad k_z^2 = A_{55,\pm}(E,n), \tag{4.84}$$

where

$$A_{55}(E,n) = (2\Psi_{17})^{-1}\left[-\Psi_{18}(E,n) + \left[\Psi_{18}^2(E,n) - 4\Psi_{17}\Psi_{19}(E,n)\right]^{\frac{1}{2}}\right],$$

$$\Psi_{17} = \left[\frac{I(\bar{a})^4}{16} + \frac{\bar{\lambda}_0\bar{\upsilon}(\bar{a})^4}{16}\right],$$

$$\Psi_{18}(E,n) = \left[\frac{I(\bar{a})^4 eB}{4\hbar}\left(n+\frac{1}{2}\right) + \Psi_{15}(E,n)\frac{\bar{\upsilon}(\bar{a})^2}{4} - \Psi_{16}(E,n)\frac{\bar{\lambda}_0(\bar{a})^2}{4}\right],$$

$$\Psi_{20} = \frac{\bar{\upsilon}(\bar{a})^2}{4},\quad \Psi_{21}(E,n) = [\Psi_{16}(E,n) + \Psi_{15}(E,n)]$$

$$\Psi_{19}(E,n) = \left(\Psi_{15}(E,n)\Psi_{16}(E,n) - \frac{I(\bar{a})^4 e^2 B^2\left(n+\frac{1}{2}\right)^2}{4\hbar^2}\right)$$
$$\Psi_{15}(E,n) = \left[E + \frac{\bar{\lambda}_0(\bar{a})^2 eB}{2\hbar}\left(n+\frac{1}{2}\right) - \frac{\bar{l}(\bar{a})^2 eB}{2\hbar}\left(n+\frac{1}{2}\right)\right]$$

and

$$\Psi_{16}(E,n) = \left[E + \bar{\delta}_0 - \frac{\bar{\upsilon}(\bar{a})^2 eB}{2\hbar}\left(n+\frac{1}{2}\right) - \frac{\bar{n}(\bar{a})^2 eB}{2\hbar}\left(n+\frac{1}{2}\right)\right]$$

The EEM for this case can be written as

$$m^*(E_{\mathrm{FB}},n) = \frac{\hbar^2}{2}A'_{55,\pm}(E_{\mathrm{FB}},n). \tag{4.85}$$

The electron concentration can be expressed as

$$n_0 = \left(\frac{eBg_v}{2\,\pi^2\,\hbar}\right)\sum_{n=0}^{n_{\max}}[Y_{47}(E_{FB},n) + Z_{47}(E_{FB},n)], \tag{4.86}$$

where

$$Y_{47}(E_{FB},n) = [\sqrt{A_{55,+}(E_{FB},n)} + \sqrt{A_{55,-}(E_{FB},n)}]$$

and

$$Z_{47}(E_{FB},n) = \sum_{r=1}^{s_0} L_B(r)[Y_{47}(E_{FB},n)]$$

4.2.10 The EEM in Bismuth Telluride Under Magnetic Quantization

In the presence of a quantizing magnetic field $\vec{B}$ along k_x direction, the magneto-dispersion relation of the carriers in Bi_2Te_3 can be written following (1.128) as

$$E(1+\alpha E) = \bar{\omega}_1 k_x^2 + \hbar\omega_{31}\left(n+\frac{1}{2}\right), \tag{4.87}$$

where

$$\omega_{31} = \frac{eB}{M_{31}}, \quad M_{31} = \frac{m_0}{\left[\bar{\alpha}_{22}\bar{\alpha}_{23} - \frac{(\bar{\alpha}_{23})^2}{4}\right]^{\frac{1}{2}}}$$

$$\text{Therefore,} \quad k_x^2 = \frac{E(1+\alpha E) - \hbar\omega_{31}\left(n+\frac{1}{2}\right)}{\bar{\omega}_1}.$$

The EEM can be expressed as

$$m^*(E_{FB}) = \frac{\hbar^2}{2}\left[\frac{1+2\alpha E_{FB}}{\bar{\omega}_1}\right]. \tag{4.88}$$

The electron concentration can be expressed as

$$n_0 = \left(\frac{eBg_v}{\pi^2\,\hbar}\right)\sum_{n=0}^{n_{\max}}[Y_{48}(E_{FB},n) + Z_{48}(E_{FB},n)], \tag{4.89}$$

where

$$Y_{48}(E_{FB},n) = \left\{\frac{1}{\bar{\omega}_1}\left[E_{FB}(1+\alpha E_{FB}) - \left(n+\frac{1}{2}\right)\hbar\omega_{31}\right]\right\}^{1/2}$$

and

$$Z_{48}(E_{FB}, n) = \sum_{r=1}^{s_0} Z_B(r)[Y_{48}(E_{FB}, n)].$$

4.2.11 The EEM in Germanium Under Magnetic Quantization

(a) The model of Cardona et al.

The dispersion relation of the conduction electrons in $n - Ge$ in accordance with the model of Cardona et al. in the presence of quantizing magnetic field $\vec{B}$ along z-direction can be written following (1.135) as

$$E(1 + \alpha E) = \hbar\omega_\perp \left(n + \frac{1}{2}\right) + \frac{\hbar^2 k_z^2}{2m_{||}^*} + 2\alpha E \left(\frac{\hbar^2 k_z^2}{2m_{||}^*}\right) - \alpha \left(\frac{\hbar^2 k_z^2}{2m_{||}^*}\right)^2, \quad (4.90)$$

where $\omega_\perp = \frac{eB}{m_\perp^*}$, $m_{||}^*$ and $m_\perp^*$ are the longitudinal and transverse effective masses along <111> direction at the edge of the conduction band respectively.

$$\text{Therefore,} \quad k_z^2 = \frac{2m_{||}^*}{\hbar^2} A_{69}(E, n), \quad (4.91)$$

where

$$A_{69}(E, n) = (2\alpha)^{-1} \left[1 + 2\alpha E - \left[1 + 4\alpha \left(n + \frac{1}{2}\right) \hbar\omega_\perp\right]^{\frac{1}{2}}\right].$$

The EEM can be written as

$$m^*(E_{FB}, n) = m_{||}^* A'_{69}(E_{FB}, n). \quad (4.92)$$

The electron concentration can be expressed as

$$n_0 = \left(\frac{eBg_v}{\pi^2 \hbar}\right) \sum_{n=0}^{n_{max}} [Y_{49}(E_{FB}, n) + Z_{49}(E_{FB}, n)], \quad (4.93)$$

where

$$Y_{49}(E_{FB}, n) = \frac{\sqrt{2m_{||}^*}}{\hbar} [\sqrt{A_{69}(E_{FB}, n)}]$$

and

$$Z_{49}(E_{FB}, n) = \sum_{r=1}^{s_0} L_B(r)[Y_{49}(E_{FB}, n)].$$

(b) The model of Wang and Ressler

The magneto-dispersion law in $n - Ge$ in accordance with the model of Wang and Ressler can be written following (1.143) as

$$k_z^2 = \frac{2m_{\|}^*}{\hbar^2}[A_{71}(E,n)], \tag{4.94}$$

where

$$A_{71}(E,n) = \left[\Psi_{24}(n) - \frac{1}{2\bar{e}_1}[\Psi_{25}(n) - 4\bar{e}_1 E]^{\frac{1}{2}}\right],$$

$$\Psi_{24}(n) = (2\bar{e}_1)^{-1}\left[1 - \bar{d}_1\left(n + \frac{1}{2}\right)\hbar\omega_\perp\right]$$

and

$$\Psi_{25}(n) = \left[\left\{1 - \bar{d}_1\left(n + \frac{1}{2}\right)\hbar\omega_\perp\right\}^2 + \bar{e}_1\left\{\left(n + \frac{1}{2}\right)\hbar\omega_\perp - \bar{c}_1\left\{\left(n + \frac{1}{2}\right)\hbar\omega_\perp\right\}^2\right\}\right].$$

The EEM is given by

$$m^*(E_{\mathrm{FB}}, n) = m_{\|}^* A'_{71}(E_{\mathrm{FB}}, n). \tag{4.95}$$

The electron concentration can be expressed as

$$n_0 = \left(\frac{eBg_v}{\pi^2\hbar}\right)\sum_{n=0}^{n_{\max}}[Y_{50}(E_{\mathrm{FB}}, n) + Z_{50}(E_{\mathrm{FB}}, n)], \tag{4.96}$$

where

$$Y_{50}(E_{\mathrm{FB}}, n) = \frac{\sqrt{2m_{\|}^*}}{\hbar}[\sqrt{A_{71}(E_{\mathrm{FB}}, n)}]$$

and

$$Z_{50}(E_{\mathrm{FB}}, n) = \sum_{r=1}^{s_0} L_B(r)[Y_{50}(E_{\mathrm{FB}}, n)].$$

4.2.12 The EEM in Gallium Antimonide Under Magnetic Quantization

The magneto-dispersion relation is given by

$$k_z^2 = \frac{2m_c}{\hbar^2}\left[I_{16}(E) - \left(n+\frac{1}{2}\right)\hbar\omega_c\right], \tag{4.97a}$$

where $I_{16}(E)$ has been defined in (1.153),

(4.97a) can be expressed as

$$k_z^2 = \frac{2m_c}{\hbar^2}\left[A_{73}(E_{FB}, n)\right], \tag{4.97b}$$

where $[A_{73}(E_{FB}, n)] = \left[I_{16}(E) - (n+\frac{1}{2})\hbar\omega_c\right]$

The EEM can be expressed as

$$m^*(E_{FB}) = m_c I'_{16}(E_{FB}). \tag{4.98}$$

The electron concentration can be expressed as

$$n_0 = \left(\frac{eBg_v}{\pi^2\hbar}\right)\sum_{n=0}^{n_{\max}}\left[Y_{501}(E_{FB}, n) + Z_{501}(E_{FB}, n)\right], \tag{4.99}$$

where

$$Y_{501}(E_{FB}, n) = \frac{\sqrt{2m_c}}{\hbar}\left[\sqrt{A_{73}(E_{FB}, n)}\right]$$

and

$$Z_{501}(E_{FB}, n) = \sum_{r=1}^{s_0} L_B(r)\left[Y_{501}(E_{FB}, n)\right]$$

4.2.13 The EEM in II–V Semiconductors Under Magnetic Quantization

The dispersion relation of the holes are given by [76–78]

$$E = \theta_1 k_x^2 + \theta_2 k_y^2 + \theta_3 k_z^2 + \delta_4 k_x \mp \left[\{\theta_5 k_x^2 + \theta_6 k_y^2 + \theta_7 k_z^2 + \delta_5 k_x\}^2 + G_3^2 k_y^2 + \Delta_3^2\right]^{\frac{1}{2}} \pm \Delta_3, \tag{4.100a}$$

where k_x, k_y and k_z are expressed in the units of $10^{10}\ \text{m}^{-1}$,

$$\theta_1 = \frac{1}{2}(a_1 + b_1),\quad \theta_2 = \frac{1}{2}(a_2 + b_2),\quad \theta_3 = \frac{1}{2}(a_3 + b_3),\quad \delta_4 = \frac{1}{2}(A + B),$$
$$\theta_5 = \frac{1}{2}(a_1 - b_1),\quad \theta_6 = \frac{1}{2}(a_2 - b_2),\quad \theta_7 = \frac{1}{2}(a_3 - b_3),\quad \delta_5 = \frac{1}{2}(A - B),$$

$$a_i (i = 1, 2, 3, 4), b_i, A, B, G_3 \text{ and } \Delta_3 \text{ are system cons tan ts}$$

The magneto-dispersion law in II–V semiconductors in the presence of a magnetic field $\vec{B}$ along k_y direction can be written as

$$k_y^2 = A_{75,\pm}(E,n), \tag{4.100b}$$

where

$$A_{75,\pm}(E,n) = \left[I_{35}E + I_{36,\pm}(n) \pm \sqrt{E^2 + EI_{38,\pm}(n) + I_{39,\pm}(n)}\right], \quad I_{35} = \frac{\theta_2}{(\theta_2^2 - \theta_5^2)},$$

$$I_{36,\pm}(n) = \frac{I_{33,\pm}(n)}{2(\theta_2^2 - \theta_5^2)}, \; I_{38,\pm}(n) = (4\theta_5^2)^{-1}\left[4\theta_2 I_{33,\pm}(n) + 8\theta_2^2 I_{31,\pm}(n) - \theta_5^2 I_{31,\pm}(n)\right],$$

$$I_{39,\pm}(n) = (4\theta_5^2)^{-1}\left[I_{33,\pm}^2(n) + 4\theta_2^2 I_{34,\pm}(n) - 4\theta_5^2 I_{34,\pm}(n)\right],$$
$$I_{33,\pm}(n) = \left[G_3^2 + 2\theta_5 I_{32}(n) - 2\theta_2 I_{31,\pm}(n)\right],$$

$$I_{34,\pm}(n) = \left[I_{32}^2(n) + \Delta_3^2 - I_{31,\pm}(n)\right], \quad I_{31,\pm}(n) = \left[\left(n + \frac{1}{2}\right)\hbar\omega_{31} - \frac{\delta_4^2}{4\theta_1} \pm \Delta_3\right],$$

$$I_{32}(n) = \left[\left(n + \frac{1}{2}\right)\hbar\omega_{32} - \frac{\delta_5^2}{4\theta_5}\right], \quad \omega_{31} = \frac{eB}{\sqrt{M_{31}M_{32}}}, \quad \omega_{32} = \frac{eB}{\sqrt{M_{33}M_{34}}},$$

$$M_{31} = \frac{\hbar^2}{2\theta_1}, \quad M_{32} = \frac{\hbar^2}{2\theta_3}, \quad M_{33} = \frac{\hbar^2}{2\theta_5} \quad \text{and} \quad M_{34} = \frac{\hbar^2}{2\theta_7}.$$

The EEM is given by

$$m^*(E_{FB}, n) = \frac{\hbar^2}{2} A'_{75,\pm}(E_{FB}, n). \tag{4.101}$$

The electron concentration can be expressed as

$$n_0 = \left(\frac{eBg_v}{2\pi^2\hbar}\right)\sum_{n=0}^{n_{max}}[Y_{51}(E_{FB}, n) + Z_{51}(E_{FB}, n)], \tag{4.102}$$

where

$$Y_{51}(E_{FB}, n) = [\sqrt{A_{75,+}(E_{FB}, n)} + \sqrt{A_{75,-}(E_{FB}, n)}]$$

and

$$Z_{51}(E_{FB}, n) = \sum_{r=1}^{s_0} L_B(r)[Y_{51}(E_{FB}, n)]$$

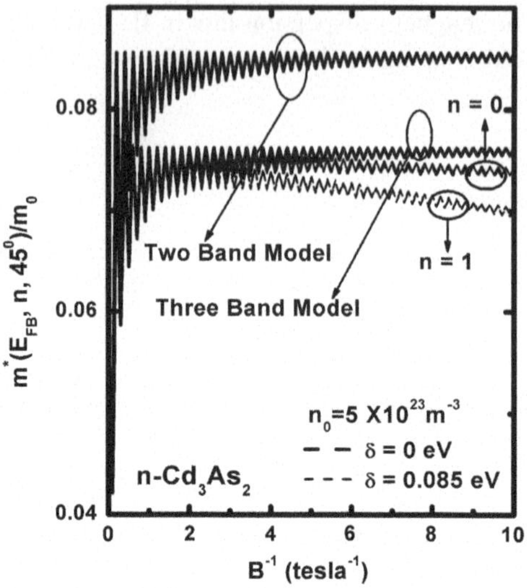

Fig. 4.1 Plot of the magnetic quantum number-dependent EEM as function of inverse magnetic field for n-Cd_3As_2 considering (4.2) both in the presence and absence of crystal field-splitting constant. The magnetic field has been oriented at 45° to the k_z axis. The graphs for the *three- and two-band models* of Kane have also been exhibited

4.3 Results and Discussion

Using (4.2) and (4.5) together with the energy band constants as given in Table 1.1 we have plotted the EEM in n-Cd_3As_2 and $CdGeAs_2$ as functions of inverse magnetic field for the first two magnetic subbands in Figs. 4.1 and 4.2 respectively. For the purpose of self-assessment, in the same figures, we have also plotted the effect of absence of the crystal field splitting together with the simplified three- and two-band models of Kane. From these figures, it appears that the EEM is an oscillatory function of the inverse quantizing magnetic field. The magnetic field has been tilted to an angle of 45° to the k_z direction in both the figures. The oscillatory dependence is due to the crossing over of the Fermi level by the Landau subbands in steps resulting in successive reduction the number of occupied Landau levels as the magnetic field is increased. For each coincidence of a Landau level, with the Fermi level, there would be a discontinuity in the density-of-states function resulting in a peak of oscillation.

Thus the peaks should occur whenever the Fermi energy is a multiple of energy separation between the two consecutive Landau levels and it may be noted that the origin of oscillations in the EEM is the same as that of the Subhnikov-de Hass oscillations. With increase in magnetic field, the amplitude of the oscillation increases and, ultimately, at very large values of the magnetic field, the conditions for the quantum limit is reached when the EEM is found to decrease monotonically with increase in magnetic field. Further, in this case we see that the EEM is a strong function of the subband quantum number n.

For this reason, we has also plotted the EEM for the next higher subband $n = 1$. It thus appears that the increasing the index decreases the EEM for lower values of

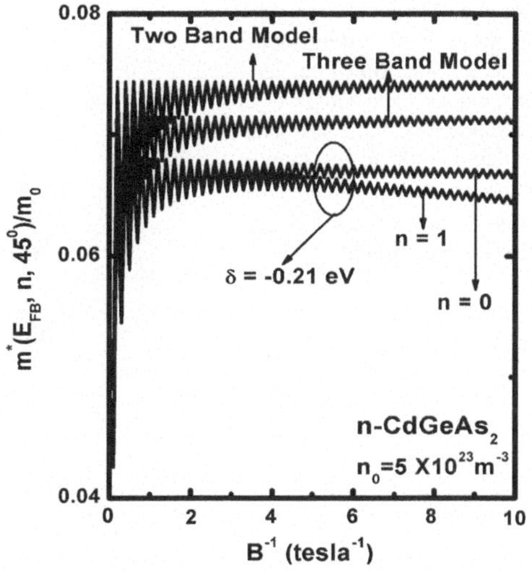

Fig. 4.2 Plot of the magnetic quantum number-dependent EEM as function of inverse magnetic field for n-CdGeAs$_2$ considering (4.2). The magnetic field has been oriented at 45° to the k_z axis. The plots for the *three- and two-band models* of Kane have also been exhibited

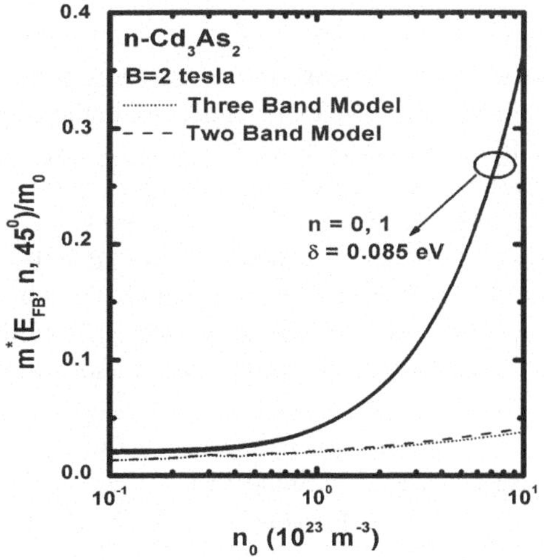

Fig. 4.3 Plot of the magnetic quantum number-dependent EEM as function of carrier degeneracy for n-Cd$_3$As$_2$. The magnetic field has been oriented at 45° to the k_z axis. The plots for the *three- and two-band models* of Kane have also been exhibited

the field. However at higher field, the difference between them diminishes and all the respective band models tend to coincide with each other which stand out to be a remarkable mathematical simplicity in deriving the analytical expressions of the EEM. The presence of the isotropic spin orbit splitting constant in the three-band model of Kane changes the value of the EEM as compared with the corresponding two-band model.

Figures 4.3 and 4.4 exhibit the variation of the EEM on the carrier degeneracy in both the aforementioned materials. Oscillatory dependences are exhibited in the case of the equivalent three- and the two-band model of Kane, the deviation among which for both the materials are almost zero. Further, we also see that there is almost

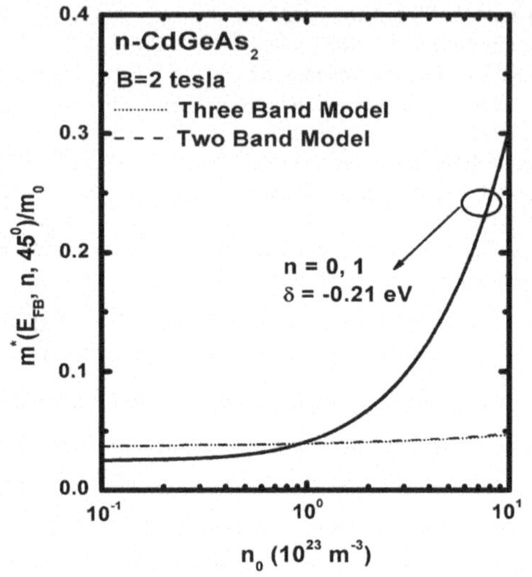

Fig. 4.4 Plot of the magnetic quantum number-dependent EEM as function of carrier degeneracy for n-CdGeAs$_2$. The magnetic field has been oriented at 45° to the k_z axis. The plots for *three- and two-band models* of Kane have also been exhibited

no significant change in the variation of the subband index from $n = 1$ to $n = 2$ in both the cases. An exponential rise in the EEM can be observed beyond 10^{23} m^{-3} for both the materials. In case of Cd$_3$As$_2$, we see that decreasing the carrier degeneracy converges the EEM from all the band models to a unique value. Incidentally, this is not the case of CdGeAs$_2$. There is a crossing over of the EEM near to the concentration zone of 10^{23} m^{-3} which overestimates the numerical result. In addition, the EEM exhibits different numerical values for both the materials, the rate of variations of which are different due to the influence of the energy band constants in accordance with all the types of band models and follow the same trend as shown in Figs. 4.3 and 4.4.

The dependency of the EEM on the angular orientation of the quantizing magnetic field has been exhibited in Figs. 4.5 and 4.6 in both n-Cd$_3$As$_2$ and n-CdGeAs$_2$ respectively. It appears that the EEM exhibits a periodic variation increasing θ from 0° up to 120°.

The effect of the crystal field splitting constant in both the cases has been exhibited for relative assessment. It appears that the influence of the crystal field splitting constant on the EEM for Cd$_3$As$_2$ is relatively insignificant, while there appears a cross-over regime in the EEM in CdGeAs$_2$ around 50°. In the later case, the crystal field constant tends to reduce the EEM beyond 50° which exhibits the influence of δ.

For the three- and the two-band models of Kane, the EEM becomes independent of θ, since the dispersion relation of the bulk materials in accordance with the said band models is spherical in constant energy wave vector space, whereas the generalized band model represents the ellipsoid of revolution in the same space.

Figures 4.7, 4.8, 4.9, 4.10, and 4.11 exhibit the variation of the EEM on the quantizing magnetic field for n-InAs, n-GaAs, n-InSb, n-Hg$_{1-x}$Cd$_x$Te and

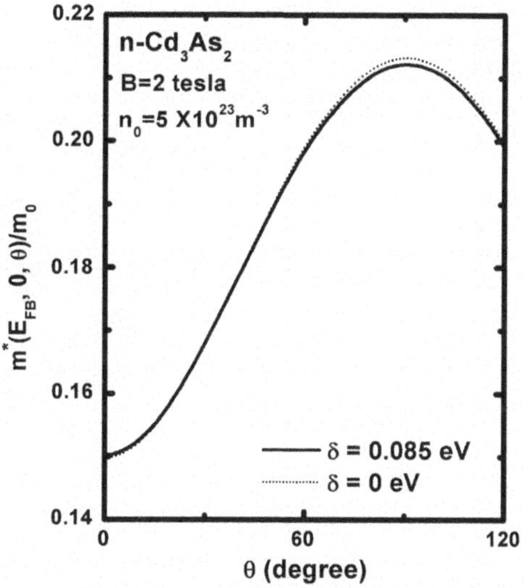

Fig. 4.5 Plot of the lowest magnetic quantum number-dependent EEM as function of orientation of the magnetic field in n-Cd_3As_2 both in the presence and absence of the crystal field-splitting constant

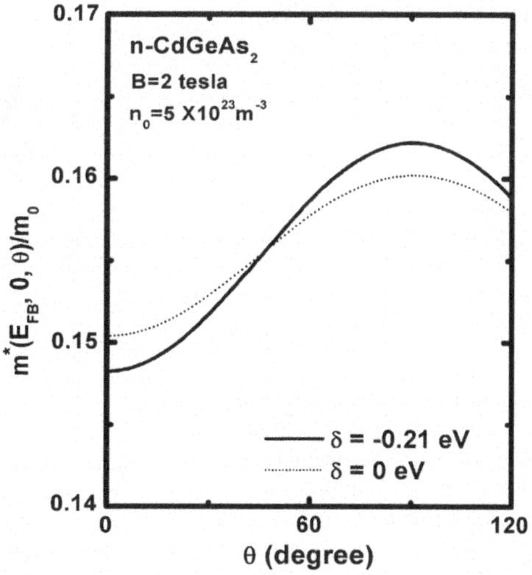

Fig. 4.6 Plot of the lowest magnetic quantum number-dependent EEM as function of angular dependency of the magnetic field in n-$CdGeAs_2$ both in the presence and absence of the of crystal field-splitting constant

$In_{1-x}Ga_xAs_yP_{1-y}$ lattice matched to InP in accordance with the three- and two-band models of Kane respectively. The variations of the EEM are periodic and independent of the subband index number with the quantizing magnetic field and the influence of the energy band constants on the EEM in accordance with all the band models is apparent from the said figures.

Figures 4.12, 4.13, 4.14, 4.15 and 4.16 exhibit the concentration dependence of the periodic EEM for all the respective aforementioned materials.

It appears from Figs. 4.12, 4.13, 4.14, 4.15, and 4.16 that the periodic oscillatory numerical values of the EEM is greatest for the quaternary materials while the least

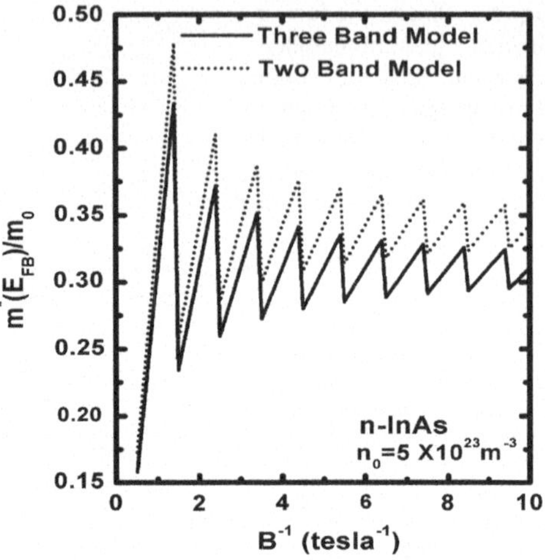

Fig. 4.7 Plot of the EEM as function of inverse magnetic field for n-InAs considering the *three and two band models* of Kane

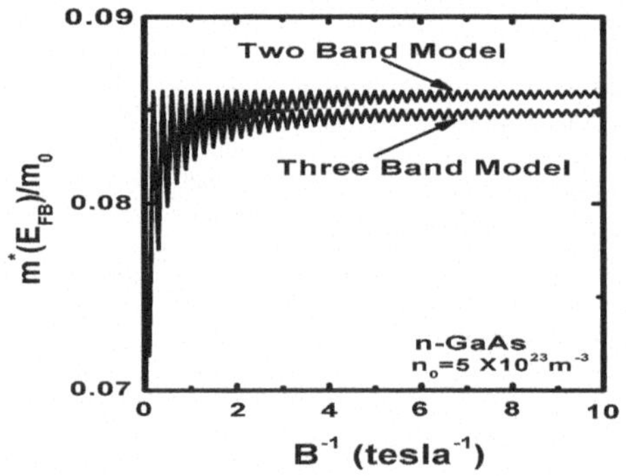

Fig. 4.8 Plot of the EEM as function of inverse magnetic field for n-GaAs considering the *three- and two-band models* of Kane

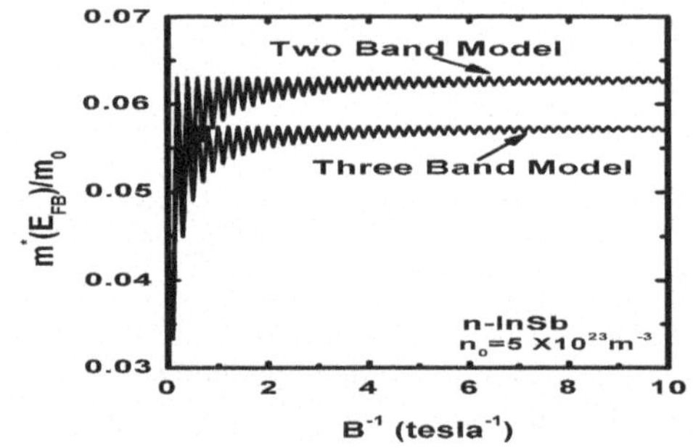

Fig. 4.9 Plot of the EEM as function of inverse magnetic field for n-InSb considering the *three- and two-band models* of Kane

for InSb for all types of variables in accordance with all types of band models of III–V, ternary and quaternary materials. In Fig. 4.17, we have plotted the variation of the EEM as function of alloy composition in HgCdTe and InGaAsP lattice matched to InP. It appears that the EEM increases with the alloy fraction in an almost linear

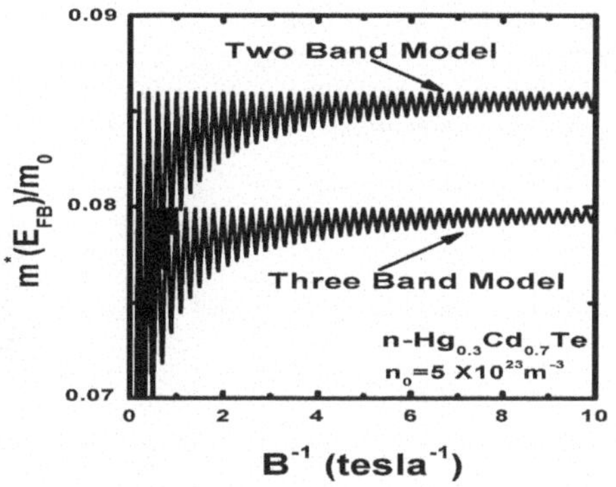

Fig. 4.10 Plot of the EEM as function of inverse magnetic field for n-$Hg_{0.3}Cd_{0.7}Te$ considering the three- and two-band models of Kane

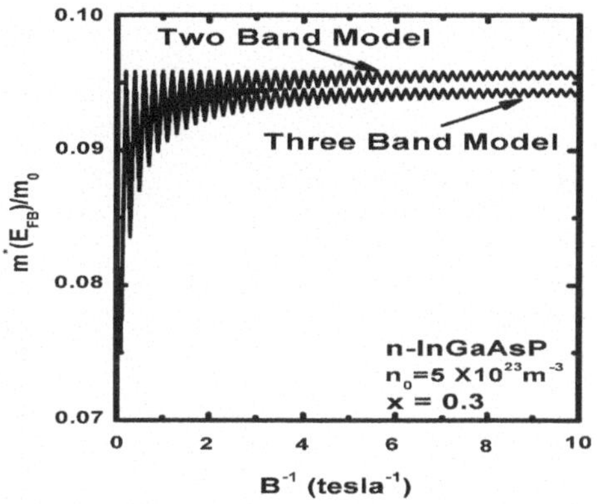

Fig. 4.11 Plot of the EEM as function of inverse magnetic field for n-$In_{1-x}Ga_xAs_yP_{1-y}$ lattice matched to InP considering the *three- and two-band models* of Kane

way. The result of the EEM arising due to the difference in the band structure also appears to be extremely less.

The numerical computations for the models according to Stillman (Eq. (4.27)) and Palik (Eq. (4.32)) have been left as an exercise for the reader. Also, from Eq. (4.36), we see that the EEM in II–VI material like CdS remains invariant with the magnetic field, hence we have not exhibited this.

Using Eqs. (4.39) and (4.42) for McClure–Choi model, (4.48) and (4.50) for the model of Cohen and (4.55) and (4.56) for the model of Lax, we have plotted the EEM for Bi as functions of inverse quantizing magnetic field and carrier degeneracy as shown in Figs. 4.18 and 4.19 respectively considering the first two magnetic subbands for models of McClure–Choi.

From Fig. 4.18, it appears that the effect of the energy band structure namely due to the model of Cohen and the Lax on the EEM almost coincides with each other. However, the quadratic nonlinear energy dispersion relation of McClure and Choi tends to increase the EEM. It appears that the increase in the magnetic subband index increases the EEM in this case as compared with that of Figs. 4.1 and 4.2 for nonlinear tetragonal materials. This increase of the EEM for the present case results due to the presence of the respective dominant energy spectrum parameters. As the

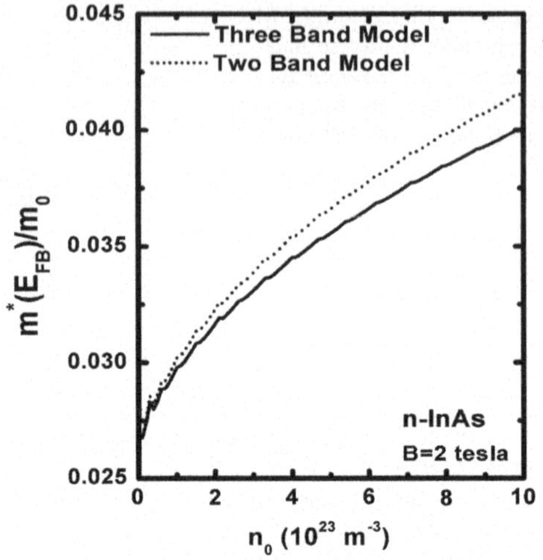

Fig. 4.12 Plot of the EEM as function of carrier degeneracy for n-InAs considering the *three- and two-band models* of Kane

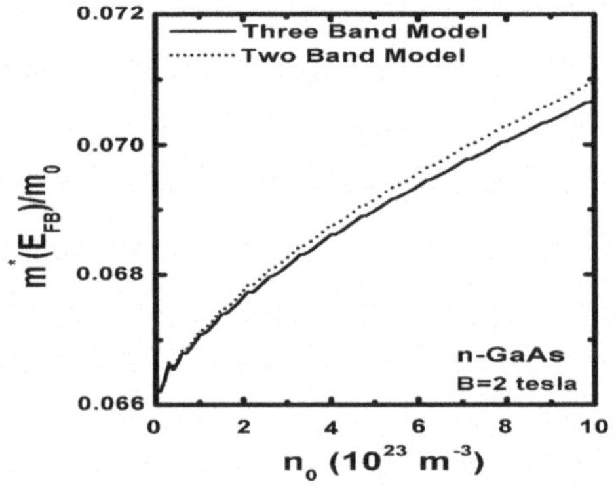

Fig. 4.13 Plot of the EEM as function of carrier degeneracy for n-GaAs considering the *three- and two-band models* of Kane

magnetic field increases, we see that with increase in the subband index the EEM exhibits a sharp discontinuity and can become a negative quantity, thus questioning the validity of the McClure and Choi model in the beyond-10 tesla zone.

The variation of the EEM on the carrier degeneracy for Bi in Fig. 4.19 is rather slow over $0.1–0.5 \times 10^{23}\ m^{-3}$ zone.

Figure 4.20 exhibits the variation of the EEM against the quantizing magnetic field for IV–VI materials considering PbTe as an example using the dispersion relation provided by the Dimmock model at the lowest quantizing subband. In the same figure we have demonstrated the variation of the EEM for stressed InSb for the first two lowest subbands. Large oscillations are exhibited for PbTe case as compared with that of the stressed case, where we have considered the stress to be composed of all the diagonal and off-diagonal strain components as given in Table 1.1. It appears the deviation of the EEM from its ground state value is almost zero when the angular dependency is 45°. To exhibit this difference, we have further plotted the EEM at the

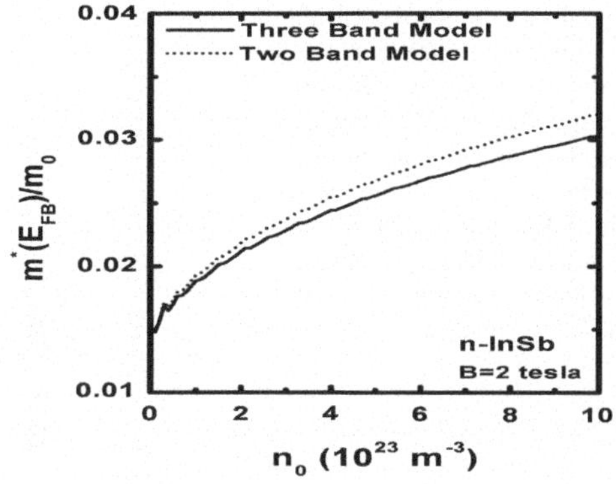

Fig. 4.14 Plot of the EEM as function of carrier degeneracy for n-InSb considering the *three- and two-band models* of Kane

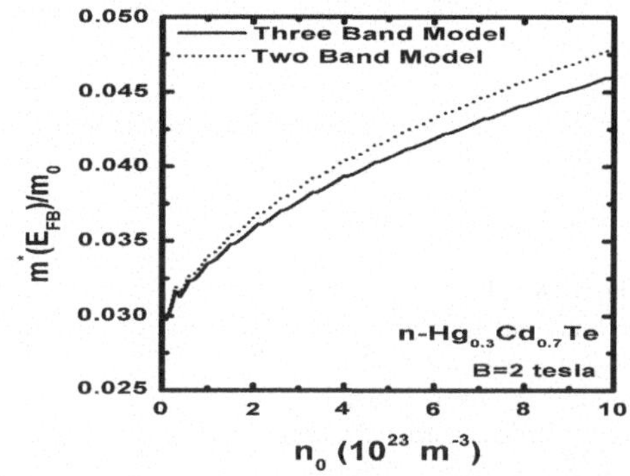

Fig. 4.15 Plot of the EEM as function of carrier degeneracy for n-$Hg_{0.3}Cd_{0.7}Te$ considering the *three- and two-band models* of Kane

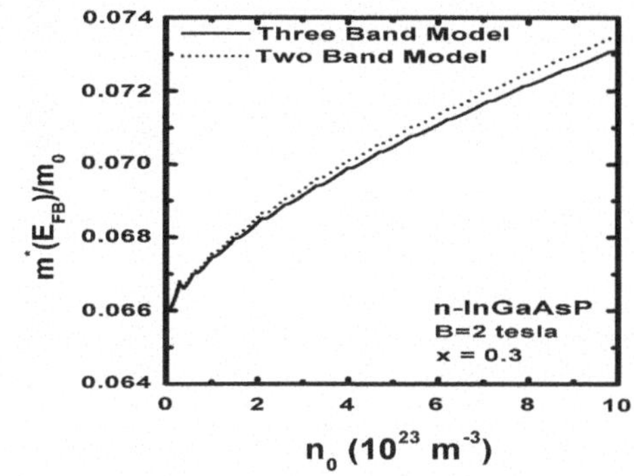

Fig. 4.16 Plot of the EEM as function of carrier degeneracy for n-$In_{1-x}Ga_xAs_yP_{1-y}$ lattice matched to InP considering the *three- and two-band models* of Kane

lowest two subbands as function of the angle of orientation of the field in Fig. 4.21. It appears that the EEM exhibits periodical variation over the entire angular range as shown in the same figure with the deviation between the subband values at the two minima and the mid angular zone (Fig. 4.21).

Figure 4.22 exhibits the EEM in Te, GaP, $PtSb_2$, Bi_2Te_3, GaSb and as function of quantizing magnetic field at the lowest subband level. The usual periodical oscillatory

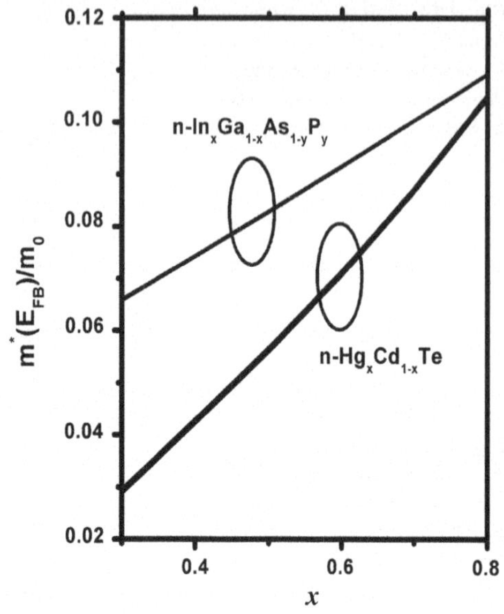

Fig. 4.17 Plot of the EEM as function of alloy composition for n-$Hg_xCd_{1-x}Te$ and n-$In_{1-x}Ga_xAs_yP_{1-y}$ lattice matched to InP considering the *three- and two-band models* of Kane at a quantizing magnetic field of 2 Tesla and carrier degeneracy of 5×10^{23} m^{-3}

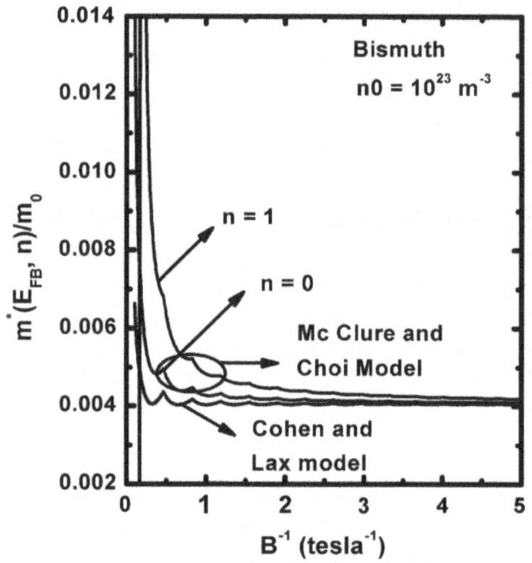

Fig. 4.18 Plot of the EEM as function of inverse magnetic field for Bi considering the *energy band models* of McClure and Choi, Cohen and Lax respectively

nature is exhibited for all the said materials with Bi_2Te_3 to exhibit the highest EEM numerical values. In case of Ge, we see from (4.92) that the Cardona et al. model and Wang et al. register a subband index-dependent EEM. We leave the reader to carry out investigation of the EEM using both the models for Ge, other allied models for IV–VI, together with that for II–VI materials.

We wish to note that the effect of electron spin has not been considered in obtaining the oscillatory plots. The peaks in all the figures would increase in number with decrease in amplitude if spin splitting term is included in the respective numerical

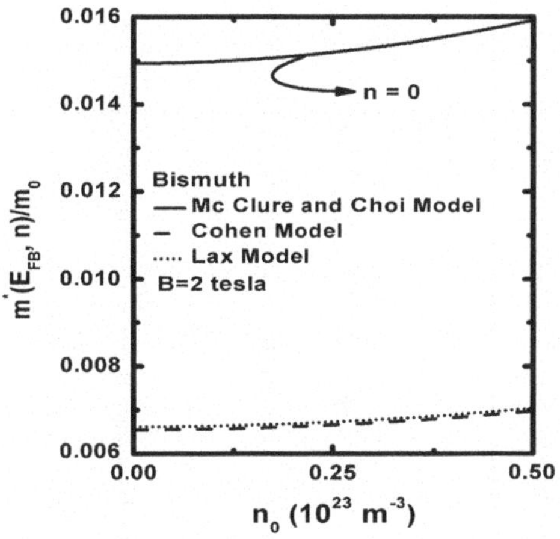

Fig. 4.19 Plot of the EEM as function of carrier degeneracy for Bi considering the *energy band models* of McClure and Choi, Cohen and Lax respectively

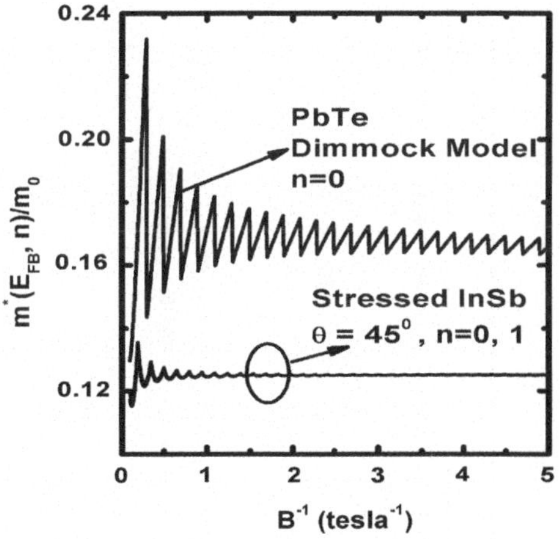

Fig. 4.20 Plot of the EEM as function of quantizing magnetic field for PbTe and stressed InSb at the *two lowest magnetic subbands*

computations. Though the effects of collisions are usually small at low temperatures, the sharpness of the amplitude of the oscillatory plots would somewhat be reduced by collision broadening. Nevertheless, the present analysis would remain valid since the effects of collision broadening can usually be taken into account by an effective increase in temperature. Although in a more rigorous statement the many-body effects should be considered along with the self-consistent procedure, the simplified analysis as presented exhibits the basic qualitative features of the EEM in this under the magnetic quantization with reasonable accuracy. For the purpose of condensed presentation, the carrier statistics and the EEM in different materials as considered in this chapter have been presented in Table 4.1.

Table 4.1 The carrier statistics and the EEM under magnetic quantization from nonlinear optical, III–V, II–VI, Bismuth, IV–VI, stressed materials, Te, n-GaP, $PtSb_2$, Bi_2Te_3, n-Ge, GaSb and II–V materials

Type of materials	The carrier statistics	The EEM
1. Nonlinear optical materials	In accordance with the generalized electron dispersion relation as given by (4.1a), $n_0 = \frac{g_v \lvert e\rvert B}{2\pi^2\hbar}\sum_{n=0}^{n_{max}}[T_{33}(n, E_{FB}) + T_{34}(n, E_{FB})]$ (4.5)	On the basis of (4.1a), $m^*_{k_{z1}}(E_{FB}n, \theta) = \left(\frac{\hbar^2}{2}\right)\left\{\left[\frac{[\psi_1(E_{FB})]' - [\bar{A}\pm(n, E_{FB,\theta})]'}{a_0(E_{FB,\theta})}\right] - \left[\frac{[a_0(E_{FB,\theta})]'}{a_0^2(E_{FB,\theta})}\right][\psi_1(E_{FB}) - \bar{A}\pm(n, E_{FB,\theta})]\right\}$ (4.2)
2. III–V materials, the conduction electrons of which can be defined by five types of energy wave vector dispersion relations as described in the column beside	(a) Three-band model of Kane: In accordance with the three-band model of Kane (4.6), which is the special case of (4.1a) $n_0 = \frac{g_v \lvert e\rvert B\sqrt{2m_c}}{2\pi^2\hbar^2}\sum_{n=0}^{n_{max}}[T_{35}(n, E_{FB}) + T_{36}(n, E_{FB})]$ (4.9)	The basis of (4.6), $m^*_{k_z}(E_{FB}) = m_c\left[\{I_{11}(E_{FB})\}' \pm \frac{\lvert e\rvert B\hbar\Delta}{6m_c}\left[(E_{FB} + E_g + \frac{2}{3}\Delta)\right]^{-2}\right]$ (4.7)
	(b) Two-band model of Kane: In accordance with the two-band model of Kane (4.11), $n_0 = \frac{g_v \lvert e\rvert B\sqrt{2m_c}}{2\pi^2\hbar^2}\sum_{n=0}^{n_{max}}[T_{39}(n, E_{FB}) + T_{310}(n, E_{FB})]$ (4.14)	On the basis of (4.11) $m^*_{k_z}(E_{FB}) = m_c[1 + 2\alpha E_{FB}]$ (4.12)
	(c) The model of Stillman et al.: In accordance with the model of Stillman et al. (4.25), $n_0 = \frac{\sqrt{2m_c}eBg_v}{\pi^2\hbar^2}\sum_{n=0}^{n_{max}}[Y_{33}(E_{FB}, n) + Z_{33}(E_{FB}, n)]$ (4.28)	On the basis of (4.25), $m^*_{k_z}(E_{FB}) = m_c[I'_{11}(E_{FB})]$ (4.27)
	(d) The model of Palik et al.: In accordance with the model of Palik et al. (4.31), $n_0 = \frac{eBg_v}{2\pi^2 h}\sum_{n=0}^{n_{max}}[Y_{34}(E_{FB}, n) + Z_{34}(E_{FB}, n)]$ (4.33)	On the basis of (4.31) $m^*_{k_z}(E_{FB}) = m_c[A'_{35,\pm}(E_{FB}, n)]$ (4.32)

(continued)

Table 4.1 (continued)

Type of materials	The carrier statistics	The EEM
3. Bismuth, the carriers of which can be defined by five types of energy band models as described in the column beside	(a) The McClure and Choi model: In accordance with (4.38), $n_0 = \frac{g_v \lvert e \rvert B\sqrt{2m_3}}{2\pi^2\hbar^2}\sum_{n=0}^{n_{\max}}[T_{313}(n, E_{FB}) + T_{314}(n, E_{FB})]$ (4.42)	On the basis of (4.38) $m^*_{k_z}(n, E_{FB}) = m_3\left[\left[1-\frac{\alpha}{2}\left(n+\frac{1}{2}\right)\hbar\omega(E_{FB})\right]^{-1} \times\left[1+2\alpha E_{FB}-\left(n+\frac{1}{2}\right)\hbar\omega'(E_{FB}) -\frac{1}{2}(n^2+n+1)\alpha\hbar^2\omega(E_{FB})\omega'(E_{FB})\right] +\frac{\alpha\left(n+\frac{1}{2}\right)\hbar\omega'(E_{FB})}{2}\left[1-\frac{\alpha\left(n+\frac{1}{2}\right)\hbar\omega(E_{FB})}{2}\right]^{-2} \times\left[E_{FB}(1+\alpha E_{FB})-\left(n+\frac{1}{2}\right)\hbar\omega(E_{FB}) -\frac{\alpha\hbar^2\omega^2(E_{FB})}{4}(n^2+1+n)\pm\frac{1}{2}g^*\mu_0 B\right]\right]$ (4.39)
	(b) The Cohen Model: In accordance with (4.43), $n_0 = \frac{g_v \lvert e \rvert B\sqrt{2m_3}}{2\pi^2\hbar^2}\sum_{n=0}^{n_{\max}}[T_{319}(n, E_{FB}) + T_{320}(n, E_{FB})]$ (4.50)	$m^*_{k_z}(n, E_{FB}) = m_3\left[2\alpha E_{FB}+1-\left(n+\frac{1}{2}\right)\hbar\omega'(E_{FB}) -\frac{3}{4}\alpha\hbar^2\omega(E_{FB})\omega'(E_{FB})\left(n^2+n+\frac{1}{2}\right)\right]$ (4.48)
	(c) The Lax Model: $n_0 = \frac{eBg_v\sqrt{2m_3}}{2\pi^2\hbar^2}\sum_{n=0}^{n_{\max}}[Y_{40}(E_{FB}, n) + Z_{40}(E_{FB}, n)]$ (4.56)	On the basis of (4.53) $m^*(E_{FB}) = m_3(1+2\alpha E_{FB})$ (4.55)

(continued)

Table 4.1 (continued)

Type of materials	The carrier statistics	The EEM
4. IV–VI materials, the carriers of which can be defined by the model of Dimmock, Bangert & Kastner and Foley & Landenberg	(a) The Dimmock Model On the basis of (4.59) $$n_0 = \left(\frac{eBg_v}{\pi^2 \hbar}\right) \sum_{n=0}^{n_{\max}} [Y_{41}(E_{FB}, n) + Z_{41}(E_{FB}, n)] \quad (4.61)$$	On the basis of (4.59) $$m^*(E_{FB}, n) = \frac{\hbar^2}{2} A'_{42}(E_{FB}, n) \quad (4.60)$$
	(b) The Bangert & Kastner Model In accordance with (4.64) $$n_0 = \left(\frac{eBg_v}{\pi^2 \hbar}\right) \sum_{n=0}^{n_{\max}} [Y_{42}(E_{FB}, n) + Z_{42}(E_{FB}, n)] \quad (4.66)$$	On the basis of (4.64) $$m^*(E_{FB}, n) = \frac{\hbar^2}{2} A'_{44}(E_{FB}, n) \quad (4.65)$$
	(c) The Foley & Landenberg Model In accordance with (4.68) $$n_0 = \left(\frac{eBg_v}{\pi^2 \hbar}\right) \sum_{n=0}^{n_{\max}} [Y_{43}(E_{FB}, n) + Z_{43}(E_{FB}, n)] \quad (4.70)$$	On the basis of (4.68) $$m^*(E_{FB}, n) = \frac{\hbar^2}{2} A'_{46}(E_{FB}, n) \quad (4.69)$$
5. Stressed materials, as defined by the model of Seiler et al.	In accordance with (4.71), $$n_0 = \frac{g_v \lvert e \rvert B}{\pi^2 \hbar} \sum_{n=0}^{n_{\max}} [T_{327}(n, E_{FB}) + T_{328}(n, E_{FB})] \quad (4.74)$$	On the basis of (4.71), $$m^*_{k'_z}(n, E_{FB}) = \frac{\hbar^2}{2} \left[-\{I_3(n, E_{FB})\}' I_2(n, E_{FB}) + (1 - I_3(n, E_{FB}))\{I_2(n, E_{FB})\}' \right] \quad (4.72)$$

(continued)

Table 4.1 (continued)

Type of materials	The carrier statistics	The EEM
6. Tellurium the conduction electrons of which can be defined by the model of Bouat et al.	In accordance with (4.76), $n_0 = \left(\frac{eBg_v}{2\pi^2\hbar}\right)\sum_{n=0}^{n_{\max}}[Y_{45}(E_{FB},n) + Z_{45}(E_{FB},n)]$ (4.78)	On the basis of (4.76) $m^*(E_{FB},n) = \frac{\hbar^2}{2}A'_{50,\pm}(E_{FB},n)$ (4.77)
7. n-GaP as described by the Rees model	In accordance with (4.80), $n_0 = \left(\frac{eBg_v}{2\pi^2\hbar}\right)\sum_{n=0}^{n_{\max}}[Y_{46}(E_{FB},n) + Z_{46}(E_{FB},n)]$ (4.82)	On the basis of (4.80) $m^*(E_{FB},n) = \frac{\hbar^2}{2}A'_{52,\pm}(E_{FB},n)$ (4.81)
8. $PtSb_2$, as defined by the Emtage model	In accordance with (4.84), $n_0 = \left(\frac{eBg_v}{2\pi^2\hbar}\right)\sum_{n=0}^{n_{\max}}[Y_{47}(E_{FB},n) + Z_{47}(E_{FB},n)]$ (4.86)	On the basis of (4.84) $m^*(E_{FB},n) = \frac{\hbar^2}{2}A'_{55,\pm}(E_{FB},n)$ (4.85)
9. Bi_2Te_3, which follows the model of Stordeur et al.	In accordance with (4.87), $n_0 = \left(\frac{eBg_v}{\pi^2\hbar}\right)\sum_{n=0}^{n_{\max}}[Y_{48}(E_{FB},n) + Z_{48}(E_{FB},n)]$ (4.89)	On the basis of (4.87) $m^*(E_{FB}) = \frac{\hbar^2}{2}\left[\frac{1+2\alpha E_{FB}}{\bar{\omega}_1}\right]$ (4.88)

(continued)

Table 4.1 (continued)

Type of materials	The carrier statistics	The EEM
10. n-Ge, the conduction electrons of which can be defined by two types of energy band models as described in the column of carrier statistics	(a) In accordance with the model of Cardona et al. (4.91), $$n_0 = \left(\frac{eBg_v}{\pi^2 \hbar}\right) \sum_{n=0}^{n_{\max}} [Y_{49}(E_{FB}, n) + Z_{49}(E_{FB}, n)] \quad (4.93)$$ (b) In accordance with the model of Wang and Ressler (4.94), $$n_0 = \left(\frac{eBg_v}{\pi^2 \hbar}\right) \sum_{n=0}^{n_{\max}} [Y_{50}(E_{FB}, n) + Z_{50}(E_{FB}, n)] \quad (4.96)$$	On the basis of (4.91), $$m^*(E_{FB}, n) = m_{\parallel}^* A'_{69}(E_{FB}, n) \quad (4.92)$$ On the basis of (4.94) $$m^*(E_{FB}, n) = m_{\parallel}^* A'_{71}(E_{FB}, n) \quad (4.95)$$
11. Gallium Antimonide, the carriers of which can be defined by the model of Mathur et al.	In accordance with (4.97), $$n_0 = \left(\frac{eBg_v}{\pi^2 \hbar}\right) \sum_{n=0}^{n_{\max}} [Y_{501}(E_{FB}, n) + Z_{501}(E_{FB}, n)] \quad (4.99)$$	On the basis of (4.97), $$m^*(E_{FB}) = m_c I'_{16}(E_{FB}) \quad (4.98)$$
12. II–V materials, as defined by the model of Yamada	In accordance with (4.100), $$n_0 = \left(\frac{eBg_v}{2\pi^2 \hbar}\right) \sum_{n=0}^{n_{\max}} [Y_{51}(E_{FB}, n) + Z_{51}(E_{FB}, n)] \quad (4.102)$$	On the basis of (4.100) $$m^*(E_{FB}, n) = \frac{\hbar^2}{2} A'_{75,\pm}(E_{FB}, n) \quad (4.101)$$

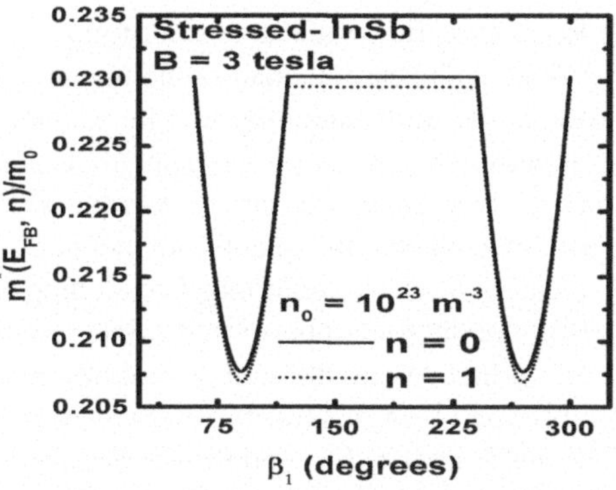

Fig. 4.21 Plot of the EEM as function of angular orientation of the magnetic field for stressed InSb at the *two lowest magnetic subbands*

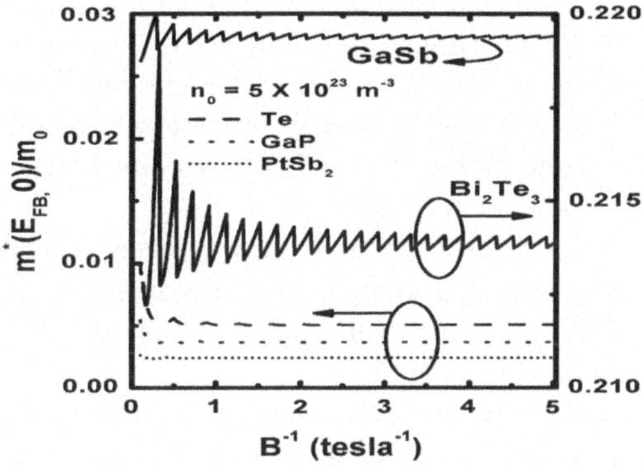

Fig. 4.22 Plot of the EEM as function quantizing magnetic field for Bi_2Te_3, GaSb, Te, GaP and $PtSb_2$ at the *lowest magnetic subband*

4.4 Open Research Problems

R.4.1

(a) Investigate the EEM, EAM, DEM, CEM, CoEM, FREM and OEM in all the bulk semiconductors as considered in this chapter in the absence of any field.
(b) Investigate the same set of masses as defined in (R 4.1) in the presence of an arbitrarily oriented quantizing magnetic field including broadening and the electron spin (applicable under magnetic quantization) for all the bulk semiconductors whose unperturbed carrier energy spectra are defined in Chap. 1.

R.4.2 Investigate the same set of masses as defined in (R 1.1) in the presence of quantizing magnetic field under an arbitrarily oriented (a) non-uniform electric field and (b) alternating electric field respectively for all the semiconductors whose unperturbed carrier energy spectra are defined in Chap. 1 by including spin and broadening respectively.

R.4.3 Investigate the same set of masses as defined in (R 1.1) under an arbitrarily oriented alternating quantizing magnetic field by including broadening and the electron spin for all the semiconductors whose unperturbed carrier energy spectra as defined in Chap. 1.

R.4.4 Investigate the same set of masses as defined in (R 1.1) under an arbitrarily oriented alternating quantizing magnetic field and crossed alternating electric field by including broadening and the electron spin for all the semiconductors whose unperturbed carrier energy spectra as defined in Chap. 1.
R.4.5 Investigate the same set of masses as defined in (R 1.1) under an arbitrarily oriented alternating quantizing magnetic field and crossed alternating non-uniform electric field by including broadening and the electron spin whose for all the semiconductors unperturbed carrier energy spectra as defined in Chap. 1.
R.4.6 Investigate the same set of masses as defined in (R 1.1) in the presence and absence of an arbitrarily oriented quantizing magnetic field under exponential, Kane, Halperin, Lax and Bonch-Bruevich band tails [70] for all the semiconductors whose unperturbed carrier energy spectra as defined in Chap. 1 by including spin and broadening (applicable under magnetic quantization).
R.4.7 Investigate the same set of masses as defined in (R 1.1) in the presence of an arbitrarily oriented quantizing magnetic field for all the semiconductors as defined in (R 4.6) under an arbitrarily oriented (a) non-uniform electric field and (b) alternating electric field respectively whose unperturbed carrier energy spectra as defined in Chap. 1.
R.4.8 Investigate the same set of masses as defined in (R 1.1) under an arbitrarily oriented alternating quantizing magnetic field by including broadening and the electron spin for all semiconductors whose unperturbed carrier energy spectra as defined in Chap. 1.
R.4.9 Investigate the same set of masses as defined in (R 1.1) under an arbitrarily oriented alternating quantizing magnetic field and crossed alternating electric field by including broadening and the electron spin for all the semiconductors whose unperturbed carrier energy spectra as defined in Chap. 1.
R.4.10 Investigate all the appropriate problems of this chapter after proper modifications introducing new theoretical formalisms for functional, negative refractive index, macro molecular, organic, and magnetic materials.
R.4.11 Investigate all the appropriate problems of this chapter for *p*-InSb, *p*-CuCl and stressed semiconductors having diamond structure valence bands whose dispersion relations of the carriers in bulk semiconductors are given by Cunningham [79], Yekimov et al. [80] and Roman et al. [81] respectively.
R.4.12 Investigate all the problems of this chapter by removing all the mathematical approximations and establishing the respective appropriate uniqueness conditions.

References

1. N. Miura, *Physics of Semiconductors in High Magnetic Fields*, Series on Semiconductor Science and Technology (Oxford University Press, USA, 2007)
2. K.H.J Buschow, F.R. de Boer, *Physics of Magnetism and Magnetic Materials* (Springer, New York, 2003)
3. D. Sellmyer, R. Skomski (ed.), *Advanced Magnetic Nanostructures* (Springer, New York, 2005)

4. J.A.C. Bland, B. Heinrich (ed.), *Ultrathin Magnetic Structures III: Fundamentals of Nanomagnetism (Part 3)* (Springer, Germany, 2005)
5. B.K. Ridley, *Quantum Processes in Semiconductors* 4th edn. (Oxford publications, Oxford, 1999)
6. J.H. Davies, *Physics of Low Dimensional Semiconductors* (Cambridge University Press, UK, 1998)
7. S. Blundell, *Magnetism in Condensed Matter* Oxford Master Series in Condensed Matter Physics (Oxford University Press, USA, 2001)
8. C. Weisbuch, B. Vinter, *Quantum Semiconductor Structures: Fundamentals and Applications* (Academic Publishers, USA, 1991)
9. D. Ferry, *Semiconductor Transport* (CRC, USA, 2000)
10. M. Reed (ed.), *Semiconductors and Semimetals: Nanostructured Systems* (Academic Press, USA, 1992)
11. T. Dittrich, *Quantum Transport and Dissipation* (Wiley, Germany, 1998)
12. A.Y. Shik, *Quantum Wells: Physics and Electronics of Twodimensional Systems* (World Scientific, USA, 1997)
13. K.P. Ghatak, M. Mondal, Zietschrift fur naturforschung A **41a**, 881 (1986)
14. K.P. Ghatak, M. Mondal, J. Appl. Phys. **62**, 922 (1987)
15. K.P. Ghatak, S.N. Biswas, Phys. Stat. Sol. (b) **140**, K107 (1987)
16. K.P. Ghatak, M. Mondal, J. Mag. Mag. Mat. **74**, 203 (1988)
17. K.P. Ghatak, M. Mondal, Phys. Stat. Sol. (b) **139**, 195 (1987)
18. K.P. Ghatak, M. Mondal, Phys. Stat. Sol. (b) **148**, 645 (1988)
19. K.P. Ghatak, B. Mitra, A. Ghoshal, Phys. Stat. Sol. (b) **154**, K121 (1989)
20. K.P. Ghatak, S.N. Biswas, J. Low Temp. Phys. **78**, 219 (1990)
21. K.P. Ghatak, M. Mondal, Phys. Stat. Sol. (b) **160**, 673 (1990)
22. K.P. Ghatak, B. Mitra, Phys. Letts. A **156**, 233 (1991)
23. K.P. Ghatak, A. Ghoshal, B. Mitra, Nouvo Cimento D **13D**, 867 (1991)
24. K.P. Ghatak, M. Mondal, Phys. Stat. Sol. (b) **148**, 645 (1989)
25. K.P. Ghatak, B. Mitra, Int. J. Elect. **70**, 345 (1991)
26. K.P. Ghatak, S.N. Biswas, J. Appl. Phys. **70**, 299 (1991)
27. K.P. Ghatak, A. Ghoshal, Phys. Stat. Sol. (b) **170**, K27 (1992)
28. K.P. Ghatak, Nouvo Cimento D **13D**, 1321 (1992)
29. K.P. Ghatak, B. Mitra, Int. J. Elect. **72**, 541 (1992)
30. K.P. Ghatak, S.N. Biswas, Nonlinear Opt. **4**, 347 (1993)
31. K.P. Ghatak, M. Mondal, Phys. Stat. Sol. (b) **175**, 113 (1993)
32. K.P. Ghatak, S.N. Biswas, Nonlinear Opt. **4**, 39 (1993)
33. K.P. Ghatak, B. Mitra, Nouvo Cimento **15D**, 97 (1993)
34. K.P. Ghatak, S.N. Biswas, Nanostruct. Mater. **2**, 91 (1993)
35. K.P. Ghatak, M. Mondal, Phys. Stat. Sol. (b) **185**, K5 (1994)
36. K.P. Ghatak, B. Goswami, M. Mitra, B. Nag, Nonlinear Opt. **16**, 9 (1996)
37. K.P. Ghatak, M. Mitra, B. Goswami, B. Nag, Nonlinear Opt. **16**, 167 (1996)
38. K.P. Ghatak, D.K. Basu, B. Nag, J. Phys. Chem. Sol. **58**, 133 (1997)
39. K.P. Ghatak, B. Nag, Nanostruct. Mater. **10**, 923 (1998)
40. D. Roy Choudhury, A.K. Choudhury, K.P. Ghatak, A.N. Chakravarti, Phys. Stat. Sol. (b) **98**, K141 (1980)
41. A.N. Chakravarti, K.P. Ghatak, A. Dhar, S. Ghosh, Phys. Stat. Sol. (b) **105**, K55 (1981)
42. A.N. Chakravarti, A.K. Choudhury, K.P. Ghatak, Phys. Stat. Sol. (a) **63**, K97 (1981)
43. A.N. Chakravarti, A.K. Choudhury, K.P. Ghatak, S. Ghosh, A. Dhar, Appl. Phys. **25**, 105 (1981)
44. A.N. Chakravarti, K.P. Ghatak, G.B. Rao, K.K. Ghosh, Phys. Stat. Sol. (b) **112**, 75 (1982)
45. A.N. Chakravarti, K.P. Ghatak, K.K. Ghosh, H.M. Mukherjee, Phys. Stat. Sol. (b) **116**, 17 (1983)
46. M. Mondal, K.P. Ghatak, Phys. Stat. Sol. (b) **133**, K143 (1984)
47. M. Mondal, K.P. Ghatak, Phys. Stat. Sol. (b) **126**, K47 (1984)
48. M. Mondal, K.P. Ghatak, Phys. Stat. Sol. (b) **126**, K41 (1984)

49. M. Mondal, K.P. Ghatak, Phys. Stat. Sol. (b) **129**, K745 (1985)
50. M. Mondal, K.P. Ghatak, Phys. Scr. **31**, 615 (1985)
51. M. Mondal, K.P. Ghatak, Phys. Stat. Sol. (b) **135**, 239 (1986)
52. M. Mondal, K.P. Ghatak, Phys. Stat. Sol. (b) **93**, 377 (1986)
53. M. Mondal, K.P. Ghatak, Phys. Stat. Sol. (b) **135**, K21 (1986)
54. M. Mondal, S. Bhattacharyya, K.P. Ghatak, Appl. Phys. A **42A**, 331 (1987)
55. S.N. Biswas, N. Chattopadhyay, K.P. Ghatak, Phys. Stat. Sol. (b) **141**, K47 (1987)
56. B. Mitra, K.P. Ghatak, Phys. Stat. Sol. (b) **149**, K117 (1988)
57. B. Mitra, A. Ghoshal, K.P. Ghatak, Phys. Stat. Sol. (b) **150**, K67 (1988)
58. M. Mondal, K.P. Ghatak, Phys. Stat. Sol. (b) **147**, K179 (1988)
59. M. Mondal, K.P. Ghatak, Phys. Stat. Sol. (b) **146**, K97 (1988)
60. B. Mitra, A. Ghoshal, K.P. Ghatak, Phys. Stat. Sol. (b) **153**, K209 (1989)
61. B. Mitra, K.P. Ghatak, Phys. Letts. **142A**, 401 (1989)
62. B. Mitra, A. Ghoshal, K.P. Ghatak, Phys. Stat. Sol. (b) **154**, K147 (1989)
63. B. Mitra, K.P. Ghatak, Sol. State Elect. **32**, 515 (1989)
64. B. Mitra, A. Ghoshal, K.P. Ghatak, Phys. Stat. Sol. (b) **155**, K23 (1989)
65. B. Mitra, K.P. Ghatak, Phys. Letts. **135A**, 397 (1989)
66. B. Mitra, K.P. Ghatak, Phys. Letts. A **146A**, 357 (1990)
67. B. Mitra, K.P. Ghatak, Phys. Stat. Sol. (b) **164**, K13 (1991)
68. S.N. Biswas, K.P. Ghatak, Int. J. Elect. **70**, 125 (1991)
69. P.R. Wallace, Phys. Stat. Sol. (b) **92**, 49 (1979)
70. B.R. Nag, *Electron Transport in Compound Semiconductors*, Springer Series in Solid-State Sciences, vol. 11 (Springer, Germany, 1980)
71. K.P. Ghatak, S. Bhattacharya, D. De, *Einstein Relation in Compound Semiconductors and Their Nanostructures*, Springer Series in Materials Science, vol. 116 (Springer, Germany, 2009)
72. C.C. Wu, C.J. Lin, J. Low, Temp. Phys. **57**, 469 (1984)
73. M.H. Chen, C.C. Wu, C.J. Lin, J. Low, Temp. Phys. **55**, 127 (1984)
74. E. Bangert, P. Kastner, Phys. Stat. Sol (b) **61**, 503 (1974)
75. G.M.T. Foley, P.N Langenberg. Phys. Rev. B **15B**, 4850 (1977)
76. M. Singh, P.R. Wallace, S.D. Jog, E. Arushanov, J. Phys. Chem. Solids **45**, 409 (1984)
77. Y. Yamada, Phys. Soc. Japan **35**, 1600 (1973)
78. Y. Yamada, Phys. Soc. Japan **37**, 606 (1974)
79. R.W. Cunningham, Phys. Rev. **167**, 761 (1968)
80. A.I. Yekimov, A.A. Onushchenko, A.G. Plyukhin, Al, L. Efros, J. Expt. Theor. Phys. **88**, 1490 (1985)
81. B.J. Roman, A.W. Ewald, Phys. Rev. B **5**, 3914 (1972)

Chapter 5
The EEM in Nanowires of Non-Parabolic Semiconductors

5.1 Introduction

It is well known that in nanowires (NWs), the restriction of the motion of the carriers along two directions may be viewed as carrier confinement by two infinitely deep one dimensional (1D) rectangular potential wells, along any two orthogonal directions leading to quantization of the wave vectors along the said directions, allowing 1D carrier transport [1–3]. With the help of modern fabrication techniques, such one dimensional quantized structures have been experimentally realized and enjoy an enormous range of important applications in the realm of nanoscience. They have generated much interest in the analysis of nanostructured devices for investigating their electronic, optical, and allied properties [4–11]. Examples of such new applications are based on the different transport properties of ballistic charge carriers which include nanoresistors [12–14], resonant tunneling diodes and band filters [15, 16], nanoswitches [17], nanosensors [18, 19], nanologic gates [20, 21], nanotransistors and subtuners [22, 23], heterojunction [24], high-speed digital networks [25–27], high-frequency microwave circuits [28], optical modulators [29], optical switching systems [30], and other nanoscale devices. In this chapter, we shall study the EEM in NWs of non-parabolic semiconductors having different band structures.

In Sect. 5.2.1, the EEM in NWs of nonlinear optical semiconductors has been investigated. In Sect. 5.2.2, the EEM in NWs of III–V, ternary and quaternary semiconductors has been studied in accordance with the said band models and the simplified results for wide-gap materials having parabolic energy bands under certain limiting conditions have further been demonstrated as a special case and thus confirming the compatibility test. The Sect. 5.2.3 contains the investigation of the EEM in NWs of II–VI compounds. In Sect. 5.2.4, the EEM in NWs of Bi has been formulated in accordance with the aforementioned energy band models for the purpose of relative assessment. Besides, under certain limiting conditions all the results for all the models of 1D systems are reduced to the well-known result of the EEM in

S. Bhattacharya and K. P. Ghatak, *Effective Electron Mass in Low-Dimensional Semiconductors*, Springer Series in Materials Science 167,
DOI: 10.1007/978-3-642-31248-9_5,

NWs of wide-gap materials. This above statement exhibits the compatibility test of our theoretical analysis. In Sect. 5.2.5, the EEM in NWs of IV–VI semiconductors has been studied taking PbTe, PbSe, and PbS as examples. The EEM in NWs of stressed compounds (taking stressed n-InSb as an example) has been investigated in Sect. 5.2.6. The Sect. 5.2.7 contains the investigation of EEM in NWs of Tellurium. In Sect. 5.2.8, the EEM in NWs of n-GaP has been studied. The Sect. 5.2.9 explores the EEM in NWs of $PtSb_2$. In Sect. 5.2.10, the EEM in NWs of Bi_2Te_3 has been considered. In Sect. 5.2.11, the EEM has been studied in NWs of Ge. In Sect. 5.2.12, the EEM in NWs of GaSb has been studied. In Sect. 5.2.13, we shall study the EEM in NWs of II–V semiconductors. The Sect. 5.2.14 explores the EEM in carbon nanotubes, a very important quantum material in nanotechnology. The Sect. 5.3 contains the result and discussions pertaining to this chapter. The last Sect. 5.4 contains open research problems.

5.2 Theoretical Background

5.2.1 The EEM in Nanowires of Nonlinear Optical Semiconductors

For two-dimensional (2D) quantizations along x and y directions, (1.2) assumes the form

$$k_z^2 = A_{11}(E, n_x, n_y) \tag{5.1}$$

where

$$A_{11}(E, n_x, n_y) = [f_2(E)]^{-1}[\gamma(E) - \phi_1(n_x, n_y) f_1(E)],$$
$$\phi_1(n_x, n_y) \equiv \left(\frac{n_x \pi}{d_x}\right)^2 + \left(\frac{n_y \pi}{d_y}\right)^2,$$

$n_x = (1, 2, 3, \ldots), n_y = (1, 2, 3, \ldots)$ are the size quantum numbers along x and y directions, respectively and d_x and d_y are the nanothickness along x and y directions, respectively.

The quantized sub-band energy (E_{11}) is given by

$$\gamma(E_{11}) = f_1(E_{11})\phi_1(n_x, n_y). \tag{5.2}$$

The EEM can be expressed as

$$m^*(E_{F1D}, n_x, n_y) = \frac{\hbar^2}{2} A'_{11}(E_{F1D}, n_x, n_y) \tag{5.3}$$

where E_{F1D} is the Fermi energy in the presence of 2D quantization as measured from the edge of the conduction band in the vertically upward direction in the absence of any quantization.

$$A'_{11}(E_{F1D}, n_x, n_y) = \left[\frac{-A_{11}(E_{F1D}, n_x, n_y) f'_2(E_{F1D})}{f_2(E_{F1D})} + [f_2(E_{F1D})]^{-1}[\gamma'(E_{F1D}) - f'_1(E_{F1D})\phi_1(n_x, n_y)]\right],$$

$$f'_2(E_{F1D}) = \left[[\hbar^2 E_{g0}(E_{g0} + \Delta_\parallel)]\left[2m^*_\parallel\left(E_{g0} + \frac{2}{3}\Delta_\parallel\right)\right]^{-1} \times \left[2E_{F1D} + 2E_{g0} + \frac{2}{3}\Delta_\parallel\right]\right],$$

$$f'_1(E_{F1D}) = \left[\left[\hbar^2 E_{g0}(E_{g0} + \Delta_\perp)\right]\left[2m^*_\perp\left(E_{g0} + \frac{2}{3}\Delta_\perp\right)\right]^{-1} \times \left[2E_{F1D} + 2E_{g0} + \frac{2}{3}\Delta_\parallel + \delta\right]\right]$$

and

$$\gamma'(E_{F1D}) = \left[\frac{\gamma(E_{F1D})(2E_{F1D} + E_{g0})}{E_{F1D}(E_{F1D} + E_{g0})} + [E_{F1D}(E_{F1D} + E_{g0})][2E_{F1D} + 2E_{g0} + \Delta_\parallel + \delta]\right].$$

Thus, we observe that the EMM is the function of both the size quantum numbers (n_x and n_y) and the Fermi energy due to the combined influence of the crystal filed splitting constant and the anisotropic spin–orbit splitting constants, respectively. The density-of-states function per sub-band ($N_{1D}(E)$) is given by,

$$N_{1D}(E) = \frac{g_v}{\pi}\left[\frac{\{\psi_1(E) - \psi_2(E)\phi(n_x, n_y)\}}{\psi_3(E)}\right]^{-1/2} [\psi_3(E)]^{-2}[\psi_3(E)\{\{\psi_1(E)\}' - \{\psi_2(E)\}'\phi(n_x, n_y)\} - \{\psi_1(E) - \psi_2(E)\phi(n_x, n_y)\}\{\psi_3(E)\}']. \tag{5.4}$$

The electron concentration per unit length can be expressed as

$$n_0 = \frac{2g_v}{\pi}\sum_{n_x=1}^{n_{x\max}}\sum_{n_y=1}^{n_{y\max}}[B_{11}(E_{F1D}, n_x, n_y) + B_{12}(E_{F1D}, n_x, n_y)] \tag{5.5}$$

where

$$B_{11}(E_{F1D}, n_x, n_y) = [A_{11}(E_{F1D}, n_x, n_y)]^{1/2},$$

$$B_{12}(E_{F1D}, n_x, n_y) = \sum_{r=1}^{r_0} Z_{1D}(r)\left[B_{11}(E_{F1D}, n_x, n_y)\right],$$

and

$$Z_{1D}(r) = 2(k_B T)^{2r}(1 - 2^{1-2r})\xi(2r)\frac{\partial^{2r}}{\partial E_{F1D}^{2r}}.$$

5.2.2 The EEM in Nanowires of III–V Semiconductors

The dispersion relation of the conduction electrons of III–V compounds are described by the models of Kane (both three and two bands) [31, 32], Stillman et al. [33], Newson and Kurobe [34] and Palik et al. [35] respectively. For the purpose of complete and coherent presentation, the EEM in NWs of III–V semiconductors have also been investigated in accordance with the aforementioned different dispersion relations for the purpose of relative comparison as follows:

(a) Under the substitutions $\delta = 0$, $\Delta_\parallel = \Delta_\perp = \Delta$ and $m_\parallel^* = m_\perp^* = m_c$ (5.1) assumes the form

$$\frac{\hbar^2 k_z^2}{2m_c} = I_{11}(E) - \frac{\hbar^2}{2m_c}\phi(n_x, n_y) \tag{5.6}$$

Using (5.6), the EMM along k_z direction for this case can be written as

$$m^*(E_{F1D}) = m_c\,\{I_{11}(E_{F1D})\}' \tag{5.7}$$

where

$$I'_{11}(E_{F1D}) = \left[I_{11}(E_{F1D})\left[\frac{1}{E_{F1D}} + \frac{1}{E_{F1D} + E_{g0}} + \frac{1}{E_{F1D} + E_{g0} + \Delta} - \frac{1}{E_{F1D} + E_{g0} + (2/3)\Delta}\right]\right].$$

It is worth noting that the EMM in this case is a function of Fermi energy alone and is independent of size quantum number.

The sub-band energy $(E_{n_{xy2}})$ can be written as

$$I_{11}(E_{n_{xy2}}) = \frac{\hbar^2}{2m_c}\phi(n_x, n_y). \tag{5.8}$$

The 1D carrier concentration can thus be written as

$$n_{1D} = \frac{2g_v}{\pi}\left(\frac{2m_c}{\hbar^2}\right)^{1/2} \sum_{n_x=1}^{n_{x\max}} \sum_{n_y=1}^{n_{y\max}} \left[T_{63}(E_{F1D}, n_x, n_y) + T_{64}(E_{F1D}, n_x, n_y)\right] \tag{5.9}$$

where

$$T_{63}(E_{F1D}, n_x, n_y) \equiv \left[I_{11}(E_{F1D}) - \frac{\hbar^2}{2m_c}\phi(n_x, n_y)\right]^{1/2}$$

and

$$T_{64}(E_{F1D}, n_x, n_y) \equiv \sum_{r=1}^{s} Z_{1D}(r) T_{63}(E_{F1D}, n_x, n_y).$$

(b) Under the inequalities $\Delta \gg E_{g0}$ or $\Delta \ll E_{g0}$, (5.6) assumes the form

$$E(1 + \alpha E) = \frac{\hbar^2}{2m_c}\phi(n_x, n_y) + \frac{\hbar^2 k_z^2}{2m_c}. \tag{5.10}$$

The EMM along k_z direction can be written as

$$m^*(E_{F1D}) = m_c(1 + 2\alpha E_{F1D}). \tag{5.11}$$

Thus, we observe that the EMM in the present case is a function of Fermi energy only due to the presence of band non-parabolicity.

For NWs, whose energy band structures for the corresponding bulk semiconductors obey the two-band model of Kane, the density-of-states function per sub-band assumes the form

$$N_{1D}(E) = \frac{g_v}{\pi}\left(\frac{2m_c}{\hbar^2}\right)^{1/2} \frac{(1 + 2\alpha E)}{\left[E(1 + \alpha E) - \frac{\hbar^2}{2m_c}\phi(n_x, n_y)\right]^{1/2}}. \tag{5.12}$$

In this case the sub-band energy ($E_{n_{xy3}}$) can be expressed as

$$\frac{\hbar^2}{2m_c}\phi(n_x, n_y) = E_{n_{xy3}}(1 + \alpha E_{n_{xy3}}). \tag{5.13}$$

The use of (5.12) leads to the expression of the 1D electron statistics as

$$n_{1D} = \frac{2g_v}{\pi}\left(\frac{2m_c}{\hbar^2}\right)^{1/2} \sum_{n_x=1}^{n_{x\max}} \sum_{n_y=1}^{n_{y\max}} \left[T_{65}(E_{F1D}, n_x, n_y) + T_{66}(E_{F1D}, n_x, n_y)\right] \tag{5.14}$$

where

$$T_{65}(E_{F1D}, n_x, n_y) \equiv \left[E_{F1D}(1 + \alpha E_{F1D}) - \frac{\hbar^2}{2m_c}\phi(n_x, n_y)\right]^{1/2}$$

and

$$T_{66}(E_{F1D}, n_x, n_y) \equiv \sum_{r=1}^{s} Z_{1D}(r) T_{65}(E_{F1D}, n_x, n_y).$$

Under the condition, $\alpha E_{F1D} \ll 1$, the expressions of the 1D electron statistics can be written as

$$\begin{aligned} n_{1D} = {} & \frac{2g_v\sqrt{2m_c\pi k_B T}}{h} \sum_{n_x=1}^{n_{x\max}} \sum_{n_y=1}^{n_{y\max}} \frac{1}{\sqrt{i_1}} \\ & \times \left[\left(1 + \frac{3}{2}\alpha i_2\right) F_{-1/2}(\eta_6) + \frac{3}{4}\alpha k_B T F_{1/2}(\eta_6)\right] \end{aligned} \tag{5.15}$$

where

$$i_1 \equiv \left[1 + \alpha \frac{\hbar^2}{2m_c}\phi(n_x, n_y)\right], i_2 \equiv \left(\frac{\hbar^2}{2m_c}\right)\phi(n_x, n_y)(i_1)^{-1}$$

and

$$\eta_6 \equiv (E_{F_1} - i_2)/k_B T.$$

(c) Under the condition $\alpha \to 0$, the expression of n_{1D} for NWs of isotropic parabolic energy bands can be written from (5.15) as

$$\begin{aligned} n_{1D} &= \frac{2g_v\sqrt{2\pi m_c k_B T}}{h} \sum_{n_x=1}^{n_{x\max}} \sum_{n_y=1}^{n_{y\max}} [F_{-1/2}(\eta_7)], \\ \eta_7 &\equiv \left(\frac{1}{k_B T}\right)[E_{F1D} - \{(\hbar^2/2m_c)\phi(n_x, n_y)\}] \end{aligned} \tag{5.16}$$

(d) *The model of Stillman et al.*

In accordance with the model of Stillman et al. [33], the electron dispersion law of NWs of 1D III–V materials assumes the form

$$k_z^2 = A_{14}(E, n_x, n_y) \tag{5.17}$$

where

$$A_{14}(E, n_x, n_y) = \left[\frac{2m_c}{\hbar^2}\{I_{12}(E)\} - \phi_1(n_x, n_y)\right].$$

The EEM in this case assume the from

$$m^*(E_{F1D}) = m_c I'_{12}(E_{F1D}) \tag{5.18}$$

where

$$I'_{12}(E_{F1D}) = \left(\frac{a_{11}a_{12}}{2}\right)(1 - a_{12}E_{F1D})^{-1/2}.$$

The quantized sub-band energy (E_{14}) is given by

$$I_{12}(E_{14}) = \left[\frac{\hbar^2}{2m_c}\right]\phi_1(n_x, n_y). \tag{5.19}$$

The electron concentration per unit length can be expressed as

$$n_0 = \frac{2g_v}{\pi}\sum_{n_x=1}^{n_{x\max}}\sum_{n_y=1}^{n_{y\max}}\left[B_{17}(E_{F1D}, n_x, n_y) + B_{18}(E_{F1D}, n_x, n_y)\right] \tag{5.20}$$

where

$$B_{17}(E_{F1D}, n_x, n_y) = \left[A_{14}(E_{F1D}, n_x, n_y)\right]^{1/2}$$

and

$$B_{18}(E_{F1D}, n_x, n_y) = \sum_{r=1}^{r_0} Z_{1D}(r)[B_{17}(E_{F1D}, n_x, n_y)].$$

(e) *The model of Newson and Kurobe*
(f) In accordance with the model of Newson and Kurobe [34], the electron dispersion law in this case assumes the form

$$E = a_{13}k_z^4 + \left[\frac{\hbar^2}{2m_c} + a_{14}k_s^2\right]k_z^2 + \frac{\hbar^2}{2m_c}k_s^2 + a_{14}k_x^2k_y^2 + a_{13}\left(k_x^4 + k_y^4\right) \tag{5.21}$$

where a_{13} is the non-parabolicity constant, $a_{14}(\equiv 2a_{13} + a_{15})$ and a_{15} is known as the warping constant.

The 1D E-k_z relation can be expressed as

$$k_z^2 = A_{15}(E, n_x, n_y) \tag{5.22}$$

where

$$A_{15}(E, n_x, n_y) = (2a_{13})^{-1}[-\overline{L_1}(n_x, n_y) + [\{\overline{L_1}(n_x, n_y)\}^2 - 4a_{13}[\overline{L_2}(n_x, n_y) - E]^{1/2}],$$

$$\overline{L_1}(n_x, n_y) = \frac{\hbar^2}{2m_c} + a_{14}\left[\left(\frac{n_x\pi}{d_x}\right)^2 + \left(\frac{n_y\pi}{d_y}\right)^2\right] \quad \text{and}$$

$$\overline{L_2}(n_x, n_y) = \left[\frac{\hbar^2}{2m_c}\phi_1(n_x, n_y) + a_{14}\left(\frac{n_x\pi}{d_x}\cdot\frac{n_y\pi}{d_y}\right)^2 + a_{13}\left[\left(\frac{n_x\pi}{d_x}\right)^4 + \left(\frac{n_y\pi}{d_y}\right)^4\right]\right].$$

The EEM can be written from (5.22) as

$$m^*(E_{F1D}, n_x, n_y) = \frac{\hbar^2}{2}A'_{15}(E_{F1D}, n_x, n_y) \tag{5.23}$$

where

$$A'_{15}(E_{F1D}, n_x, n_y) = [\{\overline{L_1}(n_x, n_y)\}^2 - 4a_{13}[\overline{L_2}(n_x, n_y) - E_{F1D}]^{-1/2}].$$

The mass is a function of quantum numbers in addition to Fermi energy due to band non-parabolicity.

The quantized sub-band energy (E_{16}) is given by

$$E_{16} = \overline{L_2}(n_x, n_y). \tag{5.24}$$

The electron concentration per unit length can be written as

$$n_0 = \frac{2g_v}{\pi}\sum_{n_x=1}^{n_{x\max}}\sum_{n_y=1}^{n_{y\max}}\left[B_{19}(E_{F1D}, n_x, n_y) + B_{20}(E_{F1D}, n_x, n_y)\right] \tag{5.25}$$

where

$$B_{19}(E_{F1D}, n_x, n_y) = [A_{15}(E_{F1D}, n_x, n_y)]^{1/2}$$

and

$$B_{20}(E_{F1D}, n_x, n_y) = \sum_{r=1}^{r_0} Z_{1D}(r)[B_{19}(E_{F1D}, n_x, n_y)].$$

(g) *Model of Palik et al.*

The energy spectrum of the conduction electrons in NWs of III–V semiconductors up to the fourth order in effective mass theory, taking into account the interactions of heavy hole, light hole, and the split-off holes can be expressed in accordance with the model of Palik et al. [35] and following (1.36) as

The 1D E-k_z relation can be written as

$$k_z^2 = A_{16}(E, n_x, n_y) \tag{5.26}$$

where

$$A_{16}(E, n_x, n_y) = \left[\frac{2m_c}{\hbar^2}\{I_{13}(E)\} - \phi_1(n_x, n_y)\right].$$

The EEM can be written from (5.26) as

$$m^*(E_{F1D}) = m_c I'_{13}(E_{F1D}) \tag{5.27}$$

where

$$I'_{13}(E_{F1D}) = 2\bar{b}_{12}\bar{B}_{11}\left[(\bar{a}_{12})^2 - 4E_{F1D}\bar{B}_{11}\right]^{-1/2}.$$

The electron concentration per unit length can be expressed as

$$n_0 = \frac{2g_v}{\pi}\sum_{n_x=1}^{n_{x\max}}\sum_{n_y=1}^{n_{y\max}}\left[B_{21}(E_{F1D}, n_x, n_y) + B_{22}(E_{F1D}, n_x, n_y)\right] \tag{5.28}$$

where

$$B_{21}(E_{F1D}, n_x, n_y) = [A_{16}(E_{F1D}, n_x, n_y)]^{1/2}$$

and

$$B_{22}(E_{F1D}, n_x, n_y) = \sum_{r=1}^{r_0} Z_{1D}(r)\left[B_{21}(E_{F1D}, n_x, n_y)\right].$$

5.2.3 The EEM in Nanowires of II–VI Semiconductors

The 1D dispersion relation for NWs of II–VI semiconductors can be expressed following (1.42) as

$$E = b'_0 k_z^2 + G_{3,\pm}(n_x, n_y) \tag{5.29}$$

where

$$G_{3,\pm}(n_x,n_y)\equiv\left[a_0'\left\{\left(\frac{\pi n_x}{d_x}\right)^2+\left(\frac{\pi n_y}{d_y}\right)^2\right\}\pm\bar{\lambda}_0\left\{\left(\frac{\pi n_x}{d_x}\right)^2+\left(\frac{\pi n_y}{d_y}\right)^2\right\}^{1/2}\right].$$

The EEM can be written from (5.29) as

$$m^*(E_{F1D})=m_{\parallel}^*. \tag{5.30}$$

From (5.30), it appears that the EEM is constant in this case.

The 1D electron statistics can be written as

$$n_{1D}=\frac{g_v}{\pi\sqrt{b_0'}}\sum_{n_x=1}^{n_{x\max}}\sum_{n_y=1}^{n_{y\max}}\left[t_7(E_{F1D},n_x,n_y)+t_8(E_{F1D},n_x,n_y)\right] \tag{5.31}$$

where

$$t_7(E_{F1D},n_x,n_y)\equiv[E_{F1D}-[G_{3,+}(n_x,n_y)]]^{1/2}+[E_{F1D}-[G_{3,-}(n_x,n_y)]]^{1/2}\quad\text{and}$$

$$t_8(E_{F1D},n_x,n_y)=\sum_{r=1}^{r_0}Z_{1D}(r)\left[t_7(E_{F1D},n_x,n_y)\right]$$

5.2.4 The EEM in Nanowires of Bismuth

(a) *The McClure and Choi model*

The dispersion relation of the carriers in NWs of Bi can be written in accordance with the McClure and Choi and following (1.49) as

$$E(1+\alpha E)=\left\{\frac{\hbar^2k_x^2}{2m_1}\left[1-\frac{\alpha\hbar^2}{2m_2}\left(\frac{\pi n_y}{d_y}\right)^2\right]+G_{12}+\frac{\hbar^2}{2m_2}\alpha E\left\{1-\left(\frac{m_2}{m_2'}\right)\right\}\left(\frac{\pi n_y}{d_y}\right)^2\right\} \tag{5.32}$$

where

$$G_{12}\equiv\left\{\frac{\hbar^2}{2m_2}\left(\frac{\pi n_y}{d_y}\right)^2+\frac{\hbar^2}{2m_3}\left(\frac{\pi n_z}{d_z}\right)^2+\frac{\alpha\hbar^4}{4m_2m_2'}\left(\frac{\pi n_y}{d_y}\right)^4-\frac{\alpha}{4m_2m_3}\left(\frac{\hbar^2n_yn_z\pi^2}{d_yd_z}\right)^2\right\}.$$

Following (5.32), the EEM in this case assumes the from

$$m^*(E_{F1D}, n_y) = m_1 \left\{ \left[1 - \frac{\alpha\hbar^2}{2m_2} \left(\frac{\pi n_y}{d_y} \right)^2 \right]^{-1} \right.$$
$$\left. \times \left[1 + 2\alpha E_{F1D} - \frac{\hbar^2}{2m_2} \alpha \left\{ 1 - \left(\frac{m_2}{m_2'} \right) \right\} \left(\frac{\pi n_y}{d_y} \right)^2 \right] \right\}. \tag{5.33}$$

Thus, EEM in this case is a function of Fermi energy and the size quantum number n_y due to the presence of band non-parabolicity only.

Using (5.32), the 1D electron statistics can be expressed as

$$n_{1D} = \frac{2g_v}{\pi} \frac{\sqrt{2m_1}}{\hbar} \sum_{n_y=1}^{n_{y\max}} \sum_{n_z=1}^{n_{z\max}} \left[t_{27}(E_{F1D}, n_y, n_z) + t_{28}(E_{F1D}, n_y, n_z) \right] \tag{5.34}$$

where

$$t_{27}(E_{F1D}, n_y, n_z) \equiv \left\{ \left[1 - \frac{\alpha\hbar^2}{2m_2} \left(\frac{\pi n_y}{d_y} \right)^2 \right]^{-1/2} \left[E_{F1D}(1 + \alpha E_{F1D}) \right. \right.$$
$$\left. \left. - G_{12} - \frac{\hbar^2}{2m_2} \alpha E_{F1D} \left\{ 1 - \left(\frac{m_2}{m_2'} \right) \right\} \left(\frac{\pi \hbar n_y}{d_y} \right)^2 \right]^{1/2} \right\}$$

and

$$t_{28}(E_{F1D}, n_y, n_z) \equiv \sum_{r=1}^{s_o} Z_{1D}(r) \left[t_{27}(E_{F1D}, n_y, n_z) \right].$$

(b) *The Hybrid Model*

Following (1.58), the 1D dispersion relation in this case assumes the form

$$E(1 + \alpha E) = \frac{\hbar^2 k_x^2}{2m_1} + G_{14} + \frac{\hbar^2}{2M_2} \left(\frac{\pi n_y}{d_y} \right)^2 \alpha E(1 - \gamma_0) \tag{5.35}$$

where

$$G_{14} = \left[\frac{\hbar^2}{2m_3} \left(\frac{\pi n_z}{d_z} \right)^2 + \frac{\hbar^2}{2M_2} \left(\frac{\pi n_y}{d_y} \right)^2 (1 + \bar{\delta}_0) + \frac{\alpha \gamma_0 \hbar^4}{4M_2^2} \left(\frac{\pi n_y}{d_y} \right)^4 \right].$$

Using this (5.35), the EEM can be expressed as

$$m^*(E_{F1D}, n_y) = m_1 \left\{ \left[1 + 2\alpha E_{F1D} - \frac{\hbar^2}{2M_2}\alpha \{1 - (\gamma_0)\} \left(\frac{\pi n_y}{d_y} \right)^2 \right] \right\}. \quad (5.36)$$

Thus, EEM in this case is a function of Fermi energy and the size quantum number n_y due to the presence of band non-parabolicity only.

The use of (5.35) leads to the expression for the electron concentration per unit length as

$$n_{1D} = \frac{2g_v}{\pi} \frac{\sqrt{2m_1}}{\hbar} \sum_{n_y=1}^{n_{y\max}} \sum_{n_z=1}^{n_{z\max}} \left[t_{31}(E_{F1D}, n_y, n_z) + t_{32}(E_{F1D}, n_y, n_z) \right] \quad (5.37)$$

where

$$t_{31}(E_{F1D}, n_y, n_z) \equiv \left[E_{F1D}(1 + \alpha E_{F1D}) - G_{14} - \frac{\hbar^2}{2M_2} \left(\frac{\pi n_y}{d_y} \right)^2 \alpha E_{F1D}(1 - \gamma_0) \right]^{1/2},$$

and

$$t_{32}(E_{F1D}, n_y, n_z) \equiv \sum_{r=1}^{s_o} Z_{1D}(r) \left[t_{31}(E_{F1D}, n_y, n_z) \right].$$

(c) *The Cohen model*

The 1D carrier dispersion law in this case can be written following (1.65) as

$$\alpha E^2 + E l_7 - G_{15} = \frac{\hbar^2 k_x^2}{2m_1} \quad (5.38)$$

where

$$l_7 = \left[1 - \frac{\alpha\hbar^2}{2m_2} \left(\frac{\pi n_y}{d_y} \right)^2 + \frac{\alpha\hbar^2}{2m_2'} \left(\frac{\pi n_y}{d_y} \right)^2 \right]$$

and

$$G_{15} = \left[\frac{\hbar^2}{2m_3} \left(\frac{\pi n_z}{d_z} \right)^2 + \frac{\hbar^2}{2m_2} \left(\frac{\pi n_y}{d_y} \right)^2 + \frac{\alpha\hbar^4}{4m_2 m_2'} \left(\frac{\pi n_y}{d_y} \right)^4 \right].$$

Using this (5.38), the EEM can be expressed as

$$m^*(E_{F1D}, n_y) = m_1 \{[1 + 2\alpha E_{F1D} + l_7]\}. \quad (5.39)$$

Thus, EEM in this case is a function of Fermi energy and the size quantum number n_y due to the presence of band non-parabolicity only.

The 1D electron concentration per unit length assumes the form

$$n_{1D} = \frac{2g_v}{\pi}\frac{\sqrt{2m_1}}{\hbar}\sum_{n_y=1}^{n_{y\max}}\sum_{n_z=1}^{n_{z\max}}\left[t_{35}(E_{F1D}, n_y, n_z) + t_{36}(E_{F1D}, n_y, n_z)\right] \quad (5.40)$$

where

$$t_{35}(E_{F1D}, n_y, n_z) = [\alpha E_{F1D}^2 + E_{F1D} l_7 - G_{15}]^{1/2}$$

and

$$t_{36}(E_{F1D}, n_y, n_z) = \sum_{r=1}^{s_o} Z_{1D}(r)\left[t_{35}(E_{F1D}, n_y, n_z)\right].$$

(d) *The Lax model*

The 1D dispersion relation in this case can be expressed following (1.71) as

$$E(1 + \alpha E) = \frac{\hbar^2 k_x^2}{2m_1} + G_{16}. \quad (5.41)$$

Using (5.41), the EEM can be expressed as

$$m^*(E_{F1D}) = m_1\{[1 + 2\alpha E_{F1D}]\}. \quad (5.42)$$

Thus, EEM in this case is a function of Fermi energy and is independent of the size quantum number n_y due to the presence of band non-parabolicity only.

The 1D electron statistics is given by

$$n_{1D} = \frac{2g_v}{\pi}\frac{\sqrt{2m_1}}{\hbar}\sum_{n_y=1}^{n_{y\max}}\sum_{n_z=1}^{n_{z\max}}\left[t_{37}(E_{F1D}, n_y, n_z) + t_{38}(E_{F1D}, n_y, n_z)\right] \quad (5.43)$$

where

$$t_{37}(E_{F1D}, n_y, n_z) = [E_{F1D}(1 + \alpha E_{F1D}) - G_{16}]^{1/2}$$

and

$$t_{38}(E_{F1D}, n_y, n_z) = \sum_{r=1}^{s_o} Z_{1D}(r)[t_{37}(E_{F1D}, n_y, n_z)].$$

It may be noted that under the conditions $\alpha \to 0$, $M_2' \to \infty$ and isotropic effective electron mass at the edge of the conduction band, m all models of Bismuth convert into isotropic parabolic energy bands leading to the confirmatory test.

5.2.5 The EEM in Nanowires of IV–VI Semiconductors

The 1D dispersion relation in this case in accordance with Dimmock model can be expressed from (1.83) as

$$k_z^2 = A_{23}(E, n_x, n_y) \tag{5.44}$$

where

$$A_{23}(E, n_x, n_y) = (2h_4)^{-1}[h_6(E, n_x, n_y) - [h_6^2(E, n_x, n_y) + 4h_4 h_7(E, n_x, n_y)]^{1/2}], \quad h_4 = \left[\frac{\alpha \hbar^4}{4x_3 x_6}\right],$$

$$x_3 = \frac{3m_t^- m_l^-}{2m_l^- + m_t^-}, \quad x_6 = \frac{3m_t^+ m_l^+}{2m_l^+ + m_t^+},$$

$$h_6(E, n_x, n_y) = \left[\frac{\alpha E \hbar^2}{2x_6} - \frac{\alpha \hbar^2}{2x_6}\left[\left(\frac{\pi n_x}{d_x}\right)^2 \frac{\hbar^2}{2x_1} + \left(\frac{\pi n_y}{d_y}\right)^2 \frac{\hbar^2}{2x_2}\right] - \frac{\alpha \hbar^2}{2x_3}\left[\left(\frac{\pi n_x}{d_x}\right)^2 \frac{\hbar^2}{2x_4} + \left(\frac{\pi n_y}{d_y}\right)^2 \frac{\hbar^2}{2x_5}\right] - \frac{\hbar^2}{2m_3} - \frac{(1+\alpha E)\hbar^2}{2x_3}\right],$$

$$x_1 = m_t^-, \quad x_2 = \frac{m_t^- + 2m_l^-}{3}, \quad x_4 = m_t^+, \quad x_5 = \frac{m_t^+ + 2m_l^+}{3},$$

$$m_3 = \frac{3m_t^* m_l^*}{m_t^* + 2m_l^*},$$

$$h_7(E, n_x, n_y) = \left[E(1+\alpha E) + \alpha E\left[\left(\frac{\pi n_x}{d_x}\right)^2 \frac{\hbar^2}{2x_4} + \left(\frac{\pi n_y}{d_y}\right)^2 \frac{\hbar^2}{2x_5}\right] - (1+\alpha E)\left[\left(\frac{\pi n_x}{d_x}\right)^2 \frac{\hbar^2}{2x_1} + \left(\frac{\pi n_y}{d_y}\right)^2 \frac{\hbar^2}{2x_2}\right] - \alpha\left[\left(\frac{\pi n_x}{d_x}\right)^2 \frac{\hbar^2}{2x_1} + \left(\frac{\pi n_y}{d_y}\right)^2 \frac{\hbar^2}{2x_2}\right] \times \left[\left(\frac{\pi n_x}{d_x}\right)^2 \frac{\hbar^2}{2x_4} + \left(\frac{\pi n_y}{d_y}\right)^2 \frac{\hbar^2}{2x_5}\right]\right]$$

$$-\left[\left(\frac{\pi n_x}{d_x}\right)^2\frac{\hbar^2}{2m_1}+\left(\frac{\pi n_y}{d_y}\right)^2\frac{\hbar^2}{2m_2}\right]\right],$$

$$m_1=m_t^* \quad \text{and} \quad m_2=\frac{m_t^*+2m_l^*}{3}.$$

The EEM can be written from (5.44) as

$$m^*(E_{F1D},n_x,n_y)=\frac{\hbar^2}{2}A'_{23}(E_{F1D},n_x,n_y) \tag{5.45}$$

where

$$A'_{23}(E_{F1D},n_x,n_y)$$
$$=(2h_4)^{-1}\left[h'_6-\frac{h_6(E_{F1D},n_x,n_y)h'_6+2h_4h'_7(E_{F1D},n_x,n_y)}{\left[h_6^2(E_{F1D},n_x,n_y)+4h_4h_7(E_{F1D},n_x,n_y)\right]^{1/2}}\right]$$
$$h'_6=\frac{\alpha\hbar^2}{2}\left(\frac{1}{x_6}-\frac{1}{x_3}\right)$$

and

$$h'_7(E_{F1D},n_x,n_y)=\left[1+2\alpha E_{F1D}+\alpha\left[\frac{\hbar^2}{2x_4}\left(\frac{\pi n_x}{d_x}\right)^2+\left(\frac{\pi n_y}{d_y}\right)^2\frac{\hbar^2}{2x_5}\right]\right.$$
$$\left.-\alpha\left[\frac{\hbar^2}{2x_1}\left(\frac{\pi n_x}{d_x}\right)2+\left(\frac{\pi n_y}{d_y}\right)^2\frac{\hbar^2}{2x_2}\right]\right].$$

The mass is a function of quantum numbers in addition to Fermi energy due to band non-parabolicity.

The electron concentration is given by

$$n_0=\frac{2g_v}{\pi}\sum_{n_x=1}^{n_{x\max}}\sum_{n_y=1}^{n_{y\max}}\left[B_{32}(E_{F1D},n_x,n_y)+B_{33}(E_{F1D},n_x,n_y)\right] \tag{5.46}$$

where

$$B_{32}(E_{F1D},n_x,n_y)=\left[A_{23}(E_{F1D},n_x,n_y)\right]^{1/2}$$

and

$$B_{33}(E_{F1D},n_x,n_y)=\sum_{r=1}^{r_0}Z_{1D}(r)\left[B_{32}(E_{F1D},n_x,n_y)\right].$$

5.2.6 The EEM in Nanowires of Stressed Semiconductors

The 1D dispersion relation of the carriers in stressed materials in this case can be written following (1.98) as

$$k_z^2 = A_{24}(E, n_x, n_y) \tag{5.47}$$

where

$$A_{24}(E, n_x, n_y) = \left[c^*(E)\right]^2 \left[1 - \left(\frac{\pi n_x}{d_x}\right)^2 \left[a^*(E)\right]^{-2} - \left(\frac{\pi n_y}{d_y}\right)^2 \left[b^*(E)\right]^{-2}\right].$$

The EEM can be written from (5.47) as

$$m^*(E_{F1D}, n_x, n_y) = \frac{\hbar^2}{2} A'_{24}(E_{F1D}, n_x, n_y) \tag{5.48}$$

where

$$\begin{aligned} A'_{24}(E_{F1D}, n_x, n_y) &= \left[\frac{K'_0(E_{F1D})}{L_0(E_{F1D})} - \frac{K_0(E_{F1D})L'_0(E_{F1D})}{L_0^2(E_{F1D})} \right. \\ &\quad - \left(\frac{n_x\pi}{d_x}\right)^2 \frac{\left[M'_0(E_{F1D}) + \frac{1}{2}N'_0(E_{F1D})\right]}{L_0(E_{F1D})} \\ &\quad + \left(\frac{n_x\pi}{d_x}\right)^2 \cdot \frac{L'_0(E_{F1D})}{L_0^2(E_{F1D})}\left[M_0(E_{F1D}) + \frac{1}{2}N_0(E_{F1D})\right] \\ &\quad + \left(\frac{n_y\pi}{d_y}\right)^2 \cdot \frac{L'_0(E_{F1D})}{L_0^2(E_{F1D})}\left[M_0(E_{F1D}) - \frac{1}{2}N_0(E_{F1D})\right] \\ &\quad \left. - \left(\frac{n_y\pi}{d_y}\right)^2 \cdot \frac{1}{L_0(E_{F1D})}\left[M'_0(E_{F1D}) - \frac{1}{2}N'_0(E_{F1D})\right]\right], \end{aligned}$$

$$\begin{aligned} K'_0(E_{F1D}) &= \left[\left[1 + \frac{2C_2^2\varepsilon_{xy}^2}{3\left\{E'_{g0}(E_{F1D})\right\}^2}\right]\left(\frac{3E'_{g0}(E_{F1D})}{2B_2^2}\right)\right. \\ &\quad \left. + \left[E_{F1D} - C_1\varepsilon - \frac{2C_2^2\varepsilon_{xy}^2}{3E'_{g0}(E_{F1D})}\right]\left(\frac{3}{2B_2^2}\right)\right], \end{aligned}$$

$$M_0'(E_{F1D}) = \left[\frac{(\bar{a}_0 + C_1)\varepsilon}{(E'_{g_0}(E_{F1D}))^2} - \frac{3\bar{b}_0\varepsilon_{xx}}{2(E'_{g_0}(E_{F1D}))^2} + \frac{\bar{b}_0\varepsilon}{2(E'_{g_0}(E_{F1D}))^2}\right] \text{ and}$$

$$N_0'(E_{F1D}) = -\left(\bar{d}_0\sqrt{3}\right)\frac{\varepsilon_{xy}}{(E'_{g_0}(E_{F1D}))^2}.$$

The mass is a function of quantum numbers in addition to Fermi energy due to stress. The sub-band energy E_{23} assumes the form

$$\left[\left(\frac{\pi n_x}{d_x}\right)^2 \left[a^*(E_{23})\right]^{-2} + \left(\frac{\pi n_y}{d_y}\right)^2 \left[b^*(E_{23})\right]^{-2}\right] = 1. \tag{5.49}$$

Using (5.47), the 1D electron statistics can be expressed as

$$n_{1D} = \frac{2g_v}{\pi}\sum_{n_y=1}^{n_{y\max}}\sum_{n_z=1}^{n_{z\max}}\left[B_{34}(E_{F1D}, n_y, n_z) + B_{35}(E_{F1D}, n_y, n_z)\right] \tag{5.50}$$

where

$$B_{34}(E_{F1D}, n_y, n_z) = \sqrt{A_{24}(E_{F1D}, n_x, n_y)}$$

and

$$B_{35}(E_{F1D}, n_x, n_y) = \sum_{r=1}^{s_o} Z_{1D}(r)\left[B_{35}(E_{F1D}, n_x, n_y)\right].$$

5.2.7 The EEM in Nanowires of Tellurium

From (1.105), the 1D dispersion relation can be written in accordance with the model of Bouat et al. as

$$k_z^2 = A_{25,\pm}(E, n_x, n_y) \tag{5.51}$$

where

$$A_{25,\pm}(E, n_x, n_y) = [(\psi_5(E) - \psi_6 k_s^2 \pm \psi_7[\psi_8(E) - k_s^2]^{1/2})],$$

$$k_s^2 = \phi_1(n_x, n_y),\ \psi_5(E) = \left[\frac{E}{\psi_1} + \frac{E}{2\psi_1^2}\right],\quad \psi_6 = \frac{\psi_2}{\psi_1},$$

$$\psi_7 = (2\psi_1^2)^{-1}[4\psi_3^2\psi_1\psi_2 - 4\psi_1^2\psi_4^2]^{1/2}$$

and

$$\psi_8(E) = \left[\frac{\psi_4^4 + 4E\psi_3^2\psi_1}{4\psi_3^2\psi_1\psi_2 - 4\psi_1^2\psi_4^2} \right].$$

The EEM can be written from (5.47) as

$$m^*(E_{F1D}, n_x, n_y) = \frac{\hbar^2}{2} A'_{25,\pm}(E_{F1D}, n_x, n_y) \tag{5.52}$$

where

$$A'_{25,\pm}(E_{F1D}, n_x, n_y) = \left[\frac{1}{\psi_1} \pm \frac{\psi_7}{2}[\psi_8(E_{F1D}) - \phi_1(n_x, n_y)]^{-1/2} \psi'_8(E_{F1D}) \right]$$

and

$$\psi'_8(E_{F1D}) = \left[\frac{4\psi_3^2\psi_1}{4\psi_3^2\psi_1\psi_2 - 4\psi_1^2\psi_4^2} \right].$$

Thus, the EEM is the function of the Fermi energy and the size quantum numbers which is the characteristic feature of such model.

The sub-band energies are given by

$$E_{26,\pm} = \psi_2\phi_1(n_x, n_y) \pm \psi_4 \left(\phi_1(n_x, n_y)\right)^{1/2}. \tag{5.53}$$

The electron concentration per unit length can be expressed as

$$n_0 = \frac{g_v}{\pi} \sum_{n_x=1}^{n_{x\max}} \sum_{n_y=1}^{n_{y\max}} \left[B_{36,\pm}(E_{F1D}, n_x, n_y) + \theta_{5,\pm} \right] \tag{5.54}$$

where

$$B_{36,\pm}(E_{F1D}, n_x, n_y) = \sqrt{A_{25,+}(E_{F1D}, n_x, n_y)} + \sqrt{A_{25,-}(E_{F1D}, n_x, n_y)}$$

and

$$\theta_{5,\pm} = \sum_{r=1}^{s_o} Z_{1D}(r) \left[B_{36,+}(E_{F1D}, n_x, n_y) + B_{36,-}(E_{F1D}, n_x, n_y) \right].$$

5.2.8 The EEM in Nanowires of Gallium Phosphide

The 1D dispersion relation in this case following (1.111) can be written in accordance with the model of Rees et al. as

$$k_x^2 = A_{26}(E, n_y, n_z) \tag{5.55}$$

where

$$A_{26}(E, n_y, n_z) = \Bigg[(2a^2)^{-1}\{2a(E - t_1) + D - \sqrt{[2a(E - t_1) + D]^2 - 4a^2[(E - t_1)^2 - t_2]}\} - \left(\frac{n_y \pi}{d_y}\right)^2 \Bigg],$$

t_1, D and t_2 have already been defined in connection with (1.112).

The EEM can be written from (5.55) as

$$m^*(E_{F1D}, n_y, n_z) = \frac{\hbar^2}{2} A'_{26}(E_{F1D}, n_y, n_z) \tag{5.56}$$

where

$$A'_{26}(E_{F1D}, n_y, n_z) = \frac{1}{a^2}\left[a - \frac{2a(E_{F1D} - t_1) + D - 4a^2(E_{F1D} - t_1)}{\sqrt{[2a(E_{F1D} - t_1) + D]^2 - 4a^2\left[(E_{F1D} - t_1)^2 - t_2\right]}} \right].$$

Thus, the mass is a function of Fermi energy and the size quantum numbers which are the characteristic features of such model.

The sub-band energy E_{27} can be written as

$$A_{26}(E_{27}, n_y, n_z) = 0. \tag{5.57}$$

The electron concentration per unit length can be expressed as

$$n_0 = \frac{2g_v}{\pi} \sum_{n_x=1}^{n_{x\max}} \sum_{n_y=1}^{n_{y\max}} \left[B_{38}(E_{F1D}, n_y, n_z) + B_{39}(E_{F1D}, n_y, n_z) \right] \tag{5.58}$$

where

$$B_{38}(E_{F1D}, n_y, n_z) = \sqrt{A_{26}(E_{F1D}, n_y, n_z)}$$

and

$$B_{39}(E_{F1D}, n_y, n_z) = \sum_{r=1}^{s_o} Z_{1D}(r)\left[B_{38}(E_{F1D}, n_y, n_z)\right].$$

5.2.9 The EEM in Nanowires of Platinum Antimonide

The 1D dispersion relation in this case can be written following (1.118) as

$$k_x^2 = \left[2\overline{A_9}\right]^{-1}\left[-\overline{A_{10}}(E, n_z) + \left[\overline{A_{10}}^2(E, n_z) + 4(\overline{A_9})\overline{A_{11}}(E, n_z)\right]\right]^{\frac{1}{2}} - \left(\frac{n_y \pi}{d_Y}\right)^2 \tag{5.59}$$

where

$$\overline{A_9} = (I_1 + \omega_1\omega_3),\ I_1 = I\frac{(\overline{a})^4}{16},$$

$$\overline{A_{10}}(E, n_z) = \left[\omega_3 E + \omega_1\left[E + \overline{\delta_0} - \omega_4\left(\frac{n_z\pi}{d_z}\right)^2\right] + \omega_2\omega_3\left(\frac{n_z\pi}{d_z}\right)^2 + 2I_1\left(\frac{n_z\pi}{d_z}\right)^2\right],$$

$$\omega_1 = \frac{(\overline{a})^2}{4}\left[(\overline{\lambda_0}) - \overline{l}\right],\ \omega_2 = \left(\overline{\lambda_0}\frac{(\overline{a})^2}{4}\right),\ \omega_3 = \frac{(\overline{a})^2}{4}(\overline{n} + \overline{\nu}),\ \omega_4 = \frac{(\overline{a})^2}{4}\overline{\nu}$$

and

$$\overline{A_{11}}(E, n_z) = \left[E\left[E + \overline{\delta_0} - \omega_4\left(\frac{n_z\pi}{d_z}\right)^2\right] + \omega_2\left(\frac{n_z\pi}{d_z}\right)^2\left[E + \overline{\delta_0} - \omega_4\left(\frac{n_z\pi}{d_z}\right)^2\right] - I_1\left(\frac{n_z\pi}{d_z}\right)^4\right].$$

The (1.92) can be expressed as

$$k_x^2 = A_{27}(E, n_y, n_z) \tag{5.60}$$

where

$$A_{27}(E, n_y, n_z) = \left[2\overline{A_9}\right]^{-1} \left[-\overline{A_{10}}(E, n_z) + \left[\overline{A_{10}}^2(E, n_z) + 4(\overline{A_9})\overline{A_{11}}(E, n_z)\right]\right]^{\frac{1}{2}} - \left(\frac{n_y \pi}{d_Y}\right)^2.$$

The EEM in this case can be written following (5.60) as

$$m^*(E_{F1D}, n_x, n_y) = \frac{\hbar^2}{2} A'_{27}(E_{F1D}, n_x, n_y) \tag{5.61}$$

where

$$A'_{27}(E_{F1D}, n_y, n_z) = \left[2\overline{A_9}\right]^{-1} \Bigg[- (\overline{A_{10}})' + [\overline{A_{10}}^2(E_{F1D}, n_z) + 4(\overline{A_9})\overline{A_{11}}(E_{F1D}, n_z)]^{-1/2}[\overline{A_{10}}(E_{F1D}, n_z)(\overline{A_{10}})' + 2(\overline{A_9})(\overline{A_{11}}(E_{F1D}, n_z))'] \Bigg], \; A'_{10} = (\omega_1 + \omega_3)$$

and

$$\left(\overline{A_{11}}(E_{F1D}, n_z)\right)' = \left[2E_{F1D} + \overline{\delta_0} - \omega_4 \left(\frac{n_z \pi}{d_z}\right)^2 + \omega_2 \left(\frac{n_z \pi}{d_z}\right)^2\right]. \tag{5.62}$$

Thus, the mass is a function of Fermi energy and the size quantum numbers which are the characteristic features of such model.

The electron concentration per unit length can be written as

$$n_0 = \frac{2g_v}{\pi} \sum_{n_x=1}^{n_{x\max}} \sum_{n_y=1}^{n_{y\max}} \left[B_{40}(E_{F1D}, n_x, n_y) + B_{41}(E_{F1D}, n_x, n_y)\right] \tag{5.63}$$

where

$$B_{40}(E_{F1D}, n_x, n_y) = \sqrt{A_{27}(E_{F1D}, n_x, n_y)}$$

and

$$B_{41}(E_{F1D}, n_x, n_y) = \sum_{r=1}^{s_o} Z_{1D}(r) \left[B_4 0(E_{F1D}, n_x, n_y)\right].$$

5.2.10 The EEM in Nanowires of Bismuth Telluride

The 1D electron energy spectrum following (1.127) assumes the form

$$k_x^2 = A_{28}(E, n_y, n_z) \tag{5.64}$$

where

$$A_{28}(E, n_y, n_z) = \left[E(1+\alpha E) - \bar{\omega}_2\left(\frac{n_y\pi}{d_y}\right)^2 - \bar{\omega}_3\left(\frac{n_z\pi}{d_z}\right)^2 - \bar{\omega}_4\left(\frac{n_y\pi}{d_y}\right)\left(\frac{n_z\pi}{d_z}\right)\right](\bar{\omega}_1)^{-1}.$$

The subband energy (E_{30}) can be expressed as

$$E_{30} = (2\alpha)^{-1}\left[-1 + \sqrt{1 + 4\alpha\theta_{32}(n_y, n_z)}\right] \tag{5.65}$$

where

$$\theta_{32}(n_y, n_z) = \left[\bar{\omega}_2\left(\frac{n_y\pi}{d_y}\right)^2 + \bar{\omega}_3\left(\frac{n_z\pi}{d_z}\right)^2 + \bar{\omega}_4\left(\frac{n_y\pi}{d_y}\right)\left(\frac{n_z\pi}{d_z}\right)\right].$$

The EEM in this case can be written following (5.64) as

$$m^*(E_{F1D}, n_x, n_y) = \frac{\hbar^2}{2\bar{\omega}_1}(1 + 2\alpha E_{F1D}) \tag{5.66}$$

The electron concentration per unit length is given by

$$n_0 = \frac{2g_v}{\pi}\sum_{n_z=1}^{n_{z\max}}\sum_{n_y=1}^{n_{y\max}}\left[B_{42}(E_{F1D}, n_z, n_y) + B_{43}(E_{F1D}, n_z, n_y)\right] \tag{5.67}$$

where

$$B_{42}(E_{F1D}, n_z, n_y) = \sqrt{A_{28}(E_{F1D}, n_y, n_z)}$$

and

$$B_{43}(E_{F1D}, n_z, n_y) = \sum_{r=1}^{s_o} Z_{1D}(r)\left[B_{42}(E_{F1D}, n_z, n_y)\right].$$

5.2.11 The EEM in Nanowires of Germanium

(a) The 1D electron energy spectrum for NWs of Ge in this case can be expressed following (1.134) as

$$k_y^2 = A_{29}(E, n_x, n_z) \tag{5.68}$$

where

$$A_{29}(E, n_x, n_z) = \left[\left[\gamma_{15}(E, n_z) - \left(\frac{\hbar^2}{2m_1^*}\right)\left(\frac{n_x\pi}{d_x}\right)^2\right]\left(2m_2^*/\hbar^2\right)\right],$$

$$\gamma_{15}(E, n_z) = \left[E(1+\alpha E) - (1+2\alpha E)\left(\frac{\hbar^2}{2m_3^*}\right)\left(\frac{n_z\pi}{d_z}\right)^2 + \alpha\left[\left(\frac{\hbar^2}{2m_3^*}\right)\left(\frac{n_z\pi}{d_z}\right)^2\right]^2\right],$$

$m_1^* = m_\perp^*$, $m_2^* = \frac{m_\perp^* + 2m_\parallel^*}{3}$ and $m_3^* = \frac{3m_\perp^* m_\parallel^*}{m_\perp^* + 2m_\parallel^*}$.

The quantized energy levels (E_{31}) can be expressed through the equation

$$E_{31} = (2\alpha)^{-1}\left[-\rho_{91}(n_z) + \sqrt{\rho_{91}(n_z)^2 - 4\alpha\rho_{92}(n_z)}\right] \tag{5.69}$$

where

$$\rho_{91}(n_z) = \left[1 - 2\alpha\frac{\hbar^2}{2m_3^*}\left(\frac{n_z\pi}{d_z}\right)^2\right]$$

and

$$\rho_{92}(n_z) = \left[\frac{\hbar^2}{2m_3^*}\left(\frac{n_z\pi}{d_z}\right)^2 - \alpha\left[\frac{\hbar^2}{2m_3^*}\left(\frac{n_z\pi}{d_z}\right)^2\right]^2\right].$$

The EEM in this case can be written following (5.68) as

$$m^*(E_{F1D}, n_x, n_y) = \frac{\hbar^2}{2}A_{29}'(E_{F1D}, n_x, n_y) \tag{5.70}$$

where

$$A'_{29}(E_{F1D}, n_x, n_z) = \left[\left[1 + 2\alpha E_{F1D} - \left(\alpha \frac{\hbar^2}{m_3^*}\right)\left(\frac{n_z \pi}{d_z}\right)^2\right]\left(2m_2^*/\hbar^2\right)\right].$$

For (5.70), we observe that the EEM is a function of the Fermi energy and the size quantum number due to the presence of band non-parabolicity only. The electron concentration per unit length is given by

$$n_0 = \frac{2g_v}{\pi} \sum_{n_x=1}^{n_{x\max}} \sum_{n_z=1}^{n_{z\max}} \left[B_{44}(E_{F1D}, n_x, n_z) + B_{45}(E_{F1D}, n_x, n_z)\right] \tag{5.71}$$

where

$$B_{44}(E_{F1D}, n_x, n_z) = \sqrt{A_{29}(E_{F1D}, n_x, n_z)} \text{ and}$$

$$B_{45}(E_{F1D}, n_x, n_z) = \sum_{r=1}^{s_o} Z_{1D}(r)\left[B_44(E_{F1D}, n_x, n_z)\right].$$

(b) The 1D electron energy spectrum for NWs of Ge in this case can be expressed following (1.143) as

$$k_y^2 = A_{30}(E, n_x, n_z) \tag{5.72}$$

where

$$A_{30}(E, n_x, n_z) = \left[\left[I_{29}(E, n_z) - \left(\frac{\hbar^2}{2m_1^*}\right)\left(\frac{n_x \pi}{d_x}\right)^2\right]\left(2m_2^*/\hbar^2\right)\right],$$

$$I_{29}(E, n_z) = \left[2\overline{C_1}\right]^{-1}\left[\overline{A_6}(n_z) + \left[\overline{A_6}^2(n_z) - 4\bar{C}_1 E + 4(\overline{C_1})\overline{A_5}(n_z)\right]^{\frac{1}{2}}\right]$$

$$\overline{A_5}(n_z) = \left(\frac{\hbar^2}{2m_3^*}\right)\left(\frac{n_z \pi}{d_z}\right)^2\left[1 - \bar{e}_1\left(\frac{\hbar^2}{2m_3^*}\right)\left(\frac{n_z \pi}{d_z}\right)^2\right] \text{ and}$$

$$\overline{A_6}(n_z) = \left[1 - \bar{d}_1\left(\frac{\hbar^2}{2m_3^*}\right)\left(\frac{n_z \pi}{d_z}\right)^2\right]$$

The quantized energy levels (E_{32}) can be expressed through the equation

$$E_{32} = \bar{A}_5(n_z) + \left(\frac{1}{4\bar{C}_1}\right)\left[\left[\frac{\bar{C}_1 \hbar^2}{m_1^*}\left(\frac{n_x \pi}{d_x}\right)^2\right]^2 - 2\bar{A}_6(n_z)\frac{\bar{C}_1 \hbar^2}{m_1^*}\left(\frac{n_x \pi}{d_x}\right)^2\right]. \tag{5.73}$$

The EEM in this case can be written following (5.72) as

$$m^*(E_{F1D}, n_x, n_y) = \frac{\hbar^2}{2} A'_{30}(E_{F1D}, n_x, n_y) \tag{5.74}$$

where

$$A'_{30}(E_{F1D}, n_x, n_z) = (2m_2^*/\hbar^2) I'_{29}(E_{F1D}, n_z) \text{ and}$$
$$I'_{29} E_{F1D}(, n_z) = \left[\left[\overline{A_6}^2(n_z) - 4\bar{C}_1 E_{F1D} + 4(\overline{C_1})\overline{A_5}(n_z) \right]^{-1/2} \right].$$

Thus, the mass is a function of Fermi energy and the size quantum numbers due to the presence of band non-parabolicity only.

The electron concentration per unit length is given by

$$n_0 = \frac{2g_v}{\pi} \sum_{n_x=1}^{n_{x\max}} \sum_{n_z=1}^{n_{z\max}} \left[B_{46}(E_{F1D}, n_x, n_z) + B_{47}(E_{F1D}, n_x, n_z) \right] \tag{5.75}$$

where

$$B_{46}(E_{F1D}, n_x, n_z) = \sqrt{A_30(E_{F1D}, n_x, n_z)} \text{ and}$$
$$B_{47}(E_{F1D}, n_x, n_z) = \sum_{r=1}^{s_o} Z_{1D}(r) \left[B_{46}(E_{F1D}, n_x, n_z) \right].$$

5.2.12 The EEM in Nanowires of Gallium Antimonide

The 1D electron energy spectrum for NWs of GaSb can be expressed following (1.153) as

$$k_z^2 = A_{31}(E, n_x, n_y) \tag{5.76}$$

where $A_{31}(E, n_x, n_y) = \left[\left[I_{36}(E) - \phi_1(n_x, n_y) \right] (2m_c/\hbar^2) \right]$.

The quantized energy levels (E_{33}) can be expressed through the equation

$$I_{36}(E_{33}) = \left(\frac{\hbar^2}{2m_c} \right) \phi_1(n_x, n_y). \tag{5.77}$$

The EEM in this case can be written following (5.76) as

$$m^*(E_{F1D}) = \frac{\hbar^2}{2} A'_{31}(E_{F1D}) \tag{5.78}$$

where

$$A'_{31}(E_{F1D}) = [1 - (m_c/m_0)(\bar{E}'_{g0}/2)[(\bar{E}'_{g0}/2)^2 + [((\bar{E}'_{g0})^2/2)(1 - (m_c/m_0))]$$
$$+ [(\bar{E}'_{g0}/2)(1 - (m_c/m_0))]^2$$
$$+ [E_{F1D}\bar{E}'_{g0}(1 - (m_c/m_0))]^{-1/2}](2m_c/\hbar^2)].$$

The electron concentration per unit length is given by

$$n_0 = \frac{2g_v}{\pi}\sum_{n_x=1}^{n_{x\max}}\sum_{n_y=1}^{n_{y\max}}\left[B_{48}(E_{F1D}, n_x, n_y) + B_{49}(E_{F1D}, n_x, n_y)\right] \tag{5.79}$$

where

$$B_{48}(E_{F1D}, n_x, n_y) = \sqrt{A_{31}(E_{F1D}, n_x, n_y)}$$

and

$$B_{49}(E_{F1D}, n_x, n_y) = \sum_{r=1}^{s_o} Z_{1D}(r)\left[B_{48}(E_{F1D}, n_x, n_y)\right].$$

5.2.13 The EEM in Nanowires of II–V Materials

The 1D electron energy spectrum for NWs of II–V materials can be expressed following (4.100a) as

$$k_z^2 = A_{32,\pm}(E, n_x, n_y) \tag{5.80}$$

where

$$A_{32,\pm}(E, n_x, n_y) = \alpha_{4,\pm}(n_x, n_y) + \beta_4 E \pm \left[\beta_5 E^2 + E\beta_{6,\pm}(n_x, n_y) + \beta_{7,\pm}(n_x, n_y)\right]^{\frac{1}{2}},$$

$$\alpha_{4,\pm}(n_x, n_y) = \left[2(\theta_3^2 - \theta_7^2)\right]^{-1}\left[2\theta_7\alpha_2(n_x, n_y) - 2\alpha_{1,\mp}(n_x, n_y)\theta_3\right],$$

$$\alpha_{1,\mp}(n_x, n_y) = \theta_1\left(\frac{n_x\pi}{d_x}\right)^2 + \theta_2\left(\frac{n_y\pi}{d_y}\right)^2 + \delta_4\left(\frac{n_x\pi}{d_x}\right) \mp \Delta_3,$$

$$\alpha_2(n_x, n_y) = \left[\theta_5\left(\frac{n_x\pi}{d_x}\right)^2 + \theta_6\left(\frac{n_y\pi}{d_y}\right)^2 + \delta_5\left(\frac{n_x\pi}{d_x}\right)\right],$$

$$\beta_4 = 2\theta_3\left[2(\theta_3^2 - \theta_7^2)\right]^{-1}, \beta_5 = \left[2(\theta_3^2 - \theta_7^2)\right]^{-2}\left[4\theta_7^2\right],$$

$$\beta_{6,\pm}(n_x, n_y) = \left[8\theta_3\theta_7\alpha_2(n_x, n_y) - 8\theta_7^2\alpha_{1,\mp}(n_x, n_y)\right]\left[2\left(\theta_3^2 - \theta_7^2\right)\right]^{-1},$$

$$\beta_{7,\pm}(n_x,n_y) = \left[2\left(\theta_3^2-\theta_7^2\right)\right]^{-2}\left[4\theta_7^2\alpha_2^2(n_x,n_y) - 8\theta_7\theta_3\alpha_2(n_x,n_y)\alpha_{1,\mp}(n_x,n_y)\right.$$
$$\left. +4\theta_3^2\alpha_3(n_y) + 4\theta_7^2\alpha_{1,\mp}(n_x,n_y) - 4\alpha_3(n_y)\theta_7^2\right]$$

and

$$\alpha_3(n_y) = G_3^2\left(\frac{n_y\pi}{d_y}\right)^2 + \Delta_3^2.$$

The quantized energy levels ($E_{34,\pm}$) can be expressed through the equation

$$E_{34,\pm} = \alpha_{1,\mp}(n_x,n_y) \pm \left[\alpha_2^2(n_x,n_y) + \alpha_3(n_y)\right]^{\frac{1}{2}}. \tag{5.81}$$

The EEM in this case can be written following (5.80) as

$$m^*(E_{F1D},n_x,n_y) = \frac{\hbar^2}{2}A'_{32\pm}(E_{F1D},n_x,n_y) \tag{5.82}$$

where

$$A'_{32,\pm}(E_{F1D},n_x,n_y) = \left[\beta_4 \pm \frac{1}{2}\left[2\beta_5E_{F1D} + \beta_{6,\pm}(n_x,n_y)\right].\right.$$
$$\left.\times\left[\beta_5E_{F1D}^2 + \beta_{6,\pm}(n_x,n_y)E_{F1D} + \beta_{7,\pm}(n_x,n_y)\right]^{-\frac{1}{2}}\right].$$

Thus, the mass is a function of Fermi energy and the size quantum numbers which are the characteristic features of such model.

The electron concentration per unit length is given by

$$n_0 = \frac{g_v}{\pi}\sum_{n_x=1}^{n_{x\max}}\sum_{n_y=1}^{n_{y\max}}\left[B_{49}(E_{F1D},n_x,n_y) + B_{50}(E_{F1D},n_x,n_y)\right] \tag{5.83}$$

where

$$B_{49}(E_{F1D},n_x,n_y) = [\sqrt{A_{32},+(E_{F1D},n_x,n_y)} + \sqrt{A_{32},(E_{F1D},n_x,n_y)}]$$

and

$$B_{50}(E_{F1D},n_x,n_y) = \sum_{r=1}^{s_o} Z_{1D}(r)\left[B_{49}(E_{F1D},n_x,n_y)\right].$$

5.2.14 Carbon Nanotubes

With the discovery of carbon nanotubes (CNs) in 1991 by Iijima [36], the CNs have been recognized as fascinating materials with nanometer dimensions uncovering new phenomena in the sphere of low dimensional science and technology. The significant physical properties of these nanomaterials make them ideal candidates to reveal new phenomena in nanoelectronics. The CNs find wide applications in conductive [37, 38] and high strength composites [39], chemical sensors [40], field emission displays [41, 42], hydrogen storage media [43, 44], nanotweezeres [45], nanogears [46], nanocantilever devices [47], nanomotors [48, 49], and nanoelectronic devices [50, 51]. Single walled carbon nanotubes (SWCNs) emerge to be excellent materials for single molecule electronics [52–56] such as nanotube based diodes [57, 58], single electron transistors [51, 59], random access memory cells [60], logic circuits [61], gigahertz oscillators [62–67], data storage nanodevices [68–73], nanorelay [47, 74–78], and in other low-dimensional devices. The CNs can be bespoke into a metal or a semiconductor based on the diameter and the chiral index numbers (m, n), where the integers m and n denote the number of unit vectors along two directions in the honeycomb crystal lattice of graphene [79, 80]. For armchair and zigzag nanotubes, the chiral indices are given as $m = n$ and $m = 0$, respectively [79, 80]. Another class of CN called as chiral CN has distinct integers m and n. Besides, a CN can be a metallic if $m - n = 3q$; where $q = 1, 2, 3, \ldots$ otherwise it is a semiconductor. Metallic SWCNs have received substantial attention as potential substitutions for traditional interconnect materials like Cu due to their excellent inherent electrical and thermal properties. Since the carriers are confined, in a metallic SWCN, the inclusion of the sub-band energy owing to Born–Von Karman (BVK) boundary conditions [81] for their unique band structure becomes prominent. The quantization of the motion of the carriers in such structures leads to the discontinuity in the DOS function due to van Hove singularity (VHS) [82] of the wave vectors. In this section, we shall explore the EEM in carbon nanotubes.

For (n, n) and $(n, 0)$ tubes, the energy dispersion relations are given by [79, 80]

$$E_m(k_y) = \pm t_c\sqrt{1 + 4\cos\frac{m\pi}{n}\cos\frac{k_y a_c}{2} + 4\cos^2\frac{k_y a_c}{2}},$$
$$-\frac{\pi}{a_c} < k_y < \frac{\pi}{a_c} \quad \text{and} \quad m(= 1, 2, \ldots, 2n) \tag{5.84a}$$

$$E_m(k_x) = \pm t_c\sqrt{1 + 4\cos\frac{\sqrt{3}k_x a_c}{n}\cos\frac{m\pi}{n} + 4\cos^2\frac{m\pi}{n}},$$
$$-\frac{\pi}{\sqrt{3}a_c} < k_x < \frac{\pi}{\sqrt{3}a_c} \quad \text{and} \quad m(= 1, 2, \ldots, 2n) \tag{5.84b}$$

where t_c [60] is the C–C bonding energy and m and n are the chiral indices [79, 80]. The Eqs. (5.84a) and (5.84b) are the analytic expressions throughout the entire Brillouin zone [79, 80].

Near Fermi energy the 1D $E\!-\!k_y$ relation can approximately written as [83]

$$E^2 - E_i^2 = Ak_y^2, \tag{5.84c}$$

where $E_i = \frac{|3i-m+n|}{2}|t_c|\frac{a_c}{r},\ i = 1, 2, 3, \ldots, i_{\max}$,

a_c is the nearest neighbor C–C bonding distance, r is the nano tube radius and $A = (9t_c^2 d^2/4\pi^2)$. From (5.84c), it appears that when $k_y \to 0$, $E \to E_i$.

Using this idea from (5.84a), we can write

$$4\cos\left(\frac{m\pi}{n}\right) = \frac{E_i^2}{t_c^2} - 5. \tag{5.84d}$$

By Eliminating $\cos(\frac{m\pi}{n})$ between (5.84a) and (5.84d), we can write

$$4z^2 + \left(\frac{E_i^2}{t_c^2} - 5\right)z + 1 - \frac{E_i^2}{t_c^2} = 0 \tag{5.84e}$$

where $z = \cos\left(\frac{k_y\sqrt{3}a_c}{2}\right)$.

From (5.84e), we can write that the $E\!-\!k_y$ relation for arm chair nanotube is given by

$$k_y = \frac{2}{\sqrt{3}a_c} f_1(E, m, n) \tag{5.84f}$$

where

$$f_1(E, m, n) = \cos^{-1}[\theta_1(E, m, n)]$$

and

$$\theta_1(E, m, n) = \frac{1}{8}\left[-\left(\frac{E_i^2}{t_c^2} - 5\right) + \left[\left(\frac{E_i^2}{t_c^2} - 5\right)^2 + 16\left(\frac{E^2}{t_c^2} - 1\right)\right]^{1/2}\right].$$

The EEM in this case is given by

$$m^*(E_{F_1}, m, n) = \frac{4\hbar^2}{3a_c^2} f_1(E_{F_1}, m, n)\frac{\theta_1'(E_{F_1}, m, n)}{\sqrt{1 - \theta_1^2(E_{F_1}, m, n)}} \tag{5.84g}$$

where E_{F_1} is the Fermi energy in this case and

$$\theta_1'(E_{F_1}, m, n) = \frac{4E_{F_1}}{t_c^2}\left[\left(\frac{E_i^2}{t_c^2} - 5\right)^2 + 16\left(\frac{E_{F_1}^2}{t_c^2} - 1\right)\right]^{-1/2}.$$

The electron concentration in this case can be expressed as

It appears from (5.86) and (5.87) that the effective mass in CNTs is a function of m and n in addition to Fermi energy which is the characteristic feature of such nanomaterials.

Using (5.84f), the electron statistics in this case, can be written as,

$$n_{1D} = \frac{8}{a_c\pi\sqrt{3}}\sum_{i=0}^{i_{\max}}\left[f_1(E_{F_1}, m, n) + B_{c_2}(E_{F_1}, m, n)\right] \tag{5.85}$$

where

$$B_{c2}(E_{F_1}, m, n) \equiv \sum_{r=1}^{s} Z_r f_1(E_{F_1}, m, n).$$

Similarly, the $E{-}k_y$ relation of zigzag nanotube is given by

$$k_y = \frac{2}{3a_c} f_2(E, m, n) \tag{5.86}$$

where

$$f_2(E, m, n) = \cos^{-1}[\theta_2(E, m, n)]$$

and

$$\theta_2(E, m, n) = \left[\left[\left(\frac{E^2}{t_c^2} - 1\right) - \left(\frac{E_i}{t_c} - 1\right)^2\right]\left[\frac{2E_i}{t_c} - 1\right]^{-1}\right].$$

The EEM and electron concentration for this case can respectively be expressed as

$$m^*(E_{F_2}, m, n) = \frac{4\hbar^2}{9a_c^2} f_2(E_{F_2}, m, n)\frac{\theta_2'(E_{F_2}, m, n)}{\sqrt{1 - \theta_2^2(E_{F_2}, m, n)}} \tag{5.87}$$

and

$$n_{1D} = \frac{8}{3a_c\pi}\sum_{i=0}^{i_{\max}}\left[f_2(E_{F_2}, m, n) + B_{c_3}(E_{F_2}, m, n)\right] \tag{5.88}$$

$$\theta_2'(E_{F_2}, m, n) = \left[\left(\frac{2E_{F_2}}{t_c^2}\right)\right]\left[\frac{2E_i}{t_c} - 1\right]^{-1},$$

E_{F_2} is the Fermi energy in this case, where

$$B_{c_3}(E_{F_2}, m, n) \equiv \sum_{r=1}^{s} Z_r f_2(E_{F_2}, m, n).$$

5.3 Results and Discussion

The variation of the EEM in NWs of different materials along the transport direction has been exhibited in the figures below at various conditions. Throughout our formalism, we have assumed the NWs to be of rectangular cross-sectional dimensions so that the usual "particle-in-a-box" concept can be applied along the quantized directions. The NWs are assumed to be degenerately doped with a carrier density starting from $10^8\,m^{-1}$. While deriving the closed form analytical solutions of EEM in all the materials, we have also assumed that the constants of the energy band structures of the materials are independent of thickness in the range beyond 5 nm. Generally speaking, as also will be shown in Chap. 8, that the "band-gap" is a strong function of cross-sectional dimension and its geometry, i.e., whether the cross-section is circular or triangular. For example in Si NW, we will observe in Chap. 8, that the band gap is very high and even becomes direct rather than its usual indirect nature in the zone 1–4 nm cross-sectional dimensions and beyond this, the band gap is nearly equal to its bulk value. Keeping this trend in view, we have assumed the invariant property of the material energy spectrum constants and evaluated the EEM in NWs of Cd_3As_2 and $CdGeAs_2$ as function of wire thickness along their respective transport directions in Figs. 5.1 and 5.2.

It appears from Figs. 5.1 and 5.2 that the EEM at the lowest subband in both the cases are strong cross-sectional functions of the dimensions which converge to their corresponding bulk values at larger dimensions. The effect of crystal field splitting in case of dispersion relation of Cd_3As_2 lets the asymptotic fall to be closer to the bulk value. It should be noted that all the curves have been evaluated at T = 4 K where the average thermal energy i.e., $E_F + k_B T$ is very less than that of the difference of the adjacent sub-band energies. This leads the carrier to reside in the lowest sub-bands only.

Figures 5.3, 5.4 and 5.5 exhibit the variation of the EEM with the wire thickness for III–V materials namely InAs, InSb, and GaAs in accordance with the well-known standard non-parabolic dispersion relation of Kane. The reader is expected to evaluate the EEM for other models as derived in this chapter for III–V materials.

It appears from these figures that the difference in the energy band models in predicting the EEM is almost insignificant. Hence for all practical purposes for determination of EEM, the second-order model of Kane can fit well. It should be noted that the EEM for these materials as presented here can be compared with that of the EEM in 2D systems as given in Chap. 1.

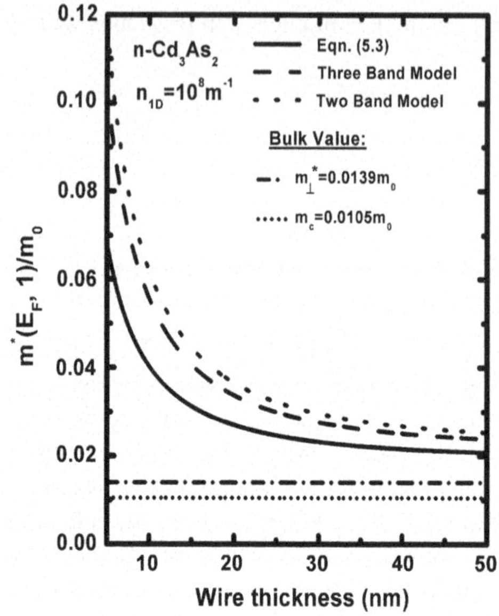

Fig. 5.1 Plot of the EEM as function of wire thickness for QWs of Cd_3As_2 considering (5.3). The plots for three- and two-band models of Kane have also been exhibited with their corresponding anisotropic bulk values as presented in Fig. 1.1

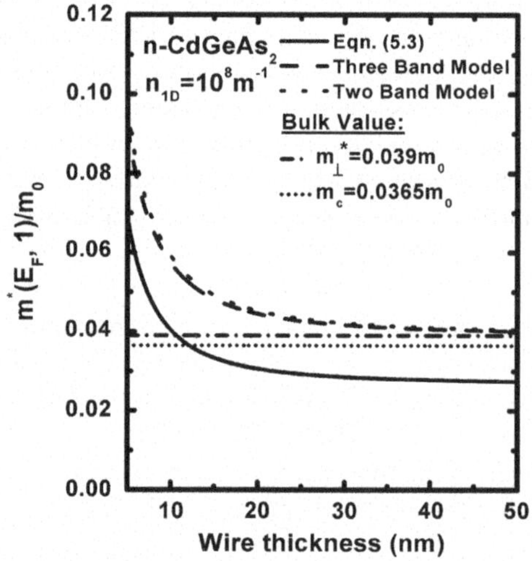

Fig. 5.2 Plot of the EEM as a function of wire thickness for QWs of $CdGeAs_2$ for all the cases of Fig. 5.1

A quick view can lead to interpret that the EEM for both the case are same. However, it should also be kept in mind about the difference in the wire thickness and carrier concentration. All the curves in this chapter have been evaluated at those concentrations for which the EEM stand close to that of their corresponding 2D systems. Further in deriving the results, we have assumed that the conduction band valley does not splits along the channel transport direction, which is a usual case with Silicon NW along [110] and [111] valleys (Chap. 8).

Figures 5.6 and 5.7 exhibit the variation of the EEM for $Hg_{1-x}Cd_xTe$ and $In_{1-x}Ga_xAs_{1-y}P_y$ considering all the aforementioned cases at x = 0.3. In Figs. 5.8,

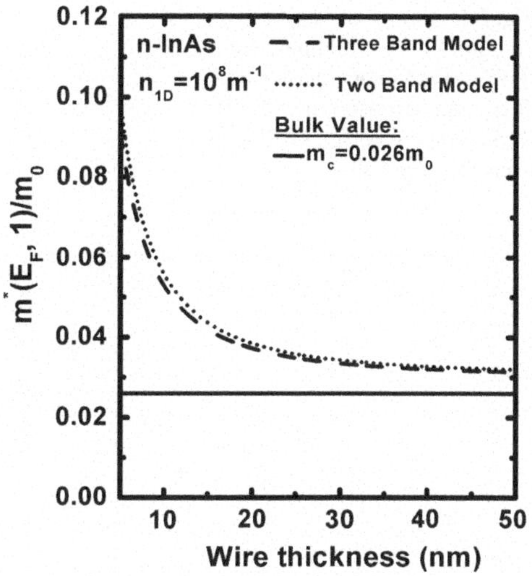

Fig. 5.3 Plot of the EEM as function of wire thickness for NWs of InAs considering the three- and two-band models of Kane with the corresponding isotropic bulk value

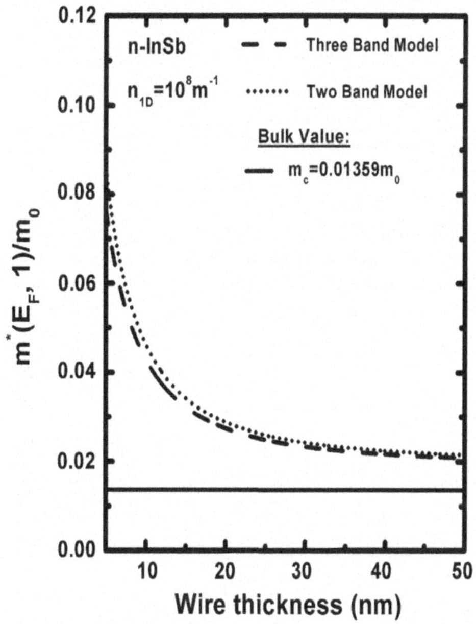

Fig. 5.4 Plot of the EEM as function of wire thickness for NWs of InSb considering the three- and two-band models of Kane with the corresponding isotropic bulk value

5.9, 5.10, 5.11, 5.12, 5.13, 5.14, we have exhibited the variation of the EEM as function of carrier degeneracy. It appears from the said figures that the EEM increases with the increase in the degeneracy. The EEM rises sharply above $10^7\ m^{-1}$ for all the materials in an exponential way due to the presence of the Fermi–Dirac probability factor in the respective carrier concentration equation.

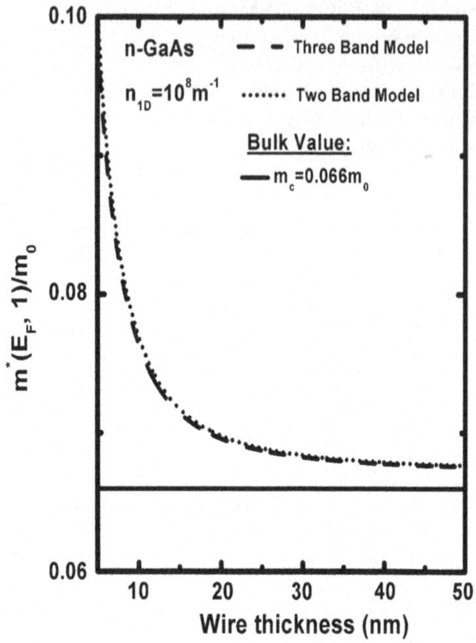

Fig. 5.5 Plot of the EEM as function of wire thickness for NWs of GaAs considering the three- and two-band models of Kane with the corresponding isotropic bulk value

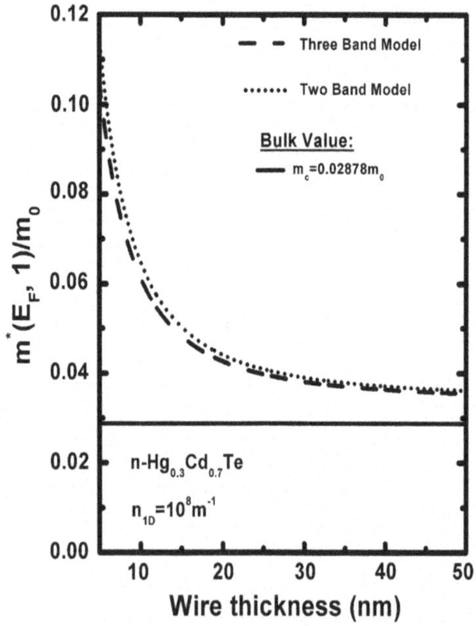

Fig. 5.6 Plot of the EEM as function of wire thickness for NWs of $Hg_{1-x}Cd_xTe$ considering the three- and two-band models of Kane with the corresponding isotropic bulk value at $x = 0.3$

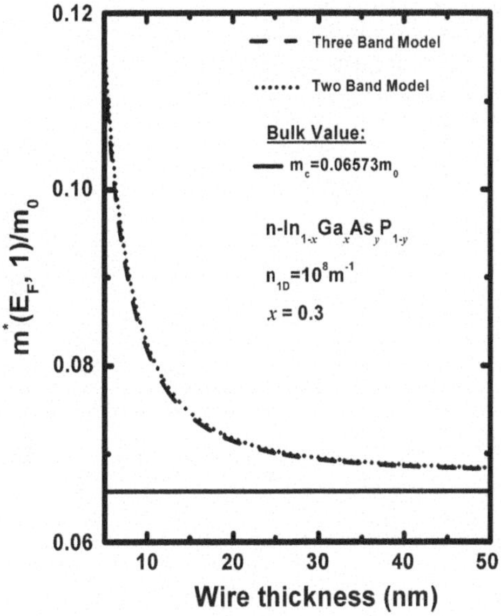

Fig. 5.7 Plot of the EEM as function of wire thickness for NWs of $In_{1-x}Ga_xAs_{1-y}P_y$ considering the three- and two-band models of Kane with the corresponding isotropic bulk value at $x = 0.3$

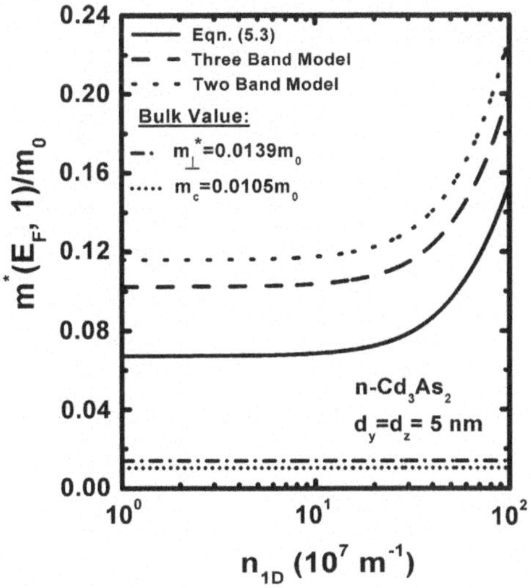

Fig. 5.8 Plot of the EEM as function of carrier degeneracy for n-Cd_3As_2 nanowire for all cases of Fig. 5.1

Figure 5.15 exhibits the variation of the EEM as function of the alloy composition in ternary and quaternary systems. It appears that as in the previous cases of quantum confinements, the EEM in this case also exhibits an increasing variation with x.

We have also plotted the variation of the bulk effective mass as x varies to present a comparative view. The influence of band non-parabolicity on the EEM in these two materials can easily be seen. In both the cases, we see that the EEM is a slow variation function of x due to the change in band gap.

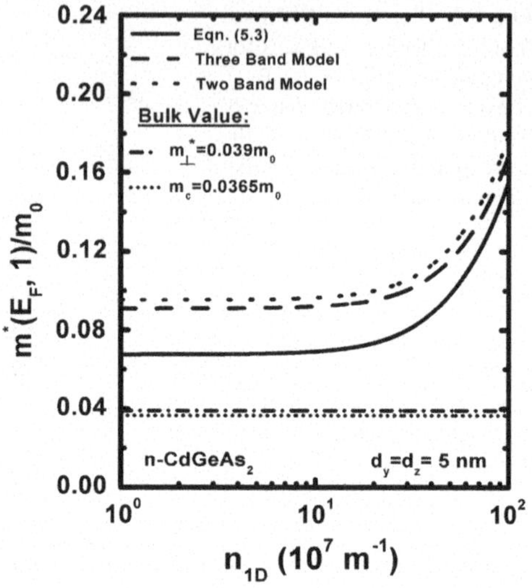

Fig. 5.9 Plot of the EEM as function of carrier degeneracy for NWs of $CdGeAs_2$ for all cases of Fig. 5.2

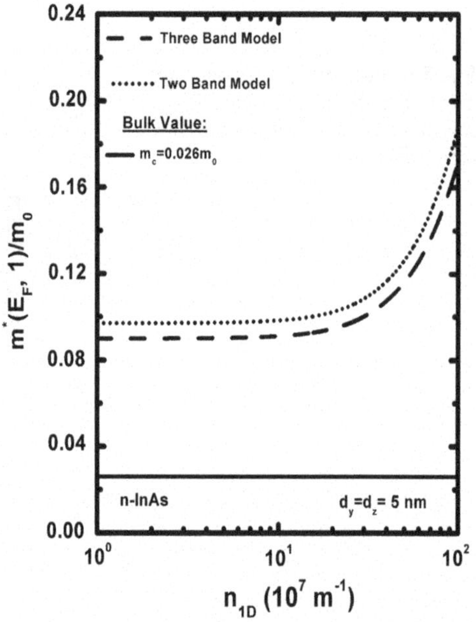

Fig. 5.10 Plot of the EEM as function of carrier degeneracy for n-InAs nanowire for all cases of Fig. 5.3

The effect of dimensionality on NW of Bismuth has been exhibited in Figs. 5.16 and 5.17 for the energy band models of McClure and Choi, Hybrid, Cohen, and Lax. It appears that as the dimension reduces in Fig. 5.16, the EEM increases which is generally accepted. However, using McClure and Choi model the EEM tends to decrease in the sub-5 nm regime there by unfolding the validity of the model in this

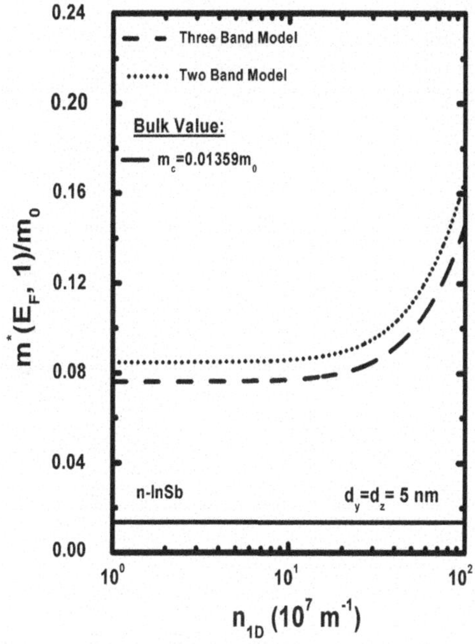

Fig. 5.11 Plot of the EEM as function of carrier degeneracy for n-InSb nanowire for all cases of Fig. 5.4

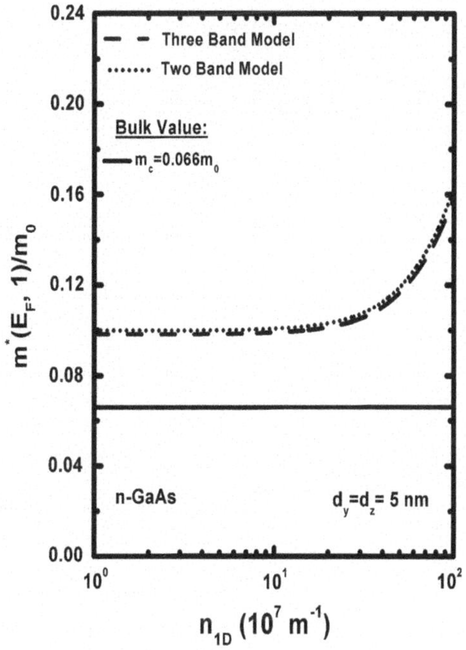

Fig. 5.12 Plot of the EEM as function of carrier degeneracy for n-GaAs nanowire for all cases of Fig. 5.5

Fig. 5.13 Plot of the EEM as function of carrier degeneracy for n-$Hg_{1-x}Cd_xTe$ nanowire for all cases of Fig. 5.6

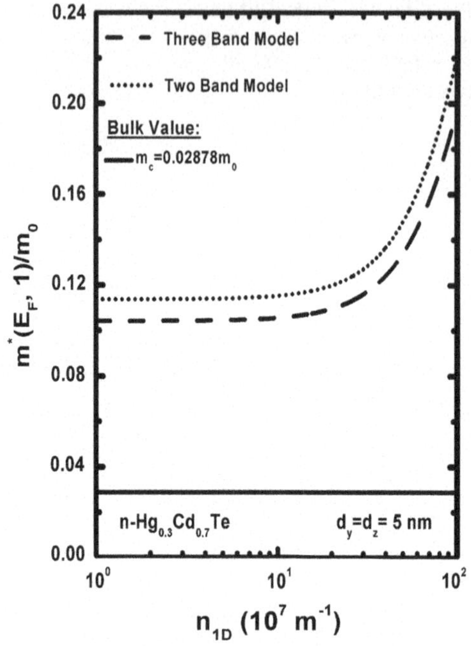

Fig. 5.14 Plot of the EEM as function of carrier degeneracy for n-$In_{1-x}Ga_xAs_{1-y}P_y$ nanowire for all cases of Fig. 5.8

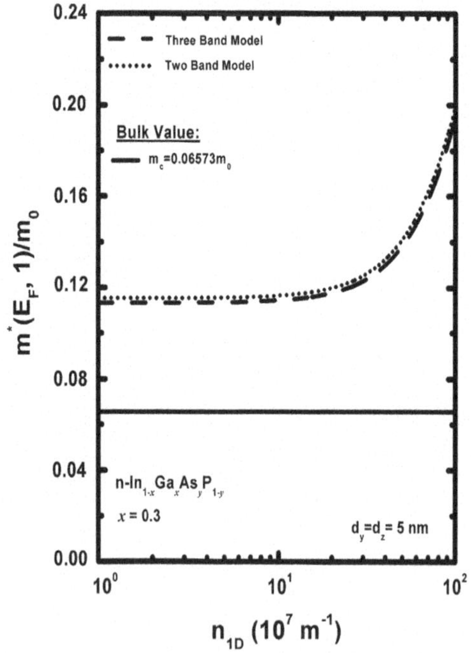

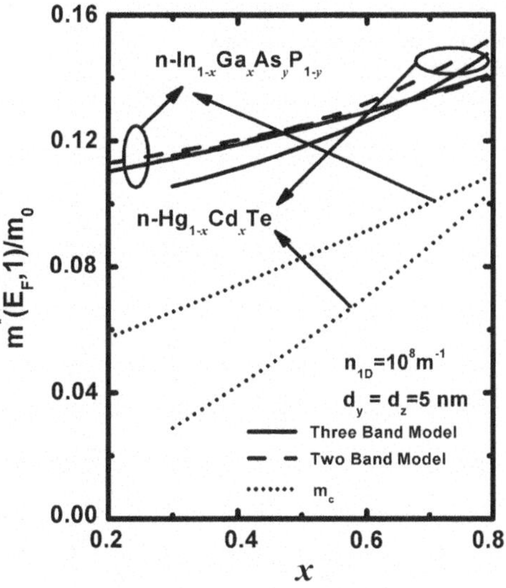

Fig. 5.15 Plot of the EEM as function of alloy composition for n-$Hg_{1-x}Cd_xTe$ and n-$In_{1-x}Ga_xAs_{1-y}P_y$ nanowires

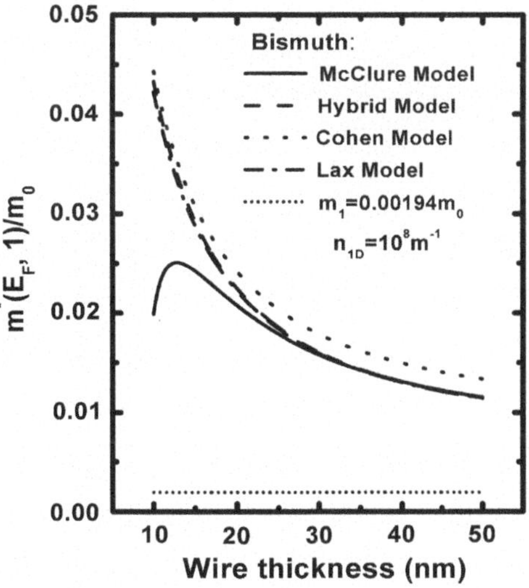

Fig. 5.16 Plot of the EEM as function of wire thickness for Bismuth nanowire

zone. However at large thicknesses, the entire model tends to their corresponding bulk value 0.00194 m_0.

The variation of the EEM as function of carrier degeneracy has further been plotted using the aforementioned band structure models. It appears from the two figures that the effect of different band structure models has significantly less deviation from one another except for model of McClure and Choi.

The EEM is found to increase almost linear with degeneracy $2 \times 10^8\,m^{-1}$ and beyond using the all the models. In case of PbTe, the EEM rises sharply with

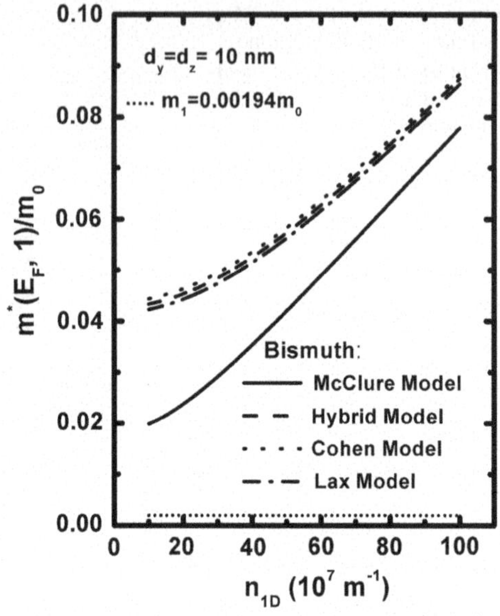

Fig. 5.17 Plot of the EEM as function of carrier degeneracy for Bismuth nanowire

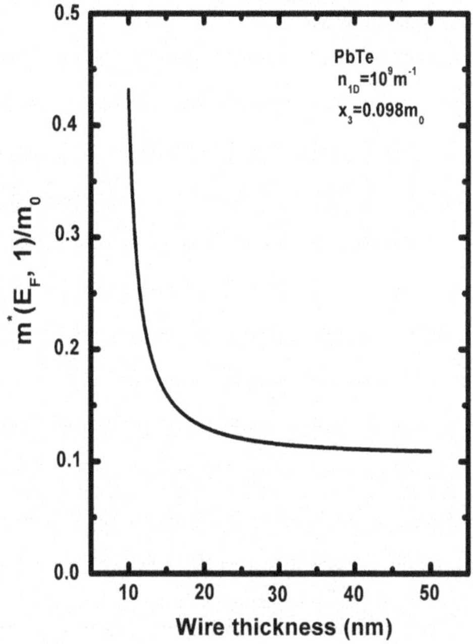

Fig. 5.18 Plot of the EEM as function of carrier degeneracy for PbTe nanowire

decreasing wire thickness below sub-15 nm from the bulk value $0.098\,m_0$ at carrier degeneracy of $10^9\,m^{-1}$, which can affect the carrier mobility strongly (Fig. 5.18).

The effect of strain on stressed InSb NWs has been exhibited in Fig. 5.19 for two different momentum matrix elements to signify its importance as dimension reduces.

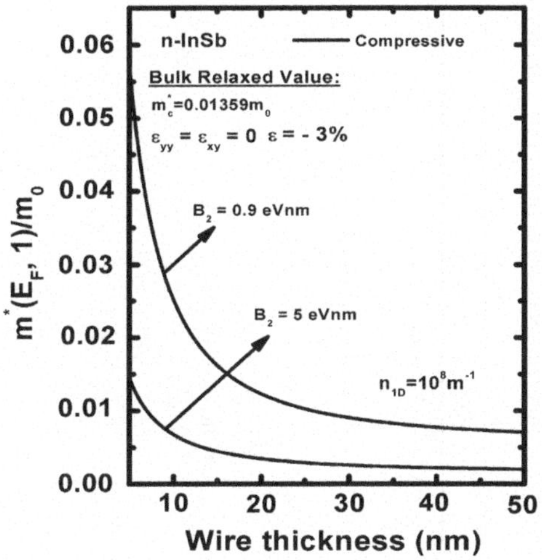

Fig. 5.19 Plot of the EEM as function of wire thickness for stressed InSb nanowire

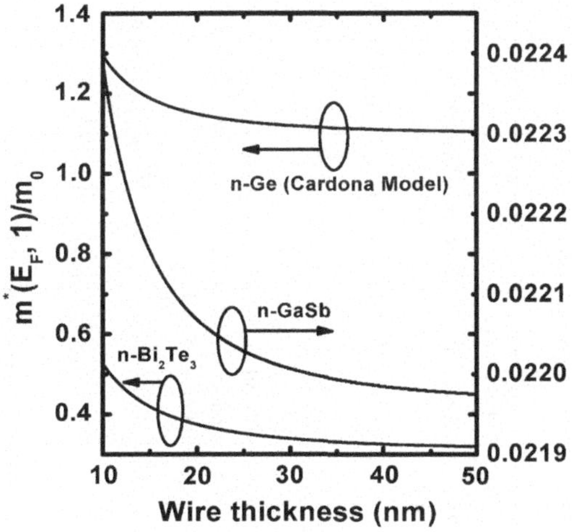

Fig. 5.20 Plot of the EEM as function of wire thickness for Ge, GaSb, and Bi_2Te_3 nanowires

A compressive strain of 3 % has been applied along x and z directions to predict the variation of the EEM along the transport direction. At this point, we could not provide the influence of strain on the energy band structure of InSb NW due to the lack in both experimental and simulation investigations. It is also not very clear about the exact value of how the momentum matrix element B_2 will change in the definition of the strain. All the parameters together with the factor $a+C_1$, as given in the Chap. 1 is expected to modulate with the applied strain orientation. However, to correlate with our strain model on InSb, we have discussed with the existing investigation of strain effects on silicon NWs in Chap. 8.

Figure 5.20 exhibits the effect of cross-sectional dimension on the EEM in n-Ge, n-GaSb, and topological insulators like n-Bi_2Te_3. We have used the Cardona model

Table 5.1 The carrier statistics and the EEM in nanowires of nonlinear optical, III–V, II–VI, Bismuth, IV–VI, stressed materials, Te, n-GaP, $PtSb_2$, $\mathbf{Bi_2Te_3}$, **n-Ge, GaSb, II–V, and carbon nanotubes**

Type of materials	The carrier statistics	The electron statistics
1. Nonlinear optical materials	In accordance with the generalized electron dispersion relation as given by (5.1), $n_0 = \frac{2g_v}{\pi}\sum_{n_x=1}^{n_{x\max}}\sum_{n_y=1}^{n_{y\max}}[B_{11}(E_{F1D}, n_x, n_y) + B_{12}(E_{F1D}, n_x, n_y)]$ (5.5)	On the basis of (5.1), $m^*(E_{F1D}, n_x, n_y) = \frac{\hbar^2}{2}A'_{11}(E_{F1D}, n_x, n_y)$ (5.3)
2. III–V materials, the conduction electrons of which can be defined by five types of energy wave vector dispersion relations as described in the column of carrier statistics	(a) Three-band model of Kane: In accordance with the three-band model of Kane (5.6), which is the special case of (5.1) $n_{1D} = \frac{2g_v}{\pi}\left(\frac{2m_c}{\hbar^2}\right)^{1/2}\sum_{n_x=1}^{n_{x\max}}\sum_{n_y=1}^{n_{y\max}}[T_{63}(E_{F1D}, n_x, n_y) + T_{64}(E_{F1D}, n_x, n_y)]$ (5.9)	On the basis of (5.6), $m^*(E_{F1D}) = m_c\{I_{11}(E_{F1D})\}'$ (5.7)
	(b) Two-band model of Kane: In accordance with the two-band model of Kane (5.10), $n_{1D} = \frac{2g_v}{\pi}\left(\frac{2m_c}{\hbar^2}\right)^{1/2}\sum_{n_x=1}^{n_{x\max}}\sum_{n_y=1}^{n_{y\max}}[T_{65}(E_{F1D}, n_x, n_y) + T_{66}(E_{F1D}, n_x, n_y)]$ (5.14)	On the basis of (5.10), $m^*(E_{F1D}) = m_c(1 + 2\alpha E_{F1D})$ (5.11)
	(c) The model of Stillman et al.: In accordance with the model of Stillman et al. (5.17), $n_0 = \frac{2g_v}{\pi}\sum_{n_x=1}^{n_{x\max}}\sum_{n_y=1}^{n_{y\max}}[B_{17}(E_{F1D}, n_x, n_y) + B_{18}(E_{F1D}, n_x, n_y)]$ (5.20)	On the basis of (5.17), $m^*(E_{F1D}) = m_c I'_{12}(E_{F1D})$ (5.18)

(continued)

Table 5.1 (continued)

Type of materials	The carrier statistics	The electron statistics
	(d) The model of Newson and Kurobe: In accordance with the model of Newson and Kurobe (5.22), $n_0 = \frac{2g_v}{\pi} \sum_{n_x=1}^{n_{x\max}} \sum_{n_y=1}^{n_{y\max}} [B_{19}(E_{F1D}, n_x, n_y) + B_{20}(E_{F1D}, n_x, n_y)]$ (5.25)	On the basis of (5.22), $m^*(E_{F1D}, n_x, n_y) = \frac{\hbar^2}{2} A'_{15}(E_{F1D}, n_x, n_y)$ (5.23)
	(e) The model of Palik et al.: In accordance with the model of Palik et al. (5.26), $n_0 = \frac{2g_v}{\pi} \sum_{n_x=1}^{n_{x\max}} \sum_{n_y=1}^{n_{y\max}} [B_{21}(E_{F1D}, n_x, n_y) + B_{22}(E_{F1D}, n_x, n_y)]$ (5.28)	On the basis of (5.26), $m^*(E_{F1D}) = m_c I'_{13}(E_{F1D})$ (5.27)
3. II–VI materials as described by Hopfield model	In accordance with (5.29), $n_{1D} = \frac{g_v}{\pi \sqrt{B_0}} \sum_{n_x=1}^{n_{x\max}} \sum_{n_y=1}^{n_{y\max}} [t_7(E_{F1D}, n_x, n_y) + t_8(E_{F1D}, n_x, n_y)]$ (5.31)	On the basis of (5.29), $m^*(E_{F1D}) = m^*_{\parallel}$ (5.30)
4. Bismuth, the carriers of which can be defined by five types of energy band models as described in the column of carrier statistics	(a) The McClure and Choi model: In accordance with (5.32), $n_{1D} = \frac{2g_v}{\pi} \frac{\sqrt{2m_1}}{\hbar} \sum_{n_y=1}^{n_{y\max}} \sum_{n_z=1}^{n_{z\max}} [t_{27}(E_{F1D}, n_y, n_z) + t_{28}(E_{F1D}, n_y, n_z)]$ (5.34)	On the basis of (5.32), $m^*(E_{F1D}, n_y) = m_1 \left\{ \left[1 - \frac{\alpha\hbar^2}{2m_2} \left(\frac{\pi n_y}{d_y}\right)^2 \right]^{-1} \times \left[1 + 2\alpha E_{F1D} - \frac{\hbar^2}{2m_2} \times \alpha \left\{ 1 - \left(\frac{m_2}{m'_2}\right) \right\} \left(\frac{\pi n_y}{d_y}\right)^2 \right] \right\}$ (5.33)

(continued)

Table 5.1 (continued)

Type of materials	The carrier statistics	The electron statistics
	(b) The Hybrid Model: In accordance with (5.35), $n_{1D} = \frac{2g_v}{\pi}\frac{\sqrt{2m_1}}{\hbar}\sum_{n_y=1}^{n_{y\max}}\sum_{n_z=1}^{n_{z\max}}[t_{31}(E_{F1D}, n_y, n_z) + t_{32}(E_{F1D}, n_y, n_z)]$ (5.37)	On the basis of (5.35), $m^*(E_{F1D}, n_y) = m_1\left[1 + 2\alpha E_{F1D} - \frac{\hbar^2}{2M_2}\alpha\{1-(\gamma_0)\}\left(\frac{\pi n_y}{d_y}\right)^2\right]$ (5.36)
	(c) The Cohen Model: In accordance with (5.38), $n_{1D} = \frac{2g_v}{\pi}\frac{\sqrt{2m_1}}{\hbar}\sum_{n_y=1}^{n_{y\max}}\sum_{n_z=1}^{n_{z\max}}[t_{35}(E_{F1D}, n_y, n_z) + t_{36}(E_{F1D}, n_y, n_z)]$ (5.40)	On the basis of (5.38), $m^*(E_{F1D}, n_y) = m_1\{[1 + 2\alpha E_{F1D} + l_7]\}$ (5.39)
	(d) The Lax Model: In accordance with (5.41), $n_{1D} = \frac{2g_v}{\pi}\frac{\sqrt{2m_1}}{\hbar}\sum_{n_y=1}^{n_{y\max}}\sum_{n_z=1}^{n_{z\max}}\left[t_{37}(E_{F1D}, n_y, n_z) + t_{38}(E_{F1D}, n_y, n_z)\right]$ (5.43)	On the basis of (5.41), $m^*(E_{F1D}) = m_1\{[1 + 2\alpha E_{F1D}]\}$ (5.42)
5. IV–VI materials, the carriers of which can be defined by the model of Dimmock	In accordance with the model of Dimmock (5.44), $n_0 = \frac{2g_v}{\pi}\sum_{n_x=1}^{n_{x\max}}\sum_{n_y=1}^{n_{y\max}}[B_{32}(E_{F1D}, n_x, n_y) + B_{33}(E_{F1D}, n_x, n_y)]$ (5.46)	On the basis of (5.44), $m^*(E_{F1D}, n_x, n_y) = \frac{\hbar^2}{2}A'_{23}(E_{F1D}, n_x, n_y)$ (5.45)
6. Stressed materials, as defined by the model of Seiler et al.	In accordance with (5.47), $n_{1D} = \frac{2g_v}{\pi}\sum_{n_y=1}^{n_{y\max}}\sum_{n_z=1}^{n_{z\max}}[B_{34}(E_{F1D}, n_y, n_z) + B_{35}(E_{F1D}, n_y, n_z)]$ (5.50)	On the basis of (5.47), $m^*(E_{F1D}, n_x, n_y) = \frac{\hbar^2}{2}A'_{24}(E_{F1D}, n_x, n_y)$ (5.48)

(continued)

Table 5.1 (continued)

Type of materials	The carrier statistics	The electron statistics
7. Tellurium, as defined by the model of Bouat et al.	In accordance with (5.51), $n_0 = \frac{g_v}{\pi} \sum_{n_x=1}^{n_{x\max}} \sum_{n_y=1}^{n_{y\max}} \times [B_{36,\pm}(E_{F1D}, n_x, n_y) + \theta_{5,\pm}]$ (5.54)	On the basis of (5.51), $m^*(E_{F1D}, n_x, n_y) = \frac{\hbar^2}{2} A'_{25,\pm}(E_{F1D}, n_x, n_y)$ (5.52)
8. n-GaP as described by the Rees model	In accordance with (5.55), $n_0 = \frac{2g_v}{\pi} \sum_{n_x=1}^{n_{x\max}} \sum_{n_y=1}^{n_{y\max}} [B_{38}(E_{F1D}, n_x, n_y) + B_{39}(E_{F1D}, n_x, n_y)]$ (5.58)	On the basis of (5.55), $m^*(E_{F1D}, n_x, n_y) = \frac{\hbar^2}{2} A'_{26}(E_{F1D}, n_x, n_y)$ (5.56)
9. $PtSb_2$, as defined by the Emtage model	In accordance with (5.60), $n_0 = \frac{2g_v}{\pi} \sum_{n_x=1}^{n_{x\max}} \sum_{n_y=1}^{n_{y\max}} [B_{40}(E_{F1D}, n_x, n_y) + B_{41}(E_{F1D}, n_x, n_y)]$ (5.63)	On the basis of (5.60), $m^*(E_{F1D}, n_x, n_y) = \frac{\hbar^2}{2} A'_{27}(E_{F1D}, n_x, n_y)$ (5.61)
10. Bi_2Te_3, which follows the model of Stordeur et al.	In accordance with (5.65), $n_0 = \frac{2g_v}{\pi} \sum_{n_z=1}^{n_{z\max}} \sum_{n_y=1}^{n_{y\max}} [B_{42}(E_{F1D}, n_z, n_y) + B_{43}(E_{F1D}, n_z, n_y)]$ (5.67)	On the basis of (5.65), $m^*(E_{F1D}, n_x, n_y) = \frac{\hbar^2}{2\bar{\omega}_1}(1 + 2\alpha E_{F1D})$ (5.66)
11. n-Ge, the conduction electrons of which can be defined by two types of energy band models as described in the column of carrier statistics	(a) In accordance with the model of Cardona et al. (5.68), $n_0 = \frac{2g_v}{\pi} \sum_{n_x=1}^{n_{x\max}} \sum_{n_z=1}^{n_{z\max}} [B_{44}(E_{F1D}, n_x, n_z) + B_{45}(E_{F1D}, n_x, n_z)]$ (5.71)	On the basis of (5.69), $m^*(E_{F1D}, n_x, n_y) = \frac{\hbar^2}{2} A'_{29}(E_{F1D}, n_x, n_y)$ (5.70)

(continued)

Table 5.1 (continued)

Type of materials	The carrier statistics	The electron statistics
	(b) In accordance with the model of Wang and Ressler (5.72), $n_0 = \frac{2g_v}{\pi} \sum_{n_x=1}^{n_{x\max}} \sum_{n_z=1}^{n_{z\max}} [B_{46}(E_{F1D}, n_x, n_z) + B_{47}(E_{F1D}, n_x, n_z)]$ (5.75)	On the basis of (5.73), $m^*(E_{F1D}, n_x, n_y) = \frac{\hbar^2}{2} A'_{30}(E_{F1D}, n_x, n_y)$ (5.74)
12. Gallium Antimonide, the carriers of which can be defined by the model of Mathur et al.	In accordance with (5.76), $n_0 = \frac{2g_v}{\pi} \sum_{n_x=1}^{n_{x\max}} \sum_{n_y=1}^{n_{y\max}} [B_{48}(E_{F1D}, n_x, n_y) + B_{49}(E_{F1D}, n_x, n_y)]$ (5.79)	On the basis of (5.77), $m^*(E_{F1D}) = \frac{\hbar^2}{2} A'_{31}(E_{F1D})$ (5.78)
13. II–V materials, as defined by the model of Yamada	In accordance with (5.80), $n_0 = \frac{g_v}{\pi} \sum_{n_x=1}^{n_{x\max}} \sum_{n_y=1}^{n_{y\max}} [B_{49}(E_{F1D}, n_x, n_y) + B_{50}(E_{F1D}, n_x, n_y)]$ (5.83)	On the basis of (5.81), $m^*(E_{F1D}, n_x, n_y) = \frac{\hbar^2}{2} A'_{32\pm}(E_{F1D}, n_x, n_y)$ (5.82)
14. Carbon nanotubes		
(a) (n,n) Nanotubes	(a) In accordance with (5.84f), $n_{1D} = \frac{8}{a_c \pi \sqrt{3}} \sum_{i=0}^{i_{\max}} \left[f_1(E_{F_1}, m, n) + B_{c_2}(E_{F_1}, m, n) \right]$ (5.85)	On the basis of (5.84f), $m^*(E_{F_1}, m, n) = \frac{4\hbar^2}{3a_c^2} f_1(E_{F_1}, m, n) \times \frac{\theta'_1(E_{F_1}, m, n)}{\sqrt{1-\theta_1^2(E_{F_1}, m, n)}}$ (5.84g)
(b) (n,0)Nanotubes	(b) In accordance with (5.86) $n_{1D} = \frac{8}{3a_c \pi} \sum_{i=0}^{i_{\max}} \left[f_2(E_{F_2}, m, n) + B_{c_3}(E_{F_2}, m, n) \right]$ (5.88)	On the basis of (5.86) $m^*(E_{F_2}, m, n) = \frac{4\hbar^2}{9a_c^2} f_2(E_{F_2}, m, n) \times \frac{\theta'_2(E_{F_2}, m, n)}{\sqrt{1-\theta_2^2(E_{F_2}, m, n)}}$ (5.87)

to evaluate the EEM in Ge NWs. The reader is left with the corresponding evaluation of the EEM for Wang et al. model of the same. It appears that the EEM in n-Ge and Bi_2Te_3 evolve from its corresponding bulk value significantly from below 15 nm. However in case of GaSb, the EEM is almost invariant. The summary of this chapter has been presented in Table 5.1.

5.4 Open Research Problems

(R5.1) Investigate the EEM, EAM, DEM, CEM, CoEM, FREM, and OEM for NWs of all the semiconductors whose unperturbed carrier energy spectra are defined in Chap. 1 by considering the presence of finite, symmetric infinite, asymmetric infinite, parabolic, finite circular, infinite circular, and annular infinite potential wells applied separately in the two different orthogonal directions.

(R5.2) Investigate the same set of masses as defined in (R5.1) when all the potentials of (R5.1) are being applied in the two different non-orthogonal directions.

(R5.3) Investigate the same set of masses as defined in (R5.1) in the presence of arbitrarily oriented non-uniform electric field.

(R5.4) Investigate the same set of masses as defined in (R5.1) to (R5.3) under an arbitrarily oriented alternating electric field.

(R5.5) Investigate the same set of masses as defined in (R5.4) all the appropriate problems of this chapter under an arbitrarily oriented alternating magnetic field by including broadening and the electron spin.

(R5.6) Investigate the same set of masses as defined in (R5.1) for the appropriate problems of this chapter under an arbitrarily oriented alternating magnetic field and crossed alternating electric field by including broadening and the electron spin for all the materials whose unperturbed carrier energy spectra are defined Chap. 1.

(R5.7) Investigate the same set of masses as defined in (R5.1) for the appropriate problems of this chapter under an arbitrarily oriented alternating magnetic field and crossed alternating non-uniform electric field by including broadening and the electron spin whose for all the materials unperturbed carrier energy spectra are defined Chap. 1.

(R5.8) Investigate the same set of masses as defined in (R5.1) in the absence of magnetic field for all the appropriate problems of this chapter under exponential, Kane, Halperin, Lax, and Bonch-Bruevich band tails [84] for all the materials whose unperturbed carrier energy spectra are defined Chap. 1.

(R5.9) Investigate the same set of masses as defined in (R5.1) in the absence of magnetic field for all the appropriate problems of this chapter for all the materials whose unperturbed carrier energy spectra are defined in Chap. 1 under an arbitrarily oriented non-uniform alternating electric field, respectively.

(R5.10) Investigate the same set of masses as defined in (R5.1) under an arbitrarily oriented alternating magnetic field by including broadening and the electron spin whose unperturbed carrier energy spectra are defined in Chap. 1.

(R5.11) Investigate the same set of masses as defined in (R5.1) for all the appropriate problems of this chapter for all the materials whose unperturbed carrier energy spectra are defined in Chap. 1 under an arbitrarily oriented alternating magnetic field and crossed alternating electric field by including broadening and the electron spin.

(R5.12) Investigate the same set of masses as defined in (R5.1) all the appropriate problems for all types of systems as discussed in this chapter for p-InSb, p-CuCl, and stressed semiconductors having diamond structure valence bands whose dispersion relations of the carriers in bulk materials are given by Cunningham [84], Yekimov et al. [85] and Roman et al. [86], respectively.

(R5.12) Investigate the same set of masses as defined in (R5.1) the influence of deep traps and surface states separately for all the appropriate problems of this chapter after proper modifications.

(R5.13) Investigate the same set of masses as defined in (R5.1) for all the appropriate problems of this chapter for multiple NWs of all the heavily doped materials as described in (R5.8).

(R5.14) Investigate the same set of masses as defined in (R5.1) all the problems of this chapter by removing all the mathematical approximations and establishing the respective appropriate uniqueness conditions.

References

1. P. Harrison, *Quantum Wells, Wires and Dots* (Wiley, New York, 2002)
2. B.K. Ridley, *Electrons and Phonons in Semiconductors Multilayers* (Cambridge University Press, Cambridge, 1997)
3. V.V. Martin, A.A. Kochelap, M.A. Stroscio, *Quantum Heterostructures* (Cambridge University Press, Cambridge, 1999)
4. C.S. Lent, D.J. Kirkner, J. Appl. Phys. **67**, 6353 (1990)
5. F. Sols, M. Macucci, U. Ravaioli, K. Hess, Appl. Phys. Lett. **54**, 350 (1980)
6. C.S. Kim, A.M. Satanin, Y.S. Joe, R.M. Cosby, Phys. Rev. B **60**, 10962 (1999)
7. S. Midgley, J.B. Wang, Phys. Rev. B **64**, 153304 (2001)
8. T. Sugaya, J.P. Bird, M. Ogura, Y. Sugiyama, D.K. Ferry, K.Y. Jang, Appl. Phys. Lett. **80**, 434 (2002)
9. B. Kane, G. Facer, A. Dzurak, N. Lumpkin, R. Clark, L. PfeiKer, K. West, Appl. Phys. Lett. **72**, 3506 (1998)
10. C. Dekker, Phys. Today **52**, 22 (1999)
11. A. Yacoby, H.L. Stormer, N.S. Wingreen, L.N. Pfeiffer, K.W. Baldwin, K.W. West, Phys. Rev. Lett. **77**, 4612 (1996)
12. Y. Hayamizu, M. Yoshita, S. Watanabe, H.A.L. PfeiKer, K. West, Appl. Phys. Lett. **81**, 4937 (2002)
13. S. Frank, P. Poncharal, Z.L. Wang, W.A. Heer, Science **280**, 1744 (1998)

14. I. Kamiya, I.I. Tanaka, K. Tanaka, F. Yamada, Y. Shinozuka, H. Sakaki, Phys. E **13**, 131 (2002)
15. A.K. Geim, P.C. Main, N. LaScala, L. Eaves, T.J. Foster, P.H. Beton, J.W. Sakai, F.W. Sheard, M. Henini, G. Hill et al., Phys. Rev. Lett. **72**, 2061 (1994)
16. A.S. Melinkov, V.M. Vinokur, Nature **415**, 60 (2002)
17. K. Schwab, E.A. Henriksen, J.M. Worlock, M.L. Roukes, Nature **404**, 974 (2000)
18. L. Kouwenhoven, Nature **403**, 374 (2000)
19. S. Komiyama, O. Astafiev, V. Antonov, H. Hirai, Nature **403**, 405 (2000)
20. E. Paspalakis, Z. Kis, E. Voutsinas, A.F. Terziz, Phys. Rev. B **69**, 155316 (2004)
21. J.H. Jefferson, M. Fearn, D.L.J. Tipton, T.P. Spiller, Phys. Rev. A **66**, 042328 (2002)
22. J. Appenzeller, C. Schroer, T. Schapers, A. Hart, A. Froster, B. Lengeler, H. Luth, Phys. Rev. B **53**, 9959 (1996)
23. J. Appenzeller, C. Schroer, J. Appl. Phys. **87**, 31659 (2002)
24. P. Debray, O.E. Raichev, M. Rahman, R. Akis, W.C. Mitchel, Appl. Phys. Lett. **74**, 768 (1999)
25. D. Kasemet, C.S. Hong, N.B. Patel, P.D. Dapkus, Appl. Phys. Lett. **41**, 912 (1982)
26. K. Woodbridge, P. Blood, E.D. Pletcher, P.J. Hulyer, Appl. Phys. Lett. **45**, 16 (1984)
27. S. Tarucha, H.O. Okamoto, Appl. Phys. Lett. **45**, 16 (1984)
28. M.I. Nathan, Appl. Phys. Lett. **47**, 1105 (1985)
29. I. Suemune, L.A. Coldren, IEEE. J. Quant. Electron. **24**, 1178 (1988)
30. J.L. Shay, J.W. Wernick, *Ternary Chalcopyrite Semiconductors-Growth, Electronic Properties and Applications* (Pergamon Press, London, 1975)
31. E.O. Kane, in *Semiconductors and Semimetals*, vol. 1, ed. by R.K. Willardson, A.C. Beer (Academic Press, New York, 1966), p. 75
32. B.R. Nag, *Electron Transport in Compound Semiconductors* (Springer, Berlin, 1980)
33. G.E. Stillman, C.M. Wolfe, J.O. Dimmock, in Semiconductors and Semimetals, vol. 12, ed. by R.K. Willardon, A.C. Beer, (Academic Press, IV, V, New York, 1977), p. 169
34. D.J. Newson, A. Karobe, Semicond. Sci. Tech. **3**, 786 (1988)
35. E.D. Palik, G.S. Picus, S. Teither, R.E. Wallis, Phys. Rev. **122**, 475 (1961)
36. S. Iijima, Nature **354**, 56 (1991)
37. V.N. Popov, Mater. Sci. Eng. R **43**, 61 (2004)
38. J. Sandler, M.S.P. Shaffer, T. Prasse, W. Bauhofer, K. Schulte, A.H. Windle, Polymer **40**, 5967 (1999)
39. D. Qian, E.C. Dickey, R. Andrews, T. Rantell, Appl. Phys. Lett. **76**, 2868 (2000)
40. J. Kong, N.R. Franklin, C.W. Zhou, M.G. Chapline, S. Peng, K.J. Cho, H.J. Dai, Science **287**, 622 (2000)
41. W.A. Deheer, A. Chatelain, D. Ugarte, Science **270**, 1179 (1995)
42. A.G. Rinzler, J.H. Hafner, P. Nikolaev, L. Lou, S.G. Kim, D. Tomanek, P. Nordlander, D.T. Olbert, R.E. Smalley, Science **269**, 1550 (1995)
43. A.C. Dillon, K.M. Jones, T.A. Bekkedahl, C.H. Kiang, D.S. Bethune, M.J. Heben, Nature **386**, 377 (1997)
44. C. Liu, Y.Y. Fan, M. Liu, H.T. Cong, H.M. Cheng, M.S. Dresselhaus, Science **286**, 1127 (1999)
45. P. Kim, C.M. Lieber, Science **286**, 2148 (1999)
46. D. Srivastava, Nanotechnology **8**, 186 (1997)
47. C. Ke, H.D. Espinosa, Appl. Phys. Lett. **85**, 681 (2004)
48. J.W. Kang, H.J. Hwang, Nanotechnology **15**, 1633 (2004)
49. J. Cumings, A. Zettl, Science **289**, 602 (2000)
50. S.J. Tans, M.H. Devoret, H.J. Dai, A. Thess, R.E. Smalley, L.J. Geerligs, C. Dekker, Nature **386**, 474 (1997)
51. S.J. Tans, A.R.M. Verschueren, C. Dekker, Nature **393**, 49 (1998)
52. P. Avouris, Acc. Chem. Res. **35**, 1026 (2002)
53. P.G. Collins, A. Zettl, H. Bando, A. Thess, R.E. Smalley, Science **278**, 100 (1997)
54. S. Saito, Science **278**, 77 (1997)
55. J.C. Charlier, Acc. Chem. Res. **35**, 1063 (2002)
56. R.H. Baughman, A.A. Zakhidov, W.A. de Heer, Science **297**, 787 (2002)
57. Z. Yao, H.W.C. Postma, L. Balents, C. Dekker, Nature **402**, 273 (1999)

58. C.W. Zhou, J. Kong, E. Yenilmez, H. Dai, Science **290**, 1552 (2000)
59. HWCh. Postma, T. Teepen, Z. Yao, M. Grifoni, C. Dekker, Science **293**, 76 (2001)
60. T. Rueckes, K. Kim, E. Joselevich, G.Y. Tseng, C.L. Cheung, C.M. Lieber, Science **289**, 94 (2000)
61. A. Bacthtold, P. Hadley, T. Nakanish, C. Dekker, Science **294**, 1317 (2001)
62. Q. Zheng, Q. Jiang, Phys. Rev. Lett. **88**, 045503 (2002)
63. Q. Zheng, J.S. Liu, Q. Jiang, Phys. Rev B **65**, 245409 (2002)
64. Y. Zhao, C.-C. Ma, G.H. Chen, Q. Jiang, Phys. Rev. Lett. **91**, 175504 (2003)
65. S.B. Legoas, V.R. Coluci, S.F. Braga, P.Z. Coura, S.O. Dantus, D.S. Galvao, Phys. Rev. Lett. **90**, 055504 (2003)
66. S.B. Legaos, V.R. Coluci, S.F. Braga, P.Z. Coura, S.O. Dantus, D.S. Galvao, Nanotechnology **15**, S184 (2004)
67. J.W. Kang, H.W. Hwang, J. Appl. Phys. **96**, 3900 (2004)
68. W.Y. Choi, J.W. Kang, H.W. Hwang, Phys. E **23**, 125 (2004)
69. H.J. Hwang, K.R. Byun, J.W. Kang, Phys. E **23**, 208 (2004)
70. J.W. Kang, H.J. Hwang, Phys. E **23**, 36 (2004)
71. J.W. Kang, H.J. Hwang, J. Appl. Phys. **73**, 4447 (2004)
72. J.W. Kang, H.J. Hwang, J. Phys. Soc. Jpn. **73**, 1077 (2004)
73. J.W. Kang, Y.W. Choi, H.J. Hwang, J. Comp. Theor. Nanosci. **1**, 199 (2004)
74. M. Dequesnes, S.V. Rotkin, N.R. Aluru, Nanotechnology **13**, 120 (2002)
75. J.M. Kinaret, T. Nord, S. Viefers, Appl. Phys. Lett. **82**, 1287 (2003)
76. L.M. Jonsson, T. Nord, J.M. Kinaret, S. Viefers, J. Appl. Phys. **96**, 629 (2004)
77. L.M. Jonsson, S. Axelsson, T. Nord, S. Viefers, J.M. Kinaret, Nanotechnology **15**, 1497 (2004)
78. S.W. Lee, D.S. Lee, R.E. Morjan, S.H. Jhang, M. Sveningsson, O.A. Nerushev, Y.W. Park, E.E.B. Campbell, Nano. Lett. **4**, 2027 (2004)
79. M. Lundstrom, J. Guo, *Nanoscale Transistors: Device Physics, Modeling and Simulation* (Springer, New York, 2006)
80. R. Saito, G. Dresselhaus, M.S. Dresselhaus, *Physical Properties of Carbon Nanotubes* (Imperial College Press, London, 1998)
81. R. Heyd, A. Charlier, E. McRae, Phys. Rev. B **55**, 6820 (1997)
82. J.W. Mintmire, C.T. White, Phys. Rev. Lett. **81**, 2506 (1998)
83. B.R. Nag, *Electron Transport in Compound Semiconductors* (Springer, Heidelberg, 1980)
84. R.W. Cunningham, Phys. Rev. **167**, 761 (1968)
85. A.I. Yekimov, A.A. Onushchenko, A.G. Plyukhin, Al.L. Efros, J. Exp. Theor. Phys. **88**, 1490 (1985)
86. B.J. Roman, A.W. Ewald, Phys. Rev. B **5**, 3914 (1972)

Part II
Influence of Light Waves on the EEM in Optoelectronic Semiconductors

Chapter 6
The EEM in Quantum Confined Optoelectronic Semiconductors in the Presence of Light Waves

6.1 Introduction

With the advent of nanophotonics, there has been a considerable interest in studying the optical processes in semiconductors and their nanostructures [1]. It appears from the literature that the investigations have been carried out on the assumption that the carrier energy spectra are invariant quantities in the presence of intense light waves, which is not fundamentally true. The physical properties of semiconductors in the presence of light waves which change the basic dispersion relation are relatively less investigated in the literature [2–4]. In this chapter, we shall study the EEM in III–V, ternary, and quaternary semiconductors and their nanostructure on the basis of newly formulated electron dispersion law under external photo excitation under different physical conditions.

In Sect. 6.2.1 of the theoretical background (Sect. 6.2), we have formulated the dispersion relation of the conduction electrons of III–V, ternary, and quaternary materials in the presence of light waves whose unperturbed electron energy spectrum is described by the three-band model of Kane. In the same section, we have studied the dispersion relations for the said materials in the presence of external photo-excitation when the unperturbed energy spectra are defined by the two band model of Kane and that of parabolic energy bands, respectively, for the purpose of relative comparison. In Sect. 6.2.2, we have derived the expressions of the electron statistics and the EEM for all the aforementioned cases. We have also investigated the EEM for the aforementioned band models in the absence of light waves consequently. In Sect. 6.3, the EEM has been numerically investigated by taking n-InAs and n-InSb as examples of III–V semiconductors, n-$Hg_{1-x}Cd_xTe$ as an example of ternary compounds and n-$In_{1-x}Ga_xAs_yP_{1-y}$ lattice matched to InP as an example of quaternary materials in accordance with the three- and two-band models of Kane together with model of parabolic energy bands, respectively, for the purpose of relative assessment.

S. Bhattacharya and K. P. Ghatak, *Effective Electron Mass in Low-Dimensional Semiconductors*, Springer Series in Materials Science 167,
DOI: 10.1007/978-3-642-31248-9_6,

6.2 Theoretical Background

6.2.1 The Formulation of the Electron Dispersion Relation in the Presence of Light Waves in III–V, Ternary and Quaternary Semiconductors

The Hamiltonian $(\hat{H})$ of an electron in the presence of light wave characterized by the vector potential $\vec{A}$ can be written following [5] as

$$\hat{H} = \left[\left|(\hat{p} + |e|\,\vec{A})\right|^2 \Big/ 2m\right] + V(\vec{r}) \tag{6.1}$$

in which, $\hat{p}$ is the momentum operator, $V(\vec{r})$is the crystal potential and m is the free electron mass. Equation (6.1) can be expressed as

$$\hat{H} = \hat{H}_0 + \hat{H}' \tag{6.2}$$

where, $\hat{H}_0 = \frac{\hat{p}^2}{2m} + V(\vec{r})$ and

$$\hat{H}' = \frac{|e|}{2m}\vec{A} \cdot \hat{p} \tag{6.3}$$

The perturbed Hamiltonian $\hat{H}'$ can be written as

$$\hat{H}' = \left(\frac{-i\hbar\,|e|}{2m}\right)(\vec{A} \cdot \nabla) \tag{6.4}$$

where $i = \sqrt{-1}$ and $\hat{p} = -i\hbar\nabla$.

The vector potential $(\vec{A})$ of the monochromatic light of plane wave can be expressed as

$$\vec{A} = A_0 \vec{\varepsilon}_s \cos(\vec{s}_0 \cdot \vec{r} - \omega t) \tag{6.5}$$

where A_0 is the amplitude of the light wave, $\vec{\varepsilon}_s$ is the polarization vector, $\vec{s}_0$ is the momentum vector of the incident photon, $\vec{r}$ is the position vector, ω is the angular frequency of light wave, and t is the time scale. The matrix element of $\hat{H}'_{nl}$ between initial state, $\psi_l(\vec{q}, \vec{r})$ and final state $\psi_n(\vec{k}, \vec{r})$ in different bands can be written as

$$\hat{H}'_{nl} = \frac{|e|}{2m}\left\langle n\vec{k}\left|\vec{A} \cdot \hat{p}\right| l\vec{q}\right\rangle \tag{6.6}$$

Using (6.4) and (6.5), we can rewrite (6.6) as

$$\hat{H}'_{nl} = \left(\frac{-i\hbar\,|e|\,A_0}{4m}\right)\vec{\varepsilon}_s \cdot \left[\left\{\left\langle n\vec{k}\left|e^{(i\vec{s}_0\cdot\vec{r})}\nabla\right| l\vec{q}\right\rangle e^{-i\omega t}\right\} + \left\{\left\langle n\vec{k}\left|e^{(-i\vec{s}_0\cdot\vec{r})}\nabla\right| l\vec{q}\right\rangle e^{i\omega t}\right\}\right] \tag{6.7}$$

The first matrix element of (6.7) can be written as

$$\left\langle n\vec{k}\left|e^{(i\vec{s}_0\cdot\vec{r})}\nabla\right| l\vec{q}\right\rangle = \int e^{\left(i\left[\vec{q}+\vec{s}_0-\vec{k}\right]\cdot\vec{r}\right)} i\vec{q}u_n^*(\vec{k},\vec{r})u_l(\vec{q},\vec{r})\mathrm{d}^3r + \int e^{\left(i\left[\vec{q}+\vec{s}_0-\vec{k}\right]\cdot\vec{r}\right)} u_n^*(\vec{k},\vec{r})\nabla u_l(\vec{q},\vec{r})\mathrm{d}^3r \tag{6.8}$$

The functions $u_n^* u_l$ and $u_n^* \nabla u_l$ are periodic. The integral over all space can be separated into a sum over unit cells times an integral over a single unit cell. It is assumed that the wavelength of the electromagnetic wave is sufficiently large so that if $\vec{k}$ and $\vec{q}$ are within the Brillouin zone, $(\vec{q} + \vec{s}_0 - \vec{k})$ is not a reciprocal lattice vector.

Therefore, we can write (6.8) as

$$\begin{aligned}\left\langle n\vec{k}|e^{(i\vec{s}_0\cdot\vec{r})}\nabla|l\vec{q}\right\rangle &= \left[\frac{(2\pi)^3}{\Omega}\right]\left\{i\vec{q}\delta(\vec{q}+\vec{s}_0-\vec{k})\delta_{nl} + \delta(\vec{q}+\vec{s}_0-\vec{k})\int_{\text{cell}} u_n^*(\vec{k},\vec{r})\nabla u_l(\vec{q},\vec{r})\mathrm{d}^3r\right\}\\ &= \left[\frac{(2\pi)^3}{\Omega}\right]\left\{\delta(\vec{q}+\vec{s}_0-\vec{k})\int_{\text{cell}} u_n^*(\vec{k},\vec{r})\nabla u_l(\vec{q},\vec{r})\mathrm{d}^3r\right\}\end{aligned} \tag{6.9}$$

where Ω is the volume of the unit cell and $\int u_n^*(\vec{k},\vec{r})u_l(\vec{q},\vec{r})\mathrm{d}^3r = \delta(\vec{q}-\vec{k})\delta_{nl} = 0$, since $n \neq l$.

The delta function expresses the conservation of wave vector in the absorption of light wave and $\vec{s}_0$ is small compared to the dimension of a typical Brillouin zone and we set $\vec{q} = \vec{k}$.

From (6.8) and (6.9), we can write,

$$\hat{H}'_{nl} = \frac{|e|\,A_0}{2m}\vec{\varepsilon}_s \cdot \hat{p}_{nl}(\vec{k})\delta(\vec{q}-\vec{k})\cos(\omega t) \tag{6.10}$$

where, $\hat{p}_{nl}(\vec{k}) = -i\hbar\int u_n^*\nabla u_l\mathrm{d}^3r = \int u_n^*(\vec{k},\vec{r})\hat{p}u_l(\vec{k},\vec{r})\mathrm{d}^3r$

Therefore, we can write

$$\hat{H}'_{nl} = \frac{|e|\,A_0}{2m}\vec{\varepsilon}\cdot\hat{p}_{nl}(\vec{k}) \tag{6.11}$$

where, $\vec{\varepsilon} = \vec{\varepsilon}_s\cos\omega t$.

When a photon interacts with a semiconductor, the carriers (i.e., electrons) are generated in the bands which are followed by the interband transitions. For example, when the carriers are generated in the valence band, the carriers then make interband transition to the conduction band. The transition of the electrons within the same band, i.e., $\hat{H}'_{nn} = \left\langle n\vec{k} \left| \hat{H}' \right| n\vec{k} \right\rangle$ is neglected. Because, in such a case, i.e., when the carriers are generated within the same bands by photons, they are lost by recombination within the aforementioned band resulting in zero carriers.

Therefore,

$$\left\langle n\vec{k} \left| \hat{H}' \right| n\vec{k} \right\rangle = 0 \tag{6.12}$$

With $n = c$ stands for conduction band and $l = v$ stand for valance band, the energy equation for the conduction electron can approximately be written as

$$I_{11}(E) = \left(\frac{\hbar^2 k^2}{2m_c}\right) + \frac{\left(\frac{|e|A_0}{2m}\right)^2 \left\langle \left| \vec{\varepsilon} \cdot \hat{p}_{cv}(\vec{k}) \right|^2 \right\rangle_{av}}{E_c(\vec{k}) - E_v(\vec{k})} \tag{6.13}$$

where, $I_{11}(E) \equiv E(aE+1)(bE+1)/(cE+1), a \equiv 1/E_{g_0}$, E_{g_0} is the un-perturbed band-gap, $b \equiv 1/(E_{g_0} + \Delta), c \equiv 1/(E_{g_0} + 2\Delta/3)$, and $\left\langle \left| \vec{\varepsilon} \cdot \hat{p}_{cv}(\vec{k}) \right|^2 \right\rangle_{av}$ represents the average of the square of the optical matrix element (OME).

For the three-band model of Kane, we can write,

$$\xi_{1k} = E_c(\vec{k}) - E_v(\vec{k}) = (E_{g_0}^2 + E_{g_0}\hbar^2 k^2/m_r)^{1/2} \tag{6.14}$$

where, m_r is the reduced mass and is given by $m_r^{-1} = (m_c)^{-1} + m_v^{-1}$, and m_v is the effective mass of the heavy hole at the top of the valance band in the absence of any field.

The doubly degenerate wave functions $u_1(\vec{k}, \vec{r})$ and $u_2(\vec{k}, \vec{r})$ can be expressed as [6, 7]

$$u_1(\vec{k}, \vec{r}) = a_{k+}\left[(is)\downarrow'\right] + b_{k+}\left[\frac{X' - iY'}{\sqrt{2}}\uparrow'\right] + c_{k+}\left[Z'\downarrow'\right] \tag{6.15}$$

and

$$u_2(\vec{k}, \vec{r}) = a_{k-}\left[(is)\uparrow'\right] - b_{k-}\left[\frac{X' + iY'}{\sqrt{2}}\downarrow'\right] + c_{k-}\left[Z'\uparrow'\right] \tag{6.16}$$

s is the s-type atomic orbital in both unprimed and primed coordinates, $\downarrow'$ indicates the spin down function in the primed coordinates,

$$a_{k\pm} \equiv \beta[E_{g_0} - (\gamma_{0k\pm})^2(E_{g_0} - \delta')^{1/2}(E_{g_0} + \delta')^{-1/2}],$$
$$\beta \equiv [(6(E_{g0} + 2\Delta/3)(E_{g0} + \Delta))/\chi]^{1/2},$$

$$\chi \equiv (6E_{g_0}^2 + 9E_{g_0}\Delta + 4\Delta^2), \quad \gamma_{0k\pm} \equiv \left[\frac{(\xi_{1k} \mp E_{g_0})}{2(\xi_{1k} + \delta')}\right]^{1/2},$$

$$\xi_{1k} \equiv E_c(\vec{k}) - E_v(\vec{k}) = E_{g_0}\left[1 + 2\left(1 + \frac{m_c}{m_v}\right)\frac{I_{11}(E)}{E_{g_0}}\right]^{1/2},$$

$\delta' \equiv (E_{g_0}^2\Delta)(\chi)^{-1}$, X', Y', and Z' are the p-type atomic orbitals in the primed coordinates, $\uparrow'$ indicates the spin-up function in the primed coordinates, $b_{k\pm} \equiv \rho\gamma_{0k\pm}$, $\rho \equiv (4\Delta^2/3\chi)^{1/2}$, $c_{k\pm} \equiv t\gamma_{0_{k\pm}}$ and $t \equiv \left[6(E_{g_0} + 2\Delta/3)^2/\chi\right]^{1/2}$.

We can, therefore, write the expression for the OME as

$$\text{OME} = \hat{p}_{cv}(\vec{k}) = \langle u_1(\vec{k},\vec{r})|\hat{p}|u_2(\vec{k},\vec{r})\rangle \tag{6.17}$$

Since the photon vector has no interaction in the same band for the study of interband optical transition, we can therefore write

$$\langle S|\hat{p}|S\rangle = \langle X|\hat{p}|X\rangle = \langle Y|\hat{p}|Y\rangle = \langle Z|\hat{p}|Z\rangle = 0$$

and $\langle X|\hat{p}|Y\rangle = \langle Y|\hat{p}|Z\rangle = \langle Z|\hat{p}|X\rangle = 0$.

There are finite interactions between the conduction band (CB) and the valance band (VB) and we can obtain

$$\langle S|\hat{P}|X\rangle = \hat{i}\cdot\hat{P} = \hat{i}\cdot\hat{P}_x$$
$$\langle S|\hat{P}|Y\rangle = \hat{j}\cdot\hat{P} = \hat{j}\cdot\hat{P}_y$$
$$\langle S|\hat{P}|Z\rangle = \hat{k}\cdot\hat{P} = \hat{k}\cdot\hat{P}_z$$

where, $\hat{i}$, $\hat{j}$, and $\hat{k}$ are the unit vectors along x, y, and z axes, respectively.

It is well known that

$$\begin{bmatrix}\uparrow'\\ \downarrow'\end{bmatrix} = \begin{bmatrix} e^{-i\phi/2}\cos(\theta/2) & e^{i\phi/2}\sin(\theta/2)\\ -e^{-i\phi/2}\sin(\theta/2) & e^{i\phi/2}\cos(\theta/2)\end{bmatrix}\begin{bmatrix}\uparrow\\ \downarrow\end{bmatrix}$$

and

$$\begin{bmatrix}X'\\ Y'\\ Z'\end{bmatrix} = \begin{bmatrix}\cos\theta\cos\phi & \cos\theta\sin\phi & -\sin\theta\\ -\sin\phi & \cos\phi & 0\\ \sin\theta\cos\phi & \sin\theta\sin\phi & \cos\theta\end{bmatrix}\begin{bmatrix}X\\ Y\\ Z\end{bmatrix}$$

Besides, the spin vector can be written as

$$\vec{S} = \frac{\hbar}{2}\vec{\sigma}, \quad \text{where,} \quad \sigma_x = \begin{bmatrix}0 & 1\\ 1 & 0\end{bmatrix}, \sigma_y = \begin{bmatrix}0 & -i\\ i & 0\end{bmatrix} \quad \text{and} \quad \sigma_z = \begin{bmatrix}1 & 0\\ 0 & -1\end{bmatrix}.$$

From above, we can write

$$\begin{aligned}\hat{p}_{CV}(\vec{k}) &= \left\langle u_1(\vec{k},\vec{r})\left|\hat{P}\right|u_2(\vec{k},\vec{r})\right\rangle \\ &= \left\langle\left\{a_{k_+}\left[(iS)\downarrow'\right]+b_{k_+}\left[\left(\frac{X'-iY'}{\sqrt{2}}\right)\uparrow'\right]+c_{k_+}\left[Z'\downarrow'\right]\right\}\left|\hat{P}\right| \right.\\ &\quad\left.\times\left\{a_{k_-}\left[(iS)\uparrow'\right]-b_{k_-}\left[\left(\frac{X'+iY'}{\sqrt{2}}\right)\downarrow'+c_{k_-}\left[Z'\uparrow'\right]\right]\right\}\right\rangle.\end{aligned}$$

Using above relations, we get

$$\begin{aligned}\hat{p}_{CV}(\vec{k}) &= \left\langle u_1(\vec{k},\vec{r})\left|\hat{P}\right|u_2(\vec{k},\vec{r})\right\rangle \\ &= \frac{b_{k_+}a_{k_-}}{\sqrt{2}}\{\langle(X'-iY')|\hat{P}|iS\rangle\langle\uparrow'|\uparrow'\rangle\}+c_{k_+}a_{k_-}\{\langle Z'|\hat{P}|iS\rangle\langle\downarrow'|\uparrow'\rangle\} \\ &\quad-\frac{a_{k_+}b_{k_-}}{\sqrt{2}}\{\langle iS|\hat{P}|(X'+iY')\rangle\langle\downarrow'|\downarrow'\rangle\}+a_{k_+}c_{k_-}\{\langle iS|\hat{P}|Z'\rangle\langle\downarrow'|\uparrow'\rangle\}\end{aligned} \tag{6.18}$$

From (6.18), we can write

$$\begin{aligned}\left\langle(X'-iY')\left|\hat{P}\right|iS\right\rangle &= \left\langle(X')\left|\hat{P}\right|iS\right\rangle-\left\langle(iY')\left|\hat{P}\right|iS\right\rangle \\ &= i\int u_{X'}^{*}\hat{P}S-\int -iu_{Y'}^{*}\hat{P}iu_X = i\left\langle X'\left|\hat{P}\right|S\right\rangle-\left\langle Y'\left|\hat{P}\right|S\right\rangle\end{aligned}$$

From the above relations, for X', Y' and Z', we get

$$\left|X'\right\rangle = \cos\theta\cos\phi\,|X\rangle+\cos\theta\sin\phi\,|Y\rangle-\sin\theta\,|Z\rangle$$

Thus,

$$\left\langle X'\left|\hat{P}\right|S\right\rangle = \cos\theta\cos\phi\left\langle X\left|\hat{P}\right|S\right\rangle+\cos\theta\sin\phi\left\langle Y\left|\hat{P}\right|S\right\rangle-\sin\theta\left\langle Z\left|\hat{P}\right|S\right\rangle = \hat{P}\hat{r}_1$$

where $\hat{r}_1 = \hat{i}\cos\theta\cos\phi+\hat{j}\cos\theta\sin\phi-\hat{k}\sin\theta$

$$\left|Y'\right\rangle = -\sin\phi\,|X\rangle+\cos\phi\,|Y\rangle+0\,|Z\rangle$$

Thus,

$$\left\langle Y'\left|\hat{P}\right|S\right\rangle = -\sin\phi\left\langle X\left|\hat{P}\right|S\right\rangle+\cos\phi\left\langle Y\left|\hat{P}\right|S\right\rangle+0\left\langle Z\left|\hat{P}\right|S\right\rangle = \hat{P}\hat{r}_2$$

where $\hat{r}_2 = -\hat{i}\sin\phi+\hat{j}\cos\phi$,
so that $\left\langle(X'-iY')\left|\hat{P}\right|S\right\rangle = \hat{P}(i\hat{r}_1-\hat{r}_2)$.

Thus,

$$\frac{a_{k_-}b_{k_+}}{\sqrt{2}}\left\langle (X'-iY')\left|\hat{P}\right|S\right\rangle\langle\uparrow'|\uparrow'\rangle=\frac{a_{k_-}b_{k_+}}{\sqrt{2}}\hat{P}(i\hat{r}_1-\hat{r}_2)\langle\uparrow'|\uparrow'\rangle \tag{6.19}$$

Now since,

$$\left\langle iS\left|\hat{P}\right|(X'+iY')\right\rangle=i\left\langle S\left|\hat{P}\right|X'\right\rangle-\left\langle S\left|\hat{P}\right|Y'\right\rangle=\hat{P}(i\hat{r}_1-\hat{r}_2)$$

We can write,

$$-\left[\frac{a_{k_+}b_{k_-}}{\sqrt{2}}\left\{\left\langle iS\left|\hat{P}\right|(X'+iY')\right\rangle\langle\downarrow'|\downarrow'\rangle\right\}\right]=-\left[\frac{a_{k_+}b_{k_-}}{\sqrt{2}}\hat{P}(i\hat{r}_1-\hat{r}_2)\langle\downarrow'|\downarrow'\rangle\right] \tag{6.20}$$

Similarly, we get

$$\left|Z'\right\rangle=\sin\theta\cos\phi\,|X\rangle+\sin\theta\sin\phi\,|Y\rangle+\cos\theta\,|Z\rangle$$

so that, $\left\langle Z'\left|\hat{P}\right|iS\right\rangle=i\left\langle Z'\left|\hat{P}\right|S\right\rangle=i\hat{P}\left\{\sin\theta\cos\phi\hat{i}+\sin\theta\sin\phi\hat{j}+\cos\theta\hat{k}\right\}=i\hat{P}\hat{r}_3$
where $\hat{r}_3=\hat{i}\sin\theta\cos\phi+\hat{j}\sin\theta\sin\phi+\hat{k}\cos\theta$.

Thus,

$$c_{k_+}a_{k_-}\left\langle Z'\left|\hat{P}\right|iS\right\rangle\langle\downarrow'|\uparrow'\rangle=c_{k_+}a_{k_-}i\hat{P}\hat{r}_3\langle\downarrow'|\uparrow'\rangle \tag{6.21}$$

Similarly, we can write,

$$c_{k_-}a_{k_+}\left\langle iS\left|\hat{P}\right|Z'\right\rangle\langle\downarrow'|\uparrow'\rangle=c_{k_-}a_{k_+}i\hat{P}\hat{r}_3\langle\downarrow'|\uparrow'\rangle \tag{6.22}$$

Therefore, we obtain

$$\begin{aligned}&\frac{a_{k_-}b_{k_+}}{\sqrt{2}}\left\{\left\langle (X'-iY')\left|\hat{P}\right|S\right\rangle\langle\uparrow'|\uparrow'\rangle\right\}-\frac{a_{k_+}b_{k_-}}{\sqrt{2}}\left\{\left\langle iS\left|\hat{P}\right|(X'+iY')\right\rangle\langle\downarrow'|\downarrow'\rangle\right\}\\&=\frac{\hat{P}}{\sqrt{2}}(-a_{k_+}b_{k_-}\langle\downarrow'|\downarrow'\rangle+a_{k_-}b_{k_+}\langle\uparrow|\uparrow'\rangle)(i\hat{r}_1-\hat{r}_2)\end{aligned} \tag{6.23}$$

Also, we can write,

$$\begin{aligned}&c_{k_+}a_{k_-}\left\langle Z'\left|\hat{P}\right|iS\right\rangle\langle\downarrow'|\uparrow'\rangle+c_{k_-}a_{k_+}\left\langle iS\left|\hat{P}\right|Z'\right\rangle\langle\downarrow'|\uparrow'\rangle\\&=i\hat{P}(c_{k_+}a_{k_-}+c_{k_-}a_{k_+})\hat{r}_3\left[\langle\downarrow'|\downarrow'\rangle\right]\end{aligned} \tag{6.24}$$

Combining (6.23) and (6.24), we find

$$\hat{p}_{CV}(\vec{k}) = \frac{\hat{P}}{\sqrt{2}}(i\hat{r}_1 - \hat{r}_2)\left\{(b_{k_+}a_{k_-})\langle\uparrow'|\uparrow'\rangle - (b_{k_-}a_{k_+})\langle\downarrow'|\downarrow'\rangle\right\} + i\hat{P}\hat{r}_3(c_{k_+}a_{k_-} - c_{k_-}a_{k_+})\langle\downarrow'|\uparrow'\rangle \tag{6.25}$$

From the above relations, we obtain,

$$\left.\begin{aligned}\uparrow' &= e^{-i\phi/2}\cos(\theta/2)\uparrow + e^{i\phi/2}\sin(\theta/2)\downarrow\\ \downarrow' &= -e^{-i\phi/2}\sin(\theta/2)\uparrow + e^{i\phi/2}\cos(\theta/2)\downarrow\end{aligned}\right\} \tag{6.26}$$

Therefore,

$$\begin{aligned}\langle\downarrow'|\uparrow'\rangle_x = &-\sin(\theta/2)\cos(\theta/2)\langle\uparrow|\uparrow\rangle_x + e^{-i\phi}\cos^2(\theta/2)\langle\downarrow|\uparrow\rangle_x\\ &- e^{i\phi}\sin^2(\theta/2)\langle\uparrow|\downarrow\rangle_x + \sin(\theta/2)\cos(\theta/2)\langle\downarrow|\downarrow\rangle_x\end{aligned} \tag{6.27}$$

But we know from above that

$$\langle\uparrow|\uparrow\rangle_x = 0,\quad \langle\downarrow|\uparrow\rangle = \frac{1}{2},\quad \langle\downarrow|\uparrow\rangle_x = \frac{1}{2}\quad \text{and}\quad \langle\downarrow|\downarrow\rangle_x = 0$$

Thus, from Eq. (6.27), we get

$$\begin{aligned}\langle\downarrow'|\uparrow'\rangle_x &= \frac{1}{2}\left[e^{-i\phi}\cos^2(\theta/2) - e^{i\phi}\sin^2(\theta/2)\right]\\ &= \frac{1}{2}\left[(\cos\phi - i\sin\phi)\cos^2(\theta/2) - (\cos\phi + i\sin\phi)\sin^2(\theta/2)\right]\\ &= \frac{1}{2}\left[\cos\phi\cos\theta - i\sin\phi\right]\end{aligned} \tag{6.28}$$

Similarly, we obtain

$$\langle\downarrow'|\uparrow'\rangle_y = \frac{1}{2}\left[i\cos\phi + \sin\phi\cos\theta\right]\quad \text{and}\quad \langle\downarrow'|\uparrow'\rangle_z = \frac{1}{2}\left[-\sin\theta\right]$$

Therefore,

$$\begin{aligned}\langle\downarrow'|\uparrow'\rangle &= \hat{i}\langle\downarrow'|\uparrow'\rangle_x + \hat{j}\langle\downarrow'|\uparrow'\rangle_y + \hat{k}\langle\downarrow'|\uparrow'\rangle_z\\ &= \frac{1}{2}\left\{(\cos\theta\cos\phi - i\sin\phi)\hat{i} + (i\cos\phi + \sin\phi\cos\theta)\hat{j} - \sin\theta\hat{k}\right\}\\ &= \frac{1}{2}\left[\left\{(\cos\theta\cos\phi)\hat{i} + (\sin\phi\cos\theta)\hat{j} - \sin\theta\hat{k}\right\} + i\left\{-\hat{i}\sin\phi + \hat{j}\cos\phi\right\}\right]\\ &= \frac{1}{2}\left[\hat{r}_1 + i\hat{r}_2\right] = -\frac{1}{2}i\left[i\hat{r}_1 - \hat{r}_2\right]\end{aligned}$$

Similarly, we can write

$$\langle\uparrow'|\uparrow'\rangle = \frac{1}{2}\left[\hat{i}\sin\theta\cos\phi + \hat{j}\sin\theta\sin\phi + \hat{k}\cos\theta\right] = \frac{1}{2}\hat{r}_3 \quad \text{and} \quad \langle\downarrow'|\downarrow'\rangle = -\frac{1}{2}\hat{r}_3$$

Using the above results and following (6.25) we can write

$$\begin{aligned}
\hat{p}_{CV}(\vec{k}) &= \frac{\hat{P}}{\sqrt{2}}(i\hat{r}_1 - \hat{r}_2)\left\{(a_{k_}b_{k_+})\langle\uparrow'|\uparrow'\rangle - (b_{k_}a_{k_+})\langle\downarrow'|\downarrow'\rangle\right\} \\
&\quad + i\hat{P}\hat{r}_3\left\{(c_{k_+}a_{k_} - c_{k_}a_{k_+})\langle\downarrow'|\uparrow'\rangle\right\} \\
&= \frac{\hat{P}}{2}\hat{r}_3(i\hat{r}_1 - \hat{r}_2)\left\{\left(\frac{a_{k_}b_{k_+}}{\sqrt{2}} + \frac{b_{k_}a_{k_+}}{\sqrt{2}}\right)\right\} \\
&\quad + \frac{\hat{P}}{2}\hat{r}_3(i\hat{r}_1 - \hat{r}_2)\left\{(c_{k_+}a_{k_} + c_{k_}a_{k_+})\right\}
\end{aligned}$$

Thus,

$$\hat{p}_{CV}(\vec{k}) = \frac{\hat{P}}{2}\hat{r}_3(i\hat{r}_1 - \hat{r}_2)\left\{a_{k_+}\left(\frac{b_{k\_}}{\sqrt{2}} + c_{k\_}\right) + a_{k\_}\left(\frac{b_{k_+}}{\sqrt{2}} + c_{k_+}\right)\right\} \tag{6.29}$$

We can write that,

$$|\hat{r}_1| = |\hat{r}_2| = |\hat{r}_3| = 1, \quad \text{also,} \quad \hat{P}\hat{r}_3 = \hat{P}_x\sin\theta\cos\phi\hat{i} + \hat{P}_y\sin\theta\sin\phi\hat{j} + \hat{P}_z\cos\theta\hat{k}$$

where, $\hat{P} = \left\langle S\middle|\hat{P}\middle|X\right\rangle = \left\langle S\middle|\hat{P}\middle|Y\right\rangle = \left\langle S\middle|\hat{P}\middle|Z\right\rangle$,

$$\left\langle S\middle|\hat{P}\middle|X\right\rangle = \int u_C^*(0,\vec{r})\hat{P}u_{VX}(0,\vec{r})\mathrm{d}^3r = \hat{P}_{CVX}(0) \quad \text{and} \quad \left\langle S\middle|\hat{P}\middle|Z\right\rangle = \hat{P}_{CVZ}(0)$$

Thus,

$$\hat{P} = \hat{P}_{CVX}(0) = \hat{P}_{CVY}(0) = \hat{P}_{CVZ}(0) = \hat{P}_{CV}(0)$$

where, $\hat{P}_{CV}(0) \equiv \int u_c^*(0,\vec{r})\hat{P}u_V(0,\vec{r})\mathrm{d}^3r \equiv \hat{P}$.

For a plane polarized light wave, we have the polarization vector $\vec{\varepsilon}_s = \hat{k}$, when the light wave vector is traveling along the z-axis. Therefore, for a plane polarized light wave, we have considered $\vec{\varepsilon}_s = \hat{k}$.

Then, from (6.29) we get

$$(\vec{\varepsilon}\cdot\hat{p}_{CV}(\vec{k})) = \vec{k}\cdot\frac{\hat{P}}{2}\hat{r}_3(i\hat{r}_1 - \hat{r}_2)\left[A(\vec{k}) + B(\vec{k})\right]\cos\omega t \tag{6.30}$$

and

$$\left.\begin{aligned} A(\vec{k}) &= a_{k_-}\left(\frac{b_{k_+}}{\sqrt{2}} + c_{k_+}\right) \\ B(\vec{k}) &= a_{k+}\left(\frac{b_{k_-}}{\sqrt{2}} + c_{k-}\right) \end{aligned}\right\} \tag{6.31}$$

Thus,

$$\begin{aligned} \left|\vec{\varepsilon}\cdot\hat{p}_{cv}(\vec{k})\right|^2 &= \left|\hat{k}\cdot\frac{\hat{P}}{2}\hat{r}_3\right|^2 \left|i\hat{r}_1 - \hat{r}_2\right|^2 \left[A(\vec{k}) + B(\vec{k})\right]^2 \cos^2\omega t \\ &= \frac{1}{4}\left|\hat{P}_z\cos\theta\right|^2 \left[A(\vec{k}) + B(\vec{k})\right]^2 \cos^2\omega t \end{aligned} \tag{6.32}$$

So, the average value of $\left|\vec{\varepsilon}\cdot\hat{p}_{cv}(\vec{k})\right|^2$ for a plane polarized light wave is given by

$$\begin{aligned} \left\langle\left|\vec{\varepsilon}\cdot\hat{p}_{cv}\left(\vec{k}\right)\right|^2\right\rangle_{av} &= \frac{2}{4}\left|\hat{P}_z\right|^2\left[A(\vec{k}) + B(\vec{k})\right]^2 \left(\int_0^{2\pi} d\phi \int_0^{\pi} \cos^2\theta\sin\theta d\theta\right)\left(\frac{1}{2}\right) \\ &= \frac{2\pi}{3}\left|\hat{P}_z\right|^2\left[A(\vec{k}) + B(\vec{k})\right]^2 \end{aligned} \tag{6.33}$$

where $\left|\hat{P}_z\right|^2 = (\frac{1}{2})\left|\vec{k}\cdot\hat{p}_{cv}(0)\right|^2$ and

$$\left|\vec{k}\cdot\hat{p}_{cv}(0)\right|^2 = \frac{m^2}{4m_r}\frac{E_{g_0}\left(E_{g_0}+\Delta\right)}{\left(E_{g_0}+\frac{2}{3}\Delta\right)} \tag{6.34}$$

We shall express $A(\vec{k})$ and $B(\vec{k})$ in terms of constants of the energy spectra in the following way:

Substituting $a_{k_\pm}$, $b_{k_\pm}$, $c_{k_\pm}$ and $\gamma_{0k\pm}$ in $A(\vec{k})$ and $B(\vec{k})$ in (6.31) we get

$$A(\vec{k}) = \beta\left(t + \frac{\rho}{\sqrt{2}}\right)\left\{\left(\frac{E_{g_0}}{E_{g_0}+\delta'}\right)\gamma_{0k_+}^2 - \gamma_{0k_+}^2\gamma_{0k_-}^2\left(\frac{E_{g_0}-\delta'}{E_{g_0}+\delta'}\right)\right\}^{1/2} \tag{6.35}$$

$$B(\vec{k}) = \beta\left(t + \frac{\rho}{\sqrt{2}}\right)\left\{\left(\frac{E_{g_0}}{E_{g_0}+\delta'}\right)\gamma_{0k_-}^2 - \gamma_{0k_+}^2\gamma_{0k_-}^2\left(\frac{E_{g_0}-\delta'}{E_{g_0}+\delta'}\right)\right\}^{1/2} \tag{6.36}$$

in which,

$$\gamma_{0k_+}^2 \equiv \frac{\xi_{1k} - E_{g_0}}{2\left(\xi_{1k}+\delta'\right)} \equiv \frac{1}{2}\left[1 - \left(\frac{E_{g_0}+\delta'}{\xi_{1k}+\delta'}\right)\right]$$

and

$$\gamma_{0k_-}^2 \equiv \frac{\xi_{1k} + E_{g0}}{2\left(\xi_{1k} + \delta'\right)} \equiv \frac{1}{2}\left[1 + \left(\frac{E_{g0} - \delta'}{\xi_{1k} + \delta'}\right)\right]$$

Substituting $x \equiv \xi_{1k} + \delta'$ in $\gamma_{0k_\pm}^2$, we can write,

$$A(\vec{k}) = \beta\left(t + \frac{\rho}{\sqrt{2}}\right)\left\{\left(\frac{E_{g0}}{E_{g0} + \delta'}\right)\frac{1}{2}\left(1 - \frac{E_{g_0} + \delta'}{x}\right)\right.$$
$$\left. -\frac{1}{4}\left(\frac{E_{g0} - \delta'}{E_{g0} + \delta'}\right)\left(1 - \frac{E_{g0} + \delta'}{x}\right)\left(1 + \frac{E_{g0} - \delta'}{x}\right)\right\}^{1/2}$$

Thus,

$$A(\vec{k}) = \frac{\beta}{2}\left(t + \frac{\rho}{\sqrt{2}}\right)\left\{1 - \frac{2a_0}{x} + \frac{a_1}{x^2}\right\}^{1/2}$$

where $a_0 \equiv (E_{g0}^2 + \delta'^2)(E_{g0} + \delta')^{-1}$ and $a_1 \equiv (E_{g0} - \delta')^2$.

After tedious algebra, one can show that

$$A(\vec{k}) = \frac{\beta}{2}\left(t + \frac{\rho}{\sqrt{2}}\right)(E_{g0} - \delta')\left[\frac{1}{\xi_{1k} + \delta'} - \frac{1}{E_{g0} + \delta'}\right]^{1/2}$$
$$\times\left[\frac{1}{\xi_{1k} + \delta'} - \frac{(E_{g0} + \delta')}{(E_{g0} - \delta')^2}\right]^{1/2} \tag{6.37}$$

Similarly, from (6.36), we can write,

$$B(\vec{k}) = \beta\left(t + \frac{\rho}{\sqrt{2}}\right)\left\{\left(\frac{E_{g0}}{E_{g0} + \delta'}\right)\frac{1}{2}\left(1 + \frac{E_{g0} - \delta'}{x}\right)\right.$$
$$\left. -\frac{1}{4}\left(\frac{E_{g0} - \delta'}{E_{g0} + \delta'}\right)\left(1 - \frac{E_{g0} + \delta'}{x}\right)\left(1 + \frac{E_{g0} - \delta'}{x}\right)\right\}^{1/2}$$

so that, finally we get,

$$B(\vec{k}) = \frac{\beta}{2}\left(t + \frac{\rho}{\sqrt{2}}\right)\left(1 + \frac{E_{g0} - \delta'}{\xi_{1k} + \delta'}\right) \tag{6.38}$$

Using (6.33), (6.34), (6.37), and (6.38), we can write

$$\left(\frac{|e|\,A_0}{2m}\right)^2 \frac{\left\langle\left|\vec{\varepsilon}\cdot\hat{p}_{cv}(\vec{k})\,\right|^2\right\rangle_{av}}{E_c(\vec{k}) - E_v(\vec{k})} = \left(\frac{|e|\,A_0}{2m}\right)^2 \frac{2\pi}{3}\left|\vec{k}\cdot\hat{p}_{cv}(0)\,\right|^2 \frac{\beta^2}{4}\left(t + \frac{\rho}{\sqrt{2}}\right)^2$$

$$\frac{1}{\xi_{1k}}\left\{\left(1+\frac{E_{g_0}-\delta'}{\xi_{1k}+\delta'}\right)+(E_{g_0}-\delta')\left[\frac{1}{\xi_{1k}+\delta'}-\frac{1}{E_{g_0}+\delta'}\right]^{1/2}\right.$$
$$\left.\times\left[\frac{1}{\xi_{1k}+\delta'}-\frac{E_{g_0}+\delta'}{(E_{g_0}-\delta')^2}\right]^{1/2}\right\}^2 \tag{6.39}$$

Following Nag [8], it can be shown that

$$A_0^2=\frac{I\lambda^2}{2\pi^2c^3\sqrt{\varepsilon_{sc}\varepsilon_0}} \tag{6.40}$$

where I is the light intensity of wavelength λ, ε_0 is the permittivity of free space, and c is the velocity of light. Thus, the simplified electron energy spectrum in III–V, ternary, and quaternary materials in the presence of light waves can approximately be written as

$$\frac{\hbar^2k^2}{2m^*}=\beta_0(E,\lambda) \tag{6.41}$$

where $\beta_0(E,\lambda)\equiv[I_{11}(E)-\theta_0(E,\lambda)]$,

$$\theta_0(E,\lambda)\equiv\frac{|e|^2}{96m_r\pi c^3}\frac{I\lambda^2}{\sqrt{\varepsilon_{sc}\varepsilon_0}}\frac{E_{g_0}(E_{g_0}+\Delta)}{(E_{g_0}+\frac{2}{3}\Delta)}\frac{\beta^2}{4}\left(t+\frac{\rho}{\sqrt{2}}\right)^2\frac{1}{\phi_0(E)}$$
$$\left\{\left(1+\frac{E_{g_0}-\delta'}{\phi_0(E)+\delta'}\right)+(E_{g_0}-\delta')\left[\frac{1}{\phi_0(E)+\delta'}-\frac{1}{E_{g_0}+\delta'}\right]^{1/2}\right.$$
$$\left.\left[\frac{1}{\phi_0(E)+\delta'}-\frac{E_{g_0}+\delta'}{(E_{g_0}-\delta')^2}\right]^{1/2}\right\}^2$$

and $\phi_0(E)\equiv E_{g_0}\left(1+2\left(1+\frac{m_c}{m_v}\right)\frac{I_{11}(E)}{E_{g_0}}\right)^{1/2}$.

Thus, under the limiting condition $\vec{k}\to 0$, from (6.41), we observe that $E\neq 0$ and is positive. Therefore, in the presence of external light waves, the energy of the electron does not tend to zero when $\vec{k}\to 0$, whereas for the unperturbed three-band model of Kane, $I_{11}(E)=\left[\hbar^2k^2/(2m_c)\right]$ in which $E\to 0$ for $\vec{k}\to 0$. As the conduction band is taken as the reference level of energy, therefore, the lowest positive value of E for $\vec{k}\to 0$ provides the increased bandgap (ΔE_g) of the semiconductor due to photon excitation. The values of the increased bandgap can be obtained by computer iteration processes for various values of I and λ, respectively.

Special Cases:

(1) For the two-band model of Kane, we have $\Delta\to 0$. Under this condition, $I_{11}(E)\to E(1+aE)=\frac{\hbar^2k^2}{2m_c}$. Since, $\beta\to 1$, $t\to 1$, $\rho\to 0$, $\delta'\to 0$ for $\Delta\to 0$, from Eq. (6.41), we can write the energy spectrum of III–V, ternary and quaternary materials in the presence of external photo-excitation whose

unperturbed conduction electrons obey the two-band model of Kane as

$$\frac{\hbar^2 k^2}{2m_c} = \tau_0(E, \lambda) \tag{6.42}$$

where $\tau_0(E, \lambda) \equiv E(1 + aE) - B_0(E, \lambda)$,

$$B_0(E, \lambda) \equiv \frac{|e|^2 I\lambda^2 E_{g_0}}{384\pi c^3 m_r \sqrt{\varepsilon_{sc}\varepsilon_0}} \frac{1}{\phi_1(E)} \left\{ \left(1 + \frac{E_{g_0}}{\phi_1(E)}\right) + E_{g_0}\left[\frac{1}{\phi_1(E)} - \frac{1}{E_{g_0}}\right]\right\}^2,$$

$$\phi_1(E) \equiv E_{g_0}\left\{1 + \frac{2m_c}{m_r} aE(1 + aE)\right\}^{1/2}.$$

(2) For relatively wide bandgap semiconductors, one can write, $a \to 0$, $b \to 0$, $c \to 0$ and $I_{11}(E) \to E$.
Thus, from (6.42), we get,

$$\frac{\hbar^2 k^2}{2m_c} = \rho_0(E, \lambda) \tag{6.43}$$

$$\rho_0(E, \lambda) \equiv E - \frac{|e|^2 I\lambda^2}{96\pi c^3 m_r \sqrt{\varepsilon_{sc}\varepsilon_0}} \left[1 + \left(\frac{2m_c}{m_r}\right) aE\right]^{-3/2} \tag{6.44}$$

6.2.2 The Formulation of the EEM in the Presence of Light Waves in III–V, Ternary, and Quaternary Semiconductors

The EEM can, in general [6, 7], be written as

$$m_c(E_F) = \left[(\hbar k) \Big/ \left(\frac{1}{\hbar} \frac{\partial E}{\partial k} \right) \right] \Bigg|_{E=E_F} = \hbar^2 k \left. \frac{\partial k}{\partial E} \right|_{E=E_F} \tag{6.45}$$

where E_F is the Fermi energy in the present case.
Using (6.41) and (6.45) we get,

$$m_c(E_F) = m_c \left[I'_{11}(E_F) - \theta'_0(E_F, \lambda) \right] \tag{6.46}$$

where the primes indicate the differentiation of the differentiable functions with respect to E_F. It appears then that the formulation of the EEM requires an expression of electron statistics, which, in turn, is determined by the density-of-states function. Using (6.41), the density-of-states function for III–V, ternary, and quaternary materials in the presence of light waves whose unperturbed conduction electrons obey the three-band model of Kane can be expressed as

$$D_0(E) = 4\pi \left(\frac{2m_c}{h^2}\right)^{3/2} g_v \sqrt{\beta_0(E,\lambda)}\beta_0'(E,\lambda) \tag{6.47}$$

where $\beta_0'(E,\lambda) = \frac{\partial}{\partial E}[\beta_0(E,\lambda)]$.

Combining (6.47) with the Fermi-Dirac occupation probability factor and using the generalized Sommerfeld's lemma [9], the electron concentration can be written as

$$n_0 = (3\pi^2)^{-1} \left(\frac{2m_c}{\hbar^2}\right)^{3/2} g_v [M_1(E_F,\lambda) + N_1(E_F,\lambda)] \tag{6.48}$$

where $M_1(E_F,\lambda) \equiv [\beta_0(E_F,\lambda)]^{3/2}$, $N_1(E_F,\lambda) = \sum_{r=1}^{s} L(r) M_1(E_F,\lambda)$ and $L(r) \equiv \left[2(k_BT)^{2r}(1-2^{1-2r})\xi(2r)\right]\left(\frac{\partial^{2r}}{\partial E^{2r}}\right)\Big|_{E=E_F}$.

The expressions of EEM and n_0, for III–V, ternary, and quaternary materials in the presence of light waves whose unperturbed conduction electrons obey the two-band model of Kane can be expressed as

$$m^*(E_F) = m_c\left[(1+2\alpha E_F) - B_0'(E_F,\lambda)\right] \tag{6.49}$$

$$n_0 = (3\pi^2)^{-1} \left(\frac{2m_c}{\hbar^2}\right)^{3/2} g_v [M_2(E_F,\lambda) + N_2(E_F,\lambda)] \tag{6.50a}$$

where $M_2(E_F,\lambda) \equiv [\omega_0(E_F,\lambda)]^{3/2}$ and $N_2(E_F,\lambda) = \sum_{r=1}^{s} L(r) M_2(E_F,\lambda)$.

The expression of EEM for III–V, ternary, and quaternary materials in the presence of light waves whose unperturbed conduction electrons obey the parabolic energy bands can be expressed as

$$m^*(E_F) = m_c\left[\rho_0'(E_F,\lambda)\right] \tag{6.50b}$$

$$n_0 = (3\pi^2)^{-1} \left(\frac{2m_c}{\hbar^2}\right)^{3/2} g_v [M_3(E_F,\lambda) + N_3(E_F,\lambda)] \tag{6.50c}$$

where $M_3(E_F,\lambda) \equiv [\rho_0(E_F,\lambda)]^{3/2}$ and $N_3(E_F,\lambda) = \sum_{r=1}^{s} L(r) M_3(E_F,\lambda)$.

In the absence of external photo-excitations, the expressions of the EEM and n_0 in accordance with the three band model of Kane assume the forms

$$m^*(E_{F_0}) = m_c\left[I_{11}'(E_{F_0})\right] \tag{6.51}$$

$$n_0 = g_v(3\pi^2)^{-1} \left(\frac{2m_c}{\hbar^2}\right)^{3/2} \left[M_4(E_{F_0}) + N_4(E_{F_0})\right] \tag{6.52}$$

where E_{F_0} is the Fermi energy in the absence of photo-excitation, $M_4(E_{F_0}) \equiv \left[I_{11}(E_{F_0})\right]^{3/2}$ and $N_4(E_{F_0}) = \sum_{r=1}^{s} L(r) M_4(E_{F_0})$.

In accordance with the two-band model of Kane, the corresponding expressions of the EEM and n_0 are given by

$$m^*(E_{F_0}) = m_c\left[(1 + 2\alpha E_{F_0})\right] \tag{6.53}$$

$$n_0 = (3\pi^2)^{-1}\left(\frac{2m_c}{\hbar^2}\right)^{3/2} g_v\left[M_5(E_{F_0}) + N_5(E_{F_0})\right] \tag{6.54}$$

where $M_5(E_{F_0}) \equiv \left[E_{F_0}(1 + \alpha E_{F_0})\right]^{3/2}$ and $N_5(E_{F_0}) = \sum_{r=1}^{s} L(r) M_5(E_{F_0})$.

Under the constraints $\Delta \gg E_{g_0}$ or $\Delta \ll E_{g_0}$ together with the condition $\alpha E_{F_0} \ll 1$, the (6.54) assumes the form

$$n_0 = N_c g_v\left[F_{1/2}(\eta) + \left(\frac{15\alpha k_B T}{4}\right) F_{3/2}(\eta)\right] \tag{6.55}$$

where $\eta = \frac{E_{F_0}}{k_B T}$ for relatively wide gap materials $E_{g_0} \to \infty$ and the (6.53) and (6.55) get simplified as to the well-known results [8] as

$$m^*(E_F) = m_c \tag{6.56}$$

and

$$n_0 = N_c g_v F_{1/2}(\eta) \tag{6.57}$$

6.3 Results and Discussion

Using the appropriate equations and the values of the energy band constants from Table 1.1, we have plotted in Figs. 6.1, 6.2, 6.3 and 6.4, the EEM as functions of electron concentration at $T = 4.2$ K by taking n-InAs, n-InSb, n-$Hg_{1-x}Cd_xTe$ and n-$In_{1-x}Ga_xAs_yP_{1-y}$ lattice matched to InP as examples of III–V, ternary, and quaternary materials which are used for the purpose of numerical computations in accordance with the perturbed three- and two-band models of Kane and that of perturbed parabolic energy bands, respectively. In Figs. 6.5, 6.6, 6.7, and 6.8 we have plotted the EEM as a function of intensity. In Figs. 6.9, 6.10, 6.11, and 6.12 we have plotted the EEM as a function of wavelength. In Figs. 6.13 and 6.14, we have plotted the EEM as function of the alloy composition for ternary and quaternary materials, respectively.

From Fig. 6.1 it appears that the EEM increases with the increasing electron concentration for n-InAs and the numerical values of the EEM in the presence of light waves in accordance with all the band models are relatively larger than that of the

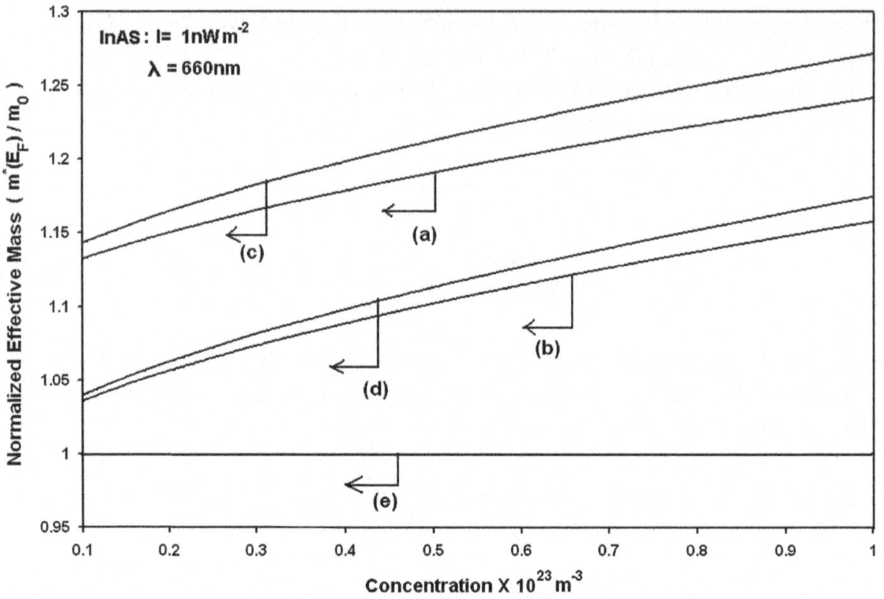

Fig. 6.1 Plot of the normalized EEM as a function of electron concentration for *n*-InAs in the presence of light waves in which the *curves* (*a*) and (*c*) represent the three and two band models of Kane, respectively. The *curves* (*b*) and (*d*) exhibit the same variation in the absence of external photo-excitation. The *curve* (*e*) represents the parabolic energy band model both in the presence and in the absence of the external photo-excitation

same in the absence of the external photo-excitation excluding curve (e), where for parabolic energy band the EEM is numerically concentration invariant due to large bandgap. The reason behind such behavior is the fact that the Fermi energy is the monotonic increasing function of electron concentration and the EEM increases monotonically with increasing Fermi energy both in the presence and absence of light waves. By comparing the plot (a) with plot (c) in Fig. 6.1 we observe that the presence of spin–orbit splitting in curve (a) decreases the value of the EEM as compared with curve (c) in the whole range of carrier degeneracy as considered here. For relatively low values of n_0, the curves (a), (b), (c), and (d) exhibit converging tendency whereas they differ with each other for relatively higher values of carrier degeneracy. The curve (e) of Fig. 6.1 represents the EEM both in the presence and absence of external light waves for the relatively wide bandgap model which is independent of doping. Plots (a) and (c) of Fig. 6.2 diverge for relatively low values of n_0, intersect each other for a particular zone of concentration, and then exhibit small difference although both of them increase with increasing degeneracy. The numerical values of the EEM for the curves (a), (b), (c), and (d) of *n*-InSb as given in Fig. 1.1b are greater as compared with the same for *n*-InAs as given in Fig. 6.1, although the nature of the curve(e) is same for both Figs. 6.1 and 6.2, respectively.

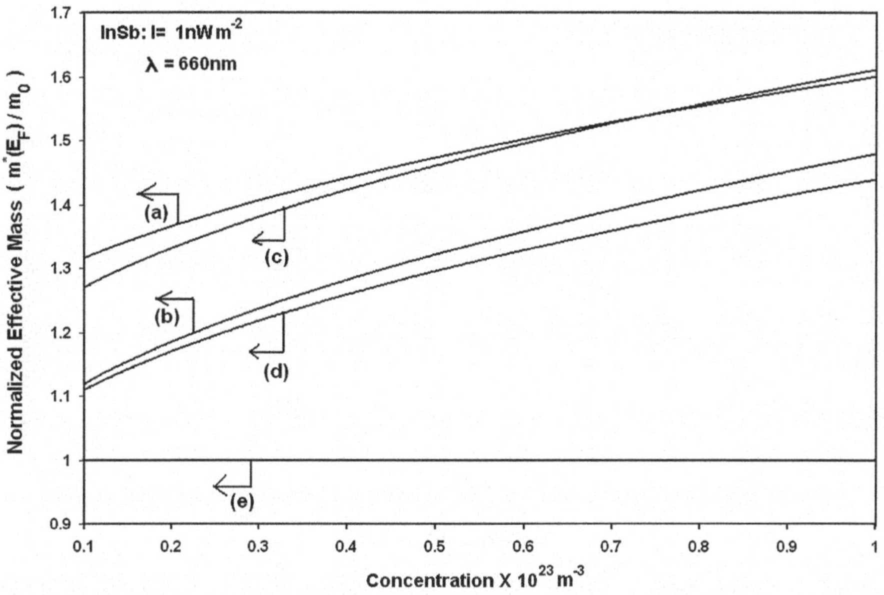

Fig. 6.2 Plot of the normalized EEM as a function of electron concentration for n-InSb in the presence of light waves in which the *curves* (a) and (c) represent the three- and two-band models of Kane, respectively. The *curves* (b) and (d) exhibit the same variation in the absence of external photo-excitation. The *curve* (e) represents the parabolic energy band model both in the presence and in the absence of the external photo-excitation

The curves (a) and (c) explore the fact that the influence of the spin–orbit splitting on the EEM for n-$Hg_{1-x}Cd_xTe$ in the presence of light waves decreases significantly the same mass as compared with the perturbed two-band model of Kane and the two curves exhibit wide difference with each other for relatively low values of doping. In the absence of light wave, the effect of Δ on the EEM is much less. It appears by comparing the Figs. 6.1, 6.2 and 6.3 that the EEM for n-$Hg_{1-x}Cd_xTe$ in the presence of external photo-excitation is much more as that of n-InSb and n-InAs, respectively. From plots (a) and (c) of Fig. 6.4, it appears for $In_{1-x}Ga_xAs_yP_{1-y}$ lattice matched to InP that both the curves maintain constant wide difference under photo-excitation with respect to electron concentration in the whole range of electron degeneracy as considered here. The influence of Δ on the EEM in the absence of external light waves for three- and two-band models of Kane is very small as evident from the curves (b) and (d), although the EEM increases with n_0 as usual. The influence of the energy band constants on the EEM is apparent from all the plots of Figs. 6.1, 6.2, 6.3 and 6.4 and the numerical values of the EEM is greatest for ternary alloys and the least for quaternary systems under light waves. The curves (a) and (c) of Fig. 6.5 explore that the EEM increases with increasing light intensity for n-InAs in the presence of light waves for both perturbed three- and two-band models of Kane, whereas for perturbed parabolic energy bands the EEM is intensity invariant.

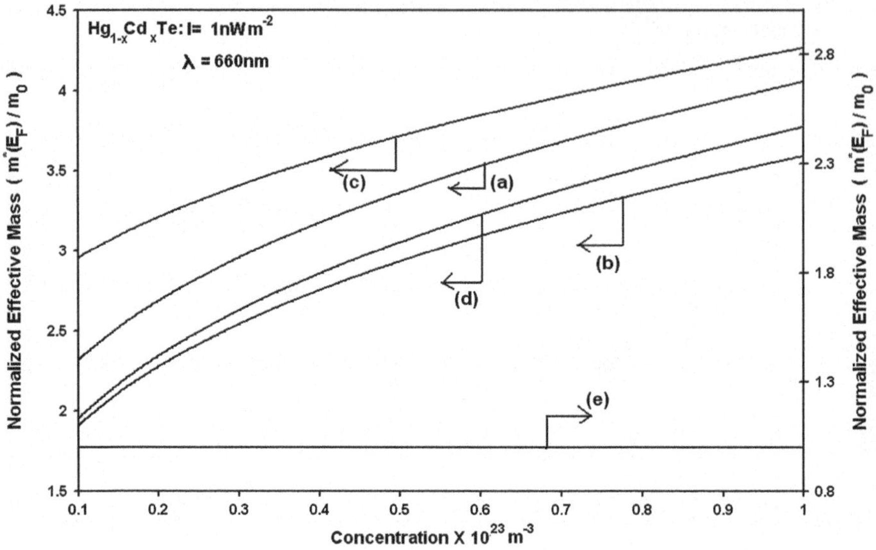

Fig. 6.3 Plot of the normalized EEM as a function of electron concentration for n-$Hg_{1-x}Cd_xTe$ in the presence of light waves in which the *curves* (*a*) and (*c*) represent the three- and two-band models of Kane, respectively. The *curves* (*b*) and (*d*) exhibit the same variation in the absence of external photo-excitation. The *curve* (*e*) represents the parabolic energy band model both in the presence and in the absence of the external photo-excitation

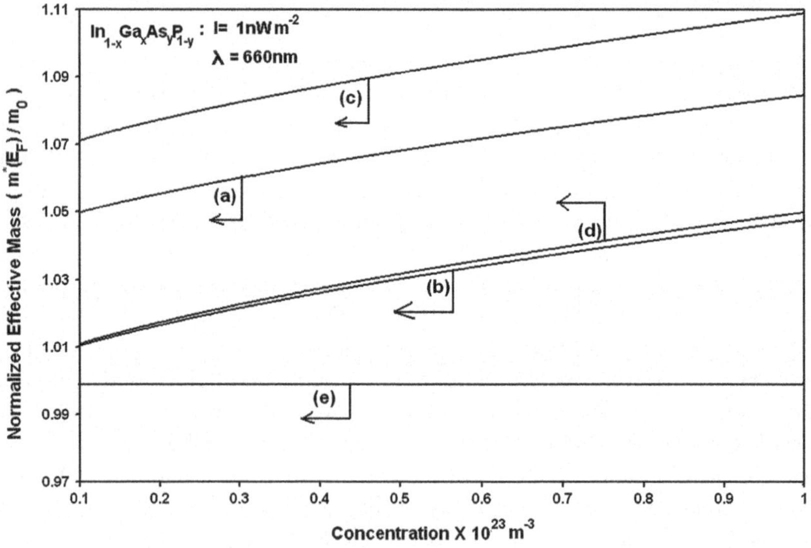

Fig. 6.4 Plot of the normalized EEM as a function of electron concentration for $In_{1-x}Ga_xAs_yP_{1-y}$ lattice matched to InP in the presence of light waves in which the *curves* (*a*) and (*c*) represent the three- and two-band models of Kane, respectively. The *curves* (*b*) and (*d*) exhibit the same variation in the absence of external photo-excitation. The *curve* (*e*) represents the parabolic energy band model both in the presence and in the absence of the external photo-excitation

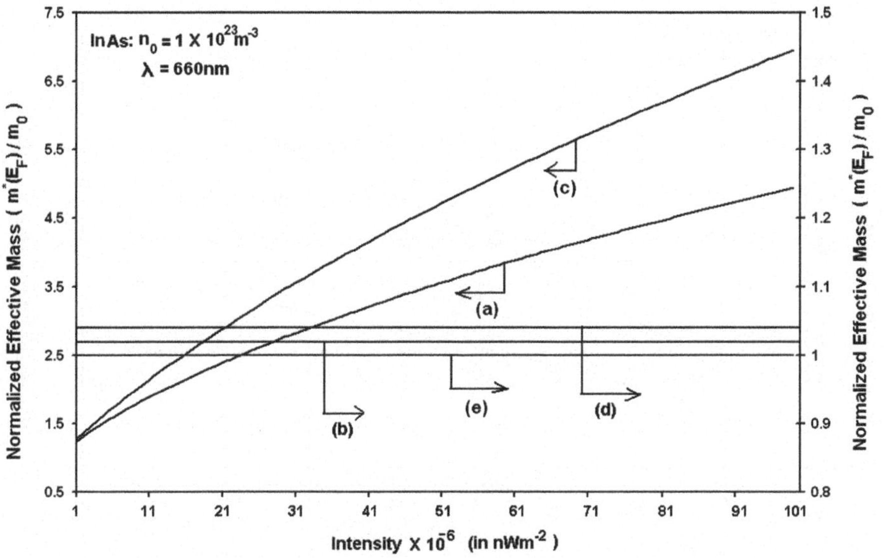

Fig. 6.5 Plot of the normalized EEM as a function of light intensity for *n*-InAs in the presence of light waves in which the *curves* (*a*) and (*c*) represent the three- and two-band models of Kane, respectively. The *curves* (*b*) and (*d*) exhibit the same variation in the absence of external photo-excitation. The *curve* (*e*) represents the parabolic energy band model both in the presence and in the absence of the external photo-excitation

The reason behind such behavior is that the Fermi energy increases with increasing light intensity and the EEM is the function of Fermi energy. For perturbed parabolic energy bands, due to very large bandgap the EEM is numerically independent of light intensity. In the absence of light waves, the EEM is naturally independent of I which is apparent from the curves (b) and (d) of Fig. 6.5. The curves (a) and (c) of Fig. 6.6 for *n*-InSb reflects the fact that the influence of spin–orbit splitting increases rapidly with increasing intensity and for low values of light intensity the EEM decreases for both perturbed three- and two-band models of Kane, whereas for higher values of I the EEM increases significantly.

From the plot (a) of Fig. 6.7 one can infer that the EEM for *n*-$Hg_{1-x}Cd_xTe$ in the presence of light waves increases with increasing I, in a more or less linear fashion in accordance with the perturbed three-band model of Kane whereas from plot (c), one observes that the EEM on the basis of the perturbed two-band model of Kane is greater as compared with plot (a) in the whole range of I. For low values of I, the curves (a) and (c) exhibit converging tendency. It is important to note that with respect to light intensity, the numerical values of the EEM in the absence of light waves, in the case of *n*-$Hg_{1-x}Cd_xTe$ are greater for both types of band models (curves (b) and (d)) as compared with that of (a) and (c) when $I \neq 0$. It appears from the plots (a) and (c) of Fig. 6.8 that the EEM increases with I for both types of perturbed band models with different small curvatures and the difference in numerical values

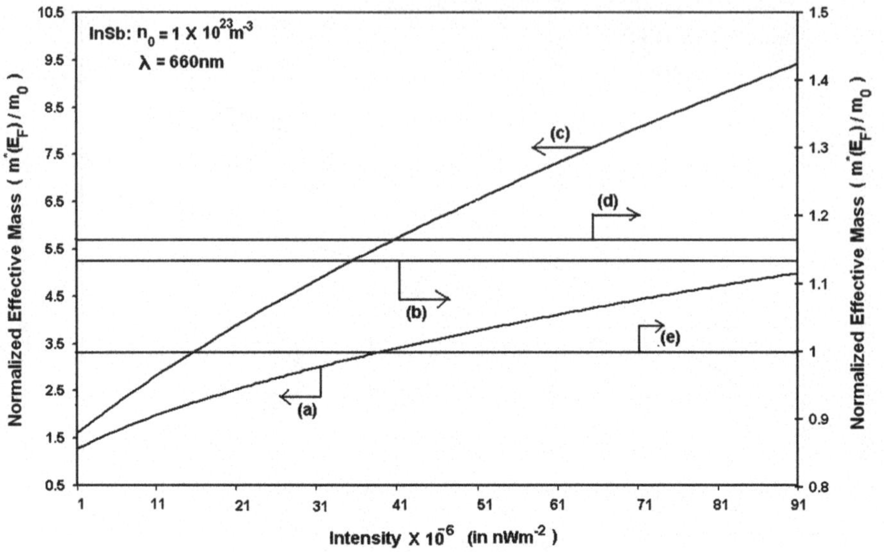

Fig. 6.6 Plot of the normalized EEM as a function of light intensity for n-InSb in the presence of light waves in which the *curves* (a) and (c) represent the three- and two-band models of Kane, respectively. The *curves* (b) and (d) exhibit the same variation in the absence of external photo-excitation. The *curve* (e) represents the parabolic energy band model both in the presence and in the absence of the external photo-excitation

of EEM increases with increasing in I, although they converge to a particular value for $I = 10^{-6}\,\mathrm{nWm}^{-2}$.

The curves (a) and (c) of Fig. 6.9 exhibit the fact that the EEM for n-InAs increases with increasing wavelength in the presence of light waves for both perturbed three- and two-band models of Kane since the Fermi energy increases with increasing wave length and the EEM is the function of Fermi energy. For perturbed parabolic energy bands, the EEM is numerically independent of wave length due to large bandgap. The curves (a) and (c) maintain wide difference with increasing wave length and the spin–orbit splitting decreases the EEM in the whole range of λ. The curves (b) and (d) exhibit the same variation in the absence of photo excitation and is independent of wave length of the incident light.

From the plots (a) and (c) of Fig. 6.10 for n-InAs, one can infer that the EEM increases with increasing wavelength in the presence of light waves for both the cases and they exhibit the diverging tendency for relatively low values of λ whereas for higher values of the wavelength exhibit the converging nature. From plots (b) and (d), we observe that the numerical values of the EEM for both three and two band model of Kane are much larger as compared with the same under photo-excitation for relatively low values of λ. The influence of Δ on the EEM for perturbed three-band model as observed in Fig. 6.10 is less as compared with the same as given in plot (a) of Fig. 6.9.

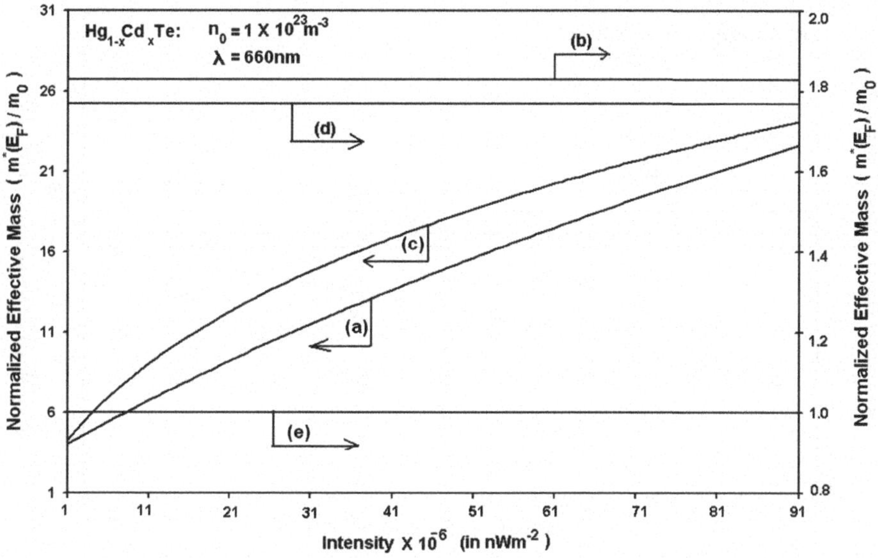

Fig. 6.7 Plot of the normalized EEM as a function of light intensity for n-$Hg_{1-x}Cd_xTe$ in the presence of light waves in which the *curves* (*a*) and (*c*) represent the three- and two-band models of Kane, respectively. The *curves* (*b*) and (*d*) exhibit the same variation in the absence of external photo-excitation. The *curve* (*e*) represents the parabolic energy band model both in the presence and in the absence of the external photo-excitation

The curves (b) and (c) for n-$Hg_{1-x}Cd_xTe$ of Fig. 1.3c increase with increasing λ and after intersection they exhibit the wide difference with each other. The spin–orbit splitting decreases the EEM remarkably for relatively higher values of λ.

From Fig. 6.12 we can write that the EEM in this case is much less when compared with the same as given I, the Figs. 6.9, 6.10 and 6.11 in the presence of external photo-excitation. From plots (a) and (c) of Fig. 6.12, it appears for $In_{1-x}Ga_xAs_yP_{1-y}$ lattice matched to InP that both the curves maintain wide difference under photo-excitation with respect to λ in the whole range of wave length as considered here. As the wave length increases, from plots (a) and (c) of Fig. 6.12 we infer that the difference also increases. The influence of Δ on the EEM in the absence of external light waves for three and two band models of Kane is very small as evident from the curves (b) and (d).

It appears that the EEM increases as the wavelength shifts from violet to red. The influence of light is immediately apparent from the plots in the Figs. 6.5, 6.6, 6.7, 6.8, 6.9, 6.10, 6.11 and 6.12 since the EEM depends strongly on I and λ for the three- and the two-band model of Kane which is in direct contrast with that for the bulk specimens of the said compounds in the absence of external photo-excitation. The variations of the EEM in the Figs. 6.5, 6.6, 6.7, 6.8, 6.9, 6.10, 6.11, and 6.12 reflect the direct signature of the light wave on the band structure-dependent physical properties of semiconductors in general in the presence of

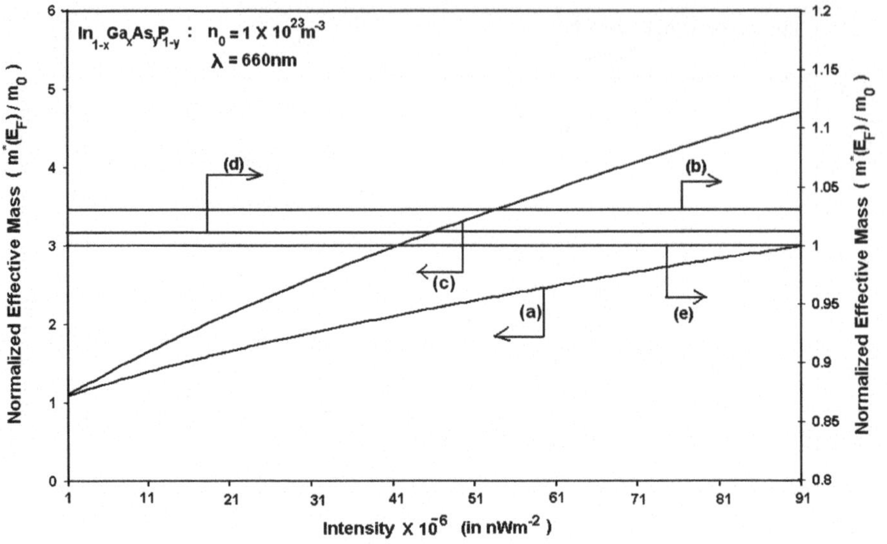

Fig. 6.8 Plot of the normalized EEM as a function of light intensity for $In_{1-x}Ga_xAs_yP_{1-y}$ lattice matched to InP in the presence of light waves in which the *curves* (*a*) and (*c*) represent the three- and two-band models of Kane, respectively. The *curves* (*b*) and (*d*) exhibit the same variation in the absence of external photo-excitation. The *curve* (*e*) represents the parabolic energy band model both in the presence and in the absence of the external photo-excitation

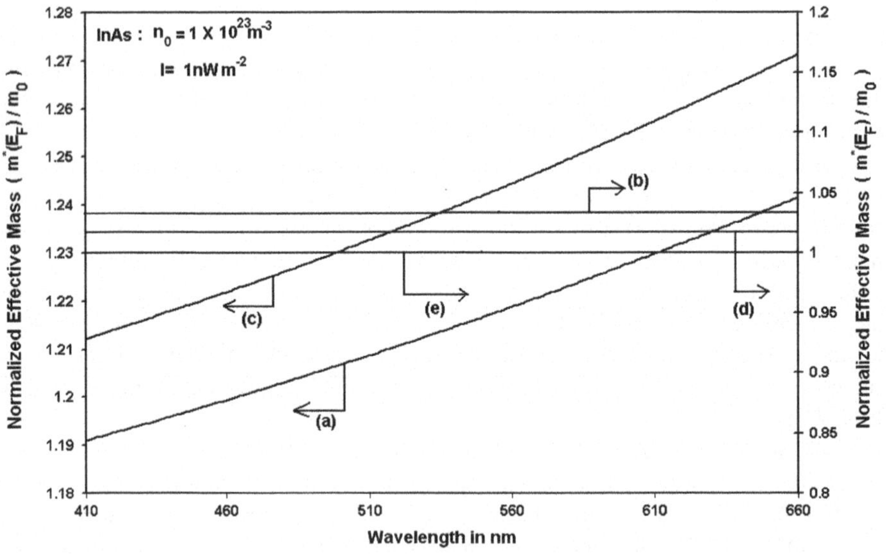

Fig. 6.9 Plot of the normalized EEM as a function of wavelength for *n*-InAs in the presence of light waves in which the curves (*a*) and (*c*) represent the three- and two-band models of Kane, respectively. The *curves* (*b*) and (*d*) exhibit the same variation in the absence of external photo-excitation. The *curve* (*e*) represents the parabolic energy band model both in the presence and in the absence of the external photo-excitation

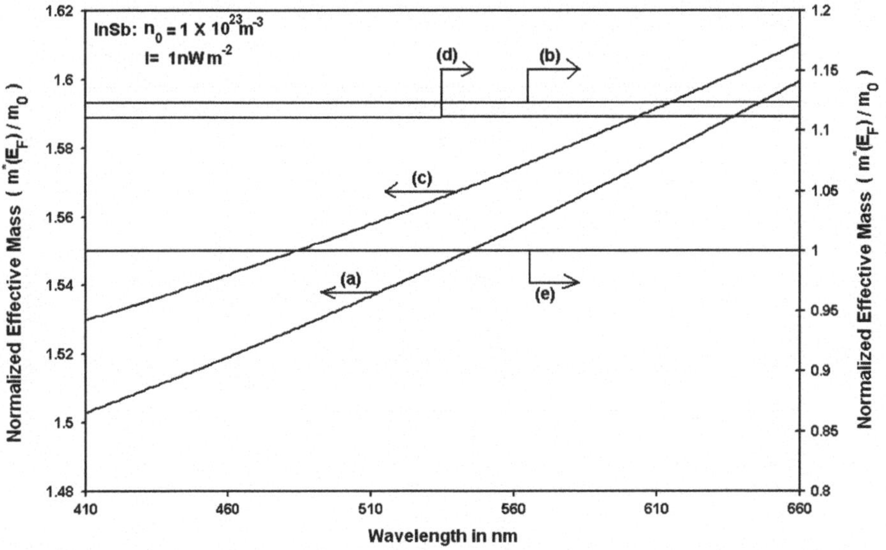

Fig. 6.10 Plot of the normalized EEM as a function of wavelength for n-InSb in the presence of light waves in which the *curves* (*a*) and (*c*) represent the three and two band models of Kane, respectively. The *curves* (*b*) and (*d*) exhibit the same variation in the absence of external photo-excitation. The *curve* (*e*) represents the parabolic energy band model both in the presence and in the absence of the external photo-excitation

external photo-excitation and the photon-assisted transport for the corresponding photonic devices. The numerical values of the EEM in the presence of the light waves are larger than that of the same in the absence of light wave for both the three- and the two-band model of Kane. Although, the EEM tends to increase with the intensity and the wavelength but the rate of increase is totally band structure dependent. It appears that the numerical values of the EEM are greatest for ternary materials and least for quaternary compounds.

In Figs. 6.13 and 6.14, the EEM has been plotted as a function of alloy composition for n-$Hg_{1-x}Cd_xTe$ and n-$In_{1-x}Ga_xAs_yP_{1-y}$ lattice matched to InP respectively in which all the cases of Figs. 6.1, 6.2, 6.3 and 6.4 have further been plotted for the purpose of relative comparison. The Fermi energy decreases with increasing alloy composition and the EEM is a function of Fermi energy. From Fig. 6.13, we can write that the EEM in ternary compounds decreases with increasing alloy composition. The numerical values of EEM in the presence of light waves are greater for both the models as appears from the plots (a), (b), (c) and (d). As alloy composition increases, the EEM for all the cases exhibit the converging tendency. The plots of the Fig. 6.13 are valid for $x > 0.17$, since for $x < 0.17$, the bandgap becomes negative in n-$Hg_{1-x}Cd_xTe$ leading to semi-metallic state. The plots of the Fig. 6.14 exhibit the variation of the EEM with y for n-$In_{1-x}Ga_xAs_yP_{1-y}$ lattice matched to InP. As the Fermi energy increases with the y, from the curves (a), (b), (c) and (d) of Fig. 6.14

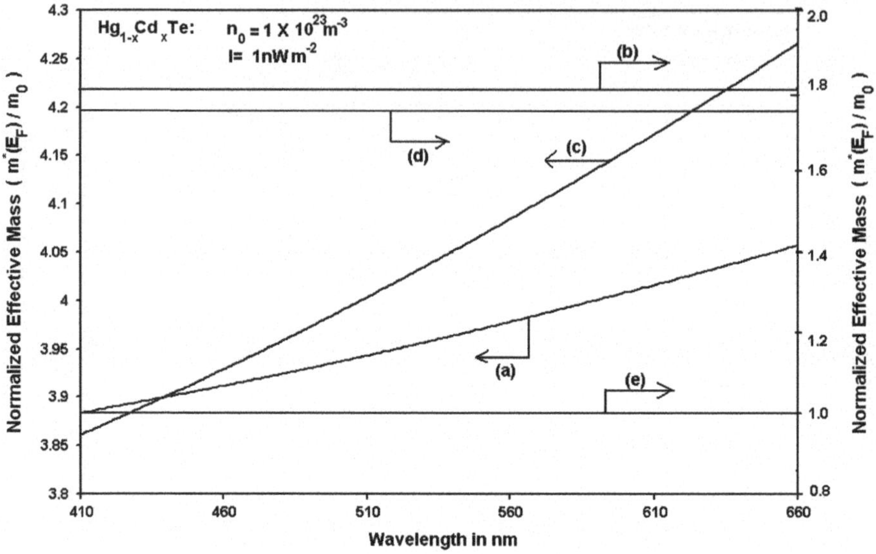

Fig. 6.11 Plot of the normalized EEM as a function of wavelength for n-$Hg_{1-x}Cd_xTe$ in the presence of light waves in which the *curves* (a) and (c) represent the three and two band models of Kane, respectively. The *curves* (b) and (d) exhibit the same variation in the absence of external photo-excitation. The *curve* (e) represents the parabolic energy band model both in the presence and in the absence of the external photo-excitation

we observe that the EEM increases with increasing y. These four plots also exhibit the fact that the influence of the spin–orbit splitting constant in the presence of light waves is much greater as compared with the same in the absence of photo excitation.

The theoretical results as presented here will be useful in determining the mobility even for relatively wide gap compounds whose energy band structures can be approximated by the parabolic energy bands both in the presence and absence of light waves. It is worth remarking that our basic Eq. (6.41) covers various materials having different energy band structures. In this section, the concentration, alloy composition, light intensity, and the wavelength dependencies of EEM in bulk specimens of n-InAs, n-InSb, n-$Hg_{1-x}Cd_xTe$ and n-$In_{1-x}Ga_xAs_yP_{1-y}$ lattice matched to InP have been studied. Thus, we have covered a wide class of optoelectronic and allied compounds whose energy band structures are defined by the three- and two-band models of Kane in the absence of photon field. Under certain limiting conditions, all the results of the EEM for different materials having various band structures lead to the well-known expression of the EEM for degenerate compounds having parabolic energy band. This indirect test not only exhibits the mathematical compatibility of our formulation but also shows the fact that our simple analysis is a more generalized one, since one can obtain the corresponding results for the relatively wide gap materials having parabolic energy bands under certain limiting conditions from our present derivation.

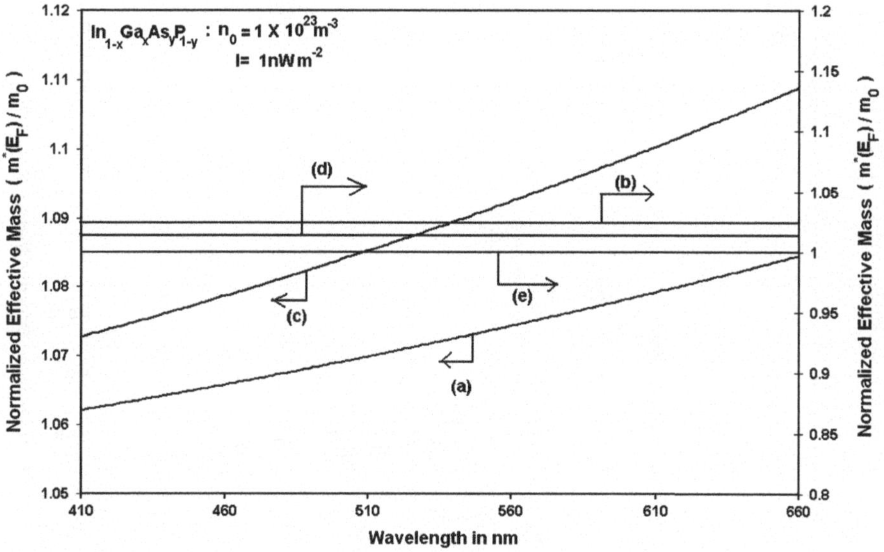

Fig. 6.12 Plot of the normalized EEM as a function of wavelength for $In_{1-x}Ga_xAs_yP_{1-y}$ lattice matched to InP in the presence of light waves in which the *curves* (*a*) and (*c*) represent the three and two band models of Kane, respectively. The *curves* (*b*) and (*d*) exhibit the same variation in the absence of external photo-excitation. The *curve* (*e*) represents the parabolic energy band model both in the presence and in the absence of the external photo-excitation

It is worth remarking that the influence of an external photo-excitation is to change radically the original band structure of the material. Because of this change, the photon field causes to increase the bandgap of semiconductor. Our method is not at all related to the DOS technique as used in the literature [10]. From the **E**-**k** dispersion relation, we can obtain the DOS, but the DOS technique as used in the literature [10] cannot provide the **E**-**k** dispersion relation. Therefore, our study is more fundamental than those of the existing literature because the Boltzman transport equation, which controls the study of the charge transport properties of semiconductor devices, can be solved if and only if the **E**-**k** dispersion relation is known. We wish to note that we have not considered the many body effects in this simplified theoretical formalism due to the lack of availability in the literature of proper analytical techniques for including them for the generalized systems as considered in this book. Our simplified approach will be useful for the purpose of comparison when methods of tackling the formidable problem after inclusion of the many body effects for the present generalized systems appear. The inclusion of the said effects would certainly increase the accuracy of the results, although the qualitative features of the EEM discussed in this book would not change in the presence of the aforementioned effects. Since the experimental results in the present case are not available in the literature to the best of our knowledge, we cannot compare our generalized theoretical analysis with

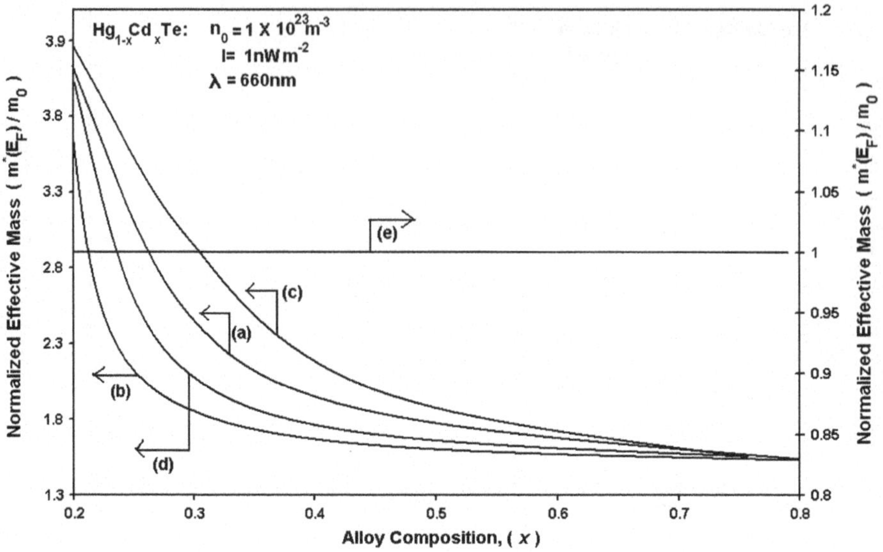

Fig. 6.13 Plot of the normalized EEM as a function of alloy composition for $Hg_{1-x}Cd_xTe$ in the presence of light waves in which the *curves* (*a*) and (*c*) represent the three- and two-band models of Kane, respectively. The *curves* (*b*) and (*d*) exhibit the same variation in the absence of external photo-excitation. The *curve* (*e*) represents the parabolic energy band model both in the presence and absence of the external photo-excitation

the corresponding experimental data. Our formalism will be useful in probing the band structure when the experimental results for our generalized systems would appear. It is worth remarking in this context that from our simplified theory, under certain limiting conditions, gets transformed to the well-known result of the EEM for wide gap materials having parabolic energy bands. We have not considered other types of optoelectronic materials and other external variables in order to keep the presentation brief. Besides, the influence of energy band models and the various band constants on the EEM for different materials can also be studied from all the figures of this book. The numerical results presented in this book would be different for other materials but the nature of variation would be unaltered. The theoretical results as given here would be useful in analyzing various other experimental data related to this phenomenon. Finally, we can write that this theory can be used to investigate the gate capacitance of nanoscale transistors, the carrier contribution to the elastic constants, the Debye screening length, the magnetic susceptibilities, the Burstien Moss shift, plasma frequency, the Hall coefficient, the specific heat, and other different transport coefficients of modern semiconductor devices operated in the presence of light waves.

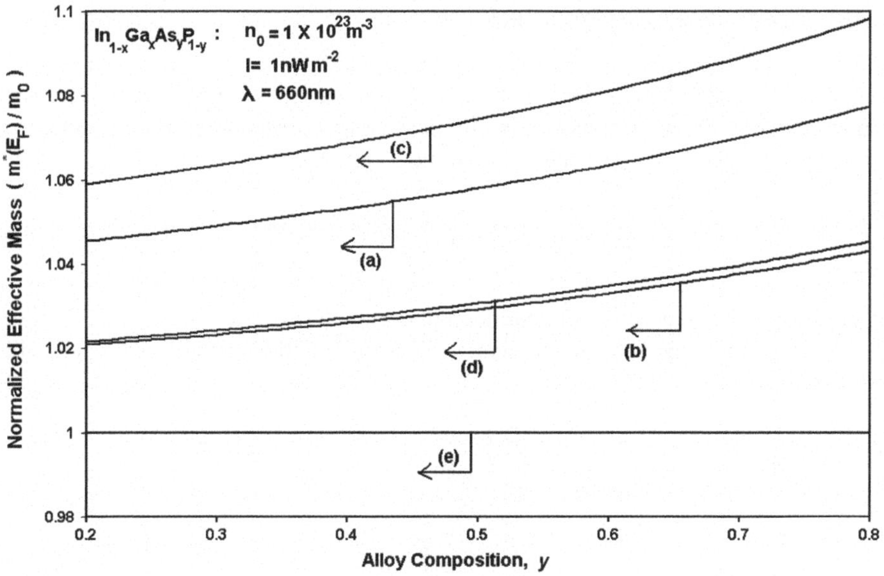

Fig. 6.14 Plot of the normalized EEM as a function of alloy composition for $In_{1-x}Ga_xAs_{1-y}P_y$ lattice matched to InP in the presence of light waves in which the *curves* (*a*) and (*c*) represent the three and two band models of Kane, respectively. The *curves* (*b*) and (*d*) exhibit the same variation in the absence of external photo-excitation. The *curve* (*e*) represents the parabolic energy band model both in the presence and in the absence of the external photo-excitation

6.4 The Formulation of the EEM in the Presence of Quantizing Magnetic Field Under External Photo Excitation in III–V, Ternary, and Quaternary Materials

6.4.1 Introduction

It is well known that the band structure of electronic materials can be dramatically changed by applying the external fields [11]. The effects of the quantizing magnetic field on the band structure of compound semiconductors have already been discussed in Chap. 4. In Sect. 6.4.2 of theoretical background, we study the effective electron mass in III–V, ternary, and quaternary semiconductors under magnetic quantization both in the presence and absence of external light waves. The Sect. 6.5 explores results and discussions in this context.

6.4.2 Theoretical Background

The magneto-dispersion law, in the absence of spin, for III–V, ternary, and quaternary semiconductors, in the presence of photo-excitation, whose unperturbed conduction electrons obey the three-band model of Kane, is given by

$$\beta_0(E,\lambda) = \left(n+\frac{1}{2}\right)\hbar\omega_0 + \frac{\hbar^2 k_z^2}{2m_c} \tag{6.58}$$

where n is the Landau quantum number and $\omega_0 = eB/m_c$.

Using (6.58), the density-of-states function in the present case can be expressed as

$$D_B(E,\lambda) = \frac{g_v|e|\sqrt{2m_c}}{2\pi^2\hbar^2}\sum_{n=0}^{n_{\max}}\left[\{\beta_0(E,\lambda)\}'\left\{\beta_0(E,\lambda) - \left(n+\frac{1}{2}\right)\hbar\omega_0\right\}^{-1/2} H(E-E_{n_{l1}})\right] \tag{6.59}$$

in which $E_{n_{l1}}$ is the positive lowest root of the equation.

$$\beta_0(E_{n_{l1}},\lambda) = \left(n+\frac{1}{2}\right)\hbar\omega_0 \tag{6.60}$$

The EEM in this case assumes the form

$$m^*(E_{F_{BL}},\lambda) = \hbar^2 k_z \left.\frac{\partial k_z}{\partial E}\right|_{E=E_{F_{BL}}} = m_c\left\{\beta_0(E_{F_{BL}},\lambda)\right\}' \tag{6.61}$$

where $E_{F_{BL}}$ is the Fermi energy under quantizing magnetic field in the presence of light waves as measured from the edge of the conduction band in the vertically upward direction in the absence of any quantization. Combining Eq. (6.59) with the Fermi-Dirac occupation probability factor and using the generalized Sommerfeld's lemma, the electron concentration can be written as

$$n_0 = \frac{g_v|e|B\sqrt{2m_c}}{\pi^2\hbar^2}\sum_{n=0}^{n_{\max}}\left[M_{13}(E_{F_{BL}},B,\lambda) + N_{13}(E_{F_{BL}},B,\lambda)\right] \tag{6.62}$$

where

$$M_{13}(E,B,\lambda) \equiv \left[\beta_0(E,\lambda) - \left(n+\frac{1}{2}\right)\hbar\omega_0\right]^{1/2}$$

and

$$N_{13}(E_{F_{BL}}, B, \lambda) \equiv \sum_{r=1}^{s} L(r) M_{13}(E_{F_{BL}}, B, \lambda).$$

The magneto-dispersion law in the absence of spin, for III–V, ternary, and quaternary semiconductors, in the presence of photo-excitation, whose unperturbed conduction electrons obey the two-band model of Kane, is given by

$$\tau_0(E, \lambda) = \left(n + \frac{1}{2}\right)\hbar\omega_0 + \frac{\hbar^2 k_z^2}{2m_c} \tag{6.63}$$

Using (6.63), the density-of-states function in this case can be written as

$$D_B(E, \lambda) = \frac{g_v |e| \sqrt{2m_c}}{2\pi^2\hbar^2} \sum_{n=0}^{n_{\max}} \left[\{\tau_0(E, \lambda)\}' \left\{ \tau_0(E, \lambda) - \left(n + \frac{1}{2}\right)\hbar\omega_0 \right\}^{-1/2} \times H(E - E_{n_{l2}}) \right] \tag{6.64}$$

where $E_{n_{l2}}$ is the Landau subband energies and can be expressed as

$$\tau_0(E_{n_{l2}}, \lambda) = \left(n + \frac{1}{2}\right)\hbar\omega_0 \tag{6.65}$$

The EEM assumes the form

$$m^*(E_{F_{BL}}, \lambda) = m_c \left\{ \tau_0(E_{F_{BL}} \lambda) \right\}' \tag{6.66}$$

Thus, the electron concentration can be written as

$$n_0 = \frac{g_v |e| B \sqrt{2m_c}}{\pi^2\hbar^2} \sum_{n=0}^{n_{\max}} \left[M_{14}(E_{F_{BL}}, B, \lambda) + N_{14}(E_{F_{BL}}, B, \lambda) \right] \tag{6.67}$$

where

$$M_{14}(E_{F_{BL}}, B, \lambda) \equiv \left[\tau_0(E_{F_{BL}}, \lambda) - \left(n + \frac{1}{2}\right)\hbar\omega_0 \right]^{1/2}$$

and

$$N_{14}(E_{F_{BL}}, B, \lambda) \equiv \sum\nolimits_{r=1}^{s} L(r) M_{14}(E_{F_{BL}}, B, \lambda).$$

The magneto-dispersion law in the absence of spin, for III–V, ternary, and quaternary semiconductors, in the presence of photo-excitation, whose unperturbed conduction electrons obey the parabolic energy bands, is given by

$$\rho_0(E,\lambda) = \left(n+\frac{1}{2}\right)\hbar\omega_0 + \frac{\hbar^2 k_z^2}{2m_c} \tag{6.68}$$

Using (6.68), the density-of-states function in this case can be written as

$$D_B(E,\lambda) = \frac{g_v\,|e|\,\sqrt{2m_c}}{2\pi^2\hbar^2}\sum_{n=0}^{n_{\max}} \times \left[\{\rho_0(E,\lambda)\}'\left\{\rho_0(E,\lambda) - \left(n+\frac{1}{2}\right)\hbar\omega_0\right\}^{-1/2} \times H(E - E_{nl3})\right]$$

where E_{nl3} is the Landau subband energies and is given by

$$\rho_0(E_{nl3},\lambda) = \left(n+\frac{1}{2}\right)\hbar\omega_0 \tag{6.69}$$

The EEM assumes the form

$$m^*(E_{F_{BL}},\lambda) = m_c\left\{\rho_0(E_{F_{BL}},\lambda)\right\}' \tag{6.70}$$

Thus, the electron concentration in this case can be written as

$$n_0 = \frac{g_v\,|e|\,B\sqrt{2m_c}}{\pi^2\hbar^2}\sum_{n=0}^{n_{\max}}\left[M_{15}(E_{F_{BL}},B,\lambda) + N_{15}(E_{F_{BL}},B,\lambda)\right] \tag{6.71}$$

where

$$M_{15}(E_{F_{BL}},B,\lambda) \equiv \left[\rho_0(E_{F_{BL}},\lambda) - \left(n+\frac{1}{2}\right)\hbar\omega_0\right]^{1/2}$$

and

$$N_{15}(E_{F_{BL}},B,\lambda) \equiv \sum_{r=1}^{s} L(r) M_{15}(E_{F_{BL}},B,\lambda).$$

6.5 Results and Discussion

Using the values of the energy band constants from Table 1.1, we have plotted the EEM along the direction of z as functions of 1/B, electron concentration, intensity and wave length (as shown in Figs. 6.15, 6.16, 6.17 and 6.18) at $T = 4.2\,\text{K}$ by taking n-InSb and n-InAs which are used for the purpose of numerical computations in accordance with the perturbed three [using (6.61) and (6.62)] and two [using (6.66) and (6.67)]-band models of Kane and that of perturbed parabolic [using (6.70) and (6.71)] energy bands respectively. It appears from Fig. 6.15 that the EEM is an oscillatory function of inverse quantizing magnetic field. The oscillatory dependence is due to the crossing over of the Fermi level by the Landau subbands in steps resulting in successive reduction the number of occupied Landau levels as the magnetic field is increased. For each coincidence of a Landau level, with the Fermi level, there would be a discontinuity in the density-of-states function resulting in a peak of oscillation. Thus the peaks should occur whenever the Fermi energy is a multiple of energy separation between the two consecutive Landau levels and it may be noted that the origin of oscillations in the EEM is the same as that of the Shubnikov–de Haas (SdH) oscillations. With increase in magnetic field, the amplitude of the oscillation will increase and, ultimately, at very large values of the magnetic field, the conditions for the quantum limit will be reached (neglecting magnetic freeze out) when the EEM will be found to decrease monotonically with increase in magnetic field. In Fig. 6.16, the concentration dependence of the magneto-EEM has been plotted for all the cases of Fig. 6.15 for both n-InSb and n-InAs. The EEM again shows oscillatory dependence with different numerical values exhibiting the signature of the SdH effect. Although the rate of variations are different, the influence of the energy band constants in accordance with all the type of the band models is apparent from the Figs. One can observe from Fig. 6.17 that the EEM has a steady increase with the increase of the light intensity although the same EEM increases sharply with the increase in wavelength in different ways, as appears from Figs. 6.17 and 6.18, respectively. The nature of variations in all the cases depends strongly on the energy spectrum constants of the respective materials and the external physical conditions. It should be noted that the numerical value of the EEM in the presence of light waves is relatively much higher even at smaller value of the magnetic field, than that in the absence of the magnetic field. Such a high value in the EEM can cause a drastic effect by reducing the electron mobility under the application of a quantized magnetic field and the contribution of the oscillatory mass or the oscillatory mobility would be is more important.

6.6 The Formulation of the EEM for the Ultrathin Films of III–IV, Ternary and Quaternary Semiconductors Under External Photo-Excitation

6.6.1 Introduction

It is well known that the concept of reduction of symmetry of the wave-vector space and its consequence can unlock the physics of low dimensional structures.

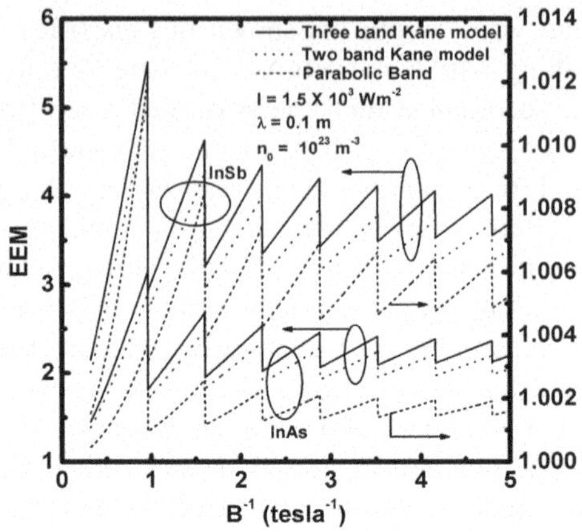

Fig. 6.15 Plot of the normalized EEM as a function of inverse magnetic field for n-InSb and n-InAs in the presence of light waves in accordance with the three, the two band models of Kane and the parabolic energy band model in the presence of external photo-excitation

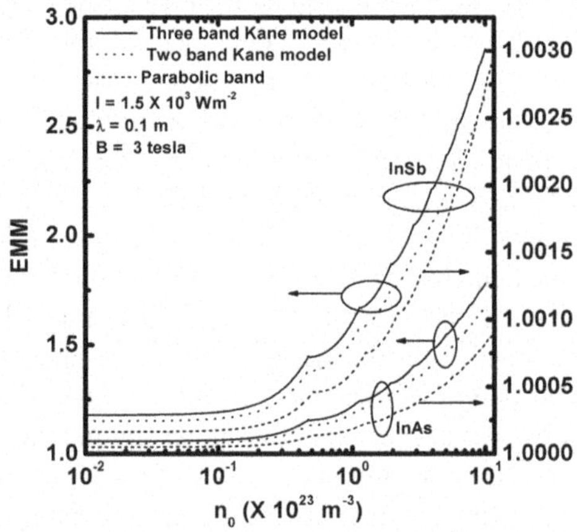

Fig. 6.16 Plot of the normalized EEM as a function of carrier concentration for n-InSb and n-InAs in the presence of light waves in accordance with the three, the two band models of Kane and the parabolic energy band model in the presence of external photo-excitation

In Sect. 6.6.2 of theoretical background, we shall study the EEM in ultrathin films of III–V, ternary and quaternary semiconductors both in the presence and absence of external light waves. The Sect. 6.7 contains result and discussions in this context.

6.6.2 Theoretical Background

The 2D electron energy spectrum in ultra-thin films of III–V, ternary and quaternary materials, whose unperturbed band structure is defined by the three-band model of Kane, in the presence of light waves can be expressed following (1.41)

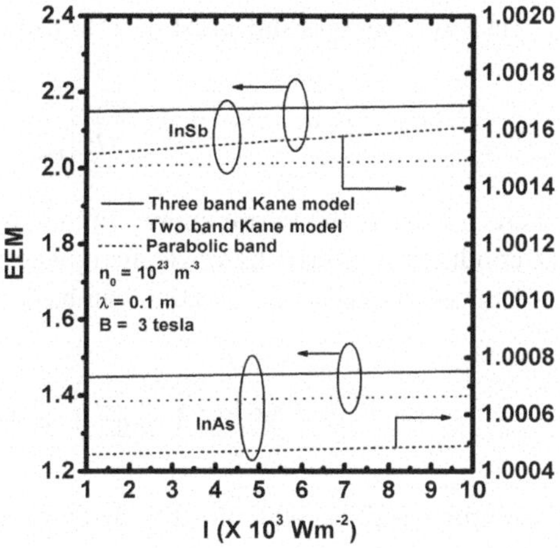

Fig. 6.17 Plot of the normalized EEM as a function of light intensity for n-InSb and n-InAs in the presence of light waves in accordance with the three, the two band models of Kane and the parabolic energy band model in the presence of external photo-excitation

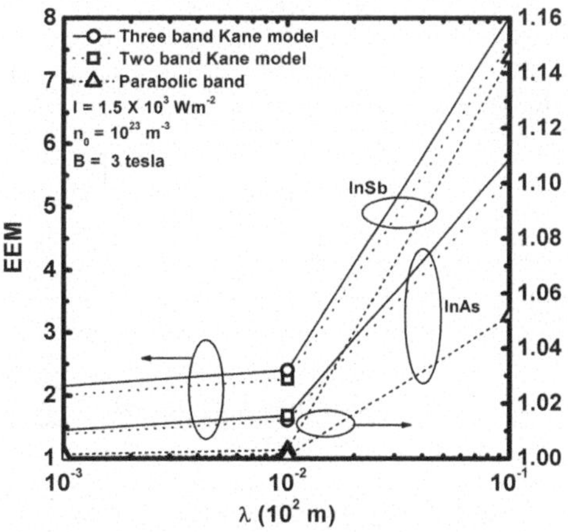

Fig. 6.18 Plot of the normalized EEM as a function of light wavelength for n-InSb and n-InAs in the presence of light waves in accordance with the three, the two-band models of Kane and the parabolic energy band model in the presence of external photo-excitation

$$\frac{\hbar^2 k_s^2}{2m_c} + \frac{\hbar^2}{2m_c}\left(\frac{n_z \pi}{d_z}\right)^2 = \beta_0(E, \lambda) \tag{6.72}$$

The subband energies $(E_{n_{l7}})$ can be written as

$$\beta_0(E_{n_{l7}}, \lambda) = \frac{\hbar^2}{2m_c}(n_z \pi / d_z)^2 \tag{6.73}$$

The expression of the EEM in this case is given by

$$m^*(E_{F2DL}, n_z, \lambda) = \hbar^2 k_s \left.\frac{\partial k_s}{\partial E}\right|_{E=E_{F2DL}} = m_c \{\beta_0(E_{F2DL}, \lambda)\}' \tag{6.74}$$

where E_{F2DL} is the Fermi energy in the present case as measured from the edge of the conduction band in the vertically upward direction in absence of any quantization.

The density-of-states function can be written as

$$N_{2D}(E, \lambda) = \left(\frac{m^* g_v}{\pi \hbar^2}\right) \sum_{n_z=1}^{n_{z\max}} [\beta_0(E, \lambda)]' H(E - E_{n_{l7}}) \tag{6.75}$$

Combining (6.75), with the Fermi-Dirac occupation probability factor, the two-dimensional electron concentration can be expressed as

$$n_{2D} = \frac{m_c g_v}{\pi \hbar^2} \sum_{n_z=1}^{n_{z\max}} \left[M_{18}(n_z, E_{F2DL}, \lambda) + N_{18}(n_z, E_{F2DL}, \lambda)\right] \tag{6.76}$$

where

$$M_{18}(n_z, E_{F2DL}, \lambda) \equiv \left[\beta_0(E_{F2DL}, \lambda) - \frac{\hbar^2}{2m_c}\left(\frac{n_z \pi}{d_z}\right)^2\right],$$

and

$$N_{18}(n_z, E_{F2DL}, \lambda) \equiv \sum_{r=1}^{s} L(r) M_{18}(n_z, E_{F2DL}, \lambda).$$

Using (1.42), the expressions for the 2D dispersion relation, the subband energies, the EEM, the density of states function and the electron concentration for ultra-thin films of III–V, ternary and quaternary semiconductors whose unperturbed band structure is defined by the two band model of Kane, can respectively be written in the presence of photo-excitation as

$$\frac{\hbar^2 k_s^2}{2m_c} + \frac{\hbar^2}{2m_c}\left(\frac{n_z \pi}{d_z}\right)^2 = \tau_0(E, \lambda) \tag{6.77}$$

$$\tau_0(E_{n_{l8}}, \lambda) = \frac{\hbar^2}{2m_c}(n_z \pi / d_z)^2 \tag{6.78}$$

$$m^*(E_{F2DL}, n_z, \lambda) = m_c \{\tau_0(E_{F2DL}, \lambda)\}' \tag{6.79}$$

$$N_{2D}(E,\lambda) = \left(\frac{m_c g_v}{\pi\hbar^2}\right)\sum_{n_z=1}^{n_{z\max}} [\tau_0(E,\lambda)]' H(E - E_{n_{l8}}) \tag{6.80}$$

$$n_{2D} = \frac{m_c g_v}{\pi\hbar^2}\sum_{n_z=1}^{n_{z\max}} \left[M_{19}(n_z, E_{F2DL},\lambda) + N_{19}(n_z, E_{F2DL},\lambda)\right] \tag{6.81}$$

where

$$M_{19}(n_z, E_{F2DL},\lambda) \equiv \left[\tau_0(E_{F2DL},\lambda) - \frac{\hbar^2}{2m^*}\left(\frac{n_z\pi}{d_z}\right)^2\right],$$

and

$$N_{19}(n_z, E_{F2DL},\lambda) \equiv \sum_{r=1}^{s} L(r) M_{18}(n_z, E_{F2DL},\lambda).$$

Using (1.43), the expressions for the 2D dispersion relation, the subband energies, the EEM, the density of states function and the electron concentration for ultra-thin films of III–V, ternary, and quaternary semiconductors, whose unperturbed band structure is defined by the parabolic energy bands, can respectively be written in the presence of photo-excitation as

$$\frac{\hbar^2 k_s^2}{2m_c} + \frac{\hbar^2}{2m_c}\left(\frac{n_z\pi}{d_z}\right)^2 = \rho_0(E,\lambda) \tag{6.82}$$

$$\rho_0(E_{n_{l9}},\lambda) = \frac{\hbar^2}{2m_c}(n_z\pi/d_z)^2 \tag{6.83}$$

$$m^*(E_{F2DL}, n_z,\lambda) = m_c\{\rho_0(E_{F2DL},\lambda)\}' \tag{6.84}$$

$$N_{2D}(E,\lambda) = \left(\frac{m_c g_v}{\pi\hbar^2}\right)\sum_{n_z=1}^{n_{z\max}} [\rho_0(E,\lambda)]' H(E - E_{n_{l9}}) \tag{6.85}$$

$$n_{2D} = \frac{m_c g_v}{\pi\hbar^2}\sum_{n_z=1}^{n_{z\max}} \left[M_{20}(n_z, E_{F2DL},\lambda) + N_{20}(n_z, E_{F2DL},\lambda)\right] \tag{6.86}$$

where

$$M_{20}(n_z, E_{F2DL},\lambda) \equiv \left[\rho_0(E_{F2DL},\lambda) - \frac{\hbar^2}{2m_c}\left(\frac{n_z\pi}{d_z}\right)^2\right],$$

$$N_{20}(n_z, E_{F2DL},\lambda) \equiv \sum_{r=1}^{s} L(r) M_{20}(n_z, E_{F2DL},\lambda).$$

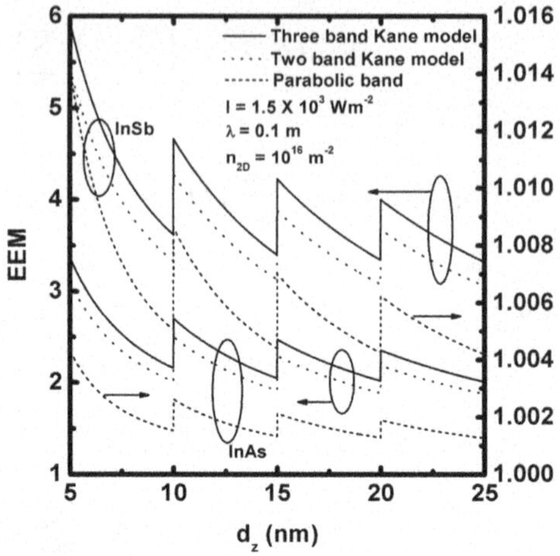

Fig. 6.19 Plot of the normalized EEM as a function of film thickness for *n*-InSb and *n*-InAs in the presence of light waves in accordance with the three- and the two-band models of Kane together with the parabolic energy band model in the presence of external photo-excitation

6.7 Results and Discussion

Using the values of the energy band constants from Table 1.1, we have plotted the EEM in the k_s plane as functions of film thickness, surface electron concentration, intensity and wavelength at $T = 4.2\,\mathrm{K}$ by taking ultra-thin films of ternary materials which are used for the purpose of numerical computations in accordance with the perturbed three [using (6.74) and (6.76)], two [using (6.79) and (6.81)], band models of Kane and that of perturbed parabolic energy bands [using (6.84) and (6.86)], as shown in Figs. 6.19, 6.20, 6.21 and 6.22, respectively. The influence of carrier confinement in 2D under the presence of an external photo-excitation on the behavior of EEM can be understood from the Figs. 6.19, 6.20, 6.21 and 6.22. The effect of quantum confinement is immediately apparent from all the curves of Fig. 6.19, since, the 2D EEM depend strongly on the nano-thickness, which is in direct contrast with the corresponding bulk specimens which is also the direct signature of quantum confinement. It appears from the said figures that the EEM in this case decreases with the increasing film thickness in a step-like manner as considered here although the numerical values vary widely and determined by the constants of the energy spectra. The oscillatory dependence is due to the crossing over of the Fermi level by the size quantized levels. For each coincidence of a size quantized level with the Fermi level, there would be a discontinuity in the density-of-states function resulting in a peak of oscillations. With large values of film thickness, the height of the steps decreases and the EEM decreases with increasing film thickness in non-oscillatory manner and exhibit monotonic decreasing dependence. The height of step size and the rate of decrement are totally dependent on the band structure. The influence of energy band nonparabolicity is immediately apparent by the comparing the curves of the said figures. The energy band non-parabolicity and the spin orbit splitting constant significantly enhances the numerical values of the

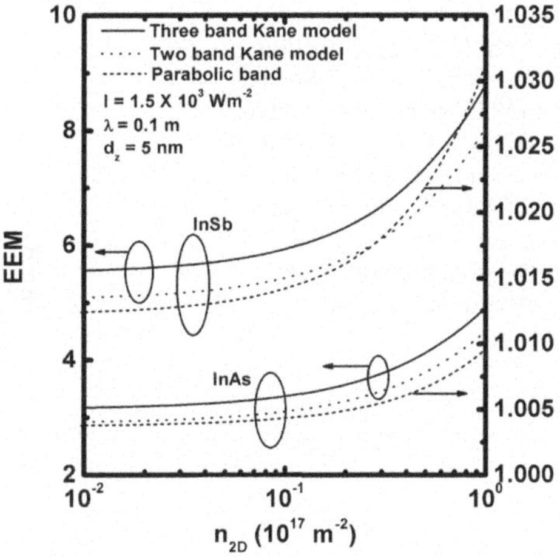

Fig. 6.20 Plot of the normalized EEM as a function of surface electron concentration for n-InSb and n-InAs in the presence of light waves in accordance with the three- and the two-band models of Kane together with the parabolic energy band model in the presence of external photo-excitation

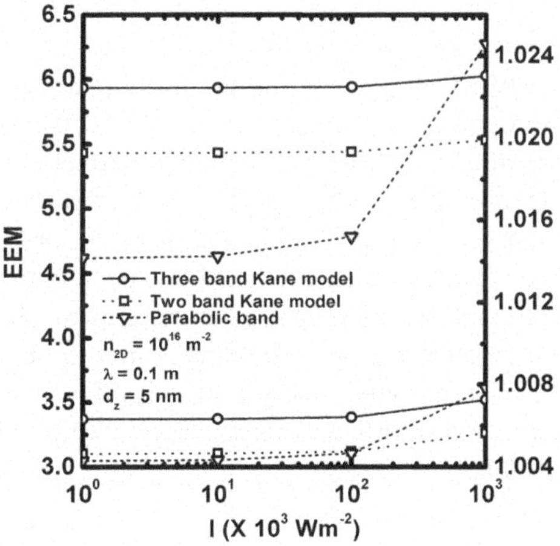

Fig. 6.21 Plot of the normalized EEM as a function of light intensity for n-InSb and n-InAs in the presence of light waves in accordance with the three- and the two-band models of Kane together with the parabolic energy band model in the presence of external photo-excitation

EEM both the cases of the materials. The numerical values of the EEM in accordance with the three band model of Kane are different as compared with the corresponding two-band model, which reflects that fact that the presence of the spin orbit splitting constant changes the magnitude of the EEM. It may be noted that the presence of the band nonparabolicity in accordance with the two-band model of Kane further changes the peaks of the oscillatory EEM for all cases of quantum confinements.

In Fig. 6.20, we have plotted the EEM as a function of surface electron concentration per unit area for all cases of Fig. 6.19. It appears that the EEM increases with increasing carrier degeneracy and also reflects the signature of the 1D confinement through the non-linear dependence with the 2D electron statistics. Since, most of the electrons at low temperatures occupies the lowest subband level, we have plotted

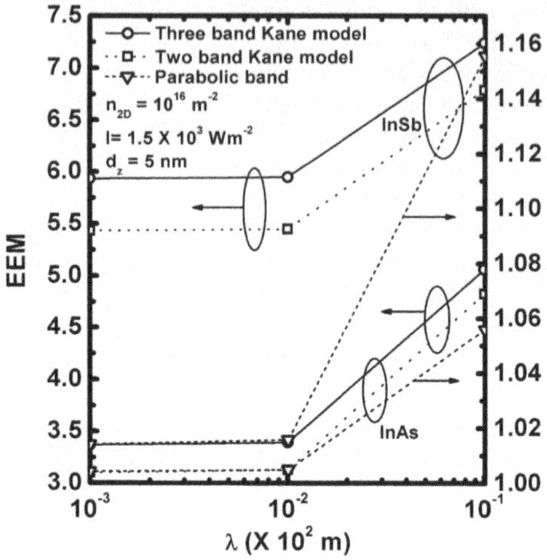

Fig. 6.22 Plot of the normalized EEM as a function of wavelength for n-InSb and n-InAs in the presence of light waves in accordance with the three- and the two-band models of Kane together with the parabolic energy band model in the presence of external photo-excitation

the EEM by considering the lowest subband energy in Figs. 6.20, 6.21 and 6.22. If more subbands were considered, the oscillatory dependence will be less and less prominent with increasing carrier concentration and ultimately, for bulk specimens of the same material, the EEM will be found to increase continuously with increasing electron concentration in a non-oscillatory manner. The effects of the light intensity and wavelength on the EEM has been exhibited in Figs. 6.21 and 6.22, respectively in the regime of very low temperatures. The EEM increases with both the variables with different slopes. It appears that the EEM in the case of parabolic energy band varies extremely slowly both with the light intensity and wavelength, although the sharp and significant variations are exhibited for both three- and the two-band energy models.

6.8 Investigation of the EEM in n-Channel Inversion Layers of III–V, Ternary and Quaternary Semiconductors Under External Photo-Excitation

6.8.1 Introduction

In Sect. 6.8.2 of theoretical background, we shall study the EEM for both weak and strong electric field limits in n-channel inversion layers of III–V, ternary and quaternary semiconductors both in the presence and absence of external light waves. The Sect. 6.9 contains result and discussions in this context.

6.8.2 Theoretical Background

In the presence of a surface electric field F_s along z direction and perpendicular to the surface, the (1.41) assumes the form

$$\frac{\hbar^2 k^2}{2m_c} = \beta_0(E - |e|\, F_s z, \lambda) \tag{6.87}$$

where, for this chapter, E represents the electron energy as measured from the edge of the conduction band at the surface in the vertically upward direction.

Using (6.87) and (3.2), the 2D electron dispersion relation in n-channel inversion layers of III–V, ternary and quaternary semiconductors (whose unperturbed electrons obey the three band model of Kane) in the presence of light waves under the condition of weak electric field limit, assumes the form

$$\beta_0(E, \lambda) = \frac{\hbar^2 k_s^2}{2m_c} + S_i \left[\frac{\hbar\, |e|\, F_s\, [\beta_0(E, \lambda)]'}{\sqrt{2m_c}}\right]^{2/3} \tag{6.88}$$

The EEM can be expressed as

$$m^*(E_{FiwL}, i) = \hbar^2 k_s \left.\frac{\partial k_s}{\partial E}\right|_{E=E_{FiwL}} = m_c\, [P_{3L}(E_{FiwL}, i)]| \tag{6.89}$$

where E_{FiwL} is the Fermi energy under the weak electric field limit as measured from the edge of the conduction band at the surface in the vertically upward direction in the presence of light waves and

$$P_{3L}(E, i) \equiv \left\{[\beta_0(E, \lambda)]' - \left\{\frac{2}{3} S_i \left[\frac{\hbar\, |e|\, F_s}{\sqrt{2m_c}}\right]^{2/3} \{[\beta_0(E, \lambda)]'\}^{-1/3} [\beta_0(E, \lambda)]''\right\}\right\}.$$

Thus, one can observe that the EEM is a function of the subband index, surface electric field, the Fermi energy and the other spectrum constants due to the combined influence of E_g, Δ and λ.

The subband energy ($E_{n_{iw2L}}$) in this case can be obtained from (6.88) as

$$\beta_0(E_{n_{iw2L}}, \lambda) = S_i \left[\frac{\hbar\, |e|\, F_s \left[\beta_0(E_{n_{iw2L}}, \lambda)\right]'}{\sqrt{2m_c}}\right]^{2/3} \tag{6.90}$$

The 2D total density-of-states function in weak electric field limit can be expressed as

$$N_{2D_i}(E) = \frac{m_c g_v}{\pi \hbar^2} \sum_{i=0}^{i_{\max}} \left[P_{3L}(E, i) H(E - E_{n_{iw2L}})\right] \tag{6.91}$$

Using (6.91) and the occupation probability, the n_{2D_w} in the present case can be written as

$$n_{2D_w} = \frac{g_v m_c}{\pi \hbar^2} \sum_{i=0}^{i_{\max}} [P_{4wL}(E_{FiwL}, i) + Q_{4wL}(E_{FiwL}, i)] \tag{6.92}$$

where

$$P_{4wL}(E_{FiwL}, i) \equiv \left\{ \beta_0(E_{FiwL}, \lambda) - S_i \left[\frac{\hbar e F_s [\beta_0(E_{FiwL}, \lambda)]'}{\sqrt{2m_c}} \right]^{2/3} \right\}$$

and

$$Q_{4L}(E_{FiwL}, i) \equiv \sum_{r=1}^{s} \{L(r) [P_{4L}(E_{FiwL}, i)]\}.$$

Using (6.87) and (3.2), the 2D electron dispersion relation in n-channel inversion layers of III–V, ternary, and quaternary semiconductors in the presence of light waves under the condition of strong electric field limit, assumes the form

$$\left[\beta_0(E, \lambda) - \left\{ \frac{|e| \, F_s \hbar}{\sqrt{2m_c}} \left(\frac{2\sqrt{2}(S_i)^{3/2}}{3} \right) \sqrt{[\beta_0(E, \lambda)]''} \right\} \right] = \frac{\hbar^2 k_s^2}{2m_c} \tag{6.93}$$

The EEM can be expressed as

$$m^*(E_{FisL}, i) = \hbar^2 k_s \left. \frac{\partial k_s}{\partial E} \right|_{E=E_{FisL}} = m_c \, [P_{5L}(E, i)]|_{E=E_{FisL}} \tag{6.94}$$

where E_{FisL} is the Fermi energy under the strong electric field limit as measured from the edge of the conduction band at the surface in the presence of light waves and

$$P_{5L}(E, i) \equiv \left\{ \{\beta_0(E, \lambda)\}' - \left\{ \frac{|e| F_s \hbar}{\sqrt{2m_c}} \left(\frac{\sqrt{2}(S_i)^{3/2}}{3} \right) ([\beta_0(E, \lambda)]'')^{-1/2} [\beta_0(E, \lambda)]''' \right\} \right\}.$$

Thus, one can observe that the EEM is a function of the subband index, surface electric field, the Fermi energy and the other spectrum constants due to the combined influence of E_g, Δ and λ.

The subband energy ($E_{n_{is2L}}$) in this case can be obtained from (6.93) as

$$\beta_0(E_{n_{is2L}}, \lambda) - \left\{ \frac{|e| \, F_s \hbar}{\sqrt{2m_c}} \left(\frac{2\sqrt{2}(S_i)^{3/2}}{3} \right) \sqrt{[\beta_0(E_{n_{is2L}}, \lambda)]''} \right\} = 0 \tag{6.95}$$

The total 2D density-of-states function under the strong electric field limit can be expressed as

$$N_{2D_i}(E) = \frac{m_c g_v}{\pi \hbar^2} \sum_{i=0}^{i_{\max}} \left[P_{5L}(E, i) H(E - E_{n_{is2L}}) \right] \quad (6.96)$$

Using (6.96) and the Fermi-Dirac occupation probability factor, the n_{2D_s} in the present case under the strong electric field can be written as

$$n_{2D_s} = \frac{g_v m^*}{\pi \hbar^2} \sum_{i=0}^{i_{\max}} \left[P_{6sL}(E_{FisL}, i) + Q_{6sL}(E_{FisL}, i) \right] \quad (6.97)$$

where

$$P_{6sL}(E_{FisL}, i) \equiv \left\{ \beta_0(E_{FisL}, \lambda) - \left[\frac{2\sqrt{2}}{3} (S_i)^{3/2} \frac{\hbar\, |e|\, F_s \sqrt{[\beta_0(E_{FisL}, \lambda)]''}}{\sqrt{2m_c}} \right] \right\},$$

$$Q_{6sL}(E_{FisL}, i) \equiv \sum_{r=1}^{s} \{ L(r) \left[P_{6sL}(E_{FisL}, i) \right] \}.$$

Again for this section the (1.42) assumes the form

$$\frac{\hbar^2 k^2}{2m_c} = \tau_0(E - |e|\, F_s z, \lambda) \quad (6.98)$$

Using (6.97) and (3.2), the 2D electron dispersion relation in n-channel inversion layers of III–V, ternary, and quaternary semiconductors (whose unperturbed electrons obey the two-band model of Kane) in the presence of light waves under the condition of weak electric field limit, assumes the form

$$\tau_0(E, \lambda) = \frac{\hbar^2 k_s^2}{2m_c} + S_i \left[\frac{\hbar\, |e|\, F_s\, [\tau_0(E, \lambda)]'}{\sqrt{2m_c}} \right]^{2/3} \quad (6.99)$$

The EEM can be expressed as

$$m^*(E_{FiwL2}, i) = m_c \left[P_{3L2}(E_{FiwL2}, i) \right] | \quad (6.100)$$

E_{FiwL2} is the Fermi energy under the weak electric field limit as measured from the edge of the conduction band at the surface in the vertically upward direction in the presence of light waves in this case and

$$P_{3L2}(E, i) \equiv \left\{ [\tau_0(E, \lambda)]' - \left\{ \frac{2}{3} S_i \left[\frac{\hbar\, |e|\, F_s}{\sqrt{2m_c}} \right]^{2/3} \{[\tau_0(E, \lambda)]'\}^{-1/3} [\tau_0(E, \lambda)]'' \right\} \right\}.$$

Thus, one can observe that the EEM is a function of the subband index, surface electric field, the Fermi energy, and the other spectrum constants due to the combined influence of E_g and λ.

The subband energy $(E_{n_{iw2L2}})$ in this case can be obtained from (6.99) as

$$\tau_0(E_{n_{iw2L2}}, \lambda) = S_i \left[\frac{\hbar |e| F_s \left[\tau_0(E_{n_{iw2L2}}, \lambda)\right]'}{\sqrt{2m_c}} \right]^{2/3} \tag{6.101}$$

The 2D total density-of-states function in weak electric field limit can be expressed as

$$N_{2D_i}(E) = \frac{m_c g_v}{\pi \hbar^2} \sum_{i=0}^{i_{\max}} \left[P_{3L2}(E, i) H(E - E_{n_{iw2L2}}) \right] \tag{6.102}$$

Using (6.102) and the occupation probability, the n_{2D_w} in the present case can be written as

$$n_{2D_w} = \frac{g_v m_c}{\pi \hbar^2} \sum_{i=0}^{i_{\max}} \left[P_{4wL2}(E_{FiwL2}, i) + Q_{4wL2}(E_{FiwL2}, i) \right] \tag{6.103}$$

where

$$P_{4wL2}(E_{FiwL2}, i) \equiv \left\{ \tau_0(E_{FiwL2}, \lambda) - S_i \left[\frac{\hbar e F_s \left[\tau_0(E_{FiwL2}, \lambda)\right]'}{\sqrt{2m_c}} \right]^{2/3} \right\}$$

and

$$Q_{4L2}(E_{FiwL2}, i) \equiv \sum_{r=1}^{s} \{ L(r) \left[P_{4L2}(E_{FiwL2}, i) \right] \}.$$

Using the appropriate equations, the 2D electron dispersion relation in n-channel inversion layers of III–V, ternary, and quaternary semiconductors (whose unperturbed electrons obey the two-band model of Kane) in the presence of light waves under the condition of strong electric field limit, assumes the form

$$\left[\tau_0(E, \lambda) - \left\{ \frac{|e| F_s \hbar}{\sqrt{2m_c}} \left(\frac{2\sqrt{2}(S_i)^{3/2}}{3} \right) \sqrt{\left[\tau_0(E, \lambda)\right]''} \right\} \right] = \frac{\hbar^2 k_s^2}{2m_c} \tag{6.104}$$

The EEM can be expressed as

$$m^*(E_{FisL2}, i) = m_c \left[P_{5L2}(E, i) \right] \big|_{E = E_{FisL2}} \tag{6.105}$$

where E_{FisL2} is the Fermi energy under the strong electric field limit as measured from the edge of the conduction band at the surface in the presence of light waves in the present case and

$$P_{5L2}(E,i) \equiv \left\{ \{\tau_0(E,\lambda)\}' - \left\{ \frac{|e|\,F_s\hbar}{\sqrt{2m_c}} \right. \right.$$
$$\left. \left. \times \left(\frac{\sqrt{2}(S_i)^{3/2}}{3} \right) \left([\tau_0(E,\lambda)]'' \right)^{-1/2} [\tau_0(E,\lambda)]''' \right\} \right\}.$$

Thus, one can observe that the EEM is a function of the subband index, surface electric field, the Fermi energy and the other spectrum constants due to the combined influence of E_g and λ.

The subband energy ($E_{n_{is2L2}}$) in this case can be obtained from (6.104) as

$$\tau_0(E_{n_{is2L2}},\lambda) - \left\{ \frac{|e|\,F_s\hbar}{\sqrt{2m_c}} \left(\frac{2\sqrt{2}(S_i)^{3/2}}{3} \right) \sqrt{[\tau_0(E_{n_{is2L2}},\lambda)]''} \right\} = 0 \qquad (6.106)$$

The total 2D density-of-states function under the strong electric field limit can be expressed as

$$N_{2D_i}(E) = \frac{m_c g_v}{\pi\hbar^2} \sum_{i=0}^{i_{\max}} \left[P_{5L2}(E,i) H(E - E_{n_{is2L2}}) \right] \qquad (6.107)$$

Using (6.107) and the Fermi-Dirac occupation probability factor, the n_{2D_s} in the present case under the strong electric field can be written as

$$n_{2D_s} = \frac{g_v m_c}{\pi\hbar^2} \sum_{i=0}^{i_{\max}} \left[P_{6sL2}(E_{FisL2},i) + Q_{6sL2}(E_{FisL2},i) \right] \qquad (6.108)$$

where

$$P_{6sL2}(E_{FisL2},i) \equiv \left\{ \tau_0(E_{FisL2},\lambda) - \left[\frac{2\sqrt{2}}{3}(S_i)^{3/2} \frac{\hbar\,|e|\,F_s\sqrt{[\tau_0(E_{FisL2},\lambda)]''}}{\sqrt{2m_c}} \right] \right\}$$

and

$$Q_{6sL2}(E_{FisL2},i) \equiv \sum_{r=1}^{s} \{ L(r) \left[P_{6sL2}(E_{FisL2},i) \right] \}.$$

Besides for this section the (1.43) assumes the form

$$\frac{\hbar^2 k^2}{2m_c} = \rho_0(E - |e| F_s z, \lambda) \tag{6.109}$$

Using (6.109) and (3.2), the 2D electron dispersion relation in n-channel inversion layers of III–V, ternary and quaternary semiconductors (whose unperturbed electrons obey the parabolic energy bands) in the presence of light waves under the condition of weak electric field limit, assumes the form

$$\rho_0(E, \lambda) = \frac{\hbar^2 k_s^2}{2m_c} + S_i \left[\frac{\hbar |e| F_s [\rho_0(E, \lambda)]'}{\sqrt{2m^*}}\right]^{2/3} \tag{6.110}$$

The EEM can be expressed as

$$m^*(E_{FiwL1}, i) = m_c [P_{3L1}(E_{FiwL1}, i)] \tag{6.111}$$

E_{FiwL1} is the Fermi energy under the weak electric field limit as measured from the edge of the conduction band at the surface in the vertically upward direction in the presence of light waves in this case and

$$P_{3L1}(E, i) \equiv \left\{[\rho_0(E, \lambda)]' - \left\{\frac{2}{3} S_i \left[\frac{\hbar |e| F_s}{\sqrt{2m_c}}\right]^{2/3} \{[\rho_0(E, \lambda)]'\}^{-1/3} [\rho_0(E, \lambda)]''\right\}\right\}.$$

Thus, one can observe that the EEM is a function of the subband index, surface electric field, the Fermi energy and the other spectrum constants due to the combined influence of E_g and λ.

The subband energy ($E_{n_{iw2L1}}$) in this case can be obtained from (1.135) as

$$\rho_0(E_{n_{iw2L1}}, \lambda) = S_i \left[\frac{\hbar |e| F_s \left[\rho_0(E_{n_{iw2L1}}, \lambda)\right]'}{\sqrt{2m_c}}\right]^{2/3} \tag{6.112}$$

The 2D total density-of-states function in weak electric field limit can be expressed as

$$N_{2D_i}(E) = \frac{m_c g_v}{\pi \hbar^2} \sum_{i=0}^{i_{\max}} \left[P_{3L1}(E, i) H(E - E_{n_{iw2L1}})\right] \tag{6.113}$$

Using (6.113) and the occupation probability, the n_{2D_w} in the present case can be written as

$$n_{2D_w} = \frac{g_v m_c}{\pi \hbar^2} \sum_{i=0}^{i_{\max}} [P_{4wL1}(E_{FiwL1}, i) + Q_{4wL1}(E_{FiwL1}, i)] \tag{6.114}$$

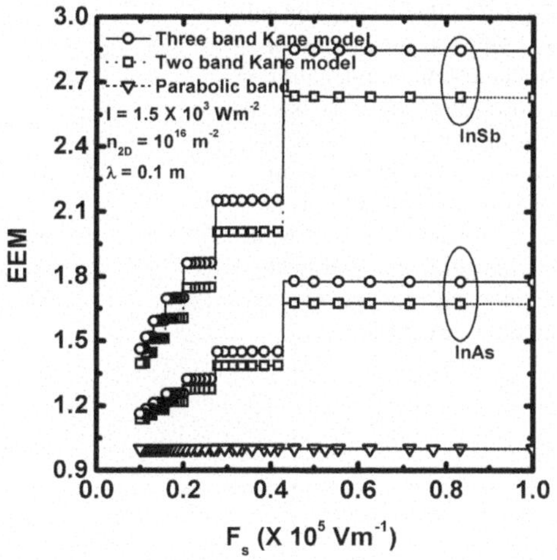

Fig. 6.23 Plot of the normalized EEM as a function of surface electric field for n-channel inversion layers of n-InSb and n-InAs in the presence of light waves in accordance with the three, the two band models of Kane and the parabolic energy band model in the presence of external photo-excitation

where

$$P_{4wL1}(E_{FiwL1}, i) \equiv \left\{ \rho_0(E_{FiwL1}, \lambda) - S_i \left[\frac{\hbar e F_s \left[\rho_0(E_{FiwL1}, \lambda)\right]'}{\sqrt{2m_c}} \right]^{2/3} \right\}$$

and

$$Q_{4wL1}(E_{FiwL1}, i) \equiv \sum_{r=1}^{s} \{L(r)\left[P_{4wL1}(E_{FiwL1}, i)\right]\}.$$

6.9 Results and Discussion

Using the values of the energy band constants from Table 1.1, we have plotted the EEM in the k_s plane for the first two subbands as functions of surface electric field, surface concentration and wavelength at $T = 4.2\,\text{K}$ by taking n-channel inversion layers of n-InSb and n-InAs semiconductors which are used for the purpose of numerical computations in accordance with the perturbed three [using (6.89) and (6.92)] and two [using (6.100) and (6.103)] band models of Kane and that of perturbed parabolic energy bands [using (6.111) and (6.114)] for weak electric field limit respectively. In Fig. 6.23, we have presented the variation of the EEM in the n-channel inversion layers of n-InSb and n-InAs as function of surface electric field in accordance with the three-band model of Kane, the two-band model of Kane and the parabolic energy bands respectively under weak electric field by considering the effect of electric subbands.

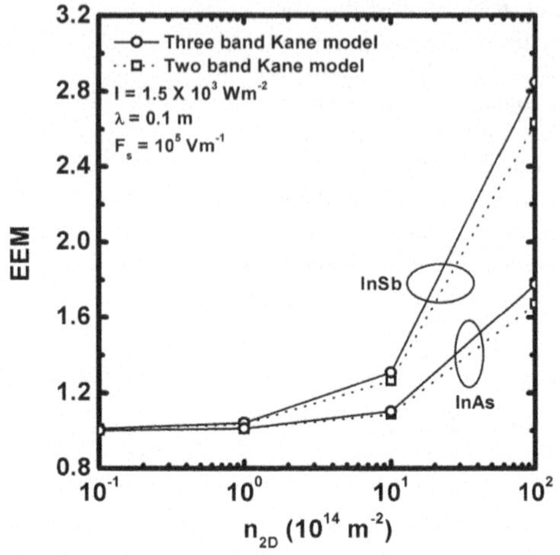

Fig. 6.24 Plot of the normalized EEM as a function of surface electron concentration for n-channel inversion layers of n-InSb and n-InAs in the presence of light waves in accordance with the three, the two-band models of Kane and the parabolic energy band model in the presence of external photo-excitation

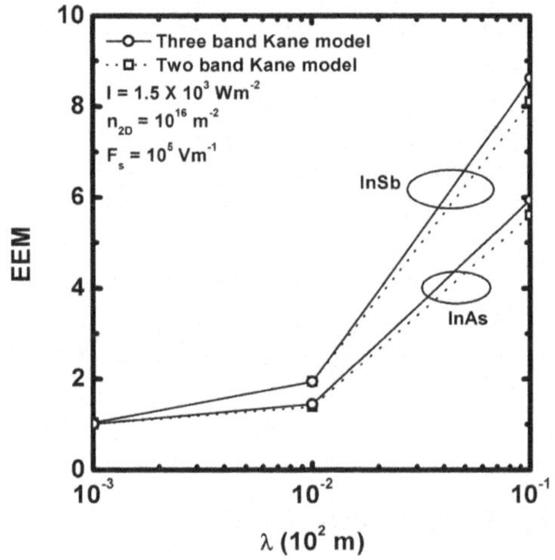

Fig. 6.25 Plot of the normalized EEM as a function of wavelength of the incident light for n-channel inversion layers of n-InSb and n-InAs in the presence of light waves in accordance with the three, the two-band models of Kane and the parabolic energy band model in the presence of external photo-excitation

It appears from Fig. 6.23 that the EEM in the n-channel inversion layers increases with increase in surface electric field for weak electric field in a step-like manner with different numerical values and the influence of the energy band constants can also be assessed from the said figures. The EEM depends on the electric subband index, surface electric field, the Fermi energy and the other spectrum constants due to the combined influence of E_g, Δ and λ which is the characteristic feature of such 2D systems under radiation. In Figs. 6.24 and 6.25, the effect of surface concentration and wavelength on the EEM for all the cases of Fig. 6.23 has been considered under the quantum limit approximation. It appears from the Figs. 6.24 and 6.25 that the EEM increases with both the variables in the straight lines segmentation fashion with different increasing slopes at particular values of surface electron concentration and light wave length. The values of EEM for InSb are greater than that of InAs. The slopes are entirely determined by the spectra constants of n channel InSb and InAs respectively. It may be noted that if the direction of application of the surface electric field applied perpendicular to the surface be taken in an arbitrary direction and not as k_z as assumed in the present work, the EEM would be different analytically for both the limits. Nevertheless, the arbitrary choice of the direction normal to the surface would not result in a change of the basic qualitative feature of the EEM in n-channel inversion layers of semiconductors in the presence of photo excitation. The approximation of the potential well at the surface by a triangular well introduces some errors, as for instance the omission of the free charge contribution to the potential. This kind of approach is reasonable if there are only few charge carriers in the inversion layer, but is responsible for an overestimation of the splitting when the inversion carrier density exceeds that of the depletion layer. It has been observed that the maximum error due to the triangular potential well is tolerable in the practical sense because for actual calculations, one need a self consistent solution which is a formidable problem, for the present generalized systems due to the non availability of the proper analytical techniques, without exhibiting a widely different qualitative behavior. The second assumption of using only two subbands in the numerical calculations is valid in the range of low temperatures, where the quantum effects become prominent. The errors which are being introduced for these assumptions are found not to be serious enough at low temperatures. We wish to note that the many body effects, the hot electron effects, the formation of band tails, arbitrary orientation of the direction of the electric quantization and the effects of surface of states have been neglected in our simplified theoretical formalism due to the lack of availability in the literature of the proper analytical techniques for including them for the generalized systems as considered in this section. The numerical computation for the corresponding cases of EEM under the strong electric field has been leftover for the readers so that they can enjoy the intricate internal computer analysis behind plots in this context. Our simplified approach will be useful for the purpose of comparison, when, the methods of tackling of the aforementioned formidable problems for the present generalized system appear and we admit the fact that the inclusion of the said effects would certainly increase the accuracy of our results.

6.10 Investigation of the EEM in *nipi* Structures of III–V, Ternary and Quaternary Semiconductors Under External Photo-Excitation

6.10.1 Introduction

In Sect. 6.10.2 of theoretical background, we shall study the EEM in nipi structures of III–V, ternary, and quaternary semiconductors both in the presence and absence of external light waves. The Sect. 6.11 contains result and discussions in this context.

6.10.2 Theoretical Background

The 2D electron dispersion relation in nipi structures of III–V, ternary, and quaternary semiconductors (whose unperturbed electrons obey the three band model of Kane) in the presence of light waves can be written as

$$\beta_0(E,\lambda) = \left(n_i + \frac{1}{2}\right)\hbar\omega_{9L}(E,\lambda) + \frac{\hbar^2 k_s^2}{2m_c} \tag{6.115}$$

where n_i $(= 0, 1, 2, \ldots)$ is the mini-band index for nipi structures, ε_{sc} is the semiconductor permittivity and

$$\omega_{9L}(E,\lambda) \equiv \left(\frac{n_0\,|e|^2}{\varepsilon_{sc}\beta_0'(E,\lambda)m_c}\right)^{1/2}.$$

The EEM in this case can be written as

$$m^*(E_{FnL},n_i) = m_c R_{82L}(E,n_i)|_{E=E_{FnL}} \tag{6.116}$$

in which, $\overline{E}_{FnL}$ is the Fermi energy in the present case as measured from the edge of the conduction band in vertically upward direction in the absence of any quantization and

$$R_{82L}(E,n_i) \equiv \left\{[\beta_0(E,\lambda)]' - \left(n_i + \frac{1}{2}\right)\hbar\,[\omega_{9L}(E,\lambda)]'\right\}.$$

From (6.116), we observe that the EEM in this case is a function of the Fermi energy, wavelength, nipi subband index and the other material constants which is the characteristic feature of nipi structures of III–V, ternary, and quaternary compounds in the presence of light waves whose bulk dispersion relation in the absence of any field is defined by the three band model of Kane.

The subband energies (E_{2niL}) can be written as

$$\beta_0(E_{2niL},\lambda) = \left(n_i + \frac{1}{2}\right)\hbar\omega_{9L}(E_{2niL},\lambda). \tag{6.117}$$

The density-of-states function in this case can be expressed as

$$N_{nipiL}(E) = \frac{m_c g_v}{\pi\hbar^2 d_0}\sum_{n_i=0}^{n_{i\max}} R_{82L}(E,n_i)H(E - E_{2niL}) \tag{6.118}$$

in which d_0 is the superlattice period.

The use of (6.118) leads to the expression of the electron concentration as

$$n_0 = \frac{m_c g_v}{\pi \hbar^2 d_0} \sum_{n_i=0}^{n_{i\max}} \left[T_{83L}(\overline{E}_{FnL}, n_i) + T_{84L}(\overline{E}_{FnL}, n_i)\right] \tag{6.119}$$

where

$$T_{83L}(\overline{E}_{FnL}, n_i) \equiv \left[\beta_0(\overline{E}_{FnL}, \lambda) - \left(n_i + \frac{1}{2}\right)\hbar\omega_{9L}(\overline{E}_{FnL}, \lambda)\right]$$

and

$$T_{84L}(\overline{E}_{FnL}, n_i) \equiv \sum\nolimits_{r=1}^{s} L(r) T_{83L}(\overline{E}_{FnL}, n_i).$$

Again, the 2D electron dispersion relation in nipi structures of III–V, ternary, and quaternary semiconductors (whose unperturbed electrons obey the two band model of Kane) in the presence of light waves can be written as

$$\tau_0(E, \lambda) = \left(n_i + \frac{1}{2}\right)\hbar\omega_{9L2}(E, \lambda) + \frac{\hbar^2 k_s^2}{2m_c} \tag{6.120}$$

where $\omega_{9L2}(E, \lambda) \equiv \left(\frac{n_0 |e|^2}{\varepsilon_{sc} \tau_0'(E, \lambda) m_c}\right)^{1/2}$.

The EEM in this case can be written as

$$m^*(E_{FnL2}, n_i) = m_c R_{82L2}(E, n_i)|_{E=E_{FnL2}} \tag{6.121}$$

in which, $R_{82L2}(E, n_i) \equiv \left\{[\tau_0(E, \lambda)]' - \left(n_i + \frac{1}{2}\right)\hbar\,[\omega_{9L2}(E, \lambda)]'\right\}$. The subband energies (E_{2niL2}) can be written as

$$\tau_0(E_{2niL2}, \lambda) = \left(n_i + \frac{1}{2}\right)\hbar\omega_{9L2}(E_{2niL2}, \lambda) \tag{6.122}$$

The density-of-states function in this case can be expressed as

$$N_{nipiL2}(E) = \frac{m_c g_v}{\pi \hbar^2 d_0} \sum_{n_i=0}^{n_{i\max}} R_{82L2}(E, n_i) H(E - E_{2niL2}) \tag{6.123}$$

The use of (6.123) leads to the expression of the electron concentration as

$$n_0 = \frac{m_c g_v}{\pi \hbar^2 d_0} \sum_{n_i=0}^{n_{i\max}} \left[T_{83L2}(\overline{E}_{FnL2}, n_i) + T_{84L2}(\overline{E}_{FnL2}, n_i)\right] \tag{6.124}$$

where $T_{83L2}(\overline{E}_{FnL2}, n_i) \equiv \left[\tau_0(\overline{E}_{FnL2}, \lambda) - \left(n_i + \frac{1}{2}\right)\hbar\omega_{9L2}(\overline{E}_{FnL2}, \lambda)\right]$.

$\overline{E}_{FnL2}$ is the Fermi energy in the present case as measured from the edge of the conduction band in vertically upward direction in the absence of any quantization and

$$T_{84L2}(\overline{E}_{FnL2}, n_i) \equiv \sum_{r=1}^{s} L(r) T_{83L2}(\overline{E}_{FnL2}, n_i).$$

Besides, the 2D electron dispersion relation in nipi structures of III–V, ternary and quaternary semiconductors (whose unperturbed electrons obey the parabolic energy bands) in the presence of light waves can be written as

$$\rho_0(E, \lambda) = \left(n_i + \frac{1}{2}\right) \hbar\omega_{9L1}(E, \lambda) + \frac{\hbar^2 k_s^2}{2m_c} \tag{6.125}$$

where $\omega_{9L1}(E, \lambda) \equiv \left(\frac{n_0 |e|^2}{\varepsilon_{sc} \rho_0'(E, \lambda) m_c}\right)^{1/2}$.

The EEM in this case can be written as

$$m^*(E_{FnL1}, n_i) = m_c R_{82L1}(E, n_i)|_{E=E_{FnL1}} \tag{6.126}$$

in which $\overline{E}_{FnL1}$ is the Fermi energy in the present case as measured from the edge of the conduction band in vertically upward direction in the absence of any quantization and

$$R_{82L2}(E, n_i) \equiv \left\{[\rho_0(E, \lambda)]' - \left(n_i + \frac{1}{2}\right) \hbar [\omega_{9L2}(E, \lambda)]'\right\}.$$

The subband energies (E_{2niL1}) can be written as

$$\rho_0(E_{2niL1}, \lambda) = \left(n_i + \frac{1}{2}\right) \hbar\omega_{9L1}(E_{2niL1}, \lambda) \tag{6.127}$$

The density-of-states function in this case can be expressed as

$$N_{nipiL1}(E) = \frac{m_c g_v}{\pi \hbar^2 d_0} \sum_{n_i=0}^{n_{i\max}} R_{82L1}(E, n_i) H(E - E_{2niL1}) \tag{6.128}$$

The use of (6.128) leads to the expression of the electron concentration as

$$n_0 = \frac{m_c g_v}{\pi \hbar^2 d_0} \sum_{n_i=0}^{n_{i\max}} \left[T_{83L1}(\overline{E}_{FnL1}, n_i) + T_{84L1}(\overline{E}_{FnL1}, n_i)\right] \tag{6.129}$$

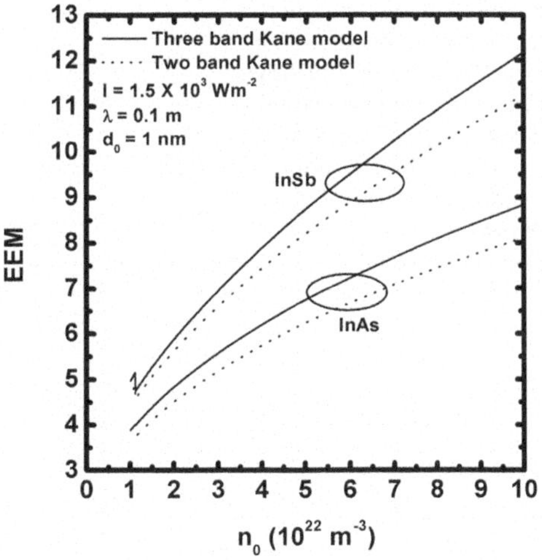

Fig. 6.26 Plot of the normalized EEM as a function of electron concentration for nipi structures of *n*-InSb and *n*-InAs in the presence of light waves in accordance with the three- and the two-band models of Kane in the presence of external photo-excitation

where

$$T_{83L1}(\overline{E}_{FnL1}, n_i) \equiv \left[\rho_0(\overline{E}_{FnL1}, \lambda) - \left(n_i + \frac{1}{2}\right) \hbar\omega_{9L1}(\overline{E}_{FnL1}, \lambda)\right]$$

and

$$T_{84L1}(\overline{E}_{FnL1}, n_i) \equiv \sum\nolimits_{r=1}^{s} L(r) T_{83L1}(\overline{E}_{FnL1}, n_i).$$

6.11 Results and Discussion

Using the values of the energy band constants from Table 1.1, we have plotted the EEM for the first two subbands as functions of wave length, intensity, thickness, and electron concentration at $T = 4.2\,\mathrm{K}$ by taking nipi structures of ternary materials which are used for the purpose of numerical computations in accordance with the perturbed three [using (6.116) and (6.119)], two [using (6.121) and (6.124)] band models of Kane and that of perturbed parabolic energy bands [using (6.126) and (6.129)] respectively.

Using the multiple subbands, one can numerically evaluate the EEM as a function of electron concentration and wavelength in nipi structures of III–V compounds by using the nipi structures of InSb and InAs as shown in Figs. 6.26 and 6.27, respectively, in accordance with three- and two-band models of Kane. The occurrence of the humps in Fig. 6.26 has been explained earlier in the context of ultrathin films. The effect of nipi structure tailoring increases the EEM to an extremely high value which severely affects the electron mobility in such structures.

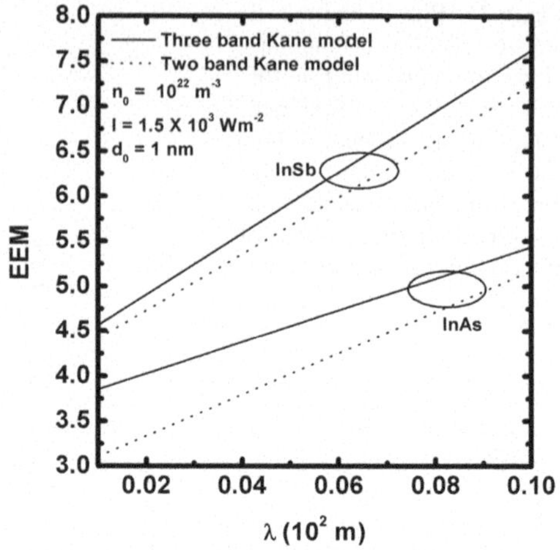

Fig. 6.27 Plot of the normalized EEM as a function of wavelength for nipi structures of *n*-InSb and *n*-InAs in the presence of light waves in accordance with the three- and the two-band models of Kane in the presence of external photo-excitation

The effect of increasing wavelength aids to increase the EEM in a linearly way, however, in case of increasing the light intensity, the tendency of increase in EEM is extremely slow. We have not considered the effect of the light intensity and wavelength on the EEM governed by the parabolic energy band due to its slow variation from the value 1.

6.12 Investigation of the EEM in *Nano Wires* of III–V, Ternary, and Quaternary Semiconductors Under External Photo-Excitation

6.12.1 Introduction

In Sect. 6.12.2 of theoretical background, we shall study the EEM in nano wires of III–V, ternary and quaternary semiconductors both in the presence and absence of external light waves. The Sect. 6.13 contains result and discussions in this context.

6.12.2 Theoretical Background

The 1D electron energy spectrum in nano wires of III–V, ternary and quaternary semiconductors, whose unperturbed band structure is defined by the three-band model of Kane, in the presence of light waves can be expressed from (1.41) as

$$\frac{\hbar^2 k_x^2}{2m_c} = \beta_0(E,\lambda) - \left[\frac{\hbar^2}{2m_c}\left(\frac{n_y \pi}{d_y}\right)^2 + \frac{\hbar^2}{2m_c}\left(\frac{n_z \pi}{d_z}\right)^2\right] \tag{6.130}$$

The subband energies ($E_{n_{l10}}$) can be expressed as

$$\beta_0(E_{n_{l10}},\lambda) = \left[\frac{\hbar^2}{2m_c}(n_y\pi/d_y)^2 + \frac{\hbar^2}{2m_c}(n_z\pi/d_z)^2\right] \tag{6.131}$$

The EEM in the free direction k_x in this case can be written from (6.130) as

$$m^*(E_{F1DL},n_y,n_z,\lambda) = m_c\,\{\beta_0(E_{F1DL},\lambda)\}' \tag{6.132}$$

where E_{F1DL} is the Fermi energy in the present case as measured from the edge of the conduction band in the vertically upward direction in absence of any quantization.

The one-dimensional density-of-states function ($N_{1D}(E,\lambda)$) is given by

$$N_{1D}(E,\lambda) = \left(\frac{g_v\sqrt{2m_c}}{\pi\hbar}\right)\sum_{n_y=1}^{n_{y\max}}\sum_{n_z=1}^{n_{z\max}}\{\beta_0(E,\lambda)\}' \times[\beta_0(E,\lambda)-\phi(n_y,n_z)]^{-1/2}H(E-E_{n_{l10}}) \tag{6.133}$$

where $\phi(n_y,n_z) = \frac{\hbar^2\pi^2}{2m_c}\left[\left(\frac{n_y}{d_y}\right)^2 + \left(\frac{n_z}{d_z}\right)^2\right]$.

Combining (6.133), with the Fermi-Dirac occupation probability factor, the one-dimensional electron concentration (n_{1D}) can thus be written as

$$n_{1D} = \frac{2g_v}{\pi}\frac{\sqrt{2m_c}}{\hbar}\sum_{n_y=1}^{n_{y\max}}\sum_{n_z=1}^{n_{z\max}}\left[M_{21}(n_y,n_z,E_{F1DL},\lambda) + N_{21}(n_y,n_z,E_{F1DL},\lambda)\right] \tag{6.134}$$

where

$$M_{21}(n_y,n_z,E_{F1DL},\lambda) \equiv \left[\beta_0(E_{F1DL},\lambda) - \phi(n_y,n_z)\right]^{1/2}$$

and

$$N_{21}(n_y,n_z,E_{F1DL},\lambda) \equiv \sum\nolimits_{r=1}^{s} L(r)M_{21}(n_y,n_z,E_{F1DL},\lambda).$$

Using (1.42), the expressions for the 1D dispersion relation, the subband energies, the EEM, the density-of-states function and the electron concentration for nano wires of III–V, ternary, and quaternary materials, whose unperturbed band structure is defined by the two-band model of Kane, can, respectively, be written in the presence of photo-excitation as

$$\frac{\hbar^2 k_x^2}{2m_c} = \{\tau_0(E,\lambda) - \phi(n_y,n_z)\} \tag{6.135}$$

$$\tau_0(E_{n_{l11}},\lambda) = \phi(n_y,n_z) \tag{6.136}$$

$$m^*(E_{F1DL},n_y,n_z,\lambda) = m_c\{\tau_0(E_{F1DL},\lambda)\}' \tag{6.137}$$

$$N_{1D}(E,\lambda) = \left(\frac{g_v\sqrt{2m_c}}{\pi\hbar}\right)\sum_{n_y=1}^{n_{y\max}}\sum_{n_z=1}^{n_{z\max}}\{\tau_0(E,\lambda)\}' \times\left[\tau_0(E,\lambda) - \phi(n_y,n_z)\right]^{-1/2} H(E - E_{n_{l11}}) \tag{6.138}$$

$$n_{1D} = \frac{2g_v}{\pi}\frac{\sqrt{2m_c}}{\hbar}\sum_{n_y=1}^{n_{y\max}}\sum_{n_z=1}^{n_{z\max}}\left[M_{22}(n_y,n_z,E_{F1DL},\lambda) + N_{22}(n_y,n_z,E_{F1DL},\lambda)\right] \tag{6.139}$$

where

$$M_{22}(n_y,n_z,E_{F1DL},\lambda) \equiv \left[\tau_0(E_{F1DL},\lambda) - \phi(n_y,n_z)\right]^{1/2}$$

and

$$N_{22}(n_y,n_z,E_{F1DL},\lambda) \equiv \sum\nolimits_{r=1}^{s} L(r) M_{22}(n_y,n_z,E_{F1DL},\lambda).$$

Using (1.43), the expressions for the 1D dispersion relation, the subband energies, the EEM, the density of states function and the electron concentration for nano wires of III–V, ternary and quaternary materials, whose unperturbed band structure is defined by the parabolic energy bands, can, respectively, be written in the presence of photo-excitation as

$$\frac{\hbar^2 k_x^2}{2m_c} = \left[\rho_0(E,\lambda) - \phi(n_y,n_z)\right] \tag{6.140}$$

$$\rho_0(E_{n_{l11}},\lambda) = \phi(n_y,n_z) \tag{6.141}$$

$$m^*(E_{F1DL},n_y,n_z,\lambda) = m_c\{\rho_0(E_{F1DL},\lambda)\}' \tag{6.142}$$

$$N_{1D}(E,\lambda) = \left(\frac{g_v\sqrt{2m_c}}{\pi\hbar}\right)\sum_{n_y=1}^{n_{y\max}}\sum_{n_z=1}^{n_{z\max}}\{\rho_0(E,\lambda)\}'$$
$$\times\left[\rho_0(E,\lambda)-\phi(n_y,n_z)\right]^{-1/2}H(E-E_{n_{l11}}) \quad (6.143)$$

$$n_{1D} = \frac{2g_v\sqrt{2m_c}}{\pi\hbar}\sum_{n_y=1}^{n_{y\max}}\sum_{n_z=1}^{n_{z\max}}\left[M_{23}(n_y,n_z,E_{F1DL},\lambda)+N_{23}(n_y,n_z,E_{F1DL},\lambda)\right] \quad (6.144)$$

where

$$M_{23}(n_y,n_z,E_{F1DL},\lambda) \equiv \left[\rho_0(E_{F1DL},\lambda)-\phi(n_y,n_z)\right]^{1/2},$$

$$N_{23}(n_y,n_z,E_{F1DL},\lambda) \equiv \sum_{r=1}^{s}L(r)M_{23}(n_y,n_z,E_{F1DL},\lambda)$$

6.13 Results and Discussion

Using the values of the energy band constants from Table 1.1, we have plotted the EEM as functions of thickness and electron concentration per unit length at $T = 4.2\,\mathrm{K}$ by taking nano wires of ternary materials which are used for the purpose of numerical computations in accordance with the perturbed three- [using (6.132) and (6.134)] and two- [using (6.137) and (6.139)] band models of Kane, respectively. In Figs. 6.28 and 6.29, we have plotted the EEM in nano wires of III–V materials as function of lateral dimension and electron concentration respectively. The effect of external photo-excitation increases the EEM significantly.

The influence of quantum confinement is immediately apparent from all the curves of Fig. 6.28 since, the 1D EEM depends strongly on the nano-thickness, which is in direct contrast with the corresponding bulk specimens. It appears from the said figures that the 1D EEM decreases with the increasing film thickness in a step-like manner as considered here although the numerical values vary widely and determined by the constants of the energy spectra. The oscillatory dependence is due to the crossing over of the Fermi level by the quantized levels. For each coincidence of a quantized level with the Fermi level, there would be a discontinuity in the density-of-states function resulting in a peak of oscillations. With large values of film thickness, the height of the steps decreases and the EEM decreases with increasing film thickness in non-oscillatory manner and exhibit monotonic decreasing dependence. The height of step size and the rate of decrement are totally dependent on the band structure. In Fig. 6.29, we have exhibited the effect of the lowest subband on the EEM when the electron concentration is varied. A direct assessment of the effect of light intensity and electron wavelength can be procured form the said figures.

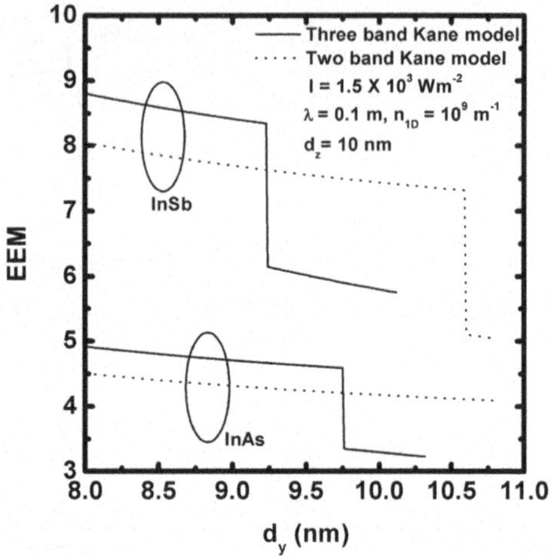

Fig. 6.28 Plot of the normalized EEM as a function of lateral film thickness field for nano wires of n-InSb and n-InAs in the presence of light waves in accordance with the three and the two band models of Kane in the presence of external photo-excitation

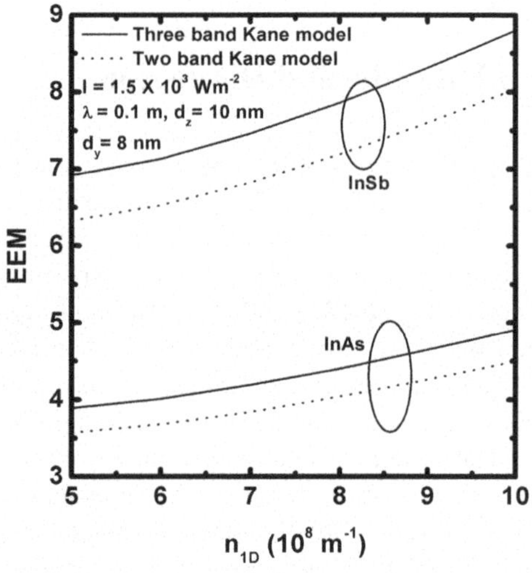

Fig. 6.29 Plot of the normalized EEM as a function of linear electron concentration for nano wires of n-InSb and n-InAs in the presence of light waves in accordance with the three- and the two-band models of Kane in the presence of external photo-excitation

6.14 The EEM in Effective Mass Superlattices of Optoelectronic Semiconductors Under Magnetic Quantization Under External Photo-Excitation

6.14.1 Introduction

In recent years, modern fabrication techniques have generated altogether a new dimension in the arena of quantum effect devices through the experimental realization of an important artificial structure known as semiconductor superlattice (SL)

by growing two similar but different semiconducting materials in alternate layers with finite thicknesses. The materials forming the alternate layers have the same kind of band structure but different energy gaps. The concept of SL was developed for the first time by Keldysh [24] and was successfully fabricated by Esaki and Tsu [25–28]. The SLs are being extensively used in thermal sensors [29, 30], quantum cascade lasers [31–33], photo-detectors [34, 35], light emitting diodes [36–39], multiplication [40], frequency multiplication [41], photo-cathodes [42, 43], thin film transistor [44], solar cells [45, 46], infrared imaging [47], thermal imaging [48, 49], infrared sensing [50], and also in other microelectronic devices.

The most extensively studied III–V SL is the one consisting of alternate layers of GaAs and $Ga_{1-x}Al_xAs$ owing to the relative easiness of fabrication. The GaAs and $Ga_{1-x}Al_xAs$ layers form the quantum wells and the potential barriers, respectively. The III–V SL's are attractive for the realization of high speed electronic and optoelectronic devices [51]. In addition to SLs with usual structure, other types of SLs such as II–VI [52], IV–VI [53] and HgTe/CdTe [54] SL's have also been investigated in the literature. The IV–VI SLs exhibit quite different properties as compared to the III–V SL due to the specific band structure of the constituent materials [55]. The epitaxial growth of II–VI SL is a relatively recent development and the primary motivation for studying the mentioned SLs made of materials with the large bandgap is in their potential for optoelectronic operation in the blue [56]. HgTe/CdTe SL's have raised a great deal of attention since 1979, when as a promising new materials for long wavelength infrared detectors and other electro-optical applications [57]. Interest in Hg-based SL's has been further increased as new properties with potential device applications were revealed. These features arise from the unique zero bandgap material HgTe and the direct bandgap semiconductor CdTe which can be described by the three band mode of Kane. The combination of the aforementioned materials with specified dispersion relation makes HgTe/CdTe SL very attractive, especially because of the tailoring of the material properties for various applications by varying the energy band constants of the SLs. In addition to it, for effective mass SLs, the electronic subbands appear continually in real space [58, 59].

In Sect. 6.14 of theoretical background, we shall study the EEM in effective mass superlattices of optoelectronic semiconductors under magnetic quantization in the presence of light waves. The Sect. 6.15 explores result and discussion for this section.

6.14.2 Theoretical Background

The dispersion relation of the conduction electrons in effective mass superlattices of III–V semiconductors (whose constituent materials obey the three band models of Kane) in the presence of external light waves can be expressed following Sasaki [58, 59] as

$$a_1 \cdot cos[c_{15}(E,\lambda,E_{g1},\Delta_1)a_0 + c_{25}(E,\lambda,E_{g2},\Delta_2)b_0] \\ - a_2 \cdot cos[c_{15}(E,\lambda,E_{g1},\Delta_1)a_0 \\ - c_{25}(E,\lambda,E_{g2},\Delta_2)b_0] = \cos(L_0 k) \quad (6.145)$$

where $L_0(\equiv a_0 + b_0)$ is the period length, a_0 and b_0 are the widths of the barrier and the well respectively,

$$a_1 = \left[\left[1+\sqrt{\frac{m_{c2}}{m_{c1}}}\right]^2 \cdot \left[4\left(\sqrt{\frac{m_{c2}}{m_{c1}}}\right)^{1/2}\right]^{-1}\right],$$

$$a_2 = \left[\left[-1+\sqrt{\frac{m_{c2}}{m_{c1}}}\right]^2 \cdot \left[4\left(\sqrt{\frac{m_{c2}}{m_{c1}}}\right)^{1/2}\right]^{-1}\right],$$

$$c_{i5}^2(E,\lambda,E_{goi},\Delta_i) = \frac{2m_{ci}}{\hbar^2}[\beta_{i50}(E,\lambda,E_{goi},\Delta_i) - k_\perp^2], \quad i = 1,2, k_\perp^2 = k_y^2 + k_z^2,$$

$$\beta_{i50}(E,\lambda,E_{goi},\Delta_i) = [I_{i50}(E,E_{goi},\Delta_i) - \theta_{i50}(E,\lambda,E_{goi},\Delta_i)]$$

$$I_{i50}(E,E_{goi},\Delta_i) = \frac{E(E+E_{goi})(E+E_{goi}+\Delta_i)(E_{goi}+\frac{2}{3}\Delta_i)}{E_{goi}(E_{goi}+\Delta_i)(E+E_{goi}+\frac{2}{3}\Delta_i)},$$

$$\theta_{i50}(E,\lambda,E_{goi},\Delta_i) = \frac{C_{i50}(\lambda,E_{goi},\Delta_i)}{\phi_{i50}(E)}\psi_{i50}^2(E),$$

$$C_{i50}(\lambda,E_{goi},\Delta_i) = \frac{e^2}{96m_{ri}\pi c^3}\frac{I_0\lambda^2}{\sqrt{\varepsilon_{sci}\varepsilon_0}}\frac{E_{g0i}(E_{g0_i}+\Delta_i)}{(E_{g0i}+\frac{2}{3}\Delta_i)}\frac{\overline{\beta}_{i50}^2}{4}\left(t_{i50}+\frac{\overline{\rho}_{i50}}{\sqrt{2}}\right)^2,$$

m_{ri} is the reduced mass and is given by $m_{ri}^{-1} = (m_{ci})^{-1} + m_{vi}^{-1}$, m_{vi} is the effective mass of the heavy hole at the top of the valance band in the absence of any field,

$$\overline{\beta}_{i50} \equiv \left[(6(E_{g0_i} + 2\Delta_i/3)(E_{g0_i} + \Delta_i))/\chi_{i50}\right]^{1/2},$$

$$\chi_{i50} = (6E_{g0_i}^2 + 9E_{g0_i}\Delta_i + 4\Delta_i^2), \ t_{i50} = \left[6(E_{g0_i} + 2\Delta_i/3)^2/\chi_{i50}\right]^{1/2},$$

$$\overline{\rho}_{i50} = (4\Delta_i^2/3\chi_{i50})^{1/2},$$

$$\phi_{i50}(E) = E_{g0_i}\left(1 + 2\left(1+\frac{m_{ci}}{m_{vi}}\right)\frac{I_{i50}(E)}{E_{g0_i}}\right)^{1/2},$$

$$\psi_{i50}(E) = \left[\left(1 + \frac{E_{goi} - \delta'_{i50}}{\phi_{i50}(E) + \delta'_{i50}}\right) + (E_{goi} - \delta'_{i50}) \times \left[\frac{1}{\phi_{i50}(E) + \delta'_{i50}} - \frac{1}{E_{goi} + \delta'_{i50}}\right]^{1/2} \times \left[\frac{1}{\phi_{i50}(E) + \delta'_{i50}} - \frac{E_{goi} + \delta'_{i50}}{(E_{goi} - \delta'_{i50})^2}\right]^{1/2}\right]$$

$$\delta'_{i50} = (E^2_{goi}\Delta_i)(\chi_{i50})^{-1}.$$

In the presence of a quantizing magnetic field B, along x-direction the magneto-energy spectrum assumes the form

$$k_x^2 = \overline{\omega}_{15}(E, \lambda, n) \tag{6.146}$$

where

$$\overline{\omega}_{15}(E, \lambda, n) = \frac{1}{L_o^2}\left[[\cos^{-1}\{\overline{f}_{15}(E, \lambda, n)\}]^2 - \frac{2eB}{\hbar}\left(n + \frac{1}{2}\right)L_o^2\right],$$

$$\overline{f}_{15}(E, \lambda, n) = [a_1 \cos[\overline{c}_{15}(E, \lambda, E_{go1}, \Delta_1, n)a_o + b_o\overline{c}_{25}(E, \lambda, E_{go2}, \Delta_2, n)] - a_2 \cos[\overline{c}_{15}(E, \lambda, E_{go1}, \Delta_1, n)a_o - b_o\overline{c}_{25}(E, \lambda, E_{go2}, \Delta_2, n)]],$$

$$\overline{c}_{15}(E, \lambda, E_{go1}, \Delta_1, n) = \left[\left(\frac{2m_{c1}}{\hbar^2}\right)[\beta_{150}(E, \lambda, E_{go1}, \Delta_1)] - \frac{2eB}{\hbar}\left(n + \frac{1}{2}\right)\right]^{1/2},$$

$$\overline{c}_{25}(E, \lambda, E_{go2}, \Delta_2, n) = \left[\left(\frac{2m_{c2}}{\hbar^2}\right)[\beta_{250}(E, \lambda, E_{go2}, \Delta_2)] - \frac{2eB}{\hbar}\left(n + \frac{1}{2}\right)\right]^{1/2}.$$

The EEM along x-direction in this case can be expressed as

$$m^*(V_0, \lambda, n) = (\hbar^2/2)\,[\overline{\omega}_{15}(V_0, \lambda, n)]' \tag{6.147}$$

where V_0 is the Fermi energy in this case and

$$\{\overline{\omega}_{15}(V_0, \lambda, n)\}' = 2\{\overline{f}_{15}(V_0, \lambda, n)\}' L_0^{-2} \cdot [\cos^{-1}[\overline{f}_{15}(V_0, \lambda, n)] \times [1 - \overline{f}_{15}(V_0, \lambda, n)]^{-1/2}],$$

$$\overline{f}_{15}(V_0, \lambda, n) = [-a_1 \sin[\overline{c}_{15}(V_0, \lambda, E_{go1}, \Delta_1, n)a_o + b_o\overline{c}_{25}(V_0, \lambda, E_{go2}, \Delta_2, n)] \times [[\overline{c}_{15}(V_0, \lambda, E_{go1}, \Delta_1, n)]'a_o + b_o[\overline{c}_{25}(V_0, \lambda, E_{go2}, \Delta_2, n)]'] + a_2 \sin[\overline{c}_{15}(V_0, \lambda, E_{go1}, \Delta_1, n)a_o + b_o\overline{c}_{25}(V_0, \lambda, E_{go2}, \Delta_2, n)] \times [[\overline{c}_{15}(V_0, \lambda, E_{go1}, \Delta_1, n)]'a_o - b_o[\overline{c}_{25}(V_0, \lambda, E_{go2}, \Delta_2, n)]']],$$

$$\{\bar{c}_{15}(V_0, \lambda, E_{go1}, \Delta_1, n)\}' = \frac{m_{c1}}{\hbar^2}[\beta'_{150}(V_0, \lambda, E_{go1}, \Delta_1)/\{\bar{c}_{15}(V_0, \lambda, E_{go1}, \Delta_1, n)\}],$$

$$\{\bar{c}_{25}(V_0, \lambda, n)\}' = \frac{m_{c2}}{\hbar^2}[\beta'_{250}(V_0, \lambda, E_{go2}, \Delta_2)/\{\bar{c}_{25}(V_0, \lambda, E_{go2}, \Delta_2, n)\}],$$

$$\beta'_{i50}(V_0, \lambda, E_{goi}, \Delta_i) = \left[I'_{i50}(V_0, E_{goi}, \Delta_i) - \theta'_{i50}(V_0, \lambda, E_{goi}, \Delta_i)\right],$$

$$I'_{i50}(V_0, E_{goi}, \Delta_i) = I_{i50}(V_0, E_{goi}, \Delta_i) \times \left[\frac{1}{V_0} + \frac{1}{V_0 + E_{goi}} + \frac{1}{V_0 + E_{goi} + \Delta_i} - \frac{1}{V_0 + E_{goi} + \frac{2}{3}\Delta_i}\right], \quad i = 1, 2,$$

$$\theta'_{i50}(V_0, \lambda, E_{goi}, \Delta_i) = \theta_{i50}(V_0, \lambda, E_{goi}, \Delta_i)\left[\frac{-\theta'_{i50}(V_0, \lambda, E_{goi}, \Delta_i)}{\theta_{i50}(V_0, \lambda, E_{goi}, \Delta_i)} + \frac{2\psi'_{i50}(V_0)}{\psi_{i50}(V_0)}\right],$$

$$\phi'_{i50}(V_0) = E_{g0i}\left(\left(1 + \frac{m_{ci}}{m_{vi}}\right)\frac{I'_{i50}(V_0, E_{goi}, \Delta_i)}{\phi_{i50}(V_0)}\right)^{1/2},$$

$$\psi'_{i50}(V_0) = \frac{-(E_{goi} - \delta'_{i50})\phi'_{i50}(V_0)}{(\phi_{i50}(V_0) + \delta'_{i50})^2}\left[1 + \frac{1}{2}\left[\frac{1}{\phi_{i50}(V_0) + \delta'_{i50}} - \frac{1}{E_{goi} + \delta'_{i50}}\right]^{-\frac{1}{2}} \times \left[\frac{1}{\phi_{i50}(V_0) + \delta'_{i50}} - \frac{(E_{goi} + \delta'_{i50})}{(E_{goi} - \delta'_{i50})^2}\right]^{1/2} + \frac{1}{2}\left[\frac{1}{\phi_{i50}(V_0) + \delta'_{i50}} - \frac{E_{goi} + \delta'_{i50}}{(E_{goi} - \delta'_{i50})^2}\right]^{-\frac{1}{2}} \cdot \left[\frac{1}{\phi_{i50}(V_0) + \delta'_{i50}} - \frac{1}{E_{goi} + \delta'_{i50}}\right]^{1/2}\right]$$

The electron concentration in this case assumes the form

$$n_o = \frac{g_v eB}{\pi^2 \hbar L_o}\sum_{n=0}^{n_{\max}}[S_{54}(V_0, \lambda, n) + T_{54}(V_0, \lambda, n)] \tag{6.148}$$

where

$$S_{54}(V_0, \lambda, n) = \left[\left[\cos^{-1}\overline{f}_{15}(V_0, \lambda, n)\right]^2 - \frac{2eB}{\hbar}\left(n + \frac{1}{2}\right)L_0^2\right]^{1/2}$$

$$T_{54}(V_0, \lambda, n) = \sum_{r=1}^{s_0} L(r)\,[S_{54}(V_0, \lambda, n)].$$

The electron concentration in this particular case when the dispersion relations of the constituent materials are defined by the perturbed two-band model of Kane can be expressed as

$$n_o = \frac{g_v eB}{\pi^2 \hbar L_o} \sum_{n=0}^{n_{\max}} [S_{55}(V_0, \lambda, n) + T_{55}(V_0, \lambda, n)] \tag{6.149}$$

where

$$S_{55}(V_0, \lambda, n) = \left[\left[\cos^{-1} \overline{f}_{151}(V_0, \lambda, n)\right]^2 - \frac{2eB}{\hbar}\left(n + \frac{1}{2}\right) L_0^2\right]^{1/2},$$

$$\overline{f}_{151}(E, \lambda, n) = [a_1 \cos[\overline{c}_{151}(E, \lambda, E_{go1}, n)a_o + b_o \overline{c}_{251}(E, \lambda, E_{go2}, n)] - a_2 \cos[\overline{c}_{151}(E, \lambda, E_{go1}, n)a_o - b_o \overline{c}_{251}(E, \lambda, E_{go2}, n)]],$$

$$\overline{c}_{151}(E, \lambda, n) = \left[\left(\frac{2m_{c1}}{\hbar^2}\right)[\omega_{150}(E, \lambda, E_{go1})] - \frac{2eB}{\hbar}\left(n + \frac{1}{2}\right)\right]^{1/2},$$

$$\overline{c}_{251}(E, \lambda, n) = \left[\left(\frac{2m_{c2}}{\hbar^2}\right)[\omega_{250}(E, \lambda, E_{go2})] - \frac{2eB}{\hbar}\left(n + \frac{1}{2}\right)\right]^{1/2}$$

$$\omega_{i50}(E, \lambda, E_{goi}) \equiv E(1 + \alpha_i E) - B_{i50}(E, \lambda), \; B_{i50}(E, \lambda) = \frac{C_{i51}(\lambda, E_{goi})\psi_{i51}^2(E)}{\phi_{i51}(E)},$$

$$C_{i51}(\lambda, E_{goi}) \equiv \frac{e^2 I_0 \lambda^2 E_{g0_i}}{384\pi c^3 m_{ri}\sqrt{\varepsilon_{sci}\varepsilon_0}}, \; \phi_{i51}(E) \equiv E_{g0i}\left\{1 + \frac{2m_{ci}}{m_{ri}}\frac{E(1 + \alpha_i E)}{E_{g0i}}\right\}^{1/2},$$

$$\alpha_i = 1/E_{goi}, \; \psi_{i51}(E) = \frac{2E_{goi}}{\phi_{i51}(E)}, \; T_{55}(E_{FB}, \lambda, n) = \sum_{r=1}^{s_0} Z_B(r)\,[S_{55}(E_{FB}, \lambda, n)].$$

The EEM in this case can be expressed as

$$m^*(V_0, \lambda, n) = (\hbar^2/2)\,[\overline{\omega}_{16}(V_0, \lambda, n)]' \tag{6.150}$$

where

$$\overline{\omega}_{16}(V_0, \lambda, n) = \frac{1}{L_o^2}\left[[\cos^{-1}\{\overline{f}_{151}(V_0, \lambda, n)\}]^2 - \frac{2eB}{\hbar}\left(n + \frac{1}{2}\right) L_o^2\right],$$

$$\{\overline{\omega}_{16}(V_0, \lambda, n)\}' = 2\{\overline{f}_{151}(V_0, \lambda, n)\}' L_0^{-2} \cdot [\cos^{-1}[[\overline{f}_{151}(V_0, \lambda, n)] [1 - \overline{f}_{151}(V_0, \lambda, n)]^{-1/2}]],$$

$$\begin{aligned}\overline{f}_{151}(V_0,\lambda,n) &= [-a_1\sin[\overline{c}_{151}(V_0,\lambda,E_{go1},n)a_o+b_o\overline{c}_{251}(V_0,\lambda,E_{go2},n)]\\ &\quad\times[[\overline{c}_{151}(V_0,\lambda,E_{go1},n)]'a_o+b_o[\overline{c}_{251}(V_0,\lambda,E_{go2},n)]']\\ &\quad+a_2\sin[\overline{c}_{151}(V_0,\lambda,E_{go1},n)a_o+b_o\overline{c}_{251}(V_0,\lambda,E_{go2},n)]\\ &\quad\times[[\overline{c}_{151}(V_0,\lambda,E_{go1},n)]'a_o-b_o[\overline{c}_{251}(V_0,\lambda,E_{go2},n)]']],\end{aligned}$$

$$\begin{aligned}\{\overline{c}_{151}(V_0,\lambda,E_{go1},n)\}' &= \tfrac{m_{c1}}{\hbar^2}[\omega_{150}(V_0,\lambda,E_{go1})]'/\{\overline{c}_{151}(V_0,\lambda,E_{go1},n)\},\\ \{\overline{c}_{251}(V_0,\lambda,E_{go2},n)\}' &= \tfrac{m_{c2}}{\hbar^2}[\omega_{250}(V_0,\lambda,E_{go1})]'/\{\overline{c}_{251}(V_0,\lambda,E_{go2},n)\},\\ \omega'_{i50}(V_0,\lambda,E_{goi}) &= \left[1+2\alpha_i V_0-B'_{i50}(V_0,\lambda)\right],\\ B'_{i50}(V_0,\lambda) &= \left[\tfrac{-B_{i50}(V_0,\lambda)\phi'_{i51}(V_0)}{\phi_{i51}(V_0)}+\tfrac{2B_{i50}(V_0,\lambda)\psi'_{i51}(V_0)}{\psi_{i51}(V_0)}\right],\\ \psi'_{i51}(V_0) &= \left[\tfrac{-\psi_{i51}(V_0)}{\phi_{i51}(V_0)}\phi'_{i51}(V_0)\right],\ \phi'_{i51}(V_0)=\tfrac{m_{ci}}{m_{ri}}\cdot\tfrac{E_{goi}(1+2\alpha_i V_0)}{\phi_{i51}(V_0)}.\end{aligned}$$

The electron concentration when the dispersion relations of the constituent materials are defined by the perturbed parabolic energy bands can be expressed as

$$n_o=\frac{g_v eB}{\pi^2\hbar L_o}\sum_{n=0}^{n_{\max}}[S_{56}(V_0,\lambda,n)+T_{56}(V_0,\lambda,n)] \tag{6.151}$$

where

$$\begin{aligned}S_{56}(V_0,\lambda,n) &= \left[\left[\cos^{-1}\overline{f}_{152}(V_0,\lambda,n)\right]^2-\frac{2eB}{\hbar}\left(n+\frac{1}{2}\right)L_0^2\right]^{1/2}\\ \overline{f}_{152}(E,\lambda,n) &= [a_1\cos[\overline{c}_{152}(E,\lambda,E_{g1},n)a_o+b_o\overline{c}_{252}(E,\lambda,E_{g2},n)]\\ &\quad-a_2\cos[\overline{c}_{152}(E,\lambda,E_{g1},n)a_o-b_o\overline{c}_{252}(E,\lambda,E_{g2},n)]]\\ \overline{c}_{152}(E,\lambda,E_{g1},n) &= \left[\left(\frac{2m_{c1}}{\hbar^2}\right)[\rho_{150}(E,\lambda,E_{go1})]-\frac{2eB}{\hbar}\left(n+\frac{1}{2}\right)\right]^{1/2},\\ \overline{c}_{252}(E,\lambda,E_{g2},n) &= \left[\left(\frac{2m_{c2}}{\hbar^2}\right)[\rho_{250}(E,\lambda,E_{go2})]-\frac{2eB}{\hbar}\left(n+\frac{1}{2}\right)\right]^{1/2},\\ \rho_{i50}(E,E_{goi},\lambda) &= E-C_{i52}(\lambda)\left[1+\left(\frac{2m_{ci}}{m_{ri}}\right)\left(\frac{E}{E_{goi}}\right)\right]^{-3/2},\\ C_{i52}(\lambda) &\equiv \frac{e^2I_0\lambda^2}{96\pi c^3 m_{ri}\sqrt{\varepsilon_{sci}\varepsilon_0}},\ \alpha_i=\frac{1}{E_{goi}},\\ T_{56}(E_{FB},\lambda,n) &= \sum_{r=1}^{S_0} Z_B(r)\,[S_{56}(E_{FB},\lambda,n)]\,.\end{aligned}$$

The EEM in this case can be written as

$$m^*(V_0, \lambda, n) = (\hbar^2/2)\,[\overline{\omega}_{17}(V_0, \lambda, n)]' \tag{6.152}$$

where

$$\begin{aligned}
\overline{\omega}_{17}(V_0, \lambda, n)' &= 2\{\overline{f}_{152}(V_0, \lambda, n)\}' L_0^{-2} \cdot [\cos^{-1}[\overline{f}_{152}(V_0, \lambda, n)] \\
&\quad \times [1 - [\overline{f}_{152}(V_0, \lambda, n)]^{-1/2}]], \\
\overline{f}_{152}(V_0, \lambda, n) &= [-a_1 \sin[\overline{c}_{152}(V_0, \lambda, E_{g1}, n)a_o + b_o\overline{c}_{252}(V_0, \lambda, E_{g2}, n)] \\
&\quad \times [[\overline{c}_{152}(V_0, \lambda, E_{g1}, n)]' a_o + b_o[\overline{c}_{252}(V_0, \lambda, E_{g2}, n)]'] \\
&\quad + a_2 \sin[\overline{c}_{152}(V_0, \lambda, E_{g1}, n)a_o + b_o\overline{c}_{252}(V_0, \lambda, E_{g2}, n)] \\
&\quad \times [[\overline{c}_{152}(V_0, \lambda, E_{g1}, n)]' a_o - b_o[\overline{c}_{252}(V_0, \lambda, E_{g2}, n)]']],
\end{aligned}$$

$$\{\overline{c}_{152}(V_0, \lambda, E_{g1}, n)\}' = \tfrac{m_{c1}}{\hbar^2}[\rho_{150}(V_0, \lambda)]'/\{\overline{c}_{152}(V_0, \lambda, E_{g1}, n)\}$$

$$\{\overline{c}_{252}(V_0, \lambda, E_{g2}, n)\}' = \tfrac{m_{c2}}{\hbar^2}[\rho_{250}(V_0, \lambda)]'/\{\overline{c}_{252}(V_0, \lambda, E_{g2}, n)\},$$

$$\rho'_{i50}(E, E_{goi}, \lambda) = 1 + C_{i52}(\lambda) \cdot \left(\tfrac{3m_{ci}}{m_{ri}}\right)\left[1 + \left(\tfrac{2m_{ci}}{m_{ri}}\right)\left(\tfrac{E}{E_{goi}}\right)\right]^{\frac{-5}{2}}$$

6.15 Results and Discussion

Using the values of the energy band constants from Table 1.1, we have plotted the EEM as functions of 1/B and electron concentration at $T = 4.2\,\mathrm{K}$ by taking effective mass super lattices of optoelectronic materials under magnetic quantization in accordance with the perturbed three [using (6.147) and (6.148)], two [using (6.149) and (6.150)] band models of Kane and that of perturbed parabolic energy bands [using (6.151) and (6.152)], respectively. In Figs. 6.30 and 6.31, we have plotted the effect of magnetic field and carrier concentration on the EEM of effective mass superlattices in GaAs/AlGaAs structures. The effect of SdH oscillations has been exhibited in this case for multi sub-band generation. In this case, the EEM is a subband index dependent and we have plotted the EEM by considering the lowest subband index. In both the figures we see that the effect of the external photo-excitation on the EEM dominated by the parabolic energy law does not tend to modulate with either of the variable compared to the bulk value of the EEM for InAs and InSb, we find that the EEM in GaAs/AlGaAs structures are extremely low and therefore the mobility in superlattices are very large as compared with the value of the mobility of the constituent materials which is very important from the application point of view for modern devices made of superlattices.

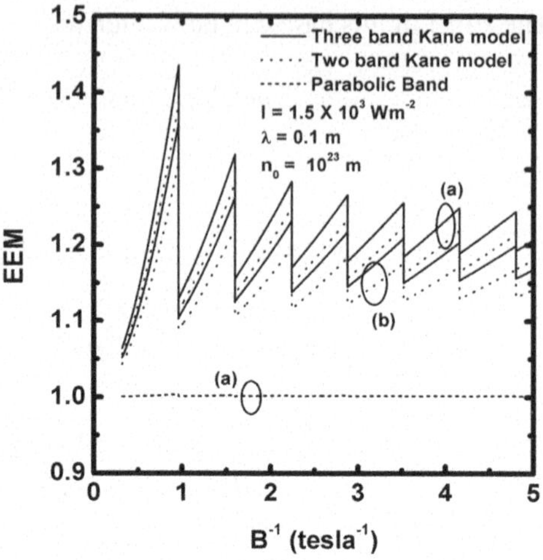

Fig. 6.30 Plot of the normalized EEM for the lowest subband index as a function of inverse magnetic field of effective mass superlattices of GaAs/AlGaAs in the presence of light waves in accordance with the three, the two-band models of Kane and parabolic energy band model in the presence of external photo-excitation

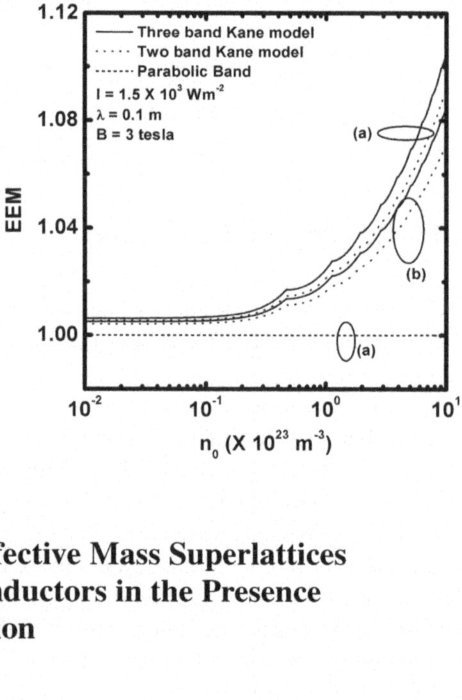

Fig. 6.31 Plot of the normalized EEM for the lowest subband index as a function of carrier concentration of effective mass superlattices of GaAs/AlGaAs in the presence of light waves in accordance with the three, the two-band models of Kane and parabolic energy band model in the presence of external photo-excitation

6.16 The EEM in Nanowire Effective Mass Superlattices of Optoelectronic Semiconductors in the Presence of External Photo-Excitation

6.16.1 Introduction

We shall study the EEM in nano wire effective mass superlattices of optoelectronic semiconductors in the presence of photo excitation in Sect. 6.16.2 of theoretical background and the Sect. 6.17 explores the result and discussions pertaining to Sect. 6.16.2.

6.16.2 Theoretical Background

The dispersion relation of the conduction electrons for nano wire effective mass superlattices in accordance with the perturbed three-band model of Kane in the presence of light waves is given by

$$k_x^2 = \overline{\omega}_{19}(E, \lambda, n_y, n_z) \tag{6.153}$$

where

$$\overline{\omega}_{19}(E, \lambda, n_y, n_z) = \left[\frac{1}{L_0^2}\left[\cos^{-1}\{\overline{f}_3(E, \lambda, n_y, n_z)\}\right]^2 - H(n_y, n_z)\right]$$

$$\begin{aligned}\overline{f}_3(E, \lambda, n_y, n_z) = [&a_1 \cos[\overline{e}_1(E, \lambda, E_{g_1}, \Delta_1, n_y, n_z)a_o \\ &+ b_o\overline{e}_2(E, \lambda, E_{g_2}, \Delta_2, n_y, n_z)] \\ &- a_2 \cos[\overline{e}_1(E, \lambda, E_{g_1}, \Delta_1, n_y, n_z)a_o \\ &- b_o\overline{e}_2(E, \lambda, E_{g_2}, \Delta_2, n_y, n_z)]],\end{aligned}$$

$$\overline{e}_i^2(E, \lambda, E_{goi}, \Delta_i, n_y, n_z) = \left[\left(\frac{2m_{c_i}}{\hbar^2}\right)[\beta_{i50}(E, \lambda, E_{goi}, \Delta_i)] - H(n_y, n_z)\right]$$

and

$$H(n_y, n_z) = \left[\left(\frac{n_y\pi}{d_y}\right)^2 + \left(\frac{n_z\pi}{d_z}\right)^2\right].$$

The expression of the electron concentration in this case can be written as

$$n_0 = \frac{2g_v}{\pi}\sum_{n_y=1}^{n_{y\max}}\sum_{n_z=1}^{n_{z\max}}[\overline{Q}_{23}(V_1, \lambda, n_y, n_z) + \overline{Q}_{24}(V_1, \lambda, n_y, n_z)] \tag{6.154}$$

where V_1 is the Fermi energy in this case,

$$\overline{Q}_{23}(V_1, \lambda, n_y, n_z) = \sqrt{\overline{\omega}_{19}(V_1, \lambda, n_y, n_z)},$$

$$\overline{Q}_{24}(V_1, \lambda, n_y, n_z) = \sum_{R=1}^{R=R_0} L(R)\overline{Q}_{23}(V_1, \lambda, n_y, n_z) \quad \text{and}$$

$$L(R) = 2(k_BT)^{2R}(1 - 2^{1-2R})\xi(2R)\frac{\partial^{2R}}{\partial V_1^{2R}}.$$

The EEM along the x-direction in this case can be expressed as

$$m^*(V_1, \lambda, n_y, n_z) = (\hbar^2/2)\left[\overline{\omega}_{19}(V_1, \lambda, n_y, n_z)\right]' \tag{6.155}$$

where

$$[\overline{\omega}_{19}(V_1,\lambda,n_y,n_z)]' = \left[1-\overline{f}_3^2(V_1,\lambda,n_y,n_z)\right]^{-1/2}[2[\overline{f}_3(V_1,\lambda,n_y,n_z)]'] \left[\cos^{-1}\{\overline{f}_3(V_1,\lambda,n_y,n_z)\}\right],$$

$$\begin{aligned}[\overline{f}_3(V_1,\lambda,n_y,n_z)]' &= -a_1\sin\left[a_0\overline{e}_1(V_1,\lambda,E_{g_1},\Delta_1,n_y,n_z)\right.\\ &\left.+b_o\overline{e}_2(V_1,\lambda,E_{g_2},\Delta_2,n_y,n_z)\right]\cdot\left[a_0[\overline{e}_1(V_1,\lambda,E_{g_1},\Delta_1,n_y,n_z)]'\right.\\ &\left.+b_o[\overline{e}_2(V_1,\lambda,E_{g_2},\Delta_2,n_y,n_z)]'\right]+a_2\sin\left[a_0\overline{e}_1(V_1,\lambda,E_{g_1},\Delta_1,n_y,n_z)\right.\\ &\left.-b_o\overline{e}_2(V_1,\lambda,E_{g_2},\Delta_2,n_y,n_z)\right]\cdot\left[a_0[\overline{e}_1(V_1,\lambda,E_{g_1},\Delta_1,n_y,n_z)]\right.\\ &\left.-b_o[\overline{e}_2(V_1,\lambda,E_{g_2},\Delta_2,n_y,n_z)]'\right]\end{aligned}$$

and $\left[\overline{e}_i(V_1,\lambda,E_{goi},\Delta_i,n_y,n_z)\right]' = \dfrac{m_{ci}\beta_{i50}(V_1,\lambda,E_{goi},\Delta_i)}{\hbar^2\overline{e}_i(V_1,\lambda,E_{goi},\Delta_i,n_y,n_z)}$.

In accordance with the perturbed two-band model of Kane the electron concentration per unit length is given by,

$$n_0 = \frac{2g_v}{\pi}\sum_{n_y=1}^{n_{y\max}}\sum_{n_z=1}^{n_{z\max}}\left[\overline{Q}_{25}(V_1,\lambda,n_y,n_z)+\overline{Q}_{26}(V_1,\lambda,n_y,n_z)\right] \tag{6.156}$$

where

$$\overline{Q}_{25}(V_1,\lambda,n_y,n_z) = \left[\sqrt{\overline{\omega}_{20}(V_1,\lambda,n_y,n_z)}\right],$$

$$\overline{Q}_{26}(V_1,\lambda,n_y,n_z) = \sum_{R=1}^{R=R_0} L(R)\left[\overline{Q}_{25}(V_1,\lambda,n_y,n_z)\right],$$

$$\overline{\omega}_{20}(V_1,\lambda,n_y,n_z) = \left[\frac{1}{L_0^2}\left[\cos^{-1}\overline{f}_4(V_1,\lambda,n_y,n_z)\right]^2 - H(n_y,n_z)\right],$$

$$\begin{aligned}\overline{f}_4(V_1,\lambda,n_y,n_z) &= \left[a_1\cos\left[a_0\overline{g}_1(V_1,\lambda,E_{g1},n_y,n_z)-b_0\overline{g}_2(V_1,\lambda,E_{g2},n_y,n_z)\right]\right.\\ &\left.-a_2\cos\left[a_0\overline{g}_1(V_1,\lambda,E_{g1},n_y,n_z)-b_0\overline{g}_2(V_1,\lambda,E_{g2},n_y,n_z)\right]\right]\end{aligned}$$

and $\overline{g}_i^2(V_1,\lambda,E_{goi},n_y,n_z) = \left[\frac{2m_{ci}}{\hbar^2}\omega_{i50}(V_1,\lambda,E_{goi}) - H(n_y,n_z)\right]$.

The EEM in this case can be expressed as

$$m^*(V_1,\lambda,n_y,n_z) = (\hbar^2/2)\left[\overline{\omega}_{20}(V_1,\lambda,n_y,n_z)\right]' \tag{6.157}$$

where

$$[\overline{\omega}_{20}(V_1,\lambda,n_y,n_z)]' = \frac{2[\overline{f}_4(V_1,\lambda,n_y,n_z)]'\left[\cos^{-1}\overline{f}_4(V_1,\lambda,n_y,n_z)\right]}{\sqrt{1-\overline{f}_4^2(V_1,\lambda,n_y,n_z)}},$$

$$\begin{aligned}[\overline{f}_4(V_1,\lambda,n_y,n_z)]' &= -a_1\sin[a_0\overline{g}_1(V_1,\lambda,E_{g1},n_y,n_z)+b_0\overline{g}_2(V_1,\lambda,E_{g2},n_y,n_z)]\\ &\times\left[a_0[\overline{g}_1(V_1,\lambda,E_{g1},n_y,n_z)]'+b_0[\overline{g}_2(V_1,\lambda,E_{g2},n_y,n_z)]'\right]\\ &-a_2\sin\left[a_0\overline{g}_1(V_1,\lambda,E_{g1},n_y,n_z)-b_0\overline{g}_2(V_1,\lambda,E_{g2},n_y,n_z)\right]\\ &\times\left[a_0[\overline{g}_1(V_1,\lambda,E_{g1},n_y,n_z)]'-b_0[\overline{g}_2(V_1,\lambda,E_{g2},n_y,n_z)]'\right],\end{aligned}$$

and $[\overline{g}_i(V_1,\lambda,E_{goi},n_y,n_z)]' = \dfrac{m_{ci}[\omega_{i50}(V_1,\lambda,E_{goi})]'}{\hbar^2\overline{g}_i(V_1,\lambda,E_{goi},n_y,n_z)}$.

In accordance with the perturbed parabolic energy bands, the electron concentration per unit length is given by,

$$n_0 = \frac{2g_v}{\pi}\sum_{n_y=1}^{n_{y\max}}\sum_{n_z=1}^{n_{z\max}}\left[\overline{Q}_{251}(V_1,\lambda,n_y,n_z)+\overline{Q}_{261}(V_1,\lambda,n_y,n_z)\right] \tag{6.158}$$

where

$$\overline{Q}_{251}(V_1,\lambda,n_y,n_z) = \left[\sqrt{\overline{\omega}_{201}(V_1,\lambda,n_y,n_z)}\right],$$

$$\overline{Q}_{261}(V_1,\lambda,n_y,n_z) = \sum_{R=1}^{R=R_0}L(R)\left[\overline{Q}_{251}(V_1,\lambda,n_y,n_z)\right],$$

$$\overline{\omega}_{201}(V_1,\lambda,n_y,n_z) = \left[\frac{1}{L_0^2}\left[\cos^{-1}\overline{f}_{41}(V_1,\lambda,n_y,n_z)\right]^2 - H(n_y,n_z)\right],$$

$$\begin{aligned}&\overline{f}_{41}(V_1,\lambda,n_y,n_z)\\ &\quad=[a_1\cos[a_0\overline{g}_{11}(V_1,\lambda,E_{g1},n_y,n_z)-b_0\overline{g}_{21}(V_1,\lambda,E_{g2},n_y,n_z)]\\ &\qquad-a_2\cos[a_0\overline{g}_{11}(V_1,\lambda,E_{g1},n_y,n_z)-b_0\overline{g}_{21}(V_1,\lambda,E_{g2},n_y,n_z)]]\end{aligned}$$

and $\overline{g}_{i1}^2(V_1,\lambda,E_{goi},n_y,n_z) = \left[\frac{2m_{ci}}{\hbar^2}\rho_{i50}(V_1,E_{goi},\lambda) - H(n_y,n_z)\right]$.

The EEM in this case can be expressed as

$$m^*(V_1,\lambda,n_y,n_z) = (\hbar^2/2)\left[\overline{\omega}_{201}(V_1,\lambda,n_y,n_z)\right]' \tag{6.159}$$

where

$$[\overline{\omega}_{201}(V_1,\lambda,n_y,n_z)]' = \frac{2[\overline{f}_{41}(V_1,\lambda,n_y,n_z)]'\left[\cos^{-1}\overline{f}_{41}(V_1,\lambda,n_y,n_z)\right]}{\sqrt{1-\overline{f}_{41}^{2}(V_1,\lambda,n_y,n_z)}},$$

$$\begin{aligned}
&[\overline{f}_{41}(V_1,\lambda,n_y,n_z)]' \\
&\quad = -a_1 \sin\left[a_0\overline{g}_{11}(V_1,\lambda,E_{g1},n_y,n_z) + b_0\overline{g}_{21}(V_1,\lambda,E_{g2},n_y,n_z)\right] \\
&[a_0[\overline{g}_{11}(V_1,\lambda,E_{g1},n_y,n_z)]' + b_0[\overline{g}_{21}(V_1,\lambda,E_{g2},n_y,n_z)]'] \\
&- a_2 \sin\left[a_0\overline{g}_{11}(V_1,\lambda,E_{g1},n_y,n_z) - b_0\overline{g}_{21}(V_1,\lambda,E_{g2},n_y,n_z)\right] \\
&\left[a_0[\overline{g}_{11}(V_1,\lambda,E_{g1},n_y,n_z)]' - b_0[\overline{g}_{21}(V_1,\lambda,E_{g2},n_y,n_z)]'\right],
\end{aligned}$$

and $[g_{i1}(V_1,\lambda,E_{goi},n_y,n_z)]' = \dfrac{m_{ci}[\rho_{i50}(V_1,E_{goi},\lambda)]'}{\hbar^2\overline{g}_{i1}(V_1,\lambda,E_{goi},n_y,n_z)}$.

6.17 Results and Discussion

Using the values of the energy band constants from Table 1.1, we have plotted the EEM for the $n_y = 1$ and $n_z = 1$ as a function of the film thickness at $T = 4.2\,\text{K}$ by taking nano wire effective mass super lattices of optoelectronic materials in accordance with the perturbed three [using (6.154) and (6.155)], two [using (6.156) and (6.157)] band models of Kane and that of perturbed parabolic energy bands [using (6.158) and (6.159)] respectively in Fig. 6.32. Figure 6.32 exhibits the variation of EEM in the nano wire effective mass superlattices of GaAs/AlGaAs by considering the quantum limit approximation. The EEM is greatest for the lowest sub-bands and for higher sub-bands the numerical values of the EEMs will be less. It appears that the EEM in such structure decreases with the increase in the film thickness in a non-linear way for the three- and the two-band energy models. Although, it appears that the EEM in case of parabolic energy band is linear; however, it depends on the photo-excitation factor, which makes the slow variation of EEM with both intensity and wavelength.

6.18 The EEM in Superlattices of Optoelectronic Semiconductors with Graded Interfaces Under Magnetic Quantization in the Presence of External Photo-Excitation

6.18.1 Introduction

We note that all the aforementioned SLs have been proposed with the assumption that the interfaces between the layers are sharply defined, of zero thickness, i.e., devoid

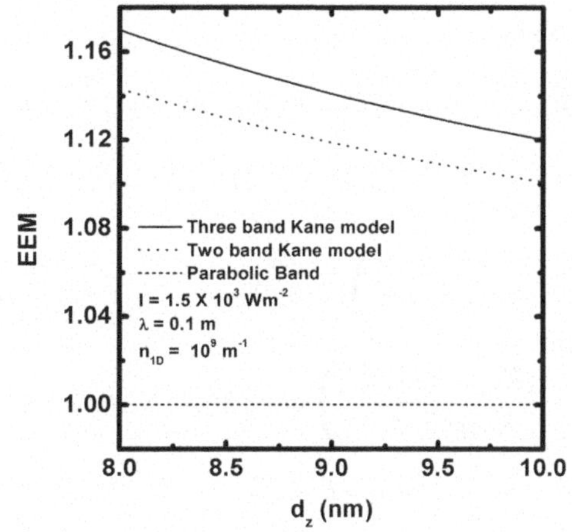

Fig. 6.32 Plot of the normalized EEM ($n_y = 1$ and $n_z = 1$) as a function of lateral film thickness of nano wire effective mass superlattices of GaAs/AlGaAs in the presence of light waves in accordance with the three, the two-band models of Kane and parabolic energy band model in the presence of external photo-excitation

of any interface effects. The SL potential distribution may be then considered as a one dimensional array of rectangular potential wells. The aforementioned advanced experimental techniques may produce SLs with physical interfaces between the two materials crystallo-graphically abrupt; adjoining their interface will change at least on an atomic scale. As the potential form changes from a well (barrier) to a barrier (well), an intermediate potential region exists for the electrons. The influence of finite thickness of the interfaces on the electron dispersion law is very important, since; the electron energy spectrum governs the electron transport in SLs [60, 61].

We shall study the EEM in superlattices of optoelectronic semiconductors with graded interfaces under magnetic quantization in the presence of photo excitation in Sect. 6.18.2 of theoretical background and the Sect. 6.19 explores the result and discussions pertaining to Sect. 6.18.2.

6.18.2 Theoretical Background

The energy spectrum in superlattices of III–V compounds with graded interfaces in the presence of light waves whose constituent materials are defined by perturbed three-band model of Kane can be written following [60, 61] as

$$\cos(L_o k) = \frac{1}{2}\Phi_{115}(E, k_s) \tag{6.160}$$

where

$$\Phi_{115}(E,k_s) = [2\cosh\{X_{215}(E,k_s)\}\cos\{Y_{215}(E,k_s)\} + \varepsilon_{215}(E,k_s)\sinh\{X_{215}(E,k_s)\}\sin\{Y_{215}(E,k_s)\} + \Delta_{21}\left[\left(\frac{K_{215}^2(E,k_s)}{K_{225}(E,k_s)} - 3K_{225}(E,k_s)\right)\cosh\{X_{215}(E,k_s)\}\sin\{Y_{215}(E,k_s)\} + \left(3K_{215}(E,k_s) - \frac{\{K_{225}(E,k_s)\}^2}{K_{215}(E,k_s)}\right)\sinh\{X_{215}(E,k_s)\}\cos\{Y_{215}(E,k_s)\}\right]$$
$$+ \Delta_{21}\left[2\left(\{K_{215}(E,k_s)\}^2 - \{K_{225}(E,k_s)\}^2\right)\cosh\{X_{215}(E,k_s)\}\cos\{Y_{215}(E,k_s)\} + \frac{1}{12}\left[\frac{5\{K_{225}(E,k_s)\}^3}{K_{215}(E,k_s)} + \frac{5\{K_{215}(E,k_s)\}^3}{K_{225}(E,k_s)} - 34K_{225}(E,k_s)K_{215}(E,k_s)\right]\sinh\{X_{215}(E,k_s)\}\sin\{Y_{215}(E,k_s)\}\right]\Bigg]$$

$$X_{215}(E,k_s) = K_{215}(E,k_s)[a_0 - \Delta_{21}],$$

$$K_{215}(E,k_s) \equiv \left[-\frac{2m_{c2}}{\hbar^2}\beta_{150}(E-\overline{V_0},\lambda,E_{go2},\Delta_2) + k_s^2\right]^{1/2},$$

$$\varepsilon(E,k_s) \equiv \left[\frac{K_{215}(E,k_s)}{K_{225}(E,k_s)} - \frac{K_{225}(E,k_s)}{K_{215}(E,k_s)}\right], \quad k_s^2 = k_x^2 + k_y^2$$

$$Y_{215}(E,k_s) = K_{225}(E,k_s)[b_0 - \Delta_{21}] \quad \text{and}$$

$$K_{225}(E,k_s) = \left[\frac{2m_{c1}\beta_{150}(E,\lambda,E_{go1},\Delta_1)}{\hbar^2} - k_s^2\right]^{1/2}.$$

In the presence of a quantizing magnetic field B along z-direction, the simplified magneto-dispersion relation can be written as

$$k_z^2 = \omega_{215}(E,\lambda,n) \tag{6.161}$$

where

$$\omega_{215}(E,\lambda,n) = \left[\frac{1}{L_0^2}\left[\cos^{-1}\left[\frac{1}{2}f_{215}(E,\lambda,n)\right]\right]^2 - \frac{2|e|B}{\hbar}\left(n+\frac{1}{2}\right)\right],$$

$$f_{215}(E,\lambda,n) = [2\cosh\{M_{215}(n,E)\}\cos\{N_{215}(n,E)\} + Z_{215}(n,E)\sinh\{M_{215}(n,E)\}\sin\{N_{215}(n,E)\}$$

$$+\Delta_{21}\left[\left(\frac{\{I_{215}(n,E)\}^2}{I_{225}(n,E)}-3I_{225}(n,E)\right)\cosh\{M_{215}(n,E)\}\right.$$

$$\times\sin\{N_{215}(n,E)\}+\left(3I_{215}(n,E)-\frac{\{I_{225}(n,E)\}^2}{I_{215}(n,E)}\right)$$

$$\times\sinh\{M_{215}(n,E)\}\cos\{N_{215}(n,E)\}]+\Delta_{21}$$

$$\times\left[2(\{I_{215}(n,E)\}^2-\{I_{225}(n,E)\}^2)\cosh\{M_{215}(n,E)\}\cos\{N_{215}(n,E)\}\right.$$

$$+\frac{1}{12}\left(\frac{5\{I_{225}(n,E)\}^3}{I_{215}(n,E)}+\frac{5\{I_{215}(n,E)\}^3}{I_{225}(n,E)}-\{34I_{225}(n,E)I_{215}(n,E)\}\right)$$

$$\left.\sinh\{M_{215}(n,E)\}\sin\{N_{215}(n,E)\}]\right],$$

$$Z_{215}(n,E)\equiv\left[\frac{I_{215}(n,E)}{I_{225}(n,E)}-\frac{I_{225}(n,E)}{I_{215}(n,E)}\right],\ M_{215}(n,E)=I_{215}(n,E)[a_0-\Delta_{21}],$$

$$I_{215}(n,E)=\left[-\frac{2m_{c2}}{\hbar^2}\beta_{250}(E-\overline{V_0},\lambda,E_{go2},\Delta_2)+\frac{2|e|B}{\hbar}\left(n+\frac{1}{2}\right)\right]^{1/2}$$

$$N_{215}(n,E)=I_{225}(n,E)[b_0-\Delta_{21}]\quad\text{and}$$

$$I_{225}(n,E)\equiv\left[\frac{2m_{c1}}{\hbar^2}\beta_{150}(E,\lambda,E_{go1},\Delta_1)-\left\{\frac{2|e|B}{\hbar}\left(n+\frac{1}{2}\right)\right\}\right]^{1/2}.$$

The electron concentration is given by

$$n_o=\frac{g_veB}{\pi^2\hbar}\left[\sum_{n=0}^{n_{\max}}[\overline{Q}_{27}(V_2,\lambda,n)+\overline{Q}_{28}(V_2,\lambda,n)]\right]\tag{6.162}$$

where $\overline{Q}_{27}(V_2,\lambda,n)=[\omega_{215}(V_2,\lambda,n)]^{1/2}$. V_2 is the Fermi energy in the present case,

$$\overline{Q}_{28}(V_2,\lambda,n)=\sum_{R=1}^{R=R_0}L(R)\left[\overline{Q}_{27}(V_2,\lambda,n)\right]\quad\text{and}$$

$$L(R)=2(k_BT)^{2R}(1-2^{1-2R})\xi(2R)\frac{\partial^{2R}}{\partial V_2^{2R}}.$$

The EEM along the z-direction in this case can be expressed as

$$m^*(V_2,\lambda,n)=(\hbar^2/2)[\omega_{215}(V_2,\lambda,n)]'\tag{6.163}$$

where

$$\omega'_{215}(V_2,\lambda,n)=\big[2M'_{215}(n,V_2)\sinh\{M_{215}(n,V_2)\}\cos\{N_{215}(n,V_2)\}$$
$$+Z_{215}(n,V_2)M'_{215}(n,V_2)\cosh\{M_{215}(n,V_2)\}\sin\{N_{215}(n,V_2)\}$$
$$-2N'_{215}(n,V_2)\sin\{N_{215}(n,V_2)\}\cosh\{M_{215}(n,V_2)\}$$
$$+Z'_{215}(n,V_2)\sinh\{M_{215}(n,V_2)\}\sin\{N_{215}(n,V_2)\}$$
$$+Z_{215}(n,V_2)N'_{215}(n,V_2)\cos\{N_{215}(n,V_2)\}\sinh\{M_{215}(n,V_2)\}$$
$$+\Delta_{21}\left[\left(\frac{\{2I_{215}(n,V_2)I'_{215}(n,V_2)\}}{I_{225}(n,V_2)}-\frac{\{I^2_{215}(n,V_2)I'_{225}(n,V_2)\}}{I^2_{225}(n,V_2)}-3I'_{225}(n,V_2)\right)\right.$$
$$\times\cosh\{M_{215}(n,V_2)\}\sin\{N_{215}(n,V_2)\}+\left(-3I_{225}(n,V_2)+\frac{\{I_{215}(n,V_2)\}^2}{I_{225}(n,V_2)}\right)$$
$$\times\big\{M'_{215}(n,V_2)\sinh\{M_{215}(n,V_2)\}\sin\{N_{215}(n,V_2)\}$$
$$+N'_{215}(n,V_2)\cosh\{M_{215}(n,V_2)\}\cos\{N_{215}(n,V_2)\}\big\}$$
$$+\left(\frac{-\{2I_{225}(n,V_2)I'_{225}(n,V_2)\}}{I_{215}(n,V_2)}+\frac{\{I^2_{225}(n,V_2)I'_{215}(n,V_2)\}}{I^2_{215}(n,V_2)}+3I'_{215}(n,V_2)\right)$$
$$\times\sinh\{M_{215}(n,V_2)\}\cos\{N_{215}(n,V_2)\}+\left(+3I_{215}(n,V_2)-\frac{\{I_{225}(n,V_2)\}^2}{I_{215}(n,V_2)}\right)$$
$$\times\big\{M'_{215}(n,V_2)\cosh\{M_{215}(n,V_2)\}\cos\{N_{215}(n,V_2)\}$$
$$\left.-N'_{215}(n,V_2)\sin\{N_{215}(n,V_2)\}\sinh\{M_{215}(n,V_2)\}\big\}\right]$$
$$+\Delta_{21}\big[4(\{I_{215}(n,V_2)I'_{215}(n,V_2)\}$$
$$-\big\{I_{225}(n,V_2)I'_{225}(n,V_2)\big\})\cosh\{M_{215}(n,V_2)\}\cos\{N_{215}(n,V_2)\}$$
$$+2(\{I_{215}(n,V_2)\}^2-\{I_{225}(n,V_2)\}^2)M'_{215}(n,V_2)$$
$$\times\sinh\{M_{215}(n,V_2)\}\cos\{N_{215}(n,V_2)\}$$
$$-N'_{215}(n,V_2)\cosh\{M_{215}(n,V_2)\}\sin\{N_{215}(n,V_2)\}$$
$$+\frac{1}{12}\left(\frac{15\left\{I^2_{225}(n,V_2)\right\}I'_{225}(n,V_2)}{I_{215}(n,V_2)}-\frac{5\{I_{225}(n,V_2)\}^3I'_{215}(n,V_2)}{I^2_{215}(n,V_2)}\right.$$
$$+\frac{15\left\{I^2_{215}(n,V_2)\right\}I'_{215}(n,V_2)}{I_{225}(n,V_2)}-\frac{5\{I_{215}(n,V_2)\}^3I'_{225}(n,V_2)}{I^2_{225}(n,V_2)}$$
$$-\big\{34I'_{225}(n,V_2)I_{215}(n,V_2)\big\}$$
$$\left.-34I_{225}(n,V_2)I'_{215}(n,V_2)\right)\sinh\{M_{215}(n,V_2)\}\sin\{N_{215}(n,V_2)\}$$
$$+\left(\frac{5\{I_{225}(n,V_2)\}^3}{I_{215}(n,V_2)}+\frac{5\{I_{215}(n,V_2)\}^3}{I_{225}(n,V_2)}-\{34I_{225}(n,V_2)I_{215}(n,V_2)\}\right)$$
$$\big\{M'_{215}(n,V_2)\cosh\{M_{215}(n,V_2)\}\sin\{N_{215}(n,V_2)\}$$
$$\left.+N'_{215}(n,V_2)\sinh\{M_{215}(n,V_2)\}\cos\{N_{215}(n,V_2)\}\big\}\big]\right],$$

$$M'_{215}(n, V_2) = I'_{215}(n, V_2)\left[a_0 - \Delta_{21}\right], I'_{215}(V_2, n) = \frac{m_{c_2}\beta'_{250}(V_2, \lambda, E_{go2}, \Delta_2)}{-\hbar^2 I_{215}(V_2, n)},$$

$$I'_{225}(V_2, n) = \frac{m_{c1}\beta'_{150}(V_2, \lambda, E_{go1}, \Delta_1)}{\hbar^2 I_{225}(V_2, n)},$$

$$I_{215}(V_2, n) = \left[\frac{2eB}{\hbar}\left(n + \frac{1}{2}\right) - \frac{2m_{c2}}{\hbar^2}\beta_{250}(V_2, \lambda, E_{go2}, \Delta_2)\right]^{1/2},$$

$$N_{215}(V_2, n) = I_{225}(V_2, n)\left[b_0 - \Delta_{21}\right],$$

$$I_{225}(V_2, n) = \left[\frac{2m_{c1}\beta_{150}(V_2, \lambda, E_{go1}, \Delta_1)}{\hbar^2} - \frac{2eB}{\hbar}\left(n + \frac{1}{2}\right)\right]^{1/2},$$

$$Z'_{215}(V_2, n) = \left[\frac{-Z_{215}(V_2, n)I'_{215}(V_2, n)}{I_{215}(V_2, n)} - \frac{Z_{215}(V_2, n)I'_{225}(V_2, n)}{I_{225}(V_2, n)}\right.$$
$$\left.+(I_{215}(V_2, n)I_{225}(V_2, n))^{-1}\left[2I'_{215}(V_2, n)I_{215}(V_2, n) - 2I'_{225}(V_2, n)I_{225}(V_2, n)\right]\right].$$

For perturbed two-band model of Kane the forms of the electron concentration and the EEM remain same where

$$I_{215}(V_2, n) = \left[\frac{2eB}{\hbar}\left(n + \frac{1}{2}\right) - \frac{2m_{c2}}{\hbar^2}\omega_{250}(V_2, \lambda, E_{go2})\right]^{1/2},$$

$$I_{225}(V_2, n) = \left[\frac{2m_{c1}\omega_{150}(V_2, \lambda, E_{go1})}{\hbar^2} - \frac{2eB}{\hbar}\left(n + \frac{1}{2}\right)\right]^{1/2},$$

$$I'_{215}(V_2, n) = \frac{m_{c2}\omega'_{250}(V_2, \lambda, E_{go2})}{\hbar^2 I_{215}(V_2, n)}, \quad \text{and}$$

$$I'_{225}(V_2, n) = \left[\frac{m_{c1}\omega'_{150}(V_2, \lambda, E_{go1})}{\hbar^2 I_{225}(V_2, n)}\right].$$

For perturbed parabolic energy bands, the forms of the electron concentration, and the EEM remain same where

$$I_{215}(V_2, n) = \left[\frac{2eB}{\hbar}\left(n + \frac{1}{2}\right) - \frac{2m_{c2}}{\hbar^2}\rho_{250}(V_2, \lambda, E_{go2})\right]^{1/2},$$

$$I_{225}(V_2, n) = \left[\frac{2m_{c1}\rho_{150}(V_2, \lambda, E_{go1})}{\hbar^2} - \frac{2eB}{\hbar}\left(n + \frac{1}{2}\right)\right]^{1/2},$$

$$I'_{215}(V_2, n) = \frac{m_{c2}\rho'_{250}(V_2, \lambda, E_{go2})}{\hbar^2 I_{215}(V_2, n)} \quad \text{and}$$

$$I'_{225}(V_2, n) = \left[\frac{m_{c1}\rho'_{150}(V_2, \lambda, E_{go1})}{\hbar^2 I_{225}(V_2, n)}\right].$$

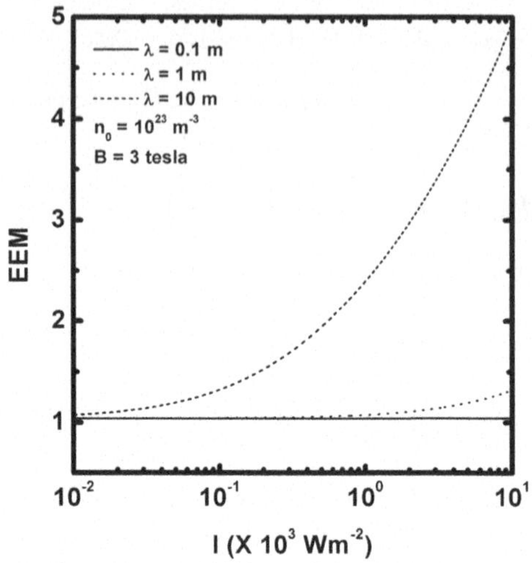

Fig. 6.33 Plot of the normalized EEM as a function of light intensity for superlattices of GaAs/AlGaAs with graded interfaces in the presence of light waves in accordance with the two-band model of Kane in the presence of external photo-excitation

6.19 Results and Discussion

Using the values of the energy band constants from Table 1.1, we have plotted the EEM for the first subband as functions of light intensity as shown in Fig. 6.33 at $T = 4.2$ K by taking superlattices of optoelectronic materials with graded interfaces under magnetic quantization in accordance with the perturbed three [using (6.162) and (6.163)] band model of Kane. It appears that with the increase in the light intensity, the EEM increases sharply as the wavelength varies for the present case.

With the incorporation of different subbands, discontinuous behavior in the EEM would be expected due to the generation of the Landau subbands. Incidentally, in this case, we have limited ourselves with the lowest energy subband at low temperatures where the electrons will be mostly occupied for prominent quantum effects.

6.20 The EEM in Quantum Wire Superlattices of Optoelectronic Semiconductors with Graded Interfaces in the Presence of External Photo-Excitation

6.20.1 Introduction

We shall study the EEM in quantum wire superlattices of optoelectronic semiconductors with graded interfaces in the presence of photo excitation in Sect. 6.20.2 of theoretical background and the Sect. 6.21 explores the result and discussions pertaining to Sect. 6.20.2.

6.20.2 Theoretical Background

The dispersion relation in accordance with the perturbed three-band model of Kane, in this case, is given by

$$k_x^2 = \omega_{225}(E, \lambda, n_y, n_z) \tag{6.164}$$

where

$$\omega_{225}(E, \lambda, n_y, n_z) = \left[\frac{1}{L_0^2}\left[\cos^{-1}\frac{1}{2} f_{135}(E, \lambda, n_y, n_z)\right]^2 - H(n_y, n_z)\right],$$

$$f_{135}(E, \lambda, n_y, n_z) = \Big[2\cosh\{M_{315}(n_y, n_z, E)\}\cos\{N_{315}(n_y, n_z, E)\}$$
$$+ Z_{315}(n_y, n_z, E)\sinh\{M_{315}(n_y, n_z, E)\}\sin\{N_{315}(n_y, n_z, E)\}$$
$$+ \Delta_{21}\Bigg[\left(\frac{\{I_{315}(n_y, n_z, E)\}^2}{I_{325}(n_y, n_z, E)} - 3I_{325}(n_y, n_z, E)\right)\cosh\{M_{315}(n_y, n_z, E)\}$$
$$\times \sin\{N_{315}(n_y, n_z, E)\} + \left(3I_{315}(n_y, n_z, E) - \frac{\{I_{325}(n_y, n_z, E)\}^2}{I_{315}(n_y, n_z, E)}\right)$$
$$\times \sinh\{M_{315}(n_y, n_z, E)\}\cos\{N_{315}(n_y, n_z, E)\}\Bigg] + \Delta_{21}\Big[2(\{I_{315}(n_y, n_z, E)\}^2$$
$$- \{I_{325}(n_y, n_z, E)\}^2)\cosh\{M_{315}(n_y, n_z, E)\}\cos\{N_{315}(n_y, n_z, E)\}$$
$$+ \frac{1}{12}\left(\frac{5\{I_{325}(n_y, n_z, E)\}^3}{I_{315}(n_y, n_z, E)} + \frac{5\{I_{315}(n_y, n_z, E)\}^3}{I_{325}(n_y, n_z, E)} - \{34 I_{325}(n_y, n_z, E)\right.$$
$$\left.\times I_{315}(n_y, n_z, E)\}\right)\sinh\{M_{315}(n_y, n_z, E)\}\sin\{N_{315}(n_y, n_z, E)\}\Big]\Big],$$

$$Z_{315}(n_y, n_z, E) \equiv \left[\frac{I_{315}(n_y, n_z, E)}{I_{325}(n_y, n_z, E)} - \frac{I_{325}(n_y, n_z, E)}{I_{315}(n_y, n_z, E)}\right],$$

$$M_{315}(n_y, n_z, E) = I_{315}(n_y, n_z, E)\,[a_0 - \Delta_{21}],$$

$$I_{315}(n_y, n_z, E) = \left[-\frac{2m_{c2}}{\hbar^2}\beta_{250}(E - \overline{V_0}, \lambda, E_{go2}, \Delta_2) + H(n_y, n_z)\right]^{1/2},$$

$$N_{315}(n_y, n_z, E) = I_{325}(n_y, n_z, E)\,[b_0 - \Delta_{21}] \quad \text{and}$$

$$I_{325}(n_y, n_z, E) \equiv \left[\frac{2m_{c1}}{\hbar^2}\beta_{150}(E, \lambda, E_{go1}, \Delta_1) - H(n_y, n_z)\right]^{1/2}.$$

The electron concentration per unit length is given by

$$n_0 = \frac{2g_v}{\pi}\sum_{n_y=1}^{n_{y\max}}\sum_{n_z=1}^{n_{z\max}}\left[\overline{Q}_{29}(V_3, \lambda, n_y, n_z) + \overline{Q}_{30}(V_3, \lambda, n_y, n_z)\right] \tag{6.165}$$

where

$$\overline{Q}_{29}(V_3, \lambda, n_y, n_z) = \left[\sqrt{\omega_{225}(V_3, \lambda, n_y, n_z)}\right],$$

$$\overline{Q}_{30}(V_3, \lambda, n_y, n_z) = \sum_{R=1}^{R=R_0} L(R)\left[\overline{Q}_{29}(V_3, \lambda, n_y, n_z)\right],$$

$$L(R) = 2(k_B T)^{2R}(1 - 2^{1-2R})\xi(2R)\frac{\partial^{2R}}{\partial V_3^{2R}}$$

and V_3 is the Fermi energy in the present case.

The EEM along x-direction in this case can be expressed as

$$m^*(V_3, \lambda, n_y, n_z) = (\hbar^2/2)\left[\omega_{225}(V_3, \lambda, n_y, n_z)\right]' \tag{6.166}$$

where

$$\omega'_{225}(V_3, \lambda, n_y, n_z) = \frac{2f'_{135}(V_3, \lambda, n_y, n_z)\left[\cos^{-1}\{\frac{1}{2}f_{135}(V_3, \lambda, n_y, n_z)\}\right]}{\sqrt{4 - f_{135}^2(V_3, \lambda, n_y, n_z)}},$$

$$\begin{aligned}
f'_{135}(V_3, \lambda, n_y, n_z) = &\Big[2M'_{315}(n_y, n_z, V_3)\sinh\left\{M_{315}(n_y, n_z, V_3)\right\} \\
&\times \cos\left\{N_{315}(n_y, n_z, V_3)\right\} + Z_{315}(n_y, n_z, V_3)M'_{315}(n_y, n_z, V_3) \\
&\times \cosh\left\{M_{315}(n_y, n_z, V_3)\right\} \cdot \sin\left\{N_{315}(n_y, n_z, V_3)\right\} - 2N'_{315}(n_y, n_z, V_3) \\
&\times \sin\left\{N_{315}(n_y, n_z, V_3)\right\}\cosh\left\{M_{315}(n_y, n_z, V_3)\right\} + Z'_{315}(n_y, n_z, V_3) \\
&\times \sinh\{M_{315}(n_y, n_z, V_3)\}\sin\{N_{315}(n_y, n_z, V_3)\} + Z_{315}(n_y, n_z, V_3)N'_{315}(n_y, n_z, V_3) \\
&\times \cos\{N_{315}(n_y, n_z, V_3)\}\sinh\left\{M_{315}(n_y, n_z, V_3)\right\} \\
&+ \Delta_{21}\Bigg[\Bigg(\frac{\{2I_{315}(n_y, n_z, V_3)I'_{315}(n_y, n_z, V_3)\}}{I_{325}(n_y, n_z, V_3)} - \frac{\{I_{315}^2(n_y, n_z, V_3)I'_{325}(n_y, n_z, V_3)\}}{I_{325}^2(n_y, n_z, V_3)} \\
&- 3I'_{325}(n_y, n_z, V_0)\Bigg) \cdot \cosh\left\{M_{315}(n_y, n_z, V_3)\right\}\sin\left\{N_{315}(n_y, n_z, V_3)\right\} \\
&+ \left(-3I_{325}(n_y, n_z, V_3) + \frac{\{I_{315}^2(n_y, n_z, V_{30})\}}{I_{325}(n_y, n_z, V_3)}\right)\{M'_{315}(n_y, n_z, V_3) \\
&\times \sinh\left\{M_{315}(n_y, n_z, V_3)\right\}\sin\left\{N_{315}(n_y, n_z, V_3)\right\} \\
&+ N'_{315}(n_y, n_z, V_3)\cosh\left\{M_{315}(n_y, n_z, V_3)\right\}\cos\left\{N_{315}(n_y, n_z, V_3)\right\}\Big\} \\
&+ \Bigg(\frac{-\{2I_{325}(n_y, n_z, V_3)I'_{325}(n_y, n_z, V_3)\}}{I_{315}(n_y, n_z, V_3)} + \frac{\{I_{325}^2(n_y, n_z, V_3)I'_{315}(n_y, n_z, V_3)\}}{I_{315}^2(n_y, n_z, V_3)}
\end{aligned}$$

$$+ 3I'_{315}(n_y, n_z, V_3)\Big) \sinh\left\{M_{315}(n_y, n_z, V_3)\right\} \cdot \cos\left\{N_{315}(n_y, n_z, V_3)\right\}$$

$$+\left(+3I_{315}(n_y, n_z, V_3) - \frac{\left\{I_{325}(n_y, n_z, V_3)\right\}^2}{I_{315}(n_y, n_z, V_3)}\right)\{M'_{315}(n_y, n_z, V_3)$$

$$\times \cosh\left\{M_{315}(n_y, n_z, V_3)\right\} \cos\left\{N_{315}(n_y, n_z, V_3)\right\}$$

$$-N'_{315}(n_y, n_z, V_3) \sin\left\{N_{315}(n_y, n_z, V_3)\right\} \sinh\left\{M_{315}(n_y, n_z, V_3)\right\}\}\Big]$$

$$+ \Delta_{21}\Big[4(\left\{I_{315}(n_y, n_z, V_3) I'_{315}(n_y, n_z, V_3)\right\}$$

$$- \left\{I_{325}(n_y, n_z, V_3) I'_{325}(n_y, n_z, V_3)\right\}) \cosh\left\{M_{315}(n_y, n_z, V_3)\right\}$$

$$\times \cos\{N_{315}(n_y, n_z, V_3)\} + 2(\{I_{315}(n_y, n_z, V_3)\}^2 - \{I_{325}(n_y, n_z, V_3)\}^2)$$

$$\times \{M'_{315}(n_y, n_z, V_3) \sinh\{M_{315}(n_y, n_z, V_3)\} \cos\{N_{315}(n_y, n_z, V_3)\}$$

$$- N'_{315}(n_y, n_z, V_3) \cosh\{M_{315}(n_y, n_z, V_3)\} \sin\{N_{315}(n_y, n_z, V_3)\}\}$$

$$+ \frac{1}{12}\left(\frac{15\{I^2_{325}(n_y, n_z, V_3)\} I'_{325}(n_y, n_z, V_3)}{I_{315}(n, V_3)}\right.$$

$$- \frac{5\{I_{325}(n_y, n_z, V_3)\}^3 I'_{315}(n_y, n_z, V_3)}{I^2_{315}(n_y, n_z, V_3)} + \frac{15\{I^2_{315}(n_y, n_z, V_3)\} I'_{315}(n_y, n_z, V_3)}{I_{325}(n_y, n_z, V_3)}$$

$$- \frac{5\{I_{315}(n_y, n_z, V_3)\}^3 I'_{325}(n_y, n_z, V_3)}{I^2_{325}(n_y, n_z, V_3)} - \{34 I'_{325}(n_y, n_z, V_3) I_{315}(n_y, n_z, V_3)\}$$

$$\left. - 34 I_{325}(n_y, n_z, V_3) I'_{315}(n_y, n_z, V_3)\right) \sinh\{M_{315}(n_y, n_z, V_3)\}$$

$$\times \sin\{N_{315}(n_y, n_z, V_3)\}$$

$$+ \left(\frac{5\{I_{325}(n_y, n_z, V_3)\}^3}{I_{315}(n_y, n_z, V_3)} + \frac{5\{I_{315}(n_y, n_z, V_3)\}^3}{I_{325}(n_y, n_z, V_3)}\right.$$

$$\left. - \{34 I_{325}(n_y, n_z, V_3) I_{315}(n_y, n_z, V_3)\}\right)$$

$$\times \{M'_{315}(n_y, n_z, V_3) \cosh\{M_{315}(n_y, n_z, V_3)\} \sin\{N_{315}(n_y, n_z, V_3)\}$$

$$+ N'_{315}(n_y, n_z, V_3) \sinh\{M_{315}(n_y, n_z, V_3)\} \cos\{N_{315}(n_y, n_z, V_3)\}\}\Big]\Big],$$

$$M'_{315}(n_y, n_z, V_3) = I'_{315}(n_y, n_z, V_3)\,[a_0 - \Delta_{21}],$$

$$I'_{315}(V_3, n_y, n_z) = \frac{m_{c2} \beta'_{250}(V_3, \lambda, E_{go2}, \Delta_2)}{\hbar^2 I_{315}(V_3, n_y, n_z)},$$

$$N'_{315}(n_y, n_z, V_3) = I'_{325}(n_y, n_z, V_3)\,[b_0 - \Delta_{21}],$$

$$I'_{325}(V_3, n_y, n_z) = \frac{m_{c1} \beta'_{150}(V_3, \lambda, E_{go2}, \Delta_2, n_y, n_z)}{\hbar^2 I_{325}(V_3, n_y, n_z)},$$

$$I_{315}(n_y, n_z, V_3) = \left[-\frac{2m_{c2}}{\hbar^2}\beta_{250}(V_3, \lambda, E_{go2}, \Delta_2) + H(n_y, n_z)\right]^{1/2},$$

$$N_{315}(n_y, n_z, V_3) = I_{325}(n_y, n_z, V_3)\,[b_0 - \Delta_{21}],$$

$$I_{325}(n_y, n_z, V_3) = \left[\frac{2m_{c1}}{\hbar^2}\beta_{150}(V_3, \lambda, E_{go1}, \Delta_1) - H(n_y, n_z)\right]^{1/2} \quad \text{and}$$

$$Z'_{315}(V_3, n_y, n_z) = \left[\frac{-Z_{315}(V_3, n_y, n_z)I'_{315}(V_3, n_y, n_z)}{I_{315}(V_3, n_y, n_z)} - \frac{Z_{315}(V_3, n_y, n_z)I'_{325}(V_3, n_y, n_z)}{I_{325}(V_3, n_y, n_z)} + (I_{315}(V_3, n_y, n_z)I_{325}(V_3, n_y, n_z))^{-1} [2I'_{315}(V_3, n_y, n_z)I_{315}(V_3, n_y, n_z) - 2I'_{325}(V_3, n_y, n_z)I_{325}(V_3, n_y, n_z)]\right].$$

For perturbed two-band model of Kane, the form of electron concentration per unit length and the field emitted current remain same where

$$I_{315}(n_y, n_z, V_3) = \left[H(n_y, n_z) - \frac{2m_{c2}}{\hbar^2}\omega_{250}(V_3, \lambda, E_{go2})\right]^{1/2},$$

$$I'_{315}(V_3, n_y, n_z) = \frac{m_{c2}\omega'_{250}(V_3, \lambda, E_{go2})}{\hbar^2 I_{315}(V_3, n_y, n_z)},$$

$$I_{325}(n_y, n_z, V_3) = \left[-H(n_y, n_z) + \frac{2m_{c1}}{\hbar^2}\omega_{150}(V_3, \lambda, E_{go1})\right]^{1/2},$$

$$I'_{325}(V_3, n_y, n_z) = \frac{m_{c1}\omega'_{150}(V_3, \lambda, E_{go1})}{\hbar^2 I_{325}(V_3, n_y, n_z)}.$$

For perturbed parabolic energy bands, the form of electron concentration per unit length and the field emitted current remain same where

$$I_{315}(n_y, n_z, V_3) = \left[H(n_y, n_z) - \frac{2m_{c2}}{\hbar^2}\rho_{250}(V_3, \lambda, E_{go2})\right]^{1/2},$$

$$I'_{315}(V_3, n_y, n_z) = \frac{m_{c2}\rho'_{250}(V_3, \lambda, E_{go2})}{\hbar^2 I_{315}(V_3, n_y, n_z)},$$

$$I_{325}(n_y, n_z, V_3) = \left[-H(n_y, n_z) + \frac{2m_{c1}}{\hbar^2}\rho_{150}(V_3, \lambda, E_{go1})\right]^{1/2},$$

$$I'_{325}(V_3, n_y, n_z) = \frac{m_{c1}\rho'_{150}(V_3, \lambda, E_{go1})}{\hbar^2 I_{325}(V_3, n_y, n_z)}.$$

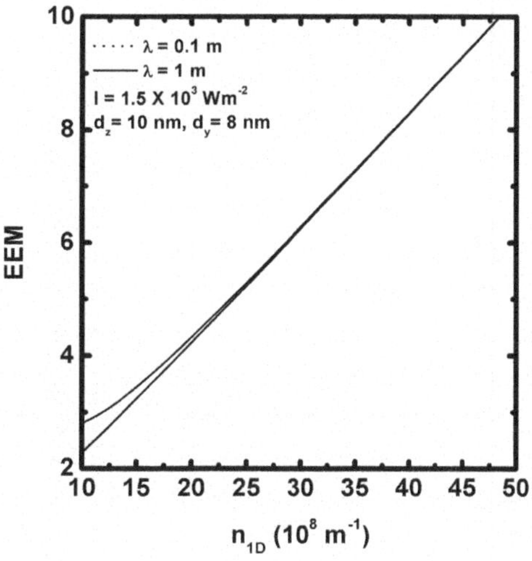

Fig. 6.34 Plot of the normalized EEM as a function of linear electron concentration of quantum wire superlattices of GaAs/AlGaAs with graded interfaces in the presence of light waves in accordance with the two-band model of Kane in the presence of external photo-excitation

6.21 Results and Discussion

Using the values of the energy band constants from Table 1.1, we have plotted the EEM for $n_y = 1$ and $n_z = 1$ as a function of electron concentration at $T = 4.2\,\mathrm{K}$ by taking quantum wire super lattices of optoelectronic materials with graded interfaces in accordance with the perturbed two-band model of Kane. In Fig. 6.34, we have plotted the EEM as function of electron concentration per unit length in quantum wires of GaAs/AlGaAs superlattices with graded interfaces. It appears from Fig. 6.34 that the effect of a single subband linearly increases the EEM for low value of wavelength.

It may be noted that with the increase in the wavelength, the EEM tends to coincide with that of the lower wavelength values at higher carrier concentration. For the purpose of condensation, the electron statistics and the EEM for this chapter has been presented in Table 6.1.

6.22 Open Research Problems

(R6.1) Investigate the EEM, EAM, DEM, CEM, CoEM, FREM, and OEM for the bulk materials whose respective dispersion relations of the carriers in the absence of any field is given in Chap. 1 in the presence of strong light waves which change the original band structure and consider its effect in the subsequent: study in each case.

(R6.2) Investigate the same set of masses as defined in (R6.1) in the presence of an arbitrarily oriented non-uniform light waves for all the materials as considered (R6.1).

Table 6.1 The EEM and the electron statistics for different quantized optoelectronic materials

Type of materials	EEM	The electron statistics
1. Opto-electronic and III–V materials under light waves		
(a) Perturbed three band model of Kane	$m^*(E_F) = m^*\left[\gamma'(E_F) - \theta_0'(E_F,\lambda)\right]$ (6.46)	$n_0 = g_v(3\pi^2)^{-1}\left(\frac{2m^*}{\hbar^2}\right)^{3/2}\left[M_1(E_F,\lambda) + N_1(E_F,\lambda)\right]$ (6.48)
(b) Perturbed two band model of Kane	$m^*(E_F) = m^*\left[(1+2\alpha E_F) - B_0'(E_F,\lambda)\right]$ (6.49)	$n_0 = g_v(3\pi^2)^{-1}\left(\frac{2m^*}{\hbar^2}\right)^{3/2}\left[M_2(E_F,\lambda) + N_2(E_F,\lambda)\right]$ (6.50a)
(c) Perturbed parabolic energy bands	$m^*(E_F) = m^*\left[\rho_0'(E_F,\lambda)\right]$ (6.50b)	$n_0 = g_v(3\pi^2)^{-1}\left(\frac{2m^*}{\hbar^2}\right)^{3/2}\left[M_3(E_F,\lambda) + N_3(E_F,\lambda)\right]$ (6.50c)
2. Bulk specimens of III–V and optoelectronic materials in the absence of light waves		
(a) Three band model of Kane	$m^*(E_{F_0}) = m^*\left[I_{\|}'(E_{F_0})\right]$ (6.51)	$n_0 = g_v(3\pi^2)^{-1}\left(\frac{2m^*}{\hbar^2}\right)^{3/2}\left[M_4(E_{F_0}) + N_4(E_{F_0})\right]$ (6.52)
(b) Two band model of Kane	$m^*(E_{F_0}) = m^*\left[(1+2\alpha E_{F_0})\right]$ (6.53)	$n_0 = g_v(3\pi^2)^{-1}\left(\frac{2m^*}{\hbar^2}\right)^{3/2}\left[M_5(E_{F_0}) + N_5(E_{F_0})\right]$ (6.54)
3. Opto-electronic and III–V materials under magnetic quantization in the presence of light waves		
(a) Perturbed three band model of Kane	$m^*(E_{F_{BL}},\lambda) = m^*\left\{\beta_0(E_{F_{BL}},\lambda)\right\}'$ (6.61)	$n_0 = \frac{g_v\|e\|B\sqrt{2m^*}}{\pi^2\hbar^2}\sum_{n=0}^{n_{\max}}[M_{13}(E_{F_{BL}},B,\lambda) + N_{13}(E_{F_{BL}},B,\lambda)]$ (6.64)
(b) Perturbed two band model of Kane	$m^*(E_{F_{BL}},\lambda) = m^*\left\{\tau_0(E_{F_{BL}}\lambda)\right\}'$ (6.66)	$n_0 = \frac{g_v\|e\|B\sqrt{2m^*}}{\pi^2\hbar^2}\sum_{n=0}^{n_{\max}}[M_{14}(E_{F_{BL}},B,\lambda) + N_{14}(E_{F_{BL}},B,\lambda)]$ (6.69)
(c) Perturbed parabolic energy bands	$m^*(E_{F_{BL}},\lambda) = m^*\left\{\rho_0(E_{F_{BL}},\lambda)\right\}'$ (6.70)	$n_0 = \frac{g_v\|e\|B\sqrt{2m^*}}{\pi^2\hbar^2}\sum_{n=0}^{n_{\max}}[M_{15}(E_{F_{BL}},B,\lambda) + N_{15}(E_{F_{BL}},B,\lambda)]$ (6.74)

(Continued)

Table 6.1 (Continued)

Type of materials	EEM	The electron statistics
4. Ultra thin films of opto-electronic and III–V materials under light waves		
(a) Perturbed three band model of Kane	$m^*(E_{F2DL}, n_z, \lambda) = m^*\{\beta_0(E_{F2DL}, \lambda)\}'$ (6.74)	$n_{2D} = \frac{m^* g_v}{\pi\hbar^2} \sum_{n_z=1}^{n_{z\max}} \left[M_{18}(n_z, E_{F2DL}, \lambda) + N_{18}(n_z, E_{F2DL}, \lambda) \right]$ (6.76)
(b) Perturbed two band model of Kane	$m^*(E_{F2DL}, n_z, \lambda) = m^*\{\tau_0(E_{F2DL}, \lambda)\}'$ (6.79)	$n_{2D} = \frac{m^* g_v}{\pi\hbar^2} \sum_{n_z=1}^{n_{z\max}} \left[M_{19}(n_z, E_{F2DL}, \lambda) + N_{19}(n_z, E_{F2DL}, \lambda) \right]$ (6.81)
(c) Perturbed parabolic energy bands	$m^*(E_{F2DL}, n_z, \lambda) = m^*\{\rho_0(E_{F2DL}, \lambda)\}'$ (6.84)	$n_{2D} = \frac{m^* g_v}{\pi\hbar^2} \sum_{n_z=1}^{n_{z\max}} \left[M_{20}(n_z, E_{F2DL}, \lambda) + N_{20}(n_z, E_{F2DL}, \lambda) \right]$ (6.86)
5. n-channel inversion layers of III–V and optoelectronic materials in the presence of light waves under low electric field limit		
(a) Perturbed three-band model of Kane	$m^*(E_{FiwL}, i) = m^* \left[P_{3L}(E_{FiwL}, i) \right]\|$ (6.89)	$n_{2D_w} = \frac{g_v m^*}{\pi\hbar^2} \sum_{i=0}^{i_{\max}} \left[P_{4wL}(E_{FiwL}, i) + Q_{4wL}(E_{FiwL}, i) \right]$ (6.92)
(b) Perturbed two band model of Kane	$m^*(E_{FiwL2}, i) = m^* \left[P_{3L2}(E_{FiwL2}, i) \right]\|$ (6.100)	$n_{2D_w} = \frac{g_v m^*}{\pi\hbar^2} \sum_{i=0}^{i_{\max}} \left[P_{4wL2}(E_{FiwL2}, i) + Q_{4wL2}(E_{FiwL2}, i) \right]$ (6.103)
(c) Perturbed parabolic energy bands	$m^*(E_{FiwL1}, i) = m^* \left[P_{3L1}(E_{FiwL1}, i) \right]\|^-$ (6.111)	$n_{2D_w} = \frac{g_v m^*}{\pi\hbar^2} \sum_{i=0}^{i_{\max}} \left[P_{4wL1}(E_{FiwL1}, i) + Q_{4wL1}(E_{FiwL1}, i) \right]$ (6.114)

(Continued)

Table 6.1 (Continued)

Type of materials	EEM	The electron statistics
6. *n*-channel inversion layers of III–V and optoelectronic materials in the presence of light waves under high electric field limit		
(a) Perturbed three-band model of Kane	$m^*(E_{FisL}, i) = m^*\,[P_{5L}(E, i)]\|_{E=E_{FisL}}$ (6.94)	$n_{2D_s} = \frac{g_v m^*}{\pi\hbar^2}\sum_{i=0}^{i_{\max}}\left[P_{6sL}(E_{FisL}, i) + Q_{6sL}(E_{FisL}, i)\right]$ (6.97)
(b) Perturbed two-band model of Kane	$m^*(E_{FisL2}, i) = m^*\,[P_{5L2}(E, i)]\|_{E=E_{FisL2}}$ (6.105)	$n_{2D_s} = \frac{g_v m^*}{\pi\hbar^2}\sum_{i=0}^{i_{\max}}\left[P_{6sL2}(E_{FisL2}, i) + Q_{6sL2}(E_{FisL2}, i)\right]$ (6.108)
7. Nipi structures of opto-electronic and III–V materials in the presence of light waves		
(a) Perturbed three band model of Kane	$m^*(E_{FnL}, n_i) = m^* R_{82L}(E, n_i)\|_{E=E_{FnL}}$ (6.116)	$n_0 = \frac{m^* g_v}{\pi\hbar_0^{2d}}\sum_{n_i=0}^{n_{i\max}}\left[T_{83L}(\overline{E}_{FnL}, n_i) + T_{84L}(\overline{E}_{FnL}, n_i)\right]$ (6.119)
(b) Perturbed two-band model of Kane	$m^*(E_{FnL2}, n_i) = m^* R_{82L2}(E, n_i)\|_{E=E_{FnL2}}$ (6.121)	$n_0 = \frac{m^* g_v}{\pi\hbar^2 d_0}\sum_{n_i=0}^{n_{i\max}}\left[T_{83L2}(\overline{E}_{FnL2}, n_i) + T_{84L2}(\overline{E}_{FnL2}, n_i)\right]$ (6.124)
(c) Perturbed parabolic energy bands	$m^*(E_{FnL1}, n_i) = m^* R_{82L1}(E, n_i)\|_{E=E_{FnL1}}$ (6.126)	$n_0 = \frac{m^* g_v}{\pi\hbar^2 d_0}\sum_{n_i=0}^{n_{i\max}}\left[T_{83L1}(\overline{E}_{FnL1}, n_i) + T_{84L1}(\overline{E}_{FnL1}, n_i)\right]$ (6.129)

(Continued)

Table 6.1 (Continued)

Type of materials	EEM	The electron statistics
8. Quantum wires of opto-electronic and III–V materials in the presence of light waves		
(a) Perturbed three-band model of Kane	$m^*(E_{F1DL}, n_y, n_z, \lambda) = m^*\{\beta_0(E_{F1DL}, \lambda)\}'$ (6.132)	$n_{1D} = \frac{2g_v}{\pi}\frac{\sqrt{2m^*}}{\hbar}\sum_{n_y=1}^{n_{y\max}}\sum_{n_z=1}^{n_{z\max}}[M_{21}(n_y, n_z, E_{F1DL}, \lambda) + N_{21}(n_y, n_z, E_{F1DL}, \lambda)]$ (6.134)
(b) Perturbed two-band model of Kane	$m^*(E_{F1DL}, n_y, n_z, \lambda) = m^*\{\tau_0(E_{F1DL}, \lambda)\}'$ (6.137)	$n_{1D} = \frac{2g_v}{\pi}\frac{\sqrt{2m^*}}{\hbar}\sum_{n_y=1}^{n_{y\max}}\sum_{n_z=1}^{n_{z\max}}[M_{22}(n_y, n_z, E_{F1DL}, \lambda) + N_{22}(n_y, n_z, E_{F1DL}, \lambda)]$ (6.139)
(c) Perturbed parabolic energy bands	$m^*(E_{F1DL}, n_y, n_z, \lambda) = m^*\{\rho_0(E_{F1DL}, \lambda)\}'$ (6.142)	$n_{1D} = \frac{2g_v\sqrt{2m^*}}{\pi\hbar}\sum_{n_y=1}^{n_{y\max}}\sum_{n_z=1}^{n_{z\max}}[M_{23}(n_y, n_z, E_{F1DL}, \lambda) + N_{23}(n_y, n_z, E_{F1DL}, \lambda)]$ (6.144)
9. Effective mass superlattices of opto-electronic and III–V materials in the presence of external light waves		
(a) Perturbed three-band model of Kane	$m^*(V_0, \lambda, n) = (\hbar^2/2)\,[\overline{\omega}_{15}(V_0, \lambda, n)]'$ (6.147)	$n_o = \frac{g_v eB}{\pi^2\hbar L_o}\sum_{n=0}^{n_{\max}}[S_{54}(V_0, \lambda, n) + T_{54}(V_0, \lambda, n)]$ (6.148)
(b) Perturbed two-band model of Kane	$m^*(V_0, \lambda, n) = (\hbar^2/2)\,[\overline{\omega}_{16}(V_0, \lambda, n)]'$ (6.150)	$n_o = \frac{g_v eB}{\pi^2\hbar L_o}\sum_{n=0}^{n_{\max}}[S_{55}(V_0, \lambda, n) + T_{55}(V_0, \lambda, n)]$ (6.149)
(c) Perturbed parabolic energy bands	$m^*(V_0, \lambda, n) = (\hbar^2/2)\,[\overline{\omega}_{17}(V_0, \lambda, n)]'$ (6.152)	$n_o = \frac{g_v eB}{\pi^2\hbar L_o}\sum_{n=0}^{n_{\max}}[S_{56}(V_0, \lambda, n) + T_{56}(V_0, \lambda, n)]$ (6.151)
10. Quantum wire effective mass superlattices of opto-electronic and III–V materials in the presence of external light waves		

(Continued)

Table 6.1 (Continued)

Type of materials	EEM	The electron statistics
(a) Perturbed three-band model of Kane	$m^*(V_1, \lambda, n_y, n_z) = (\hbar^2/2)\left[\overline{\omega}_{19}(V_1, \lambda, n_y, n_z)\right]'$ (6.155)	$n_0 = \frac{2g_v}{\pi} \sum_{n_y=1}^{n_{y\max}} \sum_{n_z=1}^{n_{z\max}} [\overline{Q}_{23}(V_1, \lambda, n_y, n_z) + \overline{Q}_{24}(V_1, \lambda, n_y, n_z)]$ (6.154)
(b) Perturbed two-band model of Kane	$m^*(V_1, \lambda, n_y, n_z) = (\hbar^2/2)\left[\overline{\omega}_{20}(V_1, \lambda, n_y, n_z)\right]'$ (6.157)	$n_0 = \frac{2g_v}{\pi} \sum_{n_y=1}^{n_{y\max}} \sum_{n_z=1}^{n_{z\max}} [\overline{Q}_{25}(V_1, \lambda, n_y, n_z) + \overline{Q}_{26}(V_1, \lambda, n_y, n_z)]$ (6.156)
(c) Perturbed parabolic energy bands	$m^*(V_1, \lambda, n_y, n_z) = (\hbar^2/2)\left[\overline{\omega}_{201}(V_1, \lambda, n_y, n_z)\right]'$ (6.159)	$n_0 = \frac{2g_v}{\pi} \sum_{n_y=1}^{n_{y\max}} \sum_{n_z=1}^{n_{z\max}} [\overline{Q}_{251}(V_1, \lambda, n_y, n_z) + \overline{Q}_{261}(V_1, \lambda, n_y, n_z)]$ (6.158)
11. Superlattices of opto electronic and III–V materials with graded interfaces under magnetic quantization in the presence of external light waves		
(a) Perturbed three-band model of Kane	$m^*(V_2, \lambda, n) = (\hbar^2/2)\,[\omega_{215}(V_2, \lambda, n)]'$ (6.163)	$n_o = \frac{g_v eB}{\pi^2 \hbar}\left[\sum_{n=0}^{n_{\max}} [\overline{Q}_{27}(V_2, \lambda, n) + \overline{Q}_{28}(V_2, \lambda, n)]\right]$ (6.162)
12. Quantum wire superlattices of opto-electronic and III–V materials with graded interfaces in the presence of external light waves		
(a) Perturbed three-band model of Kane	$m^*(V_3, \lambda, n_y, n_z) = (\hbar^2/2)\left[\omega_{225}(V_3, \lambda, n_y, n_z)\right]'$ (6.166)	$n_0 = \frac{2g_v}{\pi} \sum_{n_y=1}^{n_{y\max}} \sum_{n_z=1}^{n_{z\max}} [\overline{Q}_{29}(V_3, \lambda, n_y, n_z) + \overline{Q}_{30}(V_3, \lambda, n_y, n_z)]$ (6.165)

(R6.3) Investigate the same set of masses as defined in (R6.1) in the presence of an arbitrarily oriented non-quantizing alternating non-uniform light waves for all the cases of (R6.1).

(R6.4) Investigate the same set of masses as defined in (R6.1) for the heavily doped materials in the presence of Gaussian, exponential, Kane, Halperin, Lax, and Bonch-Bruevich types of band tails for all materials whose unperturbed carrier energy spectra are defined in (R6.1).

(R6.5) Investigate the same set of masses as defined in (R6.1) for all the materials in the presence of arbitrarily oriented non-quantizing non-uniform light waves for all the appropriate cases of problem (R6.4).

(R6.6) Investigate the same set of masses as defined in (R6.1) for all the materials in the presence of arbitrarily oriented non-quantizing alternating light waves for all the appropriate cases of problem (R6.4).

(R6.7) Investigate the same set of masses as defined in (R6.1) for the negative refractive index, organic, magnetic, and other advanced optical materials in the presence of arbitrarily oriented light waves.

(R6.8) Investigate the same set of masses as defined in (R6.1) in the presence of alternating non-quantizing light waves for all the problems of (R6.7).

(R6.9) Investigate the same set of masses as defined in (R6.1) for all the quantum confined materials (i.e., multiple quantum wells and wires) whose unperturbed carrier energy spectra are defined in (R6.1) in the presence of arbitrary oriented quantizing magnetic field by including the effects of spin and broadening respectively.

(R6.10) Investigate the same set of masses as defined in (R6.1) in the presence of an additional arbitrarily oriented alternating quantizing magnetic field, respectively, for all the problems of (R6.9).

(R6.11) Investigate the same set of masses as defined in (R6.1) in the presence of arbitrarily oriented alternating quantizing magnetic field and arbitrary oriented non-quantizing non-uniform light waves, respectively, for all the problems of (R6.9).

(R6.12) Investigate the same set of masses as defined in (R6.1) in the presence of arbitrary oriented alternating non- uniform quantizing magnetic field and additional arbitrary oriented non-quantizing alternating light waves respectively for all the problems of (R6.9).

(R6.13) Investigate the same set of masses as defined in (R6.1) in the presence of arbitrary oriented and crossed quantizing magnetic and electric fields respectively for all the problems of (R6.9).

(R6.14) Investigate the same set of masses as defined in (R6.1) for all the appropriate low-dimensional systems of this chapter in the presence of finite potential wells.

(R6.15) Investigate the same set of masses as defined in (R6.1) for all the appropriate low-dimensional systems of this chapter in the presence of parabolic potential wells.

(R6.16) Investigate the same set of masses as defined in (R6.1) for all the appropriate systems of this chapter forming quantum rings.

(R6.17) Investigate the same set of masses as defined in (R6.1) for all the above appropriate problems in the presence of elliptical Hill and quantum square rings respectively.

(R6.18) Investigate the same set of masses as defined in (R6.1) for multiple wall carbon nano-tubes.

(R6.19) Investigate the same set of masses as defined in (R6.1) for multiple wall carbon nano-tubes in the presence of non-quantizing non-uniform alternating light waves.

(R6.20) Investigate the same set of masses as defined in (R6.1) for multiple wall carbon nano-tubes in the presence of non-quantizing non-uniform alternating magnetic field.

(R6.21) Investigate the same set of masses as defined in (R6.1) for multiple wall carbon nano-tubes in the presence of crossed electric and quantizing magnetic fields.

(R6.22) Investigate the same set of masses as defined in (R6.1) for heavily doped semiconductor nano-tubes for all the materials whose unperturbed carrier dispersion laws are defined in Chap. 1.

(R6.23) Investigate the same set of masses as defined in (R6.1) for heavily doped semiconductor nanotubes in the presence of non-quantizing alternating light waves for all the materials whose unperturbed carrier dispersion laws are defined in Chap. 1.

(R6.24) Investigate the same set of masses as defined in (R6.1) for heavily doped semiconductor nanotubes in the presence of non-quantizing alternating magnetic field for all the materials whose unperturbed carrier dispersion laws are defined in Chap. 1.

(R6.25) Investigate the same set of masses as defined in (R6.1) for heavily doped semiconductor nano-tubes in the presence of non-uniform light waves for all the materials whose unperturbed carrier dispersion laws are defined in Chap. 1.

(R6.26) Investigate the same set of masses as defined in (R6.1) for heavily doped semiconductor nanotubes in the presence of alternating quantizing magnetic fields for all the materials whose unperturbed carrier dispersion laws are defined in Chap. 1.

(R6.27) Investigate the same set of masses as defined in (R6.1) for heavily doped semiconductor nanotubes in the presence of crossed electric and quantizing magnetic fields for all the materials whose unperturbed carrier dispersion laws are defined in Chap. 1.

(R6.28) Investigate the same set of masses as defined in (R6.1) for all the appropriate nipi structures of the materials whose unperturbed carrier energy spectra are defined in Chap. 1.

(R6.29) Investigate the same set of masses as defined in (R6.1) for all the appropriate nipi structures of the materials whose unperturbed carrier energy spectra are defined in Chap. 1, in the presence of an arbitrarily oriented non-quantizing non-uniform additional electric field.

(R6.30) Investigate the same set of masses as defined in (R6.1) for all the appropriate nipi structures of the materials whose unperturbed carrier energy spectra are defined in Chap. 1 in the presence of non-quantizing alternating additional magnetic field.

(R6.31) Investigate the same set of masses as defined in (R6.1) for all the appropriate nipi structures of the materials whose unperturbed carrier energy spectra are defined in Chap. 1 in the presence of quantizing alternating additional magnetic field.

(R6.32) Investigate the same set of masses as defined in (R6.1) for all the appropriate nipi structures of the materials whose unperturbed carrier energy spectra are defined in Chap. 1 in the presence of crossed electric and quantizing magnetic fields.

(R6.33) Investigate the same set of masses as defined in (R6.1) for heavily doped nipi structures for all the appropriate cases of all the above problems.

(R6.34) Investigate the same set of masses as defined in (R6.1) for the appropriate accumulation layers of all the materials whose unperturbed carrier energy spectra are defined in Chap. 1 in the presence of crossed electric and quantizing magnetic fields by considering electron spin and broadening of Landau levels.

(R6.35) Investigate the same set of masses as defined in (R6.1) for quantum confined III–V, II–VI, IV–VI, HgTe/CdTe effective mass superlattices together with short period, strained layer, random, Fibonacci, poly-type and sawtooth superlattices.

(R6.36) Investigate the same set of masses as defined in (R6.1) in the presence of quantizing magnetic field, respectively, for all the cases of (R6.35).

(R6.37) Investigate the same set of masses as defined in (R6.1) in the presence of non-quantizing non-uniform additional electric field, respectively, for all the cases of (R6.35).

(R6.38) Investigate the same set of masses as defined in (R6.1) in the presence of non-quantizing alternating light waves, respectively, for all the cases of (R6.35).

(R6.39) Investigate the same set of masses as defined in (R6.1) in the presence of crossed electric and quantizing magnetic fields, respectively, for all the cases of (R6.35).

(R6.40) Investigate the same set of masses as defined in (R6.1) for heavily doped quantum confined superlattices for all the problems of (R6.35).

(R6.41) Investigate the same set of masses as defined in (R6.1) in the presence of quantizing non-uniform magnetic field, respectively, for all the cases of (R6.40).

(R6.42) Investigate the same set of masses as defined in (R6.1) in the presence of crossed electric and quantizing magnetic fields, respectively, for all the cases of (R6.40).

(R6.43) Investigate the same set of masses as defined in (R6.1) for all the systems in the presence of strain.

(R6.44) Investigate all the problems of this chapter by removing all the mathematical approximations and establishing the respective appropriate uniqueness conditions.

References

1. P.K. Basu, *Theory of Optical Process in Semiconductors, Bulk and Microstructures* (Oxford University Press, Oxford, 1997)
2. K.P. Ghatak, S. Bhattacharya, S. Bhowmik, R. Benedictus, S. Chowdhury, J. Appl. Phys. **103**, 094314 (2008)
3. K.P. Ghatak, S. Bhattacharya, J. Appl. Phys. **102**, 073704 (2007)
4. K.P. Ghatak, S. Bhattacharya, S.K. Biswas, A. De, A.K. Dasgupta, Phys. Scr. **75**, 820 (2007)
5. K. Seeger, *Semiconductor Physics*, 6th edn. (Springer, Berlin, 1997)
6. B.R. Nag, *Electron Transport in Compound Semiconductors* (Springer, Berlin, 1980)
7. E.O. Kane, in *Semiconductors and Semimetals*, vol. 1, ed. by R.K. Willardson, A.C. Beer (Academic Press, New York, 1966), p. 75
8. B.R. Nag, *Physics of Quantum Well Devices* (Kluwer Academic, Dodrecht, 2000)
9. R.K. Pathria, *Statistical Mechanics*, 2nd edn. (Butterworth-Heinmann, Oxford, 1996)
10. M. Abramowitz, I.A. Stegun, *Handbook of Mathematical Functions* (Dover, New York, 1965)
11. J.N. Schulman, Y.C. Chang, Phys. Rev. B **24**, 4445 (1981)
12. N. Miura, *Physics of Semiconductors in High Magnetic Fields*, Series on Semiconductor Science and Technology (Oxford University Press, Oxford, 2007)
13. K.H.J. Buschow, F.R. de Boer, *Physics of Magnetism and Magnetic Materials* (Springer, New York, 2003)
14. D. Sellmyer, R. Skomski (eds.), *Advanced Magnetic Nanostructures* (Springer, New York, 2005)
15. J.A.C. Bland, B. Heinrich (eds.), *Ultrathin Magnetic Structures III: Fundamentals of Nanomagnetism* (Pt. 3) (Springer, Berlin, 2005)
16. B.K. Ridley, *Quantum Processes in Semiconductors* (Oxford Publications, Oxford, 1999)
17. J.H. Davies, *Physics of Low Dimensional Semiconductors* (Cambridge University Press, Cambridge, 1998)
18. S. Blundell, *Magnetism in Condensed Matter*, Oxford Master Series in Condensed Matter Physics (Oxford University Press, Oxford, 2001)
19. C. Weisbuch, B. Vinter, *Quantum Semiconductor Structures: Fundamentals and Applications* (Academic Publishers, New York, 1991)
20. D. Ferry, *Semiconductor Transport* (CRC Press, Boca Raton, 2000)
21. M. Reed (ed.), *Semiconductors and Semimetals: Nanostructured Systems* (Academic Press, New York, 1992)
22. T. Dittrich, *Quantum Transport and Dissipation* (Wiley-VCH Verlag GmbH, Weinheim, 1998)
23. A.Y. Shik, *Quantum Wells: Physics and Electronics of Two dimensional Systems* (World Scientific, Hackensack, 1997)
24. L.V. Keldysh, Sov. Phys. Solid State **4**, 1658 (1962)
25. L. Esaki, R. Tsu, IBM J. Res. Develop. **14**, 61 (1970)
26. G. Bastard, *Wave Mechanics Applied to Heterostructures* (Editions de Physique, Les Ulis, 1990)
27. E.L. Ivchenko, G. Pikus, *Superlattices and Other Heterostructures* (Springer, Berlin, 1995)
28. R. Tsu, *Superlattices to Nanoelectronics* (Elsevier, Amsterdam, 2005)
29. P. Fürjes, C. Dücs, M. Ádám, J. Zettner, I. Bársony, Superlattices Microstruct. **35**, 455 (2004)
30. T. Borca-Tasciuc, D. Achimov, W.L. Liu, G. Chen, H.-W. Ren, C.-H. Lin, S.S. Pei, Microscale Thermophys. Eng. **5**, 225 (2001)

31. B.S. Williams, Nat. Photonics **1**, 517 (2007)
32. A. Kosterev, G. Wysocki, Y. Bakhirkin, S. So, R. Lewicki, F. Tittel, R.F. Curl, Appl. Phys. B **90**, 165 (2008)
33. M.A. Belkin, F. Capasso, F. Xie, A. Belyanin, M. Fischer, A. Wittmann, J. Faist, Appl. Phys. Lett. **92**, 201101 (2008)
34. G.J. Brown, F. Szmulowicz, R. Linville, A. Saxler, K. Mahalingam, C.-H. Lin, C.H. Kuo, W.Y. Hwang, IEEE Photonics Technol. Lett. **12**, 684 (2000)
35. H.J. Haugan, G.J. Brown, L. Grazulis, K. Mahalingam, D.H. Tomich, Phys. E Low-Dimens. Syst. Nanostruct. **20**, 527 (2004)
36. S.A. Nikishin, V.V. Kuryatkov, A. Chandolu, B.A. Borisov, G.D. Kipshidze, I. Ahmad, M. Holtz, H. Temkin, Jpn. J. Appl. Phys. **42**, L1362 (2003)
37. Y.-K. Su, H.-C. Wang, C.-L. Lin, W.-B. Chen, S.-M. Chen, Jpn. J. Appl. Phys. **42**, L751 (2003)
38. C.H. Liu, Y.K. Su, L.W. Wu, S.J. Chang, R.W. Chuang, Semicond. Sci. Technol. **18**, 545 (2003)
39. S.-B. Che, I. Nomura, A. Kikuchi, K. Shimomura, K. Kishino, Phys. Stat. Sol. (b) **229**, 1001 (2002)
40. C.P. Endes, F. Lewen, T.F. Giesen, S. Schleemer, D.G. Paveliev, Y.I. Koschurinov, V.M. Ustinov, A.E. Zhucov, Rev. Sci. Instrum. **78**, 043106 (2007)
41. F. Klappenberger, K.F. Renk, P. Renk, B. Rieder, Y.I. Koshurinov, D.G. Pavelev, V. Ustinov, A. Zhukov, N. Maleev, A. Vasilyev, Appl. Phys. Lett. **84**, 3924 (2004)
42. X. Jin, Y. Maeda, T. Saka, M. Tanioku, S. Fuchi, T. Ujihara, Y. Takeda, N. Yamamoto, Y. Nakagawa, A. Mano, S. Okumi, M. Yamamoto, T. Nakanishi, H. Horinaka, T. Kato, T. Yasue, T. Koshikawa, J. Cryst. Growth **310**, 5039 (2008)
43. X. Jin, N. Yamamoto, Y. Nakagawa, A. Mano, T. Kato, M. Tanioku, T. Ujihara, Y. Takeda, S. Okumi, M. Yamamoto, T. Nakanishi, T. Saka, H. Horinaka, T. Kato, T. Yasue, T. Koshikawa, Appl. Phys. Express **1**, 045002 (2008)
44. B.H. Lee, K.H. Lee, S. Im, M.M. Sung, Org. Electron. **9**, 1146 (2008)
45. P.-H. Wu, Y.-K. Su, I.-L. Chen, C.-H. Chiou, J.-T. Hsu, W.-R. Chen, Jpn. J. Appl. Phys. **45**, L647 (2006)
46. A.C. Varonides, Renew. Energy **33**, 273 (2008)
47. M. Walther, G. Weimann, Phys. Stat. Sol. (b) **203**, 3545 (2006)
48. R. Rehm, M. Walther, J. Schmitz, J. Fleißner, F. Fuchs, J. Ziegler, W. Cabanski, Opto-Electron. Rev. **14**, 19 (2006)
49. R. Rehm, M. Walther, J. Scmitz, J. Fleissner, J. Ziegler, W. Cabanski, R. Breiter, Electron. Lett. **42**, 577 (2006)
50. G.J. Brown, F. Szmulowicz, H. Haugan, K. Mahalingam, S. Houston, Microelectron. J. **36**, 256 (2005)
51. K.V. Vaidyanathan, R.A. Jullens, C.L. Anderson, H.L. Dunlap, Solid State Electron. **26**, 717 (1983)
52. B.A. Wilson, IEEE. J. Quantum Electron. **24**, 1763 (1988)
53. M. Krichbaum, P. Kocevar, H. Pascher, G. Bauer, IEEE. J. Quantum Electron. **24**, 717 (1988)
54. J.N. Schulman, T.C. McGill, Appl. Phys. Lett. **34**, 663 (1979)
55. H. Kinoshita, T. Sakashita, H. Fajiyasu, J. Appl. Phys. **52**, 2869 (1981)
56. L. Ghenin, R.G. Mani, J.R. Anderson, J.T. Cheung, Phys. Rev. B **39**, 1419 (1989)
57. C.A. Hoffman, J.R. Mayer, F.J. Bartoli, J.W. Han, J.W. Cook, J.F. Schetzina, J.M. Schubman, Phys. Rev. B. **39**, 5208 (1989)
58. H. Sasaki, Phys. Rev. B **30**, 7016 (1984)
59. V.A. Yakovlev, Sov. Phys. Semicond. **13**, 692 (1979)
60. H.X. Jiang, J.Y. Lin, J. Appl. Phys. **61**, 624 (1987)
61. G.M.T. Foley, P.N. Langenberg, Phys. Rev. B **15B**, 4850 (1977)

Part III
Influence of Intense Electric Field on the EEM in Optoelectronic Semiconductors

Chapter 7
The EEM in the Presence of Intense Electric Field

7.1 Introduction

With the advent of modern nanodevices, there has been considerable interest in studying the electric field-induced processes in semiconductors having different band structures. It appears from the literature that the studies have been made on the assumption that the carrier dispersion laws are invariant quantities in the presence of intense electric field, which is not fundamentally true. In this chapter, we shall study the EEM in quantum confined optoelectronic semiconductors under strong electric field. In Sect. 7.2.1, the EEM in the bulk specimens said compounds under strong electric field has been investigated. In Sect. 7.2.2, the EEM in the presence of an arbitrarily oriented quantizing magnetic field whose unperturbed electron energy spectra are, respectively, defined by the three- and two-band models of Kane together with parabolic energy bands has been studied. In Sects. 7.2.3, 7.2.4 and 7.2.5, the EEM in quantum wells, inversion layers, and nipi structures of optoelectronic materials under strong electric field has been explored. Section 7.2.6 contains the investigation of the EEM in quantum wires of optoelectronic semiconductors. In Sect. 7.2.7, the EEM in field effective mass superlattices of optoelectronic semiconductors in the presence of strong electric field under magnetic quantization has been studied. In Sect. 7.2.7 we have investigated the EEM in quantum wire effective mass superlattices of Kane-type semiconductors. In Sect. 7.2.8 the EEM in superlattices of Kane- type compounds with graded interfaces under magnetic quantization has been investigated. In Sect. 7.2.9 the EEM in quantum wire superlattices of optoelectronic semiconductors with graded interfaces has been studied. Section 7.3 contains the results and discussion pertinent to this chapter. Section 7.4 presents open research problems.

S. Bhattacharya and K. P. Ghatak, *Effective Electron Mass in Low-Dimensional Semiconductors*, Springer Series in Materials Science 167,
DOI: 10.1007/978-3-642-31248-9_7, © Springer-Verlag Berlin Heidelberg 2013

7.2 Theoretical Background

7.2.1 The EEM in Bulk Optoelectronic Semiconductors Under Strong Electric Field

The $E-k$ dispersion relation in the presence of an external electric field F_s along x-axis for III-V, ternary and quaternary materials whose unperturbed energy band structures are defined by the well-known three band model of Kane can be expressed as [1]

$$\frac{k_x^2}{\frac{2m_c}{\hbar^2}\left[\frac{I_{11}(E)}{1+\Phi(E,F)}\right]}+\frac{k_y^2}{\frac{2m_c}{\hbar^2}I_{11}(E)}+\frac{k_z^2}{\frac{2m_c}{\hbar^2}I_{11}(E)}=1 \tag{7.1}$$

where,

$$\Phi(E,F)=\left(\frac{m_c}{2m_r^2}\right)\left[F^2\hbar^2E_g^2(E_g-\delta')^2\right]\frac{1}{\eta_1^3(E)}\frac{1}{(\eta_1(E)+\delta')^4},$$
$$\times\left[P\left(\frac{\eta_1(E)+E_g}{\eta_1(E)-E_g'}\right)^{1/2}+Q\left(\frac{\eta_1(E)-E_g}{\eta_1(E)+E_g}\right)^{1/2}\right]^2,$$

$$F=eF_s,\ \delta'=\frac{E_g^2\Delta}{\chi},\quad \chi=6E_g^2+9E_g.\Delta+4\Delta^2,\quad \eta_1(E)=\left[E_g^2+\frac{2m_cE_gI_{11}(E)}{m_r}\right]^{1/2},$$

$$P=\frac{r_0^2}{2}\left(\frac{E_g-\delta'}{E_g+\delta'}\right),\quad r_0=\left\{6.\frac{\left(E_g+\frac{2\Delta}{3}\right).(E_g+\Delta)}{\chi}\right\}^{1/2},$$

$$E_g'=\frac{E_g(E_g-3\delta')}{E_g+\delta'},\quad Q=t^2/2.\ \text{and}\ t=\left\{\frac{6\left(E_g+\frac{2\Delta}{3}\right)}{\chi}\right\}^{1/2}$$

In (7.1), the coefficients of k_x, k_y and k_z are not same and for this reason, this basic equation is "anisotropic" in nature together with the fact that the anisotropic dispersion relation is the ellipsoid of revolution in the k-space.

From (7.1) the expressions of the effective electron masses along x, y, and z directions can, respectively, be written as

$$m_x^*(E,F)=\hbar^2k_x\left.\frac{\partial k_x}{\partial E}\right|_{k_y=0,k_z=0}$$
$$=m_c\left[1+\Phi(E,F)\right]^{-2}\left[\left[1+\Phi(E,F)\right]\left[I_{11}(E)\right]'-\left[I_{11}(E)\right]\Phi'(E,F)\right] \tag{7.2}$$

$$m_y^*(E,F) = \hbar^2 k_y \left.\frac{\partial k_y}{\partial E}\right|_{k_x=0,k_z=0} = m_c\,[I_{11}(E)]' \tag{7.3}$$

$$m_z^*(E,F) = \hbar^2 k_z \left.\frac{\partial k_z}{\partial E}\right|_{k_x=0,k_y=0} = m_c\,[I_{11}(E)]' \tag{7.4}$$

where,

$$\begin{aligned}\Phi'(E,F) = &-\Phi(E,F)(\eta_1(E))'\left[\left(\frac{3}{\eta_1(E)}+\frac{4}{\eta_1(E)+\delta'}\right)\right.\\ &+\left[P\left(\frac{\eta_1(E)+E_g}{\eta_1(E)-(E_g)'}\right)^{1/2}+Q\left(\frac{\eta_1(E)-E_g}{\eta_1(E)+E_g}\right)^{1/2}\right]^{-1}\\ &\left[P\left\{\frac{1}{\sqrt{\eta_1(E)+E_g}\sqrt{\eta_1(E)-(E_g)'}}-\frac{(\eta_1(E)+E_g)^{1/2}}{(\eta_1(E)-(E_g)')^{3/2}}\right\}\right.\\ &\left.\left.+Q\left\{\frac{1}{\sqrt{\eta_1^2(E)-E_g^2}}-\frac{(\eta_1(E)-E_g)^{1/2}}{(\eta_1(E)+E_g)^{3/2}}\right\}\right]\right],\end{aligned}$$

$(\eta_1(E))' = (\eta_1(E))^{-1}\left(\frac{m_c E_g}{m_r}\right)[I_{11}(E)]'$ and $[I_{11}(E)]'$ has already been defined in Chap. 1.

It may be noted from (7.2) that the effective mass along x- direction is a function of both electron energy and electric field, respectively, whereas from (7.3) and (7.4) we can infer the expressions of the effective masses along y and z directions are same and they depend on the electron energy only. Thus, in the presence of an electric field, the mass anisotropy for Kane-type semiconductors depends both on electron energy and electric field, respectively.

Therefore, it appears that the study of the EEM at the Fermi level requires an expression of the electron concentration which inturn needs the expression for the DOS function. The DOS function in this case can be written as

$$N(E) = 4\pi g_v\left(\frac{2m_c}{h^2}\right)^{3/2}\left[\frac{\sqrt{I_{11}(E)}\,[I_{11}(E)]'}{\sqrt{1+\Phi(E,F)}}-\frac{\Phi'(E,F)}{3}\left\{\frac{I_{11}(E)}{1+\Phi(E,F)}\right\}^{3/2}\right] \tag{7.5}$$

Combining (7.5) with the Fermi Dirac occupation probability factor and applying the generalized Sommerfeld's lemma, the electron statistics in this case assumes the form

$$n_0 = C_{7L}\,[M_{1L}(E_{FF},F)+N_{1L}(E_{FF},F)] \tag{7.6a}$$

where

$$C_{7F} = \frac{g_v}{3\pi^2}\left(\frac{2m_c}{\hbar^2}\right)^{3/2}, \quad M_{1F}(E_{FF}, F) = \frac{[I_{11}(E_{FF})]^{3/2}}{\sqrt{1+\Phi(E_{FF}, F)}},$$

E_{FF} is the Fermi energy in the presence of electric field as measured from the edge of the conduction band in the vertically upward direction in the absence of any field,

$$N_{1F}(E_{FL}, F) = \sum_{r=1}^{s} X_{rF}\,[M_{1F}(E_{FF}, F)]$$

$$\text{and } X_{rF} = 2(k_B T)^{2r}(1-2^{1-2r})\zeta(2r)\frac{\partial^{2r}}{\partial E_{FF}^{2r}},$$

Under the condition $\Delta \to 0$, the (7.1) assumes the form

$$\frac{k_x^2}{\frac{2m_c}{\hbar^2}\left[\frac{\gamma_0(E)}{1+\Phi_1(E,F)}\right]} + \frac{k_y^2}{\frac{2m_c}{\hbar^2}\gamma_0(E)} + \frac{k_z^2}{\frac{2m_c}{\hbar^2}\gamma_0(E)} = 1 \tag{7.6b}$$

where $\gamma_0(E) = E(1+\alpha E)$ and $\Phi_1(E, F) = \frac{\hbar^2 F^2}{4m_r E_g^2 \gamma_0(E)}\left[1 + \frac{2m_c}{m_r}\frac{\gamma_0(E)}{E_g}\right]^{-5/2}$.

Equation (7.8) represents the electron energy spectrum of III-V, ternary and quaternary materials in the presence of an external electric field whose unperturbed band structures are defined by the two-band model of Kane.

From (7.5) the expressions of the effective electron masses along x, y, and z directions can, respectively, be written as

$$m_x^*(E_{FF}, F) = m_c\,[1+\Phi_1(E_{FF}, F)]^{-2}\,[[1+\Phi_1(E_{FF}, F)]\,[1+2\alpha E_{FF}] - [E_{FF}(1+\alpha E_{FF})]\,\Phi_1'(E_{FF}, F)] \tag{7.6c}$$

$$m_y^*(E_{FF}, F) = m_c\,[1+2\alpha E_{FF}] \tag{7.7}$$

$$m_z^*(E_{FF}, F) = m_c\,[1+2\alpha E_{FF}] \tag{7.8}$$

where

$$\Phi_1'(E_{FF}, F) = -[(1+2\alpha E_{FF})\Phi_1(E_{FF}, F)]\left[1+5\left[1+\frac{2m_c}{m_r}\frac{E_{FF}(1+\alpha E_{FF})}{E_g}\right]^{-1}\frac{m_c}{m_r}\frac{1}{E_g}\right],$$

The DOS function assumes the form

$$N(E) = 4\pi g_v \left(\frac{2m_c}{h^2}\right)^{3/2} \left[\frac{\sqrt{\gamma_0(E)}\,[1+2\alpha E]}{\sqrt{1+\Phi_1(E,F)}} - \frac{\Phi_1'(E,F)}{3}\left\{\frac{E(1+\alpha E)}{1+\Phi_1(E,F)}\right\}^{3/2}\right] \tag{7.9}$$

Combining (7.9) with the Fermi Dirac occupation probability factor and applying the generalized Sommerfeld's lemma, the electron statistics in this case can be written as

$$n_0 = C_{7L}\,[M_{2L}(E_{FF},F) + N_{2L}(E_{FF},F)] \tag{7.10}$$

where

$$C_{7F} = \frac{g_v}{3\pi^2}\left(\frac{2m_c}{\hbar^2}\right)^{3/2}, \quad M_{2F}(E_{FF},F) = \frac{[E_{FF}(1+\alpha E_{FF})]^{3/2}}{\sqrt{1+\Phi_1(E_{FF},F)}}$$

and

$$N_{2F}(E_{FL},F) = \sum_{r=1}^{s} X_{rF}\,[M_{2F}(E_{FF},F)]$$

7.2.2 The Magneto EEM in Optoelectronic Semiconductors Under Strong Electric Field

The electron dispersion law in the presence of an arbitrarily oriented quantizing magnetic field B which makes an angle Θ with k_x axis and lies in the k_x, k_z plane can be formulated in the following simplified way in this case:

The area of cross-section of the ellipsoid $\frac{x^2}{a^2}+\frac{y^2}{b^2}+\frac{z^2}{c^2}=1$ by the plane $lx+my+nz=p$ is given by [2]

$$A = \frac{\pi abc}{(a^2l^2+b^2m^2+c^2n^2)^{1/2}}\left[1-\frac{p^2}{(a^2l^2+b^2m^2+c^2n^2)}\right] \tag{7.11}$$

In our case, the ellipsoid of the revolution can be written from equation (7.1) as $\frac{k_x^2}{a_7^2(E,F)}+\frac{k_y^2}{b_7^2(E)}+\frac{k_z^2}{b_7^2(E)}=1$ and the equation of the plane is $k_z\sin\theta + k_x \cos\theta = k_z'$

where $a_7^2(E,F) = \frac{2m_c}{\hbar^2}\left(\frac{I_{11}(E)}{1+\Phi(E,F)}\right)$ and $b_7^2(E) = \frac{2m_c}{\hbar^2}I_{11}(E)$,

Therefore, the use of (7.11) leads to the expression for the area of cross-section in this case as

$$A(E,k_{z1}) = \left[\pi a_7(E)b_7^2(E)\right]\left[(b_7^2(E)\sin^2\theta)+(a_7^2(E,F)\cos^2\theta)\right]^{-1/2} \times\left[1-\frac{(k_z')^2}{\left[(b_7^2(E)\sin^2\theta)+(a_7^2(E,F)\cos^2\theta)\right]}\right] \tag{7.12}$$

The Landau area quantization rule is given by [6]

$$A(E, k_{z1}) = \frac{2\pi\,|e|\,B}{\hbar}\left(n + \frac{1}{2}\right) \tag{7.13}$$

Therefore, combining (7.12) and (7.13), the dispersion relation of the conduction electrons in optoelectronic materials under electric field can be written in presence of an arbitrarily oriented quantizing magnetic field B whose unperturbed electron energy spectrum is defined by the three band model of Kane as

$$(k_z')^2 = A_7(E, F, \theta) - \frac{2eB}{\hbar}\left(n + \frac{1}{2}\right) B_7(E, F, \theta) \tag{7.14}$$

where $A_7(E, F, \theta) = \left[a_7^2(E, F)\cos^2\theta + b_7^2(E)\sin^2\theta\right]$
and $B_7(E, F, \theta) = \dfrac{\left[a_7^2(E, F)\cos^2\theta + b_7^2(E)\sin^2\theta\right]^{3/2}}{a_7(E, F) b_7^2(E)}$

Using (7.14) the EEM can be expressed as

$$m^*(E_{FFB}, n, F, \theta) = \tfrac{\hbar^2}{2}\left[A_7'(E_{FFB}, F, \theta) - \tfrac{2eB}{\hbar}\left(n + \tfrac{1}{2}\right) B_7'(E_{FFB}, F, \theta)\right] \tag{7.15}$$

where $A_7'(E_{FFB}, F, \theta) = \Big[2a_7(E_{FFB}, F) a_7'(E_{FFB}, F)\cos^2\theta + 2b_7(E_{FFB})$
$b_7'(E_{FFB})\sin^2\theta\Big]$,

$$a_7'(E_{FFB}, F) = \Big[[a_7(E_{FFB}, F)]^{-1}\left(\frac{m_c}{\hbar^2}\right)[1 + \Phi(E_{FFB}, F)]^{-2} \times \Big[[1 + \Phi(E_{FFB}, F)]\, I_{11}'(E_{FFB}) - \Phi'(E_{FFB}, F) I_{11}(E_{FFB})\Big]\Big]$$

$$\cdot b_7'(E_{FFB}) = \left(\frac{m_c}{\hbar^2}\right)[b_7(E_{FFB})]\, I_{11}'(E_{FF}),$$

$$b_7'(E_{FFB}, F, \theta) = \Big[-\frac{B_7(E_{FFB}, F, \theta)}{a_7(E_{FFB}, F)} a_7'(E_{FFB}, F) - \frac{2B_7(E_{FFB}, F, \theta)}{b_7(E_{FFB})} b_7'(E_{FFB})$$

$$+\frac{3}{2}\left[a_7^2(E_{FFB},F)\cos^2\theta+b_7^2(E_{FFB})\sin^2\theta\right]^{-1}$$
$$\times\left[2a_7(E_{FFB},F)a_7'(E_{FFB},F)\cos^2\theta\right.$$
$$\left.\left.+\sin^2\theta 2b_7(E_{FFB})b_7'(E_{FFB})\right]\right],$$

and E_{FFB} is the Fermi energy under magnetic quantization in this case. The DOS is given by

$$N_B(E)=\frac{eB}{2\pi^2\hbar}\sum_{n=0}^{n_{\max}}\left[\left[A_7(E,F,\theta)-\frac{2eB}{\hbar}\left(n+\frac{1}{2}\right)B_7(E,F,\theta)\right]^{-1/2}\right. \tag{7.16}$$

$$\left.\left[A_7'(E,F,\theta)-\frac{2eB}{\hbar}\left(n+\frac{1}{2}\right)B_7'(E,F,\theta)\right]H(E-E_7)\right]$$

where E_7 is the lowest positive root of the equation

$$A_7(E,F,\theta)-\frac{2eB}{\hbar}\left(n+\frac{1}{2}\right)B_7(E,F,\theta)=0 \tag{7.17}$$

The electron concentration can be written as

$$n_0=\frac{eB}{\pi^2\hbar}\sum_{n=0}^{n_{\max}}\left[M_3(E_{FFB},F,\theta,n)+N_3(E_{FFB},F,\theta,n)\right] \tag{7.18}$$

where

$$M_3(E_{FFB},F,\theta,n)=\left[A_7(E,F,\theta)-\frac{2eB}{\hbar}\left(n+\frac{1}{2}\right)B_7(E,F,\theta)\right]^{\frac{1}{2}}$$

and

$$N_3(E_{FFB},F,\theta,n)=\sum_{r=1}^{s}Z_r\left[M_3(E_{FFB},F,\theta,n)\right]$$

The electron dispersion relation in the presence of an arbitrarily oriented quantizing magnetic field, the EEM , the density-of-states function, and the electron concentration in the presence of strong electric field for the materials where unperturbed condition electron obey the two-band model of Kane can, respectively, be expressed as

$$(k_z')^2=A_{71}(E,F,\theta)-\frac{2eB}{\hbar}\left(n+\frac{1}{2}\right)B_{71}(E,F,\theta) \tag{7.19}$$

$$m^*(E_{FFB}, n, F, \theta) = \frac{\hbar^2}{2}\left[A'_{71}(E_{FFB}, F, \theta) - \frac{2eB}{\hbar}\left(n+\frac{1}{2}\right)B'_{71}(E_{FFB}, F, \theta)\right] \tag{7.20}$$

$$N_B(E) = \frac{eB}{2\pi^2\hbar}\sum_{n=0}^{n_{\max}}\left[\left[A_{71}(E, F, \theta) - \frac{2eB}{\hbar}\left(n+\frac{1}{2}\right)B_{71}(E, F, \theta)\right]^{-1/2}\right. \left.\left[A'_{71}(E, F, \theta) - \frac{2eB}{\hbar}\left(n+\frac{1}{2}\right)B'_{71}(E, F, \theta)\right]H(E - E_8)\right] \tag{7.21}$$

and

$$n_0 = \frac{eB}{\pi^2\hbar}\sum_{n=0}^{n_{\max}}\left[M_4(E_{FFB}, F, \theta, n) + N_4(E_{FFB}, F, \theta, n)\right] \tag{7.22}$$

where

$$A_{71}(E, F, \theta) = [a_{71}^2(E, F)\cos^2\theta + b_{71}^2(E)\sin^2\theta],$$

$$B_{71}(E, F, \theta) = \frac{[a_{71}^2(E, F)\cos^2\theta + b_{71}^2(E)\sin^2\theta]^{3/2}}{a_{71}(E, F)b_{71}^2(E)},$$

$$A'_{71}(E_{FFB}, F, \theta) = \left[2a_{71}(E_{FFB}, F)a'_{71}(E_{FFB}, F)\cos^2\theta + 2b_{71}(E_{FFB})b'_{71}(E_{FFB})\sin^2\theta\right],$$

$$a'_{71}(E_{FFB}, F) = \left[[a_{71}(E_{FFB}, F)]^{-1}\left(\frac{m_c}{\hbar^2}\right)[1 + \Phi_1(E_{FFB}, F)]^{-2} \times [[1 + \Phi_1(E_{FFB}, F)](1 + 2\alpha E_{FFB}) - \Phi_2(E_{FFB}, F)E_{FFB}(1 + \alpha E_{FFB})]\right].$$

$$b'_{71}(E_{FFB}) = \left(\frac{m_c}{\hbar^2}\right)[b_{71}(E_{FFB})](1 + 2\alpha E_{FFB}),$$

$$B'_{71}(E_{FFB}, F, \theta) = \left[-\frac{B_{71}(E_{FFB}, F, \theta)}{a_{71}(E_{FFB}, F)}a'_{71}(E_{FFB}, F) - \frac{2B_{71}(E_{FFB}, F, \theta)}{b_{71}(E_{FFB})}b'_{71}(E_{FFB}) + \frac{3}{2}\left[a_{71}^2(E_{FFB}, F)\cos^2\theta + b_{71}^2(E_{FFB})\sin^2\theta\right]^{-1} \times \left[2a_{71}(E_{FFB}, F)a'_{71}(E_{FFB}, F)\cos^2\theta + 2\sin^2\theta\, b_{71}(E_{FFB})b'_{71}(E_{FFB})\right]\right],$$

E_8 is the lowest positive root of the equation

$$A_{71}(E, F, \theta) - \frac{2eB}{\hbar}\left(n + \frac{1}{2}\right) B_{71}(E, F, \theta) = 0 \tag{7.23}$$

$$M_4(E_{FFB}, F, \theta, n) = \left[A_{71}(E, F, \theta) - \frac{2eB}{\hbar}\left(n + \frac{1}{2}\right) B_{71}(E, F, \theta)\right]^{\frac{1}{2}}$$

and

$$N_4(E_{FFB}, F, \theta, n) = \sum_{r=1}^{s} Z_r\left[M_4(E_{FFB}, F, \theta, n)\right]$$

7.2.3 *The EEM in UFs of Optoelectronic Semiconductors Under Strong Electric Field*

In the presence of quantization along x direction we can write

$$k_s^2 = b_7^2(E)\left[1 - \left(\frac{n_x \pi}{d_x}\right)^2 \frac{1}{a_7^2(E, F)}\right] \tag{7.24}$$

where $k_s^2 = \mathrm{k}_y^2 + \mathrm{k}_z^2$, $\mathrm{n}_x = 1, 2, 3 \ldots\ldots$ is the size quantum number and d_x is the thickness along x direction.
The density-of-states function is given by

$$N(E) = \frac{2g_v}{(2\pi)^2} \sum_{n_x=1}^{n_{x\max}} \left[2b_7(E)b_7'(E)\left[1 - \left(\frac{n_x\pi}{d_x}\right)^2 (a_7(E, F))^{-2}\right.\right.$$
$$\left.\left. + \frac{2b_7^2(E)}{a_7^2(E, F)} a_7'(E, F)\left(\frac{n_x\pi}{d_x}\right)^2\right]\right] H(E - E_9) \tag{7.25}$$

where E_9 is the root of the equation

$$1 = \left(\frac{n_x\pi}{d_x}\right)^2 (a_7(E_9, F))^{-2} \tag{7.26a}$$

The EEM is given by

$$m^*(E_{FFS}, F, n_x) = \frac{\hbar^2}{2}\phi_{100} \tag{7.26b}$$

where E_{FFS} is the Fermi Energy in this case and

$$\phi_{100} = \left[2b_7(E)b_7'(E)\left\{1-\left(\frac{n_x\pi}{d_x}\right)^2 \frac{1}{a_7^2(E_{FFS},F)}\right\} + \frac{2b_7^2(E_{FFS})}{a_7^3(E_{FFS},F)}a_7'(E_{FFS},F)\left(\frac{n_x\pi}{d_x}\right)^2\right]$$

The electron concentration per unit area is given by

$$n_0 = \frac{g_v}{2\pi}\sum_{n_x=1}^{n_{x\max}}\left[M_5(E_{FFS},F,n_x)+N_5(E_{FFS},F,n_x)\right] \tag{7.27}$$

where

$$M_5(E_{FFS},F,n_x) = b_7^2(E_{FFS})\left[1-\left(\frac{n_x\pi}{d_x}\right)^2\frac{1}{a_7^2(E_{FFS},F)}\right]$$

and

$$N_5(E_{FFS},F,n_x) = \sum_{r=1}^{s} Z_r[M_5(E_{FFS},F,n_x)].$$

The electron dispersion relation to the effective electron mass, the density-of-states function and the surface electron concentration in ultrathin films in the presence of strong electron field where unperturbed conduction electrons obey the two-band model of Kane can, respectively, be expressed as

$$k_s^2 = b_{71}^2(E)\left[1-\left(\frac{n_x\pi}{d_x}\right)^2\frac{1}{a_{71}^2(E,F)}\right] \tag{7.28}$$

$$m^*(E_{FS},F,n) = \hbar^2\phi_{101} \tag{7.29}$$

where,

$$\phi_{101} = \left[2b_{71}(E_{FFS})b_{71}'(E_{FFS})\left\{1-\left(\frac{n_x\pi}{d_x}\right)^2\frac{1}{a_{71}^2(E_{FFS},F)}\right\} + \frac{2b_{71}^2(E_{FFS})}{a_{71}^3(E_{FFS},F)}a_{71}'(E_{FFS},F)\left(\frac{n_x\pi}{d_x}\right)^2\right]$$

$$N(E) = \frac{2g_v}{(2\pi)^2}\sum_{n_x=1}^{n_{x\max}}\left[2b_{71}(E_{FFS})b_{71}'(E_{FFS})\left\{1-\left(\frac{n_x\pi}{d_x}\right)^2\frac{1}{a_{71}^2(E_{FFS},F)}\right\} + \frac{2b_{71}^2(E_{FFS})}{a_{71}^3(E_{FFS},F)}a_{71}'(E_{FFS},F)\left(\frac{n_x\pi}{d_x}\right)^2\right]H(E-E_{10}) \tag{7.30}$$

where,

E_{10} is the lowest positive root of the equation

$$1 = \left(\frac{n_x \pi}{d_x}\right)^2 \frac{1}{a_{71}^2(E_{10}, F)} \tag{7.31}$$

$$n_0 = \frac{g_v}{2\pi} \sum_{n_x=1}^{n_{x\max}} [M_6(E_{FFS}, F, n_x) + N_6(E_{FFS}, F, n_x)] \tag{7.32}$$

where

$$M_6(E_{FFS}, F, n_x) = b_{71}^2(E_{FFS}) \left[1 - \left(\frac{n_x \pi}{d_x}\right)^2 \frac{1}{a_{71}^2(E_{FFS}, F)}\right]$$

and

$$N_6(E_{FFS}, F, n_x) = \sum_{r=1}^{s} Z_r [M_6(E_{FFS}, F, n_x)]$$

7.2.4 The EEM in NIPI Structures of Optoelectronic Semiconductors Under Strong Electric Field

The dispersion relation of the conduction electrons in NIPI structures of optoelectronic semiconductors under strong electric field whose unperturbed conduction electrons obey the three-band model of Kane can be written as

$$k_y^2 + k_z^2 = b_7^2(E) \left[1 - \frac{2\hbar a_7'(E)}{a_7(E)} \left(n_i + \frac{1}{2}\right)\right] \tag{7.33}$$

where, $n_i = 1, 2, 3 \ldots$ is the mini-band index in this structure

The density of states function is given by

$$N(E) = \frac{g_v}{\pi} \sum_{n_i=0}^{n_{i\max}} \left[2b_7(E) b_7'(E) \left[1 - \frac{2\hbar a_7'(E)}{a_7(E)} \left(n_i + \frac{1}{2}\right)\right]\right.$$
$$\left. + b_7^2(E) \left(n_i + \frac{1}{2}\right) 2\hbar \left\{\frac{-a_7''(E)}{a_7(E)} + \frac{(a_7'(E))^2}{a_7^2(E)}\right\}\right] H(E - E_{11}) \tag{7.34}$$

where, E_{11} is the lowest positive root of the equation

$$1 = \frac{2\hbar a_7'(E_{11})}{a_7(E_{11})} \left(n_i + \frac{1}{2}\right) \tag{7.35}$$

The EEM is given by

$$m^*(\bar{E}_{Fn}, F, n_i) = \hbar^2 \phi_{102} \tag{7.36}$$

where, $\bar{E}_{Fn}$ is the Fermi energy in this case and

$$\phi_{102} = \left[b_7(\bar{E}_{Fn}) b_7'(\bar{E}_{Fn}) \left[1 - \frac{2\hbar a_7'(\bar{E}_{Fn})}{a_7(\bar{E}_{Fn})} \left(n_i + \frac{1}{2} \right) \right] \right.$$
$$\left. + b_7^2(\bar{E}_{Fn}) \left(n_i + \frac{1}{2} \right) \hbar \left\{ \frac{-a_7''(\bar{E}_{Fn})}{a_7(\bar{E}_{Fn})} + \frac{(a_7'(\bar{E}_{Fn}))^2}{a_7^2(\bar{E}_{Fn})} \right\} \right]$$

The electron concentration can be expressed as

$$n_0 = \frac{g_v}{2\pi} \sum_{n_i=1}^{n_{i\max}} \left[M_7(\bar{E}_{Fn}, F, n_i) + N_7(\bar{E}_{Fn}, F, n_i) \right] \tag{7.37}$$

where

$$M_7(\bar{E}_{Fn}, F, n_i) = b_7^2(\bar{E}_{Fn}) \left[1 - \frac{2\hbar a_7'(\bar{E}_{Fn})}{a_7(\bar{E}_{Fn})} \left(n_i + \frac{1}{2} \right) \right]$$

and

$$N_7(\bar{E}_{Fn}, F, n_i) = \sum_{r=1}^{s} Z_r \left[M_7(\bar{E}_{Fn}, F, n_i) \right]$$

The electron dispersion relation, the effective electron mass, the density-of-states function, and the surface electron concentration in NIPI structures in the presence of strong electron field where unperturbed conduction electrons obey the two-band model of Kane can, respectively, be expressed as

$$k_y^2 + k_z^2 = b_{71}^2(E) \left[1 - \frac{2\hbar a_{71}'(E)}{a_{71}(E)} \left(n_i + \frac{1}{2} \right) \right] \tag{7.38}$$

$$m^*(\bar{E}_{Fn}, F, n_i) = \hbar^2 \phi_{103} \tag{7.39}$$

where,

$$\phi_{103} = \left[b_{71}(\bar{E}_{Fn}) b_{71}'(\bar{E}_{Fn}) \left[1 - \frac{2\hbar a_{71}'(\bar{E}_{Fn})}{a_{71}(\bar{E}_{Fn})} \left(n_i + \frac{1}{2} \right) \right] \right.$$
$$\left. + b_{71}^2(\bar{E}_{Fn}) \left(n_i + \frac{1}{2} \right) \hbar \left\{ \frac{-a_{71}''(\bar{E}_{Fn})}{a_{71}(\bar{E}_{Fn})} + \frac{(a_{71}'(\bar{E}_{Fn}))^2}{a_{71}^2(\bar{E}_{Fn})} \right\} \right]$$

$$N(E) = \frac{g_v}{\pi} \sum_{n_i=0}^{n_{i\max}} \left[2b_{71}(E)b'_{71}(E) \left[1 - \frac{2\hbar a'_{71}(E)}{a_{71}(E)} \left(n_i + \frac{1}{2} \right) \right] \right.$$
$$\left. + b_{71}^2(E) \left(n_i + \frac{1}{2} \right) 2\hbar \left\{ \frac{-a''_{71}(E)}{a_{71}(E)} + \frac{(a'_{71}(E))^2}{a_{71}^2(E)} \right\} \right] H(E - E_{12}) \tag{7.40}$$

E_{12} is the lowest positive root of the equation

$$1 = \frac{2\hbar a'_{71}(E_{12})}{a_{71}(E_{12})} \left(n_i + \frac{1}{2} \right) \tag{7.41}$$

$$n_0 = \frac{g_v}{2\pi} \sum_{n_i=1}^{n_{i\max}} \left[M_8(\bar{E}_{Fn}, F, n_i) + N_8(\bar{E}_{Fn}, F, n_i) \right] \tag{7.42}$$

where

$$M_8(\bar{E}_{Fn}, F, n_i) = b_{71}^2(\bar{E}_{Fn}) \left[1 - \frac{2\hbar a'_{71}(\bar{E}_{Fn})}{a_{71}(\bar{E}_{Fn})} \left(n_i + \frac{1}{2} \right) \right]$$

and

$$N_8(\bar{E}_{Fn}, F, n_i) = \sum_{r=1}^{s} Z_r \left[M_8(\bar{E}_{Fn}, F, n_i) \right]$$

7.2.5 The EEM in n-Channel Inversion Layers of Optoelectronic Semiconductors

(a) The 2D dispersion relation of the conduction electrons in n-channel inversion layers of optoelectronic semiconductors under weak electric field limit whose unperturbed conduction electrons obey the three-band model of Kane can be written as

$$k_s^2 = J_7(E, F, i) \tag{7.43}$$

where

$$J_7(E, F, i) = \left[\frac{t_{17}(E, F) - S_i [eF_s t'_{17}(E, F)]^{2/3}}{t_{27}(E, F) - \frac{2}{3} S_i (eF_s)^{2/3} \frac{t'_{27}(E, F)}{[t'_{17}(E, F)]^{1/3}}} \right], \quad t_{17}(E, F) = a_7^2(E, F),$$

$$t_{27}(E, F) = \frac{t_{17}(E, F)}{b_7^2(E)}, \quad t'_{17}(E, F) = 2a_7(E, F)a'_7(E, F),$$

$$t'_{27}(E,F) = \left[\frac{t'_{17}(E,F)}{b_7^2(E)} - \frac{2t_{17}(E,F)b'_7(E)}{b_7^3(E)}\right]$$

$$a'_7(E,F) = \frac{m_c}{\hbar^2 a_7(E,F)}[1+\phi(E,F)]^{-2}\left[(1+\phi(E,F))I'_{11}(E) - I_{11}(E)\phi'(E,F)\right]$$

The EEM at the Fermi level (E_{Fs}) in the present case is given by

$$m^*(E_{Fs},F,i) = \frac{\hbar^2}{2} J'_7(E,F,i)\big|_{E=E_{Fs}} \tag{7.44}$$

where

$$J'_7(E,F,i) = [\alpha_7(E,F,i)\beta'_7(E,F,i) - \alpha'_7(E,F,i)\beta_7(E,F,i)]\{\alpha_7(E,F,i)\}^{-2},$$

$$\alpha_7(E,F,i) = [t_{17}(E,F) - S_i[eF_s t'_{17}(E,F)]^{2/3}],$$

$$\beta_7(E,F,i) = \left[t_{27}(E,F) - \frac{2}{3}S_i(eF_s)^{2/3}\frac{t'_{27}(E,F)}{[t'_{17}(E,F)]^{2/3}}\right],$$

$$\alpha'_7(E,F,i) = [t'_{17}(E,F) - S_i(eF_s)^{2/3}\frac{2}{3}[t'_{17}(E,F)]^{-1/3}[t''_{17}(E,F)]],$$

$$t''_{17}(E,F) = [2[a'_7(E,F)]^2 + 2a_7(E,F)a''_7(E,F)],$$

$$a'_7(E,F) = \frac{m_c}{\hbar^2 a_7(E,F)}[1+\phi(E,F)]^{-2}\left[(1+\phi(E,F))I'_{11}(E) - I_{11}(E)\phi'(E,F)\right],$$

$$\begin{aligned} a''_7(E,F) = {} & \frac{-a'_7(E,F)}{a_7(E,F)} - 2[1+\phi(E,F)]^{-1}\phi'(E,F)a'_7(E,F) \\ & + \frac{m_c}{\hbar^2 a_7(E,F)}[1+\phi(E,F)]^{-2}\left[I''_{11}(E) - I_{11}(E)\phi''(E,F)\right], \end{aligned}$$

$$\begin{aligned} I''_{11}(E) = {} & \left[\frac{(I'_{11}(E))^2}{I_{11}(E)} - 2I_{11}(E)\left[\frac{1}{E^2} + \frac{1}{(E+E_g)^2} + \frac{1}{(E+E_g+\Delta)^2}\right.\right. \\ & \left.\left. - \frac{1}{(E+E_g+\frac{2}{3}\Delta)^2}\right]\right], \end{aligned}$$

$$\begin{aligned} \beta'_7(E,F,i) = {} & t'_{27}(E,F) - \left\{\frac{2}{3}S_i(eF_s)^{2/3}\frac{t''_{27}(E,F)}{[t'_{17}(E,F)]^{1/3}}\right\} \\ & + \frac{2}{3}S_i(eF_s)^{2/3}\frac{\{t'_{27}(E,F)t''_{17}(E,F)\}}{3[t'_{17}(E,F)]^{4/3}}, \end{aligned}$$

$$\begin{aligned} t''_{27}(E,F) = {} & \left[\frac{t''_{17}(E,F)}{b_7^2(E)} - 4\frac{t'_{17}(E,F)b'_7(E)}{b_7^3(E)} + 6\frac{t_{17}(E,F)[b'_7(E)]^2}{b_7^4(E)}\right. \\ & \left. - \frac{2t_{17}(E,F)b''_7(E)}{b_7^3(E)}\right], \end{aligned}$$

$$b'_7(E) = \frac{m_c}{\hbar^2}\frac{I'_{11}(E)}{b_7(E)}, \quad b''_7(E) = \frac{1}{b_7(E)}\left[\frac{m_c}{\hbar^2}I'_{11}(E) - (b_7(E))^2\right]$$

$$\Phi''(E,F) = \left[\frac{(\Phi'(E,F))^2}{\Phi(E,F)} + \frac{\eta_1''(E)\Phi'(E,F)}{\eta_1'(E)} - \Phi(E,F)(\eta_1(E))'\right.$$

$$\times\left[\left(\frac{-3}{\eta_1(E))^2} + \frac{-4}{(\eta_1(E)+\delta')^2}\right) - \left[P\left(\frac{\eta_1(E)+E_g}{\eta_1(E)-(E_g)'}\right)^{1/2}\right.\right.$$

$$\left.\left.+Q\left(\frac{\eta_1(E)-E_g}{\eta_1(E)+E_g}\right)^{1/2}\right]^{-2} + \frac{\eta_1'(E)}{2}\right.$$

$$\times\left[P\left\{\frac{1}{\sqrt{\eta_1(E)+E_g}\sqrt{\eta_1(E)-(E_g)'}} - \frac{(\eta_1(E)+E_g)^{1/2}}{(\eta_1(E)-(E_g)')^{3/2}}\right\}\right.$$

$$\left.\left.+Q\left\{\frac{1}{\sqrt{\eta_1^2(E)-E_g^2}} - \frac{(\eta_1(E)-E_g)^{1/2}}{(\eta_1(E)+E_g)^3/2}\right\}\right]\right] + \left[-\Phi(E,F)(\eta_1(E))'\right.$$

$$\times\left[\left(\frac{3}{\eta_1(E)} + \frac{4}{\eta_1(E)+\delta'}\right) + \left[P\left(\frac{\eta_1(E)+E_g}{\eta_1(E)-(E_g)'}\right)^{1/2}\right.\right.$$

$$\left.+Q\left(\frac{\eta_1(E)-E_g}{\eta_1(E)+E_g}\right)^{1/2}\right]^{-1} \frac{\eta_1'(E)}{2}\left[P\left\{\frac{-1}{(\eta_1(E)+E_g)^{3/2}(\eta_1(E)-E_g')^{1/2}}\right.\right.$$

$$-\frac{1}{(\eta_1(E)-E_g)^{3/2}(\eta_1(E)+E_g')^{1/2}}$$

$$\left.-\frac{(\eta_1(E)+E_g)^{-1/2}}{(\eta_1(E)-E_g)^{3/2}} + 3(\eta_1(E)-E_g')(\eta_1(E)+E_g)^{1/2}\right\}$$

$$-Q\left\{\frac{2\eta_1(E)}{[(\eta_1(E))^2+E_g^2]^{3/2}} + \frac{(\eta_1(E)-E_g)^{-1/2}}{(\eta_1(E)+E_g)^{3/2}}\right.$$

$$\left.\left.\left.-3\frac{(\eta_1(E)-E_g)^{1/2}}{(\eta_1(E)+E_g)^{5/2}}\right\}\right]\right]$$

$$\eta_1''(E) = (\eta_1(E))^{-1}\left[\frac{m_c E_g}{m_r} I_{11}'(E) - (\eta_1'(E))^2\right].$$

The electron concentration is given by

$$n_0 = \frac{g_v}{2\pi}\sum_{i=0}^{i_{\max}}[J_7(E_{\mathrm{Fs}},F,i) + H_7(E_{\mathrm{Fs}},F,i)] \tag{7.45}$$

where $H_7(E_{\mathrm{Fs}},F,i) = \sum_{r=1}^{s} Z_r[J_7(E_{\mathrm{Fs}},F,i)]$

(b) The 2D dispersion relation of the conduction electrons in n-channel inversion layers of optoelectronic semiconductors under weak electric field limit whose unperturbed conduction electrons obey the two-band model of Kane can be written as

$$k_s^2 = J_{71}(E,F,i) \tag{7.46}$$

where

$$J_{71}(E,F,i)=\left[\frac{t_{171}(E,F)-S_i[eF_s t'_{171}(E,F)]^{2/3}}{t_{271}(E,F)-\frac{2}{3}S_i(eF_s)^{2/3}\frac{t'_{271}(E,F)}{[t'_{171}(E,F)]^{1/3}}}\right],$$

$$t_{171}(E,F)=a_{71}^2(E,F),\quad t_{271}(E,F)=\frac{t_{171}(E,F)}{b_{71}^2(E)},$$

$$t'_{171}(E,F)=2a_{71}(E,F)a'_{71}(E,F),$$

$$a'_{71}(E,F)=[1+\phi_1(E,F)]^{-1}\left[-a_{71}(E,F)\phi_2(E,F)+\left(\frac{2m_c}{\hbar^2}\right)(1+2\alpha E)\right]$$

$$t'_{271}(E,F)=\left[\frac{t'_{171}(E,F)}{b_{71}^2(E)}-\frac{2t_{171}(E,F)b'_{71}(E)}{b_{71}^3(E)}\right],$$

$$b'_{71}(E)=\left(\frac{m_c}{\hbar^2}\right)(1+2\alpha E)[b_{71}(E)]^{-1}$$

The EEM at the Fermi level (E_{Fs}) in the present case is given by

$$m^*(E_{\mathrm{Fs}},F,i)=\frac{\hbar^2}{2}J'_{71}(E,F,i)\big|_{E=E_{\mathrm{Fs}}} \tag{7.47}$$

where

$$J'_{71}(E,F,i)=[\alpha_{71}(E,F,i)\beta'_{71}(E,F,i)-\alpha'_{71}(E,F,i)\beta_{71}(E,F,i)]\{\alpha_{71}(E,F,i)\}^{-2},$$

$$\alpha_{71}(E,F,i)=[t_{171}(E,F)-S_i[eF_s t'_{171}(E,F)]^{2/3}],$$

$$\beta_{71}(E,F,i)=[t_{271}(E,F)-\frac{2}{3}S_i(eF_s)^{2/3}\frac{t'_{271}(E,F)}{[t'_{171}(E,F)]^{1/3}}],$$

$$\alpha'_{71}(E,F,i)=[t'_{171}(E,F)-S_i(eF_s)^{2/3}\frac{2}{3}[t'_{171}(E,F)]^{-1/3}[t''_{171}(E,F)]],$$

$$\beta'_{71}(E,F,i)=t'_{271}(E,F)-\left\{\frac{2}{3}S_i(eF_s)^{2/3}\frac{t''_{271}(E,F)}{[t'_{171}(E,F)]^{1/3}}\right\}+\frac{2}{9}S_i(eF_s)^{2/3}\frac{\{t'_{271}(E,F)t''_{171}(E,F)\}}{[t'_{171}(E,F)]^{4/3}}$$

$$t''_{271}(E,F)=\left[\frac{t''_{171}(E,F)}{b_{71}^2(E)}-4\frac{t'_{171}(E,F)b'_{71}(E)}{b_{71}^3(E)}+6\frac{t_{171}(E,F)[b'_{71}(E)]^2}{b_{71}^4(E)}-\frac{2t_{171}(E,F)b''_{71}(E)}{b_{71}^3(E)}\right]$$

$$t''_{171}(E,F)=[2[a'_{71}(E,F)]^2+2a_{71}(E,F)a''_{71}(E,F)]$$

$$b''_{71}(E) = [b_{71}(E)]^{-1}\left[(b'_{71}(E))^2 - \frac{2m_c\alpha}{\hbar^2}\right]$$

$$a''_{71}(E,F) = [1+\phi_1(E,F)]^{-1}\Bigg[-a'_{71}(E,F)\phi_2(E,F) + \frac{4m_c\alpha}{\hbar^2} - a_{71}(E,F)\phi_3(E,F)\Bigg],$$

$$\begin{aligned}\phi_3(E,F) = {} & \Bigg[\Bigg[\frac{-\phi_2(E,F)(1+2\alpha E)}{\gamma_7(E)} + \frac{\phi_1(E,F)(1+2\alpha E)^2}{\gamma_7^2(E)} - \frac{2\alpha\phi_1(E,F)}{\gamma_7(E)}\Bigg] \\ & \times \left[1+\frac{5}{2}\left\{1+\left(\frac{2m_c}{m_r E_g}\right)E(1+\alpha E)\right\}^{-1}\left(\frac{2m_c}{m_r E_g}\right)\gamma_7(E)\right] \\ & \times \left[\frac{-\phi_1(E,F)(1+2\alpha E)}{\gamma_7(E)}\right]\left[-\frac{5}{2}\left\{1+\left(\frac{2m_c}{m_r E_g}\right)E(1+\alpha E)\right\}^{-2}\right. \\ & \times \left(\frac{2m_c}{m_r E_g}\right)^2(1+2\alpha E)\gamma_7(E) + \left(\left(\frac{5m_c}{m_r E_g}\right)(1+2\alpha E)\right) \\ & \times \left.\left\{1+\left(\frac{2m_c}{m_r E_g}\right)E(1+\alpha E)\right\}^{-1}\right]\Bigg]\end{aligned}$$

The electron concentration per unit area in this case is given by

$$n_0 = \frac{g_v}{2\pi}\sum_{i=0}^{i_{\max}}[J_{71}(E_{\mathrm{Fs}},F,i) + H_{71}(E_{\mathrm{Fs}},F,i)] \tag{7.48}$$

where

$$H_{71}(E_{\mathrm{Fs}},F,i) = \sum_{r=1}^{s} Z_r[J_{71}(E_{\mathrm{Fs}},F,i)]$$

7.2.6 The EEM in Nano Wires of Optoelectronic Semiconductors

The one-dimensional motion of the electron for quantum wires of optoelectronic materials can be expressed as

$$G(n_y,n_z) + \frac{\hbar^2 k_x^2}{2m_c} = \beta_{11}(E,F) \tag{7.49}$$

$$G(n_y,n_z) + \frac{\hbar^2 k_x^2}{2m_c} = \overline{\beta_{12}}(E,F) \tag{7.50}$$

where

$$G(n_y, n_z) = \frac{\hbar^2\pi^2}{2m_c}\left[\left(\frac{n_y}{d_y}\right)^2 + \left(\frac{n_z}{d_z}\right)^2\right]$$

$$\beta_{11}(E, F) = \left[I_{11}(E) - \overline{C_1}.\left[\frac{I_{11}(E)}{\phi^3(E)}\right]\left[\frac{T_1^2(E)}{(\phi(E)+\delta')^4}\right]\right],$$

$$\overline{C_1} = \left[\frac{m_c(\hbar e F E_g)^2(E_g - \delta')^2}{6m_r^2}\right],$$

$$\phi^2(E) = \left[E_g^2 + E_g\frac{m_c}{m_r}I_{11}(E)\right],$$

$$T_1(E) = \left[P\left(\frac{\phi(E)+E_g}{\phi(E)-E_g'}\right)^{1/2} + Q\left(\frac{\phi(E)-E_g}{\phi(E)+E_g}\right)^{1/2}\right],$$

$$\overline{\beta_{12}}(E, F) = \left[E(1+\alpha E) - \delta_5\left[E(1+\alpha E) + \frac{m_r E_g}{2m_c}\right]^{-\frac{5}{2}}\right]$$

$$\delta_5 = \left[\frac{\hbar^2 F^2 m_r^{3/2} E_g^{1/2}}{12.(2m_c)^{5/2}}\right]$$

The EEMs in this case can be expressed as

$$m^*(E_{F1D}, F) = m_c\beta_{11}'(E_{F1D}, F) \tag{7.51}$$

$$m^*(E_{F1D}, F) = m_c\bar{\beta}_{12}'(E_{F1D}, F) \tag{7.52}$$

$$\beta_{11}'(E_{F1D}, F) = \left[\gamma'(E_{F1D}) - \frac{\bar{C}_1\gamma'(E_{F1D})}{\phi^3(E_{F1D})}\cdot\frac{T_1^2(E_{F1D})}{[\phi(E_{F1D})+\delta']^4} + \frac{2\bar{C}_1\gamma(E_{F1D})T_1(E_{F1D})T_1'(E_{F1D})}{\phi^3(E_{F1D})[\phi(E_{F1D})+\delta']^4} - \frac{4\bar{C}_1\gamma(E_{F1D})T_1^2(E_{F1D})\phi'(E_{F1D})}{\phi^3(E_{F1D})[\phi(E_{F1D})+\delta']^5} + \frac{4\bar{C}_1\gamma(E_{F1D})T_1^2(E_{F1D})\phi'(E_{F1D})}{\phi^3(E_{F1D})[\phi(E_{F1D})+\delta']^5}\right],$$

$$\gamma'(E_{F1D}) = \gamma(E_{F1D})\left[\frac{1}{E_{F1D}} + \frac{1}{E_{F1D}+E_g} + \frac{1}{E_{F1D}+E_g+\Delta} - \frac{1}{E_{F1D}+E_g+\frac{2}{3}\Delta}\right],$$

$$\phi'(E_{F1D}) = \left[\frac{E_g m_c\gamma'(E_{F1D})}{2m_r\phi(E_{F1D})}\right], \quad \phi'(E_{F1D}) = \left(\frac{\phi'(E_{F1D})}{2}\right)$$

$$\times \frac{2QE_g[\phi(E_{F1D}) - E_g]^{-1/2}}{[\phi(E_{F1D}) + E_g]^{3/2}}$$

$$- \frac{(E'_g + E_g)P[\phi(E_{F1D}) + E_g]^{-1/2}}{[\phi(E_{F1D}) - E'_g]^{3/2}} \text{ and } [\bar{\beta}_{12}(E_{F1D}, F)]'$$

$$= (1 + 2\alpha(E_{F1D}) \times \left[1 + \frac{5}{2}\delta_5\left[E_{F1D}(1 + \alpha E_{F1D}) + \frac{m_r E_g}{2m_c}\right]^{\frac{-7}{2}}\right]$$

The electron concentration per unit length assumes the forms

$$n_0 = \frac{2g_v}{\pi}\sum_{n_z=1}^{n_{z\max}}\sum_{n_y=1}^{n_{y\max}}[Q_{15}(E_{F1D}, F, n_y, n_z) + Q_{16}(E_{F1D}, F, n_y, n_z)] \quad (7.53)$$

$$n_0 = \frac{2g_v}{\pi}\sum_{n_z=1}^{n_{z\max}}\sum_{n_y=1}^{n_{y\max}}[Q_{17}(E_{F1D}, F, n_y, n_z) + Q_{18}(E_{F1D}, F, n_y, n_z)] \quad (7.54)$$

where

$$Q_{15}(E_{F1D}, F, n_y, n_z) = \sqrt{\omega_{15}(E_{F1D}, F, n_y, n_z)}, \quad \omega_{15}(E_{F1D}, F, n_y, n_z)$$
$$= \frac{2m_c}{\hbar^2}[\beta_{11}(E_{F1D}, F) - G(n_y, n_z)],$$

$$Q_{16}(E_{F1D}, F, n_y, n_z) = \sum_{R=1}^{R_0} Z(R_{1D})[Q_{15}(E_{F1D}, F, n_y, n_z)], \; Z(R_{1D})$$
$$= 2(k_B T)^{2R}(1 - 2^{1-2R})\xi(2R)\frac{\partial^{2R}}{\partial E_{F1D}^{2R}},$$

E_{F1D} is the Fermi energy for one-dimensional system in the present case as measured from the edge of the conduction band in vertically upward direction in absence of any quantization,

$$Q_{17}(E_{F1D}, F, n_y, n_z) = \sqrt{\omega_{16}(E_{F1D}, F, n_y, n_z)},$$

$$\omega_{16}(E_{F1D}, F, n_y, n_z) = \frac{2m_c}{\hbar^2}[\bar{\beta}_{12}(E_{F1D}, F) - G(n_y, n_z)],$$

$$Q_{18}(E_{F1D}, F, n_y, n_z) = \sum_{R=1}^{R_0} Z(R_{1D})[Q_{17}(E_{F1D}, F, n_y, n_z)].$$

7.2.7 The EEM in Effective Mass Superlattices of Optoelectronic Semiconductors Under Magnetic Quantization

The dispersion relation of the conduction electrons in effective mass superlattices of optoelectronic semiconductors can be expressed following Sasaki [3] as

$$\begin{aligned} & a_1 \times \cos[c_1(E,F,E_{g1},\Delta_1)a_0 + c_2(E,F,E_{g2},\Delta_2)b_0] \\ & \quad - a_2 \times \cos[c_1(E,F,E_{g1},\Delta_1)a_0 - c_2(E,F,E_{g2},\Delta_2)b_0] = \cos(L_0 k) \quad (7.55) \end{aligned}$$

where

$$a_1 = \left[\left[1+\sqrt{\frac{m_{c2}}{m_{c1}}}\right]^2 \times \left[4\left(\sqrt{\frac{m_{c2}}{m_{c1}}}\right)^{1/2}\right]^{-1}\right],$$

$$a_2 = \left[\left[-1+\sqrt{\frac{m_{c2}}{m_{c1}}}\right]^2 \times \left[4\left(\sqrt{\frac{m_{c2}}{m_{c1}}}\right)^{1/2}\right]^{-1}\right],$$

$$c_i^2(E,F,E_{gi},\Delta_i) = \frac{2m_{ci}}{\hbar^2}[\beta_{1i}(E,F,E_{gi},\Delta_i) - k_\perp^2], \quad i=1,2,$$

$$k_\perp^2 = k_y^2 + k_z^2, \quad \beta_{1i}(E,F,E_{gi},\Delta_i) = \left[I_{11}(E,E_{gi},\Delta_i) - \left[\frac{L(E_{gi},\Delta_i,m_{ri})I_{11}(E,E_{gi},\Delta_i)T_i^2(E,E_{gi},\Delta_i)}{\phi_i^3(E,E_{gi},\Delta_i)[\phi_i(E,E_{gi},\Delta_i)+\delta_i']^4}\right]\right],$$

$$I_{11}(E,E_{gi},\Delta_i) = \frac{E(E+E_{gi})(E+E_{gi}+\Delta_i)(E_{gi}+\frac{2}{3}\Delta_i)}{E_{gi}(E_{gi}+\Delta_i)(E+E_{gi}+\frac{2}{3}\Delta_i)},$$

$$L(E_{gi},\Delta_i,m_{ri}) = \frac{(\hbar eF)^2(E_{gi}-\delta_i')^2 m_{ci}}{6m_{ri}^2}, \quad \delta_i' = \frac{(E_{gi})^2\Delta_i}{\chi_i},$$

$$\chi_i = [6(E_{gi})^2 + 9E_{gi}.\Delta_i + 4\Delta_i^2], \quad \frac{1}{m_{ri}} = \left(\frac{1}{m_{ci}} + \frac{1}{m_{vi}}\right),$$

$$T_i(E,E_{gi},\Delta_i) = \left[P_i\left[\frac{\{\phi_i(E,E_{gi},\Delta_i)+E_{gi}\}}{\{\phi_i(E,E_{gi},\Delta_i)-E'_{gi}\}}\right]^{1/2} + Q_i\left[\frac{\{\phi_i(E,E_{gi},\Delta_i)-E_{gi}\}}{\{\phi_i(E,E_{gi},\Delta_i)+E_{gi}\}}\right]^{1/2}\right]$$

$$P_i = \frac{r_{0i}^2}{2}\left(\frac{E_{gi}-\delta_i'}{E_{gi}+\delta_i'}\right), \quad r_{oi}^2 = \left[6\left(E_{gi}+\frac{2}{3}\Delta_i\right)(E_{gi}+\Delta_i)\right][\chi_i]^{-1}$$

$$\phi_i^2(E,E_{gi},\Delta_i) = [E_{gi}^2 + E_{gi}(m_{ci}/m_{ri})I_{11}(E,E_{gi},\Delta_i)],$$

$$E'_{gi} = \left[\frac{E_{gi}(E_{gi}-3\delta_i')}{(E_{gi}+\delta_i')}\right], \quad Q_i = \frac{t_i^2}{2} \text{ and } t_i^2 = \left[\frac{6(E_{g_i}+\frac{2}{3}\Delta_i)}{\chi_i}\right].$$

In the presence of a quantizing magnetic field B, along z-direction the magneto-energy spectrum assumes the form

$$k_z^2 = \omega_{17}(E, F, n) \tag{7.56}$$

where

$$\omega_{17}(E, F, n) = \frac{1}{L_o^2}[\cos^{-1}\{f_1(E, F, n)\}]^2 - \frac{2eB}{\hbar}\left(n + \frac{1}{2}\right),$$

$$f_1(E, F, n) = [a_1 \cos[c_1(E, F, E_{g_1}, \Delta_1, n)a_o + b_o c_2(E, F, E_{g_2}, \Delta_2, n)] \\ - a_2 \cos[c_1(E, F, E_{g_1}, \Delta_1, n)a_o - b_o c_2(E, F, E_{g_2}, \Delta_2, n)]]$$

and

$$c_i^2(E, F, E_{g_i}, \Delta_i, n) = \left[\left(\frac{2m_{c_i}}{\hbar^2}\right)[\beta_{1_i}(E, F, E_{g_i}, \Delta_i)] - \frac{2eB}{\hbar}\left(n + \frac{1}{2}\right)\right].$$

The EEM in this case can be written as

$$m^*(E_{FB}, F, n) = (\hbar^2/2)\omega'_{17}(E_{FB}, F, n) \tag{7.57}$$

E_{FB} is the Fermi energy in this case,

$$\omega'_{17}(E_{FB}, F, n) = [2f_1'(E_{FB}, F, n)\cos^{-1}[f_1(E_{FB}, F, n)][1 - f_1^2(E_{FB}, F, n)]^{-1/2}]$$

$$f_1'(E_{FB}, F, n) = [-a_1 \sin[c_1(E_{FB}, F, E_{g_1}, \Delta_1, n)a_o + b_o c_2(E_{FB}, F, E_{g_2}, \Delta_2, n)] \\ \times [c_1'(E_{FB}, F, E_{g_1}, \Delta_1, n)a_o + b_o c_2'(E_{FB}, F, E_{g_2}, \Delta_2, n)] \\ + a_2 \sin[c_1(E_{FB}, F, E_{g_1}, \Delta_1, n)a_o - b_o c_2(E_{FB}, F, E_{g_2}, \Delta_2, n_o)]. \\ \times [c_1'(E_{FB}, F, E_{g_1}, \Delta_1, n)a_o - b_o c_2'(E_{FB}, F, E_{g_2}, \Delta_2, n_o)]]$$

$$c_i'(E_{FB}, F, E_{g_i}, \Delta_i) = \left(\frac{m_{c_i}}{\hbar^2}\right)[c_i(E_{FB}, F, E_{g_i}, \Delta_i)]^{-1}[\beta_{1i}'(E_{FB}, F, E_{g_i}, \Delta_i)]$$

$$\beta_{1i}'(E_{FB}, F, E_{g_i}, \Delta_i) = [I_{11}'(E_{FB}, E_{g_i}, \Delta_i) \\ - \frac{L(E_{g_i}, \Delta_i, m_{r_i})I_{11}'(E_{FB}, E_{g_i}, \Delta_i)T_i^2(E_{FB}, E_{g_i}, \Delta_i)}{\phi_i^3(E_{FB}, E_{g_i}, \Delta_i)[\phi_i(E_{FB}, E_{g_i}, \Delta_i) + \delta_i']^4} \\ - \frac{2L(E_{g_i}, \Delta_i, m_{r_i})I_{11}(E_{FB}, E_{g_i}, \Delta_i)T_i(E_{FB}, E_{g_i}, \Delta_i)T_i'(E_{FB}, E_{g_i}, \Delta_i)}{\phi_i^3(E_{FB}, E_{g_i}, \Delta_i)[\phi_i(E_{FB}, E_{g_i}, \Delta_i) + \delta_i']^4} \\ + \frac{3L(E_{g_i}, \Delta_i, m_{r_i})I_{11}(E_{FB}, E_{g_i}, \Delta_i)T_i^2(E_{FB}, E_{g_i}, \Delta_i)\phi_i'(E_{FB}, E_{g_i}, \Delta_i)}{\phi_i^4(E_{FB}, E_{g_i}, \Delta_i)[\phi_i(E_{FB}, E_{g_i}, \Delta_i) + \delta_i']^4} \\ + \frac{4L(E_{g_i}, \Delta_i, m_{r_i})I_{11}(E_{FB}, E_{g_i}, \Delta_i)T_i^2(E_{FB}, E_{g_i}, \Delta_i)\phi_i'(E_{FB}, E_{g_i}, \Delta_i)}{\phi_i^3(E_{FB}, E_{g_i}, \Delta_i)[\phi_i E_{FB}, E_{g_i}, \Delta_i + \delta_i']^5}$$

$$I'_{11}(E_{FB}, E_{g_i}, \Delta_i) = I_{11}(E_{FB}, E_{g_i}, \Delta_i)\left[\frac{1}{E_{FB}} + \frac{1}{E_{FB}+E_{g_i}} + \frac{1}{E_{FB}+E_{g_i}+\Delta_i}\right.$$

$$\left. - \frac{1}{E_{FB}+E_{g_i}+\frac{2}{3}\Delta_i}\right] \phi'_i(E_{FB}, E_{g_i}, \Delta_i) = \frac{E_{g_i} m_{c_i} I'_{11}(E_{FB}, E_{g_i}, \Delta_i)}{2m_{r_i}\phi_i(E_{FB}, E_{g_i}, \Delta_i)}$$

$$T'_i(E_{FB}, E_{g_i}, \Delta_i) = \left[\frac{\phi'_i(E_{FB}, E_{g_i}, \Delta_i)}{2}\right]\left[\frac{2E_{g_i} Q_i[\phi_i(E_{FB}, E_{g_i}, \Delta_i) - E_{g_i}]^{-1/2}}{[\phi_i(E_{FB}, E_{g_i}, \Delta_i) - E_{g_i}]^{3/2}}\right.$$

$$\left. -(E'_{g_i} + E_{g_i})P_i[\phi_i(E_{FB}, E_{g_i}, \Delta_i) - E_{g_i}]^{-1/2}[\phi_i(E_{FB}, E_{g_i}, \Delta_i) - E'_{g_i}]^{-3/2}\right]$$

The electron concentration assumes the form

$$n_o = \frac{g_v eB}{\pi^2 \hbar L_o}\left[\sum_{n=0}^{n_{\max}} [Q_{19}(E_{FB}, F, n) + Q_{20}(E_{FB}, F, n)]\right] \quad (7.58)$$

where

$$Q_{19}(E_{FB}, F, n) = [\omega_{17}(E_{FB}, F, n)]^{1/2}$$

and

$$Q_{20}(E_{FB}, F, n) = \sum_{R=1}^{R=R_o} Z(R)[Q_{19}(E_{FB}, F, n)]$$

The electron concentration and the EEM in this case when the dispersion relations of the constituent materials are defined by the perturbed two-band model of Kane can, respectively, be expressed as

$$n_o = \frac{g_v eB}{\pi^2 \hbar L_o}\left[\sum_{n=0}^{n_{\max}} [Q_{21}(E_{FB}, F, n) + Q_{22}(E_{FB}, F, n)]\right] \quad (7.59)$$

and

$$m^*(E_{FB}, F, n) = (\hbar^2/2)\omega'_{18}(E_{FB}, F, n) \quad (7.60)$$

where

$$Q_{21}(E_{FB}, F, n) = [\omega_{18}(E_{FB}, F, n)]^{1/2},$$

$$Q_{22}(E_{FB}, F, n) = \sum_{R=1}^{R=R_o} Z(R)[Q_{21}(E_{FB}, F, n)]$$

$$\omega_{18}(E_{FB}, F, n) = \left[\frac{1}{L_o^2}[\cos^{-1}\{f_2(E_{FB}, F, n)\}]^2 - \frac{2eB}{\hbar}\left(n + \frac{1}{2}\right)\right]$$

$$f_2(E_{FB}, F, n) = [a_1 \cos[D_1(E_{FB}, F, E_{g_1}, n)a_o + b_o D_2(E_{FB}, F, E_{g_2}, n)] - a_2 \cos[D_1(E_{FB}, F, E_{g_1}, n)a_o - b_o D_2(E_{FB}, F, E_{g_2}, n)]]$$

$$D_i^2(E_{FB}, F, E_{gi}, n) = \left[\frac{2m_{ci}}{\hbar^2}\rho_{1i}(E_{FB}, F, E_{gi}) - \frac{2eB}{\hbar}\left(n + \frac{1}{2}\right)\right],$$

$$\rho_{1i}(E_{FB}, F, E_{gi}) = \left[E_{FB}(1+\alpha_i E_{FB}) - \delta_{5i}\left[E_{FB}(1+\alpha_i E_{FB}) + \frac{m_{ri}E_{gi}}{2m_{ci}}\right]^{-\frac{5}{2}}\right],$$

$$\delta_{5i} = \left[\frac{(\hbar e F)^2 m_{ri}^{3/2}(E_{gi})^{1/2}}{12(2m_{ci})^{5/2}}\right]$$

$$\omega'_{18}(E_{FB}, F, n) = [2 f'_2(E_{FB}, F, n)\cos^{-1}[f_2(E_{FB}, F, n)] \times [1 - f_2^2(E_{FB}, F, n)]^{-1/2}]$$

$$f_2{}'(E_{FB}, F, n) = [-a_1 \sin[D_1(E_{FB}, F, E_{g_1}, n)a_o + b_o D_2(E_{FB}, F, E_{g_2}, n)] \times [D_1{}'(E_{FB}, F, E_{g_1}, n)a_o + b_o D'_2(E_{FB}, F, E_{g_2}, n)] + a_2 \sin[D_1(E_{FB}, F, E_{g_1}, n)a_o - b_o D_2(E_{FB}, F, E_{g_2}, n_o)]. \times [D'_1(E_{FB}, F, E_{g_1}, n)a_o - b_o D'_2(E_{FB}, F, E_{g_2}, n_o)]],$$

$$D'_i(E_{FB}, F, E_{gi}, n) = \left[\frac{m_{ci}\rho'_{1i}(E_{FB}, F, E_{gi})}{\hbar^2 D_i(E_{FB}, F, E_{gi}, n)}\right]$$

and

$$\rho'_{1i}(E_{FB}, F, E_{gi}) = \left[1 + \frac{2E_{FB}}{E_{gi}}\right]\left[1 + \frac{5}{2}\delta_{5i}\left[E_{FB}(1 + \alpha_i E_{FB}) + \frac{m_{ri}E_{gi}}{2m_{ci}}\right]^{-7/2}\right]$$

7.2.8 The EEM in Nano Wire Effective Mass Superlattices of Optoelectronic Semiconductors

The dispersion relation of the conduction electrons for nanowire effective mass superlattices in accordance with the perturbed three-band model of Kane is given by

$$k_x^2 = \omega_{19}(E, F, n_y, n_z) \tag{7.61}$$

where

$$\omega_{19}(E, F, n_y, n_z) = \left[\frac{1}{L_0^2}\left[\cos^{-1}\{f_3(E, F, n_y, n_z)\}\right]^2 - H(n_y, n_z)\right]$$

$$f_3(E, F, n_y, n_z) = [a_1 \cos[e_1(E, F, E_{g_1}, \Delta_1, n_y, n_z)a_o$$

$$+ b_o e_2(E, F, E_{g_2}, \Delta_2, n_y, n_z)]$$
$$- a_2 \cos[e_1(E, F, E_{g_1}, \Delta_1, n_y, n_z) a_o$$
$$- b_o e_2(E, F, E_{g_2}, \Delta_2, n_y, n_z)]]$$

$$e_i^2(E, F, E_{g_i}, \Delta_i, n_y, n_z) = \left[\left(\frac{2m_{c_i}}{\hbar^2}\right)[\beta_{1_i}(E, F, E_{g_i}, \Delta_i)] - H(n_y, n_z)\right]$$

$$\text{and } H(n_y, n_z) = \left[\left(\frac{n_y \pi}{d_y}\right)^2 + \left(\frac{n_z \pi}{d_z}\right)^2\right]$$

The expression of the electron concentration in this case can be written as

$$n_0 = \frac{2g_v}{\pi}\phi_{104} \tag{7.62}$$

where

$$\phi_{104} = \sum_{n_y=1}^{n_{y\max}} \sum_{n_z=1}^{n_{z\max}} [Q_{23}(E_{FIDEMSL}, F, n_y, n_z) + Q_{24}(E_{FIDEMSL}, F, n_y, n_z)],$$

$$Q_{23}(E_{FIDEMSL}, F, n_y, n_z) = \sqrt{\omega_{19}(E_{FIDEMSL}, F, n_y, n_z)},$$

$$Q_{24}(E_{FIDEMSL}, F, n_y, n_z) = \sum_{R=1}^{R=R_0} Z(R_{IDEMSL}) Q_{23}(E_{FIDEMSL}, F, n_y, n_z),$$

$E_{FIDEMSL}$ is the Fermi energy in the present case and

$$Z(R_{IDEMSL}) = 2(k_B T)^{2R}(1 - 2^{1-2R})\xi(2R)\frac{\partial^{2R}}{\partial E_{FIDEMSL}}.$$

The EEM in this case can be expressed as

$$m^*(E_{FIDEMSL}, F, n_y, n_z) = (\hbar^2/2)\omega'_{19}(E_{FIDEMSL}, F, n_y, n_z) \tag{7.63}$$

where

$$\omega'_{19}(E_{FIDEMSL}, F, n_y, n_z) = \left[1 - f_3^2(E_{FIDEMSL}, F, n_y, n_z)\right]^{-1/2}$$
$$\times \left[2 f_3'(E_{FIDEMSL}, F, n_y, n_z)\right]$$
$$\times \left[\cos^{-1}\left\{f_3(E_{FIDEMSL}, F, n_y, n_z)\right\}\right],$$

$$f_3'(E_{FIDEMSL}, F, n_y, n_z) = -a_1 \sin[a_0 e_1(E_{FIDEMSL}, F, E_{g_1}, \Delta_1, n_y, n_z)$$

$$\begin{aligned}
&+ b_o e_2(E_{FIDEMSL}, F, E_{g_2}, \Delta_2, n_y, n_z)]. \\
&\times \left[a_0 e_1'(E_{FIDEMSL}, F, E_{g_1}, \Delta_1, n_y, n_z)\right. \\
&\left.+ b_o e_2'(E_{FIDEMSL}, F, E_{g_2}, \Delta_2, n_y, n_z)\right] \\
&+ a_2 \sin\left[a_0 e_1(E_{FIDEMSL}, F, E_{g_1}, \Delta_1, n_y, n_z)\right. \\
&\left.- b_o e_2(E_{FIDEMSL}, F, E_{g_2}, \Delta_2, n_y, n_z)\right]. \\
&\times \left[a_0 e_1'(E_{FIDEMSL}, F, E_{g_1}, \Delta_1, n_y, n_z)\right. \\
&\left.- b_o e_2'(E_{FIDEMSL}, F, E_{g_2}, \Delta_2, n_y, n_z)\right]
\end{aligned}$$

and

$$e_i'(E_{FIDEMSL}, F, E_{gi}, \Delta_i, n_y, n_z) = \frac{m_{ci}\beta_{1i}(E_{FIDEMSL}, F, E_{gi}, \Delta_i)}{\hbar^2 e_i(E_{FIDEMSL}, F, E_{gi}, \Delta_i, n_y, n_z)}.$$

In accordance with the perturbed two-band model of Kane the electron concentration per unit length is given by,

$$n_0 = \frac{2g_v}{\pi}\phi_{105} \tag{7.64}$$

where,

$$\phi_{105} = \sum_{n_y=1}^{n_{y\max}} \sum_{n_z=1}^{n_{z\max}} \left[Q_{25}(E_{FIDEMSL}, F, n_y, n_z) + Q_{26}(E_{FIDEMSL}, F, n_y, n_z)\right],$$

$$Q_{25}(E_{FIDEMSL}, F, n_y, n_z) = \left[\sqrt{\omega_{20}(E_{FIDEMSL}, F, n_y, n_z)}\right],$$

$$Q_{26}(E_{FIDEMSL}, F, n_y, n_z) = \sum_{R=1}^{R=R_0} Z(R_{IDEMSL})\left[Q_{25}(E_{FIDEMSL}, F, n_y, n_z)\right],$$

$$\omega_{20}(E_{FIDEMSL}, F, n_y, n_z) = \left[\frac{1}{L_0^2}\left[\cos^{-1} f_4(E_{FIDEMSL}, F, n_y, n_z)\right]^2 - H(n_y, n_z)\right],$$

$$\begin{aligned}
f_4(E_{FIDEMSL}, F, n_y, n_z) = \Big[&a_1 \cos[a_0 g_1(E_{FIDEMSL}, F, E_{g1}, n_y, n_z) \\
&- b_0 g_2(E_{FIDEMSL}, F, E_{g2}, n_y, n_z)] \\
&- a_2 \cos\left[a_0 g_1(E_{FIDEMSL}, F, E_{g1}, n_y, n_z)\right. \\
&\left.- b_0 g_2(E_{FIDEMSL}, F, E_{g2}, n_y, n_z)\right]\Big]
\end{aligned}$$

and

$$g_i^2(E_{FIDEMSL}, F, E_{gi}, n_y, n_z) = \left[\frac{2m_{ci}}{\hbar^2}\theta_{1i}(E_{FIDEMSL}, F, E_{gi}) - H(n_y, n_z)\right]$$

The EEM in this case can be expressed as

$$m^*(E_{FIDEMSL}, F, n_y, n_z) = (\hbar^2/2)\omega'_{20}(E_{FIDEMSL}, F, n_y, n_z) \tag{7.65}$$

where

$$\begin{aligned}
&\omega'_{20}(E_{FIDEMSL}, F, n_y, n_z) \\
&\quad = \frac{2f'_4(E_{FIDEMSL}, F, n_y, n_z)\left[\cos^{-1} f_4(E_{FIDEMSL}, F, n_y, n_z)\right]}{\sqrt{1 - f_4^2(E_{FIDEMSL}, F, n_y, n_z)}},
\end{aligned}$$

$$\begin{aligned}
&f'_4(E_{FIDEMSL}, F, n_y, n_z) \\
&\quad = -a_1 \sin\left[a_0 g_1(E_{FIDEMSL}, F, E_{g1}, n_y, n_z)\right. \\
&\qquad \left.+ b_0 g_2(E_{FIDEMSL}, F, E_{g2}, n_y, n_z)\right] \\
&\qquad \times \left[a_0 g'_1(E_{FIDEMSL}, F, E_{g1}, n_y, n_z) + b_0 g'_2(E_{FIDEMSL}, F, E_{g2}, n_y, n_z)\right] \\
&\qquad - a_2 \sin\left[a_0 g_1(E_{FIDEMSL}, F, E_{g1}, n_y, n_z) - b_0 g_2(E_{FIDEMSL}, F, E_{g2}, n_y, n_z)\right] \\
&\qquad \times \left[a_0 g_1(E_{FIDEMSL}, F, E_{g1}, n_y, n_z) - b_0 g_2(E_{FIDEMSL}, F, E_{g2}, n_y, n_z)\right],
\end{aligned}$$

and

$$g'_i(E_{FIDEMSL}, F, n_y, n_z) = \frac{m_{ci}\theta'_{1i}(E_{FIDEMSL}, F, E_{gi})}{\hbar^2 g_i(E_{FIDEMSL}, F, E_{gi}, n_y, n_z)}.$$

7.2.9 The EEM in Superlattices of Optoelectronic Semiconductors with Graded Interfaces Under Magnetic Quantization

The energy spectrum in superlattices of optoelectronic compounds with graded interfaces in the presence of electric field whose constituent materials are defined by perturbed three-band model of Kane can be written following [4] as

$$\cos(L_o k) = \frac{1}{2}\Phi_{11}(E, k_s) \tag{7.66}$$

where

$$\Phi_{11}(E, k_s) = \Big[2\cosh\{X_{21}(E, k_s)\}\cos\{Y_{21}(E, k_s)\}$$

$$
\begin{aligned}
&+\varepsilon_{21}(E,k_s)\sinh\{X_{21}(E,k_s)\}\sin\{Y_{21}(E,k_s)\}\\
&+\Delta_{21}\left[\left(\frac{K_{21}^2(E,k_s)}{K_{22}(E,k_s)}-3K_{22}(E,k_s)\right)\cosh\{X_{21}(E,k_s)\}\right.\\
&\times\sin\{Y_{21}(E,k_s)\}+\left(3K_{21}(E,k_s)-\frac{\{K_{22}(E,k_s)\}^2}{K_{21}(E,k_s)}\right)\\
&\left.\times\sinh\{X_{21}(E,k_s)\}\cos\{Y_{21}(E,k_s)\}\right]\\
&+\Delta_{21}\left[2(\{K_{21}(E,k_s)\}^2-\{K_{22}(E,k_s)\}^2)\right.\\
&\times\cosh\{X_{21}(E,k_s)\}\cos\{Y_{21}(E,k_s)\}\\
&+\frac{1}{12}\left[\frac{5\{K_{22}(E,k_s)\}^3}{K_{21}(E,k_s)}+\frac{5\{K_{21}(E,k_s)\}^3}{K_{22}(E,k_s)}\right.\\
&\left.\left.\left.-34K_{22}(E,k_s)K_{21}(E,k_s)\right]\sinh\{X_{21}(E,k_s)\}\sin\{Y_{21}(E,k_s)\}\right]\right],
\end{aligned}
$$

$$X_{21}(E,k_s)=K_{21}(E,k_s)\left[a_0-\Delta_{21}\right],$$

$$K_{21}(E,k_s)\equiv\left[-\frac{2m_{c2}}{\hbar^2}\beta_{012}(E-\overline{V_0},F,E_{g2},\Delta_2)+k_s^2\right]^{1/2}$$

$$
\begin{aligned}
\beta_{012}(E-\overline{V_0},F,E_{g2},\Delta_2)&=[\gamma(E-\overline{V_0},E_{g2},\Delta_2)\\
&-\frac{L(E_{g2},\Delta_2,m_{r2})\gamma(E-\overline{V_0},E_{g2},\Delta_2)T_2^2(E-\overline{V_0},E_{g2},\Delta_2)}{\phi_2^3(E-\overline{V_0},E_{g2},\Delta_2)[\phi_2(E-\overline{V_0},E_{g2},\Delta_2)+\delta_2']^4}],
\end{aligned}
$$

$$\varepsilon(E,k_s)\equiv\left[\frac{K_1(E,k_s)}{K_2(E,k_s)}-\frac{K_2(E,k_s)}{K_1(E,k_s)}\right],$$

$$k_s^2=k_x^2+k_y^2,\ Y_{21}(E,k_s)=K_{22}(E,k_s)\left[b_0-\Delta_{21}\right]\text{ and}$$

$$K_{22}(E,k_s)=\left[\frac{2m_{c1}\beta_{11}(E,F,E_{g1},\Delta_1)}{\hbar^2}-k_s^2\right]^{1/2}.$$

In the presence of a quantizing magnetic field B along z-direction, the simplified magneto-dispersion relation can be written as

$$k_z^2=\omega_{21}(E,F,n)\tag{7.67}$$

where, $\omega_{21}(E,F,n)=\left[\frac{1}{L_0^2}[\cos^{-1}[\frac{1}{2}f_{11}(E,F,n)]]^2-\frac{2|e|B}{\hbar}\left(n+\frac{1}{2}\right)\right]$,

$$f_{11}(E,F,n)=[2\cosh\{M_{21}(n,E)\}\cos\{N_{21}(n,E)\}+Z_{21}(n,E)\sinh\{M_{21}(n,E)\}$$
$$\times\sin\{N_{21}(n,E)\}+\Delta_{21}\left[\left(\frac{\{I_{21}(n,E)\}^2}{I_{22}(n,E)}-3I_{22}(n,E)\right)\cosh\{M_{21}(n,E)\}\right.$$
$$\left.\times\sin\{N_{21}(n,E)\}+\left(3I_{21}(n,E)-\frac{\{I_{22}(n,E)\}^2}{I_{21}(n,E)}\right)\sinh\{M_{21}(n,E)\}\cos\{N_{21}(n,E)\}\right]$$
$$+\Delta_{21}\left[2(\{I_{21}(n,E)\}^2-\{I_{22}(n,E)\}^2)\cosh\{M_{21}(n,E)\}\cos\{N_{21}(n,E)\}\right.$$
$$+\frac{1}{12}\left(\frac{5\{I_{22}(n,E)\}^3}{I_{21}(n,E)}+\frac{5\{I_{21}(n,E)\}^3}{I_{22}(n,E)}-\{34I_{22}(n,E)I_{21}(n,E)\}\right)$$
$$\times\sinh\{M_{21}(n,E)\}\sin\{N_{21}(n,E)\}]]$$

$$Z_{21}(n,E)\equiv\left[\frac{I_{21}(n,E)}{I_{22}(n,E)}-\frac{I_{22}(n,E)}{I_{21}(n,E)}\right],\ M_{21}(n,E)=I_{21}(n,E)[a_0-\Delta_{21}],$$

$$I_{21}(n,E)=\left[-\frac{2m_{c2}}{\hbar^2}\beta_{012}(E-\overline{V_0},F,E_{g2},\Delta_2)+\frac{2|e|B}{\hbar}\left(n+\frac{1}{2}\right)\right]^{1/2}$$

$N_{21}(n,E)=I_{22}(n,E)[b_0-\Delta_{21}]$ and $I_{22}(n,E)$

$$\equiv\left[\frac{2m_{c1}}{\hbar^2}\beta_{11}(E,F,E_{g1},\Delta_1)-\left\{\frac{2|e|B}{\hbar}\left(n+\frac{1}{2}\right)\right\}\right]^{1/2}.$$

The electron concentration is given by

$$n_o=\frac{g_v eB}{\pi^2\hbar}\phi_{106} \tag{7.68}$$

where,

$$\phi_{106}=[\sum_{n=0}^{n_{\max}}[Q_{27}(E_{FBGISL},F,n)+Q_{28}(E_{FBGISL},F,n)]],$$

$$Q_{27}(E_{FBGISL},F,n)=[\omega_{21}(E_{FBGISL},F,n)]^{1/2},$$

E_{FBGISL} is the Fermi energy in the present case,

$$Q_{28}(E_{FBGISL},F,n)=\sum_{R=1}^{R=R_0}Z(R_{BGISL})[Q_{27}(E_{FBGISL},F,n)]$$

and

$$Z(R_{BGISL})=2(k_BT)^2(1-2^{1-2R})\xi(2R)\frac{\partial^{2R}}{\partial E_{FBGISL}}.$$

The EEM in this case can be expressed as

$$m^*(E_{FBGISL}, F, n) = (\hbar^2/2)\omega'_{21}(E_{FBGISL}, F, n) \tag{7.69}$$

where

$$\omega_{21}'(E_{FBGISL}, F, n) = \frac{f_{11}'(E_{FBGISL}, F, n)\cos^{-1}\left[\frac{1}{2}f_{11}(E_{FBGISL}, F, n)\right]}{\sqrt{1-\frac{1}{4}f_{11}^2(E_{FBGISL}, F, n)}},$$

$$\begin{aligned}
f'_{11}(E_{FBGISL}, F, n) = {} & \big[2M'_{21}(n, E_{FBGISL})\sinh\{M_{21}(n, E_{FBGISL})\}\cos\{N_{21}(n, E_{FBGISL})\} \\
& + Z_{21}(n, E_{FBGISL})M'_{21}(n, E_{FBGISL})\cosh\{M_{21}(n, E_{FBGISL})\} \\
& \times \sin\{N_{21}(n, E_{FBGISL})\} - 2N'_{21}(n, E_{FBGISL}) \\
& \times \sin\{N_{21}(n, E_{FBGISL})\}\cosh\{M_{21}(n, E_{FBGISL})\} \\
& + Z'_{21}(n, V_0)\sinh\{M_{21}(n, E_{FBGISL})\}\sin\{N_{21}(n, E_{FBGISL})\} \\
& + Z_{21}(n, E_{FBGISL})N'_{21}(n, E_{FBGISL}) \\
& \times \cos\{N_{21}(n, E_{FBGISL})\}\sinh\{M_{21}(n, E_{FBGISL})\} \\
& + \Delta_{21}\left[\left(\frac{\{2I_{21}(n, E_{FBGISL})I'_{21}(n, E_{FBGISL})\}}{I_{22}(n, E_{FBGISL})}\right.\right. \\
& \qquad - \frac{\left\{I_{21}^2(n, E_{FBGISL})I'_{22}(n, E_{FBGISL})\right\}}{I_{22}^2(n, E_{FBGISL})} \\
& \qquad \left.\left. -3I'_{22}(n, E_{FBGISL})\right)\right] \\
& \times \cosh\{M_{21}(n, E_{FBGISL})\}\sin\{N_{21}(n, E_{FBGISL})\} \\
& + \left(-3I_{22}(n, E_{FBGISL}) + \frac{\{I_{21}(n, E_{FBGISL})\}^2}{I_{22}(n, E_{FBGISL})}\right) \\
& \times\{M'_{21}(n, E_{FBGISL})\sinh\{M_{21}(n, E_{FBGISL})\}\}\sin\{N_{21}(n, E_{FBGISL})\} \\
& +\{N'_{21}(n, E_{FBGISL})\cosh\{M_{21}(n, E_{FBGISL})\}\cos\{N_{21}(n, E_{FBGISL})\} \\
& + \left(\frac{-\{2I_{22}(n, E_{FBGISL})I'_{22}(n, E_{FBGISL})\}}{I_{21}(n, E_{FBGISL})}\right. \\
& \qquad + \frac{\left\{I_{22}^2(n, E_{FBGISL})I'_{21}(n, E_{FBGISL})\right\}}{I_{21}^2(n, E_{FBGISL})} \\
& \qquad \left. +3I'_{21}(n, E_{FBGISL})\right)
\end{aligned}$$

$$
\begin{aligned}
&\times \sinh\{M_{21}(n, E_{FBGISL})\}\cos\{N_{21}(nE_{FBGISL})\}\\
&\quad + \left(+3I_{21}(n, E_{FBGISL}) - \frac{\{I_{22}(n, E_{FBGISL})\}^2}{I_{21}(n, E_{\mathrm{FBGISL}})}\right)\\
&\quad \times \{M'_{21}(n, E_{FBGISL})\cosh\{M_{21}(n, E_{FBGISL})\}\}\cos\{N_{21}(n, E_{FBGISL})\}\\
&\quad - N'_{21}(n, E_{FBGISL})\sin\{N_{21}(n, E_{FBGISL})\}\sinh\{M_{21}(n, E_{FBGISL})\}\}]\\
&\quad + \Delta_{21}\left[4(\{I_{21}(n, E_{FBGISL})I'_{21}(n, E_{FBGISL})\}\right.\\
&\quad \left.-\{I_{22}(n, E_{FBGISL})I'_{22}(n, E_{FBGISL})\})\right]\cdot\\
&\quad \times \cosh\{M_{21}(n, E_{FBGISL})\}\cos\{N_{21}(n, E_{FBGISL})\}\\
&\quad + 2(\{I_{21}(n, E_{FBGISL})\}^2 - \{I_{22}(n, E_{FBGISL})\}^2)\\
&\quad \times \{M'_{21}(n, E_{FBGISL})\sinh\{M_{21}(n, E_{FBGISL})\}\cos\{N_{21}(n, E_{FBGISL})\}\\
&\quad - N'_{21}(n, E_{FBGISL})\cosh\{M_{21}(n, E_{FBGISL})\}\sin\{N_{21}(n, E_{FBGISL})\}\}\\
&\quad + \frac{1}{12}\left(\frac{15\left\{I_{22}^2(n, E_{FBGISL})\right\}I'_{22}(n, E_{FBGISL})}{I_{21}(n, E_{FBGISL})}\right.\\
&\qquad\qquad - \frac{5\{I_{22}(n, E_{FBGISL})\}^3 I'_{21}(n, E_{FBGISL})}{I_{21}^2(n, E_{FBGISL})}\\
&\qquad\qquad + \frac{15\left\{I_{21}^2(n, E_{FBGISL})\right\}I'_{21}(n, E_{FBGISL})}{I_{22}(n, E_{FBGISL})}\\
&\qquad\qquad - \frac{5\{I_{21}(n, E_{FBGISL})\}^3 I'_{22}(n, E_{FBGISL})}{I_{22}^2(n, E_{FBGISL})}\\
&\qquad\qquad - \left\{34I'_{22}(n, E_{FBGISL})I_{21}(n, E_{FBGISL})\right\}\\
&\qquad\qquad - 34I_{22}(n, E_{FBGISL})I'_{21}(n, E_{FBGISL}))\\
&\qquad\qquad \times \sinh\{M_{21}(n, E_{FBGISL})\}\sin\{N_{21}(n, E_{FBGISL})\}\\
&\quad + \left(\frac{5\{I_{22}(n, E_{FBGISL})\}^3}{I_{21}(n, E_{FBGISL})} + \frac{5\{I_{21}(n, E_{FBGISL})\}^3}{I_{22}(n, E_{FBGISL})}\right.\\
&\quad \left. - \{34I_{22}(n, E_{FBGISL})I_{21}(n, E_{FBGISL})\}\right)\\
&\quad \{M'_{21}(n, E_{FBGISL})\cosh\{M_{21}(n, E_{FBGISL})\}\sin\{N_{21}(n, E_{FBGISL})\}\\
&\qquad + N'_{21}(n, E_{FBGISL})\sinh\{M_{21}(n, E_{FBGISL})\}\cos\{N_{21}(n, E_{FBGISL})\}\}]],
\end{aligned}
$$

$$
\begin{aligned}
M'_{21}(n, E_{FBGISL}) &= I'_{21}(n, E_{FBGISL})\left[a_0 - \Delta_{21}\right],\\
I'_{21}(E_{FBGISL}, n) &= \frac{m_{c2}\beta'_{012}(E_{FBGISL} - \bar{V}_0, F, E_{g2}, \Delta_2)}{-\hbar^2 I_{21}(E_{FBGISL}, n)},
\end{aligned}
$$

$$
\begin{aligned}
\beta'_{012}(E_{FBGISL} - \bar{V}_0, F, E_{g2}, \Delta_2) = \Bigg[& (I'_{11}(E_{FBGISL} - \bar{V}_0, E_{g2}, \Delta_2)) \\
& - \frac{L(E_{g_2}, \Delta_2, m_{r2}) I'_{11}(E_{FBGISL} - \bar{V}_0, E_{g2}, \Delta_i) T_i^2(E_{FBGISL} - \bar{V}_0, E_{g2}, \Delta_2)}{\phi_i^3(E_{FBGISL} - \bar{V}_0, E_{g_2}, \Delta_2)[\phi_2(E_{FBGISL} - \bar{V}_0, E_{g_2}, \Delta_2) + \delta'_2]^4} \\
& - \frac{2L(E_{g_2}, \Delta_2, m_{r_2}) I_{11}(E_{FBGISL} - \bar{V}_0, E_{g_2}, \Delta_2) T_2(E_{FBGISL} - \bar{V}_0, E_{g_2}, \Delta_2) T'_2(E_{FBGISL} - \bar{V}_0, E_{g_2}, \Delta_2)}{\phi_2^3(E_{FBGISL} - \bar{V}_0, E_{g_2}, \Delta_2)[\phi_2(E_{FBGISL} - \bar{V}_0, E_{g_2}, \Delta_2) + \delta'_2]^4} \\
& + \frac{3L(E_{g_2}, \Delta_2, m_{r_2}) I_{11}(E_{FBGISL} - \bar{V}_0, E_{g_2}, \Delta_2) T_2^2(E_{FBGISL} - \bar{V}_0, E_{g_2}, \Delta_2) \phi'_2(E_{FBGISL} - \bar{V}_0, E_{g_2}, \Delta_2)}{\phi_i^4(E_{FBGISL} - \bar{V}_0, E_{g_2}, \Delta_2)[\phi_2(E_{FBGISL} - \bar{V}_0, E_{g_2}, \Delta_2) + \delta'_2]^4} \\
& + \frac{4L(E_{g_2}, \Delta_2, m_{r_2}) I_{11}(E_{FBGISL} - \bar{V}_0, E_{g_2}, \Delta_2) T_2^2(E_{FBGISL} - \bar{V}_0, E_{g_2}, \Delta_2) \phi'_2(E_{FBGISL} - \bar{V}_0, E_{g_2}, \Delta_2)}{\phi_2^3(E_{FBGISL} - \bar{V}_0, E_{g_2}, \Delta_2)[\phi_2(E_{FBGISL} - \bar{V}_0, E_{g_2}, \Delta_2) + \delta'_2]^5} \Bigg]
\end{aligned}
$$

$$
\begin{aligned}
\gamma'(E_{FBGISL} - \bar{V}_0, E_{g_2}, \Delta_2) = {} & I_{11}(E_{FBGISL} - \bar{V}_0, E_{g_2}, \Delta_2) \\
& \times \Bigg[\frac{1}{E_{FBGISL} - \bar{V}_0} + \frac{1}{E_{FBGISL} - \bar{V}_0 + E_{g2}} \\
& + \frac{1}{E_{FBGISL} - \bar{V}_0 + E_{g2} + \Delta_2} \\
& - \frac{1}{E_{FBGISL} - \bar{V}_0 + E_{g2} + \frac{2}{3}\Delta_2} \Bigg],
\end{aligned}
$$

$$
\begin{aligned}
& I_{11}(E_{FBGISL} - \bar{V}_0, E_{g_2}, \Delta_2) \\
& = \frac{(E_{FBGISL} - \bar{V}_0)(E_{FBGISL} - \bar{V}_0 + E_{g2})(E_{FBGISL} - \bar{V}_0 + E_{g2} + \Delta_2)\left(E_{g2} + \frac{2}{3}\Delta_2\right)}{E_{g_2}(E_{g_2} + \Delta_2)\left(E_{FBGISL} - \bar{V}_0 + E_{g_2} + \frac{2}{3}\Delta_2\right)},
\end{aligned}
$$

$$
\begin{aligned}
L(E_{g_2}, \Delta_2, m_{r_2}) &= \frac{(\hbar e F)^2 (E_{g2} - \delta'_2)^2 m_{c2}}{6 m_{r2}^2}, \\
\delta'_2 &= \frac{(E_{g2})^2 \Delta_2}{\chi_2}, \\
\chi_2 &= [6(E_{g2})^2 + 9 E_{g2}.\Delta_2 + 4\Delta_2^2], \\
\frac{1}{m_{r2}} &= \left(\frac{1}{m_{c2}} + \frac{1}{m_{v2}}\right),
\end{aligned}
$$

$$
\begin{aligned}
T_2(E_{FBGISL} - \bar{V}_0, E_{g_2}, \Delta_2) = \Bigg[& P_2 \left[\frac{\{\phi_2(E_{FBGISL} - \bar{V}_0, E_{g_2}, \Delta_2) + E_{g2}\}}{\{\phi_i(E_{FBGISL} - \bar{V}_0, E_{g_2}, \Delta_2) - E'_{g_2}\}} \right]^{1/2} \\
& + Q_2 \left[\frac{\{\phi_2(E_{\mathrm{FBGISL}} - \bar{V}_0, E_{g_2}, \Delta_2) - E_{g2}\}}{\{\phi_2(E_{FBGISL} - \bar{V}_0, E_{g_2}, \Delta_2) + E_{g2}\}} \right]^{1/2} \Bigg]
\end{aligned}
$$

$$P_2 = \frac{r_{02}^2}{2}\left(\frac{E_{g2} - \delta_2'}{E_{g_2} + \delta_2'}\right), \; r_{o2}^2 = \left[6\left(E_{g2} + \frac{2}{3}\Delta_2\right)(E_{g_2} + \Delta_2)\right][\chi_2]^{-1},$$

$$\phi_2^2(E_{FBGISL} - \bar{V}_0, E_{g_2}, \Delta_2) = [E_{g_2}^2 + E_{g_2}(m_{c_2}/m_{r_2})I_{11}(E_{FBGISL} - \bar{V}_0, E_{g_2}, \Delta_2)],$$

$$E'_{g_2} = \left[\frac{E_{g_2}(E_{g_2} - 3\delta_2')}{(E_{g_2} + \delta_2')}\right], \; Q_2 = \frac{t_2^2}{2}, \; t_2^2 = \left[\frac{6\left(E_{g_2} + \frac{2}{3}\Delta_2\right)}{\chi_2}\right],$$

$$\phi_2'(E_{FBGISL} - \bar{V}_0, E_{g_2}, \Delta_2) = \frac{E_{g_2} m_{c_2} I_{11}'(E_{FBGISL} - \bar{V}_0, E_{g_2}, \Delta_2)}{2m_{r_2}\phi_2(E_{FBGISL} - \bar{V}_0, E_{g_2}, \Delta_2)},$$

$$\begin{aligned} T_2'(E_{FBGISL} - \bar{V}_0, E_{g_2}, \Delta_2) = {} & \left[\frac{\phi_2'(E_{FBGISL} - \bar{V}_0, E_{g_2}, \Delta_2)}{2}\right] \\ & \times \left[\frac{2E_{g_2} Q_2[\phi_2(E_{FBGISL} - \bar{V}_0, E_{g_2}, \Delta_2) - E_{g2}]^{-1/2}}{[\phi_2(E_{FBGISL} - \bar{V}_0, E_{g_2}, \Delta_2) - E_{g_2}]^{3/2}}\right. \\ & -(E'_{g_2} + E_{g_2})P_2[\phi_2(E_{FBGISL} - \bar{V}_0, E_{g_2}, \Delta_2) - E_{g_2}]^{-1/2} \\ & \left.\times[\phi_2(E_{FBGISL} - \bar{V}_0, E_{g_2}, \Delta_2) - E'_{g_2}]^{-3/2}\right], \end{aligned}$$

$$I_{22}'(E_{FBGISL}, n) = \frac{m_{c1}\beta_{11}'(E_{FBGISL}, F, E_{g1}, \Delta_1)}{\hbar^2 I_{22}(E_{FBGISL}, n)},$$

$$I_{21}(E_{FBGISL}, n) = \left[\frac{2eB}{\hbar}\left(n + \frac{1}{2}\right) - \frac{2m_{c2}}{\hbar^2}\beta_{012}(E_{FBGISL} - \bar{V}_0, F, E_{g2}, \Delta_2)\right]^{\frac{1}{2}},$$

$$N_{21}(E_{FBGISL}, n) = I_{22}(E_{FBGISL}, n)\,[b_0 - \Delta_{21}],$$

$$I_{22}(E_{FBGISL}, n) = \left[\frac{2m_{c1}\beta_{11}(E_{FBGISL}, F, E_{g1}, \Delta_1)}{\hbar^2} - \frac{2eB}{\hbar}\left(n + \frac{1}{2}\right)\right]^{\frac{1}{2}},$$

$$\begin{aligned} Z_{21}'(E_{FBGISL}, n) = {} & \left[\frac{-Z_{21}(E_{FBGISL}, n)I_{21}'(E_{FBGISL}, n)}{I_{21}(V_0, n)}\right. \\ & - \frac{Z_{21}(E_{FBGISL}, n)I_{22}'(E_{FBGISL}, n)}{I_{22}(V_0, n)} \\ & + (I_{21}(E_{FBGISL}, n)I_{22}(E_{FBGISL}, n))^{-1} \\ & \times\left[2I_{21}'(E_{FBGISL}, n)I_{21}(E_{FBGISL}, n)\right. \\ & \left.\left.-2I_{22}'(E_{FBGISL}, n)I_{22}(E_{FBGISL}, n)\right]\right] \end{aligned}$$

For perturbed two-band model of Kane the forms of the electron concentration and field emitted current density remain same where

$$I_{21}(E, n) = \left[\frac{2eB}{\hbar}\left(n + \frac{1}{2}\right) - \frac{2m_{c2}}{\hbar^2}\rho_{012}(E - \bar{V}_0, F, E_{g2})\right]^{\frac{1}{2}},$$

$$\rho_{012}(E-\bar{V}_0,F,E_{g2}) = \Bigg[(E-\bar{V}_0)\left[1+\alpha_2(E-\bar{V}_0)\right] - \delta_{52}\left[(E-V_0)\left[1+\alpha_2(E-\bar{V}_0)\right]+\frac{m_{r2}E_{g2}}{2m_{c2}}\right]^{5/2}\Bigg]$$

$$\delta_{52} = \left[\frac{(\hbar eF)^2 m_{r2}^{3/2}(E_{g2})^{1/2}}{12(2m_{c2})^{5/2}}\right],$$

$$I_{22}(E,n) = \left[\frac{2m_{c1}\rho_{11}(E,F,E_{g1})}{\hbar^2} - \frac{2eB}{\hbar}\left(n+\frac{1}{2}\right)\right]^{\frac{1}{2}},$$

$$\rho_{11}(E,F,E_{g1}) = \left[E(1+\alpha_1 E) - \delta_{51}\left[E(1+\alpha_1 E)+\frac{m_{r1}E_{g1}}{2m_{c1}}\right]^{\frac{5}{2}}\right],$$

$$\alpha_i = \frac{1}{E_{gi}},$$

$$\delta_{51} = \left[\frac{(\hbar eF)^2 m_{r1}^{3/2}(E_{g1})^{1/2}}{12(2m_{c1})^{5/2}}\right],$$

$$I'_{21}(E_{FBGISL},n) = \frac{m_{c2}\rho'_{012}(E_{FBGISL}-\bar{V}_0,F,E_{g2})}{\hbar^2 I_{21}(E_{FBGISL},n)},$$

$$\rho'_{012}(E_{FBGISL}-\bar{V}_0,F,E_{g2}) = \Bigg[\left[1+2\alpha_2(E_{FBGISL}-\bar{V}_0)\right] \times\left[1-\frac{5}{2}\delta_{52}(E_{FBGISL}-\bar{V}_0)\right. \times \left.\left[1+\alpha_2(E_{FBGISL}-\bar{V}_0)\right]+\frac{m_{r2}E_{g2}}{2m_{c2}}\right]^{3/2}\Bigg]$$

$$I'_{22}(E_{FBGISL},n) = \left[\frac{2m_{c1}\rho'_{11}(E_{FBGISL},F,E_{g1})}{\hbar^2 I_{22}(E_{FBGISL},n)}\right] \text{ and}$$

$$\rho'_{11}(E_{FBGISL},F,E_{g1}) = \Bigg[\left[1+2\alpha_2(E_{FBGISL})\right] \times\left[1-\frac{5}{2}\delta_{51}(E_{FBGISL})\left[1+\alpha_1(E_{FBGISL})\right]+\frac{m_{r1}E_{g1}}{2m_{c1}}\right]^{3/2}\Bigg].$$

7.2.10 The EEM in Quantum Wire Superlattices of Optoelectronic Semiconductors with Graded Interfaces

The dispersion relation in accordance with the perturbed three-band model of Kane, in this case is given by

$$k_x^2 = \omega_{22}(E,F,n_y,n_z) \tag{7.70}$$

where

$$\omega_{22}(E,F,n_y,n_z) = \left[\frac{1}{L_0^2}\left[\cos^{-1}\frac{1}{2}f_{13}(E,F,n_y,n_z)\right]^2 - H(n_y,n_z)\right],$$

$$f_{13}(E,F,n_y,n_z) = \left[2\cosh\left\{M_{31}(n_y,n_z,E)\right\}\cos\left\{N_{31}(n_y,n_z0,E)\right\} + Z_{31}(n_y,n_z,E)\sinh\left\{M_{31}(n_y,n_z,E)\right\}\right.$$

$$\sin\left\{N_{31}(n_y,n_z,E)\right\} + \Delta_{21}\left[\left(\frac{\left\{I_{31}(n_y,n_z,E)\right\}^2}{I_{32}(n_y,n_z,E)} - 3I_{32}(n_y,n_z,E)\right)\cosh\left\{M_{31}(n_y,n_z,E)\right\}\right.$$

$$\sin\left\{N_{31}(n_y,n_z,E)\right\} + \left(3I_{31}(n_y,n_z,E) - \frac{\left\{I_{32}(n_y,n_z,E)\right\}^2}{I_{31}(n_y,n_z,E)}\right)$$
$$\times\sinh\left\{M_{31}(n_y,n_z,E)\right\}\cos\left\{N_{31}(n_y,n_z,E)\right\}]$$
$$+\Delta_{21}\left[2\left(\left\{I_{31}(n_y,n_z,E)\right\}^2 - \left\{I_{32}(n_y,n_z,E)\right\}^2\right)\cosh\left\{M_{31}(n_y,n_z,E)\right\}\right.$$
$$\times\cos\left\{N_{31}(n_y,n_z,E)\right\}$$
$$+\frac{1}{12}\left(\frac{5\left\{I_{32}(n_y,n_z,E)\right\}^3}{I_{31}(n_y,n_z,E)} + \frac{5\left\{I_{31}(n_y,n_z,E)\right\}^3}{I_{32}(n_y,n_z,E)} - \left\{34I_{32}(n_y,n_z,E)I_{31}(n_y,n_z,E)\right\}\right)$$
$$\sinh\left\{M_{31}(n_y,n_z,E)\right\}\sin\left\{N_{31}(n_y,n_z,E)\right\}]]$$

$Z_{31}(n_y,n_z,E) \equiv \left[\frac{I_{31}(n_y,n_z,E)}{I_{32}(n_y,n_z,E)} - \frac{I_{32}(n_y,n_z,E)}{I_{31}(n_y,n_z,E)}\right]$, $M_{31}(n_y,n_z,E) = I_{31}(n_y,n_z,E)\,[a_0 - \Delta_{21}]$,
$I_{31}(n_y,n_z,E) = \left[-\frac{2m_{c2}}{\hbar^2}\beta_{012}(E - \overline{V_0},F,E_{g2},\Delta_2) + H(n_y,n_z)\right]^{1/2}$,

$$N_{31}(n_y,n_z,E) = I_{32}(n_y,n_z,E)\,[b_0 - \Delta_{21}] \text{ and}$$

$$I_{32}(n_y,n_z,E) \equiv \left[\frac{2m_{c1}}{\hbar^2}\beta_{11}(E,F,E_{g1},\Delta_1) - H(n_y,n_z)\right]^{1/2}$$

The electron concentration per unit length is given by

$$n_0 = \frac{2g_v}{\pi}\phi_{107} \tag{7.71}$$

where

$$\phi_{107} = \sum_{n_y=1}^{n_{y\max}}\sum_{n_z=1}^{n_{z\max}}\left[Q_{29}(E_{FQWGISL},F,n_y,n_z) + Q_{30}(E_{FQWGISL},F,n_y,n_z)\right],$$

$$Q_{29}(E_{FQWGISL},F,n_y,n_z) = \left[\sqrt{\omega_{22}(E_{FQWGISL},F,n_y,n_z)}\right],$$
$$Q_{30}(E_{FQWGISL},F,n_y,n_z) = \sum_{R=1}^{R=R_0} Z(R_{FQWGISL})\left[Q_{29}(E_{FQWGISL},F,n_y,n_z)\right],$$

$$Z(R_{FQWGSL}) = 2(k_BT)^{2R}(1-2^{1-2R})\xi(2R)\frac{\partial^{2R}}{\partial E_{FQWGSL}}$$

and E_{FQWGSL} is the Fermi energy in the present case.

The EEM in this case can be expressed as

$$m^*(E_{FQWGSL}, F, n_y, n_z) = (\hbar^2/2)\omega'_{22}(E_{FQWGSL}, F, n_y, n_z) \tag{7.72}$$

where

$$\omega'_{22}(E_{FQWGSL}, F, n_y, n_z) = \frac{2f'_{13}(E_{FQWGSL}, F, n_y, n_z)\left[\cos^{-1}\left\{\frac{1}{2}f_{13}(E_{FQWGSL}, F, n_y, n_z)\right\}\right]}{\sqrt{4 - f_{13}^2(E_{FQWGSL}, F, n_y, n_z)}},$$

$$\begin{aligned}
&f'_{13}(E_{FQWGSL}, F, n_y, n_z) = \left[2M'_{31}(n_y, n_z, E_{FQWGSL})\right.\\
&\times \sinh\left\{M_{31}(n_y, n_z, E_{FQWGSL})\right\} \cos\left\{N_{31}(n_y, n_z, E_{FQWGSL})\right\}\\
&+Z_{31}(n_y, n_z, E_{FQWGSL})M'_{31}(n_y, n_z, E_{FQWGSL}) \cosh\left\{M_{31}(n_y, n_z, E_{FQWGSL})\right\}\\
&\times \sin\left\{N_{31}(n_y, n_z, V_0)\right\} - 2N'_{31}(n_y, n_z, V_0) \sin\left\{N_{31}(n_y, n_z, V_0)\right\} \cosh\left\{M_{31}(n_y, n_z, V_0)\right\}\\
&+Z'_{31}(n_y, n_{z,} V_0) \sinh\left\{M_{31}(n_y, n_z, V_0)\right\} \sin\left\{N_{31}(n_y, n_{z,} V_0)\right\}\\
&+Z_{31}(n_y, n_z, E_{FQWGSL})N'_{31}(n_y, n_z, E_{FQWGSL}) \cos\left\{N_{31}(n_y, n_z, E_{FQWGSL})\right\}\\
&\times \sinh\left\{M_{31}(n_y, n_{z,} E_{FQWGSL})\right\}\\
&+\Delta_{21}\left[(\left\{2I_{31}(n_y, n_z, E_{FQWGSL})I'_{31}(n_y, n_z, E_{FQWGSL})\right\} (I_{32}(n_y, n_z, E_{FQWGSL}))^{-1}\right.\\
&-\left\{I_{31}^2(n_y, n_z, E_{FQWGSL})I'_{32}(n_y, n_z, E_{FQWGSL})\right\}\\
&\left.\times\ (I_{32}^2(n_y, n_z, E_{FQWGSL}))^{-1} - 3I'_{32}(n_y, n_z, E_{FQWGSL}))\right]\cdot
\end{aligned}$$

$$\begin{aligned}
&\cosh\left\{M_{31}(n_y, n_z, E_{FQWGSL})\right\} \sin\left\{N_{31}(n_y, n_z, E_{FQWGSL})\right\}\\
&+(-3I_{32}(n_y, n_z, E_{FQWGSL}) + \left\{I_{31}^2(n_y, n_z, E_{FQWGSL})\right\} (I_{32}(n_y, n_z, E_{FQWGSL}))^{-1})\\
&\{M'_{31}(n_y, n_z, E_{FQWGSL}) \sinh\left\{M_{31}(n_y, n_z, E_{FQWGSL})\right\} \sin\left\{N_{31}(n_y, n_z, E_{FQWGSL})\right\}\\
&+\{N'_{31}(n_y, n_z, E_{FQWGSL}) \cosh\left\{M_{31}(n_y, n_z, E_{FQWGSL})\right\} \cos\left\{N_{31}(n_y, n_z, E_{FQWGSL})\right\}\\
&+\left(\frac{-\left\{2I_{32}(n_y, n_z, E_{FQWGSL})I'_{32}(n_y, n_z, E_{FQWGSL})\right\}}{I_{31}(n_y, n_z, E_{FQWGSL})}\right.\\
&\qquad\left.+\frac{\left\{I_{32}^2(n_y, n_z, E_{FQWGSL})I'_{31}(n_y, n_z, E_{FQWGSL})\right\}}{I_{31}^2(n_y, n_z, E_{FQWGSL})} + 3I'_{31}(n_y, n_z, E_{FQWGSL})\right)\cdot\\
&\sinh\left\{M_{31}(n_y, n_z, E_{FQWGSL})\right\} \cos\left\{N_{31}(n_y, n_z, E_{FQWGSL})\right\}\\
&+\left(+3I_{31}(n_y, n_z, E_{FQWGSL}) - \frac{\left\{I_{32}(n_y, n_z, E_{FQWGSL})\right\}^2}{I_{31}(n_y, n_z, E_{FQWGSL})}\right)\\
&\times \{M'_{31}(n_y, n_z, E_{FQWGSL}) \cosh\left\{M_{31}(n_y, n_z, E_{FQWGSL})\right\}\cdot\\
&\cos\left\{N_{31}(n_y, n_z, E_{FQWGSL})\right\} - N'_{31}(n_y, n_z, E_{FQWGSL}) \sin\left\{N_{31}(n_y, n_z, E_{FQWGSL})\right\}\\
&\times \sinh\left\{M_{31}(n_y, n_z, E_{FQWGSL})\right\}\}]\\
&+\Delta_{21}\left[4\left(\left\{I_{31}(n_y, n_z, E_{FQWGSL})I'_{31}(n_y, n_z, E_{FQWGSL})\right\}\right.\right.\\
&\left.-\left\{I_{32}(n_y, n_z, E_{FQWGSL})I'_{32}(n_y, n_z, E_{FQWGSL})\right\}\right)\cdot
\end{aligned}$$

$$\cosh\{M_{31}(n_y, n_z, E_{FQWGSL})\}\cos\{N_{31}(n_y, n_z, E_{FQWGSL})\}$$
$$+2\left(\{I_{31}(n_y, n_z, E_{FQWGSL})\}^2 - \{I_{32}(n_y, n_z, E_{FQWGSL})\}^2\right)$$
$$\times\{M'_{31}(n_y, n_z, E_{FQWGSL})\sinh\{M_{31}(n_y, n_z, E_{FQWGSL})\}$$

$$\cos\{N_{31}(n_y, n_z, E_{FQWGSL})\} - N'_{31}(n_y, n_z, E_{FQWGSL})$$
$$\times\cosh\{M_{31}(n_y, n_z, E_{FQWGSL})\}\sin\{N_{31}(n_y, n_z, E_{FQWGSL})\}\}$$
$$+\frac{1}{12}\left(\frac{15\{I_{32}^2(n_y,n_z,E_{FQWGSL})\}I'_{32}(n_y,n_z,E_{FQWGSL})}{I_{21}(n,E_{FQWGSL})} - \frac{5\{I_{32}(n_y,n_z,E_{FQWGSL})\}^3 I'_{31}(n_y,n_z,E_{FQWGSL})}{I_{31}^2(n_y,n_z,E_{FQWGSL})}\right.$$
$$+\frac{15\{I_{31}^2(n_y,n_z,E_{FQWGSL})\}I'_{31}(n_y,n_z,E_{FQWGSL})}{I_{32}(n_y,n_z,E_{FQWGSL})} - \frac{5\{I_{31}(n_y,n_z,E_{FQWGSL})\}^3 I'_{32}(n_y,n_z,E_{FQWGSL})}{I_{32}^2(n_y,n_z,E_{FQWGSL})}$$
$$-\{34I'_{32}(n_y, n_z, E_{FQWGSL})I_{31}(n_y, n_z, E_{FQWGSL})\}$$
$$\left.-34I_{32}(n_y, n_z, E_{FQWGSL})I'_{31}(n_y, n_z, E_{FQWGSL})\right)\cdot$$

$$\sinh\{M_{31}(n_y, n_z, E_{FQWGSL})\}\sin\{N_{31}(n_y, n_z, E_{FQWGSL})\}$$
$$+\left(\frac{5\{I_{32}(n_y, n_z, E_{FQWGSL})\}^3}{I_{31}(n_y, n_z, E_{FQWGSL})} + \frac{5\{I_{31}(n_y, n_z, E_{FQWGSL})\}^3}{I_{32}(n_y, n_z, E_{FQWGSL})}\right.$$
$$\left.-\{34I_{32}(n_y, n_z, E_{FQWGSL})I_{31}(n_y, n_z, E_{FQWGSL})\}\right)$$

$$\{M'_{31}(n_y, n_z, E_{FQWGSL})\cosh\{M_{31}(n_y, n_z, E_{FQWGSL})\}\sin\{N_{31}(n_y, n_z, E_{FQWGSL})\}$$
$$+N'_{31}(n_y, n_z, E_{FQWGSL})\sinh\{M_{31}(n_y, n_z, E_{FQWGSL})\}\cos\{N_{31}(n_y, n_z, E_{FQWGSL})\}\}]],$$
$$M'_{31}(n_y, n_z, E_{FQWGSL}) = I'_{31}(n_y, n_z, E_{FQWGSL})[a_0 - \Delta_{21}],$$
$$I'_{31}(E_{FQWGSL}, n_y, n_z) = \frac{m_{c2}\beta'_{012}(E_{FQWGSL} - \bar{V}_0, F, E_{g2}, \Delta_2)}{-\hbar^2 I_{31}(E_{FQWGSL}, n_y, n_z)},$$
$$N'_{31}(n_y, n_z, E_{FQWGSL}) = I'_{32}(n_y, n_z, E_{FQWGSL})[b_0 - \Delta_{21}],$$
$$I'_{32}(E_{FQWGSL}, n_y, n_z) = \frac{m_{c1}\beta'_{11}(E_{FQWGSL}, F, E_{g2}, \Delta_2, n_y, n_z)}{\hbar^2 I_{32}(E_{FQWGSL}, n_y, n_z)},$$
$$I_{31}(n_y, n_z, E_{FQWGSL}) = \left[-\frac{2m_{c2}}{\hbar^2}\beta_{012}(E_{FQWGSL} - \overline{V_0}, F, E_{g2}, \Delta_2) + H(n_y, n_z)\right]^{1/2},$$
$$N_{31}(n_y, n_z, E_{FQWGSL}) = I_{32}(n_y, n_z, E_{FQWGSL})[b_0 - \Delta_{21}],$$

$$I_{32}(n_y, n_z, E_{FQWGSL}) = \left[\frac{2m_{c1}}{\hbar^2}\beta_{012}(E_{FQWGSL}, F, E_{g1}, \Delta_1) - H(n_y, n_z)\right]^{1/2} \text{ and}$$

$$Z'_{31}(E_{FQWGSL}, n_y, n_z) = \left[\frac{-Z_{31}(E_{FQWGSL}, n_y, n_z)I'_{31}(E_{FQWGSL}, n_y, n_z)}{I_{31}(E_{FQWGSL}, n_y, n_z)}\right.$$
$$-\frac{Z_{31}(E_{FQWGSL}, n_y, n_z)I'_{32}(E_{FQWGSL}, n_y, n_z)}{I_{32}(V_0, n_y, n_z)}$$
$$+(I_{31}(E_{FQWGSL}, n_y, n_z)I_{32}(E_{FQWGSL}, n_y, n_z))^{-1}$$
$$\times\left[2I'_{31}(E_{FQWGSL}, n_y, n_z)I_{31}(E_{FQWGSL}, n_y, n_z)\right.$$
$$\left.\left.-2I'_{32}(E_{FQWGSL}, n_y, n_z)I_{32}(E_{FQWGSL}, n_y, n_z)\right]\right]$$

For perturbed two band model of Kane, the form of electron concentration per unit length and the EEM remain same where

$$I_{31}(n_y, n_z, E) = \left[H(n_y, n_z) - \frac{2m_{c2}}{\hbar^2}\rho_{012}(E - \bar{V}_0, F, E_{g2}) \right]^{1/2},$$

$$I'_{31}(E_{FQWGSL}, n_y, n_z) = \frac{m_{c2}\rho'_{012}(E_{FQWGSL} - \bar{V}_0, F, E_{g2})}{\hbar^2 I_{31}(E_{FQWGSL}, n_y, n_z)},$$

$$I_{32}(n_y, n_z, E_{FQWGSL}) = \left[-H(n_y, n_z) + \frac{2m_{c1}}{\hbar^2}\rho_{11}(E_{FQWGSL}, F, E_{g1}) \right]^{1/2},$$

and $I'_{32}(E_{FQWGSL}, n_y, n_z) = \frac{m_{c1}\rho'_{11}(E_{FQWGSL}, F, E_{g1})}{\hbar^2 I_{32}(E_{FQWGSL}, n_y, n_z)}$.

7.3 Results and Discussion

The effect of an intense electric field on the EEM of the III-V materials has been exhibited in Figs. 7.1 and 7.2 by taking n-InSb as an example. Using (7.2) and (7.6c), the variation of the EEM along k_x direction has been demonstrated in Fig. 7.1 at different field strengths. It appears that the EEM at higher field strength becomes a linear function of carrier concentration, thus exhibiting an exponential dependency of the Fermi energy on the degeneracy. However, at lower field strengths, the EEM approaches the corresponding bulk variation. It also appears from the same figure that the EEM at different field strengths converges at higher carrier degeneracy zone rather than at non-degeneracy zone. This variation is opposite to that exhibited in Fig. 7.2 for the EEM along the rest other two directions using Eqs. (7.3) and (7.4). It appears that the EEM departs from their bulk isotropic value to almost 4 times, a significant increment due to the carrier degeneracy, where the electric field changes the EEM at the higher degeneracy level to almost 33%.

The effect of film thickness under the presence of a strong electric field in quantum wells of n-InSb has been exhibited in Fig. 7.3 for the two lowest subbands due to the size quantization. Quantized variations in EEM are exhibited at both the subband levels which marks a quantum number-dependent EEM. It appears that as the field strength increases, the response of the EEM increases however at slow rate. A close inspection reveals that for a particular subband the EEM has a tendency to decrease as also exhibited in Fig. 1.4 of Chap. 1. However, the presence of the electric field raises the subbands when the former crosses with the Fermi energy. The negative values of the EEM at the second subband in Fig. 7.3 is of interest and specify the validity region of the band structure formalism at such high field. It appears that the EEM becomes negative in the sub-15 nm film thickness and thus questioning the validity of the quantum number- dependent EEM in this regime.

We have already written a lot and still we have to move an infinitely long path in the direction to reach the creative knowledge temple. We leave all the computer programming and related graphs together with the inside physics for all the remaining materials of this chapter to the able shoulders of our readers whom we believe are creatively fur superior than that of us. As usual, for the last time the summary of this chapter has been presented in Table 7.1.

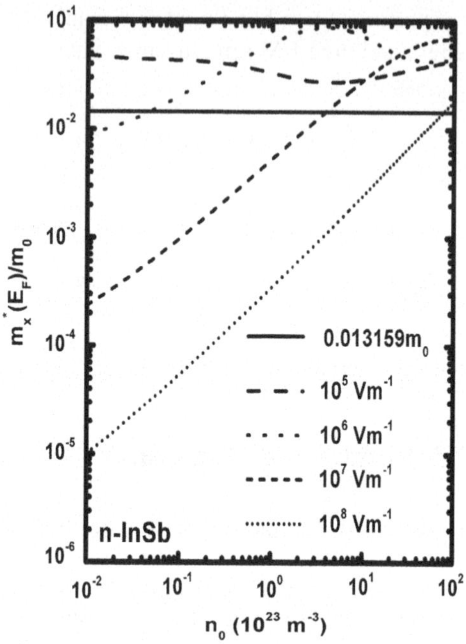

Fig. 7.1 Plot of the EEM along the k_x direction as function of carrier concentration in bulk n-InSb at different field strengths

7.4 Open Research Problems

(R.7.1) Investigate the EEM, EAM, DEM, CEM, CoEM, FREM, and OEM for the bulk materials whose respective dispersion relations of the carriers in the absence of any field is given in Chap. 1 in the presence of intense electric field which change the original band structure and consider its effect in the subsequent: study in each case.

(R7.2) Investigate the same set of masses as defined in (R7.1) in the presence of an arbitrarily oriented non-uniform light waves for all the materials as considered R7.1.

(R7.3) Investigate the same set of masses as defined in (R7.1) in the presence of an arbitrarily oriented non-quantizing alternating non-uniform electric field for all the cases of R7.1.

(R7.4) Investigate the same set of masses as defined in (R7.1) for the heavily doped materials in the presence of Gaussian, exponential, Kane, Halperin, Lax, and Bonch-Bruevich types of band tails for all materials whose unperturbed carrier energy spectra are defined in R7.1

(R7.5) Investigate the same set of masses as defined in (R7.1) for all the materials in the presence of arbitrarily oriented non-quantizing non-uniform electric field for all the appropriate cases of problem R7.4.

(R7.6) Investigate the same set of masses as defined in (R7.1) for all the materials in the presence of arbitrarily oriented non-quantizing alternating electric field for all the appropriate cases of problem R7.4.

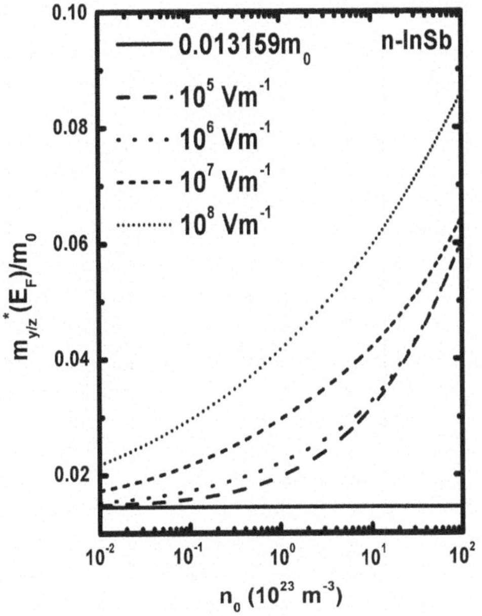

Fig. 7.2 Plot of the EEM along the k_y and k_z directions as function of carrier concentration in bulk n-InSb at different field strengths

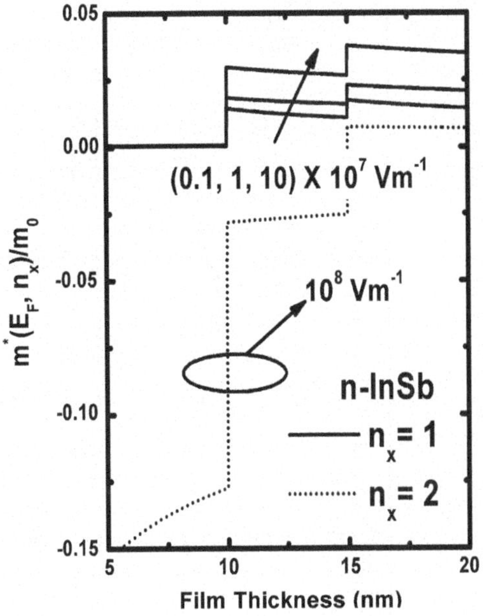

Fig. 7.3 Plot of the EEM at the lowest two subbands as function of film thickness in quantum wells of n-InSb at different field strengths and at an extreme carrier degeneracy of 10^{18} m^{-2}

Table 7.1 The EEM and the electron statistics for different quantized of optoelectronic materials as considered in this chapter in the presence of intense electric field

Type of materials	EEM	The electron statistics
1. Bulk opto-electronic and III–V materials under intense electric field:		
a) Perturbed three band model of Kane	$m_x^*(E,F) = m_c\,[1+\Phi(E,F)]^{-2}$ $\left[[1+\Phi(E,F)]\,[I_{11}(E)]' - [I_{11}(E)]\,\Phi'(E,F)\right]$ (7.2) $m_y^*(E,F) = m_c\,[I_{11}(E)]'$ (7.3) $m_z^*(E,F) = m_c\,[I_{11}(E)]'$ (7.4)	$n_0 = C_{7L}\,[M_{1L}(E_{FF},F) + N_{1L}(E_{FF},F)]$ (7.6a)
b) Perturbed two band model of Kane	$m_x^*(E_{FF},F) = m_c\,[1+\Phi_1(E_{FF},F)]^{-2}$ $[[1+\Phi_1(E_{FF},F)]\,[1+2\alpha E_{FF}]$ $-[E_{FF}(1+\alpha E_{FF})]\,\Phi_1'(E_{FF},F)]$ (7.6c) $m_y^*(E_{FF},F) = m_c\,[1+2\alpha E_{FF}]$ (7.7) $m_z^*(E_{FF},F) = m_c\,[1+2\alpha E_{FF}]$ (7.8)	$n_0 = C_{7L}\,[M_{2L}(E_{FF},F) + N_{2L}(E_{FF},F)]$ (7.10)
2.The magneto EEM in optoelectronic and III–V semiconductors under strong electric field		
a) Perturbed three band model of Kane	$m^*(E_{FFB},n,F,\theta) = \frac{\hbar^2}{2}$ $\left[A_7'(E_{FFB},F,\theta) - \frac{2eB}{\hbar}\left(n+\frac{1}{2}\right)B_7'(E_{FFB},F,\theta)\right]$ (7.15)	$n_0 = \frac{eB}{\pi^2\hbar}\sum_{n=0}^{n_{\max}}[M_3(E_{FFB},F,\theta,n) + N_3(E_{FFB},F,\theta,n)]$ (7.18)
b) Perturbed two band model of Kane	$m^*(E_{FFB},n,F,\theta) = \frac{\hbar^2}{2}$ $\left[A_{71}'(E_{FFB},F,\theta) - \frac{2eB}{\hbar}\left(n+\frac{1}{2}\right)B_{71}'(E_{FFB},F,\theta)\right]$ (7.20)	$n_0 = \frac{eB}{\pi^2\hbar}\sum_{n=0}^{n_{\max}}[M_4(E_{FFB},F,\theta,n) + N_4(E_{FFB},F,\theta,n)]$ (7.22)
3. The EEM in quantum wells of optoelectronic semiconductors under strong electric field:		

(continued)

Table 7.1 (continued)

Type of materials	EEM	The electron statistics
a) Perturbed three-band model of Kane	$m^*(E_{FFS}, F, n_x) = \frac{\hbar^2}{2}\phi_{100}$ (7.26b)	$n_0 = \frac{g_v}{2\pi} \sum_{n_x=1}^{n_{x\max}} [M_5(E_{FFS}, F, n_x) + N_5(E_{FFS}, F, n_x)]$ (7.27)
b)Perturbed two band model of Kane	$m^*(E_{FS}, F, n) = \hbar^2\phi_{101}$ (7.29)	$n_0 = \frac{g_v}{2\pi} \sum_{n_x=1}^{n_{x\max}} [M_6(E_{FFS}, F, n_x) + N_6(E_{FFS}, F, n_x)]$ (7.32)
4. The EEM in NIPI structures of optoelectronic semiconductors under strong electric field:		
a) Perturbed three band model of Kane	$m^*(\bar{E}_{Fn}, F, n_i) = \hbar^2\phi_{102}$ (7.36)	$n_0 = \frac{g_v}{2\pi} \sum_{n_i=1}^{n_{i\max}} \left[M_7(\bar{E}_{Fn}, F, n_i) + N_7(\bar{E}_{Fn}, F, n_i)\right]$ (7.37)
b) Perturbed two-band model of Kane	$m^*(\bar{E}_{Fn}, F, n_i) = \hbar^2\phi_{103}$ (7.39)	$n_0 = \frac{g_v}{2\pi} \sum_{n_i=1}^{n_{i\max}} \left[M_8(\bar{E}_{Fn}, F, n_i) + N_8(\bar{E}_{Fn}, F, n_i)\right]$ (7.42)
5.The EEM in n-channel inversion layers of optoelectronic semiconductors:		
a) Perturbed three-band model of Kane	$m^*(E_{Fs}, F, i) = \frac{\hbar^2}{2} J_7'(E, F, i)\big\vert_{E=E_{Fs}}$ (7.44)	$n_0 = \frac{g_v}{2\pi} \sum_{i=0}^{i_{\max}} [J_7(E_{Fs}, F, i) + H_7(E_{Fs}, F, i)]$ (7.45)
b) Perturbed two-band model of Kane	$m^*(E_{Fs}, F, i) = \frac{\hbar^2}{2} J_{71}'(E, F, i)\big\vert_{E=E_{Fs}}$ (7.47)	$n_0 = \frac{g_v}{2\pi} \sum_{i=0}^{i_{\max}} [J_{71}(E_{Fs}, F, i) + H_{71}(E_{Fs}, F, i)]$ (7.48)
6.The EEM in nanowires of optoelectronic semiconductors:		
a) Perturbed three-band model of Kane	$m^*(E_{F1D}, F) = m_c\beta_{11}'(E_{F1D}, F)$ (7.51)	$n_0 = \frac{2g_v}{\pi} \sum_{n_z=1}^{n_{z\max}} \sum_{n_y=1}^{n_{y\max}} [Q_{15}(E_{F1D}, F, n_y, n_z) + Q_{16}(E_{F1D}, F, n_y, n_z)]$ (7.53)
	$m^*(E_{F1D}, F) = m_c\bar{\beta}_{12}'(E_{F1D}, F)$ (7.52)	$n_0 = \frac{2g_v}{\pi} \sum_{n_z=1}^{n_{z\max}} \sum_{n_y=1}^{n_{y\max}} [Q_{17}(E_{F1D}, F, n_y, n_z) + Q_{18}(E_{F1D}, F, n_y, n_z)]$ (7.54)

(continued)

Table 7.1 (continued)

Type of materials	EEM	The electron statistics
7.The EEM in effective mass superlattices of optoelectronic semiconductors under magnetic quantization :		
a) Perturbed three-band model of Kane	$m^*(E_{\mathrm{FB}}, F, n) = (\hbar^2/2)\omega'_{17}(E_{\mathrm{FB}}, F, n)$ (7.57)	$n_o = \frac{g_v eB}{\pi^2 \hbar L_o}[\sum_{n=0}^{n_{\max}}[Q_{19}(E_{\mathrm{FB}}, F, n) + Q_{20}(E_{\mathrm{FB}}, F, n)]]$ (7.58)
b) Perturbed three-band model of Kane	$m^*(E_{\mathrm{FB}}, F, n) = (\hbar^2/2)\omega'_{18}(E_{\mathrm{FB}}, F, n)$ (7.60)	$n_o = \frac{g_v eB}{\pi^2 \hbar L_o}[\sum_{n=0}^{n_{\max}}[Q_{21}(E_{\mathrm{FB}}, F, n) + Q_{22}(E_{\mathrm{FB}}, F, n)]]$ (7.59)
8.The EEM in nanowire effective mass superlattices of optoelectronic semiconductors:		
a) Perturbed three band model of Kane	$m^*(E_{FIDEMSL}, F, n_y, n_z) = (\hbar^2/2)\omega'_{19}(E_{FIDEMSL}, F, n_y, n_z)$ (7.63)	$n_0 = \frac{2g_v}{\pi}\phi_{104}$ (7.62)
b) Perturbed three-band model of Kane	$m^*(E_{FIDEMSL}, F, n_y, n_z) = (\hbar^2/2)\omega'_{20}(E_{FIDEMSL}, F, n_y, n_z)$(7.65)	$n_0 = \frac{2g_v}{\pi}\phi_{105}$ (7.64)
9.The EEM in superlattices of optoelectronic semiconductors with graded interfaces under magnetic quantization:		
a) Perturbed three-band model of Kane	$m^*(E_{FBGISL}, F, n) = (\hbar^2/2)\omega'_{21}(E_{FBGISL}, F, n)$ (7.69)	$n_o = \frac{g_v eB}{\pi^2 \hbar}\phi_{106}$ (7.68)
10.The EEM in quantum wire superlattices of optoelectronic semiconductors with graded interfaces :		
a) Perturbed three-band model of Kane	$m^*(E_{FQWGSL}, F, n_y, n_z) = (\hbar^2/2)\omega'_{22}(E_{FQWGSL}, F, n_y, n_z)$ (7.72)	$n_0 = \frac{2g_v}{\pi}\phi_{107}$ (7.71)

(R7.7) Investigate the same set of masses as defined in (R7.1) for the negative refractive index, organic, magnetic, and other advanced optical materials in the presence of arbitrarily oriented electric field.
(R7.8) Investigate the same set of masses as defined in (R7.1) in the presence of alternating non-quantizing electric field for all the problems of R7.7.
(R7.9) Investigate the same set of masses as defined in (R7.1) for all the quantum confined materials (i.e, multiple quantum wells and wires) whose unperturbed carrier energy spectra are defined in R7.1 in the presence of arbitrary oriented quantizing magnetic field by including the effects of spin and broadening respectively.
(R7.10) Investigate the same set of masses as defined in (R7.1) in the presence of an additional arbitrarily oriented alternating quantizing magnetic field, respectively, for all the problems of R7.9.
(R7.11) Investigate the same set of masses as defined in (R7.1) in the presence of arbitrarily oriented alternating quantizing magnetic field and arbitrary oriented non-quantizing non-uniform electric field, respectively, for all the problems of R7.9.
(R7.12) Investigate the same set of masses as defined in (R7.1) in the presence of arbitrary oriented alternating non- uniform quantizing magnetic field and additional arbitrary oriented non-quantizing alternating electric field respectively for all the problems of R7.9.
(R7.13) Investigate the same set of masses as defined in (R7.1) in the presence of arbitrary oriented and crossed quantizing magnetic and electric fields respectively for all the problems of R7.9.
(R7.14) Investigate the same set of masses as defined in (R7.1) for all the appropriate low-dimensional systems of this chapter in the presence of finite potential wells.
(R7.15) Investigate the same set of masses as defined in (R7.1) for all the appropriate low-dimensional systems of this chapter in the presence of parabolic potential wells.
(R7.16) Investigate the same set of masses as defined in (R7.1) for all the appropriate systems of this chapter forming quantum rings.
(R7.17) Investigate the same set of masses as defined in (R7.1) for all the above appropriate problems in the presence of elliptical Hill and quantum square rings respectively.
(R7.18) Investigate the same set of masses as defined in (R7.1) for multiple wall carbon nano-tubes. .
(R7.19) Investigate the same set of masses as defined in (R7.1) for multiple wall carbon nano-tubes in the presence of non-quantizing non-uniform alternating light waves.
(R7.20) Investigate the same set of masses as defined in (R7.1) for multiple wall carbon nanotubes in the presence of non-quantizing non-uniform alternating magnetic field.
(R7.21) Investigate the same set of masses as defined in (R7.1) for multiple wall carbon nanotubes in the presence of crossed electric and quantizing magnetic fields.
(R7.22) Investigate the same set of masses as defined in (R7.1) for heavily doped semiconductor nano-tubes for all the materials whose unperturbed carrier dispersion laws are defined in Chap. 1.

(R7.23) Investigate the same set of masses as defined in (R7.1) for heavily doped semiconductor nanotubes in the presence of non-quantizing alternating light waves for all the materials whose unperturbed carrier dispersion laws are defined in Chap. 1.

(R7.24) Investigate the same set of masses as defined in (R7.1) for heavily doped semiconductor nanotubes in the presence of non-quantizing alternating magnetic field for all the materials whose unperturbed carrier dispersion laws are defined in Chap. 1.

(R7.25) Investigate the same set of masses as defined in (R7.1) for heavily doped semiconductor nanotubes in the presence of non-uniform light waves for all the materials whose unperturbed carrier dispersion laws are defined in Chap. 1.

(R7.26) Investigate the same set of masses as defined in (R7.1) for heavily doped semiconductor nano-tubes in the presence of alternating quantizing magnetic fields for all the materials whose unperturbed carrier dispersion laws are defined in Chap. 1.

(R7.27) Investigate the same set of masses as defined in (R7.1) for heavily doped semiconductor nano-tubes in the presence of crossed electric and quantizing magnetic fields for all the materials whose unperturbed carrier dispersion laws are defined in Chap. 1.

(R7.28) Investigate the same set of masses as defined in (R7.1) for all the appropriate nipi structures of the materials whose unperturbed carrier energy spectra are defined in Chap. 1.

(R7.29) Investigate the same set of masses as defined in (R7.1) for all the appropriate nipi structures of the materials whose unperturbed carrier energy spectra are defined in Chap. 1, in the presence of an arbitrarily oriented non-quantizing non-uniform additional electric field.

(R7.30) Investigate the same set of masses as defined in (R7.1) for all the appropriate nipi structures of the materials whose unperturbed carrier energy spectra are defined in Chap. 1 in the presence of non-quantizing alternating additional magnetic field.

(R7.31) Investigate the same set of masses as defined in (R7.1) for all the appropriate nipi structures of the materials whose unperturbed carrier energy spectra are defined in Chap. 1. in the presence of quantizing alternating additional magnetic field.

(R7.32) Investigate the same set of masses as defined in (R7.1) for all the appropriate nipi structures of the materials whose unperturbed carrier energy spectra are defined in Chap. 1 in the presence of crossed electric and quantizing magnetic fields.

(R7.33) Investigate the same set of masses as defined in (R7.1) for heavily doped nipi structures for all the appropriate cases of all the above problems.

(R7.34) Investigate the same set of masses as defined in (R7.1) for the appropriate accumulation layers of all the materials whose unperturbed carrier energy spectra are defined in Chap. 1 in the presence of crossed electric and quantizing magnetic fields by considering electron spin and broadening of Landau levels.

(R7.35) Investigate the same set of masses as defined in (R7.1) for quantum confined III-V, II-VI, IV-VI, HgTe/CdTe effective mass super-lattices together with short period, strained layer, random, Fibonacci, poly-type and sawtooth super-lattices

(R7.36) Investigate the same set of masses as defined in (R7.1) in the presence of quantizing magnetic field, respectively, for all the cases of R7.35.

(R7.37) Investigate the same set of masses as defined in (R7.1) in the presence of non-quantizing non-uniform additional electric field, respectively, for all the cases of R7.35.

(R7.38) Investigate the same set of masses as defined in (R7.1) in the presence of non-quantizing alternating electric field, respectively, for all the cases of R7.35

(R7.39) Investigate the same set of masses as defined in (R7.1) in the presence of crossed electric and quantizing magnetic fields, respectively, for all the cases of R7.35.

(R7.40) Investigate the same set of masses as defined in (R7.1) for heavily doped quantum confined superlattices for all the problems of R7.35.

(R7.41) Investigate the same set of masses as defined in (R7.1) in the presence of quantizing non-uniform magnetic field, respectively, for all the cases of R7.40.

(R7.42) Investigate the same set of masses as defined in (R7.1) in the presence of crossed electric and quantizing magnetic fields respectively for all the cases of R7.40.

(R7.43) Investigate the same set of masses as defined in (R7.1) for all the systems in the presence of strain.

(R7.44) Investigate all the problems of this chapter by removing all the mathematical approximations and establishing the respective appropriate uniqueness conditions.

References

1. P.K. Chakraborty, S. Choudhury, K.P. Ghatak, Physica B **387**, 333 (2007)
2. K.P. Ghatak, S. Bhattacharya, D. De, *Einstein Relation in Compound Semiconductors and Their Nanostructures*, Springer Series in Materials Science, **116** (Springer-Verlag, Germany, 2009)
3. H. Sasaki, Phys. Rev. B **30**, 7016 (1984)
4. H.X. Jiang, J.Y. Lin, J. Appl. Phys. **61**, 624 (1987)

Chapter 8
Applications and Brief Review of Experimental Results

8.1 Introduction

In this monograph, we have investigated many aspects of the effective masses of the carriers based on the dispersion relations of the semiconductor nanostructures of different technologically important quantum confined materials having different band structures. In this chapter, we shall discuss few applications in this context in Sect. 8.2 and also present a very brief review of the experimental investigations in Sect. 8.3. The Sect. 8.4 contains the single experimental open research problem.

8.2 Applications

The investigations as presented in this monograph find nine different applications in the realm of modern quantum effect devices.

8.2.1 Thermoelectric Power:

In recent years, with the advent of Quantum Hall Effect (QHE) [1, 2], there has been considerable interest in studying the thermoelectric power under strong magnetic field (TPSM) in various types of nanostructured materials having quantum confinement of their charge carriers in one, two, and three dimensions of the respective wave vector space leading to different carrier energy spectra [3–44]. The classical TPSM equation is valid only under the condition of carrier non-degeneracy, being independent of carrier concentration and reflects the fact that the signature of the band structure of any material is totally absent in the same.

S. Bhattacharya and K. P. Ghatak, *Effective Electron Mass in Low-Dimensional Semiconductors*, Springer Series in Materials Science 167,
DOI: 10.1007/978-3-642-31248-9_8, © Springer-Verlag Berlin Heidelberg 2013

Zawadzki [9] demonstrated that the TPSM for electronic materials having degenerate electron concentration is essentially determined by their respective energy band structures. It has, therefore, different values in different materials and changes with the doping, magnitude of the reciprocal quantizing magnetic field under magnetic quantization, quantizing electric field as in inversion layers, nanothickness as in quantum wells, wires and dots, with superlattice period as in quantum confined semiconductor superlattices with graded interfaces having various carrier energy spectra and also in other types of field assisted nanostructured materials.

The magnitude of the thermoelectric power G can be written as [10]

$$G = \frac{1}{|e|\, T n_0} \int_{-\infty}^{\infty} (E - E_{\mathrm{F}}) R(E) \left[-\frac{\partial f_0}{\partial E} \right] \mathrm{d}E, \tag{8.1}$$

where $R(E)$ is the total number of states. The (8.1) can be written under the condition of carrier degeneracy [4] as

$$G = \left(\frac{\pi^2 k_B^2 T}{3\, |e|\, n_0} \right) \left(\frac{\partial n_0}{\partial E_{\mathrm{F}}} \right). \tag{8.2}$$

For inversion layers, heavily doped semiconductors and their nanostructures and the nipi superlattices, under the condition of electric quantum limit, (8.1) assumes the form

$$G = \left(\frac{\pi^2 k_B^2 T}{3\, |e|\, \bar{n}_0} \right) \left[\frac{\mathrm{d}\bar{n}_0}{d(\bar{E}_{F0} - \bar{E}_0)} \right]. \tag{8.3}$$

Thus, we can use the carrier statistics for different low dimensional materials to investigate the TPSM in such compounds and for the purpose of completeness we present few results of TPSM for bulk specimens as written below:

(i) Nonlinear optical materials and Cd_3As_2

The electron concentration of bulk specimens in this case can be expressed following [1, 2] as

$$n_0 = g_v (3\pi^2)^{-1} \left[M_{1a}(E_{F_b}) + N_{1a}(E_{F_b}) \right], \tag{8.4}$$

where $M_{1a}(E_{\mathrm{F}_b}) \equiv \left[\frac{[\gamma(E_{\mathrm{F}_b})]^{\frac{3}{2}}}{f_1(E_{\mathrm{F}_b})\sqrt{f_2(E_{\mathrm{F}_b})}} \right]$, E_{F_b} is the Fermi energy as measured from the edge of the conduction band in the vertically upward direction in the absence of any quantization, $N_{1a}(E_{\mathrm{F}_b}) \equiv \sum_{r=1}^{s} Z_{1a}(r) M_{1a}(E_{\mathrm{F}_b})$

and $Z_{1a}(r) \equiv \left[2(k_B T)^{2r} (1 - 2^{1-2r}) \xi(2r) \right] \left[\frac{\partial^{2r}}{\partial E_{\mathrm{F}_b}^{2r}} \right]$.

Using (8.4) and (8.2), the TPSM in this case is given by

$$G_0 = \frac{\pi^2 k_B^2 T}{3e} \left[M'_{1a}(E_{F_b}) + N'_{1a}(E_{F_b})\right] \left[M_{1a}(E_{F_b}) + N_{1a}(E_{F_b})\right]^{-1}. \quad (8.5)$$

(ii) III–V materials

(a) *Three band model of Kane*

In accordance with this model the electron concentration can be expressed as

$$n_0 = \frac{g_v}{3\pi^2}\left(\frac{2m^*}{\hbar^2}\right)^{3/2}\left[\bar{M}_A(E_{F_b}) + \bar{N}_A(E_{F_b})\right], \quad (8.6)$$

where

$$\bar{M}_A(E_{F_b}) = \left[\frac{E_{F_b}(E_{F_b}+E_g)(E_{F_b}+E_g+\Delta)\left(E_g+\frac{2}{3}\Delta\right)}{E_g\left(E_g+\Delta\right)\left(E_{F_b}+E_g+\frac{2}{3}\Delta\right)}\right]^{3/2}$$

and

$$\bar{N}_A(E_{F_b}) = \sum_{r=1}^{s} 2(k_B T)^{2r}(1-2^{1-2r})\zeta(2r)\frac{\partial^{2r}}{\partial E_{F_b}^{2r}}\left[\bar{M}_A(E_{F_b})\right]$$

Using (8.6) and (8.2), the TPSM in this case can be written as

$$G_0 = \left(\frac{\pi^2 k_B^2 T}{3e}\right)\left[\frac{(\bar{M}_A(E_{F_b}))' + (\bar{N}_A(E_{F_b}))'}{\bar{M}_A(E_{F_b}) + \bar{N}_A(E_{F_b})}\right] \quad (8.7)$$

(b) *The model of Stillman et al.*

The expression of electron concentration in this case can be written as

$$n_0 = \frac{g_v}{3\pi^2}\left(\frac{2m^*}{\hbar^2}\right)^{3/2}\left[M_{A_{10}}(E_{F_b}) + N_{A_{10}}(E_{F_b})\right], \quad (8.8)$$

where

$$M_{A_{10}}(E_{F_b}) = [I_{11}(E_{F_b})]^{3/2}$$

and

$$N_{A_{10}}(E_{F_b}) = \sum_{r=1}^{s} 2(k_B T)^{2r}(1-2^{1-2r})\zeta(2r)\frac{\partial^{2r}}{\partial E_{F_b}^{2r}}\left[M_{A_{10}}(E_{F_b})\right]$$

Using (8.8) and (8.2), the TPSM can be expressed as

$$G_0 = \left(\frac{\pi^2 k_B^2 T}{3e}\right)\left[\frac{M'_{A_{10}}(E_{F_b}) + N'_{A_{10}}(E_{F_b})}{M_{A_{10}}(E_{F_b}) + N_{A_{10}}(E_{F_b})}\right] \quad (8.9)$$

(c) *The model of Palik et al.*
In accordance with this model the electron concentration can be expressed as

$$n_0 = \frac{g_v}{3\pi^2}\left(\frac{2m^*}{\hbar^2}\right)^{3/2}\left[\bar{M}_{12A_b}(E_{F_b}) + \bar{N}_{12A_b}(E_{F_b})\right], \quad (8.10)$$

where

$$\bar{M}_{12A_b}(E_{F_b}) = \left[I_{12}(E_{F_b})\right]^{3/2}$$

and

$$\bar{N}_{12A_B}(E_{F_b}) = \sum_{r=1}^{s} 2(k_B T)^{2r}(1 - 2^{1-2r})\zeta(2r)\frac{\partial^{2r}}{\partial E_{F_b}^{2r}}\left[\bar{M}_{12A_b}(E_{F_b})\right]$$

Using (8.2) and (8.10), the TPSM in this case can be written as

$$G_0 = \left(\frac{\pi^2 k_B^2 T}{3e}\right)\left[\frac{(\bar{M}_{12A_b}(E_{F_b}))' + (\bar{N}_{12A_b}(E_{F_b}))'}{\bar{M}_{12A_b}(E_{F_b}) + \bar{N}_{12A_b}(E_{F_b})}\right] \quad (8.11)$$

(d) *Model of Johnson and Dicley*
The expressions of the electron concentration and the TPSM for this model are given by

$$n_0 = \frac{g_v}{3\pi^2}\left[M_{13A_b}(E_{F_b}) + N_{13A_b}(E_{F_b})\right] \quad (8.12)$$

and

$$G_0 = \left(\frac{\pi^2 k_B{}^2 T}{3e}\right)\left[\frac{M'_{13A_b}(E_{F_b}) + N'_{13A_b}(E_{F_b})}{M_{13A_b}(E_{F_b}) + N_{13A_b}(E_{F_b})}\right], \quad (8.13)$$

where

$$M_{13A_b}(E_{F_b}) = \left[\bar{e}_8\left(E_{F_b}\right)\right]^{3/2},$$

$$N_{13A_b}(E_{F_b}) = \sum_{r=1}^{s} 2(k_B T)^{2r}(1 - 2^{1-2r})\zeta(2r)\frac{\partial^{2r}}{\partial E_{F_b}{}^{2r}}\left[M_{13A_b}(E_{F_b})\right],$$

$$\bar{e}_8(E_{F_b}) = \left[(E_{g_0} + 2E_{F_b})e_7 + \frac{E_{g_0}^2}{4} e_8(E_{F_b}) \right.$$
$$\left. - \left[E_{g_0}^2 \left[e_7^2 + \frac{E_{g_0}^2}{16} e_8^2(E_{F_b}) + E_{F_b} e_7 e_8(E_{F_b}) + \frac{e_7 E_{g_0}}{2} e_8(E_{F_b}) \right] \right]^{1/2} \right] .(2e_7^2)^{-1}$$

$$e_7 = \frac{\hbar^2}{2}\left[\frac{1}{m^*} - \frac{1}{m_0}\right], \quad e_8(E_{F_b}) = \frac{2\hbar^2 \varphi_{A_1}(E_{F_b})}{E_{g_0} m^*},$$
$$\varphi_{A_1}(E_{F_b}) = \frac{(E_{g_0} + \Delta)\left(E_{F_b} + E_{g_0} + \frac{2}{3}\Delta\right)}{(E_{g_0} + \frac{2}{3}\Delta)\left(E_{F_b} + E_{g_0} + \Delta\right)}$$

(iii) *n-type Gallium Phosphide*

In this case, the electron concentration and the TPSM can, respectively, be written as

$$n_0 = \frac{2g_v}{4\pi^2}\left[M_{A_1}(E_{F_b}) + N_{A_1}(E_{F_b})\right] \tag{8.14}$$

$$G_0 = \left(\frac{\pi^2 k_B^2 T}{3e}\right)\left[\frac{M'_{A_1}(E_{F_b}) + N'_{A_1}(E_{F_b})}{M_{A_1}(E_{F_b}) + N_{A_1}(E_{F_b})}\right], \tag{8.15}$$

where

$$M_{A_1}(E_{F_b}) = \left[(t_{A_1}) \cdot (E_{F_b})\theta_-(E_{F_b}) + t_{A_2}\theta_-(E_{F_b}) - \frac{t_{A_3}(\theta_-(E_{F_b}))^3}{3} \right.$$
$$- \frac{t_{A_4}\theta_-(E_{F_b})}{2}\left[(\theta_-(E_{F_b}))^2 + t_{A_5}(E_{F_b})\right]^{\frac{1}{2}}$$
$$\left. + \frac{t_{A_4} t_{A_5}(E_{F_b})}{2} \ln\left|\frac{\theta_-(E_{F_b}) + \sqrt{(\theta_-(E_{F_b}))^2 + t_{A_5}(E_{F_b})}}{\sqrt{t_{A_5}(E_{F_b})}}\right| \right]$$

$$t_{A_1} = \frac{1}{a}, \quad a = \left(\frac{\hbar^2}{2m_\perp^*} + A\frac{\hbar^2}{2m_\parallel^*}\right), \quad b = \frac{\hbar^2}{2m_\parallel^*}, \quad c = \frac{\hbar^2 k_0^2}{m_\parallel^{*2}}, D = |V_G|^2,$$
$$t_{A_2} = \left[\frac{g_1}{2a^2}\right], \quad t_{A_3} = \left(\frac{b}{a}\right), \quad t_{A_4} = \left(\frac{\sqrt{g_3}}{2a^2}\right),$$

$$g_1 = (2aD - c), \quad g_2 = \left[4a^2b^2 + c^2 - 4acD\right], \quad g_3 = \left[4abc + 4a^2c\right]$$

$$t_{A_5}(E_{F_b}) = \left[\frac{g_2 - (4ac).(E_{F_b})}{g_3}\right], \quad t_{A_6} = (t_{A_4}^2 + 2t_{A_2}t_{A_3}), \quad t_{A_7} = (2t_{A_1}t_{A_3}),$$

$$t_{A_8} = \left[t_{A_4}^4 + 4t_{A_4}^2 t_{A_2} t_{A_3} + (4t_{A_3}^2 t_{A_4}^2 g_2/g_3)\right],$$
$$t_{A_9} = \left[4t_{A_1} t_{A_3} t_{A_4}^2 + 8t_{A_1} t_{A_2} t_{A_3}^2 - (16t_{A_3}^2 t_{A_4}^2 ac/g_3)\right]$$

and $\theta_-(E_{F_b}) = (t_{A_3}.\sqrt{2})^{-1}\left[t_{A_6} + (E_{F_b}).(t_{A_7}) - (t_{A_8} + (t_{A_9}).(E_{F_b}))^{1/2}\right]$

(iv) *II–VI materials*

The expressions of electron concentration and the TPSM for II–VI materials assume the forms

$$n_0 = \frac{1}{2}\left(\frac{k_B T}{\pi b_0'}\right)^{3/2}\left(\frac{b_0'}{a_0'}\right)\left[F_{\frac{1}{2}}(\eta) + \frac{\bar{\lambda}_0^2}{2a_0' k_B T} F_{\frac{-1}{2}}(\eta)\right] \tag{8.16}$$

$$G_0 = \left(\frac{\pi^2 k_B}{3e}\right)\left[\frac{F_{\frac{-1}{2}}(\eta) + (\bar{\lambda}_0^2/2a_0' k_B T) F_{\frac{-3}{2}}(\eta)}{F_{\frac{1}{2}}(\eta) + (\bar{\lambda}_0^2/2a_0' k_B T) F_{\frac{-1}{2}}(\eta)}\right] \tag{8.17}$$

(v) Stressed Materials

In this case electron concentration and TPSM assume the forms

$$n_0 = g_v(3\pi^2)^{-1}\left[M_{A_2}(E_{F_b}) + N_{A_2}(E_{F_b})\right] \tag{8.18}$$

$$G_0 = \left(\frac{\pi^2 k_B^2 T}{3e}\right)\left[\frac{M_{A_2}'(E_{F_b}) + N_{A_2}'(E_{F_b})}{M_{A_2}(E_{F_b}) + N_{A_2}(E_{F_b})}\right], \tag{8.19}$$

where

$$M_{A_2}(E_{F_b}) = \left[a^*(E_{F_b}) b^*(E_{F_b}) c^*(E_{F_b})\right]$$

and

$$N_{A_2}(E_{F_b}) = \sum_{r=1}^{s} 2(k_B T)^{2r}(1 - 2^{1-2r})\zeta(2r)\frac{\partial^{2r}}{\partial E_{F_b}^{2r}}\left[M_{A_2}(E_{F_b})\right].$$

(vi) *IV–VI Semiconductors*

(a) *Bangert and Kästner model*

In this case electron concentration and the TPSM can, respectively, be expressed as

$$n_0 = \left(\frac{g_v}{3\pi^2}\right)\left[M_{A_3}(E_{F_b}) + N_{A_3}(E_{F_b})\right] \tag{8.20}$$

and

$$G_0 = \left(\frac{\pi^2 k_B^2 T}{3e}\right)\left[\frac{M_{A_3}'(E_{F_b}) + N_{A_3}'(E_{F_b})}{M_{A_3}(E_{F_b}) + N_{A_3}(E_{F_b})}\right], \tag{8.21}$$

where

$$M_{A_3}(E_{F_b}) = \left[\tau_A(E_{F_b})\right]^{3/2}\left[\bar{F}_1(E_{F_b})\sqrt{\bar{F}_2(E_{F_b})}\right]^{-1}, \quad \tau_A(E_{F_b}) = 2E_{F_b}$$

and

$$N_{A_3}(E_{F_b}) = \sum_{r=1}^{s} 2(k_B T)^{2r}(1-2^{1-2r})\zeta(2r)\frac{\partial^{2r}}{\partial E_{F_b}^{2r}}\left[M_{A_3}(E_{F_b})\right].$$

(b) *Cohen Model*
In this case electron concentration and the TPSM can, respectively, be written as

$$n_0 = \left(\frac{g_v\sqrt{m_1 m_3}}{\pi^2\hbar}\right)\left[M_{A_3}(E_{F_b}) + N_{A_3}(E_{F_b})\right] \tag{8.22}$$

$$G_0 = \left(\frac{\pi^2 k_B^2 T}{3e}\right)\left[\frac{M'_{A_3}(E_{F_b}) + N'_{A_3}(E_{F_b})}{M_{A_3}(E_{F_b}) + N_{A_3}(E_{F_b}}\right], \tag{8.23}$$

where

$$M_{A_3}(E_{F_b}) = \tau_{A_1}(E_{F_b})\left[E_{F_b}(1+\alpha E_{F_b}) - \frac{\tau_{A_1}^4(E_{F_b})}{20m_2m_2'} + \frac{\alpha E_{F_b}\tau_{A_1}^2(E_{F_b})}{6m_2'} - \frac{\tau_{A_1}^2(E_{F_b})(1+\alpha E_{F_b})}{6m_2}\right]$$

$$\tau_{A_1}(E_{F_b}) = \left[\frac{\alpha}{2m_2m_2'}\right]^{-1/2}\left[-\left[\frac{1+\alpha E_{F_b}}{2m_2} - \frac{\alpha E_{F_b}}{2m_2'}\right] + \left[\left[\frac{1+\alpha E_{F_b}}{2m_2} - \frac{\alpha E_{F_b}}{2m_2'}\right]^2 + \frac{\alpha E_{F_b}(1+\alpha E_{F_b})}{m_2m_2'}\right]^{1/2}\right]^{1/2}$$

and

$$N_{A_3}(E_{F_b}) = \sum_{r=1}^{s} 2(k_B T)^{2r}(1-2^{1-2r})\zeta(2r)\frac{\partial^{2r}}{\partial E_{F_b}^{2r}}\left[M_{A_3}(E_{F_b})\right].$$

(c) *Dimmock Model*
In this case electron concentration and the TPSM assume the forms

$$n_0 = \left(\frac{g_v}{2\pi^2}\right)\left[M_{A_4}(E_{F_b}) + N_{A_4}(E_{F_b})\right] \tag{8.24}$$

$$G_0 = \left(\frac{\pi^2 k_B^2 T}{3e}\right)\left[\frac{M'_{A_4}(E_{F_b}) + N'_{A_4}(E_{F_b})}{M_{A_4}(E_{F_b}) + N_{A_4}(E_{F_b})}\right], \tag{8.25}$$

where

$$M_{A_4}(E_{F_b}) = \left[\alpha_5 J_{A_1}(E_{F_b}) - \alpha_3(E_{F_b})\bar{\tau}_{A_1}(E_{F_b}) - \frac{\alpha_4}{3}\left[\bar{\tau}_{A_1}(E_{F_b})\right]^3\right],$$

$$\alpha_5 = \left[\frac{2m_t^+ m_t^-}{\alpha \hbar^2}\omega_{A_1}\right],$$

$$\omega_{A_1} = \left[\frac{\alpha^2}{16}\left[\frac{1}{m_t^- m_l^+} + \frac{1}{m_l^- m_t^+}\right]^2 - \frac{\alpha^2}{4m_l^+ m_t^- m_l^- m_t^+}\right],$$

$$J_{A_1}(E_{F_b}) = \frac{A_A(E_{F_b})}{3}\left[-(A_A^2(E_{F_b}) + B_A^2(E_{F_b}))E(\lambda, q) + 2B_A^2(E_{F_b})F(\lambda, q)\right] + \frac{\bar{\tau}_{A_1}(E_{F_b})}{3}$$

$$\left[(\bar{\tau}_{A_1}(E_{F_b}))^2 + A_A^2(E_{F_b}) + 2B_A^2(E_{F_b})\right]\left[A_A^2(E_{F_b}) + \bar{\tau}_{A_1}^2(E_{F_b})\right]^{1/2} \times \left[B_A^2(E_{F_b}) + \bar{\tau}_{A_1}^2(E_{F_b})\right]^{-1/2}$$

$$\lambda = \tan^{-1}\frac{\bar{\tau}_A(E_{F_b})}{B_A(E_{F_b})}, \quad q = \left[\frac{\sqrt{A_A^2(E_{F_b}) - B_A^2(E_{F_b})}}{A_A(E_{F_b})}\right],$$

$$A_A(E_{F_b}) = \left[\tau_{A_2}(E_{F_b}) + \sqrt{\tau_{A_2}^2(E_{F_b}) - 4\tau_{A_3}(E_{F_b})}\right]^{1/2} \Big/ \sqrt{2},$$

$$B_A(E_{F_b}) = \left[\tau_{A_2}(E_{F_b}) - \sqrt{\tau_{A_2}^2(E_{F_b}) - 4\tau_{A_3}(E_{F_b})}\right]^{1/2} \Big/ \sqrt{2},$$

$$\tau_{A_2}(E_{F_b}) = \frac{\omega_{A_2}(E_{F_b})}{\omega_{A_1}^2}, \tau_{A_3}(E_{F_b}) = \frac{\omega_{A_3}(E_{F_b})}{\omega_{A_1}^2},$$

$$\omega_{A_2}(E_{F_b}) = \left[\frac{\alpha}{2}\left[\frac{1}{2m_t^*} - \frac{\alpha.E_{F_b}}{2m_t^+} + \frac{1+\alpha.E_{F_b}}{2m_t^-}\right].\left[\frac{1}{m_t^- m_l^+} + \frac{1}{m_l^- m_t^+}\right] - \frac{\alpha}{m_t^+ m_t^-}\left[\frac{1}{2m_l^*} + \frac{\alpha.E_{F_b}}{2m_l^+} + \frac{1+\alpha.E_{F_b}}{2m_l^-}\right]\right]$$

$$\omega_{A_3}(E_{F_b}) = \left[\frac{\alpha.E_{F_b}(1+\alpha.E_{F_b})}{m_t^+ m_t^-} + \left[\frac{1}{2m_t^*} - \frac{\alpha.E_{F_b}}{2m_t^+} + \frac{1+\alpha.E_{F_b}}{2m_t^-}\right]^2\right],$$

$$\alpha_2(E_{F_b}) = \left[\frac{1}{2m_t^*} - \frac{\alpha.E_{F_b}}{2m_t^+} + \frac{1+\alpha.E_{F_b}}{2m_t^-}\right],$$

$$\alpha_3 = \frac{\alpha\hbar^2}{4}\left[\frac{1}{m_t^- m_l^+} + \frac{1}{m_l^- m_t^+}\right],$$

$$\tau_{A_1}(E_{F_b}) = \left[\frac{2m_l^+ m_l^-}{\alpha\hbar^2}\right]^{1/2}\left[-\left[\frac{1}{2m_l^*} + \frac{1+\alpha.E_{F_b}}{m_l^-} - \frac{\alpha.E_{F_b}}{2m_l^+}\right] + \left[\left[\frac{1}{2m_l^*} + \frac{1+\alpha.E_{F_b}}{m_l^-} - \frac{\alpha.E_{F_b}}{2m_l^+}\right]^2 + \frac{\alpha.E_{F_b}(1+\alpha.E_{F_b})}{m_l^- m_l^+}\right]^{1/2}\right]^{1/2}$$

$E(\lambda, q) = \int_0^{\lambda}\left[1 - q^2\sin^2\alpha\right]^{1/2} d\alpha$ is the complete Elliptic integral of second kind,

$F(\lambda, q) = \int_0^{\lambda}\frac{d\alpha}{\sqrt{1 - q^2\sin^2\alpha}}$ is the complete Elliptic integral of first kind

and $N_{A_4}(E_{F_b}) = \sum_{r=1}^{s} 2(k_B T)^{2r}(1 - 2^{1-2r})\zeta(2r)\frac{\partial^{2r}}{\partial E_{F_b}^{2r}}\left[M_{A_4}(E_{F_b})\right]$

(d) *Foley and Langenberg Model*

In this case electron concentration and the TPSM can, respectively, be expressed as

$$n_0 = \left(\frac{2g_v}{4\pi^2}\right)\left[h_{A_6}(E_{F_b}) + h_{A_7}(E_{F_b})\right] \tag{8.26}$$

$$G_0 = \left(\frac{\pi^2 k_B^2 T}{3e}\right)\left[\frac{h'_{A_6}(E_{F_b}) + h'_{A_7}(E_{F_b})}{h_{A_6}(E_{F_b}) + h_{A_7}(E_{F_b})}\right], \tag{8.27}$$

where

$$h_{A_6}(E_{F_b}) = \left[\frac{1}{3}\delta_{A_5} h_{A_3}^3(E_{F_b}) - \delta_{A_4}(E_{F_b}) h_{A_3}(E_{F_b}) + \delta_{A_{10}} J_{A_6}(E_{F_b})\right],$$

$$\delta_{A_6} = \left[\frac{\hbar^4}{2}\left(\frac{1}{(m_\perp^+)^2} - \frac{1}{(m_\perp^-)^2}\right)\right]^{-1},$$

$$\delta_{A_4}(E_{F_b}) = \delta_{A_6}\left[\frac{\hbar^2}{2m_\perp^-}(E_{g0} + 2E_{F_b}) + P_\perp^2 + \frac{\hbar^2 E_{g0}}{2m_\perp^+}\right],$$

$$\delta_{A_7} = \hbar^8\left[\frac{1}{4(m_\perp^+ m_\parallel^+)^2} - \frac{1}{2m_\perp^+ m_\perp^- m_\parallel^+ m_\parallel^-} + \frac{1}{4(m_\perp^- m_\parallel^-)^2}\right],$$

$$\delta_{A_5} = \delta_{A6}\hbar^4\left[\frac{1}{2m_\perp^- m_\parallel^-} - \frac{1}{2m_\perp^+ m_\parallel^+}\right],$$

$$\begin{aligned}\delta_{A_8}(E_{F_b}) = \Bigg[&\frac{\hbar^6 E_{g0}}{2(m_\perp^+)^2 m_\parallel^-} + \frac{\hbar^6(E_{g0} + 2E_{F_b})}{2m_\perp^+ m_\perp^- m_\parallel^+} - \frac{\hbar^6 E_{g0}}{2m_\perp^+ m_\perp^- m_\parallel^-} - \frac{\hbar^4 P_\perp^2}{m_\perp^- m_\parallel^-}\\ &-\frac{\hbar^6(E_{g0} + 2E_{F_b})}{2(m_\perp^-)^2 m_\parallel^-} - \frac{\hbar^6(E_{g0} + 2E_{F_b})}{2m_\parallel^-(m_\perp^+)^2} - \frac{\hbar^6 E_{g0}}{2m_\parallel^+(m_\perp^+)^2}\\ &-\frac{P_\parallel^2 \hbar^4}{(m_\perp^+)^2} + \frac{\hbar^6(E_{g0} + 2E_{F_b})}{2m_\parallel^-(m_\perp^-)^2} + \frac{\hbar^6 E_{g0}}{2m_\parallel^+(m_\perp^-)^2} + \frac{\hbar^4 P_\parallel^2}{(m_\perp^+)^2}\Bigg],\end{aligned}$$

$$\begin{aligned}\delta_{A_9}(E_{F_b}) = &\left[P_\perp^4 + \frac{\hbar^4 E_{g0}^2}{4(m_\perp^+)^2} + \frac{\hbar^4(E_{g0} + 2E_{Fb})}{4(m_\perp^-)^2} + \frac{E_{g0}\hbar^2 P_\perp^2}{m_\perp^+} + \frac{\hbar^2 P_\perp^2(E_{g0} + 2E_{F_b})}{m_\perp^-}\right]\\ &+\left[\frac{E_{g0}\hbar^4(E_{g0} + 2E_{F_b})}{2m_\perp^+ m_\perp^-} + \frac{\hbar^4 {E_{F_b}}^2}{(m_\perp^+)^2} + \frac{\hbar^4 E_{F_b} E_{g0}}{(m_\perp^+)^2} - \frac{\hbar^4 {E_{F_b}}^2}{(m_\perp^-)^2} - \frac{\hbar^4 E_{F_b} E_{g0}}{(m_\perp^-)^2}\right]\end{aligned}$$

$$\delta_{A_{10}} = \delta_{A_6}(\delta_{A_7})^{1/2}, \quad \delta_{A_{11}}(E_{F_b}) = \left[\delta_{A_8}(E_{F_b})\delta_{A_7}\right],$$

$$h_{A_2}(E_{F_b}) = \frac{\hbar^2}{2}\left[\frac{2E_{F_b} + E_{g0}}{m_\parallel^-} + \frac{E_{g0}}{m_\parallel^+} + P_\parallel^2\right]$$

$$h_{A_3}(E_{F_b}) = (2h_{A_1})^{1/2}\left[\sqrt{h_{A_2}^2(E_{F_b}) + 4h_{A_1} E_{F_b}(E_{F_b} + E_{g0})} - h_{A_2}(E_{F_b})\right]^{1/2},$$

$$J_{A_6}(E_{F_b}) = \frac{h_{A_4}(E_{F_b})}{3}\Big[-E(\lambda_1, q_1)\left[h^2_{A_4}(E_{F_b}) + h^2_{A_5}(E_{F_b})\right]$$
$$+2h^2_{A_5}(E_{F_b})F(\lambda_1, q_1) + \frac{h_{A_4}(E_{F_b})}{3}$$
$$\times\left[h^2_{A_3}(E_{F_b}) + h^2_{A_4}(E_{F_b}) + 2h^2_{A_5}(E_{F_b})\right]$$
$$\times\left[(h^2_{A_4}(E_{Fb}) + h^2_{A_5}(E_{Fb}))\right] \Big/ \left[(h^2_{A_5}(E_{Fb}) + h^2_{A_3}(E_{Fb}))\right]\Big]^{1/2},$$

$$\lambda_1 = \tan^{-1}\left[h_{A_3}(E_{F_b})/h_{A_5}(E_{F_b})\right], \quad q_1 = \left[\frac{h^2_{A_4}(E_{F_b}) - h^2_{A_5}(E_{F_b})}{h_{A_4}(E_{F_b})}\right]$$

and

$$h_{A_7}(E_{F_b}) = \sum_{r=1}^{s_0} 2(k_BT)^{2r}(1-2^{1-2r})\zeta(2r)\frac{\partial^{2r}}{\partial E^{2r}_{F_b}}\left[h_{A_6}(E_{F_b})\right]$$

(vii) n-Ge

(a) *Model of Cardona et al.*

The expressions for the electron concentration and the TPSM can be written as

$$n_0 = N_{c0}\left[F_{\frac{1}{2}}(\eta) + \bar{\alpha}_2 F_{\frac{3}{2}}(\eta) - \bar{\alpha}_3 F_{\frac{7}{2}}(\eta)\right] \tag{8.28}$$

$$G_0 = \left(\frac{\pi^2 k_B}{3e}\right)\left[\frac{F_{\frac{-1}{2}}(\eta) + \bar{\alpha}_2 F_{\frac{1}{2}}(\eta) - \bar{\alpha}_3 F_{\frac{5}{2}}(\eta)}{F_{\frac{1}{2}}(\eta) + \bar{\alpha}_2 F_{\frac{3}{2}}(\eta) - \bar{\alpha}_3 F_{\frac{7}{2}}(\eta)}\right], \tag{8.29}$$

where

$$N_{c0} = 2g_v(2\pi m^*_D k_B T \big/ h^2)^{3/2}, \; m^*_D = ((m^*_\perp)^2 m^*_\parallel)^{1/3}, \; \bar{\alpha}_2 = \frac{45\alpha k_B T}{24}$$

and

$$\bar{\alpha}_3 = \frac{189}{8}\alpha(k_BT)^2\left(\frac{k_BT(m^*_\parallel)^2}{\hbar^4}\right).$$

(b) *Model of Wang and Ressler*

The expressions for the electron concentration and the TPSM assume the forms

$$n_0 = \left(\frac{m^*_\perp g_v}{\pi^2\hbar^2}\right)\left[M_{A_5}(E_{F_b}) + N_{A_5}(E_{F_b})\right] \tag{8.30}$$

$$G_0 = \left(\frac{\pi^2 k_B^2 T}{3e}\right)\left[\frac{M'_{A_5}(E_{F_b}) + N'_{A_5}(E_{F_b})}{M_{A_5}(E_{F_b}) + N_{A_5}(E_{F_b})}\right], \tag{8.31}$$

where

$$M_{A_5}(E_{F_b}) = \left[\bar{\alpha}_8 \rho_{A_1}(E_{F_b}) - \frac{\bar{\alpha}_9}{3}\rho^3_{A_1}(E_{F_b}) - \bar{\alpha}_{10} J_{A_2}(E_{F_b})\right],$$

$$\bar{\alpha}_4 = \beta_4 (2m^*_\perp \big/ \hbar^2)^2 ,$$

$$\beta_4 = 1.4\beta_5, \qquad \beta_5 = \frac{1}{4}\left(\alpha\hbar^4 \big/ (m^*_\perp)^2\right) \cdot \left(1 - \frac{m^*_\perp}{m_0}\right)^2,$$

$$\bar{\alpha}_5 = \bar{\alpha}_7 (4 m^*_\perp m^*_\parallel / \hbar^4), \quad \bar{\alpha}_7 = 0.8\beta_5,$$

$$\bar{\alpha}_6 = (0.005\beta_5)(2m^*_\parallel / \hbar^2)^2,$$

$$\bar{\alpha}_{10} = \left(\frac{1}{2\bar{\alpha}_4}\right) \cdot \left(\frac{\hbar^2}{2m^*_\parallel}\right)\left[\bar{\alpha}_5^2 - 4\bar{\alpha}_4\bar{\alpha}_6\right]^{1/2},$$

$$\bar{\alpha}_{11} = \left(\frac{2m^*_\parallel}{\hbar^2}\right)\left[\frac{4\bar{\alpha}_4 - 2\bar{\alpha}_5}{\bar{\alpha}_5^2 - 4\bar{\alpha}_4\bar{\alpha}_6}\right],$$

$$\bar{\alpha}_{12}(E_{F_b}) = \left(\frac{2m^*_\parallel}{\hbar^2}\right)^2\left[\frac{(1 - 4\bar{\alpha}_4 E_{F_b})}{\bar{\alpha}_5^2 - 4\bar{\alpha}_4\bar{\alpha}_6}\right],$$

$$\rho_{A_1}(E_{F_b}) = \frac{1}{\hbar}\left(\frac{m^*_\parallel}{\bar{\alpha}_6}\right)^{1/2}\left[\left[1 - \sqrt{1 - 4\bar{\alpha}_6(E_{F_b})}\right]^{1/2}\right],$$

$$\bar{A}^2_{A_1}(E_{F_b}) = \frac{1}{2}\left[\bar{\alpha}_{11} + \left[\bar{\alpha}^2_{11} - 4\bar{\alpha}_{12}(E_{F_b})\right]^{1/2}\right],$$

$$\bar{B}^2_{A_1}(E_{F_b}) = \frac{1}{2}\left[\bar{\alpha}_{11} - \left[\bar{\alpha}^2_{11} - 4\bar{\alpha}_{12}(E_{F_b})\right]^{1/2}\right],$$

$$J_{A_2}(E_{F_b}) = \frac{\bar{A}_{A_1}(E_{F_b})}{3}\Big[-E(\lambda_3, q_3)\left[\bar{A}^2_{A_1}(E_{F_b}) + \bar{B}^2_{A_1}(E_{F_b})\right] + 2\bar{B}^2_{A_1}(E_{F_b}) F(\lambda_3, q_3)\Big]$$

$$+\frac{\bar{A}_{A_1}(E_{F_b})}{3}\left[\bar{\rho}^2_{A_1}(E_{F_b})+\bar{A}^2_{A_1}(E_{F_b})+2\bar{B}^2_{A_1}(E_{F_b})\right]$$
$$\times\left[\frac{\bar{A}^2_{A_1}(E_{F_b})+\rho^2_{A_1}(E_{F_b})}{\bar{B}^2_{A_1}(E_{F_b})+\rho^2_{A_1}(E_{F_b})}\right]^{1/2},$$
$$\lambda_3=\tan^{-1}\frac{\rho_{A_1}(E_{F_b})}{\bar{B}_{A_1}(E_{F_b})},\quad q_3=\left[\frac{\bar{A}^2_{A_1}(E_{F_b})-\bar{B}^2_{A_1}(E_{F_b})}{\bar{A}_{A_1}(E_{F_b})}\right]$$

and

$$N_{A_5}(E_{F_b})=\sum_{r=1}^{s}2(k_BT)^{2r}(1-2^{1-2r})\zeta(2r)\frac{\partial^{2r}}{\partial E^{2r}_{F_b}}\left[M_{A_5}(E_{F_b})\right]$$

(viii) *Platinum Antimonide*

The expressions for the electron concentration and the TPSM can be written as

$$n_0=\left(\frac{g_v}{2\pi^2}\right)\left[M_{A_6}(E_{F_b})+N_{A_6}(E_{F_b})\right] \tag{8.32}$$
$$G_0=\left(\frac{\pi^2k_B^2T}{3e}\right)\left[\frac{M'_{A_6}(E_{F_b})+N'_{A_6}(E_{F_b})}{M_{A_6}(E_{F_b})+N_{A_6}(E_{F_b})}\right], \tag{8.33}$$

where

$$M_{A_6}(E_{F_b})=\left[T_{A_9}(E_{F_b})\rho_{A_2}(E_{F_b})-T_{A_{10}}(E_{F_b})\frac{\rho^3_{A_2}(E_{F_b})}{3}-T_{A_{11}}J_{A_3}(E_{F_b})\right],$$
$$T_{A_1}=[I_1+\omega_1\omega_3],$$
$$T_{A_2}(E_{F_b})=\left[-E_{F_b}\omega_3+\omega_1(E_{F_b}+\delta_0)\right],$$
$$T_{A_3}=[2I_1+\omega_2\omega_4+\omega_3\omega_2],$$
$$T_{A_4}=[I_1+\omega_2\omega_4],$$
$$T_{A_5}(E_{F_b})=\omega_2(E_{F_b}+\delta_0),$$
$$T_{A_6}(E_{F_b})=\left[E_{F_b}(E_{F_b}+\delta_0)-E_{F_b}\omega_4\right],$$
$$\bar{T}_{A_6}=\left[T^2_{A_3}-4T_{A_1}T_{A_4}\right],$$
$$T_{A_7}(E_{F_b})=\left[2T_{A_3}T_{A_2}(E_{F_b})-4T_{A_1}T_{A_5}(E_{F_b})\right],$$
$$T_{A_8}(E_{F_b})=\left[T^2_{A_2}(E_{F_b})+4T_{A_1}T_{A_6}(E_{F_b})\right],$$
$$T_{A_9}(E_{F_b})=\frac{T_{A_2}(E_{F_b})}{2T_{A_1}},\quad T_{A_{10}}=\left[T_{A_3}/2T_{A_1}\right],$$
$$T_{A_{11}}=\frac{\sqrt{\bar{T}_{A_6}}}{2T_{A_1}},\quad T_{A_{12}}(E_{F_b})=\left[T_{A_7}(E_{F_b})/\bar{T}_{A_6}\right],$$

$$T_{A_{13}}(E_{F_b}) = T_{A_8}(E_{F_b}) / \bar{T}_{A_6} ,$$

$$\rho_{A_2}(E_{F_b}) = \left[\left[T_{A_5}(E_{F_b}) - \sqrt{T_{A_5}^2(E_{F_b}) + 4T_{A_4} T_{A_6}(E_{F_b})}\right] / (2T_{A_4})\right]^{1/2} ,$$

$$A_{A_3}^2(E_{F_b}) = \frac{1}{2}\left[T_{A_{12}}(E_{F_b}) + \sqrt{T_{A_{12}}^2(E_{F_b}) - 4T_{A_{13}}(E_{F_b})}\right],$$

$$B_{A_3}^2(E_{F_b}) = \frac{1}{2}\left[T_{A_{12}}(E_{F_b}) - \sqrt{T_{A_{12}}^2(E_{F_b}) - 4T_{A_{13}}(E_{F_b})}\right],$$

$$\begin{aligned} J_{A_3}(E_{F_b}) = {} & \frac{\rho_{A_2}(E_{F_b})}{3} [[A_{A_3}^2(E_{F_b}) + B_{A_3}^2(E_{F_b})] \\ & \times E(\eta_1, t_1) - \left[A_{A_3}^2(E_{F_b}) - B_{A_3}^2(E_{F_b})\right] F(\eta_1, t_1)] \\ & + \frac{\rho_{A_2}(E_{F_b})}{3} \left[\left[A_{A_3}^2(E_{F_b}) - \rho_{A_2}^2(E_{F_b})\right]\left[B_{A_3}^2(E_{F_b}) - \rho_{A_2}^2(E_{F_b})\right]\right]^{1/2} \end{aligned}$$

$$\eta_1 = \tan^{-1}\left[\rho_{A_2}(E_{F_b}) / B_{A_3}(E_{F_b})\right], \quad t_1 = \left[B_{A_3}(E_{F_b}) / A_{A_3}(E_{F_b})\right]$$

and

$$N_{A_6}(E_{F_b}) = \sum_{r=1}^{s} 2(k_B T)^{2r}(1 - 2^{1-2r})\zeta(2r) \frac{\partial^{2r}}{\partial E_{F_b}^{2r}} \left[M_{A_6}(E_{F_b})\right].$$

(ix) *n-GaSb*
In accordance of model of Mathur and Jain, the electron concentration and the TPSM can be expressed as

$$n_0 = \frac{g_v}{3\pi^2}\left(\frac{2m^*}{\hbar^2}\right)^{3/2} \left[\delta_{A_2}(E_{F_b}) + \delta_{A_3}(E_{F_b})\right] \tag{8.34}$$

$$G_0 = \left(\frac{\pi^2 k_B^2 T}{3e}\right)\left[\frac{\delta'_{A_2}(E_{F_b}) + \delta'_{A_3}(E_{F_b})}{\delta_{A_2}(E_{F_b}) + \delta_{A_3}(E_{F_b})}\right], \tag{8.35}$$

where

$$\delta_{A_2}(E_{F_b}) = \left[\delta_{A_1}(E_{F_b})\right]^{3/2},$$

$$\delta_{A_1}(E_{F_b}) = \left[E_{F_b} + E_{g_1} - \frac{m^*}{m_0}\frac{E_{g_1}}{2} - \left[\left(\frac{E_{g_1}}{2}\right)^2 + \left[\frac{E_{g_1}}{2}\left(1 - \frac{m^*}{m_0}\right)\right]^2 + \frac{(E_{g_1})^2}{2}\left(1 - \frac{m^*}{m_0}\right) + E_{F_b}E_{g_1}\left(1 - \frac{m^*}{m_0}\right)\right]^{1/2}\right]$$

and

$$\delta_{A_3}(E_{F_b}) = \sum_{r=1}^{s} 2(k_B T)^{2r}(1 - 2^{1-2r})\zeta(2r)\frac{\partial^{2r}}{\partial E_{F_b}^{2r}}\left[\delta_{A_2}(E_{F_b})\right]$$

(x) *n-Te*

The electron concentration and TPSM in n-Te in accordance with the model of Bouat et al. can be written as

$$n_0 = \frac{g_v}{3\pi^2}\left[M_{A_9}(E_{F_b}) + N_{A_9}(E_{F_b})\right] \tag{8.36}$$

$$G_0 = \left(\frac{\pi^2 k_B^2 T}{3e}\right)\left[\frac{M'_{A_9}(E_{F_b}) + N'_{A_9}(E_{F_b})}{M_{A_9}(E_{F_b}) + N_{A_9}(E_{F_b})}\right] \tag{8.37}$$

$$M_{A_9}(E_{F_b}) = \left[3\psi_5(E_{F_b})\Gamma_3(E_{F_b}) - \psi_6\Gamma_3^3(E_{F_b})\right], \quad \psi_5(E_{F_b}) = \left[\frac{E_{F_b}}{\psi_2} + \frac{\psi_4^2}{2\psi_2^2}\right],$$

$$\Gamma_3(E_{F_b}) = [2\psi_1]^{-1}\left[\sqrt{\psi_3^2 + 4\psi_1 E_{F_b}} - \psi_3\right], \quad \psi_6 = (\psi_1/\psi_2),$$

$$N_{A_9}(E_{F_b}) = \sum_{r=1}^{s} 2(k_B T)^{2r}(1 - 2^{1-2r})\zeta(2r)\frac{\partial^{2r}}{\partial E_{F_b}^{2r}} \times \left[M_{A_9}(E_{F_b})\right], \psi_1 = A_6, \quad \psi_2 = A_7, \quad \psi_3^2 = A_8$$

and $\psi_4^2 = A_9$

(xi) *Bismuth*

(a) *McClure and Choi model*

The electron concentration and TPSM in Bi in accordance with this model can be written as

$$n_0 = \left(\frac{g_v}{4\pi^3}\right)h_{A_8}\left[h_{A_{10}}(E_{F_b}) + h_{A_{11}}(E_{F_b})\right] \tag{8.38}$$

$$G_0 = \left(\frac{\pi^2 k_B^2 T}{3e}\right)\left[\frac{h'_{A_{10}}(E_{F_b}) + h'_{A_{11}}(E_{F_b})}{h_{A_{10}}(E_{F_b}) + h_{A_{11}}(E_{F_b})}\right], \tag{8.39}$$

where

$$h_{A_8} = \frac{4\pi^2\sqrt{m_1 m_3}}{\hbar^2 \theta_{A_4}}, \quad \theta_{A_4} = \frac{\alpha\hbar^2}{2m_2}, \quad \theta_{A_3} = \frac{\alpha\hbar^4}{4m_2 m'_2},$$

$$\theta_{A_2}(E_{F_b}) = (\alpha E_{F_b}\hbar^4/2m_2)\left[1 - \frac{m_2}{m'_2}\right], \quad \theta^2_{A_5} = \frac{1}{\theta_{A_4}},$$

$$h_{A_{10}}(E_{F_b}) = \left[\frac{h_{A_9}(E_{F_b})}{2\theta_{A_5}} \ln\left|\frac{\theta_{A_5} + \bar{h}_{A_4}(E_{F_b})}{\theta_{A_5} - \bar{h}_{A_4}(E_{F_b})}\right| + (\theta_{A_5}(E_{F_b}) + \theta_{A_3}\theta^2_{A_5})\bar{h}_{A_4}(E_{F_b}) + \frac{\theta_{A_3}}{3}\left[\bar{h}_{A_4}(E_{F_b})\right]^3\right],$$

$$h_{A_9}(E_{F_b}) = \left[E_{F_b}(1 + \alpha E_{F_b}) - \theta_{A_5}(E_{F_b})\theta^2_{A_5} - \theta_{A_3}\theta^4_{A_5}\right],$$

$$\bar{h}_{A_4}(E_{F_b}) = \frac{\sqrt{2m_2 m'_2}}{\sqrt{\alpha}\hbar^2}\left[\frac{-\alpha E_{F_b}\hbar^2}{2m_2}\left(1 - \frac{m_2}{m'_2}\right) + \left[\frac{\alpha^2 E^2_{F_b}\hbar^4}{4m_2^2}\left(1 - \frac{m_2}{m'_2}\right)^2 + \frac{\alpha E(1 + \alpha E_{F_b})\hbar^4}{m_2 m'_2}\right]^{1/2}\right]^{1/2}$$

and

$$h_{A_{11}}(E_{F_b}) = \sum_{r=1}^{s} 2(k_B T)^{2r}(1 - 2^{1-2r})\zeta(2r)\frac{\partial^{2r}}{\partial E^{2r}_{F_b}}\left[h_{A_{10}}(E_{F_b})\right]$$

(b) *Hybrid model*
In accordance with Hybrid model, the expressions for n_0 and G_0 are given by

$$n_0 = \left(\frac{g_v}{2\pi^2}\right)\left[h_{A_{12}}(E_{F_b}) + h_{A_{13}}(E_{F_b})\right] \tag{8.40}$$

$$G_0 = \left(\frac{\pi^2 k_B^2 T}{3e}\right)\left[\frac{h'_{A_{12}}(E_{F_b}) + h'_{A_{13}}(E_{F_b})}{h_{A_{12}}(E_{F_b}) + h_{A_{13}}(E_{F_b})}\right], \tag{8.41}$$

where

$$h_{A_{12}}(E_{F_b}) = \left[E_{F_b}(1+\alpha E_{F_b}) - \frac{L_{A_1}(E_{F_b})\hbar^2 I_{A_4}^2(E_{F_b})}{6M_2} - \frac{L_{A_2}\hbar^4 I_{A_5}^5(E_{F_b})}{20M_2^2 E_{g_0}}\right],$$

$$L_{A_1}(E_{F_b}) = \left[1 + L_{A_3} + \alpha E_{F_b}(1 - L_{A_2})\right],\ L_{A_3} = M_2/m_2,\ L_{A_2} = M_2/M_2'$$

$$I_{A_4}(E_{F_b}) = \left[\frac{L_{A_2}}{2E_{g_0}M_2^2}\right]^{-1/2}\left[\frac{-L_{A_1}(E_{F_b})}{2M_2} + \left[\frac{L_{A_1}^2(E_{F_b})}{4M_2^2} + \frac{L_{A_2}E_{F_b}(1+\alpha E_{F_b})}{4E_{g_0}M_2^2}\right]^{1/2}\right]^{1/2}$$

and

$$h_{A_{13}}(E_{F_b}) = \sum_{r=1}^{s} 2(k_B T)^{2r}(1 - 2^{1-2r})\zeta(2r)\frac{\partial^{2r}}{\partial E_{F_b}^{2r}}\left[h_{A_{12}}(E_{F_b})\right]$$

8.2.2 Debye Screening Length:

The Debye screening length (DSL) of the carriers in the semiconductors is a fundamental quantity, characterizing the screening of the Coulomb field of the ionized impurity centers by the free carriers. It affects many special features of the modern semiconductor devices, the carrier mobility under different mechanisms of scattering, and the carrier plasmas in semiconductors [45–58].

The DSL (L_D) can, in general, be writ ten as [48–58]

$$L_D = \left(\frac{|e|^2}{\varepsilon_{sc}}\frac{\partial n_0}{\partial E_F}\right)^{-1/2}, \tag{8.42}$$

where n_0 and E_F are applicable for bulk samples.

Using (8.42) and (8.2), one obtains

$$L_D = \left(3|e|^3 n_0 G \big/ \varepsilon_{sc}\pi^2 k_B^2 T\right)^{-1/2}. \tag{8.43}$$

Therefore, we can experimentally determine L_D by knowing the experimental curve of G versus carrier concentration at a fixed temperature. It is evident that the DSL for a system can be investigated if the functional dependence between the electron concentration and the Fermi energy of that particular material is known. For the purpose of completeness we present few results of DSL as written below:

(i) In the presence of external light waves, the DSL in optoelectronic materials whose unperturbed conduction electrons obey the three and two band models of Kane together with parabolic energy bands can, respectively, be expressed as

$$L_D = \left[\left(\frac{e^2}{3\pi^2\varepsilon_{sc}}\right)\left(\frac{2m_c}{\hbar^2}\right)^{3/2}\right]^{-1/2} \times \left[G'_{70}(E_{Fl}, \lambda, E_{g0}, \Delta) + H'_{70}(E_{Fl}, \lambda, E_{g0}, \Delta)\right]^{1/2} \quad (8.44)$$

$$L_D = \left[\left(\frac{e^2}{3\pi^2\varepsilon_{sc}}\right)\left(\frac{2m_c}{\hbar^2}\right)^{3/2}\right]^{-1/2} \times \left[G'_{71}(E_{Fl}, \lambda, E_{g0}) + H'_{71}(E_{Fl}, \lambda, E_{g0})\right]^{-1/2} \quad (8.45)$$

$$L_D = \left[\left(\frac{e^2}{3\pi^2\varepsilon_{sc}}\right)\left(\frac{2m_c}{\hbar^2}\right)^{3/2}\right]^{-1/2} \times \left[G'_{72}(E_{Fl}, \lambda, E_{g0}) + H'_{72}(E_{Fl}, \lambda, E_{g0})\right]^{-1/2}, \quad (8.46)$$

where the primes indicate the differentiation of the differentiable functions with respect to the Fermi energy, $G_{70}(E_{Fl}, \lambda, E_{g0}, \Delta) = [\beta_0(E_{Fl}, \lambda)]^{3/2}$, $H_{70}(E_{Fl}, \lambda, E_{g0}, \Delta) = \sum_{r=1}^{s} z_t(r) G_{70}(E_{Fl}, \lambda, E_{g0}, \Delta)$ $z_t(r) = 2(k_B T)^{2r}(1 - 2^{1-2r})\xi(2r)\frac{\partial^{2r}}{\partial E_{Fl}^{2r}}$, $t = l$ or F_s, E_{Fl} is the Fermi energy as measured in the presence of light waves as measured from the edge of the conduction band in the vertically upward direction in the absence of any field,

$$H_{71}(E_{Fl}, \lambda, E_{g0}) = \sum_{r=1}^{s} z_t(r) G_{71}(E_{Fl}, \lambda),$$

$$G_{71}(E_{Fl}, \lambda, E_{g0}) = [\omega_0(E_{Fl}, \lambda)]^{3/2},$$

$$G_{72}(E_{Fl}, \lambda, E_{g0}) = [\rho_0(E_{Fl}, \lambda)]^{3/2},$$

$$H_{72}(E_{Fl}, \lambda, E_{g0}) = \sum_{r=1}^{s} z_t(r) G_{72}(E_{Fl}, \lambda, E_{g0})$$

$$n_0 = \frac{1}{3\pi^2}\left(\frac{2m_c}{\hbar^2}\right)^{3/2}\left[G_{70}(E_{Fl}, \lambda, E_{g0}, \Delta) + H_{70}(E_{Fl}, \lambda, E_{g0}, \Delta)\right],$$

$$n_0 = \frac{1}{3\pi^2}\left(\frac{2m_c}{\hbar^2}\right)^{3/2}\left[G_{71}(E_{Fl}, \lambda, E_{g0}) + H_{71}(E_{Fl}, \lambda, E_{g0})\right]$$

and

$$n_0 = \frac{1}{3\pi^2}\left(\frac{2m_c}{\hbar^2}\right)^{3/2}\left[G_{72}(E_{Fl}, \lambda, E_{g0}) + H_{72}(E_{Fl}, \lambda, E_{g0})\right].$$

(ii) In the presence of intense electric field, the DSL in optoelectronic semiconductors in accordance with the perturbed three and two band models of Kane can, respectively, be expressed as

$$L_D=\left[\left(\frac{e^2}{3\pi^2\varepsilon_{sc}}\right)\left(\frac{2m_c}{\hbar^2}\right)^{3/2}\right]^{-1/2}\left[g'_{70}(E_{Fs},F)+h'_{70}(E_{Fs},F)\right]^{-1/2} \tag{8.47}$$

$$L_D=\left[\left(\frac{e^2}{3\pi^2\varepsilon_{sc}}\right)\left(\frac{2m_c}{\hbar^2}\right)^{3/2}\right]^{-1/2}\left[g'_{71}(E_{Fs},F)+h'_{71}(E_{Fs},F)\right]^{-1/2} \tag{8.48a}$$

$$g_{70}(E_{Fs},F)=[\beta(E_{Fs},F)]^{3/2},\quad h_{70}(E_{Fs},F)=\sum_{r=1}^{s}z_t(r)g_{70}(E_{Fs},F)$$

$$\beta(E_{Fs},\lambda)=\left[\frac{1+\phi(E_{Fs},F)}{3I_{11}(E_{Fs})}+\frac{2}{3I_{11}(E_{Fs})}\right]^{-1}$$

E_{Fs} is the Fermi energy as measured in the presence of intense electric field as measured from the edge of the conduction band in the vertically upward direction in the absence of any field $g_{71}(E_{Fs},F)=[\beta_1(E_{Fs},F)]^{3/2}$, $h_{71}(E_{Fs},F)=\sum_{r=1}^{s}z_t(r)g_{71}(E_{Fs},F)$,

$$\beta_1(E_{Fs},\lambda)=\left[\frac{1+\phi_1(E_{Fs},F)}{3\gamma_0(E_{Fs})}+\frac{2}{3\gamma_0(E_{Fs})}\right]^{-1}$$

$$n_0=\frac{1}{3\pi^2}\left(\frac{2m_c}{\hbar^2}\right)^{3/2}\left[g_{70}(E_{Fs},F)+h_{70}(E_{Fs},F)\right], \tag{8.48b}$$

$$n_0=\frac{1}{3\pi^2}\left(\frac{2m_c}{\hbar^2}\right)^{3/2}\left[g_{71}(E_{Fs},F)+h_{71}(E_{Fs},F)\right] \tag{8.48c}$$

In the absence of any field, the expressions for the DSL and the electron concentration for optoelectronic semiconductors whose energy band structures are defined by the unperturbed two band model of Kane, under the condition $(E_F E_{g0}^{-1}) \ll 1$, assume the well-known forms as [52]

$$L_D=\left[\frac{e^2}{\varepsilon_{sc}}N_c k_B T\left[F_{-1/2}(\eta)+\frac{15\alpha k_B T}{4}F_{1/2}(\eta)\right]\right]^{-1/2} \tag{8.49}$$

$$n_0=N_c\left[F_{1/2}(\eta)+\frac{15\alpha k_B T}{4}F_{3/2}(\eta)\right], \tag{8.50}$$

where $\eta=\dfrac{E_F}{k_B T}$.

8.2.3 Carrier Contribution to the Elastic Constants:

The knowledge of the carrier contribution to the elastic constants is important in studying the mechanical properties of the materials and has been investigated in the literature [59–81]. The electronic contribution to the second- and third-order elastic constants can be written as [59–81]

$$\Delta C_{44} = -\frac{(\overline{G_0})^2}{9}\frac{\partial n_0}{\partial E_F}, \tag{8.51}$$

and

$$\Delta C_{456} = \frac{(\overline{G_0})^3}{27}\frac{\partial^2 n_0}{\partial E_F^2}, \tag{8.52}$$

where $\overline{G_0}$ is the deformation potential constant. Thus, using (8.2), (8.51), and (8.52), we can write

$$\Delta C_{44} = \left[-n_0(\overline{G_0})^2\,|e|\,G_0\Big/(3\pi^2k_B^2T)\right] \tag{8.53}$$

and

$$\Delta C_{456} = (n_0\,|e|\,(\overline{G_0})^3G_0^2\Big/(3\pi^4k_B^3T))(1+\frac{n_0}{G_0}\frac{\partial G_0}{\partial n_0}) \tag{8.54}$$

Thus, again the experimental graph of G_0 versus n_0 allows us to determine the electronic contribution to the elastic constants for materials having arbitrary spectra. We present a few results in this context:

The expressions for ΔC_{44} and ΔC_{456} in quantum wires of nonlinear optical materials, III–V, II–VI, Bismuth, IV–VI, stressed semiconductors, Te, n-GaP, $PtSb_2$, Bi_2Te_3, n-Ge, and II–V can, respectively, be expressed as

(a) *Nonlinear optical materials*:

$$\Delta C_{44} = -\left(\frac{2(\bar{G}_0)^2g_v}{9\pi}\right)\sum_{n_x=1}^{n_{x\max}}\sum_{n_y=1}^{n_{y\max}}[B'_{11}(E_{F1D},n_x,n_y)+B'_{12}(E_{F1D},n_x,n_y)], \tag{8.55}$$

$$\Delta C_{456} = \left(\frac{2(\bar{G}_0)^3g_v}{27\pi}\right)\sum_{n_x=1}^{n_{x\max}}\sum_{n_y=1}^{n_{y\max}}\left[B''_{11}(E_{F1D},n_x,n_y)+B''_{12}(E_{F1D},n_x,n_y)\right]. \tag{8.56}$$

(b) *III–V materials:*

1. *Three band model of Kane*:

$$\Delta C_{44} = -\left(\frac{2(\bar{G}_0)^2 g_v}{9\pi}\right)\sum_{n_x=1}^{n_{x\max}}\sum_{n_y=1}^{n_{y\max}}[T'_{63}(E_{F1D}, n_x, n_y) + T'_{64}(E_{F1D}, n_x, n_y)], \tag{8.57}$$

$$\Delta C_{456} = \left(\frac{2(\bar{G}_0)^3 g_v}{27\pi}\right)\sum_{n_x=1}^{n_{x\max}}\sum_{n_y=1}^{n_{y\max}}[T''_{63}(E_{F1D}, n_x, n_y) + T''_{64}(E_{F1D}, n_x, n_y)]. \tag{8.58}$$

2. *Two band model of Kane*:

$$\Delta C_{44} = -\left(\frac{2(\bar{G}_0)^2 g_v}{9\pi}\right)\sum_{n_x=1}^{n_{x\max}}\sum_{n_y=1}^{n_{y\max}}[T'_{65}(E_{F1D}, n_x, n_y) + T'_{66}(E_{F1D}, n_x, n_y)], \tag{8.59}$$

$$\Delta C_{456} = \left(\frac{2(\bar{G}_0)^3 g_v}{27\pi}\right)\sum_{n_x=1}^{n_{x\max}}\sum_{n_y=1}^{n_{y\max}}[T''_{65}(E_{F1D}, n_x, n_y) + T''_{66}(E_{F1D}, n_x, n_y)]. \tag{8.60}$$

3. *The model of Stillman et al.:*

$$\Delta C_{44} = -\left(\frac{2(\bar{G}_0)^2 g_v}{9\pi}\right)\sum_{n_x=1}^{n_{x\max}}\sum_{n_y=1}^{n_{y\max}}[B'_{17}(E_{F1D}, n_x, n_y) + B'_{18}(E_{F1D}, n_x, n_y)], \tag{8.61}$$

$$\Delta C_{456} = \left(\frac{2(\bar{G}_0)^3 g_v}{27\pi}\right)\sum_{n_x=1}^{n_{x\max}}\sum_{n_y=1}^{n_{y\max}}[B''_{17}(E_{F1D}, n_x, n_y) + B''_{18}(E_{F1D}, n_x, n_y)]. \tag{8.62}$$

4. *The model of Newson and Kurobe:*

$$\Delta C_{44} = -\left(\frac{2(\bar{G}_0)^2 g_v}{9\pi}\right)\sum_{n_x=1}^{n_{x\max}}\sum_{n_y=1}^{n_{y\max}}[B'_{19}(E_{F1D}, n_x, n_y) + B'_{20}(E_{F1D}, n_x, n_y)], \tag{8.63}$$

$$\Delta C_{456} = \left(\frac{2(\bar{G}_0)^3 g_v}{27\pi}\right)\sum_{n_x=1}^{n_{x\max}}\sum_{n_y=1}^{n_{y\max}}[B''_{19}(E_{F1D}, n_x, n_y) + B''_{20}(E_{F1D}, n_x, n_y)]. \tag{8.64}$$

5. *The model of Palik et al.:*

$$\Delta C_{44} = -\left(\frac{2(\bar{G}_0)^2 g_v}{9\pi}\right)\sum_{n_x=1}^{n_{x\max}}\sum_{n_y=1}^{n_{y\max}}[B'_{21}(E_{F1D}, n_x, n_y) + B'_{22}(E_{F1D}, n_x, n_y)], \tag{8.65}$$

$$\Delta C_{456} = \left(\frac{2(\bar{G}_0)^3 g_v}{27\pi}\right) \sum_{n_x=1}^{n_{x\max}} \sum_{n_y=1}^{n_{y\max}} \left[B''_{21}(E_{F1D}, n_x, n_y) + B''_{22}(E_{F1D}, n_x, n_y)\right]. \tag{8.66}$$

(b) *II–VI materials:*

$$\Delta C_{44} = -\frac{(\bar{G}_0)^2 g_v}{9\pi\sqrt{B_0}} \sum_{n_x=1}^{n_{x\max}} \sum_{n_y=1}^{n_{y\max}} \left[t'_7(E_{F1D}, n_x, n_y) + t'_8(E_{F1D}, n_x, n_y)\right], \tag{8.67}$$

$$\Delta C_{456} = \frac{(\bar{G}_0)^3 g_v}{27\pi\sqrt{B_0}} \sum_{n_x=1}^{n_{x\max}} \sum_{n_y=1}^{n_{y\max}} \left[t''_7(E_{F1D}, n_x, n_y) + t''_8(E_{F1D}, n_x, n_y)\right]. \tag{8.68}$$

(c) *Bismuth*:

1. *The model of McClure and Choi*:

$$\Delta C_{44} = -\frac{2(\bar{G}_0)^2 g_v}{9\pi} \frac{\sqrt{2m_1}}{\hbar} \sum_{n_y=1}^{n_{y\max}} \sum_{n_z=1}^{n_{z\max}} \left[t'_{27}(E_{F1D}, n_y, n_z) + t'_{28}(E_{F1D}, n_y, n_z)\right], \tag{8.69}$$

$$\Delta C_{456} = \frac{2(\bar{G}_0)^3 g_v}{27\pi} \frac{\sqrt{2m_1}}{\hbar} \sum_{n_y=1}^{n_{y\max}} \sum_{n_z=1}^{n_{z\max}} \left[t''_{27}(E_{F1D}, n_y, n_z) + t''_{28}(E_{F1D}, n_y, n_z)\right]. \tag{8.70}$$

2. *Hybrid model*:

$$\Delta C_{44} = -\frac{2(\bar{G}_0)^2 g_v}{9\pi} \frac{\sqrt{2m_1}}{\hbar} \sum_{n_y=1}^{n_{y\max}} \sum_{n_z=1}^{n_{z\max}} \left[t'_{31}(E_{F1D}, n_y, n_z) + t'_{32}(E_{F1D}, n_y, n_z)\right], \tag{8.71}$$

$$\Delta C_{456} = \frac{2(\bar{G}_0)^3 g_v}{27\pi} \frac{\sqrt{2m_1}}{\hbar} \sum_{n_y=1}^{n_{y\max}} \sum_{n_z=1}^{n_{z\max}} \left[t''_{31}(E_{F1D}, n_y, n_z) + t''_{32}(E_{F1D}, n_y, n_z)\right]. \tag{8.72}$$

3. *Cohen model*:

$$\Delta C_{44} = -\frac{2(\bar{G}_0)^2 g_v}{9\pi} \frac{\sqrt{2m_1}}{\hbar} \sum_{n_y=1}^{n_{y\max}} \sum_{n_z=1}^{n_{z\max}} \left[t'_{35}(E_{F1D}, n_y, n_z) + t'_{36}(E_{F1D}, n_y, n_z)\right], \tag{8.73}$$

$$\Delta C_{456} = \frac{2(\bar{G}_0)^3 g_v}{27\pi} \frac{\sqrt{2m_1}}{\hbar} \sum_{n_y=1}^{n_{y\max}} \sum_{n_z=1}^{n_{z\max}} \left[t''_{35}(E_{F1D}, n_y, n_z) + t''_{36}(E_{F1D}, n_y, n_z)\right]. \tag{8.74}$$

4. *Lax model*:

$$\Delta C_{44} = -\frac{2(\bar{G}_0)^2 g_v}{9\pi}\frac{\sqrt{2m_1}}{\hbar}\sum_{n_y=1}^{n_{y\max}}\sum_{n_z=1}^{n_{z\max}}\left[t'_{37}(E_{F1D}, n_y, n_z) + t'_{38}(E_{F1D}, n_y, n_z)\right], \tag{8.75}$$

$$\Delta C_{456} = \frac{2(\bar{G}_0)^3 g_v}{27\pi}\frac{\sqrt{2m_1}}{\hbar}\sum_{n_y=1}^{n_{y\max}}\sum_{n_z=1}^{n_{z\max}}\left[t''_{37}(E_{F1D}, n_y, n_z) + t''_{38}(E_{F1D}, n_y, n_z)\right]. \tag{8.76}$$

(d) *IV–VI materials:*
Dimmock model:

$$\Delta C_{44} = -\frac{2(\bar{G}_0)^2 g_v}{9\pi}\sum_{n_x=1}^{n_{x\max}}\sum_{n_y=1}^{n_{y\max}}[B'_{32}(E_{F1D}, n_x, n_y) + B'_{33}(E_{F1D}, n_x, n_y)], \tag{8.77}$$

$$\Delta C_{456} = \frac{2(\bar{G}_0)^3 g_v}{27\pi}\sum_{n_x=1}^{n_{x\max}}\sum_{n_y=1}^{n_{y\max}}[B''_{32}(E_{F1D}, n_x, n_y) + B''_{33}(E_{F1D}, n_x, n_y)]. \tag{8.78}$$

(e) *Stressed materials*:

$$\Delta C_{44} = -\frac{2(\bar{G}_0)^2 g_v}{9\pi}\sum_{n_y=1}^{n_{y\max}}\sum_{n_z=1}^{n_{z\max}}\left[B'_{34}(E_{F1D}, n_y, n_z) + B'_{35}(E_{F1D}, n_y, n_z)\right], \tag{8.79}$$

$$\Delta C_{456} = \frac{2(\bar{G}_0)^3 g_v}{27\pi}\sum_{n_y=1}^{n_{y\max}}\sum_{n_z=1}^{n_{z\max}}\left[B''_{34}(E_{F1D}, n_y, n_z) + B''_{35}(E_{F1D}, n_y, n_z)\right]. \tag{8.80}$$

(f) *Tellurium*:

$$\Delta C_{44} = -\frac{(\bar{G}_0)^2 g_v}{9\pi}\sum_{n_x=1}^{n_{x\max}}\sum_{n_y=1}^{n_{y\max}}\left[B'_{36,\pm}(E_{F1D}, n_x, n_y) + \theta'_{5,\pm}\right], \tag{8.81}$$

$$\Delta C_{456} = \frac{(\bar{G}_0)^3 g_v}{27\pi}\sum_{n_x=1}^{n_{x\max}}\sum_{n_y=1}^{n_{y\max}}\left[B''_{36,\pm}(E_{F1D}, n_x, n_y) + \theta''_{5,\pm}\right]. \tag{8.82}$$

(e) *Gallium phosphide*:

$$\Delta C_{44} = -\frac{2(\bar{G}_0)^2 g_v}{9\pi}\sum_{n_x=1}^{n_{x\max}}\sum_{n_y=1}^{n_{y\max}}\left[B'_{38}(E_{F1D}, n_x, n_y) + B'_{39}(E_{F1D}, n_x, n_y)\right], \tag{8.83}$$

$$\Delta C_{456} = \frac{2(\bar{G}_0)^3 g_v}{27\pi} \sum_{n_x=1}^{n_{x\max}} \sum_{n_y=1}^{n_{y\max}} \left[B''_{38}(E_{F1D}, n_x, n_y) + B''_{39}(E_{F1D}, n_x, n_y)\right]. \tag{8.84}$$

(f) *Platinum Antimonide*:

$$\Delta C_{44} = -\frac{2(\bar{G}_0)^2 g_v}{9\pi} \sum_{n_x=1}^{n_{x\max}} \sum_{n_y=1}^{n_{y\max}} \left[B'_{40}(E_{F1D}, n_x, n_y) + B'_{41}(E_{F1D}, n_x, n_y)\right], \tag{8.85}$$

$$\Delta C_{456} = \frac{2(\bar{G}_0)^3 g_v}{27\pi} \sum_{n_x=1}^{n_{x\max}} \sum_{n_y=1}^{n_{y\max}} \left[B''_{40}(E_{F1D}, n_x, n_y) + B''_{41}(E_{F1D}, n_x, n_y)\right]. \tag{8.86}$$

(g) *Bismuth Telluride*:

$$\Delta C_{44} = -\frac{2(\bar{G}_0)^2 g_v}{9\pi} \sum_{n_z=1}^{n_{z\max}} \sum_{n_y=1}^{n_{y\max}} \left[B'_{42}(E_{F1D}, n_z, n_y) + B'_{43}(E_{F1D}, n_z, n_y)\right], \tag{8.87}$$

$$\Delta C_{456} = \frac{2(\bar{G}_0)^3 g_v}{27\pi} \sum_{n_z=1}^{n_{z\max}} \sum_{n_y=1}^{n_{y\max}} \left[B''_{42}(E_{F1D}, n_z, n_y) + B''_{43}(E_{F1D}, n_z, n_y)\right]. \tag{8.88}$$

(h) *Germanium*:

1. *The model of Cardona et al.:*

$$\Delta C_{44} = -\frac{2(\bar{G}_0)^2 g_v}{9\pi} \sum_{n_x=1}^{n_{x\max}} \sum_{n_z=1}^{n_{z\max}} \left[B'_{44}(E_{F1D}, n_x, n_z) + B'_{45}(E_{F1D}, n_x, n_z)\right], \tag{8.89}$$

$$\Delta C_{456} = \frac{2(\bar{G}_0)^3 g_v}{27\pi} \sum_{n_x=1}^{n_{x\max}} \sum_{n_z=1}^{n_{z\max}} \left[B''_{44}(E_{F1D}, n_x, n_z) + B''_{45}(E_{F1D}, n_x, n_z)\right]. \tag{8.90}$$

2. *The model of Wang and Ressler:*

$$\Delta C_{44} = -\frac{2(\bar{G}_0)^2 g_v}{9\pi} \sum_{n_x=1}^{n_{x\max}} \sum_{n_z=1}^{n_{z\max}} \left[B'_{46}(E_{F1D}, n_x, n_z) + B'_{47}(E_{F1D}, n_x, n_z)\right], \tag{8.91}$$

$$\Delta C_{456} = \frac{2(\bar{G}_0)^3 g_v}{27\pi} \sum_{n_x=1}^{n_{x\max}} \sum_{n_z=1}^{n_{z\max}} \left[B''_{46}(E_{F1D}, n_x, n_z) + B''_{47}(E_{F1D}, n_x, n_z)\right]. \tag{8.92}$$

(i) *Gallium Antimonide:*

$$\Delta C_{44} = -\frac{2(\bar{G}_0)^2 g_v}{9\pi} \sum_{n_x=1}^{n_{x\max}} \sum_{n_y=1}^{n_{y\max}} \left[B'_{48}(E_{F1D}, n_x, n_y) + B'_{49}(E_{F1D}, n_x, n_y)\right], \tag{8.93}$$

$$\Delta C_{456} = \frac{2(\bar{G}_0)^3 g_v}{27\pi} \sum_{n_x=1}^{n_{x\max}} \sum_{n_y=1}^{n_{y\max}} \left[B''_{48}(E_{F1D}, n_x, n_y) + B''_{49}(E_{F1D}, n_x, n_y)\right]. \tag{8.94}$$

(j) *II–V materials*:

$$\Delta C_{44} = -\frac{(\bar{G}_0)^2 g_v}{9\pi} \sum_{n_x=1}^{n_{x\max}} \sum_{n_y=1}^{n_{y\max}} \left[B'_{49}(E_{F1D}, n_x, n_y) + B'_{50}(E_{F1D}, n_x, n_y)\right], \tag{8.95}$$

$$\Delta C_{456} = \frac{(\bar{G}_0)^3 g_v}{27\pi} \sum_{n_x=1}^{n_{x\max}} \sum_{n_y=1}^{n_{y\max}} \left[B''_{49}(E_{F1D}, n_x, n_y) + B''_{50}(E_{F1D}, n_x, n_y)\right]. \tag{8.96}$$

8.2.4 Diffusivity-Mobility Ratio:

The diffusivity (D) tomobility (μ) ratio (DMR) of the carriers in semiconductor devices is known to be very useful [82] since the diffusion constant (a quantity often used in device analysis but whose exact experimental determination is rather difficult) can be obtained from this ratio by knowing the experimental values of the mobility. In addition, it is more accurate than any of the individual relation for the diffusivity or the mobility, which are the two widely used quantities of carrier transport of modern nanostructured materials and devices. The classical DMR equation is valid for both types of carriers. In its conventional form, it appears that, the DMR increases linearly with the temperature T being independent of the carrier concentration. This relation holds only under the condition of carrier non-degeneracy although its validity has been suggested erroneously for degenerate materials [83]. The performance of the electron devices at the device terminals and the speed of operation of modern switching transistors are significantly influenced by the degree of carrier degeneracy present in these devices [84]. The simplest way of analyzing them under degenerate condition is to use the appropriate DMR to express the performance of the devices at the device terminals and the switching speed in terms of the carrier concentration [84].

It is well known from the fundamental work of Landsberg [85–87] that the DMR for electronic materials having degenerate electron concentration is essentially determined by their respective energy band structures. This relation is useful for semiconductor homostructures [88, 89], semiconductor–semiconductor heterostructures [90, 91], metals-semiconductor heterostructures [92–95], and insulator-semiconductor heterostructures [96–99]. It has different values in different materials and varies with the doping, with the magnitude of the reciprocal quantizing magnetic field under magnetic quantization, with the quantizing electric field as in inversion layers, with the nanothickness as in quantum wells and quantum well wires, and with superlattice period as in the quantum confined superlattices of small gap semiconductors with graded interfaces having various carrier energy spectra [100–112]. It can, in general, be proved that for bulk specimens the DMR is given by [101]

$$\frac{D}{\mu} = \left(\frac{n_0}{|e|}\right) \Big/ \left(\frac{\partial n_0}{\partial E_{\mathrm{F}}}\right). \tag{8.97}$$

The electric quantum limit as in inversion layers and nipi structures refers to the lowest electric sub-band and (8.97) assumes the form [101]

$$\frac{D}{\mu} = \left(\frac{\bar{n}_0}{|e|}\right) \Big/ \left(\frac{\partial \bar{n}_0}{\partial(\bar{E}_{F0} - \bar{E}_0)}\right), \tag{8.98}$$

where $\bar{n}_0$, $\bar{E}_{F0}$,and $\bar{E}_0$ are the electron concentration, the energy of the electric sub-band and the Fermi energy in the electric quantum limit.

Using the appropriate equations one obtains

$$\frac{D}{\mu} = \left(\frac{\pi^2 k_B^2 T}{3\,|e|^2\, G}\right) \tag{8.99}$$

Thus, the DMR for degenerate materials can be determined by knowing the experimental values of G.

The suggestion for the experimental determination of the DMR for degenerate semiconductors having arbitrary dispersion laws as given by (8.99) does not contain any energy band constants. For a fixed temperature, the DMR varies inversely as G. Only the experimental values of G for any material as a function of electron concentration will generate the experimental values of the DMR for that range of n_0 for that system. Since G decreases with increasing n_0, from (8.99) one can infer that the DMR will increase with increase in n_0. This statement is the compatibility test so far as the suggestion for the experimental determination of DMR for degenerate materials is concerned.

Although the DMR has extensively been investigated in the literature [100–112] nevertheless it appears that the influence of electric field on the DMR in optoelectronic semiconductors together with its various quantum confined counterpart has yet to be reported. We present few results in this context.

(a) In the presence of intense electric field, the DMR in III–V, ternary and quaternary materials in accordance perturbed three and two band models of Kane can, respectively, be expressed as

$$\frac{D}{\mu} = \frac{1}{e}\left[\frac{M_{1L}(E_{\mathrm{FF}}, F) + N_{1L}(E_{\mathrm{FF}}, F)}{M'_{1L}(E_{\mathrm{FF}}, F) + N'_{1L}(E_{\mathrm{FF}}, F)}\right], \tag{8.100}$$

$$\frac{D}{\mu} = \frac{1}{e}\left[\frac{M_{2L}(E_{\mathrm{FF}}, F) + N_{2L}(E_{\mathrm{FF}}, F)}{M'_{2L}(E_{\mathrm{FF}}, F) + N'_{2L}(E_{\mathrm{FF}}, F)}\right]. \tag{8.101}$$

In the absence of any field, the expressions for the DMR for optoelectronic materials whose energy band structures are defined by the unperturbed two band model of Kane, under the condition $(E_{\mathrm{F}} E_{\mathrm{g0}}^{-1}) \ll 1$ assume the well-known forms as [100]

$$\frac{D}{\mu} = \frac{k_B T}{e}\left[\frac{F_{1/2}(\eta) + \frac{15\alpha k_B T}{4} F_{3/2}(\eta)}{F_{-1/2}(\eta) + \frac{15\alpha k_B T}{4} F_{1/2}(\eta)}\right] \tag{8.102}$$

(b) In the presence of intense electric field, the DMR in III–V, ternary, and quaternary materials in accordance perturbed three and two band models of Kane can, respectively, be expressed under arbitrarily oriented quantizing magnetic field as

$$\frac{D}{\mu} = \frac{1}{e}\left[\frac{\sum_{n=0}^{n_{\max}} [M_3(E_{\mathrm{FFB}}, F, \theta, n) + N_3(E_{\mathrm{FFB}}, F, \theta, n)]}{\sum_{n=0}^{n_{\max}} [M'_3(E_{\mathrm{FFB}}, F, \theta, n) + N'_3(E_{\mathrm{FFB}}, F, \theta, n))]}\right] \tag{8.103}$$

and

$$\frac{D}{\mu} = \frac{1}{e}\left[\frac{\sum_{n=0}^{n_{\max}} [M_4(E_{\mathrm{FFB}}, F, \theta, n) + N_4(E_{\mathrm{FFB}}, F, \theta, n)]}{\sum_{n=0}^{n_{\max}} [M'_4(E_{\mathrm{FFB}}, F, \theta, n) + N'_4(E_{\mathrm{FFB}}, F, \theta, n)]}\right]. \tag{8.104}$$

(c) In the presence of intense electric field, the DMR in quantum wells of optoelectronic semiconductors in accordance perturbed three and two band models of Kane can, respectively, be expressed as

$$\frac{D}{\mu} = \frac{1}{e}\left[\frac{\sum_{n_x=1}^{n_{x\max}} [M_5(E_{\mathrm{FFS}}, F, n_x) + N_5(E_{\mathrm{FFS}}, F, n_x)]}{\sum_{n_x=1}^{n_{x\max}} [M'_5(E_{\mathrm{FFS}}, F, n_x) + N'_5(E_{\mathrm{FFS}}, F, n_x)]}\right], \tag{8.105}$$

$$\frac{D}{\mu} = \frac{1}{e}\left[\frac{\sum_{n_x=1}^{n_{x\max}} [M_6(E_{\mathrm{FFS}}, F, n_x) + N_6(E_{\mathrm{FFS}}, F, n_x)]}{\sum_{n_x=1}^{n_{x\max}} [M'_6(E_{\mathrm{FFS}}, F, n_x) + N'_6(E_{\mathrm{FFS}}, F, n_x)]}\right]. \tag{8.106}$$

(d) In the presence of intense electric field, the DMR in quantum wires of III–V, ternary, and quaternary materials in accordance perturbed three and two band models of Kane can, respectively, be expressed as

$$\frac{D}{\mu} = \frac{1}{e}\left[\frac{\sum_{n_z=1}^{n_{z\max}}\sum_{n_y=1}^{n_{y\max}}[Q_{15}(E_{F1D}, F, n_y, n_z) + Q_{16}(E_{F1D}, F, n_y, n_z)]}{\sum_{n_z=1}^{n_{z\max}}\sum_{n_y=1}^{n_{y\max}}[Q'_{15}(E_{F1D}, F, n_y, n_z) + Q'_{16}(E_{F1D}, F, n_y, n_z)]}\right], \tag{8.107}$$

$$\frac{D}{\mu} = \frac{1}{e}\left[\frac{\sum_{n_z=1}^{n_{z\max}}\sum_{n_y=1}^{n_{y\max}}[Q_{17}(E_{F1D}, F, n_y, n_z) + Q_{18}(E_{F1D}, F, n_y, n_z)]}{\sum_{n_z=1}^{n_{z\max}}\sum_{n_y=1}^{n_{y\max}}[Q'_{17}(E_{F1D}, F, n_y, n_z) + Q'_{18}(E_{F1D}, F, n_y, n_z)]}\right]. \tag{8.108}$$

(e) In the presence of intense electric field, the DMR in effective mass super lattices of optoelectronic materials in accordance perturbed three and two band models of Kane can, respectively, be expressed under magnetic quantization as

$$\frac{D}{\mu} = \frac{1}{e}\left[\frac{\sum_{n=0}^{n_{\max}}[Q_{19}(E_{FB}, F, n) + Q_{20}(E_{FB}, F, n)]}{\sum_{n=0}^{n_{\max}}[Q'_{19}(E_{FB}, F, n) + Q'_{20}(E_{FB}, F, n)]}\right] \tag{8.109}$$

and

$$\frac{D}{\mu} = \frac{1}{e}\left[\frac{\sum_{n=0}^{n_{\max}}[Q_{21}(E_{FB}, F, n) + Q_{22}(E_{FB}, F, n)]}{\sum_{n=0}^{n_{\max}}[Q'_{21}(E_{FB}, F, n) + Q'_{22}(E_{FB}, F, n)]}\right]. \tag{8.110}$$

(f) In the presence of intense electric field, the DMR in quantum wire effective mass super lattices of optoelectronic materials in accordance perturbed three and two band models of Kane can, respectively, be expressed as

$$\frac{D}{\mu} = \frac{1}{e}\left[\frac{\sum_{n_z=1}^{n_{z\max}}\sum_{n_y=1}^{n_{y\max}}[Q_{23}(E_{FIDEMSL}, F, n_y, n_z) + Q_{24}(E_{FIDEMSL}, F, n_y, n_z)]}{\sum_{n_z=1}^{n_{z\max}}\sum_{n_y=1}^{n_{y\max}}[Q'_{23}(E_{FIDEMSL}, F, n_y, n_z) + Q'_{24}(E_{FIDEMSL}, F, n_y, n_z)]}\right], \tag{8.111}$$

$$\frac{D}{\mu}=\frac{1}{e}\left[\frac{\sum_{n_z=1}^{n_{z\max}}\sum_{n_y=1}^{n_{y\max}}[Q_{25}(E_{\text{FIDEMSL}},F,n_y,n_z)+Q_{26}(E_{\text{FIDEMSL}},F,n_y,n_z)]}{\sum_{n_z=1}^{n_{z\max}}\sum_{n_y=1}^{n_{y\max}}[Q'_{25}(E_{\text{FIDEMSL}},F,n_y,n_z)+Q'_{26}(E_{\text{FIDEMSL}},F,n_y,n_z)]}\right]. \tag{8.112}$$

(g) In the presence of intense electric field, the DMR in super lattices of optoelectronic materials with graded interfaces in accordance with perturbed three band model of Kane can be expressed under magnetic quantization as

$$\frac{D}{\mu}=\frac{1}{e}\left[\frac{\sum_{n=0}^{n_{\max}}[Q_{27}(E_{\text{FBGISL}},F,n)+Q_{28}(E_{\text{FBGISL}},F,n)]}{\sum_{n=0}^{n_{\max}}[Q'_{27}(E_{\text{FBGISL}},F,n)+Q'_{28}(E_{\text{FBGISL}},F,n)]}\right]. \tag{8.113}$$

(h) In the presence of intense electric field, the DMR in quantum wire super lattices of optoelectronic materials with graded interfaces in accordance with perturbed three band model of Kane can be expressed as

$$\frac{D}{\mu}=\frac{1}{e}\left[\frac{\sum_{n_z=1}^{n_{z\max}}\sum_{n_y=1}^{n_{y\max}}[Q_{29}(E_{\text{FQWGISL}},F,n_y,n_z)+Q_{30}(E_{\text{FQWGISL}},F,n_y,n_z)]}{\sum_{n_z=1}^{n_{z\max}}\sum_{n_y=1}^{n_{y\max}}[Q'_{29}(E_{\text{FQWGISL}},F,n_y,n_z)+Q'_{30}(E_{\text{FQWGISL}},F,n_y,n_z)]}\right] \tag{8.114}$$

With the advent of ultra-small devices, the influence of electric field is of crucial importance in the whole spectrum of nano-science and technology. In this particular section, we have formulated the DMR in optoelectronic semiconductors and their nanostructures in the presence of intense electric field.

8.2.5 Measurement of Band Gap in the Presence of Light Waves:

Using (6.41), (6.42), and (6.43), the normalized incremental band gap (ΔE_g) has been plotted as a function of normalized I_0 (for a given wavelength and considering red light for which $\lambda = 660\,\text{nm}$) at $T = 4.2\,\text{K}$ in Figs. 8.1 and 8.2 for n-$Hg_{1-x}Cd_xTe$ and n-$In_{1-x}Ga_xAs_yP_{1-y}$ lattice matched to InP in accordance with the perturbed three and two band models of Kane and that of perturbed parabolic energy bands respectively. In Figs. 8.3 and 8.4, the normalized incremental band gap has been plotted for the aforementioned optoelectronic compounds as a function of λ. It is worth remarking that the influence of an external photoexcitation is to change radically the original band structure of the material. Because of this change, the photon field

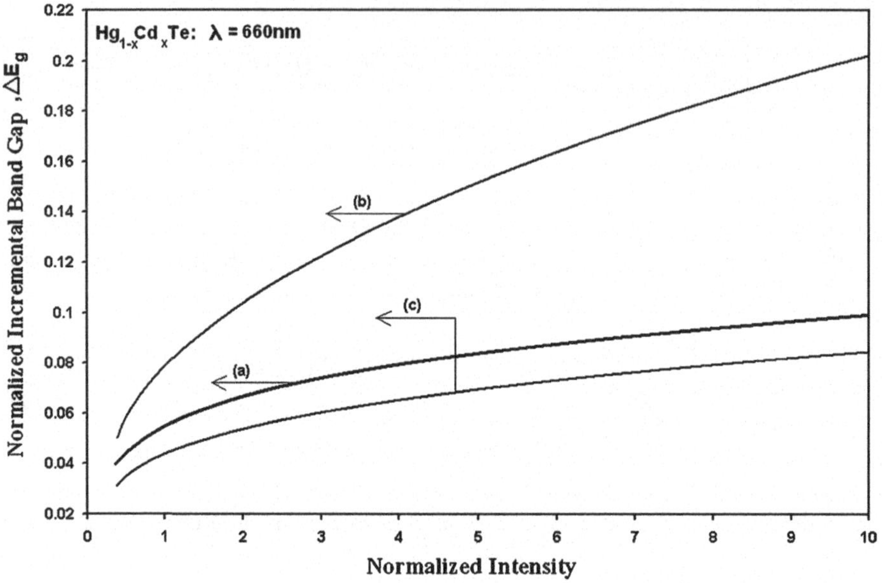

Fig. 8.1 Plots of the normalized incremental band gap (ΔE_g) for n-$Hg_{1-x}Cd_xTe$ as a function of normalized light intensity in which the curves **a** and **b** represent the perturbed three and two band models of Kane respectively. The curve **c** represents the same variation in n-$Hg_{1-x}Cd_xTe$ in accordance with the perturbed parabolic energy bands

causes to increase the band gap of semiconductors. We propose the following two experiments for the measurement of band gap of semiconductors under photoexcitation.

(A) A white light with color filter is allowed to fall on a semiconductor and the optical absorption coefficient ($\overline{\alpha}_0$) is being measured experimentally. For different colors of light, $\overline{\alpha}_0$ is measured and $\overline{\alpha}_0$ versus $\hbar\omega$ (the incident photon energy) is plotted and we extrapolate the curve such that $\overline{\alpha}_0 \to 0$ at a particular value $\hbar\omega_1$. This $\hbar\omega_1$ is the unperturbed band gap of the semiconductor. During this process, we vary the wavelength with fixed I_0. From our present study, we have observed that the band gap of the semiconductor increases for various values of λ when I_0 is fixed (from Figs. 8.3 and 8.4). This implies that the band gap of the semiconductor measured (i.e., $\hbar\omega_1 = E_g$) is not the unperturbed band gap E_{g_0} but the perturbed band gap E_g; where $E_g = E_{g_0} + \Delta E_g$, ΔE_g is the increased band gap at $\hbar\omega_1$. Conventionally, we consider this E_g as the unperturbed band gap of the semiconductor and this particular concept needs modification. Furthermore, if we vary I_0 for a monochromatic light (when λ is fixed) the band gap of the semiconductor will also change consequently (Figs. 8.1 and 8.2). Consequently, the absorption coefficient will change with the intensity of light [77]. For the overall understanding, the detailed theoretical and experimental investigations are needed in this context for various materials having different band structures.

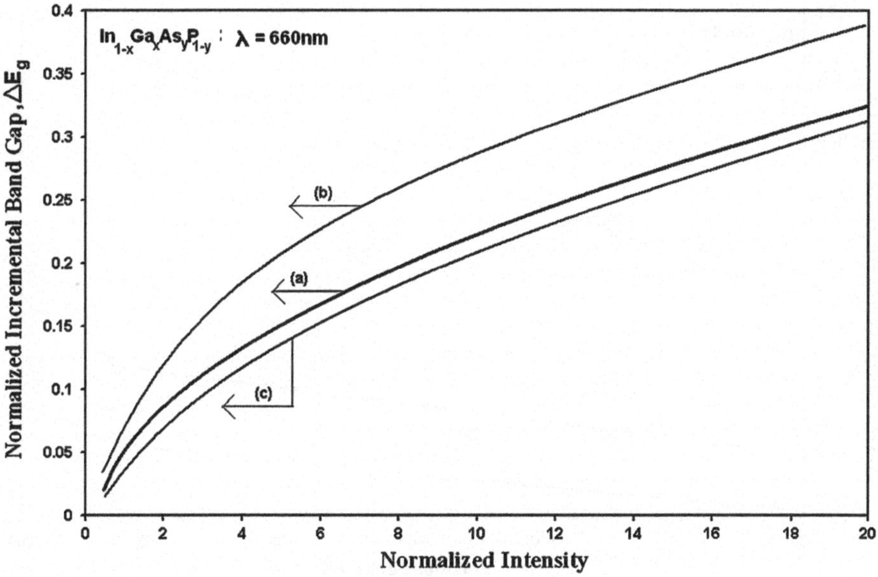

Fig. 8.2 Plots of the normalized incremental band gap (ΔE_g) for $In_{1-x}Ga_xAs_yP_{1-y}$ lattice matched to InP as a function of normalized light intensity for all cases of Fig. 8.1

(B) The conventional idea for the measurement of the band gap of the semiconductors is the fact that the minimum photon energy $h\nu$(ν is the frequency of the monochromatic light) should be equal to the band gap E_{g_0} (unperturbed) of the semiconductor, i.e.,

$$h\nu = E_{g_0}. \tag{8.115}$$

In this case, λ is fixed for a given monochromatic light and the semiconductor is exposed to a light of wavelength λ. Also the intensity of the light is fixed. From Figs. 8.3 and 8.4, we observe that the band gap of the semiconductor is not E_{g_0}(for a minimum value of $h\nu$) but E_g, the perturbed band gap. Thus, we can rewrite the above equality as

$$h\nu = E_g. \tag{8.116}$$

Furthermore, if we vary the intensity of light (Figs. 8.1 and 8.2) for the study of photoemission, the minimum photon energy should be

$$h\nu_1 = E_{g_1}, \tag{8.117}$$

where E_{g_1} is the perturbed band gap of the semiconductor due to various intensity of light when ν and ν_1 are different.

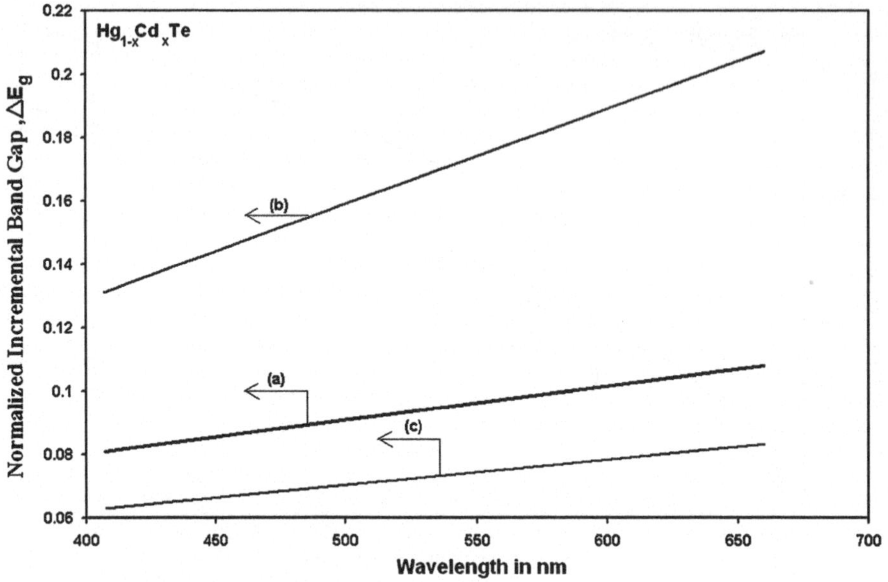

Fig. 8.3 Plots of the normalized incremental band gap (ΔE_g) for $Hg_{1-x}Cd_xTe$ as a function of wavelength for all cases of Fig. 8.1

Thus, we arrive at the following conclusions:

(a) Under different intensity of light, keeping λ fixed, the condition of band gap measurement is given by

$$h\nu_1 = E_{g_1} = E_{g_0} + \Delta E_{g_1}. \tag{8.118}$$

(b) Under different color of light, keeping the intensity fixed, the condition of band gap measurement assumes the form

$$h\nu = E_g = E_{g_0} + \Delta E_g \tag{8.119}$$

and not the conventional result as given by (8.115).

8.2.6 Diffusion Coefficient of the Minority Carriers:

This particular coefficient in quantum confined lasers can be expressed [85] as

$$D_i / D_0 = \mathrm{d}E_{\mathrm{Fi}} / \mathrm{d}E_{\mathrm{F}}, \tag{8.120}$$

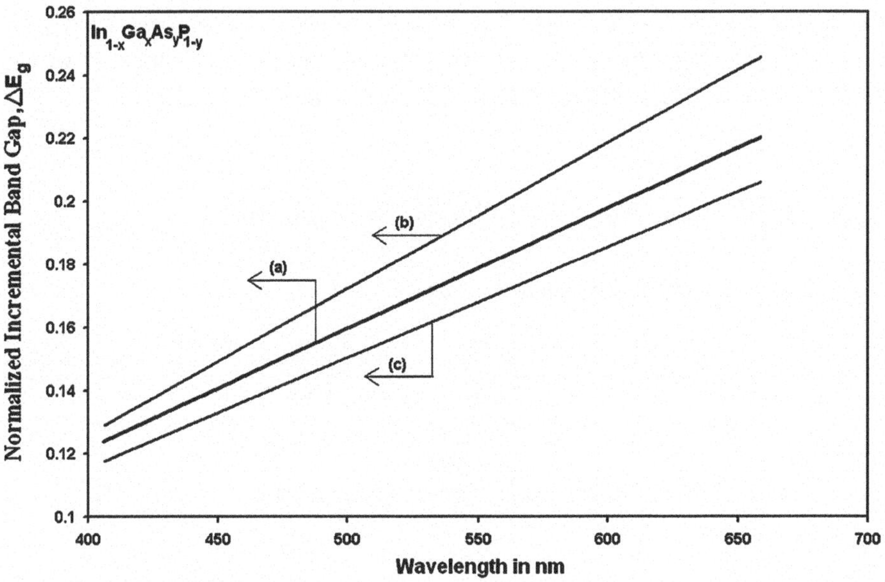

Fig. 8.4 Plots of the normalized incremental band gap (ΔE_g) for $In_{1-x}Gs_xAs_yP_{1-y}$ lattice matched to InP as a function of wavelength for all cases of Fig. 8.2

where D_i and D_0 are the diffusion coefficients of the minority carriers both in the presence and absence of quantum confinements and E_{Fi} and E_F are the Fermi energies in the respective cases. It appears then that, the formulation of the above ratio requires a relation between E_{Fi} and E_F, which, in turn, is determined by the appropriate carrier statistics. Thus, our present study plays an important role in determining the diffusion coefficients of the minority carriers of quantum confined lasers with materials having arbitrary band structures. Therefore in the investigation of the optical excitation of the optoelectronic materials which lead to the study of the ambipolar diffusion coefficients the present results contribute significantly.

8.2.7 Nonlinear Optical Response:

The nonlinear response from the optical excitation of the free carriers is given by [113]

$$Z_0 = \frac{-e^2}{\omega^2\hbar^2}\int_0^\infty \left(k_x\frac{\partial k_x}{\partial E}\right)^{-1} f(E)\, N(E)\mathrm{d}E, \tag{8.121}$$

where ω is the optical angular frequency, $N(E)$ is the density-of-states function. From the various E-k relations of different materials under different physical conditions, we can formulate the expression of $N(E)$ and from band structure we can derive the term $\left(k_x \frac{\partial k_x}{\partial E}\right)$ and thus by using the density-of-states function as formulated, we can study the Z_0 for all types of materials as considered in this monograph.

8.2.8 Third-Order Nonlinear Optical Susceptibility:

This particular susceptibility can be written as [114]

$$\chi_{\mathrm{NP}}(\omega_1, \omega_2, \omega_3) = \frac{n_0 e^4 \left\langle \varepsilon^4 \right\rangle}{24\omega_1\omega_2\omega_3(\omega_1 + \omega_2 + \omega_3)\hbar^4}, \tag{8.122}$$

where

$$n_0 \left\langle \varepsilon^4 \right\rangle = \int_0^\infty \frac{\partial^4 E}{\partial k_z^4} N(E) f(E)\, \mathrm{d}E$$

and the other notations are defined in [114]. The term $\left(\frac{\partial^4 E}{\partial k_z^4}\right)$ can be formulated by using the dispersion relations of different materials as given in appropriate sections of this monograph. Thus one can investigate the $\chi_{\mathrm{NP}}(\omega_1, \omega_2, \omega_3)$ for all materials as considered in this monograph.

8.2.9 Generalized Raman Gain:

The generalized Raman gain in optoelectronic materials can be expressed as [115]

$$R_G = \bar{I}\left(\frac{16\pi^2 c^2}{\hbar\omega \rho g \omega_s^2 n_s n_p}\right)\left(\frac{\Gamma_\rho}{\Gamma}\right)\left(\left(\frac{e^2}{mc^2}\right)^2 m^2 R^2\right), \tag{8.123}$$

where $\bar{I} = \sum_{n.t_z}\left[f_0(n, k_z \uparrow) - f_0(n, k_z \downarrow)\right]$, $f_0(n, k_z \uparrow)$ is the Fermi factor for spin-up Landau levels, $f_0(n, k_z \downarrow)$ is the Fermi factor for spin down Landau levels, n is the Landau quantum number and the other notations are defined in [115]. It appears then the formulation of R_G is determined by the appropriate derivation requires the magneto-dispersion relations. By using the different appropriate formulas as formulated in various chapters of this monograph R_G can, in general, be investigated.

8.2.10 Einstein's Photo-Electric Effect

It is well known that the Einstein's photoelectric effect occupies a singular position in the whole arena of materials science and related disciplines in general together with the fact that the photoemission from the electronic materials is also a vital physical phenomena from the viewpoint of modern optoelectronics and photoemission spectroscopy [116, 117]. The classical equation of the photoemitted current density is [118] $J = \left[4\pi e m_c g_v (k_B T)^2 / h^3\right] \exp\left[(hv - \varphi)/(k_B T)\right]$, where $h\upsilon$ and ϕ are incident photon energy along z-axis and work function respectively. The aforementioned equation is valid for both the charge carriers and in this conventional form it appears that, the photoemission changes with the effective mass, temperature, work function, and the incident photon energy, respectively. This relation holds only under the condition of carrier non-degeneracy.

The Einstein's photoemission has different values for different materials and varies with doping and with external fields which creates quantization of the wave vector space of the carriers leading to various types of quantized structures. The nature of these variations has been studied in [118–154] and some of the significant features are as follow:

1. The photoemission from bulk materials increases with the increase in doping.
2. The photoemission exhibits oscillatory dependence with inverse quantizing magnetic field because of the Shubnikov de Haas (SdH) effect.
3. The photoemission changes significantly with the magnitude of the externally applied quantizing electric field in electronic materials.
4. The photoemission from quantum confined Bismuth, nonlinear optical, III–V, II–VI, and IV–VI materials oscillate with nano-thickness in various manners which are totally band structure dependent.
5. The nature of variations is significantly influenced by the energy band constants of various materials having different band structures.
6. The photoemission has significantly different values in quantum confined semiconductor superlattices and various other quantized structures.

It is important to note that, in the methods as given in the literature, the physics of photoemission has been incorporated in the lower limit of the photoemission integral and assuming that the band structure of the bulk materials becomes an invariant quantity in the presence of photo-excitation necessary for Einstein's photoelectric effect. The basic band structure of optoelectronic materials changes in the presence of external light waves in a fundamental way, which has been incorporated mathematically through the expressions of the DOS function and the velocity along the direction of photoemission respectively in addition to the appropriate fixation of the lower limit of the photoemission integral for the purpose of investigating the Einstein's photoemission from bulk specimens of optoelectronic compounds.

The consequence of the photoelectric effect is the creation of the concept of photoelectric current density (J) which, can, in turn, be written through the photoemission integral (P_I) as [118]

$$J = \frac{\alpha_0 e}{4}(P_I), \tag{8.124}$$

where α_0 is the probability of photoemission,

$$P_I = \int_{E_0}^{\infty} N(E')v_z(E')f(E)\mathrm{d}E' \qquad (8.125)$$

in which, $E_0 \equiv W - h\upsilon$, W is the electron affinity, $E' \equiv E - E_0$, $N(E')$ is the density-of-states function at $E = E'$, $v_z(E')$ is the velocity of the emitted electron along z-axis when, $E = E'$ and $f(E)$ is the Fermi Dirac occupation probability factor

Using (6.41), $v_z(E')$ and $N(E')$ for optoelectronic materials in the presence of light waves whose unperturbed conduction electrons obey the three band model of Kane can be written as

$$v_z(E') = \frac{\sqrt{2}}{\sqrt{m_c}} \frac{[\beta_0(E',\lambda)]^{1/2}}{\beta_0'(E',\lambda)} \qquad (8.126)$$

and

$$N(E') = \left[4\pi\left(\frac{2m_c}{h^2}\right)^{3/2} \cdot g_v\right]\sqrt{\beta_0(E',\lambda)}\beta_0'(E',\lambda), \qquad (8.127)$$

where

$$\beta_0'(E,\lambda) \equiv \frac{\partial}{\partial E}[\beta_0(E,\lambda)].$$

Using (8.124–8.127), the photoemitted current density in this case can be written using the generalized Sommerfield's lemma as

$$J_L = \left[\frac{4\pi e m_c (k_B T)^2 \alpha_0 g_v}{h^3}\right]\left[\left[\frac{(1+\frac{2}{3}\alpha\Delta)}{(+\alpha\Delta)}\right]\left[2\alpha k_B T F_2(\eta_L) + \left(1 + 2\alpha E_0 + \frac{1}{3}\alpha\Delta\right)F_1(\eta_L)\right.\right.$$
$$\left.\left. + \frac{E_0 + \alpha E_0^2 + \frac{1}{3}\alpha\Delta E_0}{k_B T}F_{-1}(\eta_L) + a_0\left[\boldsymbol{ln}\left|\frac{a_0+\eta_L}{a_0}\right| + \phi_5(\eta_L)\right]\right] - \left\{\frac{B_{51}}{(k_B T)^2}\right\}I_2\right] \qquad (8.128)$$

$\eta_L \equiv (E_{F_L} - E_0)(k_B T)^{-1}$, E_{F_L} is the Fermi energy in the presence of light waves as measured from the edge of the conduction band in the absence of any field, $a_0 \equiv (E_0 + E_{g_0} + \frac{2}{3}\Delta)(k_B T)^{-1}$

$$\phi(\eta_L) \equiv \sum_{r=1}^{s_0} 2(1-2^{1-2r})\xi(2r)\frac{(-1)^{2r-1}(2r-1)!}{(a_0+\eta_L)^{2r}},$$

$$B_{51} \equiv \frac{e^2}{48 m_r \pi c^3}\frac{I_0\lambda^2}{\sqrt{\varepsilon_{sc}\varepsilon_0}}\frac{E_{g_0}(E_{g_0}+\Delta)}{(E_{g_0}+\frac{2}{3}\Delta)}\frac{\beta^2}{4}\left(t+\frac{\rho}{\sqrt{2}}\right)^2,$$

$$I_2 \equiv \int_{E_0}^{\infty} f_1(E')f(E)\mathrm{d}E'$$

and

$$f_1(E') \equiv \frac{1}{\phi_0(E')}\left\{(1+\frac{E_{g_0}-\delta'}{\phi_0(E')+\delta'})+(E_{g_0}-\delta')\left[\frac{1}{\phi_0(E')+\delta'}-\frac{1}{E_{g_0}+\delta'}\right]^{1/2}\right.$$
$$\left.\times\left[\frac{1}{\phi_0(E')+\delta'}-\frac{E_{g_0}+\delta'}{(E_{g_0}-\delta')^2}\right]^{1/2}\right\}^2.$$

The expression of J_L for optoelectronic materials in the presence of light waves whose unperturbed conduction electrons obey the two band model of Kane can be written following (6.42) as

$$J_L=\left[\frac{4\pi em_c(k_BT)^2\alpha_0g_v}{h^3}\right]\left[F_1(\eta_L)+2\alpha k_BTF_2(\eta_L)-\left(\frac{\overline{B}_{51}\alpha}{k_BT}\right)F_0(\eta_L)\right.$$
$$\left.-\left(\frac{3\overline{C}_{51}}{2}\right)k_BTF_1(\eta_L)+2C_{52}(k_BT)^2F_2(\eta_L)\right],\tag{8.129}$$

where

$$\overline{B}_{51}\equiv\frac{e^2}{192m_r\pi c^3}\frac{I_0\lambda^2E_{g_0}}{\sqrt{\varepsilon_{sc}\varepsilon_0}},\quad \overline{C}_{51}\equiv\frac{2m_c\alpha}{m_v},\quad C_{52}\equiv\left[\frac{15(\overline{C}_{51})^2}{8}-\left(\frac{3}{2}\alpha(\overline{C}_{51})\right)\right].$$

The expression of J_L for optoelectronic materials in the presence of light waves whose unperturbed conduction electrons obey the parabolic energy bands can be expressed as

$$J_L=\left[\frac{4\pi em_c(k_BT)^2\alpha_0g_v}{h^3}\right]\left[F_1(\eta_L)-\left(\frac{B_{53}}{k_BT}\right)\left[F_0(\eta_L)-3\alpha k_BT\left(1+\frac{m_c}{m_v}\right)F_1(\eta_L)\right]\right],\tag{8.130}$$

where

$$B_{53}\equiv\frac{e^2}{8m_r\pi(m_c)^2c^3}\frac{I_0\lambda^2}{\sqrt{\varepsilon_{sc}\varepsilon_0}}.$$

Special Case:

Formulation of current density for unperturbed three and two band models of Kane for optoelectronic materials:

(i) The expression of J in accordance with the unperturbed three band model of Kane assume the forms

$$J=\left[\frac{4\pi em_c(k_BT)^2\alpha_0g_v}{h^3}\right]$$

$$\times\left[\frac{(1+\frac{2}{3}\alpha\Delta)}{(1+\alpha\Delta)}\left[2\alpha k_B T F_2(\eta_0)+\left(1+2\alpha E_0+\frac{1}{3}\alpha\Delta\right)F_1(\eta_0)\right.\right.$$
$$\left.\left.+\frac{E_0+\alpha E_0^2+\frac{1}{3}\alpha\Delta E_0}{k_B T}F_{-1}(\eta_0)+a_0\left[\boldsymbol{ln}\left|\frac{\overline{a_0}+\eta_0}{\overline{a_0}}\right|+\varphi(\eta_0)\right]\right]\right], \tag{8.131}$$

where

$$a_0=\frac{E_0+E_{g_0}+\frac{2}{3}\Delta}{k_B T},\quad \overline{a_0}=\frac{2}{9}\frac{\Delta^2}{(k_B T)^2}\left(1+\frac{2}{3}\alpha\Delta\right),$$
$$\phi(\eta_0)=\sum_{r=1}^{S_0}2(1-2^{1-2r})\zeta(2r)\frac{(-1)^{2r-1}(2r-1)!}{(\overline{a_0}+\eta_0)^{2r}}.$$

(ii) In accordance with the unperturbed two band model of Kane, the corresponding expression of J is given by

$$J=\left[\frac{4\pi e m_c(k_B T)^2\alpha_0 g_v}{h^3}\right][F_1(\eta_0)+2\alpha k_B T F_2(\eta_0)] \tag{8.132}$$

For $\alpha\to 0$, (8.132) gets simplified as.

$$J=\frac{4\pi\alpha_0 e m_c g_v(k_B T)^2}{h^3}F_1(\eta_0), \tag{8.133}$$

where $\eta_0\equiv\dfrac{h\nu-\phi}{k_B T}$

Under the condition of non-degeneracy (8.133) gets transform to the well-known form as states already.

8.3 Brief Review of Simulation and Experimental Results

The experimental aspect of the effective mass is very wide and it is not possible to highlight even the major developments in a single chapter. For the purpose of condensed presentation the experimental aspect of the effective mass for different technologically important materials are given below.

Using (1.2) of Chap. 1, the density-of-states effective mass for bulk specimens of nonlinear optical materials and n-Cd_3As_2 can, respectively be expressed as

$$m_D^*=\frac{h^2}{2}(12\pi^3)^{-2/3}(E_{F_b})^{-1/3}$$
$$\times\left[\frac{3}{2}f_1(E_{F_b})\sqrt{f_2(E_{F_b})\gamma(E_{F_b})}.\gamma'(E_{F_b})-(\gamma(E_{F_b}))^{3/2}\right.$$

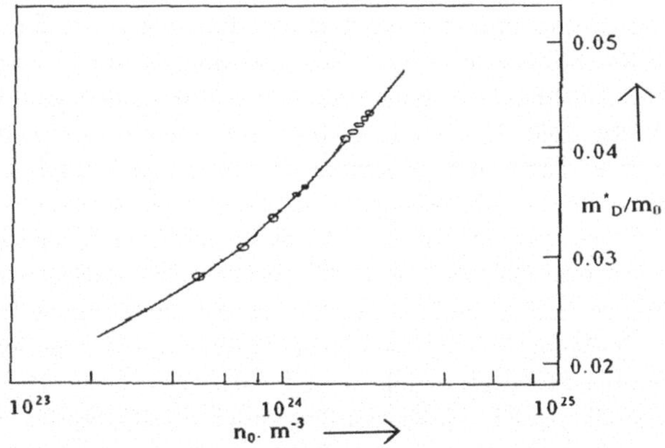

Fig. 8.5 Plot of concentration dependence of the density-of-states effective mass in bulk specimens of n-Cd_3As_2 by using (8.134) where the circular points exhibit the experimental results as given in [155]

$$[f_1'(E_{F_b})\sqrt{f_2(E_{F_b})}+\frac{f_2'(E_{F_b})f_1(E_{F_b})}{2\sqrt{f_2(E_{F_b})}}]/(f_1^2(E_{F_b})f_2(E_{F_b}))]^{3/2}. \tag{8.134}$$

Using (8.134) and (8.4) and the energy band constants of n-Cd_3As_2 from Table 1.1 of Chap. 1, the plot of the density-of-states effective mass as a function of electron concentration in bulk specimens of n-Cd_3As_2 (which is an example of tetragonal material, the conduction electrons of which obey the generalized energy-wave vector dispersion relation for nonlinear optical compounds as formulated in ((1.2) of Chap. 1) is shown in Fig. 8.5, where the circular points exhibit the experimental result [155]. It appears from the Fig. 8.5 that the density-of-states effective mass in bulk specimens of n-type Cadmium Arsenide increases with increasing electron concentration in the whole range of the carrier degeneracy as considered here and the theoretical plot is in good agreement with the experimental data as given in the above reference. It is worth remarking to note that the generalized theoretical formulation of the EEM for different materials, defined by the respective carrier energy spectrum, as formulated in Appendices together with the open research problem as given there and the consequent experimental verification in each case will constitute very important experimental study in this particular arena.

The EEM in GaN_xAs_{1-x}/GaAs quantum wells (QWs) has been investigated, and detected by cyclotron resonance technique by P. N. Hai et al. [156]. The values of EEM are $0.12m_0$ and $0.19m_0$ which are directly determined for the 70-Å-thick QWs with N composition of 1.2 % and 2.0 %, respectively. This sizable increase in the EEM is consistent with the earlier theoretical predictions based on the strong interaction of the lowest conduction band states with the upper lying band states or impurity band induced by the incorporation of N. DiVincenzo et al. [157] formulated the self-consistent effective mass theory for intralayer screening in graphite intercalation compounds. The effective mass approximation (EMA) differential equations appropriate for impurities in a graphite host are constructed and are used to solve self-consistently for the screening response surrounding a single intercalant atom. The screening cloud is found to have a very slow algebraic decay with a characteristic length of 0.38 nm which is due to both the semimetallic and two dimensional character of graphite. The transferred charge in alkali-metal-graphite intercalation compounds is distributed nearly homogeneously on a carbon plane.

Perlin et al. [158] performed infrared reflectivity and Hall effect measurements on highly conducting n-type GaN ($n \approx 6 \times 10^{19}$ cm^3) bulk crystals grown by the high-pressure high-temperature method for the purpose of the experimental determination of the EEM of GaN. Values of electron-plasma frequency and free-electron concentration were determined for each sample of the set of seven crystals. It enabled them to calculate the perpendicular EEM in the wurtzite structure of GaN as $m^* = 0.22 \pm 0.02$ m_0 and the effects of nonparabolicity together with the difference between parallel and perpendicular components of the effective mass are small and do not exceed the experimental error. The EEM has been determined by magnetophotoluminescence in as-grown and hydrogenated $GaAs_{1-x}N_x$ samples for a wide range of nitrogen concentrations (from $x < 0.01\,\%$ to $x = 1.78\,\%$) by the group of Masia et al. [159]. A modified $\mathbf{k} \cdot \mathbf{p}$ model, which takes into account hybridization effects between N cluster states and the conduction band reproduces quantitatively the experimental me values up to $x \leq 0.6\,\%$. Experimental and theoretical evidence is provided for the N complexes responsible for the nonmonotonic and initially puzzling compositional dependence of the EEM. Sewall et al. [160] investigated the experimental tests of EEM and atomistic approaches to quantum dot electronic structure. The overall symmetry of the envelope functions for the four lowest energy excitonic states in colloidal CdSe quantum dots are assigned using excitonic state-resolved pump/probe spectroscopy.

G.E. Smith [161] performed the experimental determination of the EEM's in Bismuth-Antimony alloy. It was found that the EEM's in $Bi_{95}Sb_5$ are smaller by about a factor of two than that of pure Bi and the hole masses are essentially unchanged. SdH investigations on n-InP are presented by the group of Schneider et al. [162] and EEM as a function of carrier concentration has been determined. The experiments were carried out with bulk and liquid phase epitactically grown material and carrier concentrations between $n_0 = 10^{20}\,\mathrm{m}^{-3}$ and $10^{22}\,\mathrm{m}^{-3}$ within the ranges of temperature between 2–77 K and magnetic field $B = 22$ Tesla. The experimental result agrees very well the theoretical relations. The values of the EEM in Cd_3As_2 were obtained from low temperature SdH, magneto-seeback and Hall measurements by Caron et al. [163]. The theoretical estimation of the variation of the energy gap at Γ as a function of temperature and pressure have been obtained. There is a band reversal in $Cd_{3-x}Zn_xAs_2$ and $Cd_3As_xP_{2-x}$ alloys. Bhattacharya et al. [164] evaluated the EEM in compound semiconductor films of CdS_xTe_{1-x} and CdS_xSe_{1-x} which showed bowing phenomena similar to those for optical bandgaps for the above alloy films.

B. Slomski et al. [165] has investigated the in-plane EEM of quantum well states in thin Pb films on a Bi reconstructed Si (111) surface by angle-resolved photoemission spectroscopy. It is found that this EEM is a factor of 3 lower than the unusually high values reported for Pb films grown on a Pb reconstructed Si (111) surface. Through a quantitative low-energy electron diffraction analysis the change in EEM as a function of coverage and for the different interfaces is linked to a change of about 2 % in the in-plane lattice constant. To corroborate this correlation, density functional theory calculations are performed on freestanding Pb slabs with different in-plane lattice constants. These calculations show an anomalous dependence of the EEM on the lattice constant including a change of sign for values close to the lattice constant of Si (111). This unexpected relation is due to a combination of reduced orbital overlap of the $6pz$ states and altered hybridization between the $6pz$ and the $6pxy$ derived quantum well states. Furthermore, it is shown by core-level spectroscopy that the Pb films are structurally and temporally stable at temperatures below 100 K.

The EEM's for spin-up and spin-down electrons of a partially spin-polarized Fermi liquid are theoretically different as proposed by L.M. Wei et al. [166]. They extracted the spin-up and spin-down EEM's from magneto transport measurements at different temperatures for a 2D electron gas in an $In_{0.65}Ga_{0.35}As/In_{0.52}Al_{0.48}As$ quantum well exhibiting zero-field spin splitting. Two analytical methods are used, one involving the simultaneous fitting of fast Fourier transform (FFT) spectra and the other involving inverse FFT (IFFT) analysis. Both methods confirm that the EEM's for spin-up and spin-down are different, consistent with theoretical expectations. The group of Karra et al. [167] performed Cyclotron-resonance measurements for wide (100–300 nm) modulation-doped $Al_xGa_{1-x}As$ graded parabolic quantum wells for electron areal densities $10^9/cm^2$–$2.5x10^{11}/cm^2$. A clear dependence of the cyclotron frequency on N_s is observed in the extreme quantum limit which is understood in terms of alloy effects. Self-consistent calculations that include the x dependence of the local effective mass and exchange–correlation effects in a local density approximation are in quantitative agreement with the measurements for high densities. At low densities a pinning of the cyclotron frequency is observed that is not predicted by the model.

Rößner et al. [168] reported the dependence of the effective masses on hole density in remotely doped strained Ge layers on relaxed $Si_{0.3}Ge_{0.7}$ buffers with sheet densities from $2.9x10^{11}$ cm^{-2} to $1.9x10^{12}$ cm^{-2}. The masses have been determined using temperature dependent Shubnikov–de Haas oscillations. No noticeable dependence of the mass on the magnetic field has been found. The extrapolated G point effective mass has been found to be 0.080 times the free electron mass. From the measured data the variation of the mass with kinetic energy and the shape of the topmost heavy hole subband have been calculated. The results are in good agreement with theoretical predictions. The determination of the EEM of the two-dimensional electron gas (2DEG) and nonparabolicity effects in modulation-doped $In_{0.65}Ga_{0.35}As/In_{0.52}Al_{0.48}As$ single quantum well were investigated by T. M. Kim et al. [169] by performing temperature-dependent Shubnikov-de Haas (SdH) measurements and FFT and the IFFT analyses. The result of the angular dependent SdH measurements clearly demonstrated the occupation of two subbands in the quantum wells by the 2DEGs. The EEM's determined from temperature-dependent S$\pm$dH measurements and the FFT and IFFT analyses were 0.05869 and 0.05385m_e for the first and zeroth subbands, respectively. The EEM's obtained from the S-dH measurements and the FFT and IFFT analyses measurements qualitatively satisfy the nonparabolicity behavior in the $In_{0.65}Ga_{0.35}As$ single quantum well.

The nonparabolic EEM's in InGaAs quantum wells (QWs), sandwiched by thick InAlAs barriers of 0.52-eV band offset, were studied by N. Kotera et al. [170] in normal and parallel directions to the QW plane. The normal mass was experimentally obtained by observing interband photocurrent spectra of undoped InGaAs multi-QW structures. The mass increased by more than 50 % from the bulk band edge mass, 0.041 m_0. Electron eigenenergies were calculated in QWs based on Kane's three-level band theory. The calculated 'apparent' normal mass as a function of kinetic energy up to 0.5 eV agreed well with experiments. The parallel mass in n-type modulation-doped InGaAs QWs was experimentally obtained by pulse cyclotron resonance up to 100 T. The analysis in quantizing magnetic fields, modified for 2D QWs, fits well with cyclotron energy. The 'apparent' parallel mass as a function of energy was obtained consistently. Interband optical transitions of $In_{0.53}Ga_{0.47}As/In_{0.52}Al_{0.48}As$ multi-quantum wells have been observed Tanaka et al. [171] in photocurrent spectra. Interband transitions were assigned from the spectral structures. Eigenenergies of conduction band were not proportional to the square of quantum numbers. An EEM normal to the quantum well plane was 50' %-heavier than the bulk bandedge mass of InGaAs.

The electronic structures of Bi_2Te_3 and Sb_2Te_3 were computed and related to the thermoelectric properties of Bi_2Te_3 and Sb_2Te_3 superlattices by Wang and Cagin [172]. They found that the similarity of the electronic structure of the two materials permits the Bi_2Te_3 and Sb_2Te_3 superlattices inherit high band edge degeneracy, and thus have high electrical conductivity. From the calculated EEM along the superlattice growth direction, they infer that presence of more Sb_2Te_3than Bi_2Te_3 in the superlattice leads to a smaller EEM and enhanced carrier mobility. Furthermore, their results suggest that external tensile strain parallel to the interface may further improve the thermoelectric performance of the Bi_2Te_3 and Sb_2Te_3 superlattices. Engineered energy-wave-vector dispersion relations of either electrons or holes hold great promise for realizing fundamental oscillators at terahertz frequencies if they contain sections with a negative EEM at appropriate energy levels as suggested by Gribnikov et al. [173], although, neither bulk semiconductor materials nor quantum wells or quantum wires exhibit such negative EEM sections in the dispersion relations at favorable energy levels. Therefore, the novel use of a nanostructure is proposed to create an NEM section of electrons at suitable energy levels. This structure utilizes a heterojunction with a QW channel grown perpendicular to a superlattice. At small values of the wave vector k; the electron wave function ψ resides mostly in the QW channel and, as k increases, ψ extends further into the superlattice. This spread of ψ induces an negative EEM section in the energy dispersion relation and several combinations of suitable material systems are considered by them.

Chen and Bajaj [174] have shown that the nonparabolicity and spin splitting enhance the EEM appreciably in a quantum wire. In bulk materials, these effects are usually small since electrons are near the conduction band edges. In nanowires, strong confinement puts the electrons far from the conduction-band edge giving rise to large nonparabolicity effects. They derive a simple expression for the

EEM parallel to the transport direction in the nanowire, taking into account the band nonparabolicity, anisotropy, and spin splitting. They apply their formalism to $GaAs/Al_{0.3}Ga_{0.7}As$nanowires with the conclusion that the parallel mass could be 50 % more than the bulk value for a wire width of 50 Å. The space dependence of the EEM in nanowires results in the appearance of an additional momentum dependent potential has been considered by Borovitskaya and Shur [175]. If the EEM is anisotropic (as in silicon or germanium), this effect strongly depends on the transverse mass for a given sub-band and they consider Si-Ge p-type nanowires, where the impact ionization by holes should be determined by the impact ionization rates in silicon and not in SiGe. Dacal et al. [176] investigated the conductance of 3D semiconductor nano wires considering different EEM's in the contacts and in the channel. They have shown that, with respect to the case with equal masses in the channel and in the contacts, the amplitude of the conductance oscillations increases if the EEM in the channel is larger and decreases if it is smaller than in the contacts. Effects on the density of probability are also considered and these effects of the EEM discontinuity are explained in terms of kinetic confinement and transmission coefficient modulation. Candidate materials for strained-layer EEM superlattices are investigated by Sasaki [177], and sixteen combinations of III–V semiconductor layers are presented. Among these $In_{0.69}Ga_{0.31}As/InP$ possesses the smallest lattice mismatch, 1.1 %. The electronic subband of $In_{0.69}Ga_{0.31}$ As/InP layers is calculated through the Kronig-Penny approach. The energy gap obtained for the conduction band of composite semiconductors is, for example, 52 mcV for one period of alternating layer thickness of 40Å/27Å. Maan et al. [178] investigated the far infrared radiation transmission of a highly doped InAs-GaSb superlattice as a function of the magnetic field, exhibiting helicon wave propagation. The EEM and the carrier density are determined from an analysis of the results as a function of frequency to be $0.082 \pm 0.005 m_0$ and 3.4×10^{18} cm^{-3}. The carrier density is equal to that obtained from Hall measurements. The EEM is significantly higher than the value expected from the InAs conduction band nonparabolicity ($0.063 m_0$).

Synthesizing single-walled carbon nanotubes (SWCNTs) with accurate structural control has been widely acknowledged as an exceedingly complex task culminating in the realization of CNT devices with uncertain electronic behavior. El Shabrawy et al. [179] applied a statistical approach in predicting the SWCNT band gap and EEM variation for typical uncertainties associated with the geometrical structure. They carried out the same by proposing a simulation-efficient analytical model which evaluates the bandgap of an isolated SWCNT with a specified diameter and chirality. They developed an SWCNT EEM model, which is applicable to CNTs of any chirality and diameters >1 nm. A Monte Carlo method has been adopted to simulate the bandgap and effective mass variation for a selection of structural parameter distributions. They established analytical expressions that separately specify the bandgap and EEM variability with respect to the CNT mean diameter and standard deviation which offer insight from a theoretical perspective on the optimization of diameter-related process parameters with the aim of suppressing bandgap and effective mass variation.

Effective electron mass in nanowires of Si and Ge

The study of energy band constants of Silicon nanowire (SiNW) in past few years has emerged as a building block for the next generation nano-electronic devices as it can accommodate multiple gate transistor architecture with excellent electrostatic integrity. As the experimental extraction of its various energy band constants at the nanoscale regime is an extremely challenging task, it is customary to adopt atomic level simulations, the results of which are at par with the experimental data.

In recent years, there have been extensive investigations on the variation of band gap and EEM along different channel orientations in both relaxed [180–183] and strained [184–186] respectively. The physics of SiNWs, are based on numerical methods like the first principle, pseudo-potential, semi-empirical, etc. Although there exists a large number of empirical relations of the band gap in relaxed SiNW [183, 187], there is a growing demand for the development of a physics based analytical model to standardize different energy band constants which particularly demands its application in TCAD software for predicting different electrical characteristics of novel devices like SiNW-based relaxed tunnel field effect transistors and its strained counterpart [188].

The main challenge involved in the formulation of the analytical method for these two quantities (i.e., the band gap and the EEM's) comes from the transition of the indirect energy band gap of bulk Si near X point of the Brillouin zone to direct energy band gap at Γpoint of SiNW. Due to this, the direct energy band gap starts depending on the conduction subband EEM's at the Γpoint, which in turn depends on the conduction and valance subband energies. This conduction subband energy is again dependent on the subband EEM's, thus making it a coupled relation. This results in a parallel variation of all the constants of an intrinsic Si which are entangled to each other.

An intrinsic relaxed bulk Si crystal consists of six equivalent conduction band minima located symmetrically along $\langle 100\rangle$ at a distance of approximately $k_0 = 0.815(2\pi/a_0)$ from the Γ point along X line in a 3D Brillioun zone, in which a_0 is the relaxed lattice constant of Si. The non-parabolic energy dispersion relation of the bulk conduction band electrons can then be written following the EMA formalism as [189]

$$E(1+\alpha E) = \frac{\hbar^2}{2m_l}(k_z - k_0)^2 + \frac{\hbar^2 k_x^2}{2m_t} + \frac{\hbar^2 k_y^2}{2m_t} \tag{8.135}$$

in which $m_l = 0.91m_0$ and $m_t = 0.19m_0$ are the longitudinal and transverse EEM's respectively and $\alpha = 0.5(\text{eV})^{-1}$ [189]. At this point, it should be noted that this relation is isotropic in (001) plane and fails to describe the conduction band wrapping and the subband structure correctly in (110) oriented Si films [189, 190].

In particular, to correlate a complete analytical conduction band dispersion relation with the advanced empirical tight binding model like $sp^3d^5s^*$, a two band degenerate **k.p** model should be used where a second conduction band close to the first conduction band must be taken into account, the two of which becomes degenerate just at the X point [189] as exhibited in Fig. 8.6. These are generally called as primed ($\Delta_{2'}$) and unprimed (Δ_1) bands respectively.

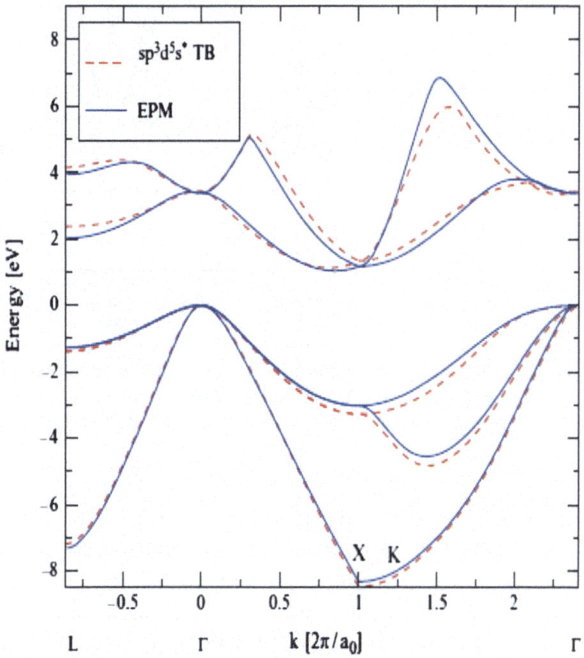

Fig. 8.6 Band structure if bulk Si from the empirical pseudo-potential (*EPM*) method (*solid*) and from the $sp^3d^5s^*$ model (*dashed*), [189]

The band structure of relaxed SiNW whose electron transport is along [107] direction is an involved task. The $sp^3d^5s^*$ model exhibits the fact that the symmetry between the six equivalent conduction band minima is now displaced due to the difference in the effective mass as a result of the quantum confinement of the carriers along y and z directions. Because of this, the six conduction band valleys are now grouped in four in-plane (Δ_4) along y and z directions and two out of plane (Δ_2) valleys along x direction (Fig. 8.7). The former is projected at the Γ point of the 1D Brillioun zone, while the later is zone folded to $k_x = \pm 0.37\pi/a_0$ [180, 191]. Due to the lighter EEM in the Δ_4 valley, the corresponding energy minimum is at a lower position than that of the Δ_2 valley, thus making the NW to be a direct band gap. This chronological transition of the energy-wave vector minimum from an indirect to a direct band gap as a result of the corresponding change from the bulk Si structure to its [107] NW depends not only on the effective masses at the band minima but also onto the subband energies along the confinement directions. This phenomena is, however, not exhibited in the simple non-parabolic EMA relation as given in (8.134), but since the electron energy in a state of the art MOSFET is of few tenths of electron-volts [192] within which the energy diagram from (8.134) and the $sp^3d^5s^*$ are almost same (Fig. 8.6) [189], one can use (8.134) safely for a simplified analytical solution of the band gap and EEM without affecting the electron transport mechanism.

As seen from Fig. 8.8, (a) represents a schematic diagram of a [107] oriented SiNW, the atomistic cross-sectional view along y and z of which is exhibited in (b).

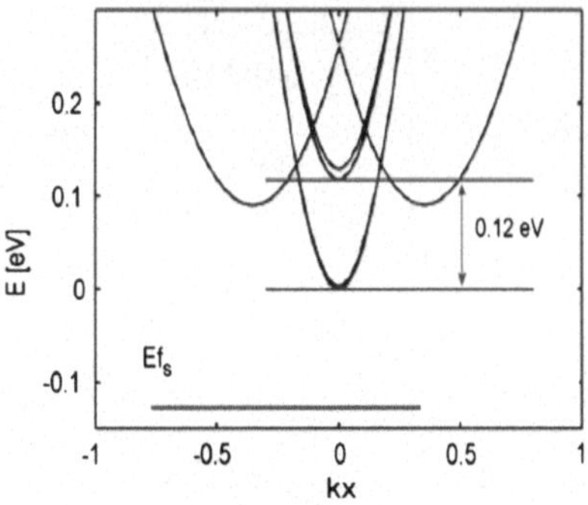

Fig. 8.7 E(k) plot for a 3-nm [107] rectangular wire exhibiting the conduction band valleys at Γ(*direct band gap*) and off-Γpoints (*in direct band gap*), where Ef_s is the source Fermi level. The valley splitting is clearly exhibited using the $sp^3d^5s^*$ model [180]

This has been carried out using the Atomistix Tool Kit (ATK) simulator [193] after a cleaved [107] fully relaxed atomic configured SiNW, the dangling bonds on the Si surface are sp^3 passivated with hydrogen atoms to dissolve any surface states in the band gap region [194]. The nearest Si–Si and Si–H bond lengths have been considered to be 0.235 nm and 0.152 nm respectively [195]. For the band structure computation, the semi-empirical extended Ḧuckel method has been instead of the usual ATK-Density Functional Theory (DFT) method. This has been used due to two main reasons: first, the DFT calculation does not provide a good estimation of the energy band gap and second, the extended Ḧuckel approach is more computationally efficient with a simultaneous good convergence [196]. The Ḧuckel basis set used for the computations were Cerda Silicon (GW Diamond) [197] and Hoffman Hydrogen having a vacuum energy level of -7.67 eV and 0 eV respectively with a Wolfsberg weighting scheme. The tolerance parameter being 10-5 with a maximum steps of 100 and a Pulay mixer algorithm [198] were used as the iteration control parameters. In addition, the k-point sampling of $1 \times 1 \times 11$ grid were used with a mesh cut-off energy of 20 Hatree. Figure 8.8c exhibits the energy band structure of the [107] SiNW for a square cross-section of width 1.5 nm. It can be seen from (c) that using the Hückel basis set, the lifting of the valley degeneracy due to the difference in EEM is not captured which has already been stated earlier. As the valley splitting energy even in room temperature is relatively small in [107] and [1, 2] SiNWs, one can ignore its contribution to the modification of the carrier transport mechanism [180, 182, 199]. Further, as the band gap for a 1.5 nm width SiNW exhibited as a direct one, one can ignore this lifting of the valley degeneracy for the present relaxed case and can concentrate on the lowest valley at the Γ point which essentially determines the band gap.

Figure 8.9 exhibits the variation of [107] relaxed SiNW band gap as function of wire width of equal thickness as exhibited by [194]. The effect of the carrier confinement along the [1, 2] and [12] directions lead to the discrete subband energy levels for both the electrons and holes. In case of valance bands, the heavy hole

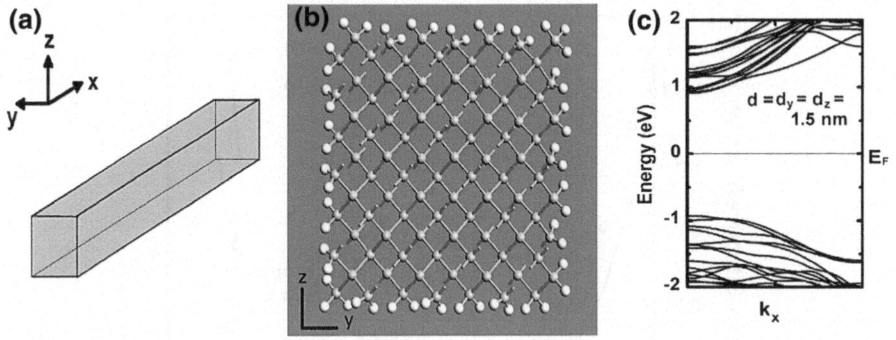

Fig. 8.8 **a** Schematic diagram of [107] oriented channel of SiNW with cross-sectional thicknesses d_y and d_z along y and z directions respectively. **b** ATK built an sp^3 Hydrogen passivated (100) SiNW plane. **c** Energy band structure exhibiting a direct band gap in a [107] Hydrogen passivated SiNW of square cross-sectional area using ATK builder which uses an extended Hückel approach [194]

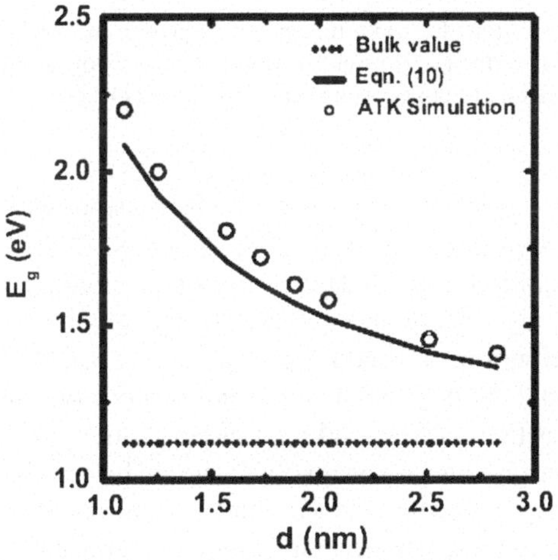

Fig. 8.9 Plot of the band gap using in relaxed [107] SiNW as function of lateral wire width $d_y = d_z = d$. The symbols are the simulation data which has been obtained by using the ATK by passivating the Si atoms at the surface of the wire using Hydrogen atoms as shown in Fig. 8.6b followed by the use of semi-empirical extended Hückel method. The line exhibits the analytical result [194]

(HH) and light hole (LH) forms separate energy subband levels due to the difference in their energies. Using this, the first subband of HH in a $1.5 \times 1.5\,\mathrm{nm}^2$ SiNW is lies about 0.1 eV below compared to that of the maxima point of the HH in case of bulk. However, for the LH subband, this is about 2.2 eV below the same. Thus, the energy band gap difference in case of SiNW should be considered from the lowest conduction subband to the lowest HH subband. Figure 8.10 exhibits the variations of the transport and subband effective mass as a function of wire width. It appears

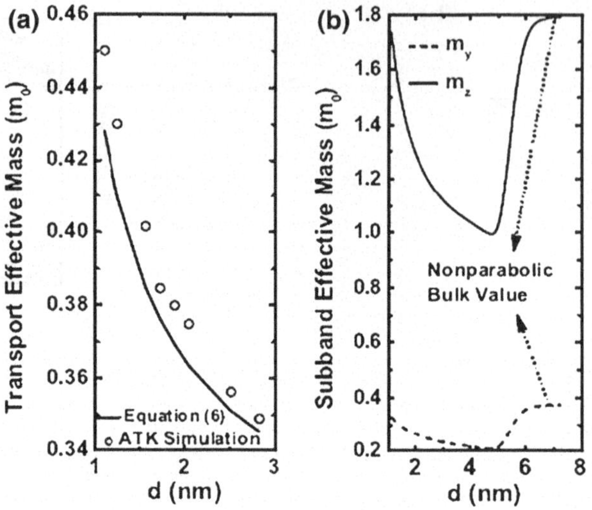

Fig. 8.10 Analytical plot of the electron **a** transport effective mass and **b** subband effective mass as function of wire thickness for [107] oriented SiNW. The symbol represents the extracted data from the energy band structure obtained using ATK simulation [194]

that transport effective mass decreases with the increase in width and as $d \to \infty$ it decreases tends to its non-parabolic bulk value which is 0.32 m_0. In case of subband effective mass, the variation is divided into two parts. Roughly below 5 nm, it appears that both m_y and m_z increase with decrease in wire thickness. This is due to the reason that the contribution of conduction band wave vector at the band minima approaches the Γ point. As the thickness increases, both the subband masses starts increasing and reaches their corresponding non-parabolic bulk effective mass which is precisely 0.38 m_0 and 1.81 m_0 respectively. It should be noted that these bulk values are measured with respect to the valance band maxima at Γ point. If the origin is shifted to k_0, the value of these masses converges to 0.19 m_0 and 0.91 m_0 respectively.

Using this approach, the maximum error between the analytical formulation and simulation data are within 3 %. The main reason behind this error is due to the complete negligence of the spin–orbit interaction between the split-off holes and HH/LH in our model. The other part of the error comes due to the omission of the At large wire cross-sections, the [107] and [118] located at Γ approach the bulk $m_t = 0.19 m_0$. The mass of the [119] wire is larger because it combines m_t and $m_l = 0.89 m_0$. As the wire dimensions shrink, the mass of the [118] wire is reduced, whereas the masses of the other two wires increase. (c) Off-Γvalley masses for the cases of the [118] and [107] wires. Both increase as the dimensions decrease. (The bulk mass values for every orientation are denoted.) The percentage change denoted is the change in the effective masses between the 1.5-nm mass value (mostly scaled wire) and the 7.1-nm wire [180].

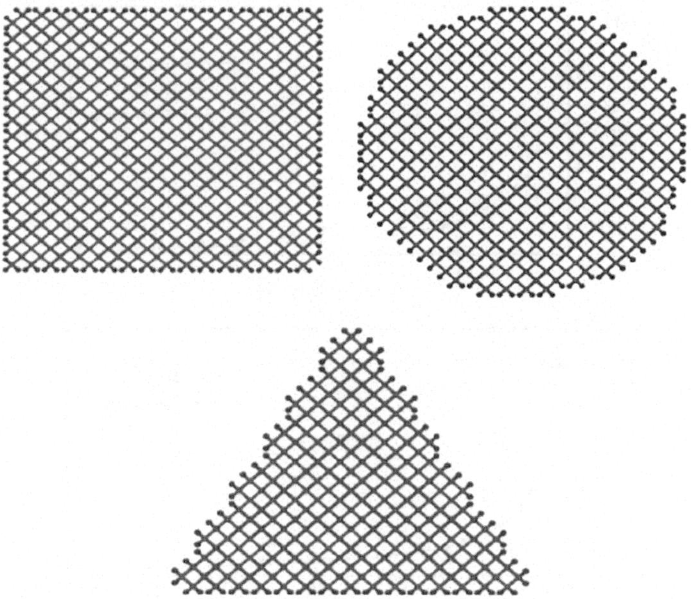

Fig. 8.11 ⟨100⟩ Silicon nanowires with different cross-sectional shapes—Square, Circular, and Triangular [200]

Interaction of the plane waves of Hydrogen on the Si-atoms due to which the band structure of ultra-small thin SiNW gets affected. The analytical model in [194] can also be compared to the band gap of circular SiNW under identical conditions. For other different cross-sectional shapes like circular and triangular [107] SiNW as shown in Fig. 8.11 following Sajjad et al. [200], the band gap in Fig. 8.12 exhibits almost zero deviations from each other when plotted against the cross-sectional area, whereas, if plotted against cross-sectional dimension, both the transport effective mass and band gap exhibits slight deviations [4, 182]. An excellent simulation observation has been studied by Lundstrom group [192] as shown in Fig. 8.13.

The EEM's m_z, m_{le}, and m_{te} have been obtained as explained in [201] from the longitudinal and the transverse masses of the bulk crystal energy dispersion by using the values $0.916m_0$ and $0.19m_0$ for the Δ valleys of the bulk silicon, $1.6m_0$ and $0.093m_0$ for the λ valleys, $0.888m_0$ and $0.194m_0$ for the Λ valleys, and $0.05m_0$ for the Γ valley of the bulk Germanium [202] (Fig. 8.14).

The validity of the parabolic EMA, which is almost universally used to describe the size and bias-induced quantization in n-MOSFETs has been exhibited by Steen, et al. [202]. In particular, the EMA results has been compared with a full-band quantization approach based on the linear combination of bulk bands (LCBB) and has been studied for the most relevant quantities for the modeling of the mobility and of the on-current of the devices, namely, the minima of the 2-D subbands, the transport masses, and the density-of-states function of carriers. The study deals with

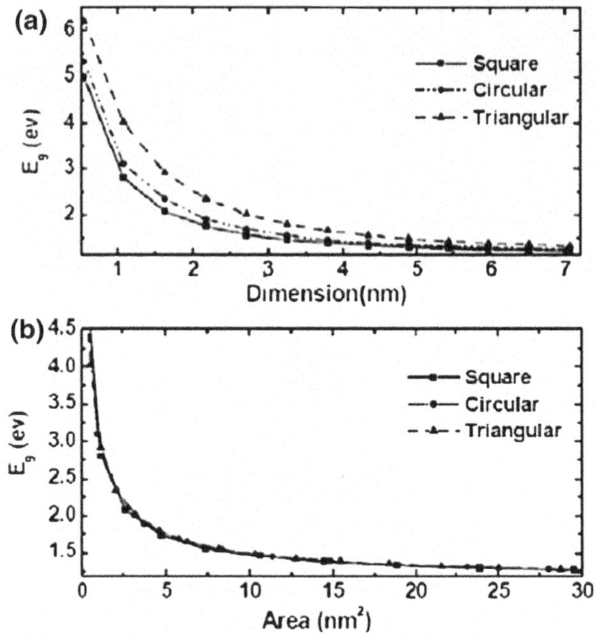

Fig. 8.12 Bandgap variation with **a** dimension, **b** cross-section area [200]

Table 8.1 Parameters of the EMA model for parameters of the EMA model for different materials and quantization directions

	Quantization directions	Valley	n_v	ΔE [eV]	m_{te}	m_{le}	m_z
Si	(001)	$D_{0.916}$	2	0	0.190	0.190	0.916
		$D_{0.19}$	4	0	0.190	0.916	0.190
	(110)	$D_{0.315}$	4	0	0.190	0.553	0.315
		$D_{0.19}$	2	0	0.190	0.916	0.190
	(111)	$D_{0.268}$	6	0	0.190	0.674	0.268
Ge	(110)	$L_{0.25}$	2	0	0.093	0.595	0.25
		$L_{0.093}$	2	0	0.093	1.60	0.093
		$D_{0.318}$	4	0.189	0.194	0.541	0.318
		$D_{0.194}$	2	0.189	0.194	0.888	0.194
		$\Gamma_{0.268}$	1	0.145	0.05	0.05	0.05

For each valley, n_v is the degeneracy, m_z is the quantization mass, and m_{le} and m_{te} are the longitudinal and transverse mass of the elliptic energy dispersion around the minimum (in unit of m_0), respectively and ΔE denotes the energy split between the valleys in the bulk semiconductor

both silicon and germanium n-MOSFETs with different crystal orientations and shows that, in most cases, the validity of the EMA is quite satisfactory. The LCBB approach is then used to calculate the values of the effective masses that help to improve the EMA accuracy. Table 8.1 exhibits the summary of the EEM at different valleys of Si and Ge.

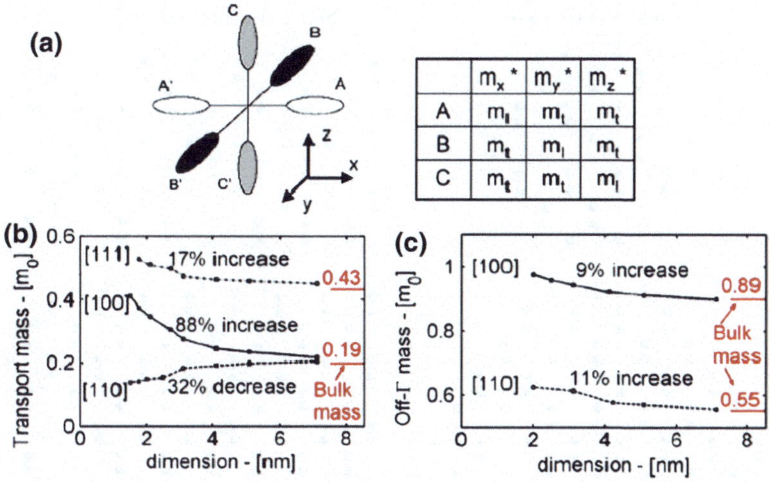

Fig. 8.13 **a** The three equivalent pairs of ellipsoids in the conduction band of Si are described by the longitudinal and transverse masses. Combining these masses results in the quantization and transport masses of nanowires under arbitrary orientations. **b** Transport masses oriented in [107, 118], and [119] versus the wire dimension as calculated from Tight Binding (TB)

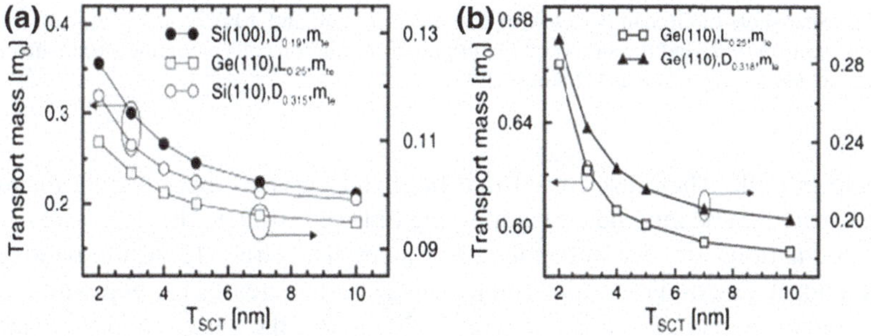

Fig. 8.14 Transverse (m_{te}) and longitudinal (m_{le}) effective masses versus the semiconductor thickness for some valleys of the Si (100) and the Ge (110) inversion layers. **a** Si (100), $D_{0.19}$, m_{te}; Ge (110), $L_{0.25}$, m_{te}; Si (110), $D_{0.315}$, *mte*. **b** Ge(110), $L_{0.25}$,m_{le}; Ge (110), $D_{0.318}$, m_{te}, in which D and L are the $\Delta$and L valleys respectively and the subscripts denotes the minima value of the wave vector space at those valleys

These effective masses exhibit a nonnegligible dependence on semiconductor thickness T_{SCT} and deviate from the values reported in Table 8.1 for very small semiconductor thicknesses [202]. The effect of uniaxial strain and quantum confinement on the effective mass of electrons and holes and band gap of Ge NWs has been demonstrated using the DFT-based first-principles simulations by Logan et al. [203] along the [118] direction as shown in Figs. 8.15, 8.16, 8.17, 8.18, and 8.19. The diameters of the nanowires being studied are up to 50 Å. As shown in [203], the Ge [118] nanowires possess a direct band gap, in contrast to the nature of an indirect

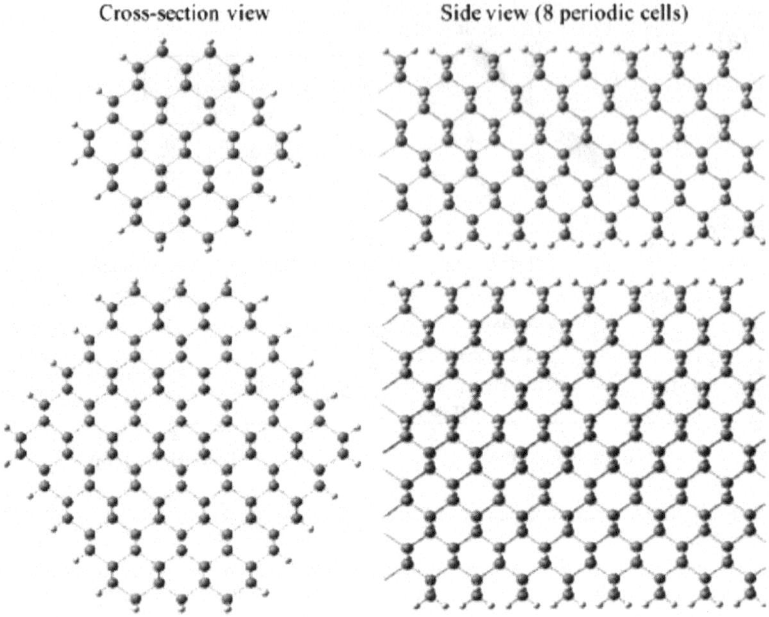

Fig. 8.15 Snapshots of Ge nanowires with size of 18 Å (*top*) and 30 Å (*bottom*) viewed from the wire cross-section (*left*) and the side (eight contiguous simulation cells along the axial z direction). *Blue dots* are Ge atoms, *white* are H atoms [203]

band gap in bulk. They discovered that the band gap and the effective masses of charge carriers can be modulated by applying uniaxial strain to the nanowires. These strain modulations are size dependent. For a smaller wire ~12 Å, the band gap is almost a linear function of strain; compressive strain increases the gap while tensile strain reduces the gap. For a larger wire (~20–50 Å), the variation in the band gap with respect to strain shows nearly parabolic behavior: compressive strain beyond $-1\,\%$ also reduces the gap. In addition, their studies showed that strain affects effective masses of the electron and hole very differently. The effective mass of the hole increases with a tensile strain while the EEM increases with a compressive strain. Our results suggested both strain and size can be used to tune the band structures of nanowires, which may help in design of future nanoelectronic devices. We also discussed our results by applying the tight-binding model.

Bulk Ge is an indirect band gap material with the conduction band minima located at L along the [119] direction. However, if the Ge nanowires are along the [118] direction, they exhibits a direct band gap at Γ[203]. In Fig. 8.16, Logan et al. presented the band structures of Ge nanowires with varied diameters. It clearly demonstrates a direct band gap—both Conduction band minimum (CBM) and valance band maximum (VBM) located at Γ.

It is also interesting to observe that the band structures are modulated by strain. For example, in Fig. 8.17a, we compared the band structures of the Ge nanowire

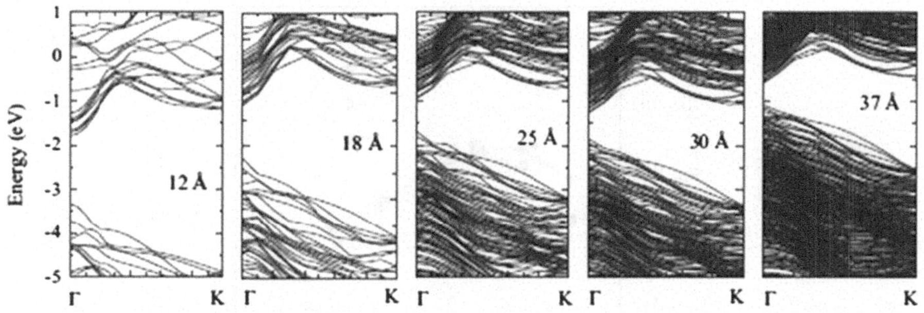

Fig. 8.16 The band structures of Ge nanowires with varied diameter along [118] direction. They show a direct band gap located at Γ [203]

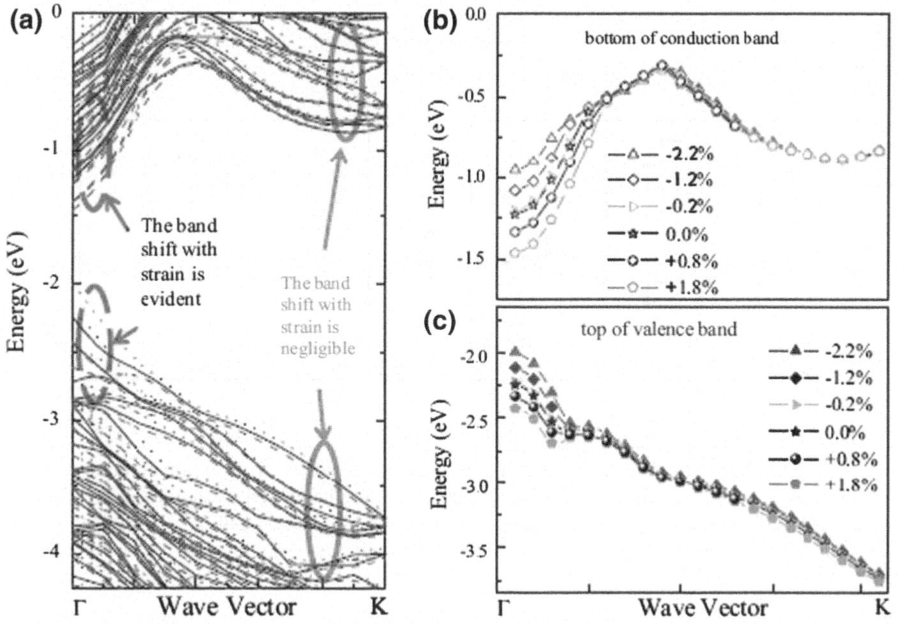

Fig. 8.17 **a** The band structures of Ge [118] nanowires with a diameter of 18 Å, with and without strain. *Black solid lines* are the band structure without strain; *red dashed lines* are under tensile uniaxial strain; *blue dotted lines* are under compressive uniaxial strain. The energy variations of the bottom of the conduction band **b** and the top of the valence band **c** in Ge nanowires of 18 Å with uniaxial strain. The uniaxial strain has a dominant effect of shifting energies on the conduction and valence bands near Γ [203]

with a diameter of 18 Å, with and without strain. Black solid lines are the band structure without strain; dashed lines are under tensile uniaxial strain; dotted lines are under compressive uniaxial strain. Generally, strain has dominant effects on the band structure near Γ (i.e., energy is shifted evidently with strain, see the dashed pink ovals), while it has negligible effects on wave vectors far away from Γ

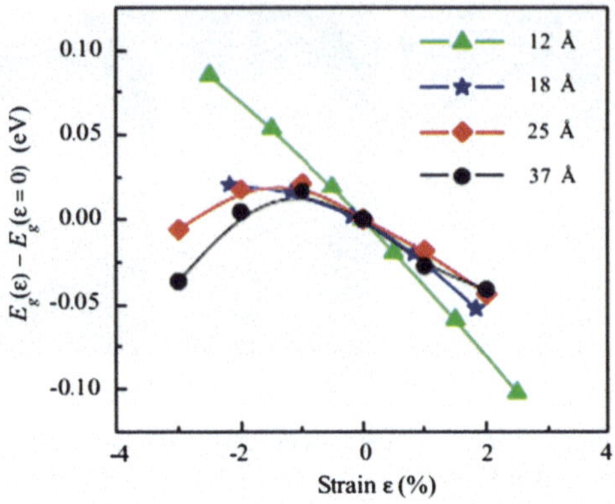

Fig. 8.18 The change in the DFT predicted band gap in Ge wires as a function of uniaxial strain at different size. Positive strain refers to uniaxial expansion while negative strain corresponds to its compression [203]

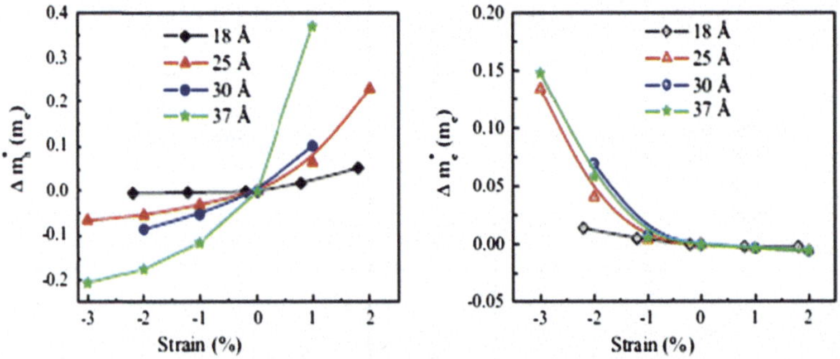

Fig. 8.19 The change in effective masses of the electron (*left*) and hole (*right*) are plotted as a function of uniaxial strain for nanowires at different size. It shows that the effective mass of the electron increases rapidly with compressive uniaxial strain, while decreasing mildly with tensile strain. Although, the effective mass of the hole reduces under compression, while enhanced dramatically with tensile strain [203]

(i.e., minimal energy shift under strain, see the solid green ovals). Most electronic properties are related to the bottom of the conduction band and the top of the valence band. Therefore, the energy variation in these two edges was particularly singled out and presented in Figs. 8.17b, c. From those two figures we can clearly see that strain modifies the energies of CBE and VBE dramatically near Γ, and has negligible energy shifts on wave vectors far away from Γ.

The influence of strain on bulk Si crystal has different effects along different directions and has been extensively studied in past few decades [204, 205]. Recently using the density functional theory, the effect of both uniaxial and biaxial strain on the band structure of a [107] oriented SiNW has been shown, where the modification of the positions of already lifted Δ_4 and Δ_2 valleys due to the quantum confinement

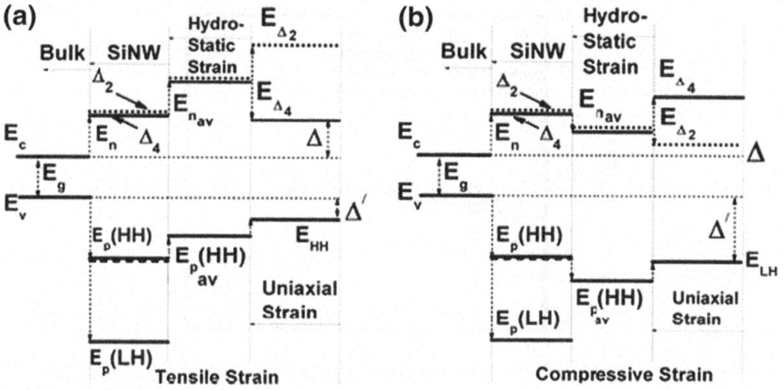

Fig. 8.20 Band alignment of the lowest conduction and valance subband using the EMA formulation under an application of a biaxial strain on [107] SiNW for **a** tensile and **b** compressive strain. Δ_2 and Δ_4 in relaxed SiNW. The average of the HH and LH subband (as shown by the *horizontal dotted line* below lowest HH subband (Ep (HH)) is assumed to coincide with the Ep (HH) for both the tensile and compressive cases due to higher effective mass of the former [194]

effects has been considered [186]. For the present quantitative analysis, we take into consideration of a uniaxial and hydrostatic strain along [107] and [1, 2] directions respectively. Figure 8.20 schematically exhibits this situation on the conduction and valance bands for both tensile (Fig. 8.20a) and compressive (Fig. 8.20b) strains on a [107] oriented SiNW. In case of a bulk Si, an application of a tensile hydrostatic strain shifts up the average energy of the conduction band with respect to its six equivalent valleys. In addition, a uniaxial strain along [107] splits this conduction band into Δ_2 and Δ_4. The position of these valleys about their bulk relaxed value, however, strictly depends whether the strain is tensile or compressive. For example, in a ⟨110⟩ uniaxial tensile strain, the position of Δ_4 is higher in energy than Δ_2 [206]. As shown in Fig. 8.20a for a relaxed SiNW, the two valleys Δ_4 (lower in energy) and Δ_2 (higher in energy) are the set of subbands as a result of $k_y = n_y\pi/d_y$ and $k_z = n_z\pi/d_z$. The average energy of these set of subbands under a tensile hydrostatic strain along [1, 2] shifts up by the same amount. The presence of a uniaxial compressive strain along [107] direction makes Δ_4 to be higher in energy than that of Δ_2 [186] as shown in Fig. 8.20b. In case of valance bands, the HH and LH split as subband energy levels in which a tensile hydrostatic strain shifts up their respective average position, while a uniaxial tensile strain shifts up the HH subbands over LH subbands (Fig. 8.20a) [194].

The effect of strain on the band gap in [107] SiNW has been exhibited in Fig. 8.21 [194]. It appears that band gap decreases as the uniaxial tensile and compressive strain increases, the rate of decrement are different due to the difference in subband energies in both the regime. It should be noted that an increase in the tensile strain decreases the energy of Δ_4 subbands while the HH subbands shifts toward the valance band maxima position of the bulk Si. This marks a reduction of the band gap as the tensile strain increases. In case of uniaxial compressive strain, it is the Δ_2 which

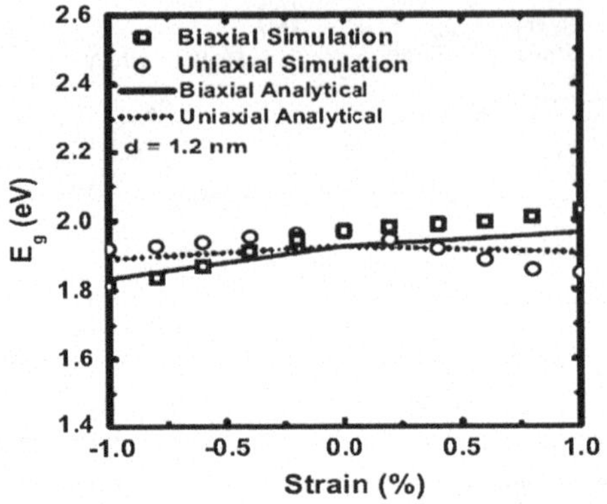

Fig. 8.21 Band gap as function of uniaxial strain along [107] direction and biaxial strain for [107] oriented SiNW. Symbols are the results of the ATK simulations [194]

shifts down and LH shifts up, thus decreasing the band gap. The scenario changes when the uniaxial strain is combined with the [1, 2] hydrostatic strain. Under this biaxial strain condition, the band gap increases along the tensile strain while decreases with compressive strain at a rate much faster than that of the corresponding uniaxial case. The $sp^3d^5s^*$ model predicts the direct to indirect transition of the band gap in a ⟨100⟩ uniaxially strained SiNW occurring inside the compressive zone [186, 207]. The reason for this is the asymmetric splitting of the six equivalent valleys in bulk Si into Δ_4 and Δ_2 due to the quantum confinement of the carriers in SiNW. Since Δ_4 lies lower at the $\Delta_2$point axis than Δ_2 which lies at higher energy at the off-Γ axis, it takes certain amount of compressive strain to bring the Δ_2 subband (at the same off-Γaxis) lower than the Δ_4. Since this confinement splitting is not arrested in EMA formalism, Fig. 8.21 exhibits that the band gap from the beginning of the compressive strain starts becoming indirect. An increase in the tensile strain decreases the Δ_4 subband at the same Γ axis, whereby the band gap remains direct always.

The variation of the transport and subband effective mass as function of strain has been exhibited in Fig. 8.22 It appears from Fig. 8.22a that with the increase in both uniaxial tensile and compressive strain, the transport effective mass follows the same rate of decrement as exhibited by its corresponding band gap variation. It appears that the subband effective mass along the z direction has larger variation due to the application of the hydrostatic strain than that of the y direction. Further above 0 strain, the effective masses are due to the direct band gap and below 0 strain, the effective masses are due to the indirect band gap [194], a reason which has already been stated earlier.

In this monograph, we have studied the EEM in quantum confined nonlinear optical, III–V, II–VI, GaP, Ge, $PtSb_2$, stressed materials, Bismuth, GaSb, IV–VI, Tellurium, II–V, Bi_2Te_3, III–V, II–VI, IV–VI, and HgTe/CdTe quantum wire superlattices with graded interfaces, III–V, II–VI, IV–VI, and HgTe/CdTe effective mass

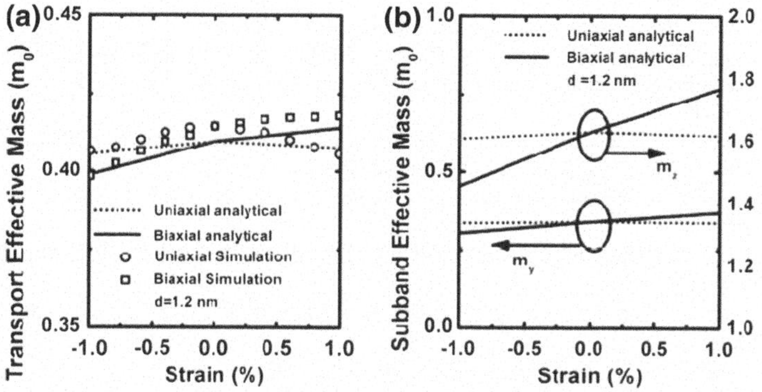

Fig. 8.22 Plot of the electron **a** transport effective mass and **b** subband effective mass as function of uniaxial and biaxial strain. The symbol represents the extracted data from the energy band structure obtained using ATK simulation [194]

superlattices under magnetic quantization, quantum confined effective mass superlattices, and superlattices of optoelectronic materials under intense electric field and light waves with graded interfaces on the basis of appropriate carrier energy spectra. Finally it may be noted that although we have considered the EEM in a plethora of quantized materials having different band structures theoretically, the detailed experimental works are still needed for an in-depth study of the EEM from such low-dimensional systems as functions of externally controllable quantities which, in turn, will add new physical phenomenon in the regime of the electron motion in nanostructured materials and related topics.

8.4 Open Research Problem

(R8.1) Investigate experimentally the EEM for all the systems as discussed in this monograph.

References

1. K.V. Klitzing, G. Dorda, M. Pepper, Phys. Rev. Lett. **45**, 494 (1980)
2. K.V. Klitzing, Rev. Mod. Phys. **58**, 519 (1986)
3. J. Hajdu, G. Landwehr, in *Strong and Ultrastrong Magnetic Fields and Their Applications*, ed. by F. Herlach (Springer, Berlin, 1985), p. 17
4. I.M. Tsidilkovskii, *Band Structure of Semiconductors* (Pergamon Press, Oxford, 1982)
5. K.P. Ghatak, S. Bhattacharya, *Thermoelectric Power in Nanostructured Materials: Strong Magnetic Fields*, Springer Series in Materials Science, **137**, (Springer, New York, 2010)
6. B. Mitra, K.P. Ghatak, Phys. Letts. A **141**, 81 (1989)
7. S.P. Zelenim, A.S. Kondratev, A.E. Kuchma, Sov. Phys. Semi. **16**, 355 (1982)

8. F.M. Peeters, P. Vasilopoulos, Phys. Rev. B **46**, 4667 (1992)
9. W. Zawadzki, in *Two-Dimensional Systems, Heterostructures and Superlattices*, Springer Series in Solid State Sciences, vol. 53, ed. by G. Bauer, F. Kuchar, H. Heinrich (Springer, Berlin, 1984)
10. B.M. Askerov, N.F. Gashimzcede, M.M. Panakhov, Sov. Phys. Sol. State **29**, 465 (1987)
11. G.P. Chuiko, Sov. Phys. Semi. **19**, 1279 (1985)
12. S. Pahari, S. Bhattacharya, K.P. Ghatak, J. Comput. Theor. Nanosci., Invited review article **6**, 2088 (2009)
13. K.P. Ghatak, S. Bhattacharya, S. Bhowmik, R. Benedictus, S. Choudhury, J. Appl. Phys. **103**, 034303 (2008)
14. K.P. Ghatak, S. Bhattacharya, S. Pahari, D. De, S. Ghosh, M. Mitra, Annalen der Physik **17**, 195 (2008)
15. K.P. Ghatak, S.N. Biswas, J. Appl. Phys. **70**, 299 (1991)
16. K.P. Ghatak, S.N. Biswas, J. Low Temp, Phys. **78**, 219 (1990)
17. K.P. Ghatak, M. Mondal, J. Appl. Phys. **65**, 3480 (1989)
18. K.P. Ghatak, B. Nag, Nanostruct. Mater. **5**, 769 (1995)
19. K.P. Ghatak, M. Mondal, Phys. Stat. Sol. (b) **185**, K5 (1994)
20. K.P. Ghatak, B. Mitra, Il Nuovo Cimento D **15**, 97 (1993)
21. K.P. Ghatak, S.N. Biswas, Phys. Stat. Solidi (b) **140**, K107 (1987)
22. K.P. Ghatak, Il Nuovo Cimento D **13**, 1321 (1991)
23. K.P. Ghatak, A. Ghoshal, Phys. Stat. Sol. (b) **170**, K27 (1992)
24. K.P. Ghatak, B. De, B. Nag, P.K. Chakraborty, Mol. Cryst. Liq. Cryst. Sci. Technol. Sect. B Nonlinear Opt. **16**, 221 (1996)
25. K.P. Ghatak, M. Mitra, B. Goswami, B. Nag, Mol. Cryst. Liq. Cryst. Sci. Technol. Sect. B Nonlinear Opt. **16**, 167 (1996)
26. K.P. Ghatak, D.K. Basu, D. Basu, B. Nag, Nuovo Cimento della Societa Italiana di Fisica D - Condensed Matter. Atomic Mol. Chem. Phys. Biophys. **18**, 947 (1996)
27. S.K. Biswas, A.R. Ghatak, A. Neogi, A. Sharma, S. Bhattacharya, K.P. Ghatak, Physica E Low-Dimension. Syst. Nanostruct. **36**, 163 (2007)
28. M. Mondal, A. Ghoshal, K.P. Ghatak, Il Nuovo Cimento D **14**, 63 (1992)
29. K.P. Ghatak, B. De, M. Mondal, S.N. Biswas, Proc. Mater. Res. Soc. Symposium **198**, 327 (1990)
30. K.P. Ghatak, Proc. SPIE Int. Soc. Opt. Eng. **1584**, 435 (1992)
31. K.P. Ghatak, B. De, Proc. Mater. Res. Soc. Symp. **234**, 55 and 59 (1991)
32. K.P. Ghatak, B. De, M. Mondal, S.N. Biswas, Proc. Mat. Res. Soc. Symp. **184**, 261 (1990)
33. K.P. Ghatak, *Influence of Band Structure on Some Quantum Processes in Tetragonal Semiconductors*, D. Eng. Thesis, (Jadavpur University, Kolkata, 1991)
34. K.P. Ghatak, S.N. Biswas, Proc. Mat. Res. Soc. Symp. **216**, 465 (1990)
35. K.P. Ghatak, M. Mondal Phys, Stat. Sol. (b) **135**, 819 (1986)
36. J. Cibert, P.M. Petroff, G.J. Dolan, S.J. Pearton, A.C. Gossard, J.H. English, Appl. Phys. Letts. **49**, 1275 (1988)
37. L.J. Singh, S. Choudhury, D. Baruah, S.K. Biswas, S. Pahari, K.P. Ghatak, Phys. B **368**, 188 (2005)
38. P.M. Petroff, A.C. Gossard, W. Wiegmann, Appl. Phys. Letts. **45**, 620 (1984)
39. J.M. Gaines, P.M. Petroff, H. Kroemer, R.J. Simes, R.S. Geels, J.H. English, J. Vac. Sci. Tech. B **6**, 1378 (1988)
40. T. Fukui, H. Saito, Appl. Phys. Letts. **50**, 824 (1987)
41. H. Sakaki, Jpn. J. Appl. Phys. **19**, L735 (1980)
42. P.M. Petroff, A.C. Gossard, R.A. Logan, W. Wiegmann, Appl. Phys. Letts. **41**, 635 (1982)
43. S. Bhattacharya, R. Sarkar, D. De, S. Mukherjee, S. Pahari, A. Saha, S. Roy, N.C. Paul, S. Ghosh, K.P. Ghatak, J. Comput. Theor. Nanosci. **6**, 112 (2009)
44. K.P. Ghatak, S. Bhattacharya, D. De, R. Sarkar, S. Pahari, A. Dey, A.K. Dasgupta, S.N. Biswas, J. Comput. Theor. Nanosci. **5**, 1345 (2008)
45. R.B. Dingle, Phil. Mag. **46**, 813 (1955)

46. D. Redfield, M.A. Afromowitz, Phil. Mag. **18**, 831 (1969)
47. H.C. Cassey, F. Stern, J. Appl. Phys. **47**, 631 (1976)
48. M. Mondal, K.P. Ghatak, Phys. Lett. **102A**, 54 (1984)
49. P.K. Chakraborty, G.C. Datta, K.P. Ghatak, Physica Scripta **68**, 368 (2003)
50. B. Mitra, D.K. Basu, B. Nag, K.P. Ghatak, Nonlinear Opt. **17**, 171 (1997)
51. K.P. Ghatak, S. Bhattacharya, J. Appl. Phys. **102**, 073704 (2007)
52. K.P. Ghatak, S. Bhattacharya, H. Saikia, D. Baruah, A. Saikia, K.M. Singh, A. Ali, S.N. Mitra, P.K. Bose, A. Sinha, J. Comput. Theor. Nanosci. **3**, 727 (2006)
53. E.O. Kane, Solid State Electron. **8**, 3 (1985)
54. T. Ando, A.H. Fowler, F. Stern, Rev. Mod. Phys. **54**, 437 (1982)
55. P.K. Basu, in *Optical Processes in Semiconductors* (Oxford University Press, New York, 2001)
56. A.N. Chakravarti, D. Mukherjee, Phys. Lett. **53A**, 403 (1975)
57. A.N. Chakravarti, S. Swaminathan, Phys. Stat. Sol. (a) **23**, K191 (1974)
58. A.N. Chakravarti, Phys. Stat. Sol (a) **25**, K 105 (1974)
59. K.P. Ghatak, S. Bhattyacharya, *Thermoelectric Power in Nanostructured Materials: Strong Magnetic Fields*, Springer Series in Materials Science, vol. 137 (Springer, Germany, 2010)
60. A.K. Sreedhar, S.C. Gupta, Phys. Rev. B **5**, 3160 (1972)
61. R.W. Keyes, IBM. J. Res. Dev. **5**, 266 (1961)
62. R.W. Keyes, Solid State Phys. **20**, 37 (1967)
63. S. Bhattacharya, S. Chowdhury, K.P. Ghatak, J. Comput. Theor. Nanosci. **3**, 423 (2006)
64. S. Choudhury, L.J. Singh, K.P. Ghatak, Physica B **365**, 5 (2005)
65. L.J. Singh, S. Choudhary, A. Mallik, K.P. Ghatak, J. Comput. Theor. Nanosci. **2**, 287 (2005)
66. K.P. Ghatak, J.Y. Siddiqui, B. Nag, Phys. Lett. A **282**, 428 (2001)
67. K.P. Ghatak, J.P. Banerjee, B. Nag, J. Appl. Phys. **83**, 1420 (1998)
68. B. Nag, K.P. Ghatak, Nonlinear Opt. **19**, 1 (1998)
69. K.P. Ghatak, B. Nag, Nanostruct. Mater. **10**, 923 (1998)
70. B. Nag, K.P. Ghatak, J. Phys. Chem. Sol. **58**, 427 (1997)
71. K.P. Ghatak, D.K. Basu, B. Nag, J. Phys. Chem. Solids **58**, 133 (1997)
72. K.P. Ghatak, J.P. Banerjee, B. Goswami, B. Nag, Nonlinear Opt. Quant. Opt. **16**, 241 (1996)
73. K.P. Ghatak, J.P. Banerjee, D. Bhattacharyya, B. Nag, Nanotechnology **7**, 110 (1996)
74. K.P. Ghatak, J.P. Banerjee, M. Mitra, B. Nag, Nonlinear Opt. **17**, 193 (1996)
75. B. Nag, K.P. Ghatak, Phys. Scr. **54**, 657 (1996)
76. K.P. Ghatak, B. Mitra, Phys. Scr. **46**, 182 (1992)
77. K.P. Ghatak, Int. J. Electron. **71**, 239 (1991)
78. K.P. Ghatak, B. De, S.N. Biswas, M. Mondal, MRS Symp. Proc. Spring Meet. **216**, 191 (1991)
79. K.P. Ghatak, B. De, MRS Symp. Proc. **226**, 191 (1991)
80. K.P. Ghatak, B. Nag, G. Majumdar, MRS Symp. Proc. **379**, 109 (1995)
81. D. Baruah, S. Choudhury, K.M. Singh, K.P. Ghatak, J. Phys. Conf. Series **61**, 80 (2007)
82. H. Kroemer, IEEE Trans. Electron. Devices **25**, 850 (1978)
83. R.W. Lade, Proc. IEEE. **51**, 743 (1964)
84. S.N. Mohammed, J. Phys. C. Solid State Phys. **13**, 2685 (1980)
85. P.T. Landsberg, Eur. J. Phys. **2**, 213 (1981)
86. P.T. Landsberg, Proc. R. Soc. A **213**, 226 (1952)
87. S.A. Hope, G. Feat, P.T. Landsberg, J. Phys. A Math. Gen. **14**, 2377 (1981)
88. C.H. Wang, A. Neugroschel, IEEE Electron. Dev. Lett. **ED-11**, 576 (1990)
89. I.-Y. Leu, A. Neugroschel, C.H. Wang, A. Neugroschel, IEEE Trans. Electron. Dev. **ED-40**, 1872 (1993)
90. F. Stengel, S.N. Mohammad, H. Morkoc, J. Appl. Phys. **80**, 3031 (1996)
91. H.J. Pan, W.C. Wang, K.B. Thai, C.C. Cheng, K.H. Yu, K.W. Lin, C.Z. Wu, W.C. Liu, Semicond. Sci. Technol. **15**, 1101 (2000)
92. S.N. Mohammad, J. Appl. Phys. **95**, 4856 (2004)
93. V.K. Arora, Appl. Phys. Lett. **80**, 3763 (2002)
94. S.N. Mohammad, J. Appl. Phys. **95**, 7940 (2004)

95. S.N. Mohammad, Phil. Mag. **84**, 2559 (2004)
96. S.N. Mohammad, J. Appl. Phys. **97**, 063703 (2005)
97. S.G. Dmitriev, YuV Markin. Semiconductors **34**, 931 (2000)
98. M. Tao, D. Park, S.N. Mohammad, D. Li, A.E. Botchkerav, H. Morkoc, Phil. Mag. B **73**, 723 (1996)
99. D.G. Park, M. Tao, D. Li, A.E. Botchkarev, Z. Fan, S.N. Mohammad, H. Morkoc, J. Vac. Sci. Technol. B **14**, 2674 (1996)
100. Z. Chen, D.G. Park, S.N. Mohammad, H. Morkoc, Appl. Phys. Lett. **69**, 230 (1996)
101. K.P. Ghatak, S. Bhattacharya, D. De, *Einstein Relation in Compound Semiconductors and Their Nanostructures*, Springer Series in Materials Science, vol. 116 (Springer, Germany, 2009)
102. K.P. Ghatak, S.N. Biswas, Nanostruct. Mater. **2**, 91 (1993)
103. K.P. Ghatak, S.N. Biswas, J. Appl. Phys. **70**, 4309 (1991)
104. K.P. Ghatak, B. Mitra, M. Mondal, Ann. der Physik **48**, 283 (1991)
105. B. Mitra, K.P. Ghatak, Physica Scripta **42**, 103 (1990)
106. B. Mitra, K.P. Ghatak, Phys. Letts. **135A**, 397 (1989)
107. M. Mondal, K.P. Ghatak, Ann. der Physik **46**, 502 (1989)
108. K.P. Ghatak, D. Bhattacharyya, Phys. Letts. A **184**, 366 (1994)
109. A.N. Charkravarti, D.P. Parui, Phys. Lett. **40A**, 113 (1972)
110. A.N. Charkravarti, D.P. Parui, Phys. Lett. **43A**, 60 (1973)
111. A.N. Charkravarti, D.P. Parui, Phys. Stat. Sol. (a) **14**, K23 (1972)
112. A.N. Charkravarti, D.P. Parui Canad, J. Phys. **51**, 451 (1973)
113. A.S. Filipchenko, I.G. Lang, D.N. Nasledov, S.T. Pavlov, L.N. Radaikine, Phys. Stat. Sol. (b) **66**, 417 (1974)
114. M. Wegener, *Extreme Nonlinear Optics* (Springer, Germany, 2005)
115. B.S. Wherreff, W. Wolland, C.R. Pidgeon, R.B. Dennis, S.D. Smith, in *Proceedings of the 12th International Conference of the Physics of the Semiconductors*, ed. by M.H. Pilkahn, R.G. Tenbner (Staffgard, 1978), p. 793
116. M. Cardona, L. Ley, in *Photoemission in Solids 1 and 2, Topics in Applied Physics*, vols. 26, 27, (Springer, Germany, 1978)
117. D.J. Lockwood, in *Light Emission in Silicon Based Materials and Devices*, vol. 2, ed. by H.S. Nalwa (Academic Press, San Diego, 2001)
118. R.K. Pathria, *Statistical Mechanics, 2nd ed* (Butterworth-Heinemann, Oxford, 1996)
119. K.P. Ghatak, S. Bhattacharya, K.M. Singh, S. Choudhury, S. Pahari, Physica B **403**, 2116 (2008)
120. K.P. Ghatak, S.N. Biswas, in *Proceedings of the Society of Photo-optical and Instrumentation Engineers*, vol. 1409, (SPIE, Nonlinear Optics II, USA, 1991), p. 28
121. K.P. Ghatak, SPIE. Process Module Metrol. USA **1594**, 110 (1992)
122. K.P. Ghatak, in *International Conference on the Application and Theory of Periodic Structures*, vol. 1545, (SPIE, USA, 1991), p. 282
123. K.P. Ghatak, M. Mondal, Solid State Electron. **31**, 1561 (1988)
124. K.P. Ghatak, M. Mondal, J. Appl. Phys. **69**, 1666 (1991)
125. K.P. Ghatak, D. Bhattacharyya, B. Nag, S.N. Biswas, J. Nonlin, Opt. Quant. Opt. **13**, 267 (1995)
126. M. Mondal, S. Banik, K.P. Ghatak, J. Low Temp, Phys. **74**, 423 (1989)
127. B. Mitra, A. Ghoshal, K.P. Ghatak, Phys. Stat. Sol. (b) **150**, K67 (1988)
128. B. Mitra, K.P. Ghatak, Physica Scripta **40**, 776 (1989)
129. K.P. Ghatak, M. Mondal, S.N. Biswas, J. Appl. Phys. **68**, 3032 (1990)
130. K.P. Ghatak, S.N. Biswas, Nonlinear Opt. **4**, 39 (1993)
131. K.P. Ghatak, S.N. Biswas, Proc. SPIE **1484**, 136 (1991)
132. K.P. Ghatak, B. De, Materials Research Society (MRS) Symposium Proceedings. MRS Spring Meet. **228**, 237 (1991)
133. K.P. Ghatak, B. Nag, G. Majumdar, MRS Symposium Proceedings. MRS Spring Meet. **379**, 85 (1995)

134. K.P. Ghatak, in *High Speed Phenomena in Photonic Materials and Optical Bistability*, vol. 1280 (SPIE, USA, 1990), p. 53
135. K.P. Ghatak, Materials and Processes Symposium Proceedings. MRS Symp. Proc. MRS Spring Meet. **216**, 469 (1990)
136. K.P. Ghatak, A. Ghoshal, S. Bhattacharyya, in *Nonlinear Optical Materials and Devices for Photonic Switching*, vol. 1216, (SPIE, USA, 1990), p. 282
137. K.P. Ghatak, in *Nonlinear Optics III*, vol. 1626 (SPIE, USA, 1992), p. 115
138. K.P. Ghatak, A. Ghoshal, B. De, in *Optoelectronic Devices and Applications*, vol. 1338, (SPIE, USA, 1990), p. 111
139. R. Houdré, C. Hermann, G. Lampel, P.M. Frijlink, Surface Sci. **168**, 538 (1986)
140. T.C. Chiang, R. Ludeke, D.E. Eastman, Phys. Rev. B. **25**, 6518 (1982)
141. S.P. Svensson, J. Kanski, T.G. Andersson, P.O. Nilsson, J. Vacuum Sci, Technol. B **2**, 235 (1984)
142. S.F. Alvarado, F. Ciccacci, M. Campagna, Appl. Phys. Letts. **39**, 615 (1981)
143. C. Majumdar, A.B. Maity, A.N. Chakravarti, Phys. Stat. Sol. (b) **140**, K7 (1987)
144. C. Majumdar, A.B. Maity, A.N. Chakravarti, Phys. Stat. Sol. (b) **141**, K35 (1987)
145. N.R. Das, K.K. Ghosh, D. Ghoshal, Phys. Stat. Sol. (b) **197**, 97 (1996)
146. C. Majumdar, A.B. Maity, A.N. Chakravarti, Phys. Stat. Sol. (b) **144**, K13 (1987)
147. N.R. Das, A.N. Chakravarti, Phys. Stat. Sol. (b) **176**, 335 (1993)
148. S. Sen, N.R. Das, A.N. Chakravarti, J. Phys. Conden. Mat. **19**, 186205 (2007)
149. N.R. Das, S. Ghosh, A.N. Chakravarti, Phys. Stat. Sol. (b) **174**, 45 (1992)
150. A.B. Maity, C. Majumdar, A.N. Chakravarti, Phys. Stat. Sol. (b) **144**, K93 (1987)
151. A.B. Maity, C. Majumdar, A.N. Chakravarti, Phys. Stat. Sol. (b) **149**, 565 (1988)
152. N.R. Das, A.N. Chakravarti, Phys. Stat. Sol. (b) **169**, 97 (1992)
153. A. Modinos, in *Field Thermionic and Secondary Electron Emission Spectroscopy* (Plenum Press, New York, 1984)
154. A.V.D. Ziel, *Solid State Physical Electronics* (Prentice Hall, Eaglewood Cliffs, 1957)
155. E.A. Aruhanov, A.F. Knyazev, A.N. Nateprov, S.I. Radautsan, Sov. Phys. Semiconductors **15**(7), 828 (1981)
156. P.N. Hai, W.M. Chen, I.A. Buyanova, H.P. Xin, C.W. Tu, Appl. Phys. Letts. **77**, 1843 (2000)
157. D.P. DiVincenzo, E.J. Mele, Phys. Rev. B **29**, 1685 (1984)
158. P. Perlin, E. Litwin-Staszewska, B. Suchanek, W. Knap, J. Camassel, T. Suski, R. Piotrzkowski, I. Grzegory, S. Porowski, E. Kaminska, J.C. Chervin, Appl. Phys. Letts. **68**, 1114 (1996)
159. F. Masia, G. Pettinari, A. Polimeni, M. Felici, A. Miriametro, M. Capizzi, A. Lindsay, S.B. Healy, E.P. O'Reilly, A. Cristofoli, G. Bais, M. Piccin, S. Rubini, F. Martelli, A. Franciosi, P.J. Klar, K. Volz, W. Stolz, Phys. Rev. B **73**, 073201 (2006)
160. S.L. Sewall, R.R. Cooney, P. Kambhampati, Appl. Phys. Letts. **94**, 243116/1–243116/3 (2009)
161. G.E. Smith, Phys. Rev. Letts. **9**, 487 (1962)
162. D. Schneider, D. Rurup, A. Plichta, H.-U. Grubert, A. Schlachetzki, K. Hansen, Z. Phys, B. **95**, 281 (1994)
163. L.G. Caron, J.-P. Jay-Gerin, M.J. Aubin, Phys. Rev. B **15**, 3879 (1977)
164. D. Bhattacharyya, R. Pal, S. Chaudhuri, A.K. Pal, Vaccum **44**, 803 (1993)
165. B. Slomski, F. Meier, J. Osterwalder, J.H. Dil, Phys. Rev B **83**, 035409 (2011)
166. L.M. Wei, K.H. Gao, X.Z. Liu, W.Z. Zhou, L.J. Cui, Y.P. Zeng, G. Yu, R. Yang, T. Lin, L.Y. Shang, S.L. Guo, N. Dai, J.H. Chu, D.G. Austing, J. Appl. Phys **110**, 063707 (2011)
167. K. Karra, M. Stopa, X. Ying, H.D. Drew, S. Das Sarma, Phys. Rev B **42**, 9732 (1990)
168. B. Roßner, G. Isella, H.V. Kanel, Appl. Phys. Letts. **82**, 754 (2003)
169. T.M. Kim, M. Jung, K.H. Yoo, J. Phys. Chem. Solids **61**, 1769 (2000)
170. N. Kotera, H. Arimoto, N. Miura, K. Shibata, Y. Ueki, K. Tanaka, H. Nakamura, T. Mishima, K. Aiki, M. Washima, Physica E **11**, 219 (2001)
171. K. Tanaka, N. Koteraa, H. Nakamura, Microelectron. Eng. **47**, 309 (1999)
172. G. Wang, T. Cagin, Appl. Phys. Letts. **89**, 152101 (2006)
173. Z.S. Gribnikov, R.R. Bashirov, H. Eisele, V.V. Mitin, G.I. Haddad, Physica E **12**, 276 (2002)
174. R. Chen, K.K. Bajaj, Phys. Rev. B **50**, 1949 (1994)

175. E. Borovitskaya, M.S. Shur, Solid-State Electron. **44**, 1293 (2000)
176. L.C.O. Dacal, E.A. de Andrada e Silva, Brazilian J. Phys. **36**, 910 (2006)
177. A. Sasaki, Surf. Sci. **174**, 624 (1986)
178. J.C. Maan, M. Altarelli, H. Sigg, P. Wyder, L.L. Chang, L. Esaki, Surf. Sci. **113**, 347 (1982)
179. K. El Shabrawy, K. Maharatna, D. Bagnall, B.M. Al-Hashimi, IEEE Trans. Nanotechnol. **9**, 184 (2010)
180. N. Neophytou, A. Paul, M.S. Lundstrom, G. Klimeck, IEEE Trans. Electron Devices **55**, 866 (2008)
181. E. Zheng, C. Rivas, R. Lake, K. Alam, T.B. Boykin, G. Klimeck, IEEE Trans. Electron. Devices **52**, 1097 (2005)
182. J. Wang, A. Rahman, A. Ghosh, G. Klimeck, M.S. Lundstrom, IEEE Trans. Electron. Devices **52**, 1589 (2005)
183. J.A. Yan, L. Yang, M.Y. Chou, Phys. Rev. B **76**, 115319 (2007)
184. M.O. Baykan, S.E. Thompson, T. Nishida, J. Appl. Phys. **108**, 092716 (2010)
185. R.N. Sajjad, K. Alam, J. Appl. Phys. **105**, 044307 (2009)
186. K.-H. Hong, J. Kim, S-H. Lee and J. K. Shin. Nano Lett. **8**, 1335 (2008)
187. P.W. Leu, B. Shan, K. Cho, Phys. Rev. B **73**, 195320 (2006)
188. International Technology Roadmap for Semiconductors, Available Online at http://www.itrs.net/Links/2009ITRS/2009Chapters-2009Tables/2009ERD.pdf
189. V. Sverdlov, in *Strain-Induced Effects in Advanced MOSFETs*, (Springer, Wein, 2011)
190. K. Uchida, A. Kinoshita, M. Saitoh, Int. Electron Devices Meeting 1019 (2006)
191. R.N. Sajjad, K. Alam, Q.D.M. Khosru, Semicond. Sci. Technol. **24**, 045023 (2009)
192. M.S. Lundstrom, J. Guo, *Nanoscale Transistors: Device Physics Modeling and Simulation* (Springer, USA, 2006)
193. Atomistix ToolKit (ATK), QuantumWise simulator, available online at: http://www.quantumwise.com/
194. R. K. Ghosh, S. Bhattacharya, S. Mahapatra, IEEE Trans. Electron Dev. In press, DOI:10.1109/TED.2012.2190737, (2012)
195. M.-F. Ng, L. Zhou, S.-W. Yang, L.Y. Sim, V.B.C. Tan, P. Wu, Phys. Rev. B **76**, 155435 (2007)
196. D. Kienle, K.H. Bevan, G.-C. Liang, L. Siddiqui, J.I. Cerda, A.W. Ghosh, J. Appl. Phys. **100**, 043715 (2006)
197. J. Cerda, F. Soria, Phys. Rev. B **61**, 7965 (2000)
198. P. Pulay, Chem. Phys. Lett. **73**, 393 (1980)
199. S. Jin, M.V. Fischetti, T.-W. Tang, IEEE Trans. Electron Devices **55**, 2886 (2008)
200. R.N. Sajjad, S. Bhowmick, Q. Khosru, IEEE Int. Conf. Electron Devices Solid-State Circ. (EDSSC) 1 (2008)
201. F. Stern, W.E. Howard, Phys. Rev. **163**, 816 (1967)
202. P.J. van der J-L, D. Steen, P. Esseni, L. Palestri, R.J.E. Selmi, Hueting. IEEE. Trans. Electron Dev. **54**, 1843 (2007)
203. P, Logan, X. Peng, Phys. Rev. B. **80**, 115322 (2009)
204. C.G.V. de Walle, R.M. Martin, Phys. Rev. B **34**, 5621 (1986)
205. J.C. Hensel, G. Feher, Phys. Rev. **129**, 10411062 (1963)
206. Y. Sun, S.E. Thompson, T. Nishida, *Strain Effect in Semiconductors: Theory and Device Applications* (Springer, Heidelberg, 2010)
207. D. Shiri, Y. Kong, A. Buin, M.P. Anantram, Appl. Phys. Lett. **93**, 073114 (2008)

Chapter 9
Conclusion and Future Research

This monograph deals with the EEM in various types of low-dimensional materials, effective mass superlattices, and superlattices with graded interfaces under different physical conditions, in the presence of quantizing magnetic field and external photo excitation and also under strong electric field altering profoundly the basic band structures which, in turn, generate pin-pointed knowledge regarding EEM in various semiconductors and their nanostructures having different carrier energy spectra. The in-depth experimental investigations covering the whole spectrum of solid state and allied science in general are extremely important to uncover the underlying physics and the related mathematics. The EEM is basically a motion-dependent phenomena and we have formulated the simplified expressions of EEM for few quantized structures together with the fact that our investigations are based on the simplified ***k.p*** formalism of solid state science without incorporating the advanced field theoretic techniques. In spite of such constraints, the role of band structure behind the curtain, which generates, in turn, new concepts are discussed throughout the text.

Finally, we present the last set of open research problems in this particular area of materials science.

(R9.1) Investigate the EEM in the presence of a quantizing magnetic field under exponential, Kane, Halperin, Lax and Bonch-Bruevich band tails [1] for all the problems of this monograph of all the materials whose unperturbed carrier energy spectra are defined in Chap. 1 by including spin and broadening effects.

(R9.2) Investigate all the appropriate problems after proper modifications introducing new theoretical formalisms for the problems as defined in (R8.1) for negative refractive index, macro molecular, nitride, and organic materials.

(R9.3) Investigate all the appropriate problems of this monograph for all types of quantum-confined p-InSb, p-CuCl, and semiconductors having diamond structure valence bands whose dispersion relations of the carriers in bulk materials are given by Cunningham [2], Yekimov et. al. [3] and Roman et. al. [4], respectively.

S. Bhattacharya and K. P. Ghatak, *Effective Electron Mass in Low-Dimensional Semiconductors*, Springer Series in Materials Science 167,
DOI: 10.1007/978-3-642-31248-9_9, © Springer-Verlag Berlin Heidelberg 2013

(R9.4) Investigate the influence of defect traps and surface states separately on the EEM, for all the appropriate problems of all the chapters after proper modifications
(R9.5) Investigate the EEM under the condition of nonequilibrium of the carrier states for all the appropriate problems of this monograph.
(R9.6) Investigate the EEM for all the appropriate problems of this monograph for the corresponding p-type semiconductors and their nanostructures.
(R9.7) Investigate the EEM for all the appropriate problems of this monograph for all types of semiconductors and their nanostructures under mixed conduction in the presence of strain.
(R9.8) Investigate the EEM for all the appropriate problems of this monograph for all types of semiconductors and their nanostructures in the presence of hot electron effects.
(R9.9) Investigate the EEM for all the appropriate problems of this monograph for all types of semiconductors and their nanostructures for nonlinear charge transport.
(R9.10) Investigate the EEM for all the appropriate problems of this monograph for all types of semiconductors and their nanostructures in the presence of strain in an arbitrary direction.
(R9.11) Investigate all the appropriate problems of this monograph for semiconductor clathrates in the presence of strain.
(R9.12) Investigate all the appropriate problems of this monograph for quasi-crystalline materials in the presence of strain.
(R9.13) Investigate all the appropriate problems of this monograph for strongly correlated electron systems in the presence of strain.
(R9.14) Investigate EEM for all the appropriate problems of this monograph for all types of transition metal silicides in the presence of strain.
(R9.15) Investigate EEM for all the appropriate problems of this monograph for all types of electrically conducting organic materials in the presence of strain.
(R9.16) Investigate EEM for all the appropriate problems of this monograph for all types of functionally graded materials in the presence of strain.
(R9.17) Investigate the EEM in all types of available super conductors in the presence of strain.
(R9.18) Investigate all the appropriate problems of this chapter in the presence of arbitrarily oriented photon field and strain.
(R9.19) Investigate all the appropriate problems of this monograph for paramagnetic semiconductors in the presence of strain.
(R9.20) Investigate all the appropriate problems of this monograph for Boron Carbides in the presence of strain.
(R9.21) Investigate all the appropriate problems of this monograph for all types of Argyrodites in the presence of strain.
(R9.22) Investigate all the appropriate problems of this monograph for layered cobalt oxides and complex chalcogenide compounds in the presence of strain.

(R9.23) Investigate all the appropriate problems of this monograph for all types of nanotubes in the presence of strain.

(R9.24) Investigate all the appropriate problems of this monograph for various types of half-Heusler compounds in the presence of strain.

(R9.25) Investigate all the appropriate problems of this monograph for various types of pentatellurides in the presence of strain.

(R9.26) Investigate all the appropriate problems of this monograph for Bi_2Te_3-Sb_2Te_3 superlattices in the presence of strain.

(R9.27) Investigate the influence of temperature-dependent energy band constants for all the appropriate problems of this monograph.

(R9.28) Investigate EEM for $Ag_{(1-x)}Cu_{(x)}TlTe$ for different appropriate physical conditions as discussed in this monograph in the presence of strain.

(R9.29) Investigate EEM for p-type SiGe under different appropriate physical conditions as discussed in this monograph in the presence of strain.

(R9.30) Investigate EEM for different metallic alloys under different appropriate physical conditions as discussed in this monograph in the presence of strain.

(R9.31) Investigate EEM for different intermetallic compounds under different appropriate physical conditions as discussed in this monograph in the presence of strain.

(R9.32) Investigate EEM for GaN under different appropriate physical conditions as discussed in this monograph in the presence of strain.

(R9.33) Investigate EEM for different disordered conductors under different appropriate physical conditions as discussed in this monograph in the presence of strain.

(R9.34) Investigate EEM for various semimetals under different appropriate physical conditions as discussed in this monograph in the presence of strain.

(R9.35) Investigate all the appropriate problems of this monograph for $Bi_2Te_{3-x}Se_x$ and $Bi_{2-x}Sb_xTe_3$ respectively in the presence of strain.

(R9.36) Investigate all the appropriate problems of this monograph for all types of skutterudites in the presence of strain.

(R9.37) Investigate all the appropriate problems of this monograph in the presence of crossed electric and quantizing magnetic fields.

(R9.38) Investigate all the appropriate problems of this monograph in the presence of crossed alternating electric and quantizing magnetic fields.

(R9.39) Investigate all the appropriate problems of this monograph in the presence of crossed electric and alternating quantizing magnetic fields.

(R9.40) Investigate all the appropriate problems of this monograph in the presence of alternating crossed electric and alternating quantizing magnetic fields.

(R9.41) Investigate all the appropriate problems of this monograph in the presence of arbitrarily oriented pulsed electric and quantizing magnetic fields.

(R9.42) Investigate all the appropriate problems of this monograph in the presence of arbitrarily oriented alternating electric and quantizing magnetic fields.

(R9.43) Investigate all the appropriate problems of this monograph in the presence of crossed inhomogeneous electric and alternating quantizing magnetic fields.

(R9.44) Investigate all the appropriate problems of this monograph in the presence of arbitrarily oriented electric and alternating quantizing magnetic fields under strain.

(R9.45) Investigate all the appropriate problems of this monograph in the presence of arbitrarily oriented electric and alternating quantizing magnetic fields under light waves.

(R9.46) Investigate all the appropriate problems of this monograph in the presence of arbitrarily oriented pulsed electric and alternating quantizing magnetic fields under light waves.

(R9.47) Investigate all the appropriate problems of this monograph in the presence of arbitrarily oriented inhomogeneous electric and pulsed quantizing magnetic fields in the presence of strain and light waves.

(R9.48) (a) Investigate the EEM for all the problems of this monograph in the presence of many body effects, strain, and arbitrarily oriented light waves, respectively.

(b) Investigate the influence of the localization of carriers for all the appropriate problems of this monograph.

(c) Investigate all the appropriate problems of this chapter for the Dirac electron.

(d) Investigate all the problems of this monograph by removing all the physical and mathematical approximations and establishing the respective appropriate uniqueness conditions.

The EEM is the consequence of motion-induced phenomena of solid state science and all the assumptions behind the said phenomena are also applicable to EEM. The formulation of EEM for all types of semiconductors and their quantum-confined counterparts after removing all the assumptions is, in general, a challenging problem. **Totally 250** open research problems have been presented in this monograph and we hope that the readers will not only solve them but also will generate new concepts, both theoretically and experimentally. Incidentally, we can easily infer how little is presented and how much more is yet to be investigated in this exciting topic which is the signature of coexistence of new physics, advanced mathematics combined with the inner fire for performing creative researches in this context from the young scientists since like Kikoin [5] we firmly believe that "A young scientist is no good if his teacher learns nothing from him and gives his teacher nothing to be proud of". In the mean time our research interest has been shifted and we are leaving this particular topic with the hope that (R9.48) alone is sufficient to draw the attention of the researchers from diverse fields and our readers are in tune with the fact that "Exposition, criticism, appreciation is the work for second-rate minds" [6].

References

1. B.R. Nag, *Electron Transport in Compound Semiconductors, Springer Series in Solid State Sciences*, vol. 11 (Springer, Berlin, 1980)
2. R.W. Cunningham, Phys. Rev. **167**, 761 (1968)
3. A.I. Yekimov, A.A. Onushchenko, A.G. Plyukhin, Al. L. Efros, J. Expt. Theor. Phys. **88**, 1490 (1985)
4. B.J. Roman, A.W. Ewald, Phys. Rev. B **5**, 3914 (1972)
5. I.K. Kikoin, *Science for Everyone: Encounters with Physicists and Physics* (Mir Publishers, Moscow, 1989), pp. 154
6. G.H. Hardy, *A Mathematician's Apology* (Cambridge University Press, Cambridge, 1990), pp. 61

Appendix A
The EEM in Compound Semiconductors and Their Nano-Structures Under Cross-Fields Configuration

A1.1 Introduction

The influence of crossed electric and quantizing magnetic fields on the transport properties of semiconductors having various band structures are relatively less investigated as compared with the corresponding magnetic quantization, although the cross-fields are fundamental with respect to the addition of new physics and the related experimental findings. It is well known that in the presence of electric field (E_o) along x-axis and the quantizing magnetic field (B) along z-axis, the dispersion relations of the conduction electrons in semiconductors become modified and for which the electron moves in both the z and y directions. The motion along y-direction is purely due to the presence of E_0 along x-axis and in the absence of electric field, the effective electron mass along y-axis tends to infinity which indicates the fact that the electron motion along y-axis is forbidden. The effective electron mass of the isotropic, bulk semiconductors having parabolic energy bands exhibits mass anisotropy in the presence of cross-fields and this anisotropy depends on the electron energy, the magnetic quantum number, the electric, and the magnetic fields respectively, although, the effective electron mass along z-axis is a constant quantity. In 1966, Zawadzki and Lax [1] formulated the electron dispersion law for III–V semiconductors in accordance with the two band model of Kane under cross-field configuration which generates the interest to study this particular topic of semiconductor science in general [2–29].

In Sect. A1.2.1 of theoretical background, the EEM in nonlinear optical materials in the presence of crossed electric and quantizing magnetic fields has been investigated by formulating the electron dispersion relation. The Sect. A1.2.2 reflects the study of the EEM in III–V, ternary, and quaternary compounds as a special case of Sect. A1.2.1. In the same section, the well-known result for the EEM in relatively wide gap materials in the absence of electric field as a limiting case has been discussed for the purpose of compatibility. The Sect. A1.2.3 contains the study of the EEM for the II–VI semiconductors in the present case.

S. Bhattacharya and K. P. Ghatak, *Effective Electron Mass in Low-Dimensional Semiconductors*, Springer Series in Materials Science 167,
DOI: 10.1007/978-3-642-31248-9, © Springer-Verlag Berlin Heidelberg 2013

In Sect. A1.2.4, the EEM under cross-field configuration in Bismuth has been investigated in accordance with the models of the McClure and Choi, the Cohen, the Lax nonparabolic ellipsoidal, and the parabolic ellipsoidal, respectively. In Sect. A1.2.5, the study of the EEM in IV-VI materials has been presented. In the Sect. A1.2.6, the EEM for the stressed Kane type semiconductors has been investigated. The Sects. A1.2.7, A1.2.8, A1.2.9, A1.2.10, A1.2.11, and A1.2.12 discuss the EEMs' in ultrathin films of the above semiconductors in the presence of cross-field configuration, respectively. The last Sect. A1.3 contains the open research problems.

A1.2 Theoretical Background

A1.2.1 Nonlinear Optical Materials

The (1.2) of Chap. 1 can be expressed as

$$U(E) = \frac{p_s^2}{2M_\perp} + \frac{p_z^2}{2M_\parallel} V(E) \tag{A1.1}$$

where

$$U(E) \equiv \left[E(1+\alpha E)\left[(E+E_g)(E+E_g+\Delta_\parallel) + \delta\left(E+E_g+\frac{2}{3}\Delta_\parallel\right) + \frac{2}{9}(\Delta_\parallel^2 - \Delta_\perp^2)\right]\right] \times \left[(E+E_g) + \left(E+E_g+\frac{2}{3}\Delta_\parallel\right) + \delta\left(E+E_g+\frac{1}{3}\Delta_\parallel\right) + \frac{1}{9}(\Delta_\parallel^2 - \Delta_\perp^2)\right]^{-1}$$

$$p_s = \hbar k_s,\ M_\parallel = \frac{m_\parallel^*(E_g + \frac{2}{3}\Delta_\parallel)}{(E_g + \Delta_\parallel)},\quad M_\perp = \frac{m_\perp^*(E_g + \frac{2}{3}\Delta_\perp)}{(E_g + \Delta_\perp)},\quad p_z = \hbar k_z$$

and

$$V(E) \equiv \left[(E+E_g)\left(E+E_g+\frac{2}{3}\Delta_\parallel\right)\right]\left[(E+E_g)\left(E+E_g+\frac{2}{3}\Delta_\parallel\right) + \delta\left(E+E_g+\frac{1}{3}\Delta_\parallel\right) + \frac{1}{9}(\Delta_\parallel^2 - \Delta_\perp^2)\right]^{-1}$$

We know that from electromagnetic theory that,

$$\vec{B} = \nabla \times \vec{A} \tag{A1.2}$$

where, $\vec{A}$ is the vector potential. In the presence of quantizing magnetic field B along z direction, the Eq. A1.2 assumes the form

$$0\hat{i}+0\hat{j}+B\hat{k}=\begin{vmatrix}\hat{i} & \hat{j} & \hat{k}\\ \dfrac{\partial}{\partial x} & \dfrac{\partial}{\partial y} & \dfrac{\partial}{\partial z}\\ A_x & A_y & A_z\end{vmatrix}$$

where $\hat{i}, \hat{j}$ and $\hat{k}$ are orthogonal triads. Thus, we can write

$$\begin{aligned}\frac{\partial A_z}{\partial y}-\frac{\partial A_y}{\partial z}&=0\\ \frac{\partial A_x}{\partial z}-\frac{\partial A_z}{\partial x}&=0\\ \frac{\partial A_y}{\partial x}-\frac{\partial A_x}{\partial y}&=B\end{aligned} \tag{A1.3}$$

This particular set of equations is being satisfied for $A_x=0$, $A_y=Bx$ and $A_z=0$.

Therefore, in the presence of the electric field E_o along x-axis and the quantizing magnetic field B along z-axis for the present case following of (A1.1) one can approximately write,

$$U(E)+|e|E_o\hat{x}\rho(E)=\frac{\hat{p}_x^2}{2M_\perp}+\frac{\left(\hat{p}_y-|e|B\hat{x}\right)^2}{2M_\perp}+\frac{\hat{p}_z^2}{2a(E)} \tag{A1.4}$$

where

$$\begin{aligned}\rho(E)\equiv\frac{\partial}{\partial E}[U(E)]=\Bigg[&\frac{U(E)(1+2\alpha E)}{E(1+\alpha E)}\\ &+U(E)\left[(E+E_g)(E+E_g+\Delta_\parallel)+\delta\left(E+E_g+\frac{2}{3}\Delta_\parallel\right)+\frac{2}{9}\left(\Delta_\parallel^2-\Delta_\perp^2\right)\right]^{-1}\\ &\times\left[2E+2E_g+\Delta_\parallel+\delta\right]-U(E)\left[(E+E_g)\left(E+E_g+\frac{2}{3}\Delta_\parallel\right)\right.\\ &\left.+\delta\left(E+E_g+\frac{1}{3}\Delta_\parallel\right)+\frac{1}{9}\left(\Delta_\parallel^2-\Delta_\perp^2\right)\right]^{-1}\times\left[2E+2E_g+\frac{2}{3}\Delta_\parallel+\delta\right]\Bigg]\end{aligned}$$

and $a(E)\equiv M_\parallel[V(E)]^{-1}$

Let us define the operator $\hat{\theta}$ as

$$\hat{\theta}=-\hat{p}_y+|e|B\hat{x}-\frac{M_\perp E_o\rho(E)}{B} \tag{A1.5}$$

Eliminating the operator $\hat{x}$, between Eqs. (A1.4) and (A1.5) the dispersion relation of the conduction electron in tetragonal semiconductors in the presence of cross-field configuration is given by

$$U(E)=\left[\left(\left(n+\frac{1}{2}\right)\hbar\omega_{01}\right)+\left(\frac{[\hbar k_z(E)]^2}{2a(E)}\right)-\left(\frac{E_o\hbar k_y\rho(E)}{B}\right)-\left(\frac{M_\perp\rho^2(E)E_o^2}{2B^2}\right)\right] \tag{A1.6}$$

where $\omega_{01} \equiv \dfrac{|e|B}{M_\perp}$.

Therefore, the EMMs along z and y directions can, respectively, be expressed as

$$m_z^*(\bar{E}_{FB}, n, E_0, B) = \left[\{a(\bar{E}_{FB})\}'\left\{\left\{U(\bar{E}_{FB}) - \left(n+\frac{1}{2}\right)\omega_{01} + \frac{M_\perp[\rho(\bar{E}_{FB})]^2 E_0^2}{2B^2}\right.\right.\right.$$
$$\left.\left.\left. + a(\bar{E}_{FB})\left[\{U(\bar{E}_{FB})\}' + \frac{M_\perp E_0^2 \rho(\bar{E}_{FB})\{\rho(\bar{E}_{FB})\}'}{B^2}\right]\right\}\right\}\right] \tag{A1.7}$$

$$m_y^*(\bar{E}_{FB}, n, E_0, B) = \left(\frac{B}{E_0}\right)^2 \left(\frac{1}{\rho(\bar{E}_{FB})}\right)\left[U(\bar{E}_{FB}) - \left(n+\frac{1}{2}\right)\hbar\omega_{01} + \frac{M_\perp[\rho(\bar{E}_{FB})]^2 E_0^2}{2B^2}\right]$$
$$\times\left[[\rho(\bar{E}_{FB})]^{-1}\left[\{U(\bar{E}_{FB})\}' + \frac{M_\perp \rho(\bar{E}_{FB})\{\rho(\bar{E}_{FB})\}' E_0^2}{B^2}\right]\right.$$
$$\left. - \left[\left[\rho(\bar{E}_{FB})\right]^{-2}\{\rho(\bar{E}_{FB})\}'\left[U(\bar{E}_{FB}) - \left(n+\frac{1}{2}\right)\hbar\omega_{01} + \frac{M_\perp[\rho(\bar{E}_{FB})]^2 E_0^2}{2B^2}\right]\right]\right] \tag{A1.8}$$

where

$$a'(\bar{E}_{FB}) = \frac{-a(\bar{E}_{FB})}{V(\bar{E}_{FB})} V'(\bar{E}_{FB}),\ V'(\bar{E}_{FB}) = \left[(\bar{E}_{FB} + E_g)(\bar{E}_{FB} + E_g + \Delta_\parallel)\right.$$
$$\left. + \delta\left(\bar{E}_{FB} + E_g + \frac{2}{3}\Delta_\parallel\right) + \frac{2}{9}\left(\Delta_\parallel^2 - \Delta_\perp^2\right)\right]^{-1}$$
$$\times\left[-(2\bar{E}_{FB} + 2E_g + \Delta_\parallel + \delta)V(\bar{E}_{FB}) + 2\bar{E}_{FB} + 2E_g + \frac{2}{3}\Delta_\parallel\right],$$

$\bar{E}_{FB}$ is the Fermi energy in the presence of cross-field configuration as measured from the edge of the conduction band in the vertically upward direction in the absence of any quantization,

$$\rho(\bar{E}_{FB}) = [U(\bar{E}_{FB})]' = \left[\frac{U(\bar{E}_{FB})(1 + 2\alpha\bar{E}_{FB})}{\bar{E}_{FB}(1 + \alpha\bar{E}_{FB})} + U(\bar{E}_{FB})\right.$$
$$\times\left[(\bar{E}_{FB} + E_g)(\bar{E}_{FB} + E_g + \Delta_\parallel) + \delta\left(\bar{E}_{FB} + E_g + \frac{2}{3}\Delta_\parallel\right) + \frac{2}{9}\left(\Delta_\parallel^2 - \Delta_\perp^2\right)\right]^{-1}$$
$$\times\left[2\bar{E}_{FB} + 2E_g + \Delta_\parallel + \delta\right] - U(\bar{E}_{FB})\left[(\bar{E}_{FB} + E_g)\left(\bar{E}_{FB} + E_g + \frac{2}{3}\Delta_\parallel\right)\right.$$
$$\left.\left. + \delta\left(\bar{E}_{FB} + E_g + \frac{1}{3}\Delta_\parallel\right) + \frac{1}{9}\left(\Delta_\parallel^2 - \Delta_\perp^2\right)\right]^{-1}\left[2\bar{E}_{FB} + 2E_g + \frac{2}{3}\Delta_\parallel + \delta\right]\right]$$

and

$$\rho'(\bar{E}_{FB}) = \left[\frac{U'(\bar{E}_{FB})(1+2\alpha\bar{E}_{FB})}{\bar{E}_{FB}(1+\alpha\bar{E}_{FB})} + \frac{2\alpha U(\bar{E}_{FB})}{\bar{E}_{FB}(1+\alpha\bar{E}_{FB})} - \frac{U(\bar{E}_{FB})(1+2\alpha\bar{E}_{FB})^2}{[\bar{E}_{FB}(1+\alpha\bar{E}_{FB})]^2}\right.$$
$$+ U'(\bar{E}_{FB})\left[(\bar{E}_{FB}+E_g)(\bar{E}_{FB}+E_g+\Delta_\parallel)+\delta\left(\bar{E}_{FB}+E_g+\frac{2}{3}\Delta_\parallel\right)+\frac{2}{9}\left(\Delta_\parallel^2-\Delta_\perp^2\right)\right]^{-1}$$
$$\times\left[2\bar{E}_{FB}+2E_g+\Delta_\parallel+\delta\right]+2U(\bar{E}_{FB})\left[(\bar{E}_{FB}+E_g)(\bar{E}_{FB}+E_g+\Delta_\parallel)+\delta\left(\bar{E}_{FB}+E_g+\frac{2}{3}\Delta_\parallel\right)+\frac{2}{9}\left(\Delta_\parallel^2-\Delta_\perp^2\right)\right]^{-1}$$
$$-U(\bar{E}_{FB})\left[2\bar{E}_{FB}+2E_g+\Delta_\parallel+\delta\right]^2\left[(\bar{E}_{FB}+E_g)(\bar{E}_{FB}+E_g+\Delta_\parallel)+\delta\left(\bar{E}_{FB}+E_g+\frac{2}{3}\Delta_\parallel\right)+\frac{2}{9}\left(\Delta_\parallel^2-\Delta_\perp^2\right)\right]^{-2}$$
$$-U'(\bar{E}_{FB})\left[(\bar{E}_{FB}+E_g)\left(\bar{E}_{FB}+E_g+\frac{2}{3}\Delta_\parallel\right)+\delta\left(\bar{E}_{FB}+E_g+\frac{1}{3}\Delta_\parallel\right)+\frac{1}{9}\left(\Delta_\parallel^2-\Delta_\perp^2\right)\right]^{-1}$$
$$\times\left[2\bar{E}_{FB}+2E_g+\frac{2}{3}\Delta_\parallel+\delta\right]-2U(\bar{E}_{FB})\left[(\bar{E}_{FB}+E_g)\left(\bar{E}_{FB}+E_g+\frac{2}{3}\Delta_\parallel\right)+\delta\left(\bar{E}_{FB}+E_g+\frac{1}{3}\Delta_\parallel\right)+\frac{1}{9}\left(\Delta_\parallel^2-\Delta_\perp^2\right)\right]^{-1}$$
$$+U(\bar{E}_{FB})\left[(\bar{E}_{FB}+E_g)\left(\bar{E}_{FB}+E_g+\frac{2}{3}\Delta_\parallel\right)+\delta\left(\bar{E}_{FB}+E_g+\frac{1}{3}\Delta_\parallel\right)+\frac{1}{9}\left(\Delta_\parallel^2-\Delta_\perp^2\right)\right]^{-1}$$
$$\left.\times\left[2\bar{E}_{FB}+2E_g+\frac{2}{3}\Delta_\parallel+\delta\right]^2\right]$$

When $E_0 \to 0$, $m_y^*(\bar{E}_{FB}, n, E_0, B) \to \infty$, which is a physically justified result. The dependence of the EMM along y direction on the Fermi energy, electric field, magnetic field, and the magnetic quantum number is an intrinsic property of cross-fields. Another characteristic feature of cross-fields is that various transport coefficients will be sampled dimension dependent. These conclusions are valid for even isotropic parabolic energy bands and cross-fields introduce the index dependent anisotropy in the effective mass.

The formulation of EEM requires the expression of the electron concentration which can, in general, be written excluding the electron spin as

$$n_o = \frac{-g_v}{L_x\pi^2}\sum_{n=0}^{n_{\max}}\int_{\bar{E}_0}^{\infty} I(E)\frac{\partial f_o}{\partial E}\mathrm{d}E \tag{A1.9}$$

where $\mathrm{L_x}$ is the sample length along x direction, $\bar{E}_0$ is determined by the equation

$$I(\bar{E}_0) = 0$$

where

$$I(E) = \int_{x_l(E)}^{x_h(E)} k_z(E)\mathrm{d}k_y \tag{A1.10}$$

in which,

$$x_l(E) \equiv \frac{-E_0 M_\perp \rho(E)}{\hbar B}$$

and

$$x_h(E) \equiv \frac{|e|BL_x}{\hbar} + x_l(E)$$

Using Eqs. (A1.6) and (A1.10), we get

$$I(E) = \frac{2}{3}\left[\frac{B\sqrt{2a(E)}}{\hbar^2 E_0 \rho(E)}\right]\left[\left[U(E) - \left(n+\frac{1}{2}\right)\frac{\hbar|e|B}{M_\perp} + |e|E_0 L_x \rho(E) - \frac{M_\perp E_0^2[\rho(E)]^2}{2B^2}\right]^{\frac{3}{2}} - \left[U(E) - \left(n+\frac{1}{2}\right)\frac{\hbar|e|B}{M_\perp} - \frac{M_\perp E_0^2[\rho(E)]^2}{2B^2}\right]^{\frac{3}{2}}\right] \tag{A1.11}$$

Combining (A1.9) and (A1.11), the electron concentration is given by

$$n_0 = \left(\frac{2g_v B\sqrt{2}}{3L_x \pi^2 \hbar^2 E_o}\right)\sum_{n=0}^{n_{\max}}[T_{41}(n, \bar{E}_{FB}) + T_{42}(n, \bar{E}_{FB})] \tag{A1.12}$$

where

$$T_{41}(n, \bar{E}_{FB}) \equiv \frac{\sqrt{a(\bar{E}_{FB})}}{\rho(\bar{E}_{FB})}\left[\left[U(\bar{E}_{FB}) - \left(n+\frac{1}{2}\right)\frac{\hbar|e|B}{M_\perp} + |e|E_0 L_x \rho(\bar{E}_{FB}) - \frac{M_\perp E_0^2[\rho(\bar{E}_{FB})]^2}{2B^2}\right]^{3/2} - \left[U(\bar{E}_{FB}) - \left(n+\frac{1}{2}\right)\frac{\hbar|e|B}{M_\perp} - \frac{M_\perp E_0^2[\rho(\bar{E}_{FB})]^2}{2B^2}\right]^{3/2}\right],$$

$$T_{42}(n, \bar{E}_{FB}) \equiv \sum_{r=1}^{s}[L(r)T_{41}(n, \bar{E}_{FB})],$$

$$L(r) = 2(k_{Bi}T)^{2r}(1 - 2^{1-2r})\xi(2r)\frac{\partial^{2r}}{\partial E_F^{2r}} \text{ and } \bar{E}_{FB}$$

is the Fermi energy in the present case.

A1.2.2 Special cases for III–V, Ternary, and Quaternary Materials

(a) Under the conditions $\delta = 0$, $\Delta_\| = \Delta_\perp = \Delta$ and $m_\|^* = m_\perp^* = m_c$, (A1.6) assumes the form

$$I_{11}(E) = \left(n+\frac{1}{2}\right)\hbar\omega_0 + \frac{[\hbar k_z(E)]^2}{2m_c} - \frac{E_0}{B}\hbar k_y\{I_{11}(E)\}' - \frac{m_c E_0^2\left[\{I_{11}(E)\}'\right]^2}{2B^2} \tag{A1.13}$$

The use of (A1.13) leads to the expressions of the EMMs along z and y directions as

$$m_z^*(\bar{E}_{FB}, n, E_0, B) = m_c\left[\{I_{11}(\bar{E}_{FB})\}'' + \frac{m_c E_0^2\{I_{11}(\bar{E}_{FB})\}'\{I_{11}(\bar{E}_{FB})\}''}{B^2}\right] \quad (A1.14)$$

$$\begin{aligned} m_y^*(\bar{E}_{FB}, n, E_0, B) = {} & \left(\frac{B}{E_0}\right)^2 \frac{1}{[\{I_{11}(\bar{E}_{FB})\}']} \\ & \times \left[I_{11}(\bar{E}_{FB}) - \left(n+\frac{1}{2}\right)\hbar\omega_0 + \frac{m_c E_0^2[\{I_{11}(\bar{E}_{FB})\}']^2}{2B^2}\right] \\ & \times \left[\frac{-\{I_{11}(\bar{E}_{FB})\}''}{[\{I_{11}(\bar{E}_{FB})\}']^2}\left[I_{11}(\bar{E}_{FB}) - \left(n+\frac{1}{2}\right)\hbar\omega_0 + \frac{m_c E_0^2[\{I_{11}(\bar{E}_{FB})\}']^2}{2B^2}\right]\right. \\ & \left. + 1 + \frac{m_c E_0^2\{I_{11}(\bar{E}_{FB})\}''}{B^2}\right] \end{aligned} \quad (A1.15)$$

The electron concentration in this case assume the forms

$$n_0 = \frac{2g_v B\sqrt{2m_c}}{3L_x\pi^2\hbar^2 E_0}\sum_{n=0}^{n_{max}}[T_{43}(n, \bar{E}_{FB}) + T_{44}(n, \bar{E}_{FB})] \quad (A1.16)$$

where

$$\begin{aligned} T_{43}(n, \bar{E}_{FB}) \equiv {} & \left[\left[I_{11}(\bar{E}_{FB}) - \left(n+\frac{1}{2}\right)\hbar\omega_0 - \frac{m_c E_0^2}{2B^2}[\{I_{11}(\bar{E}_{FB})\}']^2 + |e|E_0 L_x[\{I_{11}(\bar{E}_{FB})\}']\right]^{3/2}\right. \\ & \left. - \left[I_{11}(\bar{E}_{FB}) - \left(n+\frac{1}{2}\right)\hbar\omega_0 - \frac{m_c E_0^2}{2B^2}[\{I_{11}(\bar{E}_{FB})\}']^2\right]^{3/2}\right]\frac{1}{[\{I_{11}(\bar{E}_{FB})\}']} \end{aligned}$$

and $T_{44}(n, \bar{E}_{FB}) \equiv \sum_{r=1}^{s}[L(r)T_{43}(n, \bar{E}_{FB})]$.

(b) Under the condition $\Delta \gg E_g$, (A1.13) assumes the well-known form [1]

$$E(1+\alpha E) = \left(n+\frac{1}{2}\right)\hbar\omega_0 - \frac{E_0}{B}\hbar k_y(1+2\alpha E) - \frac{m_c E_0^2}{2B^2}(1+2\alpha E)^2 + \frac{[\hbar k_z(E)]^2}{2m_c} \quad (A1.17)$$

The use of (A1.17) leads to the expressions of the EMMs along z and y directions as

$$m_z^*(\bar{E}_{FB}, n, E_0, B) = m_c\left[(1+2\alpha\bar{E}_{FB}) + \frac{2m_c E_0^2(1+2\alpha\bar{E}_{FB})\alpha}{B^2}\right] \quad (A1.18)$$

$$m_y^*(\bar{E}_{FB}, n, E_0, B) = \left(\frac{B}{E_0}\right)^2 \frac{1}{[1+2\alpha\bar{E}_{FB}]}\left[\bar{E}_{FB}(1+\alpha\bar{E}_{FB}) - \left(n+\frac{1}{2}\right)\hbar\omega_0 + \frac{m_c E_0^2[(1+2\alpha\bar{E}_{FB})]^2}{2B^2}\right]$$
$$\times \left[\frac{-2\alpha}{[(1+2\alpha\bar{E}_{FB})]^2}\left[\bar{E}_{FB}(1+\alpha\bar{E}_{FB}) - \left(n+\frac{1}{2}\right)\hbar\omega_0 + \frac{m_c E_0^2[(1+2\alpha\bar{E}_{FB})]^2}{2B^2}\right] + 1 + \frac{2\alpha m_c E_0^2}{B^2}\right] \quad (A1.19)$$

The expressions for n_0 in this case assume the forms

$$n_0 = \frac{2g_v B\sqrt{2m_c}}{3L_x\pi^2\hbar^2 E_0}\sum_{n=0}^{n_{max}}[T_{45}(n,\bar{E}_{FB}) + T_{46}(n,\bar{E}_{FB})] \quad (A1.20)$$

where

$$T_{45}(n,\bar{E}_{FB}) \equiv \left[\left[\bar{E}_{FB}(1+\alpha\bar{E}_{FB}) - \left(n+\frac{1}{2}\right)\hbar\omega_0 + |e|E_0L_x(1+2\alpha\bar{E}_{FB}) - \frac{m_cE_0^2}{2B^2}(1+2\alpha\bar{E}_{FB})^2\right]^{3/2}\right.$$
$$\left. - \left[\bar{E}_{FB}(1+\alpha\bar{E}_{FB}) - \left(n+\frac{1}{2}\right)\hbar\omega_0 - \frac{m_cE_0^2}{2B^2}(1+2\alpha\bar{E}_{FB})^2\right]^{3/2}\right][1+2\alpha\bar{E}_{FB}]^{-1}$$

and $T_{46}(n,\bar{E}_{FB}) \equiv \sum_{r=0}^{s} L(r)[T_{45}(n,\bar{E}_{FB})]$

(c) For parabolic energy bands, $\alpha \to 0$ and we can write,

$$E = \left(n+\frac{1}{2}\right)\hbar\omega_0 + \frac{[\hbar k_z(E)]^2}{2m_c} - \frac{1}{2}m_c\left(\frac{E_0}{B}\right)^2 - \frac{E_0}{B}\hbar k_y \quad (A1.21)$$

Using Eq. (A1.21), the expressions of the EMMs along y and z directions can be written as

$$m_z^*(\bar{E}_{FB}, n, E_0, B) = m_c \quad (A1.22)$$

and

$$m_y^*(\bar{E}_{FB}, n, E_0, B) = \left(\frac{B}{E_0}\right)^2\left[\bar{E}_{FB} - \left(n+\frac{1}{2}\right)\hbar\omega_0 + \frac{m_cE_0^2}{2B^2}\right] \quad (A1.23)$$

The electron concentration in this case can, respectively, be expressed as

$$n_0 = N_c\theta g_v\left[\frac{k_BT}{|e|E_0L_x}\right]\sum_{n=0}^{n_{max}}\left[F_{\frac{1}{2}}(\eta_1) - F_{\frac{1}{2}}(\eta_2)\right] \quad (A1.24)$$

where

$$\eta_1 \equiv \frac{\bar{E}_{\text{FB}} - \overline{\theta_1}}{k_B T}, \quad \overline{\theta_1} \equiv \left[\left(n + \frac{1}{2}\right)\hbar\omega_0 + \frac{1}{2}m_c\left(\frac{E_0}{B}\right)^2 - |e|E_0 L_x\right],$$

$$\eta_2 \equiv \frac{\left(\bar{E}_{\text{FB}} - \overline{\theta_2}\right)}{k_B T}, \quad \overline{\theta_2} \equiv \overline{\theta_1} + |e|E_0 L_x$$

In the absence of electricfield $E_0 \to 0$ and the application of L' Hospital's rule transforms the (A.24) into the well-known form of electron concentration under magnetic quantization as given by (4.24b) of Chap. 4.

A1.2.3 II–VI Semiconductors

In the presence of electric field along x-axis and the quantizing magnetic field B along z-axis, from (1.42) of Chap. 1 we can write

$$\hat{E} + |e|E_0\hat{x} = \frac{\hat{p}_x^2}{2m_\perp^*} + \frac{\left(\hat{p}_y - |e|B\hat{x}^2\right)}{2m_\perp^*} + \frac{\hat{p}_z^2}{2m_\parallel^*} + D\left[\frac{\hat{p}_x^2}{2m_\perp^*} + \frac{\left(\hat{p}_y - |e|B\hat{x}\right)^2}{2m_\perp^*}\right]^{1/2} \tag{A1.25}$$

where,

$$D \equiv \pm\frac{\bar{\lambda}_0\sqrt{2m_\perp^*}}{\hbar}.$$

Let us define the operator $\widehat{\theta}$ as

$$\widehat{\theta} = -\hat{p}_y + |e|B\hat{x} - \frac{E_0 m_\perp^*}{B} \tag{A1.26}$$

Eliminating $\hat{x}$ between (A1.25) and (A1.26), one obtains

$$\hat{E} + \frac{E_0}{B}\widehat{\theta} + \frac{E_0}{B}\hat{p}_y + \left(\frac{E_0}{B}\right)^2 m_\perp^* = \frac{\hat{p}_x^2}{2m_\perp^*} + \frac{\hat{\theta}^2}{2m_\perp^*} + \frac{E_0^2 m_\perp^*}{2B^2} + \frac{\hat{\theta}E_0}{B}$$
$$+ D\left[\frac{\hat{p}_x^2}{2m_\perp^*} + \frac{\hat{\theta}^2}{2m_\perp^*} + \frac{E_0^2 m_\perp^*}{2B^2}\right]^{1/2} + \frac{\hat{p}_z^2}{2m_\parallel^*} \tag{A1.27}$$

Thus, the electron energy spectrum in this case can be expressed as

$$E = (\beta_1(n, E_0)) + \left(\frac{[\hbar k_z(E)]^2}{2m_\parallel^*}\right) - \left(\frac{E_0}{B}\hbar k_y\right) \tag{A1.28}$$

where

$$\beta_1(n, E_0) \equiv \left[\left(n+\frac{1}{2}\right)\hbar\omega_{02} - \left(\frac{E_0^2 m_\perp^*}{2B^2}\right) + D\left\{\left(n+\frac{1}{2}\right)\hbar\omega_{02} + \left(\frac{E_0^2 m_\perp^*}{2B^2}\right)\right\}^{1/2}\right]$$

and

$$\omega_{02} \equiv \frac{|e|B}{m_\perp^*}$$

The use of (A1.28) leads to the expressions of the EMMs along z and y directions as

$$m_z^*(\bar{E}_{\mathrm{FB}}, n, E_0, B) = m_\parallel^* \tag{A1.29}$$

$$m_y^*(\bar{E}_{\mathrm{FB}}, n, E_0, B) = \left(\frac{B}{E_0}\right)^2 [\bar{E}_{\mathrm{FB}} - \beta_1(n, E_0)] \tag{A1.30}$$

In this case

$$x_i = \frac{-E_0}{B\hbar} m_\perp^* + \frac{2D}{\hbar}\left(\frac{m_\perp^*}{2}\right)^{\frac{1}{2}}, \quad x_h = \frac{|e|BL_x}{\hbar} + x_l$$
$$\text{and} \quad k_z(E) = \frac{\sqrt{2m_\parallel^*}}{\hbar}\left[E - \beta_1(n, E_0) + \frac{E_0}{B}\hbar k_y\right]^{1/2} \tag{A1.31}$$

The (A1.10) for II–VI semiconductors the cross-field configuration assumes the form

$$I(E) = \int_{x_l}^{x_h} \frac{\sqrt{2m_\parallel^*}}{\hbar}\left[E - \beta_1(n, E_0) + \frac{E_0}{B}\hbar k_y\right]^{\frac{1}{2}} \mathrm{d}k_y$$

Therefore,

$$I(E) = \frac{\sqrt{2m_\parallel^*}}{\hbar}\frac{2}{3}\frac{B}{E_0\hbar}\left[\left\{E - \beta_1(n, E_0) + \frac{E_0}{B}\hbar x_h\right\}^{3/2} - \left\{E - \beta_1(n, E_0) + \frac{E_0}{B}\hbar x_l\right\}^{\frac{3}{2}}\right] \tag{A1.32}$$

The electron concentration, from (A1.9) can be expressed as

$$n_0 = -\frac{2g_v B\sqrt{2m_\parallel}}{3L_x\pi^2\hbar^2 E_o}\sum_{n=0}^{n_{\max}}\left[\int_{\theta_1}^{\infty}[E-\theta_1]^{\frac{3}{2}}\frac{\partial f_o}{\partial E}\mathrm{d}E - \int_{\theta_2}^{\infty}[E-\theta_2]^{\frac{3}{2}}\frac{\partial f_o}{\partial E}\mathrm{d}E\right] \tag{A1.33}$$

where

$$\theta_1 \equiv \beta_1(n, E_0) - \frac{E_0}{B}\hbar x_h$$

and

$$\theta_2 \equiv \beta_1(n, E_0) - \frac{E_0}{B}\hbar x_l$$

Substituting

$$\frac{E-\theta_1}{k_B T} = x_1, \quad \frac{E-\theta_2}{k_B T} = x_2, \quad \eta_3 = \frac{\bar{E}_{FB} - \theta_1}{k_B T}$$

and

$$\eta_4 = \frac{\bar{E}_{FB} - \theta_2}{k_B T},$$

from (A1.33), we can write

$$n_0 = \frac{2g_v B\sqrt{2m_{\|}^*}}{3E_0 L_x \hbar^2 \pi^2}(k_B T)^{\frac{3}{2}} \sum_{n=0}^{n_{max}} \left[\int_0^\infty \frac{x_1^{3/2}\exp(x_1 - \eta_3)}{[1+\exp(x_1-\eta_3)]^2} dx_1 - \int_0^\infty \frac{x_2^{3/2}\exp(x_2 - \eta_4)}{[1+\exp(x_2-\eta_4)]^2} dx_2 \right] \tag{A1.34}$$

Differentiating both sides of (1.22) with respect to η, one can write,

$$\Gamma(j+1)F_{j-1}(\eta) = \int_0^\infty \frac{x^j \exp(x-\eta)}{[1+\exp(x-\eta)]^2} dx \tag{A1.35}$$

Using (A1.34) and (A1.35), the electron concentration in this case can be written as

$$n_0 = \frac{g_v B\sqrt{2m_{\|}^* \pi}}{2E_0 L_x \hbar^2 \pi^2}(k_B T)^{\frac{3}{2}} \sum_{n=0}^{n_{max}} \left[F_{\frac{1}{2}}(\eta_3) - F_{\frac{1}{2}}(\eta_4) \right] \tag{A1.36}$$

A.1.2.4 Formulation of EEM in Bi

(a) **The McClure and Choi Model**

In the presence of an electric field E_0 along trigonal-axis (z direction) and the quantizing magnetic field B along bisectrix axis (y direction) from (1.49), we can write

$$E(1+\alpha E)+|e|E_0\hat{z}(1+2\alpha E)$$
$$=\frac{(\hat{p}_x-|e|B\hat{z})^2}{2m_1}+\frac{\hat{p}_y^2}{2m_2}+\frac{\hat{p}_z^2}{2m_3}+\frac{\hat{p}_y^2}{2m_2}\alpha E\left(1-\frac{m_2}{m_2'}\right)-\alpha\frac{\hat{p}_y^2\hat{p}_z^2}{4m_2m_3}+\alpha\frac{\hat{p}_y^4}{4m_2m_2'}$$
$$-\alpha\frac{(\hat{p}_x^2-|e|B\hat{z})^2\hat{p}_y^2}{4m_1m_2} \qquad (A1.37)$$

Let us define the operator $\hat{\theta}$ as

$$\hat{\theta}=|e|B\hat{z}-\hat{p}_x-\frac{m_1E_0}{B}(1+2\alpha E) \qquad (A1.38)$$

Eliminating $\hat{z}$ between (A1.37) and (A1.38), one obtains

$$E(1+\alpha E)+\frac{E_0}{B}(1+2\alpha E)\hat{\theta}+\frac{E_0}{B}(1+2\alpha E)\hat{p}_x+m_1\left(\frac{E_0}{B}\right)^2(1+2\alpha E)^2$$
$$=\left(\frac{\hat{\theta}^2}{2m_1}+\frac{\hat{p}_z^2}{2m_3}\right)+\frac{\hat{p}_y^2}{2m_2}+\frac{\hat{p}_y^2}{2m_2}\alpha E\left(1-\frac{m_2}{m_2'}\right)+\frac{\alpha\hat{p}_y^4}{4m_2m_2'}-\frac{\alpha\hat{p}_y^2}{2m_2}\left[\frac{\hat{\theta}^2}{2m_1}+\frac{\hat{p}_z^2}{2m_3}\right]$$
$$-\frac{\alpha\hat{p}_y^2m_1E_0^2}{4m_2B^2}(1+2\alpha E)^2+\frac{E_0}{B}\hat{\theta}(1+2\alpha E)+\frac{1}{2}m_1\left(\frac{E_0}{B}\right)^2(1+2\alpha E)^2 \qquad (A1.39)$$

Therefore, the required dispersion relation is given by

$$E(1+\alpha E)=\left(n+\frac{1}{2}\right)\hbar\omega_{03}+\frac{[\hbar k_y(E)]^2}{2m_2}-\frac{E_0}{B}(1+2\alpha E)\hbar k_x-\frac{1}{2}m_1\left(\frac{E_0}{B}\right)^2$$
$$\times(1+2\alpha E)^2+\frac{[\hbar k_y(E)]^2}{2m_2}\alpha E\left(1-\frac{m_2}{m_2'}\right)+\frac{\alpha[\hbar k_y(E)]^4}{4m_2m_2'}-\frac{\alpha[\hbar k_y(E)]^2}{2m_2}$$
$$\times\left(n+\frac{1}{2}\right)\hbar\omega_{03}-\frac{\alpha[\hbar k_y(E)]^2m_1E_0^2}{4m_2B^2}(1+2\alpha E)^2 \qquad (A1.40)$$

where

$$\omega_{03}\equiv\frac{|e|B}{\sqrt{m_1m_3}}.$$

When $\alpha\to 0$, from (A1.40), we can write

$$E=\left(n+\frac{1}{2}\right)\hbar\omega_{03}+\frac{[\hbar k_y(E)]^2}{2m_2}-\frac{E_0}{B}\hbar k_x-\frac{1}{2}m_1\left(\frac{E_0}{B}\right)^2 \qquad (A1.41)$$

The use of (A1.40) leads to the equations of the EMMs along x and y direction as

$$
\begin{aligned}
m_x^*(\bar{E}_{\mathrm{FB}}, n, E_0, B) = {} & \left(\frac{B}{E_0}\right)^2 (1 + 2\alpha \bar{E}_{\mathrm{FB}})^{-3} \\
& \times \left[\bar{E}_{FB}(1 + \alpha \bar{E}_{\mathrm{FB}}) - \left(n + \frac{1}{2}\right)\hbar\omega_{03} + \frac{1}{2} m_1 \left(\frac{E_0}{B}\right)^2 (1 + 2\alpha \bar{E}_{\mathrm{FB}})^2\right] \\
& \times \left[(1 + 2\alpha \bar{E}_{\mathrm{FB}})\left\{1 + 2\alpha \bar{E}_{\mathrm{FB}} + 2\alpha m_1 (1 + 2\alpha \bar{E}_{\mathrm{FB}})\left(\frac{E_0}{B}\right)^2\right\}\right. \\
& \left. - 2\alpha\left\{\bar{E}_{\mathrm{FB}}(1 + \alpha \bar{E}_{\mathrm{FB}}) - \left(n + \frac{1}{2}\right)\hbar\omega_{03} + \frac{1}{2} m_1 \left(\frac{E_0}{B}\right)^2 (1 + 2\alpha \bar{E}_{\mathrm{FB}})^2\right\}\right]
\end{aligned}
\tag{A1.42}
$$

and

$$
m_y^*(\bar{E}_{\mathrm{FB}}, n, E_0, B) = \frac{1}{4}\left[\frac{[h_4(n, \overline{E}_{\mathrm{FB}})]'}{2\sqrt{h_4(n, \overline{E}_{\mathrm{FB}})}} - [h_1(n, \overline{E}_{\mathrm{FB}})]'\right] \tag{A1.43}
$$

where

$$
\begin{aligned}
h_4(n, \bar{E}_{\mathrm{FB}}) &\equiv \left[h_1^2(n, \bar{E}_{\mathrm{FB}}) + 4h_2(n, \bar{E}_{\mathrm{FB}})\right], \\
h_1(n, \bar{E}_{\mathrm{FB}}) &\equiv \frac{4m_2 m_2'}{\alpha}\left[\frac{-\alpha}{2m_2}\left(n + \frac{1}{2}\right)\hbar\omega_{03} - \frac{\alpha m_1 E_0^2}{4m_2 B^2}(1 + 2\alpha \bar{E}_{\mathrm{FB}})^2 + \frac{1}{2m_2} + \frac{\alpha \bar{E}_{\mathrm{FB}}}{2m_2}\left(1 - \frac{m_2}{m_2'}\right)\right] \\
\text{and } h_2(n, \bar{E}_{\mathrm{FB}}) &\equiv \frac{4m_2 m_2'}{\alpha}\left[\bar{E}_{\mathrm{FB}}(1 + \alpha \bar{E}_{\mathrm{FB}}) - \left(n + \frac{1}{2}\right)\hbar\omega_{03} + \frac{1}{2} m_1 \left(\frac{E_{03}}{B}\right)^2 (1 + 2\alpha \bar{E}_{\mathrm{FB}})^2\right]
\end{aligned}
$$

In this case,

$$
x_l(E) = -\frac{m_1 E_0}{B\hbar}(1 + 2\alpha E) \quad \text{and} \quad x_h(E) = \frac{|e|BL_z}{\hbar} + x_l(E) \tag{A1.44}
$$

where, L_z is the sample length along z-direction.

The electron concentration in this case can be written as

$$
n_0 = \frac{g_v}{L_z \pi^2} \sum_{n=0}^{n_{\max}} \int_{\bar{E}_{01}}^{\infty} J(E)\left(-\frac{\partial f_o}{\partial E}\right) \mathrm{d}E \tag{A1.45}
$$

in which $\bar{E}_{01}$ is the root of the equation $J(\bar{E}_{01}) = 0$ where $J(E)$ is given by

$$
J(E) = \int_{x_l(E)}^{x_h(E)} k_y(E)\,\mathrm{d}k_x \tag{A1.46}
$$

The term $k_y(E)$ in (A1.46) satisfies the following equation

$$k_y(E) = \left[-h_1(n,E) + \sqrt{h_4(n,E) + h_5(E)k_x}\right]^{1/2} \left[(\hbar)\sqrt{2}\right]^{-1} \quad \text{(A1.47)}$$

where,

$$h_5(E) \equiv \left(\frac{16m_2m_2'}{\alpha}\right)\left[\frac{E_0}{B}(1+2\alpha E)\hbar\right]$$

Using (A1.46) and (A1.47), we get,

$$J(E) = \frac{2\sqrt{2}}{3\hbar}\left[\left[\frac{h_1(n,E)}{h_5(n,E)}\right]\left[\{-h_1(n,E)+h_7(n,E)\}^{3/2} - \{-h_1(n,E)+h_6(n,E)\}^{3/2}\right] + \left[\frac{3}{5h_5(n,E)}\right]\left[\{-h_1(n,E)+h_7(n,E)\}^{5/2} - \{-h_1(n,E)+h_6(n,E)\}^{5/2}\right]\right] \quad \text{(A1.48)}$$

where,

$$h_6(n,E) \equiv [h_4(n,E) + h_5(E)x_l(E)]^{1/2}$$

and

$$h_7(n,E) \equiv [h_4(n,E) + x_h(E)h_5(n,E)]$$

Combining (A1.45) and (A1.48), the electron concentration in this case can be written as

$$n_0 = \frac{g_v}{L_z\pi^2}\frac{2\sqrt{2}}{3\hbar}\sum_{n=0}^{n_{\max}}\left[T_{47}(n,\overline{E}_{FB}) + T_{48}(n,\overline{E}_{FB})\right] \quad \text{(A1.49)}$$

where,

$$T_{47}(n,\overline{E}_{FB}) \equiv \left(\frac{h_1(n,\overline{E}_{FB})}{h_5(n,\overline{E}_{FB})}\right)\left[[-h_1(n,\overline{E}_{FB}) + h_7(n,\overline{E}_{FB})]^{\frac{3}{2}} - \{-h_1(n,\overline{E}_{FB}) + h_6(n,\overline{E}_{FB})\}^{\frac{3}{2}}\right] + \left(\frac{3}{5h_5(n,\overline{E}_{FB})}\right)\left[[-h_1(n,\overline{E}_{FB}) + h_7(n,\overline{E}_{FB})]^{\frac{5}{2}} - [-h_1(n,\overline{E}_{FB}) + h_6(n,\overline{E}_{FB})]^{\frac{5}{2}}\right]$$

and $T_{48}(n,\overline{E}_{FB}) \equiv \sum_{r=1}^{s} L(r)\left[T_{47}(n,\overline{E}_{FB})\right]$

(b) **The Cohen Model**

In the presence of an electric field E_o along trigonal axis and the quantizing magnetic field B along bisectrix axis for this case, the (1.65) assumes the form

$$E(1+\alpha E)+|e|E_o\hat{z}(1+2\alpha E)=\frac{(\hat{p}_x-|e|B\hat{z})^2}{2m_1}+\frac{\hat{p}_z^2}{2m_3}-\frac{\alpha E\hat{p}_y^2}{2m_2'}+\frac{\hat{p}_y^2}{2m_2}\times(1+\alpha E)+\frac{\alpha\hat{p}_y^4}{4m_2m_2'} \quad \text{(A1.50)}$$

Using the same operator $\hat{\theta}$ as defined by (A1.38) and eliminating $\hat{z}$ between the (A1.50) and (A1.38), one can write

$$E(1+\alpha E)=\left(\frac{\hat{\theta}^2}{2m_1}+\frac{\hat{p}_z^2}{2m_3}\right)-\frac{E_o}{B}\hat{p}_x(1+2\alpha E)-\frac{E_0^2 m_1}{2B^2}(1+2\alpha E)^2-\frac{\alpha E\hat{p}_y^2}{2m_2'}+\frac{\hat{p}_y^2}{2m_2}(1+\alpha E)+\frac{\alpha\hat{p}_y^4}{4m_2m_2'} \quad \text{(A1.51)}$$

Thus, the electron energy spectrum can be expressed as

$$E(1+\alpha E)=\left[\left(n+\frac{1}{2}\right)\hbar\omega_{03}-\frac{E_0}{B}\hbar k_x(1+2\alpha E)-\frac{1}{2}m_1\left(\frac{E_0}{B}\right)^2(1+2\alpha E)^2-\left(\frac{\alpha E[\hbar k_y(E)]^2}{2m_2'}\right)+\frac{[\hbar k_y(E)]^2}{2m_2}(1+\alpha E)+\left[\frac{\alpha[\hbar k_y(E)]^4}{4m_2m_2'}\right]\right] \quad \text{(A1.52)}$$

The use of (A1.52) leads to the same expression of EMM along the x direction as given by (A1.42) for the McClure and Choi model and the EMM along y direction is given by

$$m_y^*(\bar{E}_{FB},n,E_0,B)=\frac{1}{4}\left[\frac{[\overline{H}_5(n,\bar{E}_{FB})]'}{2\sqrt{\overline{H}_5(n,\bar{E}_{FB})}}-[\overline{H}_1(n,\bar{E}_{FB})]'\right] \quad \text{(A1.53)}$$

where

$$\overline{H}_5(n,\bar{E}_{\mathrm{FB}})\equiv\left[\overline{H}_1^2(\bar{E}_{\mathrm{FB}})+4\overline{H}_3(n,\bar{E}_{\mathrm{FB}})\right],\quad \overline{H}_1(\bar{E}_{\mathrm{FB}})\equiv\left(\frac{4m_2m_2'}{\alpha}\right)\left[\frac{1+\alpha\bar{E}_{\mathrm{FB}}}{2m_2}-\frac{\alpha\bar{E}_{\mathrm{FB}}}{2m_2'}\right]$$

and $$\overline{H}_3(n,\bar{E}_{\mathrm{FB}})\equiv\frac{4m_2m_2'}{\alpha}\left[\bar{E}_{\mathrm{FB}}(1+\alpha\bar{E}_{\mathrm{FB}})-\left(n+\frac{1}{2}\right)\hbar\omega_{03}+\frac{1}{2}m_1\left(\frac{E_{03}}{B}\right)^2(1+2\alpha\bar{E}_{\mathrm{FB}})^2\right].$$

The term $k_y(E)$ of (A1.52) in this case can be determined from the following equation

$$[\hbar k_y(E)]^4+[\hbar k_y(E)]^2\left(\frac{4m_2m_2'}{\alpha}\right)\left[\frac{-\alpha E}{2m_2'}+\frac{1+\alpha E}{2m_2}\right]-\left[E(1+\alpha E)-\left(n+\frac{1}{2}\right)\hbar\omega_{03}+\frac{E_0}{B}\hbar k_x(1+2\alpha E)+\frac{E_0^2 m_1}{2B^2}(1+2\alpha E)^2\right]\left(\frac{4m_2m_2'}{\alpha}\right)=0 \quad \text{(A1.54)}$$

Therefore,

$$k_y(E) = \left[-\overline{H}_1(E) + \sqrt{\overline{H}_5(n,E) + \overline{H}_6(E)k_x}\right]^{1/2} \left(\hbar\sqrt{2}\right)^{-1} \tag{A1.55}$$

where

$$\overline{H}_6(E) \equiv 4\overline{H}_4(E)$$

and

$$\overline{H}_4(E) \equiv \left[\frac{E_0}{B}\hbar(1+2\alpha E)\right]\left(\frac{4m_2 m_2'}{\alpha}\right).$$

The expression of $J(E)$ in this case can be written as

$$J(E) = \frac{\sqrt{2}}{\hbar}\left[\frac{\overline{H_1}(E)}{\overline{H_6}(E)}\frac{2}{3}\left\{\left(\overline{H_8}(n,E) - \overline{H_1}(E)\right)^{3/2} - \left(\overline{H_7}(n,E) - \overline{H_1}(E)\right)^{3/2}\right\}\right.$$
$$\left. + \frac{1}{\overline{H_6}(E)}\frac{2}{5}\left[\left\{\left(\overline{H_8}(n,E) - \overline{H}_1(E)\right)\right\}^{5/2} - \left\{\left(\overline{H_7}(n,E) - \overline{H_1}(E)\right)\right\}^{5/2}\right]\right] \tag{A1.56}$$

where

$$\overline{H}_8(n,E) \equiv \left[\overline{H}_5(n,E) + \overline{H}_6(E)x_h(E)\right]^{1/2},$$
$$\overline{H}_5(n,E) \equiv \left[\left[\overline{H}_1(n,E)\right]^2 + 4\overline{H}_3(n,E)\right],$$
$$x_h(E) \equiv \frac{|e|BL_z}{\hbar} + x_l(E),$$
$$x_l(E) \equiv \frac{-E_o}{B}\frac{m_1}{\hbar}(1+2\alpha E)$$

and $\overline{H}_7(n,E) \equiv \left[\overline{H}_5(n,E) + \overline{H}_6(E)x_l(E)\right]^{1/2}.$

The expression of the electron concentration for the Cohen model in the present case is given by

$$n_0 = \frac{2g_v\sqrt{2}}{3L_z\pi^2\hbar}\sum_{n=0}^{n_{\max}}\left[T_{49}\left(n,\bar{E}_{FB}\right) + T_{410}\left(n,\bar{E}_{FB}\right)\right] \tag{A1.57}$$

where,

$$T_{49}(n,E_{\mathrm{FB}}) \equiv \left[\frac{\overline{H_1}(\bar{E}_{\mathrm{FB}})}{\overline{H_6}(\bar{E}_{\mathrm{FB}})}\left\{\left(\overline{H}_8(n,\bar{E}_{\mathrm{FB}}) - \overline{H}_1(\bar{E}_{\mathrm{FB}})\right)^{3/2} - \left(\overline{H_7}(n,\bar{E}_{\mathrm{FB}}) - \overline{H}_1(\bar{E}_{\mathrm{FB}})\right)^{3/2}\right\}\right.$$
$$\left. + \frac{1}{\overline{H_6}(\bar{E}_{\mathrm{FB}})}\frac{3}{5}\left[\left\{\left(\overline{H}_8(n,\bar{E}_{\mathrm{FB}}) - \overline{H}_1(\bar{E}_{\mathrm{FB}})\right)\right\}^{5/2} - \left\{\left(\overline{H}_7(n,\bar{E}_{\mathrm{FB}}) - \overline{H}_1(\bar{E}_{\mathrm{FB}})\right)\right\}^{5/2}\right]\right]$$

and $T_{410}(n,\bar{E}_{\mathrm{FB}}) \equiv \sum_{r=1}^{s} L(r)T_{49}(n,\bar{E}_{\mathrm{FB}})$

(c) **The Lax Model**

Under cross-field configuration from the (1.71) of Chap. 1, one can write

$$E(1+\alpha E)+|e|E_o\hat{z}(1+2\alpha E)=\frac{(p_x-|e|B\hat{z})^2}{2m_1}+\frac{\hat{p}_y^2}{2m_2}+\frac{\hat{p}_z^2}{2m_3} \tag{A1.58}$$

Using the same operator $\widehat{\theta}$ as used for McClure and Choi model, we get

$$\begin{aligned}&E(1+\alpha E)+\frac{E_0}{B}(1+2\alpha E)\widehat{\theta}+\frac{E_0}{B}(1+2\alpha E)\widehat{p}_x+m_1\left(\frac{E_0}{B}\right)^2(1+2\alpha E)^2\\&=\frac{\widehat{\theta}^2}{2m_1}+\frac{\widehat{p}_z^2}{2m_3}+\frac{m_1}{2}\left(\frac{E_0}{B}\right)^2(1+2\alpha E)^2+\frac{\widehat{\theta}E_0}{B}(1+2\alpha E)+\frac{\widehat{p}_y^2}{2m_2}\end{aligned} \tag{A1.59}$$

Therefore, the electron dispersion relation assumes the form

$$E(1+\alpha E)=\left(n+\frac{1}{2}\right)\hbar\omega_{03}-\frac{E_0}{B}(1+2\alpha E)\hbar k_x+\frac{[\hbar k_y]^2}{2m_2}-\frac{m_1}{2}\left(\frac{E_0}{B}\right)^2(1+2\alpha E)^2 \tag{A1.60}$$

The EMM along x direction in this case is given by (A1.42) and the EMM along y direction is given by

$$m_y^*(\bar{E}_{FB},n,E_0,B)=m_2\left[1+2\alpha\bar{E}_{FB}+2m_1\alpha\left(\frac{E_0}{B}\right)^2(1+2\alpha\bar{E}_{FB})\right] \tag{A1.61}$$

From (A1.60), we can write

$$k_y(E)=\frac{\sqrt{2m_2}}{\hbar}[\bar{G}_1(n,E)+\bar{G}_2(E)k_x]^{1/2}$$

where

$$\bar{G}_1(n,E)\equiv\left[E(1+\alpha E)-\left(n+\frac{1}{2}\right)\hbar\omega_{03}+\frac{m_1}{2}\left(\frac{E_0}{B}\right)^2(1+2\alpha E)^2\right]$$

and $\bar{G}_2(E)\equiv\frac{E_0}{B}(1+2\alpha E)\hbar$.

Therefore, the integral $J(E)$ in this case assumes the form

$$\begin{aligned}J(E)=&\frac{\sqrt{2m_2}}{\hbar}\frac{2}{3}\left[[\bar{G}_2(E)]^{-1}[\bar{G}_1(n,E)+\bar{G}_2(E)x_h(E)]^{3/2}\right.\\&\left.-[\bar{G}_2(E)]^{-1}[\bar{G}_1(n,E)+\bar{G}_2(E)x_l(E)]^{3/2}\right]\end{aligned} \tag{A1.62}$$

where,

$$x_l(E)\equiv-\frac{E_o}{B}\frac{m_1}{\hbar}(1+2\alpha E)$$

and

$$x_h(E) \equiv \frac{|e|BL_z}{\hbar} + x_l(E).$$

The use of (A1.45) and (A1.62), the expression of the electron concentration for the Lax model in the present case can be written as

$$n_0 = \frac{2g_v\sqrt{2m_2}}{3L_z\pi^2\hbar}\sum_{n=0}^{n_{\max}}[T_{411}(n,\bar{E}_{FB}) + T_{412}(n,\bar{E}_{FB})] \quad \text{(A1.63)}$$

where

$$T_{411}(n,\bar{E}_{\rm FB}) \equiv [\bar{G}_2(\bar{E}_{\rm FB})]^{-1}\left[[\bar{G}_1(n,\bar{E}_{\rm FB}) + \bar{G}_2(\bar{E}_{\rm FB})x_h(\bar{E}_{\rm FB})]^{3/2}\right]$$

$$-[\bar{G}_1(n,\bar{E}_{\rm FB}) + \bar{G}_2(\bar{E}_{\rm FB})x_l(\bar{E}_{\rm FB})]^{3/2}],$$

$$x_h(\bar{E}_{\rm FB}) \equiv \left(\frac{|e|BL_z}{\hbar}\right) + x_l(\bar{E}_{\rm FB}),$$

$$x_l(\bar{E}_{\rm FB}) \equiv \frac{-E_o}{B}\frac{m_1}{\hbar}(1+2\alpha\bar{E}_{\rm FB}) \quad \text{and}$$

$$T_{412}(n,\bar{E}_{\rm FB}) \equiv \sum_{r=1}^{s} L(r)T_{411}(n,\bar{E}_{\rm FB}) \, .$$

(d) **The Parabolic Ellipsoidal model**

For this model, the electron dispersion relation for the present case assumes the form

$$E = \left(n+\frac{1}{2}\right)\hbar\omega_{03} - \frac{E_0}{B}\hbar k_x + \frac{\hbar^2k_y^2}{2m_2} - \frac{m_1}{2}\left(\frac{E_0}{B}\right)^2 \quad \text{(A1.64)}$$

The EMMs along y and x directions can respectively be expressed as

$$m_y^*(\bar{E}_{\rm FB}, n, E_0, B) = m_2 \quad \text{(A1.65)}$$

$$m_x^*(\bar{E}_{\rm FB}, n, E_0, B) = \left(\frac{B}{E_0}\right)^2\left[\bar{E}_{\rm FB} - \left(n+\frac{1}{2}\right)\hbar\omega_{03} + \frac{m_1}{2}\left(\frac{E_0}{B}\right)^2\right] \quad \text{(A1.66)}$$

For this case, the electron concentration assumes the form

$$n_0 = \frac{g_vB\sqrt{2\pi m_2}(k_BT)^{3/2}}{2E_0L_z\pi^2\hbar^2}\sum_{n=0}^{n_{\max}}[F_{1/2}(\bar{e}_1) - F_{1/2}(\bar{e}_2)] \quad \text{(A1.67)}$$

where,

$$\bar{e}_1 \equiv (k_BT)^{-1}(\bar{E}_{\rm FB} - \bar{e}_3), \quad \bar{e}_3 \equiv \left[(n+1/2)\hbar\omega_{03} + \frac{m_1}{2}\left(\frac{E_0}{B}\right)^2 - |e|E_0L_z\right],$$

$$\bar{e}_2 \equiv (k_BT)^{-1}(\bar{E}_{\rm FB} - \bar{e}_4) \quad \text{and} \quad \bar{e}_4 \equiv \bar{e}_3 + |e|E_0L_z.$$

A1.2.5 IV–VI Materials

The conduction electrons of IV–VI semiconductors obey the Cohen model of bismuth and the Eqs. (A1.53) and (A1.57) should be used for the electron concentration and the EEM in this case along with the appropriate change of energy band constants.

A1.2.6 Stressed Kane Type Semiconductors

The use of (1.98) can be written as

$$(E-\alpha_1)k_x^2+(E-\alpha_2)k_y^2+(E-\alpha_3)k_z^2=t_1E^3-t_2E^2+t_3E+t_4 \qquad \text{(A1.68)}$$

where

$$\alpha_1\equiv\left[E_g-C_1\varepsilon-(\bar{a}_0+C_1)\varepsilon+\frac{3}{2}\bar{b}_0\varepsilon_{xx}-\frac{\bar{b}_0}{2}\varepsilon+\left(\sqrt{3}/2\right)\varepsilon_{xy}\bar{d}_0\right],$$

$$\alpha_2\equiv\left[E_g-C_1\varepsilon-(\bar{a}_0+C_1)\varepsilon+\frac{3}{2}\bar{b}_0\varepsilon_{xx}-\frac{\bar{b}_0}{2}\varepsilon-\left(\sqrt{3}/2\right)\varepsilon_{xy}\bar{d}_0\right],$$

$$\alpha_3\equiv\left[E_g-C_1\varepsilon-(\bar{a}_0+C_1)\varepsilon+\frac{3}{2}\bar{b}_0\varepsilon_{zz}-\frac{\bar{b}_0}{2}\varepsilon\right],\quad t_1\equiv\left(3/2B_2^2\right),$$

$$t_2\equiv\left(1/2B_2^2\right)\left[6(E_g-C_1\varepsilon)+3C_1\varepsilon\right],$$

$$t_3\equiv\left(1/2B_2^2\right)\left[3(E_g-C_1\varepsilon)^2+6C_1\varepsilon(E_g-C_1\varepsilon)-2C_2^2\varepsilon_{xy}^2\right]$$

and $$t_4\equiv\left(1/2B_2^2\right)\left[-3C_1\varepsilon(E_g-C_1\varepsilon)^2+2C_2^2\varepsilon_{xy}^2\right].$$

In the presence of quantizing magnetic field B along z direction and the electric field along x-axis, from (A1.68) one obtains

$$\frac{\hat{p}_x^2}{2M_{\parallel}(E)}+\frac{(\hat{p}_y-|e|B\hat{x})^2}{2M_{\perp}(E)}+R(E)\hat{p}_z^2=\rho_5(E)+|e|E_0\hat{x}[\rho_5(E)]' \qquad \text{(A1.69)}$$

where

$$M_{\parallel}(E)\equiv\frac{1}{2P(E)},\quad P(E)\equiv\frac{1}{\hbar^2}(E-\alpha_1),\quad M_{\perp}(E)\equiv\frac{1}{2Q(E)},\quad Q(E)\equiv\frac{1}{\hbar^2}(E-\alpha_2),$$

$$R(E)\equiv\frac{1}{\hbar^2}(E-\alpha_3)\quad\text{and}\quad\rho_5(E)\equiv\left[t_1E^3-t_2E^2+t_3E+t_4\right].$$

Let us define the operator $\hat{\theta}$ as

$$\hat{\theta}=-\hat{p}_y+|e|B\hat{x}-\frac{M_{\perp}(E)E_0[\rho_5(E)]'}{B} \qquad \text{(A1.70)}$$

Combining Eqs. (A1.69) and (A1.70), we can write

$$\frac{\hat{p}_x^2}{2M_{\|}(E)}+\frac{\hat{\theta}^2}{2M_{\perp}(E)}+\frac{M_{\perp}(E)E_0^2\{[\rho_5(E)]'\}^2}{2B^2}+\hat{\theta}\frac{E_0[\rho_5(E)]'}{B}+R(E)\hat{p}_z^2$$
$$=\rho_5(E)+\frac{E_0[\rho_5(E)]'\hat{\theta}}{B}+\frac{E_0}{B}[\rho_5(E)]'\hat{p}_y+\frac{E_0^2}{B^2}M_{\perp}(E)\{[\rho_5(E)]'\}^2 \tag{A1.71}$$

Therefore, the electron dispersion relation in stressed Kane type semiconductors in the presence of cross-field configuration can be expressed as

$$\rho_5(E)=\left(n+{}^1\!/_2\right)\hbar\bar{\omega}(E)+R(E)[\hbar k_z(E)]^2-\frac{E_0}{B}[\rho_5(E)]'\hbar k_y(E)$$
$$-\left\{\frac{M_{\perp}(E)E_0^2\{[\rho_5(E)]'\}^2}{2B^2}\right\} \tag{A1.72}$$

where

$$\bar{\omega}(E)\equiv\frac{|e|B}{\sqrt{M_{\|}(E)M_{\perp}(E)}}$$

The use of (A1.72) leads to the expressions of EMMs along z and y directions as

$$m_z^*(\bar{E}_{\mathrm{FB}},n,E_0,B)=\frac{1}{2}[R(\bar{E}_{\mathrm{FB}})]^{-2}\left\{R(\bar{E}_{\mathrm{FB}})\left[[\rho_5(\bar{E}_{\mathrm{FB}})]'-\left(n+\frac{1}{2}\right)\hbar[\bar{\omega}(\bar{E}_{\mathrm{FB}})]'\right.\right.$$
$$\left.+\frac{M_{\perp}(\bar{E}_{\mathrm{FB}})E_0^2[\rho_5(\bar{E}_{\mathrm{FB}})]'[\rho_5(\bar{E}_{\mathrm{FB}})]''}{B^2}+\frac{\{M_{\perp}(\bar{E}_{\mathrm{FB}})\}'E_0^2\{[\rho_5(\bar{E}_{\mathrm{FB}})]'\}^2}{2B^2}\right] \tag{A1.73}$$
$$\left.-\{R(\bar{E}_{\mathrm{FB}})\}'\left[\rho_5(\bar{E}_{\mathrm{FB}})-\left(n+\frac{1}{2}\right)\hbar\bar{\omega}(\bar{E}_{\mathrm{FB}})+\frac{M_{\perp}(\bar{E}_{\mathrm{FB}})E_0^2\{[\rho_5(\bar{E}_{\mathrm{FB}})]'\}^2}{2B^2}\right]\right\}$$

and

$$m_y^*(\bar{E}_{\mathrm{FB}},n,E_0,B)=\left[\frac{B}{E_0}\right]^2\{[\rho_5(\bar{E}_{\mathrm{FB}})]'\}^{-3}$$
$$\times\left[\rho_5(\bar{E}_{\mathrm{FB}})-\left(n+\frac{1}{2}\right)\hbar\bar{\omega}(\bar{E}_{\mathrm{FB}})+\frac{M_{\perp}(\bar{E}_{\mathrm{FB}})E_0^2\{[\rho_5(\bar{E}_{\mathrm{FB}})]'\}^2}{2B^2}\right]$$
$$\times\left[[\rho_5(\bar{E}_{\mathrm{FB}})]'\left[[\rho_5(\bar{E}_{\mathrm{FB}})]'-\left(n+\frac{1}{2}\right)\hbar[\bar{\omega}(\bar{E}_{\mathrm{FB}})]'+\frac{M_{\perp}(\bar{E}_{\mathrm{FB}})E_0^2[\rho_5(\bar{E}_{\mathrm{FB}})]'[\rho_5(\bar{E}_{\mathrm{FB}})]''}{B^2}\right.\right.$$
$$\left.+\frac{[M_{\perp}(\bar{E}_{\mathrm{FB}})]'E_0^2\{[\rho_5(\bar{E}_{\mathrm{FB}})]'\}^2}{2B^2}\right]-[\rho_5(\bar{E}_{\mathrm{FB}})]''$$
$$\left.\times\left[\rho_5(\bar{E}_{\mathrm{FB}})-\left(n+\frac{1}{2}\right)\hbar\bar{\omega}(\bar{E}_{\mathrm{FB}})+\frac{M_{\perp}(\bar{E}_{\mathrm{FB}})E_0^2\{[\rho_5(\bar{E}_{\mathrm{FB}})]'\}^2}{2B^2}\right]\right] \tag{A1.74}$$

For this case,

$$x_l(E) = \frac{-M_\perp(E)E_0[\rho_5(\bar{E}_{\mathrm{FB}})]'}{B}, \quad x_h(E) = \frac{|e|BL_x}{\hbar} + x_l(E) \tag{A1.75}$$

The integral $I(E)$for stressed Kane type semiconductors in the presence of crossed electric and quantizing magnetic fields assumes the form

$$I(E) = \frac{1}{\hbar\sqrt{R(E)}}\left(\frac{B}{E_0\hbar[\rho_5(\bar{E}_{\mathrm{FB}})]'}\right)\int\limits_{x_l(E)}^{x_h(E)}\left[T_5(n,E) + \frac{E_0[\rho_5(\bar{E}_{\mathrm{FB}})]'}{B}\hbar k_y\right]^{1/2} \mathrm{d}k_y \tag{A1.76}$$

where

$$T_5(n,E) \equiv \left[\rho_5(E) - \left(n+{}^1\!/_2\right)\hbar\bar{\omega}(E) + \frac{M_\perp(E)E_0^2\{[\rho_5(\bar{E}_{\mathrm{FB}})]'\}^2}{2B^2}\right].$$

From (A1.76), we get,

$$\begin{aligned} I(E) = {} & \frac{1}{\hbar\sqrt{R(E)}}\frac{B}{E_0\hbar[\rho_5(\bar{E}_{\mathrm{FB}})]'} \\ & \times\frac{2}{3}\left[\left[T_5(n,E) + \frac{E_0}{B}[\rho_5(\bar{E}_{\mathrm{FB}})]'\hbar x_h(E)\right]^{3/2} - \left[T_5(n,E) + \frac{E_0}{B}[\rho_5(\bar{E}_{\mathrm{FB}})]'\hbar x_l(E)\right]^{3/2}\right] \end{aligned} \tag{A1.77}$$

Therefore, the electron concentration can be written as

$$n_0 = \frac{2B}{3L_x\pi^2\hbar^2 E_0}\sum_{n=0}^{n_{\max}}[T_{413}(n,\bar{E}_{\mathrm{FB}}) + T_{414}(n,\bar{E}_{\mathrm{FB}})] \tag{A1.78}$$

where

$$\begin{aligned} T_{413}(n,\bar{E}_{\mathrm{FB}}) \equiv {} & \left[\frac{1}{\left(\sqrt{R(\bar{E}_{\mathrm{FB}})}\right)(\rho_5(\bar{E}_{\mathrm{FB}}))'}\right]\left[\left[T_5(n,\bar{E}_{\mathrm{FB}}) + \frac{E_0}{B}[\rho_5(\bar{E}_{\mathrm{FB}})]'\hbar x_h(\bar{E}_{\mathrm{FB}})\right]^{3/2}\right. \\ & \left. - \left[T_5(n,\bar{E}_{\mathrm{FB}}) + \frac{E_0}{B}[\rho_5(\bar{E}_{\mathrm{FB}})]'\hbar x_l(\bar{E}_{\mathrm{FB}})\right]^{3/2}\right] \end{aligned}$$

and $T_{414}(n,\bar{E}_{\mathrm{FB}}) \equiv \sum_{r=1}^{s} L(r)T_{413}(n,\bar{E}_{\mathrm{FB}})$.

A1.2.7 Ultrathin Films of Nonlinear Optical Materials

The dispersion relation of the conduction electrons in ultrathin films of nonlinear optical material in the presence of cross-field configuration can be written as

$$U(E)=\left[\left((n+\frac{1}{2})\hbar\omega_{01}\right)+\left(\frac{[\hbar]^2}{2a(E)}\right)\left(\frac{\pi n_z}{d_z}\right)^2-\left(\frac{E_o\hbar k_y\rho(E)}{B}\right)-\left(\frac{M_\perp\rho^2(E)E_o^2}{2B^2}\right)\right] \tag{A1.79}$$

From (A1.79), the EEM along k_y direction can be expressed as

$$\begin{aligned}m_y^*(e_{fA1},E_0,n,n_z)=&\left[\frac{B}{E_0\rho(e_{fA1})}\right]^2\left[\left(n+\frac{1}{2}\right)\hbar\omega_{01}+\frac{\hbar^2}{2a(e_{fA1})}\left(\frac{n_z\pi}{d_z}\right)^2-\frac{M_\perp^2\rho^2(e_{fA1})E_0^2}{2B^2}\right.\\&-U'(e_{fA1})\Big]\left[\frac{\rho'(e_{fA1})}{\rho(e_{fA1})}\left\{U(e_{fA1})+\frac{M_\perp^2\rho^2(e_{fA1})E_0^2}{2B^2}-\left(n+\frac{1}{2}\right)\hbar\omega_{01}-\frac{\hbar^2}{2a(e_{fA1})}\left(\frac{n_z\pi}{d_z}\right)^2\right\}\right.\\&\left.-U'(e_{fA1})-\frac{M_\perp^2\rho(e_{fA1})\rho'(e_{fA1})E_0^2}{B^2}-\frac{\hbar^2a'(e_{fA1})}{2a^2(e_{fA1})}\left(\frac{n_z\pi}{d_z}\right)^2\right]\end{aligned} \tag{A1.80}$$

where e_{fA1} is the Fermi energy in the present case. It appears then that the EEM is a function of the Fermi energy, Landau quantum number, size quantum number and the electric field due to the presence of electric field only. The investigation of the EEM in this case requires an expression of electron statistics which, in turn, can be written as

$$n_0=\frac{g_veB}{\pi\hbar}\sum_{n=0}^{n_{max}}\sum_{n_z=1}^{n_{zmax}}F_{-1}(\eta_{A1}) \tag{A1.81}$$

where $\eta_{A1}=\dfrac{e_{fA1}-e_{A1}}{k_BT}$, e_{fA1} is the Fermi energy in this case and e_{fA1} is the lowest positive root of the equation.

$$U(e_{A1})=\left[\left(\left(n+\frac{1}{2}\right)\hbar\omega_{01}\right)+\left(\frac{[\hbar]^2}{2a(e_{A1})}\right)\left(\frac{\pi n_z}{d_z}\right)^2-\left(\frac{M_\perp\rho^2(e_{A1})E_o^2}{2B^2}\right)\right] \tag{A1.82}$$

A1.2.8 Special Cases for Ultrathin Films of III–V, Ternary, and Quaternary Materials

(a) Under the conditions $\delta=0$, $\Delta_{\|}=\Delta_\perp=\Delta$ and $m_{\|}^*=m_\perp^*=m_c$, (A1.79) assumes the form

$$I_{11}(E) = \left(n+\frac{1}{2}\right)\hbar\omega_0 + \frac{\hbar^2}{2m_c}\left(\frac{\pi n_z}{d_z}\right)^2 - \frac{E_0}{B}\hbar k_y\{I_{11}(E)\}' - \frac{m_c E_0^2\left[\{I_{11}(E)\}'\right]^2}{2B^2} \tag{A1.83}$$

From (A1.83), the EEM assumes the form

$$m_y^*(e_{fA1}, E_0, n, n_z) = \left[\frac{B}{E_0 I_{11}'(e_{fA1})}\right]^2 \left[\left\{\left\{I_{11}'(e_{fA1}) + \frac{m_c E_0^2}{B^2} I_{11}'(e_{fA1})[I_{11}(e_{fA1})]''\right\} - \left[\frac{I_{11}''(e_{fA1})}{I_{11}'(e_{fA1})}\right]\right.\right.$$
$$\left.\times\left\{I_{11}(e_{fA1}) - \left(n+\frac{1}{2}\right)\hbar\omega_0 - \frac{\hbar^2}{2m_c}\left(\frac{n_z\pi}{d_z}\right)^2 + \frac{m_c E_0^2}{2B^2}[I_{11}'(e_{fA1})]^2\right\}\right] \tag{A1.84}$$

where

$$I_{11}'(e_{fA1}) = I_{11}(e_{fA1})\left[\frac{1}{e_{fA1}} + \frac{1}{e_{fA1}+E_g} + \frac{1}{e_{fA1}+E_g+\Delta} - \frac{1}{e_{fA1}+E_g+\frac{2}{3}\Delta}\right]$$

and
$$I_{11}''(e_{fA1}) = \left[\frac{[I_{11}'(e_{fA1})]^2}{I_{11}(e_{fA1})} - I_{11}(e_{fA1})\left[\frac{1}{(e_{fA1})^2} + \frac{1}{(e_{fA1}+E_g)^2} + \frac{1}{(e_{fA1}+E_g+\Delta)^2} - \frac{1}{\left(e_{fA1}+E_g+\frac{2}{3}\Delta\right)^2}\right]\right]$$

The electron concentration is given by

$$n_0 = \frac{g_v eB}{\pi\hbar}\sum_{n=0}^{n_{\max}}\sum_{n_z=1}^{n_{z\max}} F_{-1}(\eta_{A2}) \tag{A1.85}$$

where $\eta_{A2} = \frac{e_{fA1}-e_{A2}}{k_B T}$ and e_{A2} is the lowest positive root of the equation

$$I_{11}(e_{A2}) = \left(n+\frac{1}{2}\right)\hbar\omega_0 + \frac{\hbar^2}{2m_c}\left(\frac{\pi n_z}{d_z}\right)^2 - \frac{m_c E_0^2\left[\{I_{11}(e_{A2})\}'\right]^2}{2B^2}$$

(b) **Two band model of Kane**

Under the condition $\Delta \gg E_g$, (A1.83) assumes the form

$$E(1+\alpha E) = \left(n+\frac{1}{2}\right)\hbar\omega_0 - \frac{E_0}{B}(1+2\alpha E)\hbar k_y$$
$$+\frac{\hbar^2}{2m_c}\left(\frac{\pi n_z}{d_z}\right)^2 - \frac{m_c}{2}\left(\frac{E_0}{B}\right)^2(1+2\alpha E)^2 \tag{A1.86}$$

$$m_y^*(e_{fA1}, E_0, n, n_z) = \left[\frac{B}{E_0(1+2\alpha e_{fA1})}\right]^2 \left[\left\{\left\{(1+2\alpha e_{fA1}).\left(1+\frac{m_c E_0^2}{B^2}\right)\right\}\right\} - \left[\frac{2\alpha}{(1+2\alpha e_{fA1})}\right]\right.$$
$$\left.\times\left\{(1+\alpha e_{fA1})e_{fA1} - \left(n+\frac{1}{2}\right)\hbar\omega_0 - \frac{\hbar^2}{2m_c}\left(\frac{n_z\pi}{d_z}\right)^2 + \frac{m_c E_0^2}{2B^2}\left[(1+2\alpha e_{fA1})\right]^2\right\}\right] \tag{A1.87}$$

The electron concentration is given by

$$n_0 = \frac{g_v eB}{\pi\hbar}\sum_{n=0}^{n_{\max}}\sum_{n_z=1}^{n_{z\max}} F_{-1}(\eta_{A3}) \tag{A1.88}$$

where $\eta_{A3} = \dfrac{e_{fA1} - e_{A3}}{k_B T}$ and e_{A3} is the lowest positive root of the equation

$$e_{A3}(1+\alpha e_{A3}) = \left(n+\frac{1}{2}\right)\hbar\omega_0 + \frac{\hbar^2}{2m_c}\left(\frac{\pi n_z}{d_z}\right)^2 - \frac{m_c}{2}\left(\frac{E_0}{B}\right)^2(1+2\alpha e_{A3})^2 \tag{A1.89}$$

(c) **Parabolic Energy Bands**

The dispersion relation, the EEM, and the electron statistics for this model under this condition $\alpha \to 0$ can be written as

$$E = \left(n+\frac{1}{2}\right)\hbar\omega_0 - \frac{E_0}{B}\hbar k_y + \frac{\hbar^2}{2m_c}\left(\frac{\pi n_z}{d_z}\right)^2 - \frac{m_c}{2}\left(\frac{E_0}{B}\right)^2 \tag{A1.90}$$

$$m_y^*(e_{fA1}, E_0, n, n_z) = \left(\frac{E_0}{B}\right)^{-2}\left[e_{fA1} - \left(n+\frac{1}{2}\right)\hbar\omega_0 - \frac{\hbar^2}{2m_c}\left(\frac{\pi n_z}{d_z}\right)^2 + \frac{m_c}{2}\left(\frac{E_0}{B}\right)^2\right] \tag{A1.91}$$

$$n_0 = \frac{g_v eB}{\pi\hbar}\sum_{n=0}^{n_{\max}}\sum_{n_z=1}^{n_{z\max}} F_{-1}(\eta_{A4}) \tag{A1.92}$$

where $\eta_{A4} = \frac{e_{fA1}-e_{A4}}{k_B T}$ and e_{A4} is given by

$$e_{A4} = \left(n+\frac{1}{2}\right)\hbar\omega_0 + \frac{\hbar^2}{2m_c}\left(\frac{\pi n_z}{d_z}\right)^2 - \frac{m_c}{2}\left(\frac{E_0}{B}\right)^2 \tag{A1.93}$$

A.1.2.9 Ultrathin Films of II–VI Materials

The dispersion relation in this case in ultrathin films of II–VI semiconductors can be written as

$$E = (\beta_1(n, E_0)) + \left(\frac{[\pi n_z/d_z]^2 \hbar^2}{2m_{\parallel}^*}\right) - \left(\frac{E_0}{B}\hbar k_y\right) \tag{A1.94}$$

The EEM can be expressed as

$$m_y^*(e_{fA1}, E_0, n, n_z) = \left(\frac{E_0}{B}\right)^{-2}\left[E_{fA1} - (\beta_1(n, E_0)) - \left(\frac{[\pi n_z/d_z]^2 \hbar^2}{2m_{\parallel}^*}\right)\right] \tag{A1.95}$$

The electron concentration per unit area in this case assumes the form

$$n_0 = \frac{g_v eB}{\pi\hbar}\sum_{n=0}^{n_{\max}}\sum_{n_z=1}^{n_{z\max}} F_{-1}(\eta_{A5}) \tag{A1.96}$$

where $\eta_{A5} = \dfrac{e_{fA1} - e_{A5}}{k_B T}$ and e_{A5} is determined from the equation

$$e_{A5} = (\beta_1(n, E_0)) + \left(\frac{[\pi n_z/d_z]^2 \hbar^2}{2m_{\parallel}^*}\right) \tag{A1.97}$$

A.1.2.10 The Formulation of EEM in Ultrathin Films of Bismuth

(a) **The McClure and Choi model**

The electron dispersion law in this case assumes the form

$$\begin{aligned} E(1+\alpha E) &= \left(n+\frac{1}{2}\right)\hbar\omega_{03} + \frac{[\hbar(\pi n_y/d_y)]^2}{2m_3} - \frac{E_0}{B}(1+2\alpha E)\hbar k_x - \frac{1}{2}m_1\left(\frac{E_0}{B}\right)^2(1+2\alpha E)^2 \\ &+ \frac{[\hbar(\pi n_y/d_y)]^2}{2m_2}\alpha E\left(1-\frac{m_2}{m_2'}\right) + \frac{\alpha[\hbar(\pi n_y/d_y)]^4}{4m_2 m_2'} - \frac{\alpha[\hbar(\pi n_y/d_y)]^2}{2m_2}\left(n+\frac{1}{2}\right)\hbar\omega_{03} \\ &- \frac{\alpha[\hbar(\pi n_y/d_y)]^2 m_1 E_0^2}{4m_2 B^2}(1+2\alpha E)^2 \end{aligned} \tag{A1.98}$$

The EEM can be expressed from (A1.98) as

$$m_x^*(e_{fA1},E_0,n,n_y)=\left(\frac{E_0}{B}\right)^{-2}(1+2\alpha e_{fA1})^{-1}\left[e_{fA1}(1+\alpha e_{fA1})-\left(n+\frac{1}{2}\right)\hbar\omega_{03}-\frac{[\hbar(\pi n_y/d_y)]^2}{2m_3}\right.$$
$$+\frac{1}{2}m_1\left(\frac{E_0}{B}\right)^2(1+2\alpha e_{fA1})^2-\frac{[\hbar(\pi n_y/d_y)]^2}{2m_2}\alpha e_{fA1}\left(1-\frac{m_2}{m_2'}\right)-\frac{\alpha[\hbar(\pi n_y/d_y)]^4}{4m_2m_2'}$$
$$\left.+\frac{\alpha[\hbar(\pi n_y/d_y)]^2}{2m_2}\left(n+\frac{1}{2}\right)\hbar\omega_{03}+\frac{\alpha[\hbar(\pi n_y/d_y)]^2m_1E_0^2}{4m_2B^2}(1+2\alpha e_{fA1})^2\right]$$
$$\times\left[\frac{-2\alpha}{(1+2\alpha e_{fA1})^2}\left[e_{fA1}(1+\alpha e_{fA1})-\left(n+\frac{1}{2}\right)\hbar\omega_{03}-\frac{[\hbar(\pi n_y/d_y)]^2}{2m_3}+\frac{1}{2}m_1\left(\frac{E_0}{B}\right)^2\right.\right.$$
$$\times(1+2\alpha e_{fA1})^2-\frac{[\hbar(\pi n_y/d_y)]^2}{2m_2}\alpha e_{fA1}\left(1-\frac{m_2}{m_2'}\right)$$
$$\left.-\frac{\alpha[\hbar(\pi n_y/d_y)]^4}{4m_2m_2'}+\frac{\alpha[\hbar(\pi n_y/d_y)]^2}{2m_2}\left(n+\frac{1}{2}\right)\hbar\omega_{03}+\frac{\alpha[\hbar(\pi n_y/d_y)]^2m_1E_0^2}{4m_2B^2}(1+2\alpha e_{fA1})^2\right]$$
$$\left.+1+2\alpha m_1\left(\frac{E_0}{B}\right)^2+\alpha^2\frac{[\hbar(\pi n_y/d_y)]^2}{m_2}m_1\left(\frac{E_0}{B}\right)^2-\alpha\frac{[\hbar(\pi n_y/d_y)]^2}{2m_2}\left(1-\frac{m_2}{m_2'}\right)(1+2\alpha e_{fA1})^{-1}\right]$$
$$\text{(A1.99)}$$

The electron concentration per unit area in this case assumes the form

$$n_0=\frac{g_veB}{\pi\hbar}\sum_{n=0}^{n_{\max}}\sum_{n_z=1}^{n_{z\max}}F_{-1}(\eta_{A6}) \quad \text{(A1.100)}$$

where $\eta_{A6}=\frac{e_{fA1}-e_{A6}}{k_BT}$ and e_{A6} is the lowest positive root of the equation

$$e_{A6}(1+\alpha e_{A6})=\left(n+\frac{1}{2}\right)\hbar\omega_{03}+\frac{[\hbar(\pi n_y/d_y)]^2}{2m_3}-\frac{1}{2}m_1\left(\frac{E_0}{B}\right)^2(1+2\alpha e_{A6})^2$$
$$+\frac{[\hbar(\pi n_y/d_y)]^2}{2m_2}\alpha e_{A6}\left(1-\frac{m_2}{m_2'}\right)+\frac{\alpha[\hbar(\pi n_y/d_y)]^4}{4m_2m_2'}-\frac{\alpha[\hbar(\pi n_y/d_y)]^2}{2m_2}$$
$$\times\left(n+\frac{1}{2}\right)\hbar\omega_{03}-\frac{\alpha[\hbar(\pi n_y/d_y)]^2m_1E_0^2}{4m_2B^2}(1+2\alpha e_{A6})^2 \quad \text{(A1.102)}$$

(b) **The Cohen Model**

The electron dispersion law in this case assumes the form

$$E(1+\alpha E)=\left(n+\frac{1}{2}\right)\hbar\omega_{03}-\frac{E_0}{B}(1+2\alpha E)\hbar k_x-\frac{1}{2}m_1\left(\frac{E_0}{B}\right)^2(1+2\alpha E)^2$$
$$+\frac{[\hbar(\pi n_y/d_y)]^2}{2m_2}\alpha E\left(1-\frac{m_2}{m_2'}\right)+\frac{\alpha[\hbar(\pi n_y/d_y)]^4}{4m_2m_2'}+\frac{[\hbar(\pi n_y/d_y)]^2}{2m_2}$$
$$\text{(A1.102)}$$

The EEM can be expressed from (A1.102) as

$$m_x^*(e_{fA1},E_0,n,n_y)=\left(\frac{E_0}{B}\right)^{-2}(1+2\alpha e_{fA1})^{-2}\Bigg[e_{fA1}(1+\alpha e_{fA1})-\left(n+\frac{1}{2}\right)\hbar\omega_{03}-\frac{[\hbar(\pi n_y/d_y)]^2}{2m_2}$$
$$+\frac{1}{2}m_1\left(\frac{E_0}{B}\right)^2(1+2\alpha e_{fA1})^2-\frac{[\hbar(\pi n_y/d_y)]^2}{2m_2}\alpha e_{fA1}\left(1-\frac{m_2}{m_2'}\right)-\frac{\alpha[\hbar(\pi n_y/d_y)]^4}{4m_2m_2'}\Bigg]$$
$$\times\Bigg[\Bigg[1+2\alpha e_{fA1}+\frac{\alpha[\hbar(\pi n_y/d_y)]^2}{2m_2'}-\frac{\alpha[\hbar(\pi n_y/d_y)]^2}{2m_2}\Bigg]-\frac{2\alpha}{(1+2\alpha e_{fA1})}\Bigg[e_{fA1}(1+\alpha e_{fA1})-\left(n+\frac{1}{2}\right)\hbar\omega_{03}$$
$$-\frac{[\hbar(\pi n_y/d_y)]^2}{2m_2}+\frac{1}{2}m_1\left(\frac{E_0}{B}\right)^2(1+2\alpha e_{fA1})^2-\frac{[\hbar(\pi n_y/d_y)]^2}{2m_2}\alpha e_{fA1}\left(1-\frac{m_2}{m_2'}\right)-\frac{\alpha[\hbar(\pi n_y/d_y)]^4}{4m_2m_2'}\Bigg]\Bigg] \quad (A1.103)$$

The electron concentration per unit area in this case assumes the form

$$n_0=\frac{g_veB}{\pi\hbar}\sum_{n=0}^{n_{max}}\sum_{n_z=1}^{n_{zmax}}F_{-1}(\eta_{A7}) \quad (A1.104)$$

where $\eta_{A7}=\frac{e_{fA1}-e_{A7}}{k_BT}$ and e_{A7} is the lowest positive root of the equation

$$e_{A7}(1+\alpha e_{A7})=\left(n+\frac{1}{2}\right)\hbar\omega_{03}-\frac{1}{2}m_1\left(\frac{E_0}{B}\right)^2(1+2\alpha e_{A7})^2$$
$$+\frac{[\hbar(\pi n_y/d_y)]^2}{2m_2}\alpha e_{A7}\left(1-\frac{m_2}{m_2'}\right)+\frac{\alpha[\hbar(\pi n_y/d_y)]^4}{4m_2m_2'}+\frac{[\hbar(\pi n_y/d_y)]^2}{2m_2} \quad (A1.04)$$

(c) **The Lax Model**

The electron dispersion law in this case assumes the form

$$E(1+\alpha E)=\left(n+\frac{1}{2}\right)\hbar\omega_{03}-\frac{E_0}{B}(1+2\alpha E)\hbar k_x$$
$$-\frac{1}{2}m_1\left(\frac{E_0}{B}\right)^2(1+2\alpha E)^2+\frac{[\hbar(\pi n_y/d_y)]^2}{2m_2} \quad (A1.106)$$

The EEM can be expressed from (A1.106) as

$$m_x^*(e_{fA1},E_0,n,n_y)=\left(\frac{E_0}{B}\right)^{-2}(1+2\alpha e_{fA1})^{-2}\Bigg[e_{fA1}(1+\alpha e_{fA1})-\left(n+\frac{1}{2}\right)\hbar\omega_{03}-\frac{[\hbar(\pi n_y/d_y)]^2}{2m_2}$$
$$+\frac{1}{2}m_1\left(\frac{E_0}{B}\right)^2(1+2\alpha e_{fA1})^2\Bigg]\Bigg[1+2\alpha e_{fA1}+2\alpha m_1\left(\frac{E_0}{B}\right)^2(1+2\alpha e_{fA1})\Bigg]$$
$$\times\Bigg[1+2\alpha e_{fA1}+2\alpha m_1\left(\frac{E_0}{B}\right)^2(1+2\alpha e_{fA1})-\frac{2\alpha}{(1+2\alpha e_{fA1})}\Bigg[e_{fA1}(1+\alpha e_{fA1})-\left(n+\frac{1}{2}\right)\hbar\omega_{03}$$
$$-\frac{[\hbar(\pi n_y/d_y)]^2}{2m_2}+\frac{1}{2}m_1\left(\frac{E_0}{B}\right)^2(1+2\alpha e_{fA1})^2\Bigg]\Bigg] \quad (A1.107)$$

The electron concentration per unit area in this case assumes the form where $\eta_{A8} = \frac{e_{fA1} - e_{A8}}{k_B T}$ and e_{A8} is the lowest positive root of the equation

$$e_A(1 + \alpha e_{A8}) = \left(n + \frac{1}{2}\right)\hbar\omega_{03} - \frac{1}{2}m_1\left(\frac{E_0}{B}\right)^2(1 + 2\alpha e_{A8})^2 + \frac{[\hbar(\pi n_y/d_y)]^2}{2m_2} \tag{A1.109}$$

(d) **The Parabolic Ellipsoidal Model**

The dispersion relation, the EEM and the electron statistics for this model under this condition $\alpha \to 0$ can be written as

$$E = \left(n + \frac{1}{2}\right)\hbar\omega_0 - \frac{E_0}{B}\hbar k_y + \frac{\hbar^2}{2m_2}\left(\frac{\pi n_z}{d_z}\right)^2 - \frac{m_1}{2}\left(\frac{E_0}{B}\right)^2 \tag{A1.110}$$

$$m_y^*(e_{fA1}, E_0, n, n_z) = \left(\frac{E_0}{B}\right)^{-2}\left[e_{fA1} - \left(n + \frac{1}{2}\right)\hbar\omega_0 - \frac{\hbar^2}{2m_2}\left(\frac{\pi n_z}{d_z}\right)^2 + \frac{m_1}{2}\left(\frac{E_0}{B}\right)^2\right] \tag{A1.111}$$

$$n_0 = \frac{g_v e B}{\pi\hbar}\sum_{n=0}^{n_{\max}}\sum_{n_z=1}^{n_{z\max}} F_{-1}(\eta_{A9}) \tag{A1.112}$$

where $\eta_{A9} = \frac{e_{fA1} - e_{A9}}{k_B T}$ and e_{A9} is given by

$$e_{A9} = \left(n + \frac{1}{2}\right)\hbar\omega_0 + \frac{\hbar^2}{2m_2}\left(\frac{\pi n_z}{d_z}\right)^2 - \frac{m_1}{2}\left(\frac{E_0}{B}\right)^2 \tag{A1.113}$$

A1.2.11 Ultrathin Films of IV–VI Materials

The carriers of IV–VI materials obey the Cohen model. Thus, all the results of the Cohen model as derived earlier are perfectly valid for IV–VI materials with the change in energy band constants.

A1.2.12 Ultrathin Films of Stressed Semiconductors

The electron dispersion relation in stressed Kane type semiconductors in the presence of cross-field configuration can be expressed as

$$\rho_5(E) = \left(n + {}^1\!/\!_2\right)\hbar\bar{\omega}(E) + R(E)[\hbar\pi n_z/d_z]^2 - \frac{E_0}{B}[\rho_5(E)]'\hbar k_y(E) - (M_\perp(E)E_0^2\{[\rho_5(E)]'\}^2/2B^2) \quad \text{(A1.114)}$$

The EEM can be written as

$$m_y^*(e_{fA1}, E_0, n, n_z) = \left[\frac{B}{E_0\rho_5'(e_{fA1})}\right]^2 \left[\{\{\rho_{51}'(e_{fA1}) - (n+\frac{1}{2})\hbar[\varpi(e_{fA1})]' - R'(e_{fA1})(\hbar\pi n_z/d_z)^2\} - \left[\frac{\rho_5''(e_{fA1})}{\rho_5'(e_{fA1})}\right](\rho_{51}(e_{fA1}) - (n+\frac{1}{2})\hbar[\varpi(e_{fA1})] - R(e_{fA1})(\hbar\pi n_z/d_z)^2)\}\right] \times\left(\rho_{51}(e_{fA1}) - \left(n+\frac{1}{2}\right)\hbar[\varpi(e_{fA1})] - R(e_{fA1})(\hbar\pi n_z/d_z)^2\right) \quad \text{(A1.115)}$$

The surface electron concentration is given by

$$n_0 = \frac{g_v eB}{\pi\hbar}\sum_{n=0}^{n_{\max}}\sum_{n_z=1}^{n_{z\max}} F_{-1}(\eta_{A9}) \quad \text{(A1.116)}$$

where $\eta_{A10} = \dfrac{e_{fA1} - e_{A10}}{k_B T}$ and e_{A10} is the lowest positive root of the equation

$$\rho_5(e_{A10}) = \left(n + {}^1\!/\!_2\right)\hbar\bar{\omega}(e_{A10}) + R(e_{A10})[\hbar\pi n_z/d_z]^2 - (M_\perp(e_{A10})E_0^2\{[\rho_5(e_{A10})]'\}^2/2B^2) \quad \text{(A1.117)}$$

A1.3 Open Research Problems

R.A1.1 Investigate the EEM in the presence of an arbitrarily oriented quantizing magnetic and crossed electric fields in tetragonal semiconductors by including broadening and the electron spin. Study all the special cases for III–V, ternary, and quaternary materials in this context.

R.A1.2 Investigate the EEMs for all models of Bi, IV–VI, II–VI, and stressed Kane type compounds in the presence of an arbitrarily oriented quantizing magnetic and crossed electric fields by including broadening and electron spin.

R.A1.3 Investigate the EEM for all the materials as stated in R.2.1 of Chap. 2 in the presence of an arbitrarily oriented quantizing magnetic and crossed electric fields by including broadening and electron spin.

Appendix B
The EEM in Heavily Doped Compound Semiconductors

B1.1 Introduction

It is well known that the band tails are being formed in the forbidden zone of heavily doped semiconductors and can be explained by the overlapping of the impurity band with the conduction and valence bands [30]. Kane [31, 32] and Bonch Bruevich [33] have independently derived the theory of band tailing for semiconductors having unperturbed parabolic energy bands. Kane's model [31, 32] was used to explain the experimental results on tunneling [34] and the optical absorption edges [35, 36] in this context. Halperin and Lax [37] developed a model for band tailing applicable only to the deep tailing states. Although Kane's concept is often used in the literature for the investigation of band tailing [38, 39], it may be noted that this model [31, 32, 40] suffers from serious assumptions in the sense that the local impurity potential is assumed to be small and slowly varying in space coordinates [39]. In this respect, the local impurity potential may be assumed to be a constant. In order to avoid these approximations, we have developed in this chapter the electron energy spectra for heavily doped semiconductors for studying the EEM based on the concept of the variation of the kinetic energy [30, 39] of the electron with the local point in space coordinates. This kinetic energy is then averaged over the entire region of variation using a Gaussian type potential energy. On the basis of the *E*–*k* dispersion relation, we have obtained the electron statistics for different heavily doped materials for the purpose of numerical computation of the respective EEMs. It may be noted that, a more general treatment of many-body theory for the density-of-states of heavily doped semiconductors merges with one-electron theory under macroscopic conditions [30]. Also, the experimental results for the Fermi energy and others are the average effect of this macroscopic case. So, the present treatment of the one-electron system is more applicable to the

S. Bhattacharya and K. P. Ghatak, *Effective Electron Mass in Low-Dimensional Semiconductors*, Springer Series in Materials Science 167,
DOI: 10.1007/978-3-642-31248-9, © Springer-Verlag Berlin Heidelberg 2013

experimental point-of-view and it is also easy to understand the overall effect in such a case [41]. In a heavily doped semiconductor, each impurity atom is surrounded by the electrons, assuming a regular distribution of atoms, and it is screened independently [38, 40, 42].

The interaction energy between electrons and impurities is known as the impurity screening potential. This energy is determined by the inter-impurity distance and the screening radius, which is known as the screening length. The screening radius grows with the electron concentration and the effective mass. Furthermore, these entities are important for heavily doped materials in characterizing the semiconductor properties [43, 44] and the devices [38, 45]. The works on Fermi energy and the screening length in an n-type GaAs have already been initiated in the literature [46–48], based on Kane's model. Incidentally, the limitations of Kane's model [39], as mentioned above, are also present in their studies.

At this point, it may be noted that many band tail models are proposed using the Gaussian distribution of the impurity potential variation [31, 32, 39]. In this chapter, we have used the Gaussian band tails to obtain the exact *E-k* dispersion relations for heavily doped tetragonal, III–V, II–VI, IV–VI, and stressed Kane type compounds. Our method is not at al related with the density-of-states (DOS) technique as used in the aforementioned works. From the electron energy spectrum, one can obtain the DOS but the DOS technique, as used in the literature cannot provide the *E–k* dispersion relation. Therefore, our study is more fundamental than those in the existing literature, because the Boltzmann transport equation, which controls the study of the charge transport properties of the semiconductor devices, can be solved if and only if the *E–k* dispersion relation is known. We wish to note that the Gaussian function for the impurity potential distribution has been used by many authors. It has been widely used since 1963 when Kane first proposed it. We will also use the Gaussian distribution for the present study.

In Sect. B1.2.1, of the theoretical background, the EEM in heavily doped tetragonal materials has been investigated. The Sect. B1.2.2 contains the results for heavily doped III–V, ternary, and quaternary compounds whose undoped conduction electrons obeys the three and the two band models of Kane together with parabolic energy bands and they form the special cases of Sect. B1.2.1. The Sects. B1.2.3, B1.2.4 and B1.2.5 contain the study of the EEM for heavily doped II–VI, IV–VI and stressed Kane type semiconductors, respectively. The last Sect. B1.3 contains the open research problems.

B1.2 Theoretical Background

B1.2.1 Study of the EEM in Heavily Doped Tetragonal Materials Forming Gaussian Band Tails

The generalized unperturbed electron energy spectrum for the bulk specimens of the tetragonal materials in the absence of any doping can be expressed following (1.2) of Chap. 1 as

$$\frac{\hbar^2 k_z^2}{2m_{\|}^*}+\left(\frac{b_{\|}c_{\perp}}{b_{\perp}c_{\|}}\right)\frac{\hbar^2 k_s^2}{2m_{\perp}^*}=\left\{\frac{E(\alpha E+1)(b_{\|}E+1)}{(c_{\|}E+1)}+\frac{\alpha b_{\|}}{c_{\|}}\left[\delta E+\frac{2}{9}\left(\Delta_{\|}^2-\Delta_{\perp}^2\right)\right]\right.$$
$$\left.-\left(\frac{2}{9}\right)\frac{\alpha b_{\|}}{c_{\|}}\frac{\left(\Delta_{\|}^2-\Delta_{\perp}^2\right)}{(c_{\|}E+1)}\right\}-\left(\frac{\hbar^2 k_s^2}{2m_{\perp}^*}\right)\left\{\left(\frac{b_{\|}c_{\perp}}{b_{\perp}c_{\|}}\right)\left[\left(\frac{\delta}{2}+\frac{\Delta_{\|}^2-\Delta_{\perp}^2}{6\Delta_{\|}}\right)\frac{\alpha_{\|}}{\alpha_{\|}E+1}\right.\right.$$
$$\left.\left.+\left(\frac{\delta}{2}-\left\{\frac{\Delta_{\|}^2-\Delta_{\perp}^2}{6\Delta_{\|}}\right\}\right)\frac{c_{\|}}{c_{\|}E+1}\right]\right\} \tag{B1.1}$$

where

$$b_{\|}\equiv 1/(E_g+\Delta_{\|}),\quad c_{\perp}\equiv 1\bigg/\left(E_g+\frac{2}{3}\Delta_{\perp}\right),\quad b_{\perp}\equiv 1/(E_g+\Delta_{\perp}),$$
$$c_{\|}\equiv 1\bigg/\left(E_g+\frac{2}{3}\Delta_{\|}\right)\quad\text{and}\quad \alpha\equiv 1/E_g.$$

The Gaussian distribution *F(V)* of the impurity potential is given by [31, 32]

$$F(V)=\left(\pi\eta_g^2\right)^{-1/2}\exp\left(-V^2\big/\eta_g^2\right) \tag{B1.2}$$

where, η_g is the impurity scattering potential. It appears from (B1.2) that the variance parameter η_g is not equal to zero, but the mean value is zero. Further, the impurities are assumed to be uncorrelated and the band mixing effect has been neglected in this simplified theoretical formalism.

We have to average the Kinetic energy in the order to obtain the E-k dispersion relation in tetragonal materials including the band tailing effect. Using the (B1.1) and (B1.2), we get

$$
\begin{aligned}
&\left[\frac{\hbar^2 k_z^2}{2m_{\|}^*}\int\limits_{-\infty}^{E} F(V)\mathrm{d}V\right]+\left[\left(\frac{b_{\|}}{b_{\perp}}\frac{c_{\perp}}{c_{\|}}\right)\frac{\hbar^2 k_s^2}{2m_{\perp}^*}\int\limits_{-\infty}^{E} F(V)\mathrm{d}V\right]\\
&=\left\{\int\limits_{-\infty}^{E}\frac{(E-V)[\alpha(E-V)+1]\left[b_{\|}(E-V)+1\right]}{\left[c_{\|}(E-V)+1\right]}F(V)\mathrm{d}V\right.\\
&\quad+\frac{\alpha b_{\|}}{c_{\|}}\left[\delta\int\limits_{-\infty}^{E}(E-V)F(V)\mathrm{d}V+\frac{2}{9}\left(\Delta_{\|}^2-\Delta_{\perp}^2\right)\int\limits_{-\infty}^{E}F(V)\mathrm{d}V\right]\\
&\quad\left.-\left(\frac{2}{9}\right)\frac{\alpha b_{\|}}{c_{\|}}\left(\Delta_{\|}^2-\Delta_{\perp}^2\right)\int\limits_{-\infty}^{E}\frac{F(V)\mathrm{d}V}{\left[c_{\|}(E-V)+1\right]}\right\}\\
&\quad-\left(\frac{\hbar^2 k_s^2}{2m_{\perp}^*}\right)\left\{\left(\frac{b_{\|}}{b_{\perp}}\frac{c_{\perp}}{c_{\|}}\right)\left[\left(\frac{\delta}{2}+\frac{\Delta_{\|}^2-\Delta_{\perp}^2}{6\Delta_{\|}}\right)\alpha\int\limits_{-\infty}^{E}\frac{F(V)\mathrm{d}V}{[\alpha(E-V)+1]}\right.\right.\\
&\quad\left.\left.+\left(\frac{\delta}{2}-\frac{\Delta_{\|}^2-\Delta_{\perp}^2}{6\Delta_{\|}}\right)c_{\|}\int\limits_{-\infty}^{E}\frac{F(V)\mathrm{d}V}{\left[c_{\|}(E-V)+1\right]}\right]\right\}
\end{aligned}
\tag{B1.3}
$$

(B1.3) can be rewritten as [49–53]

$$
\begin{aligned}
&\frac{\hbar^2 k_z^2}{2m_{\|}^*}I(1)+\left(\frac{b_{\|}}{b_{\perp}}\frac{c_{\perp}}{c_{\|}}\right)\frac{\hbar^2 k_s^2}{2m_{\perp}^*}I(1)\\
&=\left\{I_3\left(c_{\|}\right)+\frac{\alpha b_{\|}}{c_{\|}}\left[\delta I(4)+\frac{2}{9}(\Delta_{\|}^2-\Delta_{\perp}^2)I(1)\right]-\left(\frac{2}{9}\right)\frac{\alpha b_{\|}}{c_{\|}}\left(\Delta_{\|}^2-\Delta_{\perp}^2\right)I_6\left(c_{\|}\right)\right\}\\
&\quad-\left(\frac{\hbar^2 k_s^2}{2m_{\perp}^*}\right)\left\{\left(\frac{b_{\|}}{b_{\perp}}\frac{c_{\perp}}{c_{\|}}\right)\left[\left(\frac{\delta}{2}+\frac{\Delta_{\|}^2-\Delta_{\perp}^2}{6\Delta_{\|}}\right)\alpha I(\alpha)+\left(\frac{\delta}{2}-\left\{\frac{\Delta_{\|}^2-\Delta_{\perp}^2}{6\Delta_{\|}}\right\}\right)c_{\|}I\left(c_{\|}\right)\right]\right\}
\end{aligned}
\tag{B1.4}
$$

where

$$
I(1)\equiv\int\limits_{-\infty}^{E}F(V)\mathrm{d}V \tag{B1.5}
$$

$$
I_3\left(c_{\|}\right)\equiv\int\limits_{-\infty}^{E}\frac{(E-V)[\alpha(E-V)+1]\left[b_{\|}(E-V)+1\right]}{\left[c_{\|}(E-V)+1\right]}F(V)\mathrm{d}V \tag{B1.6}
$$

$$
I(4)\equiv\int\limits_{-\infty}^{E}(E-V)F(V)\mathrm{d}V \tag{B1.7}
$$

$$I(\alpha) \equiv \int_{-\infty}^{E} \frac{F(V)\mathrm{d}V}{[\alpha(E-V)+1]} \tag{B1.8}$$

Let us substitute $E - V \equiv x$ and $x/\eta_g \equiv t_0$, we get from (B1.5)

$$I(1) = \left(\exp(-E^2/\eta_g^2)/\sqrt{\pi}\right) \int_0^{\infty} \exp\left[-t_0^2 + (2Et_0/\eta_g)\right]\mathrm{d}t_0$$

Thus,

$$I(1) = \left[\frac{1 + Erf(E/\eta_g)}{2}\right] \tag{B1.9}$$

From (B1.7), one can write

$$I(4) = (1/\eta_g\sqrt{\pi}) \int_{-\infty}^{E} (E-V)\exp(-V^2/\eta_g^2)\mathrm{d}V$$

$$= \frac{E}{2}[1 + Erf(E/\eta_g)] - \left\{\frac{1}{\sqrt{\pi\eta_g^2}} \int_{-\infty}^{E} V\exp\left(-E^2/\eta_g^2\right)\mathrm{d}V\right\}$$

After computing this simple integration, one obtains thus,

$$I(4) = \eta_g \exp\left(-E^2/\eta_g^2\right)(2\sqrt{\pi})^{-1} + \frac{E}{2}\left(1 + Erf(E/\eta_g)\right) = \gamma_0(E, \eta_g) \tag{B1.10}$$

From (B1.8), we can write

$$I(\alpha) = \frac{1}{\sqrt{\pi\eta_g^2}} \int_{-\infty}^{E} \frac{\exp\left(-V^2/\eta_g^2\right)\mathrm{d}V}{[\alpha(E-V)+1]} \tag{B1.11}$$

when, $V \to \pm\infty$, $\dfrac{1}{[\alpha(E-V)+1]} \to 0$ and $\exp\left(-V^2/\eta_g^2\right) \to 0$; therefore, using (B1.11) one can write

$$I(\alpha) = \frac{1}{\sqrt{\pi\eta_g^2}} \int_{-\infty}^{+\infty} \frac{\exp\left(-V^2/\eta_g^2\right)\mathrm{d}V}{[\alpha E + 1 - \alpha V]} \tag{B1.12}$$

The (B1.12) can be expressed as

$$I(\alpha) = (1/\alpha\eta_g\sqrt{\pi}) \int_{-\infty}^{\infty} \exp(-t^2)(u-t)^{-1}\mathrm{d}t \tag{B1.13}$$

where, $\frac{V}{\eta_g} \equiv t$ and $u \equiv \left(\frac{1+\alpha E}{\alpha \eta}\right)$.

It is well known that [54, 55]

$$W(Z) = (i/\pi) \int_{-\infty}^{\infty} (Z-t)^{-1} \exp(-t^2) dt \quad \text{(B1.14)}$$

in which $i = \sqrt{-1}$ and Z is, in general, a complex number. We also know [54, 55],

$$W(Z) = \exp(-Z^2) Erfc(-iZ) \quad \text{(B1.15)}$$

where

$$Erfc(Z) \equiv 1 - Erf(Z)$$

Thus

$$Erfc(-iu) = 1 - Erf(-iu)$$

Since,

$$Erf(-iu) = -Erf(iu)$$

Therefore,

$$Erfc(-iu) = 1 + Erf(iu).$$

Thus,

$$I(\alpha) = \left[-i\sqrt{\pi}/\alpha\eta_g\right] \exp(-u^2)[1 + Erf(iu)] \quad \text{(B1.16)}$$

We also know that [54]

$$Erf(x+iy) = Erf(x) + \left(\frac{e^{-x^2}}{2\pi x}\right)\left[(1-\cos(2xy)) + i\sin(2xy) + \frac{2}{\pi} e^{-x^2} \sum_{p=1}^{\infty} \frac{\exp(-p^2/4)}{(p^2+4x^2)}\right] \times \left[f_p(x,y) + i g_p(x,y) + \varepsilon(x,y)\right] \quad \text{(B1.17)}$$

where

$$f_p(x,y) \equiv [2x - 2x\cosh(py)\cos(2xy) + p\sinh(py)\sin(2xy)],$$
$$g_p(x,y) \equiv [2x\cosh(py)\sin(2xy) + p\sinh(py)\cos(2xy)],$$
$$|\varepsilon(x,y)| \approx 10^{-16}|Erf(x+iy)|$$

Substituting $x = 0$ and $y = u$ in (B1.17), one obtains,

$$Erf(iu) = \left(\frac{2i}{\pi}\right) \sum_{p=1}^{\infty} \left\{\frac{\exp(-p^2/4)}{p} \sinh(pu)\right\} \quad \text{(B1.18)}$$

Therefore, one can write

$$I(\alpha) = C_{21}(\alpha, E, \eta_g) - iD_{21}(\alpha, E, \eta_g) \tag{B1.19}$$

where

$$C_{21}(\alpha, E, \eta_g) \equiv \left[\frac{2}{\alpha\eta_g\sqrt{\pi}}\right]\exp(-u^2)\left[\sum_{p=1}^{\infty}\left\{\frac{\exp(-p^2/4)}{p}\sinh(pu)\right\}\right]$$

and

$$D_{21}(\alpha, E, \eta_g) \equiv \left[\frac{\sqrt{\pi}}{\alpha\eta_g}\exp(-u^2)\right].$$

The (B1.19) has both real and imaginary parts, and therefore $I(\alpha)$ is complex, which can also be prove by using the method of analytic continuation. The integral $I_3(c_{||})$ in (B1.6) can be written as

$$I_3(c_{||}) = \left(\frac{\alpha b_{||}}{c_{||}}\right)I(5) + \left(\frac{\alpha c_{||} + b_{||}c_{||} - \alpha b_{||}}{c_{||}^2}\right)I(4) + \frac{1}{c_{||}}\left(1 - \frac{\alpha}{c_{||}}\right)\left(1 - \frac{b_{||}}{c_{||}}\right)I(1)$$
$$- \left\{\frac{1}{c_{||}}\left(1 - \frac{\alpha}{c_{||}}\right)\left(1 - \frac{b_{||}}{c_{||}}\right)I(c_{||})\right\} \tag{B.1.20}$$

where

$$I(5) \equiv \int_{-\infty}^{E}(E - V)^2 F(V)\mathrm{d}V \tag{B1.21}$$

From (B1.21), one can write

$$I(5) = \frac{1}{\sqrt{\pi\eta_g^2}}\left[E^2\int_{-\infty}^{E}\exp\left(\frac{-V^2}{\eta_g^2}\right)\mathrm{d}V - 2E\int_{-\infty}^{E}V\exp\left(\frac{-V^2}{\eta_g^2}\right)\mathrm{d}V + \int_{-\infty}^{E}V^2\exp\left(\frac{-V^2}{\eta_g^2}\right)\mathrm{d}V\right]$$

The evaluations of the component integrals lead us to write

$$I(5) = \frac{\eta_g E}{2\sqrt{\pi}}\exp\left(\frac{-E^2}{\eta_g^2}\right) + \frac{1}{4}\left(\eta_g^2 + 2E^2\right)\left[1 + Erf\left(\frac{E}{\eta_g}\right)\right] = \theta_0(E, \eta_g) \tag{B1.22}$$

Thus, combining the aforementioned equations, $I_3(c_{||})$can be expressed as

$$I_3(c_{||}) = A_{21}(E, \eta_g) + iB_{21}(E, \eta_g) \tag{B1.23}$$

where

$$A_{21}(E,\eta) \equiv \left[\frac{\alpha b_{||}}{c_{||}}\left[\frac{\eta_g E}{2\sqrt{\pi}}\exp\left(\frac{-E^2}{\eta_g^2}\right)+\frac{1}{4}\left(\eta_g^2+2E^2\right)\left\{1+Erf\left(\frac{E}{\eta_g}\right)\right\}\right]\right.$$
$$+\left[\frac{\alpha c_{||}+b_{||}c_{||}-\alpha b_{||}}{c_{||}^2}\right]\left\{\frac{E}{2}[1+Erf(E/\eta)]+\frac{\eta_g\exp\left(-E^2/\eta_g^2\right)}{2\sqrt{\pi}}\right\}$$
$$+\frac{1}{c_{||}}\left(1-\frac{\alpha}{c_{||}}\right)\left(1-\frac{b_{||}}{c_{||}}\right)\frac{1}{2}[1+Erf(E/\eta)]$$
$$\left.-\left\{\frac{2}{c_{||}^2\eta_g\sqrt{\pi}}\left(1-\frac{\alpha}{c_{||}}\right)\left(1-\frac{b_{||}}{c_{||}}\right)\exp(-u_1^2)\right\}\left[\sum_{p=1}^{\infty}\left\{\frac{\exp(-p^2/4)}{p}\sinh(pu_1)\right\}\right]\right],$$

$$u_1 \equiv \left[\frac{1+c_{||}E}{c_{||}\eta_g}\right] \quad \text{and} \quad B_{21}(E,\eta_g) \equiv \frac{\sqrt{\pi}}{c_{||}^2\eta_g}\left(1-\frac{\alpha}{c_{||}}\right)\left(1-\frac{b_{||}}{c_{||}}\right)\exp(-u_1^2).$$

Therefore, the combination of all the appropriate equations together with the algebraic manipulations lead to the dispersion relation of the conduction electrons of heavily doped tetragonal materials forming Gaussian band tails as

$$\frac{\hbar^2 k_z^2}{2m_{||}^* T_{21}(E,\eta_g)}+\frac{\hbar^2 k_s^2}{2m_{\perp}^* T_{22}(E,\eta_g)}=1 \tag{B1.24}$$

where $T_{21}(E,\eta_g)$ and $T_{22}(E,\eta_g)$ have both real and complex parts and they are given by

$$T_{21}(E,\eta_g) \equiv \left[T_{27}(E,\eta_g)+iT_{28}(E,\eta_g)\right], \quad T_{27}(E,\eta_g) \equiv \left[\frac{T_{23}(E,\eta_g)}{T_5(E,\eta_g)}\right],$$

$$T_{23}(E,\eta_g) \equiv \left[A_{21}(E,\eta_g)+\frac{\alpha b_{||}}{c_{||}}\left[\delta\gamma_0(E,\eta_g)+\frac{1}{9}(\Delta_{||}^2-\Delta_{\perp}^2)[1+Erf(E/\eta_g)]\right]\right.$$
$$\left.-\left\{\frac{2}{9}\left(\frac{\alpha b_{||}}{c_{||}}\right)(\Delta_{||}^2-\Delta_{\perp}^2)G_{21}(c_{||},E,\eta_g)\right\}\right],$$

$$G_{21}(E,\eta_g) \equiv \frac{2}{c_{||}\eta_g\sqrt{\pi}}\exp(-u_1^2)\sum_{p=1}^{\infty}\left\{\frac{\exp(-p^2/4)}{p}\sinh(pu_1)\right\},$$

$$T_5(E,\eta_g) \equiv \frac{1}{2}\left[1+Erf(E/\eta_g)\right],$$

$$T_{28}(E,\eta_g) \equiv \left[\frac{T_{24}(E,\eta_g)}{T_5(E,\eta_g)}\right],$$

$$T_{24}(E,\eta_g) \equiv \left[B_{21}(E,\eta_g)+\frac{2}{9}\frac{\alpha b_{||}}{c_{||}}\left(\Delta_{||}^2-\Delta_{\perp}^2\right)H_{21}(c_{||},E,\eta_g)\right],$$

$$H_{21}(c_{||}, E, \eta_g) \equiv \left[\frac{\sqrt{\pi}}{\eta_g c_{||}} \exp(-u_1^2)\right],$$

$$T_{22}(E, \eta_g) \equiv [T_{29}(E, \eta_g) + iT_{30}(E, \eta_g)],$$

$$T_{29}(E, \eta_g) \equiv \frac{T_{23}(E, \eta_g)T_{25}(E, \eta_g) - T_{24}(E, \eta_g)T_{26}(E, \eta_g)}{\left[(T_{25}(E, \eta_g))^2 + (T_{26}(E, \eta_g))^2\right]},$$

$$T_{25}(E, \eta_g) \equiv \left[\left(\frac{b_{||}\, c_{\perp}}{b_{\perp}\, c_{||}}\right)\frac{1}{2}\left[1 + Erf\left(\frac{E}{\eta_g}\right)\right] + \left(\frac{b_{||}\, c_{\perp}}{b_{\perp}\, c_{||}}\right)\left(\frac{\delta}{2} + \left[\frac{\Delta_{||}^2 - \Delta_{\perp}^2}{6\Delta_{||}}\right]\right)\right.$$
$$\left.\times\, \alpha_{||} C_{21}(\alpha_{||}, E, \eta_g) + \left(\frac{b_{||} c_{\perp}}{b_{\perp}}\right)\left(\frac{\delta}{2} - \left[\frac{\Delta_{||}^2 - \Delta_{\perp}^2}{6\Delta_{||}}\right]\right) G_{21}(\alpha_{||}, E, \eta_g)\right],$$

$$C_{21}(\alpha, E, \eta_g) \equiv \left[\frac{2}{\alpha\sqrt{\pi}\eta_g} \exp(-u^2)\left[\sum_{p=1}^{\infty} \frac{\exp(-p^2/4)}{p} \sinh(pu)\right]\right],$$

$$T_{26}(E, \eta_g) \equiv \left(\frac{b_{||}\, c_{\perp}}{b_{\perp}\, c_{||}}\right)\left(\frac{\delta}{2} - \frac{\Delta_{||}^2 - \Delta_{\perp}^2}{6\Delta_{||}}\right)\alpha D_{21}(\alpha, E, \eta_g)$$
$$+ \frac{b_{||} c_{\perp}}{b_{\perp}}\left(\frac{\delta}{2} - \frac{\Delta_{||}^2 - \Delta_{\perp}^2}{6\Delta_{||}}\right) H_{21}(c_{||}, E, \eta_g),$$

and

$$T_{30}(E, \eta_g) \equiv \frac{T_{24}(E, \eta_g)T_{25}(E, \eta_g) + T_{23}(E, \eta_g)T_{26}(E, \eta_g)}{\left[(T_{25}(E, \eta_g))^2 + (T_{26}(E, \eta_g))^2\right]}.$$

From (B1.24), it appears that the energy spectrum in heavily doped tetragonal semiconductors is complex. The complex nature of the electron dispersion law in heavily doped semiconductors occurs from the existence of the essential poles in the corresponding undoped electron energy spectrum. It may be noted that the complex band structures have already been studied for bulk semiconductors and superlattices without heavy doping [56, 57] and bears no relationship with the complex electron dispersion law as indicated by (B1.24). The physical picture behind the formulation of the complex energy spectrum in heavily doped tetragonal semiconductors is the interaction of the impurity atoms in the tails with the splitting constants of the valance bands. More is the interaction causes more prominence of the complex part than the other case. When there is no heavy doping, $\eta_g \to 0$, and there is no interaction of the impurity atoms in the tails with the spin–orbit constants. As a result, there exist no complex energy spectrum and (B1.24) gets converted into (B1.2) when $\eta_g \to 0$. Besides, the complex spectra are not related to same evanescent modes in the band tails and the conduction bands.

The transverse and the longitudinal EEMs at the Fermi energy (E_{F_h}) of heavily doped tetragonal materials can be expressed respectively as

$$m_{\perp}^{*}(E_{F_h},\eta_g)=m_{\perp}^{*}\{T_{29}(E,\eta_g)\}'\Big|_{E=E_{F_h}} \tag{B1.25}$$

and

$$m_{||}^{*}(E_{F_h},\eta_g)=m_{||}^{*}\{T_{27}(E,\eta_g)\}'\Big|_{E=E_{F_h}} \tag{B1.26}$$

In the absence of band tailing effects $\eta_g \to 0$ and we get

$$m_{\perp}^{*}(E_F,0)=\frac{\hbar^2}{2}\left[\frac{\psi_2(E)\{\psi_1(E)\}'-\psi_1(E)\{\psi_2(E)\}'}{\{\psi_2(E)\}^2}\right]\Bigg|_{E=E_F} \tag{B1.27}$$

and

$$m_{||}^{*}(E_F,0)=\frac{\hbar^2}{2}\left[\frac{\psi_3(E)\{\psi_1(E)\}'-\{\psi_1(E)\}\{\psi_3(E)\}'}{\{\psi_3(E)\}^2}\right]\Bigg|_{E=E_F} \tag{B1.28}$$

Comparing the aforementioned equations, one can infer that the effective masses exist in the forbidden zone, which is impossible without the effect of band tailing. For undoped semiconductors, the effective mass in the band gap is infinity. The density-of-states function is given by

$$N_{HD}(E,\eta_g)=\frac{2g_v m_{\perp}^{*}\sqrt{2m_{||}^{*}}}{3\pi^2\hbar^3}R_{11}(E,\eta_g)\cos\left[\psi_{11}(E,\eta_g)\right] \tag{B1.29}$$

where

$$\begin{aligned}R_{11}(E,\eta_g)\equiv\Bigg[\Bigg[&\{T_{29}(E,\eta_g)\}'\sqrt{x(E,\eta_g)}+\frac{T_{29}(E,\eta_g)\{x(E,\eta_g)\}'}{2\sqrt{x(E,\eta_g)}}\\
&-\{T_{30}(E,\eta_g)\}'\sqrt{y(E,\eta_g)}-\frac{T_{30}(E,\eta_g)\{y(E,\eta_g)\}'}{2\sqrt{y(E,\eta_g)}}\Bigg]^2\\
&+\Bigg[\{T_{29}(E,\eta_g)\}'\sqrt{y(E,\eta_g)}+\frac{T_{29}(E,\eta_g)\{y(E,\eta_g)\}'}{2\sqrt{y(E,\eta_g)}}\\
&+\{T_{30}(E,\eta_g)\}'\sqrt{x(E,\eta_g)}-\frac{T_{30}(E,\eta_g)\{x(E,\eta_g)\}'}{2\sqrt{x(E,\eta_g)}}\Bigg]^2\Bigg]^{1/2},\end{aligned}$$

$$x(E,\eta_g) \equiv \frac{1}{2}\left[T_{27}(E,\eta_g) + \sqrt{\{T_{27}(E,\eta_g)\}^2 + \{T_{28}(E,\eta_g)\}^2}\right],$$

$$y(E,\eta_g) \equiv \frac{1}{2}\left[\sqrt{\{T_{27}(E,\eta_g)\}^2 + \{T_{28}(E,\eta_g)\}^2} - T_{27}(E,\eta_g)\right]$$

and
$$\psi_{11}(E,\eta_g) \equiv \tan^{-1}\left[\left[\left[\{T_{29}(E,\eta_g)\}'\sqrt{y(E,\eta_g)} + \frac{T_{29}(E,\eta_g)}{2\sqrt{y(E,\eta_g)}}\right.\right.\right.$$
$$\left. + \{T_{30}(E,\eta_g)\}'\sqrt{x(E,\eta_g)} + \frac{T_{30}\{x(E,\eta_g)\}'}{2\sqrt{x(E,\eta_g)}}\right]$$
$$\times\left[\{T_{29}(E,\eta_g)\}'\sqrt{x(E,\eta_g)} + \frac{T_{29}(E,\eta_g)\{x(E,\eta_g)\}'}{2\sqrt{x(E,\eta_g)}}\right.$$
$$\left.\left. - \{T_{30}(E,\eta_g)\}'\sqrt{y(E,\eta_g)} + \frac{T_{30}\{y(E,\eta_g)\}'}{2\sqrt{y(E,\eta_g)}}\right]^{-1}\right].$$

The oscillatory nature of the DOS for heavily doped tetragonal materials is apparent from (B1.29). For, $\psi_{11}(E,\eta_g) \geq \pi$, the cosine function becomes negative leading to the negative values of the DOS. The electrons cannot exist for the negative values of the DOS, and therefore this reason is forbidden for electrons, which indicates that in the band tail, there appears a new forbidden zone in addition to the normal band gap of the semiconductor. The use of (B1.29) the electron concentration at low temperatures can be expressed as

$$n_0 = \frac{2g_v m_\perp^* \sqrt{2m_{||}^*}}{3\pi^2\hbar^3}\left[I_{11}(E_{F_h},\eta_g)\right] \tag{B1.30}$$

where,

$$I_{11}(E_{F_h},\eta_g) \equiv \left[T_{29}(E_{F_h},\eta_g)\sqrt{x(E_{F_h},\eta_g)} - T_{30}(E_{F_h},\eta_g)\sqrt{y(E_{F_h},\eta_g)}\right].$$

For heavily doped tetragonal semiconductors, $\bar{E}_{\text{hd}}$ is the smallest negative root of the equation

$$\left[T_{27}(\bar{E}_{\text{hd}},\eta_g)T_{29}(\bar{E}_{\text{hd}},\eta_g) - T_{28}(\bar{E}_{\text{hd}},\eta_g)T_{30}(\bar{E}_{\text{hd}},\eta_g)\right] = 0 \tag{B1.31}$$

B1.2.2 Study of the EEM in Heavily Doped III–V, Ternary and Quaternary Materials Forming Gaussian Band Tails

(a) Under the conditions, $\delta = 0$, $m^*_{\|} = m^*_{\perp} = m^*$ and $\Delta_{\|} = \Delta_{\perp} = \Delta$, the electron dispersion law in this case assumes the form

$$\frac{\hbar^2 k^2}{2m^*} = T_{31}(E, \eta_g) + iT_{32}(E, \eta_g) \tag{B1.32}$$

where,

$$T_{31}(E,\eta_g) \equiv \left(\frac{2}{1+Erf(E/\eta_g)}\right)\left[\frac{\alpha b}{c}\theta_0(E,\eta_g) + \left[\frac{\alpha c + bc - \alpha b}{c^2}\right]\gamma_0(E,\eta_g)\right.$$
$$+\frac{1}{c}\left(1-\frac{\alpha}{c}\right)\left(1-\frac{b}{c}\right)\frac{1}{2}\left[1+Erf\left(\frac{E}{\eta_g}\right)\right]$$
$$\left.-\frac{1}{c}\left(1-\frac{\alpha}{c}\right)\left(1-\frac{b}{c}\right)\frac{2}{c\eta_g\sqrt{\pi}}\exp(-u_2^2)\left[\sum_{p=1}^{\infty}\frac{\exp(-p^2/4)}{p}\sinh(pu_2)\right]\right],$$
$$b \equiv \left(\frac{1}{E_g+\Delta}\right),\quad c \equiv \left(\frac{1}{E_g+\frac{2}{3}\Delta}\right),\quad u_2 \equiv \frac{1+cE}{c\eta_g}\quad \text{and}$$
$$T_{32}(E,\eta_g) \equiv \left(\frac{2}{1+Erf(E/\eta_g)}\right)\frac{1}{c}\left(1-\frac{\alpha}{c}\right)\left(1-\frac{b}{c}\right)\frac{\sqrt{\pi}}{c\eta_g}\exp(-u_2^2).$$

Thus, the complex energy spectrum occurs due to the term $T_{32}(E, \eta_g)$ and this imaginary band is quite different from the forbidden energy band.

The EEM at the Fermi level is given by

$$m^*(E_{F_h}, \eta_g) = m^*\{T_{31}(E, \eta_g)\}'\Big|_{E=E_{F_h}} \tag{B1.33}$$

Thus, the EEM in heavily doped III–V, ternary and quaternary materials exists in the band gap, which is the new attribute of the theory of band tailing. In the absence of band tailing, $\eta_g \to 0$ and the EEM assumes the form

$$m^*(E_F) = m^*\{I(E)\}'|_{E=E_F} \tag{B1.34}$$

The density-of-states function in this case assumes the form

$$N_{\mathrm{HD}}(E, \eta_g) = \frac{g_v}{3\pi^2}\left(\frac{2m^*}{\hbar^2}\right)^{3/2} R_{21}(E, \eta_g)\cos[\vartheta_{21}(E, \eta_g)] \tag{B1.35}$$

where

$$R_{21}(E,\eta_g) \equiv \left[\frac{\left[\{\alpha_{11}(E,\eta_g)\}'\right]^2}{4\alpha_{11}(E,\eta_g)} + \frac{\left[\{\beta_{11}(E,\eta_g)\}'\right]^2}{4\beta_{11}(E,\eta_g)}\right]^{1/2},$$

$$\alpha_{11}(E,\eta_g) \equiv \frac{1}{2}\left[T_{33}(E,\eta_g) + \sqrt{\{T_{33}(E,\eta_g)\}^2 + \{T_{34}(E,\eta_g)\}^2}\right],$$

$$T_{33}(E,\eta_g) \equiv \left[\{T_{31}(E,\eta_g)\}^3 - 3T_{31}(E,\eta_g)\{T_{32}(E,\eta_g)\}^2\right],$$

$$T_{34}(E,\eta_g) \equiv \left[3T_{32}(E,\eta_g)\{T_{31}(E,\eta_g)\}^2 - \{T_{32}(E,\eta_g)\}^3\right],$$

$$\beta_{11}(E,\eta_g) \equiv \frac{1}{2}\left[\sqrt{\{T_{33}(E,\eta_g)\}^2 + \{T_{34}(E,\eta_g)\}^2} - T_{33}(E,\eta_g)\right] \text{ and}$$

$$\vartheta_{21}(E,\eta_g) \equiv \tan^{-1}\left[\frac{\{\beta_{11}(E,\eta_g)\}'}{\{\alpha_{11}(E,\eta_g)\}'}\sqrt{\frac{\alpha_{11}(E,\eta_g)}{\beta_{11}(E,\eta_g)}}\right].$$

Thus, the oscillatory density-of-states function becomes negative for $\vartheta_{21}(E,\eta_g) \geq \pi$ and a new forbidden zone will appear in addition to the normal band gap. The electron concentration in the zone of low temperatures can be written as

$$n_0 = \frac{g_v}{3\pi^2}\left(\frac{2m^*}{\hbar^2}\right)^{3/2}\frac{1}{\sqrt{2}}\left[T_{33}(E_{F_h},\eta_g) + \sqrt{\{T_{33}(E_{F_h},\eta_g)\}^2 + \{T_{34}(E_{F_h},\eta_g)\}^2}\right] \tag{B1.36}$$

In this case, $\bar{E}_{\text{hd}}$ is given by

$$T_{31}(\bar{E}_{\text{hd}},\eta_g) = 0 \tag{B1.37}$$

(b) The dispersion relation in heavily doped III-V, ternary, and quaternary materials whose undoped energy spectrum obeys the two band model of Kane is given by

$$\frac{\hbar^2 k^2}{2m^*} = \gamma_2(E,\eta_g) \tag{B1.38}$$

where

$$\gamma_2(E,\eta_g) \equiv \left[\frac{2}{1 + Erf(E/\eta_g)}\right]\left[\gamma_0(E,\eta_g) + \alpha\theta_0(E,\eta_g)\right].$$

Since, the original two band Kane model is an all zero and no pole function, therefore, the heavily doped counterpart will be totally real and the complex band vanishes. The EEM in this case can be written as

$$m^*\left(E_{F_h},\eta_g\right) = m^*\{\gamma_2(E,\eta_g)\}'\Big|_{E=E_{F_h}} \quad \text{(B1.39)}$$

Thus, one again observes that the EEM in this case exists in the band gap. In the absence of band tailing, $\eta_g \to 0$ and the EEM assumes the form

$$m^*(E_F) = m^*\{1+2\alpha E\}|_{E=E_F} \quad \text{(B1.40)}$$

The density-of-states function in this case can be written as

$$N_{\mathrm{HD}}\left(E,\eta_g\right) = \frac{g_v}{2\pi^2}\left(\frac{2m^*}{\hbar^2}\right)^{3/2}\sqrt{\gamma_2(E,\eta_g)}\{\gamma_2(E,\eta_g)\}' \quad \text{(B1.41)}$$

Since, the original two band Kane model is an all zero and no pole function, therefore, the heavily doped counterpart will be totally real and the complex band vanishes.
The electron concentration at low temperatures is given by

$$n_0 = \frac{g_v}{3\pi^2}\left(\frac{2m^*}{\hbar^2}\right)^{3/2}\{\gamma_2(E_{F_h},\eta_g)\}^{3/2} \quad \text{(B1.42)}$$

In this case, $\bar{E}_{\mathrm{hd}}$ is given by

$$\gamma_2\left(\bar{E}_{\mathrm{hd}},\eta_g\right) = 0 \quad \text{(A1.43)}$$

(c) The dispersion relation in heavily doped semiconductors whose unperturbed conduction electrons obeys parabolic energy bands is given by

$$\frac{\hbar^2 k^2}{2m^*} = \gamma_3\left(E,\eta_g\right) \quad \text{(B1.44)}$$

where

$$\gamma_3\left(E,\eta_g\right) \equiv \left[\frac{2}{\left(1+Erf\left(E/\eta_g\right)\right)}\right]\gamma_0\left(E,\eta_g\right).$$

Since, the original parabolic energy band is no pole function, therefore, the heavily doped counterpart will be totally real, which is also apparent form the expression (B1.44).
The EEM in this case can be written as

$$m^*\left(E_{F_h},\eta_g\right) = m^*\{\gamma_3(E,\eta_g)\}'\Big|_{E=E_{F_h}} \tag{B1.45}$$

In the absence of band tailing, $\eta_g \to 0$ and the EEM assumes the form

$$m^*(E_F) = m^* \tag{B1.46}$$

It is well known that the EEM in undoped parabolic energy bands is a constant quantitity in general excluding cross-field configuration. But, the same mass in the corresponding heavily doped bulk counterpart is complicated functions of Fermi energy and the impurity potential together with the fact that the EEM also exists in the band gap.

The density-of-states function in this case can be written as

$$N_{\mathrm{HD}}\left(E,\eta_g\right) = \frac{g_v}{2\pi^2}\left(\frac{2m^*}{\hbar^2}\right)^{3/2}\sqrt{\gamma_3(E,\eta_g)}\{\gamma_3(E,\eta_g)\}' \tag{B1.47}$$

Since, the original parabolic energy band model is a no pole function, therefore, the heavily doped counterpart will be totally real and the complex band vanishes.The electron concentration at low temperatures is given by

$$n_0 = \frac{g_v}{3\pi^2}\left(\frac{2m^*}{\hbar^2}\right)^{3/2}\{\gamma_3(E_{F_h},\eta_g)\}^{3/2} \tag{B1.48}$$

In this case, $\bar{E}_{\mathrm{hd}}$ is given by

$$\gamma_3\left(\bar{E}_{\mathrm{hd}},\eta_g\right) = 0 \tag{B1.49}$$

B1.2.3 Study of the EEM in Heavily Doped II–VI Materials Forming Gaussian Band Tails

Using (1.42) and (A1.2), the dispersion relation of the carriers in heavily doped II–VI materials in the presence of Gaussian band tails can be expressed as

$$\gamma_3\left(E,\eta_g\right) = a_0'k_s^2 + b_0'k_z^2 \pm \bar{\lambda}_0 k_s \tag{B1.50}$$

Thus, the energy spectrum in this case is real since the corresponding undoped case as given by (1.42) is a no pole function.

The transverse and the longitudinal EEMs masses are respectively given by

$$m_\perp^*\left(E_{F_h},\eta_g\right) = m_\perp^*\{\gamma_3(E,\eta_g)\}'\left[1-\left(\frac{\bar{\lambda}_0}{\sqrt{(\bar{\lambda}_0)^2+4a_0'\gamma_3(E,\eta_g)}}\right)\right]\Bigg|_{E=E_{F_h}} \tag{B1.51}$$

and

$$m_{\|}^{*}(E_{F_h},\eta_g)=m_{\|}^{*}\{\gamma_3(E,\eta_g)\}'\Big|_{E=E_{F_h}} \tag{B1.52}$$

In the absence of band tailing effects $\eta_g \to 0$, we get

$$m_{\perp}^{*}(E_F)=m_{\perp}^{*}\left[1-\left(\frac{\bar{\lambda}_0}{\sqrt{(\bar{\lambda}_0)^2+4a_0'E}}\right)\right]\Bigg|_{E=E_F} \tag{B1.53}$$

and

$$m_{\|}^{*}(E_F)=m_{\|}^{*} \tag{B1.54}$$

Thus, the in heavily doped II–VI materials, both the transverse and the longitudinal EEM exist in the band gap.

The volume in k-space can be enclosed by the (B1.50) can be expressed as

$$V(E,\eta_g)=\frac{4\pi}{3a_0'\sqrt{b_0'}}\left[\{\gamma_3(E,\eta_g)\}^{3/2}+\frac{3}{8}\frac{(\bar{\lambda}_0)^2\sqrt{\gamma_3(E,\eta_g)}}{a_0'}\right.$$
$$\left.-\left(\frac{3}{4}\frac{\bar{\lambda}_0}{\sqrt{a_0'}}\right)\left(\gamma_3(E,\eta_g)+\frac{(\bar{\lambda}_0)^2}{4a_0'}\right)\sin^{-1}\left[\frac{\sqrt{\gamma_3(E,\eta_g)}}{\sqrt{\gamma_3(E,\eta_g)+\frac{(\bar{\lambda}_0)^2}{4a_0'}}}\right]\right] \tag{B1.55}$$

Using (B1.55), the density-of-states function in this case can be written as

$$N_{\mathrm{HD}}(E,\eta_g)=\frac{g_v}{2\pi^2a_0'\sqrt{b_0'}}\left[\{\gamma_3(E,\eta_g)\}^{1/2}\{\gamma_3(E,\eta_g)\}'+\frac{1}{8}\frac{(\bar{\lambda}_0)^2\{\gamma_3(E,\eta_g)\}'}{4a_0'\sqrt{\gamma_3(E,\eta_g)}}\right.$$
$$-\left(\frac{1}{2}\frac{\bar{\lambda}_0}{\sqrt{a_0'}}\right)\{\gamma_3(E,\eta_g)\}'\sin^{-1}\left[\frac{\sqrt{\gamma_3(E,\eta_g)}}{\sqrt{\gamma_3(E,\eta_g)+\frac{(\bar{\lambda}_0)^2}{4a_0'}}}\right]$$
$$\left.-\frac{\{\gamma_3(E,\eta_g)\}'}{2}\left(\gamma_3(E,\eta_g)+\frac{(\bar{\lambda}_0)^2}{4a_0'}\right)\left[\frac{1}{\sqrt{\gamma_3(E,\eta_g)}}-\frac{\sqrt{\gamma_3(E,\eta_g)}}{\left(\gamma_3(E,\eta_g)+\frac{(\bar{\lambda}_0)^2}{4a_0'}\right)}\right]\right] \tag{B1.56}$$

Therefore, the electron concentration in the zone of low temperatures can be expressed as

$$n_0 = \frac{g_v}{3\pi^2 a'_0 \sqrt{b'_0}} \left[\{\gamma_3(E_{F_h}, \eta_g)\}^{3/2} + \frac{3}{8} \frac{(\bar{\lambda}_0)^2 \sqrt{\gamma_3(E_{F_h}, \eta_g)}}{a'_0} - \left(\frac{3}{4} \frac{\bar{\lambda}_0}{\sqrt{a'_0}}\right) \left(\gamma_3(E_{F_h}, \eta_g) + \frac{(\bar{\lambda}_0)^2}{4a'_0}\right) \sin^{-1}\left[\frac{\sqrt{\gamma_3(E_{F_h}, \eta_g)}}{\sqrt{\gamma_3(E_{F_h}, \eta_g) + \frac{(\bar{\lambda}_0)^2}{4a'_0}}}\right]\right] \quad \text{(B1.57)}$$

In this case, $\bar{E}_{\text{hd}}$ is given by

$$\{\gamma_3(\bar{E}_{\text{hd}}, \eta_g)\} = 0 \quad \text{(B1.58)}$$

B1.2.4 Study of the EEM in Heavily Doped IV–VI Materials Forming Gaussian Band Tails

From (1.83), we can write

$$\frac{\alpha\hbar^4 k_s^4}{4m_t^+ m_t^-} + \hbar^2 k_s^2 \left[\left(\frac{1}{2m_t^*} - \frac{1}{2m_t^-}\right) + \alpha E\left(\frac{1}{2m_t^-} - \frac{1}{2m_t^+}\right) + \frac{\alpha\hbar^2 k_z^2}{4m_l^- m_t^+}\right] + \left[\left(\frac{\hbar^2 k_z^2}{2m_l^*} + \frac{\hbar^2 k_z^2}{2m_l^-}\right) + \frac{\alpha E}{2}\hbar^2 k_z^2 \left(\frac{1}{m_l^-} - \frac{1}{m_l^+}\right) + \frac{\alpha\hbar^4 k_z^4}{4m_l^+ m_l^-} - E(1 + \alpha E)\right] = 0 \quad \text{(B1.59)}$$

Using (B1.59) and (B1.2), the dispersion relation of the conduction electrons in heavily doped IV–VI materials can be expressed as

$$\frac{\alpha\hbar^4 k_s^4}{4m_t^+ m_l^-} Z_0(E, \eta_g) + \hbar^2 k_s^2 [\lambda_{71}(E, \eta_g) k_z^2 + \lambda_{72}(E, \eta_g)] + [\lambda_{73}(E, \eta_g) k_z^2 + \lambda_{74}(E, \eta_g) k_z^4 - \lambda_{75}(E, \eta_g)] = 0 \quad \text{(B1.60)}$$

where

$$Z_0(E, \eta_g) \equiv \frac{1}{2}\left[1 + Erf\left(\frac{E}{\eta_g}\right)\right], \quad \lambda_{70}(E, \eta_g) \equiv \frac{\alpha}{4m_t^+ m_t^-} Z_0(E, \eta_g),$$

$$\lambda_{71}(E, \eta_g) \equiv \left[\frac{\alpha\hbar^2}{4m_t^- m_l^+} Z_0(E, \eta_g) + \frac{\alpha\hbar^2}{4m_l^- m_t^+} Z_0(E, \eta_g)\right],$$

$$\lambda_{72}(E, \eta_g) \equiv \left[\left(\frac{1}{2m_t^*} - \frac{1}{2m_t^-}\right) Z_0(E, \eta_g) + \alpha\left(\frac{1}{2m_t^-} - \frac{1}{2m_t^+}\right)\gamma_0(E, \eta_g)\right],$$

$$\lambda_{73}(E,\eta_g) \equiv \left[\left(\frac{\hbar^2}{2m_l^*}+\frac{\hbar^2}{2m_l^-}\right)Z_0(E,\eta_g)+\frac{\alpha\hbar^2}{2}\left(\frac{1}{m_l^-}-\frac{1}{2m_l^+}\right)\gamma_0(E,\eta_g)\right],$$

$$\lambda_{74}(E,\eta_g) \equiv \frac{\alpha\hbar^4 Z_0(E,\eta_g)}{4m_l^+m_l^-}$$

and

$$\lambda_{75}(E,\eta_g) \equiv \left[\gamma_0(E,\eta_g)+\alpha\theta_0(E,\eta_g)\right].$$

Thus, the energy spectrum in this case is real since the corresponding undoped material as given by (B1.59) is a pole-less function.

The respective transverse and the longitudinal EEM in this case can be written as

$$m_\perp^*(E_{F_h},\eta_g) = \{2Z_0(E,\eta_g)\}^{-2}\left[Z_0(E,\eta_g)\left[-\{\lambda_{72}(E,\eta_g)\}'+\frac{\{\lambda_{78}(E,\eta_g)\}'}{2\sqrt{\lambda_{78}(E,\eta_g)}}\right]\right.$$
$$\left.\left.-\{Z_0(E,\eta_g)\}'.\left[-\lambda_{72}(E,\eta_g)+\sqrt{\lambda_{78}(E,\eta_g)}\right]\right]\right|_{E=E_{F_h}} \tag{B1.61}$$

where, $\lambda_{78}(E,\eta_g) \equiv [4\lambda_{70}(E,\eta_g)\lambda_{75}(E,\eta_g)]$
and

$$m_{||}^*(E_{F_h},\eta_g) = \frac{\hbar^2}{4}\left[-\{\lambda_{84}(E,\eta_g)\}'+\frac{\{\lambda_{84}(E,\eta_g)\}'\lambda_{84}(E,\eta_g)+2\{\lambda_{85}(E,\eta_g)\}'}{\sqrt{(\lambda_{84}(E,\eta_g))^2+4\lambda_{85}(E,\eta_g)}}\right]\Bigg|_{E=E_{F_h}} \tag{B1.62}$$

in which

$$\lambda_{84}(E,\eta_g) \equiv \frac{\lambda_{73}(E,\eta_g)}{\lambda_{74}(E,\eta_g)}$$

and

$$\lambda_{85}(E,\eta_g) \equiv \frac{\lambda_{75}(E,\eta_g)}{\lambda_{74}(E,\eta_g)}.$$

Thus, we can see that the both the EEMs in this case exist in the band gap.In the absence of band tailing effects $\eta_g \to 0$, we get

$$m_{\perp}^{*}(E_F) = \frac{\hbar^2}{2}\left[-\{\alpha_{11}(E)\}' + \frac{\alpha_{511}\{T_{311}(E)\}'}{2\sqrt{T_{311}(E)}}\right]\Bigg|_{E=E_F} \tag{B1.63}$$

where,

$$\alpha_{11}(E) \equiv \frac{2m_t^+ m_t^-}{\alpha\hbar^2}\alpha_{211}(E), \quad \alpha_{211}(E) \equiv \left[\frac{1}{2m_t^*} - \frac{\alpha E}{2m_t^+} + \frac{1+\alpha E}{2m_t^-}\right],$$
$$\alpha_{511} \equiv \frac{2m_t^+ m_t^-}{\alpha\hbar^2}\omega_{11},$$

$$(\omega_{11}) \equiv \left[\frac{\alpha^2}{16}\left[\frac{1}{m_t^- m_l^+} + \frac{1}{m_l^- m_t^+}\right]^2 - \frac{\alpha^2}{4m_t^- m_t^+ m_l^- m_l^+}\right]^{1/2}, \quad T_{311}(E) \equiv \frac{\omega_{311}(E)}{(\omega_{11})^2},$$
$$\omega_{311}(E) \equiv \left[\frac{\alpha E(1+\alpha E)}{m_t^+ m_t^-} + \left[\frac{1}{2m_t^*} - \left(\frac{\alpha E}{2m_t^+}\right) + \frac{(1+\alpha E)}{2m_t^-}\right]^2\right].$$

and

$$m_{||}^{*}(E_F) = \left(\frac{m_t^+ m_l^-}{\alpha}\right)\left[\left(\frac{\alpha}{2m_l^+} - \frac{\alpha}{2m_l^-}\right) + \frac{1}{2}\left\{\frac{2\left[\frac{1}{2m_l^*} + \frac{1+\alpha E}{2m_l^-} - \frac{\alpha E}{2m_l^+}\right]\left(\frac{\alpha}{2m_l^-} - \frac{\alpha}{2m_l^+}\right) + \frac{\alpha(1+2\alpha E)}{m_l^- m_l^+}}{\left[\left[\frac{1}{2m_l^*} + \frac{1+\alpha E}{2m_l^-} - \frac{\alpha E}{2m_l^+}\right]^2 + \frac{\alpha E(1+\alpha E)}{m_l^- m_l^+}\right]^{1/2}}\right\}\right]\Bigg|_{E=E_F} \tag{B1.64}$$

The volume in k-space can be enclosed by the (B1.60) can be written through the integral as

$$V(E,\eta_g) = 2\pi \int_0^{\lambda_{86}(E,\eta_g)} \left[-\left[\lambda_{79}(E,\eta_g)k_z^2 + \lambda_{80}(E,\eta_g)\right]\right.$$
$$\left. + \sqrt{\lambda_{81}(E,\eta_g)k_z^4 + \lambda_{82}(E,\eta_g)k_z^2 + \lambda_{83}(E,\eta_g)}\right]dk_z \tag{B1.65}$$

where

$$\lambda_{86}(E,\eta_g) \equiv \left[\frac{\sqrt{\left[\lambda_{84}(E,\eta_g)\right]^2 + 4\lambda_{85}(E,\eta_g)} - \lambda_{84}(E,\eta_g)}{2}\right]^{1/2},$$
$$\lambda_{79}(E,\eta_g) \equiv \frac{\lambda_{71}(E,\eta_g)}{2\hbar^2 Z_0(E,\eta_g)},$$

$$\lambda_{80}(E,\eta_g) \equiv \frac{\lambda_{72}(E,\eta_g)}{2\hbar^2 Z_0(E,\eta_g)}, \quad \lambda_{81}(E,\eta_g) \equiv \frac{\lambda_{76}(E,\eta_g)}{4\hbar^4\left[Z_0(E,\eta_g)\right]^2},$$
$$\lambda_{76}(E,\eta_g) \equiv \left[\lambda_{71}(E,\eta_g)\right]^2,$$

$$\lambda_{82}(E,\eta_g) \equiv \frac{\lambda_{77}(E,\eta_g)}{9\hbar^4[Z_0(E,\eta_g)]^2}$$

$$\lambda_{77}(E,\eta_g) \equiv [2\lambda_{71}(E,\eta_g)\lambda_{72}(E,\eta_g) - 4\lambda_{70}(E,\eta_g)\lambda_{73}(E,\eta_g) - 4\lambda_{70}(E,\eta_g)\lambda_{74}(E,\eta_g)],$$

$$\lambda_{83}(E,\eta_g) \equiv \frac{\lambda_{78}(E,\eta_g)}{9\hbar^4[Z_0(E,\eta_g)]^2} \quad \text{and} \quad \lambda_{78}(E,\eta_g) \equiv [4\lambda_{70}(E,\eta_g)\lambda_{75}(E,\eta_g)].$$

Thus,

$$V(E,\eta_g) = [\lambda_{87}(E,\eta_g)] \int_0^{\lambda_{86}(E,\eta_g)} \left[\sqrt{k_z^4 + \lambda_{88}(E,\eta_g)k_z^2 + \lambda_{89}(E,\eta_g)} - \lambda_{90}(E,\eta_g)\right] dk_z \tag{B1.66}$$

where

$$\lambda_{87}(E,\eta_g) \equiv 2\pi\sqrt{\lambda_{81}(E,\eta_g)}, \quad \lambda_{88}(E,\eta_g) \equiv \frac{\lambda_{82}(E,\eta_g)}{\lambda_{81}(E,\eta_g)},$$
$$\lambda_{89}(E,\eta_g) \equiv \frac{\lambda_{83}(E,\eta_g)}{\lambda_{81}(E,\eta_g)}$$

and

$$\lambda_{90}(E,\eta_g) \equiv 2\pi\left[\frac{\lambda_{79}(E,\eta_g)\{\lambda_{86}(E,\eta_g)\}^3}{3} + \lambda_{80}(E,\eta_g)\lambda_{89}(E,\eta_g)\right].$$

Using (B1.20), (B1.66) can be written as

$$V(E,\eta_g) = [\lambda_{87}(E,\eta_g)\lambda_{95}(E,\eta_g) - \lambda_{90}(E,\eta_g)] \tag{B1.67}$$

in which,

$$\begin{aligned}
\lambda_{95}(E,\eta_g) \equiv & \left[\frac{\lambda_{91}(E,\eta_g)}{3}[-E_i[\lambda_{93}(E,\eta_g),\lambda_{94}(E,\eta_g)]\right.\\
& \times \left[\{\lambda_{91}(E,\eta_g)\}^2 + \{\lambda_{92}(E,\eta_g)\}^2 + 2\{\lambda_{92}(E,\eta_g)\}^2 F_i[\lambda_{93}(E,\eta_g),\lambda_{94}(E,\eta_g)]\right]\\
& + \left(\frac{\{\lambda_{86}(E,\eta_g)\}}{3}\right)\left[\{\lambda_{86}(E,\eta_g)\}^2 + \{\lambda_{91}(E,\eta_g)\}^2 + 2\{\lambda_{92}(E,\eta_g)\}^2\right]\\
& \left.\times\left[\left[\{\lambda_{91}(E,\eta_g)\}^2 + \{\lambda_{86}(E,\eta_g)\}^2\right]^{1/2}\left[\{\lambda_{92}(E,\eta_g)\}^2 + \{\lambda_{86}(E,\eta_g)\}^2\right]^{-1/2}\right]\right],
\end{aligned}$$

$$\{\lambda_{91}(E,\eta_g)\}^2 \equiv \frac{1}{2}\left[\sqrt{\{\lambda_{88}(E,\eta_g)\}^2 - 4\lambda_{89}(E,\eta_g)} + \lambda_{88}(E,\eta_g)\right],$$

$E_i[\lambda_{93}(E,\eta_g), \lambda_{94}(E,\eta_g)]$

is the incomplete elliptic integral of the 2nd kind and is given by [55],

$$E_i[\lambda_{93}(E,\eta_g), \lambda_{94}(E,\eta_g)] \equiv \int_0^{\lambda_{93}(E,\eta_g)} \left[\left\{1 - \{\lambda_{94}(E,\eta_g)\}^2 \sin^2\xi\right\}^{1/2}\right] d\xi,$$

ξ is the variable of integration in this case $\lambda_{93}(E,\eta_g) \equiv \tan^{-1}\left[\frac{\lambda_{86}(E,\eta_g)}{\lambda_{92}(E,\eta_g)}\right]$,

$$\{\lambda_{92}(E,\eta_g)\}^2 \equiv \frac{1}{2}\left[\lambda_{88}(E,\eta_g) - \sqrt{\{\lambda_{88}(E,\eta_g)\}^2 - 4\lambda_{89}(E,\eta_g)}\right],$$

$$\lambda_{94}(E,\eta_g) \equiv \frac{\sqrt{\{\lambda_{91}(E,\eta_g)\}^2 - \{\lambda_{92}(E,\eta_g)\}^2}}{\lambda_{91}(E,\eta_g)},$$

$F_i[\lambda_{93}(E,\eta_g), \lambda_{94}(E,\eta_g)]$ is the incomplete elliptic integral of the 2nd kind and is given by [55],

$$F_i[\lambda_{93}(E,\eta_g), \lambda_{94}(E,\eta_g)] \equiv \int_0^{\lambda_{93}(E,\eta_g)} \left[\left\{1 - \{\lambda_{94}(E,\eta_g)\}^2 \sin^2\xi\right\}^{-1/2}\right] d\xi.$$

Using (2.3a) and (7.66), the density-of-states function is given by

$$N_{HD}(E,\eta_g) = \frac{g_v}{4\pi^3}\left[\{\lambda_{87}(E,\eta_g)\}'\lambda_{95}(E,\eta_g) + \{\lambda_{95}(E,\eta_g)\}'\lambda_{87}(E,\eta_g) - \{\lambda_{90}(E,\eta_g)\}'\right] \tag{B1.68}$$

Therefore, the electron concentration at low temperature can be expressed as

$$n_0 = \frac{g_v}{4\pi^3}[\{\lambda_{87}(E_{F_h},\eta_g)\}\lambda_{95}(E_{F_h},\eta_g) - \{\lambda_{90}(E_{F_h},\eta_g)\}] \tag{B1.69}$$

In this case, $\bar{E}_{hd}$ is given by

$$\{\lambda_{75}(\bar{E}_{hd},\eta_g)\} = 0 \tag{B1.70a}$$

B1.2.5 Study of the EEM in Heavily Doped Stressed Materials Forming Gaussian Band Tails

The use of (1.98) leads us to write

$$(E-\alpha_1)k_x^2 + (E-\alpha_2)k_y^2 + (E-\alpha_3)k_z^2 = t_1E^3 - t_2E^2 + t_3E + t_4 \tag{B1.70b}$$

where

$$\alpha_1 \equiv \left[E_g - C_1\varepsilon - (\bar{a}_0 + C_1)\varepsilon + \frac{3}{2}\bar{b}_0\varepsilon_{xx} - \frac{\bar{b}_0}{2}\varepsilon + \left(\sqrt{3}/2\right)\varepsilon_{xy}\bar{d}_0\right],$$

$$\alpha_2 \equiv \left[E_g - C_1\varepsilon - (\bar{a}_0 + C_1)\varepsilon + \frac{3}{2}\bar{b}_0\varepsilon_{xx} - \frac{\bar{b}_0}{2}\varepsilon - \left(\sqrt{3}/2\right)\varepsilon_{xy}\bar{d}_0\right],$$

$$\alpha_3 \equiv \left[E_g - C_1\varepsilon - (\bar{a}_0 + C_1)\varepsilon + \frac{3}{2}\bar{b}_0\varepsilon_{zz} - \frac{\bar{b}_0}{2}\varepsilon\right], \quad t_1 \equiv \left(3/2B_2^2\right),$$

$$t_2 \equiv \left(1/2B_2^2\right)\left[6(E_g - C_1\varepsilon) + 3C_1\varepsilon\right],$$

$$t_3 \equiv \left(1/2B_2^2\right)\left[3(E_g - C_1\varepsilon)^2 + 6C_1\varepsilon(E_g - C_1\varepsilon) - 2C_2^2\varepsilon_{xy}^2\right]$$

and $$t_4 \equiv \left(1/2B_2^2\right)\left[-3C_1\varepsilon(E_g - C_1\varepsilon)^2 + 2C_2^2\varepsilon_{xy}^2\right].$$

Using (B1.70b) and (B1.2), we can write,

$$\begin{aligned} &I(4)k^2 - T_{17}I(1)k_x^2 - T_{27}I(1)k_y^2 - T_{37}k_z^2I(1) \\ &= [q_{67}I(6) - R_{67}I(5) + V_{67}I(4) + \rho_{67}I(1)] \end{aligned} \tag{B1.70c}$$

where $T_{17} = \alpha_1$, $T_{27} = \alpha_2, T_{37} = \alpha_3$, $t_1 = q_{67}, t_2 = R_{67}$, $t_3 = V_{67}$, $t_4 = \rho_{67}$ and

$$I(6) = \int_{-\infty}^{E} (E - V)^3 F(V)\mathrm{d}V \tag{B1.71}$$

(B1.71) can be written as

$$I(6) = E^3 I(1) - 3E^2 I(7) + 3EI(8) - I(9) \tag{B1.72}$$

In which,

$$I(7) = \int_{-\infty}^{E} VF(V)\mathrm{d}V \tag{B1.73}$$

$$I(8) = \int_{-\infty}^{E} V^2F(V)\mathrm{d}V \tag{B1.74}$$

$$I(9) = \int_{-\infty}^{E} V^3F(V)\mathrm{d}V \tag{B1.75}$$

Using (B1.2) and successively (B1.73), (B1.74), and (B1.75) together with simple algebraic manipulations, one obtains

$$I(7) = \frac{-\eta_g}{2\sqrt{\pi}} \exp\left(\frac{-E^2}{\eta_g^2}\right) \tag{B1.76}$$

$$I(8) = \frac{\eta_g^2}{4}\left[1 + Erf\left(\frac{E}{\eta_g}\right)\right] \tag{B1.77}$$

and

$$I(9) = \frac{-\eta_g^3}{2\sqrt{\pi}} \exp\left(\frac{-E^2}{\eta_g^2}\right)\left[1 + \frac{E^2}{\eta_g^2}\right] \tag{B1.78}$$

Thus, (B1.72) can be written as

$$I(6) = \left[\frac{E}{2}\left[1 + Erf\left(\frac{E}{\eta_g}\right)\right]\left[E^2 + \frac{3}{2}\eta_g^2\right] + \frac{\eta_g}{2\sqrt{\pi}} \exp\left(\frac{-E^2}{\eta_g^2}\right)\left[4E^2 + \eta_g^2\right]\right] \tag{B1.79}$$

Thus, combining the appropriate equations, the dispersion relations of the conduction electrons in heavily doped stressed materials can be expressed as

$$P_{11}(E, \eta_g)k_x^2 + Q_{11}(E, \eta_g)k_y^2 + S_{11}(E, \eta_g)k_z^2 = 1 \tag{B1.80}$$

where

$$P_{11}(E, \eta_g) \equiv \left[\frac{\gamma_0(E, \eta_g) - (T_{17}/2)[1 + Erf(E/\eta_g)]}{\Delta_{14}(E, \eta_g)}\right],$$

$$\Delta_{14}(E, \eta_g) \equiv \left[q_{67}\left\{\frac{E}{2}\left[1 + Erf\left(\frac{E}{\eta_g}\right)\right]\left[E^2 + \frac{3}{2}\eta_g^2\right] + \frac{\eta_g}{2\sqrt{\pi}} \exp\left(\frac{-E^2}{\eta_g^2}\right)[4E^2 + \eta_g^2]\right\}\right.$$
$$\left. -R_{67}\theta_0(E, \eta_g) + V_{67}\gamma_0(E, \eta_g) + \frac{\rho_{67}}{2}[1 + Erf(E/\eta_g)]\right],$$

$$Q_{11}(E, \eta_g) \equiv \left[\frac{\gamma_0(E, \eta_g) - (T_{27}/2)[1 + Erf(E/\eta_g)]}{\Delta_{14}(E, \eta_g)}\right]$$

and

$$S_{11}(E, \eta_g) \equiv \left[\frac{\gamma_0(E, \eta_g) - (T_{37}/2)[1 + Erf(E/\eta_g)]}{\Delta_{14}(E, \eta_g)}\right].$$

Thus, the energy spectrum in this case is real since the corresponding undoped material as given by (1.98) is a pole-less function.

The EEMs along x, y, and z directions in this case can be written as

$$m_{xx}^{*}(E_{F_h},\eta_g)=\frac{\hbar^2}{2}\left[\left[\gamma_0(E_{F_h},\eta_g)-(T_{17}/2)\left[1+Erf(E_{F_h}/\eta_g)\right]\right]\right]^{-2}\cdot$$
$$\left[\{\Delta_{14}(E_{F_h},\eta_g)\}'\left[\gamma_0(E_{F_h},\eta_g)-(T_{17}/2)\left[1+Erf(E_{F_h}/\eta_g)\right]\right]\right]$$
$$-\Delta_{14}(E_{F_h},\eta_g)\left[\frac{1}{2}\left[1+Erf\left(\frac{E_{F_h}}{\eta_g}\right)\right]-\left\{\frac{T_{17}}{\eta_g\sqrt{\pi}}\exp\left(\frac{-E_{F_h}^2}{\eta_g^2}\right)\right\}\right]\Bigg]\tag{B1.81}$$

$$m_{yy}^{*}(E_{F_h},\eta_g)=\frac{\hbar^2}{2}\left[\left[\gamma_0(E_{F_h},\eta_g)-(T_{27}/2)[1+Erf(E_{F_h}/\eta_g)\right]\right]^{-2}$$
$$\times\left[\{\Delta_{14}(E_{F_h},\eta_g)\}'\left[\gamma_0(E_{F_h},\eta_g)-(T_{27}/2)[1+Erf(E_{F_h}/\eta_g)\right]\right]\Bigg]$$
$$-\Delta_{14}(E_{F_h},\eta_g)\left[\frac{1}{2}\left[1+Erf\left(\frac{E_{F_h}}{\eta_g}\right)\right]-\left\{\frac{T_{27}}{\eta_g\sqrt{\pi}}\exp\left(\frac{-E_{F_h}^2}{\eta_g^2}\right)\right\}\right]\Bigg]\tag{B1.82}$$

and

$$m_{zz}^{*}(E_{F_h},\eta_g)=\frac{\hbar^2}{2}\left[\left[\gamma_0(E_{F_h},\eta_g)-(T_{37}/2)[1+Erf(E_{F_h}/\eta_g)]\right]\right]^{-2}$$
$$[\{\Delta_{14}(E_{F_h},\eta_g)\}'[\gamma_0(E_{F_h},\eta_g)-(T_{37}/2)[1+Erf(E_{F_h}/\eta_g)]]]$$
$$-\Delta_{14}(E_{F_h},\eta_g)\left[\frac{1}{2}\left[1+Erf\left(\frac{E_{F_h}}{\eta_g}\right)\right]-\left\{\frac{T_{37}}{\eta_g\sqrt{\pi}}\exp\left(\frac{-E_{F_h}^2}{\eta_g^2}\right)\right\}\right]\Bigg]\tag{B1.83}$$

Thus, we can see that the EEMs in this case exist within the band gap. In the absence of band tailing effects $\eta_g \to 0$, we get

$$m_{xx}^{*}(E_F)=\hbar^2\bar{a}_0(E_F)\{\bar{a}_0(E_F)\}'\tag{B1.84}$$

$$m_{xx}^{*}(E_F)=\hbar^2\bar{b}_0(E_F)\{\bar{b}_0(E_F)\}'\tag{B1.85}$$

and

$$m_{xx}^{*}(E_F)=\hbar^2\bar{c}_0(E_F)\{\bar{c}_0(E_F)\}'\tag{B1.86}$$

The density-of-states function in this case can be written as

$$N_{\mathrm{HD}}(E,\eta_g)=\frac{g_v}{3\pi^2}\{\Delta_{15}(E,\eta_g)\}^{-2}$$
$$\left[\frac{3}{2}\{\Delta_{15}(E,\eta_g)\}\sqrt{\Delta_{14}(E,\eta_g)}\{\Delta_{14}(E,\eta_g)\}'-\{\Delta_{14}(E,\eta_g)\}^{3/2}\{\Delta_{15}(E,\eta_g)\}'\right] \tag{B1.87}$$

where

$$\Delta_{15}(E,\eta_g)\equiv\left[\left[\gamma_0(E,\eta_g)-(T_{17}/2)[1+Erf(E/\eta_g)]\right]\left[\gamma_0(E,\eta_g)\right.\right.$$
$$\left.\left.-(T_{27}/2)\left[1+Erf(E/\eta_g)\right]\right]\quad\left[\gamma_0(E,\eta_g)-(T_{37}/2)[1+Erf(E/\eta_g)]\right]\right]^{1/2}$$

Using Eq. B1.87, the electron concentration at low temperatures can be written as

$$n_0=\frac{g_v}{3\pi^2}\left[\frac{\{\Delta_{14}(E_{F_h},\eta_g)\}^{3/2}}{\Delta_{15}(E_{F_h},\eta_g)}\right] \tag{B1.88}$$

In this case, $\bar{E}_{\mathrm{hd}}$ is given by

$$\{\Delta_{14}(\bar{E}_{\mathrm{hd}},\eta_g)\}=0 \tag{B1.89}$$

B1.3 Open Research Problems

R.B1.1 Investigate the EEM for all the materials as given in problems in R. 1.1 of Chap. 1 in the presence of the Gaussian type band tails.

R.B1.2 Investigate the EEM in the presence of an arbitrarily oriented quantizing magnetic field in heavily doped tetragonal semiconductors by including broadening and the electron spin. Study all the special cases for heavily doped III–V, ternary, and quaternary materials in this context.

R.B1.3 Investigate the EEMs for heavily doped IV–VI, II–VI, and stressed Kane type compounds in the presence of an arbitrarily oriented quantizing magnetic field by including broadening and electron spin.

R.B1.4 Investigate the EEM for all the materials as stated in R.1.1 of Chap. 1 in the presence of an arbitrarily oriented quantizing magnetic field by including broadening and electron spin under the condition of heavily doping.

R.B1.5 Investigate the EEM in the presence of an arbitrarily oriented quantizing magnetic field and crossed electric fields in heavily doped tetragonal semiconductors by including broadening and the electron spin. Study all the special cases for heavily doped III–V, ternary, and quaternary materials in this context.

R.B1.6 Investigate the EEMs for heavily doped IV–VI, II–VI, and stressed Kane type compounds in the presence of an arbitrarily oriented quantizing magnetic field and crossed electric fields by including broadening and electron spin.

R.A1.7 Investigate the EEM for all the materials as stated in R.1.1 of Chap. 1 in the presence of an arbitrarily oriented quantizing magnetic field and crossed electric fields by including broadening and electron spin under the condition of heavy doping.

R.B1.8 Investigate the 2D EEM in ultrathin films of heavily doped tetragonal, III–V, II–VI, IV–VI, and stressed Kane type semiconductors.

R.B1.9 Investigate the 2D EEM for heavily doped ultrathin films of all the materials as considered in problems R.1.1.

R.B1.10 Investigate the 2D EEM in the presence of an arbitrarily oriented non-quantizing magnetic field for the ultrathin films of heavily doped tetragonal semiconductors by including the electron spin. Study all the special cases for III–V, ternary and quaternary materials in this context.

R.B1.11 Investigate the EEMs in ultrathin films of heavily doped IV–VI, II–VI and stressed Kane type compounds in the presence of an arbitrarily oriented non-quantizing magnetic field by including the electron spin.

R.B1.12 Investigate the 2D EEM for heavily doped ultrathin films of all the materials as stated in R.1.1 of Chap. 1 in the presence of an arbitrarily oriented magnetic field by including electron spin and broadening.

R.B1.13 Investigate the EEM for all the problems of R1.1 under an additional arbitrarily oriented electric field in the presence of heavy doping.

R.A1.14 Investigate the EEM for all the problems of R1.1 under the arbitrarily oriented crossed electric and magnetic fields in the presence of heavy doping.

R.B1.15 Investigate the 2D EEM for all the problems in R1.1 the presence of finite potential well under the condition of heavy doping.

R.B1.16 Investigate the 2D EEM for all the problems in R1.1 the presence of parabolic potential well under the condition heavy doping.

R.B1.17 Investigate the 2D EEM for all the problems in R1.1 the presence of circular potential well under the condition of heavy doping.

R.B1.18 Investigate the 2D EEM for accumulation layers of heavily doped tetragonal, III–V, IV–VI, II–VI, and stressed Kane type semiconductors in the presence of an arbitrary electric quantization.

R.B1.19 Investigate the 2D EEM in accumulation layers of all the materials as stated in R. 1.1 of Chap. 1 under the condition of heavy doping and in the presence of electric quantization along arbitrary direction.

R.B1.20 Investigate the 2D EEM in the presence of an arbitrarily oriented electric quantization for accumulation layers of heavily doped tetragonal semiconductors. Study all the special cases for III–V, ternary, and quaternary materials in this context.

R.B1.21 Investigate the 2D EEMs in accumulation layers of heavily doped IV–VI, II–VI, and stressed Kane type compounds in the presence of an arbitrarily oriented electric quantization.

R.B1.22 Investigate the 2D EEM in accumulation layers of all the materials as stated in R.1.1 of Chap. 1 in the presence of an arbitrarily oriented quantizing electric field under the condition of heavy doping.

R.B1.23 Investigate the 2D EEM in the presence of an arbitrarily oriented magnetic field in accumulation layers of heavily doped tetragonal semiconductors by including the electron spin. Study all the special cases for heavily doped III–V, ternary, and quaternary materials in this context.

R.A1.24 Investigate the 2D EEMs in accumulation layers of heavily doped IV–VI, II–VI, and stressed Kane type compounds in the presence of an arbitrarily oriented non-quantizing magnetic field by including the electron spin.

R.B1.25 Investigate the 2D EEM in accumulation layers of all the materials as stated in R1.1 of Chap. 1 in the presence of an arbitrarily oriented non-quantizing magnetic field by including electron spin and heavy doping.

R.B1.26 Investigate the 2D EEM in accumulation layers for all the problems from R B1.22 to R B1.26 in the presence of an additional arbitrarily oriented electric field.

R.B1.27 Investigate the 2D EEM in accumulation layers for all the problems from R B1.22 to R B1.26 in the presence of arbitrarily oriented crossed electric and magnetic fields.

R.B1.28 Investigate the 2D EEM in accumulation layers for all the problems from R B1.22 to R B1.26 in the presence of surface states.

R.B1.29 Investigate the 2D EEM in accumulation layers for all the problems from R B1.22 to R B1.26 in the presence of hot electron effects.

R.B1.30 Investigate the 2D EEM in accumulation layers for all the problems from R B1.22 to R B1.26 by including the occupancy of the electrons in various electric subbands.

R.B1.31 Investigate the 2D EEM in nipi structures of heavily doped tetragonal, III–V, II–VI, IV–VI, and stressed Kane type materials.

R.B1.32 Investigate the 2D EEM in nipi structures of all types of materials as discussed in problem R.1.1 as given in Chap. 1 under the condition of heavy doping.

R.B1.33 Investigate the 2D EEM in the presence of an arbitrarily oriented non-quantizing magnetic field for nipi structures of heavily doped tetragonal semiconductors by including the electron spin. Study all the special cases for heavily doped III–V, ternary, and quaternary materials in this context.

R.B1.34 Investigate the 2D EEMs in nipi structures of heavily doped IV–VI, II–VI, and stressed Kane type compounds in the presence of an arbitrarily oriented non-quantizing magnetic field by including the electron spin.

R.B1.35 Investigate the 2D EEM for nipi structures of all the materials as stated in R.1.1 of Chap. 1 in the presence of an arbitrarily oriented non-quantizing magnetic field by including electron spin under the condition of heavy doping.

R.B1.36 Investigate the 2D EEM for all the problems from R B1.32 to R B1.35 in the presence of an additional arbitrarily oriented non-quantizing electric field.

R.B1.37 Investigate the 2D EEM for all the problems from R B1.32 to R B1.35 in the presence of arbitrarily oriented crossed electric and magnetic fields.

R.B1.38 Investigate all the problems from R. B1.1 to R. B1.37, in the presence of light waves.

R.B1.39 Investigate all the problems from R. B1.1 up to R. B1.37 in the presence of exponential, Kane, Halperin and Lax and Bonch-Bruevich band tails [42].

R.B1.40 Investigate all the problems of this chapter by removing all the mathematical approximations and establishing the uniqueness conditions in each case.

Appendix C
The EEM in Superlattices of Heavily Doped Non-Parabolic Semiconductors

C1.1 Introduction

In recent years, modern fabrication techniques have generated altogether a new dimension in the arena of quantum effect devices through the experimental realization of an important artificial structure known as semiconductor superlattice (SL) by growing two similar but different semiconducting compounds in alternate layers with finite thicknesses. The materials forming the alternate layers have the same kind of band structure but different energy gaps. The concept of SL was developed for the first time by Keldysh [58] and was successfully fabricated by Esaki and Tsu [59–62]. The SLs are being extensively used in thermal sensors [63, 64], quantum cascade lasers [65–67], photodetectors [68, 69], light emitting diodes [70–73], multiplication [74], frequency multiplication [75], photocathodes [76, 77], thin film transistor [78], solar cells [79, 80], infrared imaging [81], thermal imaging [82, 83], infrared sensing [84], and also in other microelectronic devices.

The most extensively studied III–V SL is the one consisting of alternate layers of GaAs and $Ga_{1-x}Al_xAs$ owing to the relative easiness of fabrication. The GaAs and $Ga_{1-x}Al_xAs$ layers form the quantum wells and the potential barriers, respectively. The III–V SLs are attractive for the realization of high speed electronic and optoelectronic devices [85]. In addition to SLs with usual structure, other types of SLs such as II–VI [86], IV–VI [87], and HgTe/CdTe [88] SLs have also been investigated in the literature. The IV–VI SLs exhibit quite different properties as compared to the III–V SL due to the specific band structure of the constituent materials [89]. The epitaxial growth of II–VI SL is a relatively recent development and the primary motivation for studying the mentioned SLs made of

S. Bhattacharya and K. P. Ghatak, *Effective Electron Mass in Low-Dimensional Semiconductors*, Springer Series in Materials Science 167,
DOI: 10.1007/978-3-642-31248-9, © Springer-Verlag Berlin Heidelberg 2013

materials with the large band gap is in their potential for optoelectronic operation in the blue [89]. HgTe/CdTe SLs have raised a great deal of attention since 1979, when as a promising new materials for long wavelength infrared detectors and other electro-optical applications [60]. Interest in Hg-based SLs has been further increased as new properties with potential device applications were revealed [90, 91]. These features arise from the unique zero band gap material HgTe [92] and the direct band gap semiconductor CdTe which can be described by the three band mode of Kane [93]. The combination of the aforementioned materials with specified dispersion relation makes HgTe/CdTe SL very attractive, especially because of the tailoring of the material properties for various applications by varying the energy band constants of the SLs.

We note that all the aforementioned SLs have been proposed with the assumption that the interfaces between the layers are sharply defined, of zero thickness, i.e., devoid of any interface effects. The SL potential distribution may be then considered as a one-dimensional array of rectangular potential wells. The aforementioned advanced experimental techniques may produce SLs with physical interfaces between the two materials crystallographically abrupt; adjoining their interface will change at least on an atomic scale. As the potential form changes from a well (barrier) to a barrier (well), an intermediate potential region exists for the electrons. The influence of finite thickness of the interfaces on the electron dispersion law is very important; since, the electron energy spectrum governs the electron transport in SLs. In addition to it, for effective mass SLs [94]. The electronic subbands appear continually in real space [95].

In this chapter, we shall study the EEM under magnetic quantization in III–V, II–VI, IV–VI, HgTe/CdTe, and strained layer, heavily doped SLs with graded interfaces in Sects. C1.2.1–C1.2.5, respectively. From Sects. C1.2.6–C1.2.10, we shall investigate the same in III–V, II–VI, IV–VI, HgTe/CdTe, and strained layer, heavily doped effective mass SLs The last Sect. C1.3 contains open research problems.

C1.2 Theoretical Background

C1.2.1 Study of EEM in Heavily Doped III–V Superlattices with Graded Interfaces

The electron dispersion law in bulk specimens of the heavily doped constituent materials of III–V SLs whose undoped energy band structures are defined by three band model of Kane can be expressed as

$$\frac{\hbar^2 k^2}{2m_{cj}^*} = T_{1j}(E, \Delta_j, E_{gj}, \eta_{gj}) + iT_{2j}(E, \Delta_j, E_{gj}, \eta_{gj}) \tag{C1.1}$$

where

$$j = 1,2,$$

$$T_{ij}(E, \Delta_j, E_{gj}, \eta_{gj})$$

$$= (2/(1 + Erf(E/\eta_{gj}))[(\alpha_j b_j/c_j).\theta_0(E, \eta_{gj}) + \left[(\alpha_j c_j + b_j c_j - \alpha_j b_j)/c_j^2\right]$$

$$\times \gamma_0(E, \eta_{gj}) + \left[(1/c_j)(1 - (\alpha_j/c_j))(1 - (b_j/c_j))\frac{1}{2}[1 + Erf(E/\eta_{gj})]\right.$$

$$- (1/c_j)(1 - (\alpha_j/c_j))(1 - (b_j/c_j))(2/(c_j \eta_{gj}\sqrt{\pi})\exp\left(-u_j^2\right)$$

$$\left.\times \left[\sum_{p=1}^{\infty}(\exp(-p^2/4)/p)\sinh(pu_j)\right]\right],$$

$$b_j \equiv (E_{gj} + \Delta_j)^{-1}, \quad c_j \equiv (E_{gj} + \frac{2}{3}\Delta_j)^{-1}, \quad u_j \equiv \frac{1 + c_j E}{c_j \eta_{gj}} \quad \text{and}$$

$$T_{2j}(E, \Delta_j, E_{gj}, \eta_{gj}) \equiv \left(\frac{2}{1 + Erf(E/\eta_{gj})}\right)\frac{1}{c_j}\left(1 - \frac{\alpha_j}{c_j}\right)\left(1 - \frac{b_j}{c_j}\right)\frac{\sqrt{\pi}}{c_j \eta_{gj}}\exp\left(-u_j^2\right).$$

Therefore, the dispersion law of the electrons of heavily doped III–V SLs with graded interfaces can be expressed as

$$k_z^2 = G_8 + iH_8 \tag{C1.2}$$

where

$$G_8 = \left[\frac{C_7^2 - D_7^2}{L_0^2} - k_s^2\right], \quad C_7 = \cos^{-1}(\overline{\omega_7}),$$

$$\overline{\omega_7} = (2)^{\frac{-1}{2}}\left[(1 - G_7^2 - H_7^2) - \sqrt{(1 - G_7^2 - H_7^2)^2 + 4G_7^2}\right]^{\frac{1}{2}}$$

$$G_7 = [G_1 + (\rho_5 G_2/2) - (\rho_6 H_2/2) + (\Delta_0/2)\{\rho_6 H_2 - \rho_8 H_3 + \rho_9 H_4 - \rho_{10} H_4 + \rho_{11} H_5 - \rho_{12} H_5 + (1/12)(\rho_{12} G_6 - \rho_{14} H_6)\}],$$

$$G_1 = [(\cos(h_1))(\cosh(h_2))(\cosh(g_1))(\cos(g_2)) + (\sin(h_1))(\sinh(h_2))(\sinh(g_1))(\sin(g_2))],$$

$$\mathrm{h}_1 = e_1(b_0 - \Delta_0), \quad e_1 = 2^{\frac{-1}{2}}(\sqrt{t_1^2 + t_2^2} + t_1)^{\frac{1}{2}},$$

$$t_1 = \left[(2m_{c1}^*/\hbar^2).T_{11}(E, E_{g1}, \Delta_1, \eta_{g1}) - k_s^2\right],$$

$$t_2 = \left[(2m_{c1}^*/\hbar^2)T_{21}(E, E_{g1}, \Delta_1, \eta_{g1})\right],$$

$$\mathrm{h}_2 = e_2(b_0 - \Delta_0), \quad e_2 = 2^{\frac{-1}{2}}\left(\sqrt{t_1^2 + t_2^2} - t_1\right)^{\frac{1}{2}},$$

$$\mathrm{g}_1 = d_1(a_0 - \Delta_0), \quad \mathrm{d}_1 = 2^{\frac{-1}{2}}\left(\sqrt{x_1^2 + y_1^2} + x_1\right)^{\frac{1}{2}},$$

$$x_1 = \left[-(2m_{c2}^*/\hbar^2).T_{11}(E - V_0, E_{g2}, \Delta_2, \eta_{g2}) + k_s^2\right],$$

$$\mathrm{y}_1 = \left[(2m_{c2}^*/\hbar^2)T_{22}(E - V_0, E_{g2}, \Delta_2, \eta_{g2})\right],$$

$$\mathrm{g}_2 = d_2(a_0 - \Delta_0), \quad \mathrm{d}_2 = 2^{\frac{-1}{2}}\left(\sqrt{x_1^2 + y_1^2} - x_1\right)^{\frac{1}{2}},$$

$$\rho_5 = (\rho_3^2 + \rho_4^2)^{-1}[\rho_1\rho_3 - \rho_2\rho_4],$$

$$\rho_1 = [d_1^2 + e_2^2 - d_2^2 - e_1^2], \quad \rho_3 = [\mathrm{d}_1 e_1 + \mathrm{d}_2 e_2], \quad \rho_2 = 2[d_1 d_2 + e_1 e_2],$$

$$\rho_4 = [d_1 e_2 - e_1 d_2],$$

$$G_2 = [(\sin(h_1))(\cosh(h_2))(\sinh(g_1))(\cos(g_2)) + (\cos(h_1))(\sinh(h_2))(\cosh(g_1))(\sin(g_2))],$$

$$\rho_6 = (\rho_3^2 + \rho_4^2)^{-1}[\rho_1\rho_4 + \rho_2\rho_3],$$

$$H_2 = [(\sin(h_1))(\cosh(h_2))(\sin(g_2))(\cosh(g_1)) - (\cos(h_1))(\sinh(h_2))(\sinh(g_1))(\cos(g_2))],$$

$$\rho_7 = [(e_1^2 + e_2^2)^{-1}[e_1(d_1^2 - d_2^2) - 2d_1 d_2 e_2] - 3e_1],$$

$$G_3 = [(\sin(h_1))(\cosh(h_2))(\cosh(g_1))(\cos(g_2)) + (\cos(h_1))(\sinh(h_2))(\sinh(g_1))(\sin(g_2))],$$

$$\rho_8 = [(e_1^2 + e_2^2)^{-1}[e_2(d_1^2 - d_2^2) + 2d_1 d_2 e_1] + 3e_2],$$

$$H_3 = [(\sin(h_1))(\cosh(h_2))(\sin(g_2))(\sinh(g_1)) - (\cos(h_1))(\sinh(h_2))(\cosh(g_1))(\cos(g_2))],$$

$$\rho_9 = [(d_1^2 + d_2^2)^{-1}[d_1(e_2^2 - e_1^2) + 2e_2 d_2 e_1] + 3d_1],$$

$$G_4 = [(\cos(h_1))(\cosh(h_2))(\cos(g_2))(\sinh(g_1)) - (\sin(h_1))(\sinh(h_2))(\cosh(g_1))(\sin(g_2))],$$

$$\rho_{10} = [-(d_1^2 + d_2^2)^{-1}[d_2(-e_2^2 + e_1^2) + 2e_2 d_2 e_1] + 3d_2],$$

$$H_4 = [(cos(h_1))(\cosh(h_2))(\cosh(g_1))(\sin(g_2)) + (\sin(h_1))(\sinh(h_2))(\sinh(g_1))(\cos(g_2))],$$

$$\rho_{11} = 2[d_1^2 + e_2^2 - d_2^2 - e_1^2],$$

$$G_5 = [(\cos(h_1))(\cosh(h_2))(\cos(g_2))(\cosh(g_1)) - (\sin(h_1))(\sinh(h_2))(\sinh(g_1))(\sin(g_2))],$$

$$\rho_{12} = 4[d_1 d_2 + e_1 e_2],$$

$$H_5 = [(cos(h_1))(\cosh(h_2))(\sinh(g_1))(\sin(g_2)) + (\sin(h_1))(\sinh(h_2))(\cosh(g_1))(\cos(g_2))],$$

$$\rho_{13} = [\{5(d_1e_1^3 - 3e_1e_2^2d_1) + 5d_2(e_1^3 - 3e_1^2e_2)\}(d_1^2 + d_2^2)^{-1} + (e_1^2 + e_2^2)^{-1}\{5(e_1d_1^3 - 3d_2e_1^2d_1) + 5(d_2^3e_2 - 3d_1^2d_2e_2)\} - 34(d_1e_1 + d_2e_2)],$$

$$G_6 = [(\sin(h_1))(\cosh(h_2))(\sinh(g_1))(\cos(g_2)) + (\cos(h_1))(\sinh(h_2))(\cosh(g_1))(\sin(g_2))],$$

$$\rho_{14} = [\{5(d_1e_2^3 - 3e_2e_1^2d_1) + 5d_2(-e_1^3 + 3e_2^2e_1)\}(d_1^2 + d_2^2)^{-1} + (e_1^2 + e_2^2)^{-1}\{5(-e_1d_2^3 + 3d_1^2d_2e_1) + 5(-d_1^3e_2 + 3d_2^2d_1e_2)\} + 34(d_1e_2 - d_2e_1)],$$

$$H_6 = [(\sin(h_1))(\cosh(h_2))(\cosh(g_1))(\sin(g_2)) - (\cos(h_1))(\sinh(h_2))(\sinh(g_1))(\cos(g_2))],$$

$$H_7 = [H_1 + (\rho_5H_2/2) + (\rho_6G_2/2) + (\Delta_0/2)\{\rho_8G_3 + \rho_7H_3 + \rho_{10}G_4 + \rho_9H_4 + \rho_{12}G_5 + \rho_{11}H_5 + (1/12)(\rho_{14}G_6 + \rho_{13}H_6)\}],$$

$$H_1 = [(\sin(h_1))(\sinh(h_2))(\cosh(g_1))(\cos(g_2)) + (\cos(h_1))(\cosh(h_2))(\sinh(g_1))(\sin(g_2))],$$

$$D_7 = \sinh^{-1}(\overline{\omega_7}), \quad \mathrm{H}_8 = (2C_7D_7/L_0^2)$$

The simplified dispersion relation of heavily doped III–V superllatices with graded interfaces under magnetic quantization can be expressed as

$$k_z^2 = G_{8E,n} + iH_{8E,n} \tag{C1.3}$$

where

$$\mathrm{G}_{8E,n} = \left[\frac{C_{7E,n}^2 - D_{7E,n}^2}{L_0^2} - \left\{\frac{2eB}{\hbar}\left(n + \frac{1}{2}\right)\right\}\right], \quad C_{7E,n} = \cos^{-1}(\overline{\omega_{7E,n}}),$$

$$\overline{\omega_{7E,n}} = (2)^{\frac{-1}{2}}\left[(1 - G_{7E,n}^2 - H_{7E,n}^2) - \sqrt{(1 - G_{7E,n}^2 - H_{7E,n}^2)^2 + 4G_{7E,n}^2}\right]^{\frac{1}{2}}$$

$$G_{7E,n} = [G_{1E,n} + (\rho_{5E,n}G_{2E,n}/2) - (\rho_{6E,n}H_{2E,n}/2) + (\Delta_0/2) \times \{\rho_{6E,n}H_{2E,n} - \rho_{8E,n}H_{3E,n} + \rho_{9E,n}H_{4E,n} - \rho_{10E,n}H_{4E,n} + \rho_{11E,n}H_{5E,n} - \rho_{12E,n}H_{5E,n} + (1/12)(\rho_{12E,n}G_{6E,n} - \rho_{14E,n}H_{6E,n})\}],$$

$$G_{1E,n} = [(\cos(h_{1E,n}))(\cosh(h_{2E,n}))(\cosh(g_{1E,n}))(\cos(g_{2E,n})) + (\sin(h_{1E,n}))(\sinh(h_{2E,n}))(\sinh(g_{1E,n}))(\sin(g_{2E,n}))],$$

$$\mathrm{h}_{1E,n} = e_{1E,n}(b_0 - \Delta_0), \quad e_{1E,n} = 2^{\frac{-1}{2}}\left(\sqrt{t_{1E,n}^2 + t_2^2} + t_{1E,n}\right)^{\frac{1}{2}},$$

$$t_{1E,n} = \left[(2m_{c1}^*/\hbar^2) \cdot T_{11}(E, E_{g1}, \Delta_1, \eta_{g1}) - \left\{\frac{2eB}{\hbar}\left(n + \frac{1}{2}\right)\right\}\right],$$

$$t_2 = [(2m_{c1}^*/\hbar^2)T_{21}(E, E_{g1}, \Delta_1, \eta_{g1})],$$

$$h_{2E,n} = e_{2E,n}(b_0 - \Delta_0), \quad e_{2E,n} = 2^{\frac{-1}{2}}\left(\sqrt{t_{1E,n}^2 + t_2^2} - t_{1E,n}\right)^{\frac{1}{2}},$$

$$g_{1E,n} = d_{1E,n}(a_0 - \Delta_0), \quad d_{1E,n} = 2^{\frac{-1}{2}}\left(\sqrt{x_{1E,n}^2 + y_1^2} + x_{1E,n}\right)^{\frac{1}{2}},$$

$$x_{1E,n} = \left[-(2m_{c2}^*/\hbar^2) \cdot T_{11}(E - V_0, E_{g2}, \Delta_2, \eta_{g2}) + \left\{\frac{2eB}{\hbar}\left(n + \frac{1}{2}\right)\right\}\right],$$

$$y_1 = [(2m_{c2}^*/\hbar^2)T_{22}(E - V_0, E_{g2}, \Delta_2, \eta_{g2})], g_{2E,n} = d_{2E,n}(a_0 - \Delta_0),$$

$$d_{2E,n} = 2^{\frac{-1}{2}}\left(\sqrt{x_{1E,n}^2 + y_1^2} - x_{1E,n}\right)^{\frac{1}{2}},$$

$$\rho_{5E,n} = (\rho_{3E,n}^2 + \rho_{4E,n}^2)^{-1}[\rho_{1E,n}\rho_{3E,n} - \rho_{2E,n}\rho_{4E,n}], \rho_{1E,n} = [d_{1E,n}^2 + e_{2E,n}^2 - d_{2E,n}^2 - e_{1E,n}^2],$$

$$\rho_{3E,n} = [d_{1E,n}e_{1E,n} + d_{2E,n}e_{2E,n}],$$

$$\rho_{2E,n} = 2[d_{1E,n}d_{2E,n} + e_{1E,n}e_{2E,n}], \rho_{4E,n} = [d_{1E,n}e_{2E,n} - e_{1E,n}d_{2E,n}],$$

$$G_{2E,n} = [(\sin(h_{1E,n}))(\cosh(h_{2E,n}))(\sinh(g_{1E,n}))(\cos(g_{2E,n})) + (\cos(h_{1E,n}))(\sinh(h_{2E,n}))(\cosh(g_{1E,n}))(\sin(g_{2E,n}))],$$

$$\rho_{6E,n} = (\rho_{3E,n}^2 + \rho_{4E,n}^2)^{-1}[\rho_{1E,n}\rho_{4E,n} + \rho_{2E,n}\rho_{3E,n}],$$

$$H_{2E,n} = [(\sin(h_{1E,n}))(\cosh(h_{2E,n}))(\sin(g_{2E,n}))(\cosh(g_{1E,n})) - (\cos(h_{1E,n}))(\sinh(h_{2E,n}))(\sinh(g_{1E,n}))(\cos(g_{2E,n}))],$$

$$\rho_{7E,n} = [(e_{1E,n}^2 + e_{2E,n}^2)^{-1}[e_{1E,n}(d_{1E,n}^2 - d_{2E,n}^2) - 2d_{1E,n}d_{2E,n}e_{2E,n}] - 3e_{1E,n}],$$

$$G_{3E,n} = [(\sin(h_{1E,n}))(\cosh(h_{2E,n}))(\cosh(g_{1E,n}))(\cos(g_{2E,n})) + (\cos(h_{1E,n}))(\sinh(h_{2E,n}))(\sinh(g_{1E,n}))(\sin(g_{2E,n}))],$$

$$\rho_{8E,n} = [(e_{1E,n}^2 + e_{2E,n}^2)^{-1}[e_{2E,n}(d_{1E,n}^2 - d_{2E,n}^2) + 2d_{1E,n}d_{2E,n}e_{1E,n}] + 3e_{2E,n}],$$

$$H_{3E,n} = [(\sin(h_{1E,n}))(\cosh(h_{2E,n}))(\sin(g_{2E,n}))(\sinh(g_{1E,n})) - (\cos(h_{1E,n}))(\sinh(h_{2E,n}))(\cosh(g_{1E,n}))(\cos(g_{2E,n}))],$$

$$\rho_{9E,n} = [(d_{1E,n}^2 + d_{2E,n}^2)^{-1}[d_{1E,n}(e_{2E,n}^2 - e_{1E,n}^2) + 2e_{2E,n}d_{2E,n}e_{1E,n}] + 3d_{1E,n}],$$

$$G_{4E,n} = [(\cos(h_{1E,n}))(\cosh(h_{2E,n}))(\cos(g_{2E,n}))(\sinh(g_{1E,n})) - (\sin(h_{1E,n}))(\sinh(h_{2E,n}))(\cosh(g_{1E,n}))(\sin(g_{2E,n}))],$$

$$\rho_{10E,n} = [-(d_{1E,n}^2 + d_{2E,n}^2)^{-1}[d_{2E,n}(-e_{2E,n}^2 + e_{1E,n}^2) + 2e_{2E,n}d_{2E,n}e_{1E,n}] + 3d_{2E,n}],$$

$$H_{4E,n} = [(\cos(h_{1E,n}))(\cosh(h_{2E,n}))(\cosh(g_{1E,n}))(\sin(g_{2E,n})) + (\sin(h_{1E,n}))(\sinh(h_{2E,n}))(\sinh(g_{1E,n}))(\cos(g_{2E,n}))],$$

$$\rho_{11E,n} = 2[d_{1E,n}^2 + e_{2E,n}^2 - d_{2E,n}^2 - e_{1E,n}^2],$$

$$\begin{aligned}
G_{5E,n} &= [(\cos(h_{1E,n}))(\cosh(h_{2E,n}))(\cos(g_{2E,n}))(\cosh(g_{1E,n})) \\
&\quad - (\sin(h_{1E,n}))(\sinh(h_{2E,n}))(\sinh(g_{1E,n}))(\sin(g_{2E,n}))], \\
\rho_{12E,n} &= 4[d_{1E,n}d_{2E,n} + e_{1E,n}e_{2E,n}], \\
H_{5E,n} &= [(\cos(h_{1E,n}))(\cosh(h_{2E,n}))(\sinh(g_{1E,n}))(\sin(g_{2E,n})) \\
&\quad + (\sin(h_{1E,n}))(\sinh(h_{2E,n}))(\cosh(g_1))(\cos(g_2))], \\
\rho_{13E,n} &= [\{5(d_{1E,n}e_{1E,n}^3 - 3e_{1E,n}e_{2E,n}^2 d_{1E,n}) + 5d_{2E,n}(e_{1E,n}^3 - 3e_{1E,n}^2 e_{2E,n})\} \\
&\quad \times (d_{1E,n}^2 + d_{2E,n}^2)^{-1} + (e_{1E,n}^2 + e_{2E,n}^2)^{-1}\{5(e_{1E,n}d_{1E,n}^3 - 3d_{2E,n}e_{1E,n}^2 d_{1E,n}) \\
&\qquad + 5(d_{2E,n}^3 e_{2E,n} - 3d_{1E,n}^2 d_{2E,n}e_{2E,n})\} - 34(d_{1E,n}e_{1E,n} + d_{2E,n}e_{2E,n})], \\
G_{6E,n} &= [(\sin(h_{1E,n}))(\cosh(h_{2E,n}))(\sinh(g_{1E,n}))(\cos(g_{2E,n})) + (\cos(h_{1E,n})) \\
&\qquad (\sinh(h_{2E,n}))(\cosh(g_1))(\sin(g_2))], \\
\rho_{14E,n} &= [\{5(d_{1E,n}e_{2E,n}^3 - 3e_{2E,n}e_{1E,n}^2 d_{1E,n}) \\
&\quad + 5d_{2E,n}(-e_{1E,n}^3 + 3e_{2E,n}^2 e_{1E,n})\}(d_{1E,n}^2 + d_{2E,n}^2)^{-1} \\
&\quad + (e_{1E,n}^2 + e_{2E,n}^2)^{-1}\{5(-e_{1E,n}d_{2E,n}^3 + 3d_{1E,n}^2 d_{2E,n}e_{1E,n}) \\
&\qquad + 5(-d_{1E,n}^3 e_{2E,n} + 3d_{2E,n}^2 d_{1E,n}e_{2E,n})\} \\
&\quad + 34(d_{1E,n}e_{2E,n} - d_{2E,n}e_{1E,n})], \\
H_{6E,n} &= [(\sin(h_{1E,n}))(\cosh(h_{2E,n}))(\cosh(g_{1E,n}))(\sin(g_{2E,n})) \\
&\quad - (\cos(h_{1E,n}))(\sinh(h_{2E,n}))(\sinh(g_{1E,n}))(\cos(g_{2E,n}))], \\
H_{7E,n} &= [H_{1E,n} + (\rho_{5E,n}H_{2E,n}/2) + (\rho_{6E,n}G_{2E,n}/2) + (\Delta_0/2)\{\rho_{8E,n}G_{3E,n} \\
&\quad + \rho_{7E,n}H_{3E,n} + \rho_{10E,n}G_{4E,n} + \rho_{9E,n}H_{4E,n} \\
&\quad + \rho_{12E,n}G_{5E,n} + \rho_{11E,n}H_{5E,n} + (1/12)(\rho_{14E,n}G_{6E,n} + \rho_{13E,n}H_{6E,n})\}], \\
H_{1E,n} &= [(\sin(h_{1E,n}))(\sinh(h_{2E,n}))(\cosh(g_{1E,n}))(\cos(g_{2E,n})) \\
&\qquad + (\cos(h_{1E,n}))(\cosh(h_{2E,n}))(\sinh(g_{1E,n}))(\sin(g_{2E,n}))], \\
D_{7E,n} &= \sinh^{-1}(\overline{\omega_{7E,n}}), \quad \mathrm{H}_{8E,n} = (2C_{7E,n}D_{7E,n}/L_0^2)
\end{aligned}$$

Therefore, the EEM in this case assumes the form

$$m^*(E_{F1}, n) = \frac{\hbar^2}{2} G'_{8E,n}\Big|_{E=E_{F1}} \tag{C1.4}$$

where E_{F1} is the Fermi energy in this case and the prime denotes the differentiation with respect to E. The electron concentration is given by

$$n_0 = \frac{g_v eB}{\pi^2 \hbar} \sum_{n=0}^{n_{\max}} [\phi_{1C}(E_{F1}, n) + \phi_{2C}(E_{F1}, n)] \tag{C1.5}$$

where $\phi_{1C}(E_{F1},n) = \left[\left(G_{8E_{F1},n} + \sqrt{G^2_{8E_{F1},n} - H_{8E_{F1},n}}\right)/2\right]^{1/2}$,

$$\phi_{2C}(E_{F1},n) = \sum_{r=1}^{s} \theta_{2r,1}[\phi_{1C}(E_{F1},n)],\ \theta_{2r,i} = 2(k_BT)^{2r}(1-2^{1-2r})\zeta(2r)\frac{\partial^{2r}}{\partial E_{Fi}^{2r}}$$

and $i = 1, 2, 3, \ldots\ldots$ (C1.5)

C.1.2.2 Study of EEM in Heavily Doped II–VI Superlattices with Graded Interfaces

The electron energy spectra of the heavily doped constituent materials of II–VI SLs are given by

$$\gamma_3(E,\eta_{g1}) = \frac{\hbar^2k_s^2}{2m^*_{\perp,1}} + \frac{\hbar^2k_z^2}{2m^*_{\parallel,1}} \pm C_0k_s \quad \text{(C1.6)}$$

$$\text{and}\quad \frac{\hbar^2k^2}{2m^*_{c2}} = T_{12}(E,\Delta_2,E_{g2},\eta_{g2}) + iT_{22}(E,\Delta_2,E_{g2},\eta_{g2}) \quad \text{(C1.7)}$$

where $m^*_{\perp,1}$ and $m^*_{\parallel,1}$ are the transverse and longitudinal effective electron masses respectively at the edge of the conduction band for the first material. The energy-wave vector dispersion relation of the conduction electrons in heavily doped II–VI SLs with graded interfaces can be expressed as

$$k_z^2 = G_{19} + iH_{19} \quad \text{(C1.8)}$$

where

$$G_{19} = \left[\frac{C_{18}^2 - D_{18}^2}{L_0^2} - k_s^2\right],$$

$$C_{18} = \cos^{-1}(\omega_{18}), \omega_{18} = (2)^{\frac{-1}{2}}\left[(1 - G_{18}^2 - H_{18}^2) - \sqrt{(1 - G_{18}^2 - H_{18}^2)^2 + 4G_{18}^2}\right]^{\frac{1}{2}},$$

$$G_{18} = \frac{1}{2}[G_{11} + G_{12} + \Delta_0(G_{13} + G_{14}) + \Delta_0(G_{15} + G_{16})],$$

$$G_{11} = 2(\cos(g_1))(\cos(g_2))(\cos\gamma_{11}(E,k_s))$$

$$\gamma_{11}(E,k_s) = k_{21}(E,k_s)(b_0-\Delta_0), k_{21}(E,k_s) = \left\{\left[\gamma_3(E,\eta_{g1}) - \frac{\hbar^2 k_s^2}{2m^*_{\perp,1}} \pm C_0 k_s\right]\frac{2m^*_{\parallel,1}}{\hbar^2}\right\}^{1/2},$$

$$G_{12} = ([\Omega_1(E,k_s)(\sinh g_1)(\cos g_2) - \Omega_2(E,k_s)(\sin g_2)(\cosh g_1)](\sin\gamma_{11}(E,k_s)))$$

$$\Omega_1(E,k_s) = \left[\frac{d_1}{k_{21}(E,k_s)} - \frac{k_{21}(E,k_s)d_1}{d_1^2+d_2^2}\right] \quad \text{and} \quad \Omega_2(E,k_s) = \left[\frac{d_2}{k_{21}(E,k_s)} + \frac{k_{21}(E,k_s)d_2}{d_1^2+d_2^2}\right]$$

$$G_{13} = ([\Omega_3(E,k_s)(\cosh g_1)(\cos g_2) - \Omega_4(E,k_s)(\sinh g_1)(\sin g_2)](\sin\gamma_{11}(E,k_s)))$$

$$\Omega_3(E,k_s) = \left[\frac{d_1^2-d_2^2}{k_{21}(E,k_s)} - 3k_{21}(E,k_s)\right], \quad \Omega_4(E,k_s) = \left[\frac{2d_1d_2}{k_{21}(E,k_s)}\right]$$

$$G_{14} = ([\Omega_5(E,k_s)(\sinh g_1)(\cos g_2) - \Omega_6(E,k_s)(\sin g_1)(\cosh g_2)](\cos\gamma_{11}(E,k_s))).$$

$$\Omega_5(E,k_s) = \left[3d_1 - \frac{d_1}{d_1^2+d_2^2}k_{21}^2(E,k_s)\right], \quad \Omega_6(E,k_s) = \left[3d_2 + \frac{d_2}{d_1^2+d_2^2}k_{21}^2(E,k_s)\right]$$

$$G_{15} = ([\Omega_9(E,k_s)(\cosh g_1)(\cos g_2) - \Omega_{10}(E,k_s)(\sinh g_1)(\sin g_2)](\cos\gamma_{11}(E,k_s)))$$

$$\Omega_9(E,k_s) = [2d_1^2 - 2d_2^2 - k_{21}^2(E,k_s)], \quad \Omega_{10}(E,k_s) = [2d_1d_2],$$

$$G_{16} = ([\Omega_7(E,k_s)(\sinh g_1)(\cos g_2) - \Omega_8(E,k_s)(\sin g_1)(\cosh g_2)](\sin\gamma_{11}(E,k_s)/12)),$$

$$\Omega_7(E,k_s) = \left[\frac{5d_1}{d_1^2+d_2^2}k_{21}^3(E,k_s) + \frac{5(d_1^3-3d_2^2d_1)}{k_{21}(E,k_s)} - 34k_{21}(E,k_s)d_1\right],$$

$$\Omega_8(E,k_s) = \left[\frac{5d_2}{d_1^2+d_2^2}k_{21}^3(E,k_s) + \frac{5(d_2^3-3d_2^2d_1)}{k_{21}(E,k_s)} + 34k_{21}(E,k_s)d_2\right]$$

$$H_{18} = \frac{1}{2}[H_{11} + H_{12} + \Delta_0(H_{13}+H_{14}) + \Delta_0(H_{15}+H_{16})],$$

$$H_{11} = 2(\sinh g_1 \sin g_2 \cos\gamma_{11}(E,k_s)),$$

$$H_{12} = ([\Omega_2(E,k_s)(\sinh g_1)(\cos g_2) + \Omega_1(E,k_s)(\sin g_2)(\cosh g_1)](\sin\gamma_{11}(E,k_s))),$$

$$H_{13} = ([\Omega_4(E,k_s)(\cosh g_1)(\cos g_2) + \Omega_3(E,k_s)(\sinh g_1)(\sin g_2)](\sin\gamma_{11}(E,k_s))),$$

$$H_{14} = ([\Omega_6(E,k_s)(\sinh g_1)(\cos g_2) + \Omega_5(E,k_s)(\sin g_1)(\cosh g_2)](\cos\gamma_{11}(E,k_s))),$$

$$H_{15} = ([\Omega_{10}(E,k_s)(\cosh g_1)(\cos g_2) + \Omega_9(E,k_s)(\sinh g_1)(\sin g_2)](\cos\gamma_{11}(E,k_s))),$$

$$H_{16} = ([\Omega_8(E,k_s)(\sinh g_1)(\cos g_2) + \Omega_7(E,k_s)(\sin g_1)(\cosh g_2)](\sin\gamma_{11}(E,k_s)/12)),$$

$$\mathrm{H}_{19} = \left[\frac{2C_{18}D_{18}}{L_0^2}\right]$$

and $D_{18} = \sinh^{-1}(\omega_{18})$

The simplified dispersion relation in heavily doped II–VI superllatices with graded interfaces under magnetic quantization can be expressed as

$$k_z^2 = G_{19E,n} + iH_{19E,n} \tag{C1.9}$$

where

$$\mathrm{G}_{19E,n} = \left[\frac{C_{18E,n}^2 - D_{18E,n}^2}{L_0^2} - \left(\frac{2eB}{\hbar}\left(n+\frac{1}{2}\right)\right)\right],$$

$$C_{18E,n} = \cos^{-1}(\omega_{18E,n}), \omega_{18E,n}$$

$$= (2)^{\frac{-1}{2}}\left[(1 - G_{18E,n}^2 - H_{18E,n}^2) - \sqrt{(1 - G_{18E,n}^2 - H_{18E,n}^2)^2 + 4G_{180D}^2}\right]^{\frac{1}{2}},$$

$$G_{18E,n} = \frac{1}{2}\left[G_{11E,n} + G_{12E,n} + \Delta_0(G_{13E,n} + G_{14E,n}) + \Delta_0(G_{15E,n} + G_{16E,n})\right],$$

$$G_{11E,n} = 2(\cos(g_{1E,n}))(\cos(g_{2E,n}))(\cos\gamma_{11}(E,n)), \gamma_{11}(E,n) = k_{21}(E,n)(b_0 - \Delta_0),$$

$$k_{21}(E,n) = \left\{\left[\gamma_3(E,\eta_{g1}) - \frac{\hbar^2}{2m_{\perp,1}^*}\left\{\frac{2eB}{\hbar}\left(n+\frac{1}{2}\right)\right\} \pm C_0\left\{\frac{2eB}{\hbar}\left(n+\frac{1}{2}\right)\right\}^{1/2}\right]\frac{2m_{\|,1}^*}{\hbar^2}\right\}^{1/2},$$

$$G_{12E,n} = ([\Omega_1(E,n)(\sinh g_{1E,n})(\cos g_{2E,n}) - \Omega_2(E,n)(\sin g_{2E,n})(\cosh g_{1E,n})](\sin\gamma_{11}(E,n)))$$

$$\Omega_1(E,n) = \left[\frac{d_{1E,n}}{k_{21}(E,n)} - \frac{k_{21}(E,n)d_{1E,n}}{d_{1E,n}^2 + d_{2E,n}^2}\right], \quad \Omega_2(E,n) = \left[\frac{d_{2E,n}}{k_{21}(E,n)} + \frac{k_{21}(E,n)d_{2E,n}}{d_{1E,n}^2 + d_{2E,n}^2}\right],$$

$$G_{13E,n} = ([\Omega_3(E,n)(\cosh g_{1E,n})(\cos g_{2E,n}) - \Omega_4(E,n)(\sinh g_{1E,n})(\sin g_{2E,n})](\sin\gamma_{11}(E,n)))$$

$$\Omega_3(E,n) = \left[\frac{d_{1E,n}^2 - d_{2E,n}^2}{k_{21}(E,n)} - 3k_{21}(E,n)\right], \quad \Omega_4(E,n) = \left[\frac{2d_{1E,n}d_{2E,n}}{k_{21}(E,n)}\right],$$

$$G_{14E,n} = ([\Omega_5(E,n)(\sinh g_{1E,n})(\cos g_{2E,n}) - \Omega_6(E,n)(\sin g_{1E,n})(\cosh g_{2E,n})](\cos\gamma_{11}(E,n))).$$

$$\Omega_5(E,n) = \left[3d_{1E,n} - \frac{d_{1E,n}}{d_{1E,n}^2 + d_{2E,n}^2}k_{21}^2(E,n)\right], \quad \Omega_6(E,n) = \left[3d_{2E,n} + \frac{d_{2E,n}}{d_{1E,n}^2 + d_{2E,n}^2}k_{21}^2(E,n)\right]$$

$$G_{15E,n} = ([\Omega_9(E,n)(\cosh g_{1E,n})(\cos g_{2E,n}) - \Omega_{10}(E,n)(\sinh g_{1E,n})(\sin g_{2E,n})](\cos\gamma_{11}(E,n)))$$

$$\Omega_9(E,n) = \left[2d_{1E,n}^2 - 2d_{2E,n}^2 - k_{21}^2(E,n)\right], \quad \Omega_{10}(E,n) = [2d_{1E,n}d_{2E,n}],$$

$$G_{16E,n} = ([\Omega_7(E,n)(\sinh g_{1E,n})(\cos g_{2E,n}) - \Omega_8(E,n)(\sin g_{1E,n})(\cosh g_{2E,n})](\sin\gamma_{11}(E,n)/12)),$$

$$\Omega_7(E,n) = \left[\frac{5d_{1E,n}}{d_{1E,n}^2 + d_{2E,n}^2}k_{21}^3(E,n) + \frac{5(d_{1E,n}^3 - 3d_{2E,n}^2 d_{1E,n})}{k_{21}(E,n)} - 34k_{21}(E,n)d_{1E,n}\right]$$

$$\Omega_8(E,n) = \left[\frac{5d_{2E,n}}{d_{1E,n}^2 + d_{2E,n}^2}k_{21}^3(E,n) + \frac{5(d_{2E,n}^3 - 3d_{2E,n}^2 d_{1E,n})}{k_{21}(E,n)} + 34k_{21}(E,n)d_{2E,n}\right]$$

$$H_{18E,n} = \frac{1}{2}[H_{11E,n} + H_{12E,n} + \Delta_0(H_{13E,n} + H_{14E,n}) + \Delta_0(H_{15E,n} + H_{16E,n})],$$

$$H_{11E,n} = 2(\sinh g_{1E,n})(\sin g_{2E,n})(\cos\gamma_{11}(E,n)))$$

$$H_{12E,n} = ([\Omega_2(E,n)(\sinh g_{1E,n})(\cos g_{2E,n}) + \Omega_1(E,n)(\sin g_{2E,n})(\cosh g_{1E,n})](\sin\gamma_{11}(E,n))),$$

$$H_{13E,n} = ([\Omega_4(E,n)(\cosh g_{1E,n})(\cos g_{2E,n}) + \Omega_3(E,n)(\sinh g_{1E,n})(\sin g_{2E,n})](\sin\gamma_{11}(E,n))),$$

$$H_{14E,n} = ([\Omega_6(E,n)(\sinh g_{1E,n})(\cos g_{2E,n}) + \Omega_5(E,n)(\sin g_{1E,n})(\cosh g_{2E,n})](\cos\gamma_{11}(E,n))),$$

$$H_{15E,n} = ([\Omega_{10}(E,n)(\cosh g_{1E,n})(\cos g_{2E,n}) + \Omega_9(E,n)(\sinh g_{1E,n})(\sin g_{2E,n})](\cos\gamma_{11}(E,n))),$$

$$H_{16E,n} = ([\Omega_8(E,n)(\sinh g_{1E,n})(\cos g_{2E,n}) + \Omega_7(E,n)(\sin g_{1E,n})(\cosh g_{2E,n})](\sin\gamma_{11}(E,n)/2)),$$

$$\mathrm{H}_{19E,n} = \left[\frac{2C_{18E,n}D_{18E,n}}{L_0^2}\right] \quad \text{and} \quad D_{18E,n} = \sinh^{-1}(\omega_{18E,n})$$

Therefore, the EEM in this case assumes the form

$$m^*(E_{F2}, n) = \frac{\hbar^2}{2} G'_{19E,n}\Big|_{E=E_{F2}} \quad \text{(C1.10)}$$

where E_{F2} is the Fermi energy in this case.

The electron concentration is given by

$$n_0 = \frac{g_v eB}{\pi^2 \hbar} \sum_{n=0}^{n_{\max}} [\phi_{3C}(E_{F2}, n) + \phi_{4C}(E_{F2}, n)] \quad \text{(C1.11)}$$

where $\phi_{3C}(E_{F2}, n) = \left[\left(G_{19E_{F2},n} + \sqrt{G^2_{19E_{F2},n} - H_{19E_{F2},n}}\right)/2\right]^{1/2}$

and $\phi_{4C}(E_{F2}, n) = \sum_{r=1}^{s} \theta_{2r,2}[\phi_{3C}(E_{F2}, n)]$

C1.2.3 Study of EEM in Heavily Doped IV–VI Superlattices with Graded Interfaces

The **E–k** dispersion relation of the conduction electrons of the heavily doped constituent materials of the IV–VI SLs can be expressed as

$$k_z^2 = [2\bar{p}_{9,i}]^{-1}[-\bar{q}_{9,i}(E, k_s, \eta_{gi}) + [[\bar{q}_{9,i}(E, k_s, \eta_{gi})]^2 + 4\bar{p}_{9,i}\bar{R}_{9,i}(E, k_s, \eta_{gi})]^{\frac{1}{2}}] \quad \text{(C1.12)}$$

where

$$\bar{p}_{9,i} = (\alpha_i \hbar^4)/(4m^-_{l,i} m^+_{li}),\quad \text{i = 1,2},\quad \bar{q}_{9,i}(E, k_s, \eta_{gi}) = [(\hbar^2/2)((1/m^*_{li}) + (1/m^-_{li})) + \alpha_i(\hbar^4/4)k_s^2((1/m^+_{li} m^-_{ti}) + (1/m^+_{ti} m^-_{li})) - \alpha_i \gamma_3(E, \eta_{gi})((1/m^+_{li}) - (1/m^-_{li})]$$

and

$$\bar{R}_{9,i}(E, k_s, \eta_{gi}) = [\gamma_2(E, \eta_{gi}) + \gamma_3(E, \eta_{gi})[(\hbar^2/2)\alpha_i k_s^2((1/m^*_{ti}) - (1/m^-_{ti}))] - [(\hbar^2/2)k_s^2((1/m^*_{ti}) + (1/m^-_{ti}))] - \alpha_i(\hbar^6/4)k_s^4((1/m^+_{ti} m^-_{ti}))]$$

The electron dispersion law in heavily doped IV–VI SLs with graded interfaces can be expressed as

$$\cos(L_o k) = \frac{1}{2}\Phi_2(E, k_s) \quad \text{(C1.13)}$$

Where

$$\Phi_2(E,k_s) \equiv [2\cosh\{\beta_2(E,k_s)\}\cos\{\gamma_2(E,k_s)\} + \varepsilon_2(E,k_s)\sinh\{\beta_2(E,k_s)\}\sin\{\gamma_{22}(E,k_s)\}$$
$$+\Delta_0\left[\left(\frac{\{K_{112}(E,k_s)\}^2}{K_{212}(E,k_s)} - 3K_{212}(E,k_s)\right)\cosh\{\beta_2(E,k_s)\}\sin\{\gamma_{22}(E,k_s)\}\right.$$
$$\left.+\left(3K_{112}(E,k_s) - \frac{\{K_{212}(E,k_s)\}^2}{K_{112}(E,k_s)}\right)\sinh\{\beta_2(E,k_s)\}\cos\{\gamma_{22}(E,k_s)\}\right]$$
$$+\Delta_0\left[2\left(\{K_{112}(E,k_s)\}^2 - \{K_{212}(E,k_s)\}^2\right)\cosh\{\beta_2(E,k_s)\}\cos\{\gamma_{22}(E,k_s)\}\right.$$
$$\left.\left.+\frac{1}{12}\left[\frac{5\{K_{112}(E,k_s)\}^3}{K_{212}(E,k_s)} + \frac{5\{K_{212}(E,k_s)\}^3}{K_{112}(E,k_s)} - 34K_{212}(E,k_s)K_{112}(E,k_s)\right]\sinh\{\beta_2(E,k_s)\}\sin\{\gamma_{22}(E,k_s)\}\right]\right],$$

$$\beta_2(E,k_s) \equiv K_{112}(E,k_s)[a_0 - \Delta_0],$$
$$k_{112}^2(E,k_s) = [2\bar{p}_{9,2}]^{-1}[-\bar{q}_{9,2}(E - V_0, k_{s,}\eta_{g2})$$
$$-\left[[\bar{q}_{9,2}(E - V_0, k_{s,}\eta_{g2})]^2 + 4\bar{p}_{9,2}\bar{R}_{9,2}(E - V_0, k_{s,}\eta_{g2})\right]^{\frac{1}{2}}\Big],$$
$$\gamma_{22}(E,k_s) = K_{212}(E,k_s)[b_0 - \Delta_0],$$
$$k_{212}^2(E,k_s) = [2\bar{p}_{9,1}]^{-1}[-\bar{q}_{9,1}(E, k_{s,}\eta_{g1})$$
$$+\left[[\bar{q}_{9,1}(E, k_{s,}\eta_{g1})]^2 + 4\bar{p}_{9,1}\bar{R}_{9,1}(E, k_{s,}\eta_{g1})\right]^{\frac{1}{2}}\Big]$$

and

$$\varepsilon_2(E,k_s) \equiv \left[\frac{K_{112}(E,k_s)}{K_{212}(E,k_s)} - \frac{K_{212}(E,k_s)}{K_{112}(E,k_s)}\right].$$

The simplified dispersion relation in heavily doped IV-VI superllatices with graded interfaces under magnetic quantization can be expressed as

$$k_z^2 = \frac{1}{L_0^2}\left[\cos^{-1}\left\{\frac{1}{2}\Phi_2(E,n)\right\}\right]^2 - \frac{2eB}{\hbar}\left(n + \frac{1}{2}\right) \tag{C1.14}$$

where

$$\Phi_2(E,k_s) \equiv \Big[2\cosh\{\beta_2(E,n)\}\cos\{\gamma_2(E,n)\} + \varepsilon_2(E,n)\sinh\{\beta_2(E,n)\}\sin\{\gamma_{22}(E,n)\}$$
$$+\Delta_0\left[\left(\{K_{112}(E,n)\}^2/K_{212}(E,n) - 3K_{212}(E,n)\right)\cosh\{\beta_2(E,n)\}\sin\{\gamma_{22}(E,n)\}\right.$$
$$\left.+\left(3K_{112}(E,n) - \frac{\{K_{212}(E,n)\}^2}{K_{112}(E,n)}\right)\sinh\{\beta_2(E,n)\}\cos\{\gamma_{22}(E,n)\}\right]$$
$$+\Delta_0\left[2\left(\{K_{112}(E,n)\}^2 - \{K_{212}(E,n)\}^2\right).\cosh\{\beta_2(E,n)\}\cos\{\gamma_{22}(E,n)\}\right.$$
$$\left.\left.+\frac{1}{12}\left[\frac{5\{K_{112}(E,n)\}^3}{K_{212}(E,n)} + \frac{5\{K_{212}(E,n)\}^3}{K_{112}(E,n)} - 34K_{212}(E,n)K_{112}(E,n)\right]\sinh\{\beta_2(E,n)\}\sin\{\gamma_{22}(E,n)\}\right]\right],$$

$$\beta_2(E,n) \equiv K_{112}(E,n)[a_0 - \Delta_0],$$
$$k_{112}^2(E,n) = [2\bar{p}_{9,2n}]^{-1}[-\bar{q}_{9,2n}(E - V_0, \eta_{g2}) - \left[[\bar{q}_{9,2n}(E - V_0, \eta_{g2})]^2 + 4\bar{p}_{9,2n}\bar{R}_{9,2n}(E - V_0, \eta_{g2})]^{\frac{1}{2}}\right],$$
$$\bar{q}_{9,2n}(E - V_0, \eta_{g2}) = [(\hbar^2/2)((1/m_{l2}^*) + (1/m_{l2}^-)) + \alpha_2(\hbar^4/4)\frac{2eB}{\hbar}\left(n+\frac{1}{2}\right)((1/m_{l2}^+ m_{t2}^-) + (1/m_{t2}^+ m_{l2}^-)) - \alpha_2\gamma_3(E - V_0, \eta_{g2})((1/m_{l2}^+) - (1/m_{l2}^-)],$$

$$\bar{R}_{9,2n}(E, \eta_{g2}) = \left[\gamma_2(E - V_0, \eta_{g2}) + \gamma_3(E - V_0, \eta_{g2})[(\hbar^2/2)\alpha_2\frac{2eB}{\hbar}\left(n+\frac{1}{2}\right)((1/m_{t2}^*) - (1/m_{t2}^-))] - [(\hbar^2/2)k_{s0}^2((1/m_{t2}^*) + (1/m_{t2}^-))] - \alpha_2(\hbar^6/4) \times \left[\frac{2eB}{\hbar}\left(n+\frac{1}{2}\right)\right]^2((1/m_{t2}^+ m_{t2}^-))\right],$$
$$\gamma_2(E,n) = K_{212}(E,n)[b_0 - \Delta_0], \quad k_{212}^2(E,n) = [2\bar{p}_{9,1n}]^{-1}[-\bar{q}_{9,1n}(E, \eta_{g1}) + \left[[\bar{q}_{9,1n}(E, \eta_{g1})]^2 + 4\bar{p}_{9,1n}\bar{R}_{9,1n}(E, \eta_{g1})]^{\frac{1}{2}}\right]$$
$$\bar{q}_{9,1n}(E, \eta_{g1}) = [(\hbar^2/2)((1/m_{l1}^*) + (1/m_{l1}^-)) + \alpha_1(\hbar^4/4)\frac{2eB}{\hbar}\left(n+\frac{1}{2}\right)((1/m_{l1}^+ m_{t1}^-) + (1/m_{t1}^+ m_{l1}^-)) - \alpha_1\gamma_3(E, \eta_{g1})((1/m_{l1}^+) - (1/m_{l1}^-)],$$

$$\bar{R}_{9,1}(E, \eta_{g1}) = [\gamma_2(E, \eta_{g1}) + \gamma_3(E, \eta_{g1})[(\hbar^2/2)\alpha_1(2eB/\hbar)(n+\frac{1}{2})((1/m_{t1}^*) - (1/m_{t1}^-))] - [(\hbar^2/2)k_{s0}^2((1/m_{t1}^*) + (1/m_{t1}^-))] - \alpha_1(\hbar^6/4) \times \left((2eB/\hbar)\left(n+\frac{1}{2}\right)\right)^2((1/m_{t1}^+ m_{t1}^-))]$$

and

$$\varepsilon_2(E,n) \equiv \left[\frac{K_{112}(E,n)}{K_{212}(E,n)} - \frac{K_{212}(E,n)}{K_{112}(E,n)}\right].$$

Therefore, the EEM in this case assumes the form

$$m^*(E_{F3},n) = \left(\frac{\hbar^2}{2}\right)\left[\Phi_2'(E_{F3},n)\cos^{-1}\left[\frac{1}{2}\Phi_2(E_{F3},n)\right]\right]/\left(\sqrt{1-(1/4)\Phi_2^2(E_{F3},n)}\right) \tag{C1.15}$$

where E_{F3} is the Fermi energy in this case.

The electron concentration is given by

$$n_0 = \frac{g_v eB}{\pi^2 \hbar} \sum_{n=0}^{n_{\max}} [\phi_{5C}(E_{F3}, n) + \phi_{6C}(E_{F3}, n)]$$

$$\text{where} \quad \phi_{5C}(E_{F3}, n) = \left[\frac{1}{L_0^2}\left[\cos^{-1}\left\{\frac{1}{2}\Phi_2(E_{F3}, n)\right\}\right]^2 - \frac{2eB}{\hbar}\left(n + \frac{1}{2}\right)\right]^{1/2}$$

$$\text{and} \quad \phi_{6C}(E_{F3}, n) = \sum_{r=1}^{s} \theta_{2r,3}[\phi_{6C}(E_{F3}, n)] \tag{C1.16}$$

C1.2.4 Study of EEM in Heavily Doped HgTe/CdTe Superlattices with Graded Interfaces

The electron energy spectra of the constituent materials of HgTe/CdTe SLs are given by

$$k^2 = \left[\frac{B_{01}^2 + 4A_1E - B_{01}\sqrt{B_{01}^2 + 4A_1E}}{2A_1^2}\right] \tag{C1.17}$$

$$\text{and} \quad \frac{\hbar^2 k^2}{2m_{c2}^*} = T_{12}(E, \Delta_2, E_{g2}, \eta_{g2}) + iT_{22}(E, \Delta_2, E_{g2}, \eta_{g2}) \tag{C1.18}$$

where

$$B_{01} = \left(3|e|^2 \Big/ 128\varepsilon_{sc1}\right), \quad A_1 = (\hbar^2/2m_{c1}^*). \ \varepsilon_{sc1}$$

is the semiconductor permittivity of the first material. The energy-wave vector dispersion relation of the conduction electrons in heavily doped HgTe/CdTe SLs with graded interfaces can be expressed as

$$k_z^2 = G_{192} + iH_{192} \tag{C1.19}$$

where $G_{192} = [((C_{182}^2 - D_{182}^2)/L_0^2) - k_s^2]$,

$$C_{182} = \cos^{-1}(\omega_{182}),$$

$$\omega_{182} = (2)^{\frac{-1}{2}}\left[(1 - G_{182}^2 - H_{182}^2) - \sqrt{(1 - G_{182}^2 - H_{182}^2)^2 + 4G_{182}^2}\right]^{\frac{1}{2}},$$

$$G_{182} = \frac{1}{2}[G_{112} + G_{122} + \Delta_0(G_{132} + G_{142}) + \Delta_0(G_{152} + G_{162})],$$

$$G_{112} = 2(\cos(g_{12}))(\cos(g_{22}))(\cos\gamma_8(E, k_s))$$

$$\gamma_8(E, k_s) = k_8(E, k_s)(b_0 - \Delta_0),$$

$$k_8(E, k_s) = \left[\frac{B_{01}^2 + 4A_1E - B_{01}\sqrt{B_{01}^2 + 4A_1E}}{2A_1^2} - k_s^2\right]^{1/2},$$

$$G_{122} = ([\Omega_{12}(E,k_s)(\sinh g_{12})(\cos g_{22}) - \Omega_{22}(E,k_s)(\sin g_{22})(\cosh g_{12})](\sin \gamma_8(E,k_s)))$$

$$\Omega_{12}(E,k_s) = \left[\frac{d_{12}}{k_8(E,k_s)} - \frac{k_8(E,k_s)d_{12}}{d_{12}^2 + d_{22}^2}\right], \quad \Omega_{22}(E,k_s) = \left[\frac{d_{22}}{k_8(E,k_s)} + \frac{k_8(E,k_s)d_{22}}{d_{12}^2 + d_{22}^2}\right],$$

$$G_{132} = ([\Omega_{32}(E,k_s)(\cosh g_{12})(\cos g_{22}) - \Omega_{42}(E,k_s)(\sinh g_{12})(\sin g_{22})](\sin \gamma_8(E,k_s))),$$

$$\Omega_{32}(E,k_s) = \left[\frac{d_{12}^2 - d_{22}^2}{k_8(E,k_s)} - 3k_8(E,k_s)\right], \quad \Omega_{42}(E,k_s) = \left[\frac{2d_{12}d_{12}}{k_8(E,k_s)}\right],$$

$$G_{142} = ([\Omega_{52}(E,k_s)(\sinh g_{12})(\cos g_{22}) - \Omega_{62}(E,k_s)(\sin g_{12})(\cosh g_{22})](\cos \gamma_8(E,k_s))),$$

$$\Omega_{52}(E,k_s) = \left[3d_{12} - \frac{d_{12}}{d_{12}^2 + d_{22}^2} k_8^2(E,k_s)\right], \quad \Omega_{62}(E,k_s) = \left[3d_{22} + \frac{d_{22}}{d_{12}^2 + d_{22}^2} k_8^2(E,k_s)\right],$$

$$G_{152} = ([\Omega_{92}(E,k_s)(\cosh g_{12})(\cos g_{22}) - \Omega_{102}(E,k_s)(\sinh g_{12})(\sin g_{22})](\cos \gamma_8(E,k_s))),$$

$$\Omega_{92}(E,k_s) = [2d_{12}^2 - 2d_{22}^2 - k_8^2(E,k_s)], \quad \Omega_{102}(E,k_s) = [2d_{12}d_{22}],$$

$$G_{162} = ([\Omega_{72}(E,k_s)(\sinh g_{12})(\cos g_{22}) - \Omega_{82}(E,k_s)(\sin g_{12})(\cosh g_{22})](\sin \gamma_8(E,k_s)/12)),$$

$$\Omega_{72}(E,k_s) = \left[\frac{5d_{12}}{d_{12}^2 + d_{22}^2} k_8^3(E,k_s) + \frac{5(d_{12}^3 - 3d_{22}^2 d_{12})}{k_8(E,k_s)} - 34k_8(E,k_s)d_{12}\right],$$

$$\Omega_{82}(E,k_s) = \left[\frac{5d_{22}}{d_{12}^2 + d_{22}^2} k_8^3(E,k_s) + \frac{5(d_{22}^3 - 3d_{22}^2 d_{12})}{k_8(E,k_s)} + 34k_8(E,k_s)d_{22}\right]$$

$$H_{182} = \frac{1}{2}[H_{112} + H_{122} + \Delta_0(H_{132} + H_{142}) + \Delta_0(H_{152} + H_{162})],$$
$$H_{112} = 2(\sinh g_{12} \sin g_{22} \cos \gamma_8(E,k_s)),$$
$$H_{122} = ([\Omega_{22}(E,k_s)(\sinh g_{12})(\cos g_{22}) + \Omega_{12}(E,k_s)(\sin g_{22})(\cosh g_{12})](\sin \gamma_8(E,k_s))),$$
$$H_{132} = ([\Omega_{42}(E,k_s)(\cosh g_{12})(\cos g_{22}) + \Omega_{32}(E,k_s)(\sinh g_{12})(\sin g_{22})](\sin \gamma_8(E,k_s))),$$
$$H_{142} = ([\Omega_{62}(E,k_s)(\sinh g_{12})(\cos g_{22}) + \Omega_{52}(E,k_s)(\sin g_{12})(\cosh g_{22})](\cos \gamma_8(E,k_s))),$$
$$H_{152} = ([\Omega_{102}(E,k_s)(\cosh g_{12})(\cos g_{22}) + \Omega_{92}(E,k_s)(\sinh g_{12})(\sin g_{22})](\cos \gamma_8(E,k_s))),$$
$$H_{162} = ([\Omega_{82}(E,k_s)(\sinh g_{12})(\cos g_{22}) + \Omega_{72}(E,k_s)(\sin g_{12})(\cosh g_{22})](\sin \gamma_8(E,k_s)/12)),$$
$$\mathrm{H}_{192} = [((2C_{182}D_{182})/L_0^2)]$$

and $D_{182} = \sinh^{-1}(\omega_{182})$

The simplified dispersion relation in heavily doped HgTe/CdTe superlattices with graded interfaces under magnetic quantization can be expressed as

$$(k_z)^2 = G_{192E,n} + iH_{192E,n} \tag{C1.20}$$

where

$$\mathrm{G}_{192E,n} = \left[\frac{C^2_{182E,n} - D^2_{182E,n}}{L_0^2} - (2eB/\hbar)(n + (1/2))\right],$$

$$C_{1820D} = \cos^{-1}(\omega_{182E,n}),$$

$$\omega_{182E,n} = (2)^{\frac{-1}{2}}\left[(1 - G^2_{182E,n} - H^2_{182E,n}) - \sqrt{(1 - G^2_{182E,n} - H^2_{182E,n})^2 + 4G^2_{182E,n}}\right]^{\frac{1}{2}},$$

$$G_{182E,n} = \frac{1}{2}[G_{112E,n} + G_{122E,n} + \Delta_0(G_{132E,n} + G_{142E,n}) + \Delta_0(G_{152E,n} + G_{162E,n})]$$

$$G_{112E,n} = 2(\cos(g_{12}))(\cos(g_{22}))(\cos\gamma_8(E,n)), \quad \gamma_8(E,n) = k_8(E,n)(b_0 - \Delta_0),$$

$$k_8(E,n) = \left[\frac{B^2_{01} + 4A_1E - B_{01}\sqrt{B^2_{01} + 4A_1E}}{2A_1^2} - (2eB/\hbar)(n + (1/2))\right]^{1/2},$$

$$G_{120D} = ([\Omega_{12}(E,n)(\sinh g_{12E,n})(\cos g_{22E,n}) - \Omega_{22}(E,n)(\sin g_{22E,n})(\cosh g_{12E,n})](\sin\gamma_8(E,n))),$$

$$\Omega_{12}(E,n) = \left[\frac{d_{12E,n}}{k_8(E,n)} - \frac{k_8(E,n)d_{12E,n}}{d^2_{12E,n} + d^2_{22E,n}}\right],$$

$$\Omega_{22}(E,n) = \left[\frac{d_{22E,n}}{k_8(E,n)} + \frac{k_8(E,n)d_{22E,n}}{d^2_{12E,n} + d^2_{22E,n}}\right],$$

$$G_{1320D} = ([\Omega_{32}(E,n)(\cosh g_{12E,n})(\cos g_{22E,n}) - \Omega_{42}(E,n)(\sinh g_{12E,n})(\sin g_{22E,n})](\sin\gamma_8(E,n))),$$

$$\Omega_{32}(E,n) = \left[\frac{d^2_{12E,n} - d^2_{2E,n}}{k_8(E,n)} - 3k_8(E,n)\right], \quad \Omega_{42}(E,n) = \left[\frac{2d_{12E,n}d_{22E,n}}{k_8(E,n)}\right],$$

$$G_{1420D} = ([\Omega_{52}(E,n)(\sinh g_{12E,n})(\cos g_{22E,n}) - \Omega_{62}(E,n)(\sin g_{12E,n})(\cosh g_{22E,n})](\cos\gamma_8(E,n))),$$

$$\Omega_{52}(E,n) = \left[3d_{12E,n} - \frac{d_{12E,n}}{d^2_{12E,n} + d^2_{22E,n}}k_8^2(E,n)\right],$$

$$\Omega_{62}(E,n) = \left[3d_{22E,n} + \frac{d_{22E,n}}{d_{12E,n}^2 + d_{22E,n}^2} k_8^2(E,n)\right],$$

$$G_{1520D} = ([\Omega_{92}(E,n)(\cosh g_{12E,n})(\cos g_{22E,n}) - \Omega_{102}(E,n)(\sinh g_{12E,n})(\sin g_{22E,n})](\cos \gamma_8(E,n))),$$

$$\Omega_{92}(E,n) = [2d_{12E,n}^2 - 2d_{22E,n}^2 - k_8^2(E,n)], \quad \Omega_{102}(E,n) = [2d_{12E,n}d_{22E,n}],$$

$$G_{162E,n} = ([\Omega_{72}(E,n)(\sinh g_{12E,n})(\cos g_{22E,n}) - \Omega_{82}(E,n)(\sin g_{12E,n})(\cosh g_{22E,n})] \times (\sin \gamma_{80D}(E,n)/12)),$$

$$\Omega_{72}(E,n) = \left[\frac{5d_{12E,n}}{d_{12E,n}^2 + d_{22E,n}^2} k_8^3(E,n) + \frac{5(d_{12E,n}^3 - 3d_{22E,n}^2 d_{12E,n})}{k_8(E,n)} - 34k_8(E,n)d_{12E,n}\right],$$

$$\Omega_{82}(E,n) = \left[\frac{5d_{22E,n}}{d_{12E,n}^2 + d_{22E,n}^2} k_8^3(E,n) + \frac{5(d_{22E,n}^3 - 3d_{22E,n}^2 d_{12E,n})}{k_8(E,n)} + 34k_8(E,n)d_{22E,n}\right],$$

$$H_{182E,n} = \frac{1}{2}[H_{112E,n} + H_{122E,n} + \Delta_0(H_{132E,n} + H_{142E,n}) + \Delta_0(H_{152E,n} + H_{162E,n})],$$

$$H_{112E,n} = 2(\sinh g_{12E,n})(\sin g_{22E,n})(\cos \gamma_8(E,n))),$$

$$H_{1220D} = ([\Omega_{22}(E,n)(\sinh g_{12E,n})(\cos g_{22E,n}) + \Omega_{12}(E,n)(\sin g_{22E,n})(\cosh g_{12E,n})] \times (\sin \gamma_8(E,n))),$$

$$H_{132E,n} = ([\Omega_{42}(E,n)(\cosh g_{12E,n})(\cos g_{22E,n}) + \Omega_{32}(E,n)(\sinh g_{12E,n})(\sin g_{22E,n})] \times (\sin \gamma_8(E,n))),$$

$$H_{142E,n} = ([\Omega_{62}(E,n)(\sinh g_{12E,n})(\cos g_{22E,n}) + \Omega_{52}(E,n)(\sin g_{12E,n})(\cosh g_{22E,n})] \times (\cos \gamma_8(E,n))),$$

$$H_{1520D} = ([\Omega_{102}(E,n)(\cosh g_{12E,n})(\cos g_{22E,n}) + \Omega_{92}(E,n)(\sinh g_{12E,n})(\sin g_{22E,n})] \times (\cos \gamma_8(E,n))),$$

$$H_{162E,n} = ([\Omega_{82}(E,n)(\sinh g_{12E,n})(\cos g_{22E,n}) + \Omega_{72}(E,n)(\sin g_{12E,n})(\cosh g_{22E,n})] \times (\sin \gamma_8(E,n)/2)),$$

$$\mathrm{H}_{192E,n} = [((2C_{182E,n}D_{182E,n})/L_0^2)] \quad \text{and} \quad D_{182E,n} = \sinh^{-1}(\omega_{182E,n})$$

Therefore, the EEM in this case assumes the form

$$m^*(E_{F4},n) = \frac{\hbar^2}{2} G'_{192E,n}\Big|_{E=E_{F4}} \tag{C1.21}$$

where E_{F4} is the Fermi energy in this case.

The electron concentration is given by

$$n_0 = \frac{g_v eB}{\pi^2 \hbar} \sum_{n=0}^{n_{\max}} [\phi_{7C}(E_{F4},n) + \phi_{8C}(E_{F4},n)] \tag{C1.22}$$

where $\phi_{7C}(E_{F4},n) = \left[\left(G_{192E_{F4},n} + \sqrt{G^2_{192E_{F4},n} - H_{192E_{F4},n}}\right)/2\right]^{1/2}$

and $\phi_{8C}(E_{F4},n) = \sum_{r=1}^{s} \theta_{2r,4}[\phi_{7C}(E_{F4},n)]$

C1.2.5 Study of EEM in Heavily Doped Strained Layer Superlattices with Graded Interfaces

The dispersion relation of the conduction electrons of the constituent materials of the strained layer superlattices can be expressed as

$$[E - T_{1i}]k_x^2 + [E - T_{2i}]k_y^2 + [E - T_{3i}]k_z^2 = q_i E^3 - R_i E^2 + V_i E + \zeta_i \tag{C1.23}$$

where

$$T_{1i} = \overline{\theta}_i, \quad \overline{\theta}_i = \left[E_{gi} - C^c_{1i}\varepsilon_i - (a_i + C^c_{1i})\varepsilon_i + \frac{3}{2}b_i\varepsilon_{xxi} - \frac{b_i\varepsilon_i}{2} + \frac{\sqrt{3}d_i\varepsilon_{xyi}}{2}\right],$$

$$T_{2i} = \omega_i, \quad \omega_i = \left[E_{gi} - C^c_{1i}\varepsilon_i - (a_i + C^c_{1i})\varepsilon_i + \frac{3}{2}b_i\varepsilon_{xxi} - \frac{b_i\varepsilon_i}{2} - \frac{\sqrt{3}d_i\varepsilon_{xyi}}{2}\right],$$

$$\mathrm{T}_{3i} = \delta_i, \quad \delta_i = \left[E_{gi} - C^c_{1i}\varepsilon_i + (a_i + C^c_{1i})\varepsilon_i + \frac{3}{2}b_i\varepsilon_{zzi} - \frac{b_i\varepsilon_i}{2}\right]$$

$$R_i = q_i\left[2A_i + C^c_{1i}\varepsilon_i\right], \quad \mathrm{q}_i = \frac{3}{2B^2_{2i}}, \quad \mathrm{A}_i = E_{gi} - C^c_{1i}\varepsilon_i,$$

$$V_i = q_i\left[A_i^2 - \frac{2C^2_{2i}\varepsilon_{xyi}}{3} + 2A_i C^c_{1i}\varepsilon_i\right], \quad \zeta_i = q_i\left[\frac{2C^2_{2i}\varepsilon_{xyi}}{3} - C^c_{1i}\varepsilon_i A_i^2\right]$$

Therefore, the electron energy spectrum in heavily doped stressed materials can be written as

$$\overline{P_i}(E,\eta_{gi})k_x^2 + \overline{Q_i}(E,\eta_{gi})k_y^2 + \overline{S_i}(E,\eta_{gi})k_z^2 = 1 \tag{C1.24}$$

where $$\overline{P_i}(E,\eta_{gi}) = \frac{[\gamma_0(E,\eta_{gi}) - I_0 T_{1i}]}{\overline{\Delta_i}(E,\eta_{gi})},$$

$$\overline{\Delta_i}(E,\eta_{gi}) = \left[\frac{-q_i\eta_{gi}^3}{2\sqrt{\pi}}\exp\left(\frac{-E^2}{\eta_{gi}^2}\right)\left[1+\frac{E^2}{\eta_{gi}^2}\right] - R_i\theta_0(E,\eta_{gi}) + V_i\gamma_0(E,\eta_{gi}) + \frac{\zeta_i}{2}\left[1+Erf\left(\frac{E}{\eta_{gi}}\right)\right]\right],$$

$$I_0 = \frac{1}{2}[1+Erf(E/\eta_{gi})], \quad \overline{Q_i}(E,\eta_{gi}) = \frac{[\gamma_0(E,\eta_{gi}) - I_0 T_{2i}]}{\overline{\Delta_i}(E,\eta_{gi})}$$

and $$\overline{S_i}(E,\eta_{gi}) = \frac{[\gamma_0(E,\eta_{gi}) - I_0 T_{3i}]}{\overline{\Delta_i}(E,\eta_{gi})}$$

The energy-wave vector dispersion relation of the conduction electrons in heavily doped strained layer SLs with graded interfaces can be expressed as

$$\cos(L_0 k) = \frac{1}{2}\overline{\phi_6}(E,k_s) \tag{C1.25}$$

where

$$\begin{aligned}\overline{\phi_6}(E,k_s) = {} & [2\cosh[T_4(E,\eta_{g2})]\cos[T_5(E,\eta_{g1})]] + [T_6(E,k_s)]\sinh[T_4(E,\eta_{g2})]\sin[T_5(E,\eta_{g1})] \\ & + \Delta_0\left[\left(\frac{k_0^2(E,\eta_{g2})}{k'(E,\eta_{g1})} - 3k'(E,\eta_{g1})\right)\cosh[T_4(E,\eta_{g2})]\sin[T_5(E,\eta_{g1})]\right. \\ & \left. + \left(3k_0(E,\eta_{g2}) - \frac{k'^2(E,\eta_{g1})}{k_0(E,\eta_{g2})}\right)\sinh[T_4(E,\eta_{g2})]\cos[T_5(E,\eta_{g1})]\right] \\ & + \Delta_0[2(k_0^2(E,\eta_{g2}) - k'^2(E,\eta_{g1}))\cosh[T_4(E,\eta_{g2})]\cos[T_5(E,\eta_{g1})]] \\ & + \frac{1}{12}\left(\frac{5k_0^3(E,\eta_{g2})}{k'(E,\eta_{g1})} + \frac{5k'^3(E,\eta_{g1})}{k_0(E,\eta_{g2})} - 34k_0(E,\eta_{g2})k'(E,\eta_{g1})\right)\sinh[T_4(E,\eta_{g2})]\sin\left[T_5\left(E,\eta_{g1}\right)\right]\Big]\end{aligned}$$

$$[T_4(E,\eta_{g2})] = k_0(E,\eta_{g2})[a_0 - \Delta_0],$$

$$k_0(E,\eta_{g2}) = [\overline{S_2}(E-V_0,\eta_{g2})]^{-1/2}\left[\overline{P_2}(E-V_0,\eta_{g2})k_x^2 + \overline{Q_2}(E-V_0,\eta_{g2})k_y^2 - 1\right]^{1/2},$$

$$T_5(E,\eta_{g1}) = k'(E,\eta_{g1})[b_0 - \Delta_0],$$

$$k'(E,\eta_{g1}) = [\overline{S_1}(E,\eta_{g1})]^{-1/2}\left[1 - \overline{P_1}(E,\eta_{g1})k_x^2 - \overline{Q_1}(E,\eta_{g1})k_y^2\right]^{1/2}$$

and

$$T_6(E,k_s) = \left[\frac{k_0(E,\eta_{g2})}{k'(E,\eta_{g1})} - \frac{k'(E,\eta_{g1})}{k_0(E,\eta_{g2})}\right]$$

Therefore, the dispersion relation of the conduction electrons in heavily doped strained layer QDSLs with graded interfaces can be expressed as

$$\cos(L_0 k_0) = \frac{1}{2}\overline{\phi_6}(E,n) \tag{C1.26}$$

where

$$\begin{aligned}
\overline{\phi_6}(E,n) &= [2\cosh[T_4(E,n,\eta_{g2})]\cos[T_5(E,n,\eta_{g1})]] \\
&\quad + [T_6(E,n)]\sinh[T_4(E,n,\eta_{g2})] \\
&\quad \times \sin[T_5(E,n,\eta_{g1})] \\
&\quad + \Delta_0\left[\left(\frac{k_0^2(E,n,\eta_{g2})}{k_0'(E,n,\eta_{g1})} - 3k_0'(E,n,\eta_{g1})\right)\cosh[T_4(E,n,\eta_{g2})]\sin[T_5(E,n,\eta_{g1})]\right. \\
&\quad \left. + \left(3k_0(E,n,\eta_{g2}) - \frac{k_0'^2(E,n,\eta_{g1})}{k_0(E,n,\eta_{g2})}\right)\sinh[T_4(E,n,\eta_{g2})]\cos[T_5(E,n,\eta_{g1})]\right] \\
&\quad + \Delta_0\left[2(k_0^2(E,n,\eta_{g2}) - k_{0D}'^2(E,n,\eta_{g1}))\cosh[T_4(E,n,\eta_{g2})]\cos[T_5(E,n,\eta_{g1})]\right] \\
&\quad + \frac{1}{12}\left(\frac{5k_0^3(E,n,\eta_{g2})}{k_0'(E,n,\eta_{g1})} + \frac{5k_0'^3(E,n,\eta_{g1})}{k_0(E,n,\eta_{g2})} - 34k_0(E,n,\eta_{g2})k_0'(E,n,\eta_{g1})\right) \\
&\quad \times \sinh[T_4(E,n,\eta_{g2})]\sin[T_5(E,n,\eta_{g1})]
\end{aligned}$$

$$\begin{aligned}
[T_4(E,n,\eta_{g2})] &= k_0(E,n,\eta_{g2})[a_0 - \Delta_0], \\
k_0(E,n,\eta_{g2}) &= [\overline{S_2}(E - V_0,\eta_{g2})]^{-1/2}.\left[\left[(n+1/2)\hbar eB/(\sqrt{\rho_1(E)\rho_2(E)})\right] - 1\right]^{1/2}, \\
\rho_1(E) &= \hbar^2/(2\overline{P_2}(E - V_0,\eta_{g2})), \quad \rho_2(E) = \hbar^2/(2\overline{Q_2}(E - V_0,\eta_{g2})) \\
T_5(E,n,\eta_{g1}) &= k_0'(E,n,\eta_{g1})[b_0 - \Delta_0], \\
k_0'(E,n,\eta_{g1}) &= [\overline{S_1}(E,n,\eta_{g1})]^{-1/2}\left[1 - \left[(n+1/2)\hbar eB/(\sqrt{\rho_3(E)\rho_4(E)})\right]\right]^{1/2} \\
\rho_3(E) &= \hbar^2/(2\overline{P_1}(E,\eta_{g1})), \quad \rho_4(E) = \hbar^2/(2\overline{Q_1}(E,\eta_{g1})) \\
T_6(E,n) &= \left[\frac{k_0(E,n,\eta_{g2})}{k_0'(E,n,\eta_{g1})} - \frac{k_0'(E,n,\eta_{g1})}{k_0(E,n,\eta_{g2})}\right]
\end{aligned}$$

Therefore, the EEM in this case assumes the form

$$m^*(E_{F6},n) = \left(\frac{\hbar^2}{2}\right)\left[\bar{\Phi}_6'(E_{F6},n)\cos^{-1}\left[\frac{1}{2}\bar{\Phi}_6(E_{F6},n)\right]\right] / \left(\sqrt{1 - (1/4)\bar{\Phi}_6^2(E_{F6},n)}\right) \tag{C1.27}$$

where E_{F6} is the Fermi energy in this case.

The electron concentration is given by

$$n_0 = \frac{g_v eB}{\pi^2 \hbar} \sum_{n=0}^{n_{\max}} [\phi_{9C}(E_{F6}, n) + \phi_{10C}(E_{F6}, n)]$$

where $$\phi_{9C}(E_{F6}, n) = \left[\frac{1}{L_0^2} \left[\cos^{-1} \left\{ \frac{1}{2} \bar{\Phi}_6(E_{F6}, n) \right\} \right]^2 - \frac{2eB}{\hbar} \left(n + \frac{1}{2} \right) \right]^{1/2} \tag{C1.28}$$

and $$\phi_{8C}(E_{F6}, n) = \sum_{r=1}^{s} \theta_{2r,6} \left[\phi_{7C}(E_{F6}, n) \right]$$

C.1.2.6 Study of EEM in Heavily Doped Effective Mass III–V Superlattices

Following Sasaki [94], the electron dispersion law in III–V heavily doped effective mass superlattices (EMSLs) can be written as

$$k_x^2 = \left[\frac{1}{L_0^2} \{ \cos^{-1} (f_{21}(E, k_y, k_z)) \}^2 - k_\perp^2 \right] \tag{C1.29}$$

in which

$$\begin{aligned} f_{21}(E, k_y, k_z) &= a_1 \cos\left[a_0 C_{21}(E, k_\perp, \eta_{g1}) + b_0 D_{21}(E, k_\perp, \eta_{g2}) \right] \\ &\quad - a_2 \cos\left[a_0 C_{21}(E, k_\perp, \eta_{g1}) - b_0 D_{21}(E, k_\perp, \eta_{g2}) \right], \\ k_\perp^2 &= k_y^2 + k_z^2, \end{aligned}$$

$$a_1 = \left[\sqrt{\frac{M_2(0, \eta_{g2})}{M_1(0, \eta_{g1})}} + 1 \right]^2 \left[4 \left(\frac{M_2(0, \eta_{g2})}{M_1(0, \eta_{g1})} \right)^{1/2} \right]^{-1},$$

$$a_2 = \left[\sqrt{\frac{M_2(0, \eta_{g2})}{M_1(0, \eta_{g1})}} - 1 \right]^2 \left[4 \left(\frac{M_2(0, \eta_{g2})}{M_1(0, \eta_{g1})} \right)^{1/2} \right]^{-1},$$

$$\begin{aligned} M_i(0, \eta_{gi}) = m_{ci}^* \Bigg[& \frac{-2}{\sqrt{\pi}} T(0, \eta_{gi}) + 2 \Bigg[\frac{\alpha_i b_i}{c_i} \frac{\eta_{gi}}{\sqrt{\pi}} + \frac{1}{2} \left(\frac{\alpha_i c_i + c_i b_i - \alpha_i b_i}{c_i^2} \right) \\ & + \frac{1}{\sqrt{\pi} c_i} \left(1 - \frac{\alpha_i}{c_i} \right) \left(1 - \frac{b_i}{c_i} \right) - \frac{1}{c_i} \left(1 - \frac{\alpha_i}{c_i} \right) \left(1 - \frac{b_i}{c_i} \right) \frac{2}{c_i \eta_{gi} \sqrt{\pi}} \\ & \times \Bigg\{ \frac{-2}{c_i \eta_{gi}^2} \exp\left(\frac{-1}{c_i^2 \eta_{gi}^2} \right) \left(\sum_{p=1}^{\alpha} \left(\exp\left(\frac{-p^2}{4} \right) \right) \frac{1}{p} \sinh\left(\frac{p}{c_i \eta_{gi}} \right) \right) \\ & + \exp\left(\frac{-1}{c_i^2 \eta_{gi}^2} \right) \left(\sum_{p=1}^{\alpha} \exp\left(\frac{-p^2}{4} \right) \frac{1}{\eta_{gi}} \cosh\left(\frac{p}{c_i \eta_{gi}} \right) \right) \Bigg\} \Bigg] \Bigg], \end{aligned}$$

$$\mathrm{T}(0,\eta_{gi}) = 2\left[\frac{\alpha_i b_i}{c_i}\frac{\eta_{gi}^2}{4} + \left(\frac{\alpha_i c_i + b_i c_i - \alpha_i b_i}{c_i^2}\right)\frac{\eta_{gi}}{2\sqrt{\pi}} + \frac{1}{2c_i}\left(1-\frac{\alpha_i}{c_i}\right)\left(1-\frac{b_i}{c_i}\right)\right.$$
$$\left. - \frac{1}{c_i}\left(1-\frac{\alpha_i}{c_i}\right)\left(1-\frac{b_i}{c_i}\right)\frac{2}{c_i\eta_{gi}\sqrt{\pi}}\exp\left(\frac{-1}{c_i^2\eta_{gi}^2}\right)\sum_{p=1}^{\alpha}\frac{\exp(-p^2/4)}{p}\sinh\left(\frac{p}{c_i\eta_{gi}}\right)\right],$$

$$C_{21}(E,k_\perp,\eta_{g1}) = e_1 + ie_2, \quad D_{21}(E,k_\perp,\eta_{g2}) = e_3 + ie_4,$$

$$e_1 = \left[\left(\left(\sqrt{t_1^2+t_2^2}+t_1\right)/2\right)\right]^{\frac{1}{2}}, \quad \mathrm{e}_2 = \left[\left(\left(\sqrt{t_1^2+t_2^2}-t_1\right)/2\right)\right]^{\frac{1}{2}},$$

$$t_1 = \left[\frac{2m_{c1}^*}{\hbar^2}T_{11}(E,\Delta_1,\eta_{g1},E_{g1}) - k_\perp^2\right], \quad \mathrm{t}_2 = \frac{2m_{c1}^*}{\hbar^2}T_{21}(E,\Delta_1,\eta_{g1},E_{g1}),$$

$$e_3 = \left[\frac{\sqrt{t_3^2+t_4^2}+t_3}{2}\right]^{1/2}, \quad e_4 = \left[\frac{\sqrt{t_3^2+t_4^2}-t_3}{2}\right]^{1/2}$$

$$t_3 = \left[\frac{2m_{c2}^*}{\hbar^2}T_{12}(E,\Delta_2,\eta_{g2},E_{g2}) - k_\perp^2\right], \quad \mathrm{t}_4 = \frac{2m_{c2}^*}{\hbar^2}T_{22}(E,\Delta_2,\eta_{g2},E_{g2}),$$

Therefore, (C1.29) can be expressed as

$$k_x^2 = \delta_7 + i\delta_8 \tag{C1.30}$$

where

$$\delta_7 = \left[\frac{1}{L_0^2}(\delta_5^2 - \delta_6^2) - k_\perp^2\right], \quad \delta_5 = \cos^{-1}\mathrm{p}_5,$$

$$\mathrm{p}_5 = \left[\frac{1-\delta_3^2-\delta_4^2-\sqrt{(1-\delta_3^2-\delta_4^2)^2+4\delta_4^2}}{2}\right]^{1/2},$$

$$\delta_3 = (a_1\cos\Delta_1\cosh\Delta_2 - a_2\cos\Delta_3\cosh\Delta_4),$$
$$\delta_4 = (a_1\sin\Delta_1\sinh\Delta_2 - a_2\sin\Delta_3\sinh\Delta_4),$$
$$\Delta_1 = (a_0e_1 + b_0e_3), \quad \Delta_2 = (a_0e_2 + b_0e_4), \quad \Delta_3 = (a_0e_1 - b_0e_3),$$
$$\Delta_4 = (a_0e_2 - b_0e_4), \delta_6 = \sinh^{-1}\mathrm{p}_5 \quad \text{and} \quad \delta_8 = \left[2\delta_5\delta_6/L_0^2\right]$$

Therefore, the electron dispersion relation in heavily doped III–V QDSL assumes the form

$$(k_z)^2 = \delta_{7E,n} + i\delta_{8E,n} \tag{C1.31}$$

where

$$\delta_{7E,n} = \left[\frac{1}{L_0^2}\left(\delta_{5E,n}^2 - \delta_{6E,n}^2\right) - \left\{\frac{2eB}{\hbar}(n+\frac{1}{2})\right\}\right],\ \delta_{5E,n} = \cos^{-1} p_{5E,n},$$

$$p_{5E,n} = \left[\frac{1 - \delta_{3E,n}^2 - \delta_{4E,n}^2 - \sqrt{\left(1 - \delta_{3E,n}^2 - \delta_{4E,n}^2\right)^2 + 4\delta_{4E,n}^2}}{2}\right]^{1/2},$$

$$\delta_{3E,n} = \left(a_1 \cos\Delta_{1E,n} \cosh\Delta_{2E,n} - a_2 \cos\Delta_{3E,n} \cosh\Delta_{4E,n}\right),$$
$$\delta_{4E,n} = \left(a_1 \sin\Delta_{1E,n} \sinh\Delta_{2E,n} - a_2 \sin\Delta_{3E,n} \sinh\Delta_{4E,n}\right),$$
$$\Delta_{1E,n} = \left(a_0 e_{1E,n} + b_0 e_{3E,n}\right),\quad \Delta_{2E,n} = \left(a_0 e_{2E,n} + b_0 e_{4E,n}\right),$$
$$\Delta_{3E,n} = \left(a_0 e_{1E,n} - b_0 e_{3E,n}\right),\quad \Delta_{4E,n} = \left(a_0 e_{2E,n} - b_0 e_{4E,n}\right),$$
$$\delta_{6E,n} = \sinh^{-1} p_{5E,n} \quad \text{and} \quad \delta_{8E,n} = \left[2\delta_{5E,n}\delta_{6E,n}/L_0^2\right],$$
$$e_{1E,n} = [((\sqrt{t_{1E,n}^2 + t_2^2} + t_{1E,n})/2)]^{\frac{1}{2}},\quad e_{2E,n} = [((\sqrt{t_{1E,n}^2 + t_2^2} - t_{1E,n})/2)]^{\frac{1}{2}},$$

$$e_{3E,n} = \left[\frac{\sqrt{t_{3E,n}^2 + t_4^2} + t_{3E,n}}{2}\right]^{1/2},\quad e_{4E,n} = \left[\frac{\sqrt{t_{3E,n}^2 + t_4^2} - t_{3E,n}}{2}\right]^{1/2},$$

$$t_{1E,n} = \left[\frac{2m_{c1}^*}{\hbar^2} T_{11}\left(E, \Delta_1, \eta_{g1}, E_{g1}\right) - \frac{2eB}{\hbar}\left(n + \frac{1}{2}\right)\right],$$
$$t_{3E,n} = \left[\frac{2m_{c2}^*}{\hbar^2} T_{12}\left(E, \Delta_2, \eta_{g2}, E_{g2}\right) - \frac{2eB}{\hbar}\left(n + \frac{1}{2}\right)\right]$$

Therefore, the EEM in this case assumes the form

$$m^*(E_{F7}, n) = \frac{\hbar^2}{2}\delta'_{7E,n}\Big|_{E=E_{F7}} \tag{C1.32}$$

where E_{F7} is the Fermi energy in this case.

The electron concentration is given by

$$n_0 = \frac{g_v eB}{\pi^2\hbar}\sum_{n=0}^{n_{\max}}\left[\phi_{11}(E_{F7}, n) + \phi_{12}(E_{F7}, n)\right] \tag{C1.33}$$

where

$$\phi_{11}(E_{F7}, n) = \left[\left(\delta_{7E_{F7},n} + \sqrt{\delta_{7E_{F7},n}^2 - \delta_{8E_{F7},n}}\right)\Big/2\right]^{1/2}$$

and $\phi_{12}(E_{F7}, n) = \sum_{r=1}^{s} \theta_{2r,7}[\phi_{11}(E_{F7}, n)]$

C.1.2.7 Study of EEM in Heavily Doped Effective Mass II–VI Superlattices

Following Sasaki [95], the electron dispersion law in heavily doped II–VI EMSLs can be written as

$$k_z^2 = \Delta_{13} + i\Delta_{14}, \tag{C1.34}$$

where $\Delta_{13} = \left[\frac{1}{L_0^2}\left(\Delta_{11}^2 - \Delta_{12}^2\right) - k_s^2\right]$

$$\Delta_{11} = \cos^{-1} p_6,\ \mathrm{p}_6 = \left[\frac{1 - \Delta_9^2 - \Delta_{10}^2 - \sqrt{\left(1 - \Delta_9^2 - \Delta_{10}^2\right)^2 + 4\Delta_{10}^2}}{2}\right]^{1/2},$$

$$\Delta_9 = \left(\overline{a_1}\cos\Delta_6\cosh\Delta_7 - \overline{a_2}\cos\Delta_8\cosh\Delta_7\right),$$
$$\Delta_{10} = \left(\overline{a_1}\sin\Delta_6\sinh\Delta_7 + \overline{a_2}\sin\Delta_8\sinh\Delta_7\right),$$
$$\Delta_6 = \left[a_0 C_{22}\left(E, k_s, \eta_{g1}\right) + b_0 e_3\right], \quad \Delta_7 = b_0 e_4,$$
$$\Delta_8 = \left[a_0 C_{22}\left(E, k_s, \eta_{g1}\right) - b_0 e_3\right],$$

$$C_{22}\left(E, k_s, \eta_{g1}\right) = \left[\frac{2m_{\parallel,1}^*}{\hbar^2}\left\{\gamma_3\left(E, \eta_{g1}\right) - \frac{\hbar^2 k_s^2}{2m_{\perp,1}^*} \mp C_0 k_s\right\}\right]^{1/2},$$

$$\overline{a_1} = \left[\sqrt{\frac{M_2\left(0, \eta_{g2}\right)}{\overline{M_1}\left(0, \eta_{g1}\right)}} + 1\right]^2 \left[4\left(\frac{M_2\left(0, \eta_{g2}\right)}{\overline{M_1}\left(0, \eta_{g1}\right)}\right)^{1/2}\right]^{-1},$$

$$\overline{M_1}\left(0, \eta_{g1}\right) = m_{c1}^*\left(1 - \frac{2}{\pi}\right),$$

$$\overline{a_2} = \left[\sqrt{\frac{M_2\left(0, \eta_{g2}\right)}{\overline{M_1}\left(0, \eta_{g1}\right)}} - 1\right]^2 \left[4\left(\frac{M_2\left(0, \eta_{g2}\right)}{\overline{M_1}\left(0, \eta_{g1}\right)}\right)^{1/2}\right]^{-1}$$

$$\Delta_{12} = \cos^{-1} p_6, \quad \Delta_{14} = \frac{2\Delta_{11}\Delta_{12}}{L_0^2}$$

The electron dispersion law in heavily doped II–VI QDSL can be written as

$$\left(k_z\right)^2 = \Delta_{13E,n} + i\Delta_{14E,n}, \tag{C1.35}$$

$$\text{where} \quad \Delta_{13E,n} = \left[\frac{1}{L_0^2}\left(\Delta_{11E,n}^2 - \Delta_{12E,n}^2\right) - \left\{\frac{2eB}{\hbar}\left(n+\frac{1}{2}\right)\right\}\right]$$

$$\Delta_{11E,n} = \cos^{-1} p_{6E,n},$$

$$p_{6E,n} = \left[\frac{1-\Delta_{9E,n}^2 - \Delta_{10E,n}^2 - \sqrt{\left(1-\Delta_{9E,n}^2-\Delta_{10E,n}^2\right)^2 + 4\Delta_{10E,n}^2}}{2}\right]^{1/2},$$

$$\Delta_{9E,n} = \left(\overline{a_1}\cos\Delta_{6E,n}\cosh\Delta_{7E,n} - \overline{a_2}\cos\Delta_{8E,n}\cosh\Delta_{7E,n}\right),$$

$$\Delta_{10E,n} = \left(\overline{a_1}\sin\Delta_{6E,n}\sinh\Delta_{7E,n} + \overline{a_2}\sin\Delta_{8E,n}\sinh\Delta_{7E,n}\right),$$

$$\Delta_{6E,n} = \left[a_0 C_{22E,n}(E_{E,n},\eta_{g1}) + b_0 e_{3E,n}\right], \quad \Delta_{7E,n} = b_0 e_{4E,n},$$

$$\Delta_{8E,n} = \left[a_0 C_{22E,n}(E_{E,n},\eta_{g1}) - b_0 e_{3E,n}\right],$$

$$C_{22E,n}(E_{E,n},\eta_{g1}) = \left[\frac{2m_{\parallel,1}^*}{\hbar^2}\left\{\gamma_3(E_{E,n},\eta_{g1}) - \frac{\hbar^2}{2m_{\perp,1}^*}\left\{\frac{2eB}{\hbar}\left(n+\frac{1}{2}\right)\right\}\right.\right.$$
$$\left.\left. \mp C_0\left[\left\{\frac{2eB}{\hbar}\left(n+\frac{1}{2}\right)\right\}\right]^{1/2}\right\}\right]^{1/2},$$

$$\Delta_{12E,n} = \cos^{-1} p_{6E,n}, \quad \Delta_{14E,n} = \frac{2\Delta_{11E,n}\Delta_{12E,n}}{L_0^2},$$

Therefore, the EEM in this case assumes the form

$$m^*(E_{F8},n) = \frac{\hbar^2}{2}\Delta'_{13E,n}\Big|_{E=E_{F8}} \tag{C1.36}$$

where E_{F8} is the Fermi energy in this case.

The electron concentration is given by

$$n_0 = \frac{g_v eB}{\pi^2\hbar}\sum_{n=0}^{n_{\max}}\left[\phi_{13}(E_{F8},n) + \phi_{14}(E_{F8},n)\right] \tag{C1.37}$$

where

$$\phi_{13}(E_{F8},n) = \left[\left(\Delta_{13E_{F8},n} + \sqrt{\Delta_{13E_{F8},n}^2 - \Delta_{14E_{F8},n}}\right)\Big/2\right]^{1/2}$$

$$\text{and} \quad \phi_{14}(E_{F8},n) = \sum_{r=1}^{s}\theta_{2r,7}\left[\phi_{13}(E_{F8},n)\right]$$

C1.2.8 Study of EEM in Heavily Doped Effective Mass IV–VI Superlattices

Following Sasaki [95], the electron dispersion law in IV–VI, EMSLs can be written as

$$k_z^2 = \left[\frac{1}{L_0^2}\left\{\cos^{-1}\left(f_{23}\left(E,k_x,k_y\right)\right)\right\}^2 - k_s^2\right] \quad \text{(C1.38)}$$

where

$$f_{23}\left(E,k_x,k_y\right) = a_3\cos\left[a_0C_{23}\left(E,k_x,k_y,\eta_{g1}\right)+b_0D_{23}\left(E,k_x,k_y,\eta_{g1}\right)\right] - a_4\cos\left[a_0C_{23}\left(E,k_x,k_y,\eta_{g2}\right)-b_0D_{23}\left(E,k_x,k_y,\eta_{g2}\right)\right],$$

$$a_3 = \left[\sqrt{\frac{M_3(0,\eta_{g2})}{M_3(0,\eta_{g1})}}+1\right]^2\left[4\left(\frac{M_3(0,\eta_{g2})}{M_3(0,\eta_{g1})}\right)^{1/2}\right]^{-1},$$

$$a_4 = \left[\sqrt{\frac{M_3(0,\eta_{g2})}{M_3(0,\eta_{g1})}}-1\right]^2\left[4\left(\frac{M_3(0,\eta_{g2})}{M_3(0,\eta_{g1})}\right)^{1/2}\right]^{-1}$$

$$M_3(0,\eta_{gi}) = (4\overline{p_{9,i}})^{-1}\left[\left\{\alpha_i\left(1-\frac{2}{\pi}\right)\left(\frac{1}{m_{l,i}^+}-\frac{1}{m_{l,i}^-}\right)\right\}+\left[[\overline{q_{9,i}}(0,\eta_{gi})]^2\right.\right.$$
$$\left.\left.+(4\overline{p_{9,i}})\overline{R_{9,i}}(0,\eta_{gi})\right]^{-1/2}\left[\alpha_i\left(1-\frac{2}{\pi}\right)\left(\frac{1}{m_{l,i}^+}-\frac{1}{m_{l,i}^-}\right)\overline{q_{9,i}}(0,\eta_{gi})+2\overline{p_{9,i}}\left(1-\frac{2}{\pi}+\frac{\alpha_i\eta_{gi}}{\sqrt{\pi}}\right)\right]\right],$$

$$\overline{p_{9,i}} = \frac{\alpha_i\hbar^4}{4m_{l,i}^+m_{l,i}^-},\quad \overline{q_{9,i}}(0,\eta_{gi}) = \left[\frac{\hbar^2}{2}\left(\frac{1}{m_{l,i}^+}+\frac{1}{m_{l,i}^-}\right)-\frac{\alpha_i\eta_{gi}}{\sqrt{\pi}}\left(\frac{1}{m_{l,i}^+}-\frac{1}{m_{l,i}^-}\right)\right],$$

$$\overline{R_{9,i}}(0,\eta_{gi}) = \left[\frac{\eta_{gi}}{\sqrt{\pi}}+\frac{\alpha_i\eta_{gi}^2}{2}\right],\ C_{23}(E,k_x,k_y,\eta_{g1}) = \left[[2\overline{p_{9,1}}]^{-1}[-\overline{q_{9,1}}(E,k_x,k_y,\eta_{g1})\right.$$
$$\left.+\left[\{\overline{q_{9,1}}(E,k_x,k_y,\eta_{g1})\}^2+(4\overline{p_{9,1}})\overline{R_{9,1}}(E,k_x,k_y,\eta_{g1})\right]^{1/2}\right]\Big]^{1/2},$$

$$D_{23}(E,k_x,k_y,\eta_{g2}) = \left[[2\overline{p_{9,2}}]^{-1}[-\overline{q_{9,2}}(E,k_x,k_y,\eta_{g2})+\left[\{\overline{q_{9,2}}(E,k_x,k_y,\eta_{g2})\}^2\right.\right.$$
$$\left.\left.+(4\overline{p_{9,2}})\overline{R_{9,2}}(E,k_x,k_y,\eta_{g2})\right]^{1/2}\right]\Big]^{1/2},$$

$$\overline{q_{9,i}}(E,k_x,k_y,\eta_{gi}) = \left[\frac{\hbar^2}{2}\left(\frac{1}{m_{l,i}^*}+\frac{1}{m_{l,i}^-}\right)+\alpha_i\frac{\hbar^4}{4}k_s^2\left(\frac{1}{m_{l,i}^+m_{t,i}^-}+\frac{1}{m_{t,i}^+m_{l,i}^-}\right)\right.$$
$$\left.-\alpha_i\gamma_3(E,\eta_{gi})\left(\frac{1}{m_{l,i}^+}-\frac{1}{m_{l,i}^-}\right)\right],$$

$$\overline{R_{9,i}}(E,k_x,k_y,\eta_{gi}) = \left[\gamma_2(E,\eta_{gi})+\gamma_3(E,\eta_{gi})\alpha_i\frac{\hbar^2}{2}k_s^2\left(\frac{1}{m_{t,i}^+}-\frac{1}{m_{t,i}^-}\right)\right.$$
$$\left.-\frac{\hbar^2}{2}k_s^2\left(\frac{1}{m_{t,i}^*}-\frac{1}{m_{t,i}^-}\right)-\frac{\alpha\hbar^6}{4}\frac{k_s^4}{m_{t,i}^-m_{t,i}^+}\right],\quad a_5 = \left[\sqrt{\frac{m_2^*}{m_1^*}}+1\right]^2\left[4\left(\frac{m_2^*}{m_1^*}\right)^{1/2}\right]^{-1}$$

Therefore, the electron dispersion law in heavily doped IV–VI, EMSLs under magnetic quantization can be written as

$$(k_z)^2 = [[1/L_0^2]\{\cos^{-1}(f_{23}(E,n))\}^2 - (\frac{2eB}{\hbar}(n+\frac{1}{2}))] \tag{C1.39}$$

where

$$f_{23}(E,n) = a_3\cos\left[a_0 C_{23E,n}(E,n,\eta_{g1}) + b_0 D_{23E,n}(E,n,\eta_{g1})\right] - a_4\cos\left[a_0 C_{23E,n}(E,n,\eta_{g2}) - b_0 D_{23E,n}(E,n,\eta_{g2})\right],$$

$$C_{23}(E,n,\eta_{g1}) = \left[[2\overline{p_{9,1}}]^{-1}[-\overline{q_{9,1}}(E,n,\eta_{g1}) + \left[\{\overline{q_{9,1}}(E,n,\eta_{g1})\}^2 + (4\overline{p_{9,1}})\overline{R_{9,1}}(E,n,\eta_{g1})\right]^{1/2}\right]\Big]^{1/2},$$

$$D_{23}(E,n,\eta_{g2}) = \left[[2\overline{p_{9,2}}]^{-1}[-\overline{q_{9,2}}(E,n,\eta_{g2}) + \left[\{\overline{q_{9,2}}(E,n,\eta_{g2})\}^2 + (4\overline{p_{9,2}})\overline{R_{9,2}}(E,n,\eta_{g2})\right]^{1/2}\right]\Big]^{1/2},$$

$$\overline{q_{9,i}}(E,n,\eta_{gi}) = \left[\frac{\hbar^2}{2}\left(\frac{1}{m_{l,i}^*}+\frac{1}{m_{l,i}^-}\right) + \alpha_i\frac{\hbar^4}{4}\left(\frac{2eB}{\hbar}\left(n+\frac{1}{2}\right)\right)\left(\frac{1}{m_{l,i}^+ m_{t,i}^-}+\frac{1}{m_{t,i}^+ m_{l,i}^-}\right) - \alpha_i\gamma_3(E,\eta_{gi})\left(\frac{1}{m_{l,i}^+}-\frac{1}{m_{l,i}^-}\right)\right],$$

$$\overline{R_{9,i}}(E,n,\eta_{gi}) = \left[\gamma_2(E,\eta_{gi}) + \gamma_3(E,\eta_{gi})\alpha_i\frac{\hbar^2}{2}\left(\frac{2eB}{\hbar}\left(n+\frac{1}{2}\right)\right)\left(\frac{1}{m_{t,i}^+}-\frac{1}{m_{t,i}^-}\right) - \frac{\hbar^2}{2}\left(\frac{2eB}{\hbar}\left(n+\frac{1}{2}\right)\right)\left(\frac{1}{m_{t,i}^*}-\frac{1}{m_{t,i}^-}\right) - \frac{\alpha\hbar^6}{4}\frac{\left(\frac{2eB}{\hbar}\left(n+\frac{1}{2}\right)\right)^2}{m_{t,i}^- m_{t,i}^+}\right],$$

Therefore, the EEM in this case assumes the form

$$m^*(E_{F9},n) = \left(\frac{\hbar^2}{2}\right)\left[f_{23}'(E_{F9},n)\cos^{-1}\left[\frac{1}{2}f_{23}(E_{F9},n)\right]\right] / \left(\sqrt{1-(1/4)f_{23}^2(E_{F9},n)}\right) \tag{C1.40}$$

where E_{F9} is the Fermi energy in this case.

The electron concentration is given by

$$n_0 = \frac{g_v eB}{\pi^2 \hbar} \sum_{n=0}^{n_{\max}} \left[\phi_{15}(E_{F9}, n) + \phi_{16}(E_{F9}, n) \right] \tag{C1.41}$$

where $\phi_{15}(E_{F9}, n) = \left[\frac{1}{L_0^2} \left[\cos^{-1} \left\{ \frac{1}{2} f_{23}(E_{F9}, n) \right\} \right]^2 - \frac{2eB}{\hbar} \left(n + \frac{1}{2} \right) \right]^{1/2}$

and $\phi_{16}(E_{F9}, n) = \sum_{r=1}^{s} \theta_{2r,6} \left[\phi_{15}(E_{F9}, n) \right]$

C1.2.9 Study of EEM in Heavily Doped Effective Mass HgTe/CdTe Superlattices

Following Sasaki [95], the electron dispersion law in heavily doped HgTe/CdTe EMSLs can be written as

$$k_z^2 = \Delta_{13H} + i\Delta_{14H} \tag{C1.42}$$

where

$$\Delta_{13H} = \left[\frac{1}{L_0^2} (\Delta_{11H}^2 - \Delta_{12H}^2) - k_s^2 \right]$$

$$\Delta_{11H} = \cos^{-1} p_{6H},$$

$$p_{6H} = \left[\frac{1 - \Delta_{9H}^2 - \Delta_{10H}^2 - \sqrt{(1 - \Delta_{9H}^2 - \Delta_{10H}^2)^2 + 4\Delta_{10H}^2}}{2} \right]^{1/2},$$

$$\Delta_{9H} = (\overline{a_{1H}} \cos \Delta_{5H} \cosh \Delta_{6H} - \overline{a_{2H}} \cos \Delta_{7H} \cosh \Delta_{6H}),$$

$$\Delta_{10H} = (\overline{a_{1H}} \sin \Delta_{5H} \sinh \Delta_{6H} + \overline{a_{2H}} \sin \Delta_{7H} \sinh \Delta_{6H}),$$

$$\Delta_{5H} = \left[a_0 C_{22H}(E, k_s, \eta_{g1}) + b_0 e_3 \right], \quad \Delta_{6H} = b_0 e_4,$$

$$\Delta_{7H} = \left[a_0 C_{22H}(E, k_s, \eta_{g1}) - b_0 e_3 \right],$$

$$C_{22H}(E, k_s, \eta_{g1}) = \left[\frac{B_{01}^2 + 2A_1 E - B_{01}(B_{01}^2 + 4A_1 E)}{2A_1^2} - k_s^2 \right]^{1/2},$$

$$\overline{a_{1H}} = \left[\sqrt{\frac{M_2(0, \eta_{g2})}{m_{c1}^*}} + 1 \right]^2 \left[4 \left(\frac{M_2(0, \eta_{g2})}{m_{c1}^*} \right)^{1/2} \right]^{-1},$$

$$\overline{a_{2H}} = \left[\sqrt{\frac{M_2(0, \eta_{g2})}{m_{c1}^*}} - 1 \right]^2 \left[4 \left(\frac{M_2(0, \eta_{g2})}{m_{c1}^*} \right)^{1/2} \right]^{-1}$$

$$\Delta_{12H} = \cos^{-1} p_{6H}, \quad \Delta_{14H} = \frac{2\Delta_{11H}\Delta_{12H}}{L_0^2}$$

The electron dispersion law in heavily doped HgTe/CdTe EMSLs under magnetic quantization can be written as

$$(k_z)^2 = \Delta_{13HE,n} + i\Delta_{14HE,n} \tag{C1.43}$$

where

$$\Delta_{13HE,n} = \left[(1/L_0^2)\left(\Delta_{11HE,n}^2 - \Delta_{12HE,n}^2\right) - \frac{2eB}{\hbar}\left(n+\frac{1}{2}\right)\right]$$

$$\Delta_{11HE,n} = \cos^{-1} p_{6HE,n},$$

$$p_{6HE,n} = \left[\left(\left(1 - \Delta_{9HE,n}^2 - \Delta_{10HE,n}^2 - \sqrt{\left(1 - \Delta_{9HE,n}^2 - \Delta_{10HE,n}^2\right)^2 + 4\Delta_{10HE,n}^2}\right)/2\right)\right]^{\frac{1}{2}}$$

$$\Delta_{9HE,n} = \left(\overline{a_{1H}} \cos \Delta_{5HE,n} \cosh \Delta_{6HE,n} - \overline{a_{2H}} \cos \Delta_{7HE,n} \cosh \Delta_{6HE,n}\right),$$

$$\Delta_{10HE,n} = \left(\overline{a_{1H}} \sin \Delta_{5HE,n} \sinh \Delta_{6HE,n} + \overline{a_{2H}} \sin \Delta_{7HE,n} \sinh \Delta_{6HE,n}\right),$$

$$\Delta_{5HE,n} = \left[a_0 C_{22HE,n}\left(E_{E,n}, \eta_{g1}\right) + b_0 e_3\right], \quad \Delta_{6HE,n} = b_0 e_4,$$

$$\Delta_{7HE,n} = \left[a_0 C_{22HE,n}\left(E_{E,n}, \eta_{g1}\right) - b_0 e_3\right],$$

$$C_{22HE,n}\left(E_{E,n}, \eta_{g1}\right) = \left[\frac{B_{01}^2 + 2A_1 E_{E,n} - B_{01}\left(B_{01}^2 + 4A_1 E_{E,n}\right)}{2A_1^2} - \left[\frac{2eB}{\hbar}\left(n+\frac{1}{2}\right)\right]\right]^{1/2},$$

$$\Delta_{12HE,n} = \cos^{-1} p_{6HE,n}, \quad \Delta_{14HE,n} = \frac{2\Delta_{11HE,n}\Delta_{12HE,n}}{L_0^2}$$

Therefore, the EEM in this case assumes the form

$$m^*(E_{F10}, n) = \frac{\hbar^2}{2} \Delta'_{13HE,n}\Big|_{E=E_{F10}} \tag{C1.44}$$

where E_{F10} is the Fermi energy in this case.

The electron concentration is given by

$$n_0 = \frac{g_v eB}{\pi^2 \hbar} \sum_{n=0}^{n_{\max}} \left[\phi_{17}(E_{F10}, n) + \phi_{18}(E_{F10}, n)\right] \tag{C1.45}$$

where

$$\phi_{17}(E_{F10}, n) = \left[\left(\Delta_{13HE_{F10},n} + \sqrt{\Delta_{13HE_{F10},n}^2 - \Delta_{14HE_{F10},n}}\right)\Big/2\right]^{1/2}$$

and $$\phi_{18}(E_{F10}, n) = \sum_{r=1}^{s} \theta_{2r,7}\left[\phi_{17}(E_{F10}, n)\right]$$

C.1.2.10 Study of EEM in Heavily Doped Strained Layer Effective Mass Superlattices

The dispersion relation of the constituent materials of heavily doped III–V super lattices can be written as

$$\bar{P}_i(E,\eta_{gi})k_x^2+\bar{Q}_i(E,\eta_{gi})k_y^2+\bar{S}_i(E,\eta_{gi})k_z^2=1 \qquad \text{(C1.46)}$$

where

$$\bar{P}_i(E,\eta_{gi})=(\gamma_0(E,\eta_{gi})-I_0T_{1i})(\bar{\Delta}_i(E,\eta_{gi})^{-1},\quad I_0=(1/2)[1+Erf(E/\eta_{gi})],$$

$$T_{1i}=\left[E_{gi}-C_{1i}^c\varepsilon_i-(a_i+C_{1i}^c)\varepsilon_i+(3/2)b_i\varepsilon_{xxi}-(b_i\varepsilon_i/2)+(\sqrt{3}d_i\varepsilon_{xyi}/2)\right],$$

$$\bar{\Delta}_i(E,\eta_{gi})=[(-q_i\eta_{gi}^3/2\sqrt{\pi})\exp(-(E^2/\eta_{gi}^2))[1+(E^2/\eta_{gi}^2)]-R_i\theta_0(E,\eta_{gi})+V_i\gamma_0(E,\eta_{gi})$$

$$+(\zeta_i/2)[1+Erf(E/\eta_{gi})]],\quad q_i=(3/2B_{2i}^2),\quad R_i=q_i[2A_i+C_{1i}^c\varepsilon_i],$$

$$A_i=E_{gi}-C_{1i}^c\varepsilon_i,\ V_i=q_i[A_i^2-(2C_{2i}^2\varepsilon_{xyi}/3)+2A_iC_{1i}^c\varepsilon_i],$$

$$\zeta_i=q_i[(2C_{2i}^2\varepsilon_{xyi}/3)-C_{1i}^c\varepsilon_iA_i^2],\ \bar{Q}_i(E,\eta_{gi})=(\gamma_0(E,\eta_{gi})-I_0T_{2i})(\bar{\Delta}_i(E,\eta_{gi})^{-1},$$

$$T_{2i}=[E_{gi}-C_{1i}^c\varepsilon_i-(a_i+C_{1i}^c)\varepsilon_i+(3/2)b_i\varepsilon_{xxi}-(b_i\varepsilon_i/2)-(\sqrt{3}d_i\varepsilon_{xyi}/2)],$$

$$\bar{S}_i(E,\eta_{gi})=(\gamma_0(E,\eta_{gi})-I_0T_{3i})(\bar{\Delta}_i(E,\eta_{gi})^{-1},$$

$$T_{3i}=[E_{gi}-C_{1i}^c\varepsilon_i+(a_i+C_{1i}^c)\varepsilon_i+(3/2)b_i\varepsilon_{zzi}-(b_i\varepsilon_i/2)],$$

The electron energy spectrum in heavily doped strained layer effective mass superlattices can be written as

$$k_z^2=\left[\frac{1}{L_0^2}\left\{\cos^{-1}\left(f_{40}\left(E,k_x,k_y\right)\right)\right\}^2-k_s^2\right] \qquad \text{(C1.47)}$$

where

$$f_{40}\left(E,k_x,k_y\right)=a_{20}\cos\left[a_0C_{40}\left(E,k_x,k_y,\eta_{g1}\right)+b_0D_{40}\left(E,k_x,k_y,\eta_{g1}\right)\right]$$

$$-a_{21}\cos\left[a_0C_{40}\left(E,k_x,k_y,\eta_{g2}\right)-b_0D_{40}\left(E,k_x,k_y,\eta_{g2}\right)\right],$$

$$a_{20}=\left[\sqrt{\frac{M_{s2}(0,\eta_{g2})}{M_{s1}(0,\eta_{g1})}}+1\right]^2\left[4\left(\frac{M_{s2}(0,\eta_{g2})}{M_{s1}(0,\eta_{g1})}\right)^{1/2}\right]^{-1},$$

$$M_{si}\left(0,\eta_{gi}\right)=(\hbar/2)\rho_i(\eta_{gi})$$

$$\rho_i(\eta_{gi}) = \left[(\eta_{gi}/2\sqrt{\pi}) - (T_{3i}/2)\right]^{-2}$$
$$\times \left[\left\{(\eta_{gi}/2\sqrt{\pi}) - (T_{3i}/2)\right\}\left\{\{(V_i/2) - (R_i\eta_{gi}/\sqrt{\pi}) + (\zeta_i/\eta_{gi}\sqrt{\pi})\right\}\right.$$
$$\left. -((1/2) - (T_{3i}/\eta_{gi}\sqrt{\pi}))\left\{(\zeta_i/2) + (V_i\eta_{gi}/2\sqrt{\pi})(R_i\eta_{gi}^2/4) - \left(q_i\eta_{gi}^3/2\sqrt{\pi}\right)\right\}\right]$$

$$a_{20} = \left[\sqrt{\frac{M_{s2}(0,\eta_{g2})}{M_{s1}(0,\eta_{g1})}} + 1\right]^2 \left[4\left(\frac{M_{s2}(0,\eta_{g2})}{M_{s1}(0,\eta_{g1})}\right)^{1/2}\right]^{-1},$$

$$a_{21} = \left[\sqrt{\frac{M_{s2}(0,\eta_{g2})}{M_{s1}(0,\eta_{g1})}} - 1\right]^2 \left[4\left(\frac{M_{s2}(0,\eta_{g2})}{M_{s1}(0,\eta_{g1})}\right)^{1/2}\right]^{-1}$$

$$C_{40}(E,k_x,k_y,\eta_{g1}) = [1 - \bar{P}_1(E,\eta_{g1})k_x^2 - \bar{Q}_1(E,\eta_{g1})k_y^2]^{1/2}[\bar{S}_1(E,\eta_{g1})]^{-1/2}$$
$$D_{40}(E,k_x,k_y,\eta_{g2}) = [1 - \bar{P}_2(E,\eta_{g2})k_x^2 - \bar{Q}_2(E,\eta_{g2})k_y^2]^{1/2}[\bar{S}_2(E,\eta_{g2})]^{-1/2}$$

Therefore, the electron dispersion law in heavily doped strained layer effective mass superlattices can be expressed as [96–108]

$$(k_z)^2 = \left[\frac{1}{L_0^2}\{\cos^{-1}(f_{40}(E,n))\}^2 - \left(\frac{2eB}{\hbar}\left(n+\frac{1}{2}\right)\right)\right] \tag{C1.48}$$

where $f_{40}(E,n) = a_{20}\cos\left[a_0C_{40}(E,n,\eta_{g1}) + b_0D_{40}(E,n,\eta_{g1})\right] - a_{21}\cos\left[a_0C_{40}(E,n,\eta_{g2}) - b_0D_{40}(E,n,\eta_{g2})\right]$,

$$C_{40}(E,n,\eta_{g1}) = \left[1 - \frac{\hbar eB}{\phi_{50}(E,\eta_{g1})}\left(n+\frac{1}{2}\right)\right]^{1/2}[\bar{S}_1(E,\eta_{g1})]^{-1/2},$$

$$\phi_{50}(E,\eta_{g1}) = \sqrt{\psi_{50}(E,\eta_{g1})\psi_{51}(E,\eta_{g1})},$$

$$\psi_{50}(E,\eta_{g1}) = \frac{\hbar^2}{2\bar{P}_1(E,\eta_{g1})}, \quad \psi_{51}(E,\eta_{g1}) = \frac{\hbar^2}{2\bar{Q}_1(E,\eta_{g1})}$$

$$D_{40}(E,n,\eta_{g2}) = \left[1 - \frac{\hbar eB}{\phi_{501}(E,\eta_{g2})}\left(n+\frac{1}{2}\right)\right]^{1/2}[\bar{S}_2(E,\eta_{g2})]^{-1/2}$$

$$\phi_{501}(E,\eta_{g2}) = \sqrt{\psi_{501}(E,\eta_{g2})\psi_{511}(E,\eta_{g2})}$$

$$\psi_{501}(E,\eta_{g2}) = \frac{\hbar^2}{2\bar{P}_2(E,\eta_{g2})}, \quad \psi_{511}(E,\eta_{g2}) = \frac{\hbar^2}{2\bar{Q}_2(E,\eta_{g2})}$$

Therefore, the EEM in this case assumes the form

$$m^*(E_{F9},n) = \left(\frac{\hbar^2}{2}\right)\left[f_{40}'(E_{F11},n)\cos^{-1}\left[\frac{1}{2}f_{40}(E_{F11},n)\right]\right] \Big/ \left(\sqrt{1-(1/4)f_{40}^2(E_{F11},n)}\right) \tag{C1.49}$$

where E_{F11} is the Fermi energy in this case.

The electron concentration is given by

$$n_0 = \frac{g_v eB}{\pi^2 \hbar} \sum_{n=0}^{n_{\max}} [\phi_{19}(E_{F11}, n) + \phi_{20}(E_{F11}, n)]$$

where $\phi_{19}(E_{F11}, n) = \left[\frac{1}{L_0^2} \left[\cos^{-1} \left\{ \frac{1}{2} f_{40}(E_{F11}, n) \right\} \right]^2 - \frac{2eB}{\hbar} \left(n + \frac{1}{2} \right) \right]^{1/2}$

and $\phi_{20}(E_{F11}, n) = \sum_{r=1}^{s} \theta_{2r,6} \left[\phi_{19}(E_{F11}, n) \right]$ (C1.50)

C1.3 Open Research Problem

Investigate the concentration dependence of the effective acceleration mass (EAM), density-of-state effective mass (DEM), concentration effective mass (CEM), conductivity effective mass (CoEM), Faraday rotation effective mass (FREM), and Optical effective mass (OEM) for bulk specimens for all the materials whose carrier energy spectra are described in this book.

References

1. W. Zawadzki, B. Lax, Phys. Rev. Lett. **16**, 1001 (1966)
2. M.J. Harrison, Phys. Rev. A **29**, 2272 (1984)
3. J. Zak, W. Zawadzki, Phys. Rev. **145**, 536 (1966)
4. W. Zawadzki, Q.H. Vrehen, B. Lax, Phys. Rev. **148**, 849 (1966)
5. Q.H. Vrehen, W. Zawadzki, M. Reine, Phys. Rev. **158**, 702 (1967)
6. M.H. Weiler, W. Zawadzki, B. Lax, Phys. Rev. **163**, 733 (1967)
7. W. Zawadzki, J. Kowalski, Phys. Rev. Lett. **27**, 1713 (1971)
8. C. Chu, Phys. Rev. Lett. **41**, 653 (1978)
9. P. Hu, C.S. Ting, Phys. Rev. B **36**, 9671 (1987)
10. E.I. Butikov, A.S. Kondratev, A.E. Kuchma, Sov. Phys. Sol. State **13**, 2594 (1972)
11. K.P. Ghatak, J.P. Banerjee, B. Goswami, B. Nag, Nonlinear Opt. Quant. Opt. **16**, 241 (1996)
12. M. Mondal, K.P. Ghatak, Phys. Stat. Sol. A**133**, K67 (1986)
13. M. Mondal, N. Chattopadhyay, K.P. Ghatak, J. Low Temp. Phys. **66**, 131 (1987)
14. K.P. Ghatak, M. Mondal, Z. Phys. B **69**, 471 (1988)
15. M. Mondal, K.P. Ghatak, Phys. Lett. A **131A**, 529 (1988)
16. M. Mondal, K.P. Ghatak, Phys. Status Solidi B Germany **147**, K179 (1988)
17. B. Mitra, K.P. Ghatak, Phys. Lett. **137A**, 413 (1989)
18. B. Mitra, A. Ghoshal, K.P. Ghatak, Phys. Stat. Sol. B **154**, K147 (1989)
19. B. Mitra, K.P. Ghatak, Phys. Stat. Sol. B **164**, K13 (1991)
20. K.P. Ghatak, B. Mitra, Int. J. Electron. **70**, 345 (1991)

21. K.P. Ghatak, B. Goswami, M. Mitra, B. Nag, Nonlinear Opt. **16**, 9 (1996)
22. K.P. Ghatak, M. Mitra, B. Goswami, B. Nag, Nonlinear Opt. **16**, 167 (1996)
23. K.P. Ghatak, J.P. Banerjee, B. Goswami, B. Nag, Nonlinear Opt. Quant. Opt. **16**, 241 (1996)
24. K.P. Ghatak, D.K. Basu, B. Nag, J. Phys. Chem. Sol. **58**, 133 (1997)
25. K.P. Ghatak, N. Chattopadhyay, S. N. Biswas, Optoelectronic materials, devices, packaging and interconnects. Proc. SPIE **836,** 203 (1988)
26. K.P. Ghatak, M. Mondal, S. Bhattacharyya, SPIE **1284**, 113 (1990)
27. K.P. Ghatak, Photonic materials and optical bistability. SPIE **1280**, 53 (1990)
28. K.P. Ghatak, S.N. Biswas, Growth and characterization of materials for infrared detectors and nonlinear optical switches. SPIE **1484**, 149 (1991)
29. K.P. Ghatak, Fiber optic and laser sensors IX. SPIE **1584**, 435 (1992)
30. R.K. Willardson, A.C. Beer (eds.), *Semiconductors and Semimetal,* vol 1 (Academic, New York 1966), p. 102
31. E.O. Kane, Phys. Rev. **131**, 79 (1963)
32. E.O. Kane, Phys. Rev. **139**, 343 (1965)
33. V.L. Bonch Bruevich, Sov. Phys. Solid State **4**, 1953 (1963)
34. R.A. Logan, A.G. Chynoweth, Phys. Rev. **131**, 89 (1963)
35. C.J. Hwang, J. Appl. Phys. **40**, 3731 (1969)
36. J.I. Pankove, Phys. Rev. A **130**, 2059 (1965)
37. B.I. Halperin, M. Lax, Phys. Rev. **148**, 722 (1966)
38. R.A. Abram, G.J. Rees, B.L.H. Wilson, Adv. Phys. **27**, 799 (1978)
39. B.I. Shklovskii, A.L. Efros, *Electronics properties of doped semiconductors*, vol. 45 (Springer, Berlin, 1984)
40. E.O. Kane, Solid State Electron. **28**, 3 (1985)
41. P.K. Chakraborty, J.C. Biswas, J. Appl. Phys. **82**, 3328 (1997)
42. B.R. Nag, *Electron transport in compound semiconductors* (Springer, New York, 1980)
43. P.E. Schmid, Phys. Rev. B **23**, 5531 (1981)
44. Jr. G. E. Jellison, F. A. Modine, C. W. White, R. F. Wood and R. T. Young, Phys. Rev. Lett., 46 1414, 1981.
45. V.I. Fistul, Heavily doped semiconductors, Chapter 7 (Plenum, New York, 1969)
46. C.J. Hwang, J. Appl. Phys. **41**, 2668 (1970)
47. W. Sritrakool, H.R. Glyde, V. Sa Yakanit, Can. J. Phys **60**, 373 (1982)
48. H. Ikoma, J. Phys. Soc. Jpn. **27**, 514 (1969)
49. P.K. Chakraborty, A. Sinha, S. Bhattacharya, K.P. Ghatak, Phys. B **390**, 325 (2007)
50. K.P. Ghatak, G.C. Datta, K.P. Ghatak, Phys. Scr. **68**, 368 (2003)
51. P.K. Chakraborty, K.P. Ghatak, J. Phys. Chem. Solids **62**, 1061 (2001)
52. P.K. Chakraborty, K.P. Ghatak, Phys. Letts. A **288**, 335 (2001)
53. P.K. Chakraborty, K.P. Ghatak, Phys. D Appl. Phys. **32**, 2438 (1999)
54. M. Abramowitz, I.A. Stegun, *Handbook of Mathematical Functions with Formulas, Graphs and Mathematical Tables* (Wiley, New York, 1964)
55. I.S. Gradshteyn, I.M. Ryzhik, *Tables of Integrals, Series and Products* (Academic, New York, 1965)
56. V. Heine, Proc. Phys. Soc. **81**, 300 (1963)
57. J.N. Schulman, Y.C. Chang, Phys. Rev. B **24**, 4445 (1981)
58. L.V. Keldysh, Sov. Phys. Solid State **4**, 1658 (1962)
59. L. Esaki, R. Tsu, IBM J. Res. Dev. **14**, 61 (1970)
60. G. Bastard, *Wave Mechanics Applied to Heterostructures* (Editions de Physique, Les Ulis, France, 1990)
61. E.L. Ivchenko, G. Pikus, *Superlattices and Other Heterostructures* (Springer, Berlin, 1995)
62. R. Tsu, *Superlattices to Nanoelectronics* (Elsevier, The Netherlands, 2005)
63. P. Fürjes, Cs Dücs, M. Ádám, J. Zettner, I. Bársony, Superlattices Microstruct. **35**, 455 (2004)

64. T. Borca-Tasciuc, D. Achimov, W.L. Liu, G. Chen, H–W. Ren, C–H. Lin, S.S. Pei, Microscale Thermophys. Eng. **5**, 225 (2001)
65. B.S. Williams, Nat. Photonics **1**, 517 (2007)
66. A. Kosterev, G. Wysocki, Y. Bakhirkin, S. So, R. Lewicki, F. Tittel, R.F. Curl, Appl. Phys. B **90**, 165 (2008)
67. M.A. Belkin, F. Capasso, F. Xie, A. Belyanin, M. Fischer, A. Wittmann, J. Faist, Appl. Phys. Lett. **92**, 201101 (2008)
68. G.J. Brown, F. Szmulowicz, R. Linville, A. Saxler, K. Mahalingam, C–H. Lin, C.H. Kuo, W.Y. Hwang, IEEE Photonics Technol. Lett. **12**, 684 (2000)
69. H.J. Haugan, G.J. Brown, L. Grazulis, K. Mahalingam, D.H. Tomich, Physica E **20**, 527 (2004)
70. S.A. Nikishin, V.V. Kuryatkov, A. Chandolu, B.A. Borisov, G.D. Kipshidze, I. Ahmad, M. Holtz, H. Temkin, Jpn. J. Appl. Phys. **42**, L1362 (2003)
71. Y–K. Su, H–C. Wang, C–L. Lin, W–B. Chen, S–M. Chen, Jpn. J. Appl. Phys. 42, L751 (2003)
72. C.H. Liu, Y.K. Su, L.W. Wu, S.J. Chang, R.W. Chuang, Semicond. Sci. Technol. **18**, 545 (2003)
73. S-B. Che, I. Nomura, A. Kikuchi, K. Shimomura, K. Kishino, Phys. Status Solidi B **229**, 1001 (2002)
74. C.P. Endres, F. Lewen, T.F. Giesen, S. Schlemmer, D.G. Paveliev, Y.I. Koschurinov, V.M. Ustinov, A.E. Zhucov, Rev. Sci. Instrum. **78**, 043106 (2007)
75. F. Klappenberger, K.F. Renk, P. Renk, B. Rieder, Y.I. Koshurinov, D.G. Pavelev, V. Ustinov, A. Zhukov, N. Maleev, A. Vasilyev, Appl. Phys. Lett. **84**, 3924 (2004)
76. X. Jin, Y. Maeda, T. Saka, M. Tanioku, S. Fuchi, T. Ujihara, Y. Takeda, N. Yamamoto, Y. Nakagawa, A. Mano, S. Okumi, M. Yamamoto, T. Nakanishi, H. Horinaka, T. Kato, T. Yasue, T. Koshikawa, J. Cryst. Growth **310**, 5039 (2008)
77. X. Jin, N. Yamamoto, Y. Nakagawa, A. Mano, T. Kato, M. Tanioku, T. Ujihara, Y. Takeda, S. Okumi, M. Yamamoto, T. Nakanishi, T. Saka, H. Horinaka, T. Kato, T. Yasue, T. Koshikawa, Appl. Phys. Express **1**, 045002 (2008)
78. B.H. Lee, K.H. Lee, S. Im, M.M. Sung, Org. Electron. **9**, 1146 (2008)
79. P-H. Wu, Y-K. Su, I-L. Chen, C-H. Chiou, J-T. Hsu, W-R. Chen, Jpn. J. Appl. Phys. **45**, L647 (2006)
80. A.C. Varonides, Renewable Energy **33**, 273 (2008)
81. M. Walther, G. Weimann, Phys. Status Solidi B **203**, 3545 (2006)
82. R. Rehm, M. Walther, J. Schmitz, J. Fleiβner, F. Fuchs, J. Ziegler, W. Cabanski, Opto-Electron. Rev. **14**, 19 (2006)
83. R. Rehm, M. Walther, J. Scmitz, J. Fleissner, J. Ziegler, W. Cabanski, R. Breiter, Electron. Lett. **42**, 577 (2006)
84. G.J. Brown, F. Szmulowicz, H. Haugan, K. Mahalingam, S. Houston, Microelectron. J. **36**, 256 (2005)
85. K.V. Vaidyanathan, R.A. Jullens, C.L. Anderson, H.L. Dunlap, Solid State Electron. **26**, 717 (1983)
86. B.A. Wilson, IEEE. J. Quant. Electron. **24**, 1763 (1988)
87. M. Krichbaum, P. Kocevar, H. Pascher, G. Bauer, IEEE. J. Quant. Electron. **24**, 717 (1988)
88. J.N. Schulman, T.C. McGill, Appl. Phys. Lett. **34**, 663 (1979)
89. H. Kinoshita, T. Sakashita, H. Fajiyasu, J. Appl. Phys. **52**, 2869 (1981)
90. L. Ghenin, R.G. Mani, J.R. Anderson, J.T. Cheung, Phys. Rev. B **39**, 1419 (1989)
91. C.A. Hoffman, J.R. Mayer, F.J. Bartoli, J.W. Han, J.W. Cook, J.F. Schetzina, J.M. Schubman, Phys. Rev. B **39**, 5208 (1989)
92. V.A. Yakovlev, Sov. Phys. Semicond. **13**, 692 (1979)
93. E.O. Kane, J. Phys. Chem. Solids **1**, 249 (1957)
94. H.X. Jiang, J.Y. Lin, J. Appl. Phys. **61**, 624 (1987)
95. H. Sasaki, Phys. Rev. B **30**, 7016 (1984)

96. G.M.T. Foley, P.N. Langenberg, Phys. Rev. B **15B**, 4850 (1977)
97. N. Miura, *Physics of Semiconductors in High Magnetic Fields, Series on Semiconductor Science and Technology* (Oxford University Press, USA, 2007)
98. K.H.J Buschow, F.R. de Boer, *Physics of Magnetism and Magnetic Materials* (Springer, New York, 2003)
99. D. Sellmyer, R. Skomski (eds.), *Advanced Magnetic Nanostructures* (Springer, New York, 2005)
100. J.A.C. Bland, B. Heinrich (eds.), *Ultrathin Magnetic Structures III: fundamentals of nanomagnetism (Pt. 3)* (Springer-Verlag, Germany, 2005)
101. B.K. Ridley, *Quantum Processes in Semiconductors*, 4th edn. (Oxford publications, Oxford, 1999)
102. J.H. Davies, *Physics of Low Dimensional Semiconductors* (Cambridge University Press, UK, 1998)
103. S. Blundell, *Magnetism in Condensed Matter*, Oxford Master Series in Condensed Matter Physics (Oxford University Press, USA, 2001)
104. C. Weisbuch, B. Vinter, *Quantum Semiconductor Structures: fundamentals and applications* (Academic Publishers, USA, 1991)
105. D. Ferry, *Semiconductor Transport* (CRC, USA, 2000)
106. M. Reed (ed.), *Semiconductors and Semimetals: nanostructured systems* (Academic Press, USA, 1992)
107. T. Dittrich, *Quantum Transport and Dissipation* (Wiley-VCH Verlag GmbH, Germany, 1998)
108. A.Y. Shik, *Quantum Wells: physics and electronics of twodimensional systems* (World Scientific, USA, 1997)

Index

S. Bhattacharya and K. P. Ghatak, *Effective Electron Mass in Low-Dimensional Semiconductors*, Springer Series in Materials Science 167,
DOI: 10.1007/978-3-642-31248-9, © Springer-Verlag Berlin Heidelberg 2013

T

About the Authors

Dr. Sitangshu Bhattacharya obtained his MSc and PhD degrees in 2003 and 2009, respectively. He is the co-author of more than 50 scientific research papers in electro-thermal transport phenomena in semiconductor nanostructures in international peer reviewed journals and four research monographs among them, two from Springer series in Materials Science (Vols. 116 and 137), one from Springer Series in Nanostructure Science and Technology and one from Springer series in Solid-State Sciences (Vol. 170), respectively. His present research interest is in electro-thermal management in quantum effect devices and interconnects.

Prof. Dr. Eng. Kamakhya Prasad Ghatak is the first recipient of the degree of Doctor of Engineering of Jadavpur University in 1991 since the University inception in 1955. He is the principal co-author of more than 250 scientific research papers in international peer reviewed journals and the said four research monographs. He is the invited speaker of SPIE, MRS etc. and is the supervisor of more than two dozens of PhD candidates. His teaching interests are nonlinear circuit theory, electron devices, and nonlinear control theory. His present research interests are nano science and technology besides number theory. Presently He is the Professor and Head of the Electronics and Communication Engineering Department of National Institute of Technology, Agartala, India and for more details please visit http://www.amazon.com/Kamakhya-Prasad-Ghatak/e/B003B09OEY.

S. Bhattacharya and K. P. Ghatak, *Effective Electron Mass in Low-Dimensional Semiconductors*, Springer Series in Materials Science 167,
DOI: 10.1007/978-3-642-31248-9, © Springer-Verlag Berlin Heidelberg 2013

The manufacturer's authorised representative in the EU is Springer Nature Customer Service Centre GmbH, Europaplatz 3, 69115 Heidelberg, Germany. If you have any concerns regarding our products, please contact ProductSafety@springernature.com

Printed and bound by CPI Group (UK) Ltd, Croydon, CR0 4YY
30/09/2026
02225591-0018